ANALYSIS AND DESIGN OF SUBSTRUCTURES LIMIT STATE DESIGN

BALKEMA - Proceedings and Monographs
in Engineering, Water and Earth Sciences

Second Edition

Analysis and Design of
SUBSTRUCTURES
Limit State Design

Swami Saran

Indian Institute of Technology, Roorkee
Department of Civil Engineering
Rookee, India

CRC Press
Taylor & Francis Group
Boca Raton London New York

CRC Press is an imprint of the
Taylor & Francis Group, an **informa** business
A TAYLOR & FRANCIS BOOK

Preface to the Second Edition

In the light of feedback received from several teachers, students and colleagues of the author, the text of the book has been revised and enlarged.

Recognising the importance of the design of circular and annular rafts, a new section in the chapter on 'Shallow Foundation' has been added. An illustrative example on this is also added for lucid understanding. The chapter 15 on Reinforced Earth has been revised keeping in view of the latest knowledge available on this. A new section on 'Wall with Reinforced Backfill' has been added. More illustrative examples are given for clear understanding of the subject.

In this edition, attention has been made in revising the design on the basis of latest IS Codes.

The author gratefully acknowledge the help he has received from various organizations and individuals in making the edition up-to-date.

Dr. Swami Saran

Preface to the First Edition

This book is intended primarily for use by practising engineers who are in the business of design of foundations and for students preparing for engineering practices. It is intended to serve as a text for the course on the **Design of Substructures** or **Foundation Design** usually taught at the postgraduate level. Since the book deals largely with design aspects, complicated and involved theoretical treatment is omitted. The background theories are however generally presented in the form of concise formulae and charts.

Since 1964 the author has been actively engaged in the teaching of Foundation Engineering at both graduate and postgraduate levels at the University of Roorkee (now renamed as IIT Roorkee). The course on Design of Substructures has been offered by the author to postgraduate students for over twenty years. The author has also over thirty years experience in the practice of the art of foundation engineering during which he had the opportunity to provide consultancy in soil exploration, analysis and design of foundations for structures like multistoreyed buildings, thermal plants, cement factories, oil storage tanks, towers, chimneys etc. During this exercise, the author felt the need for a text covering comprehensively the basic aspects of geotechnical engineering, analysis and structural design of foundations and retaining structures. An attempt has been made to fill this gap. The approach has been to deal with a topic in totality, starting from the initial stage of soil exploration and testing and ending with analysis and design. The methodology of design has been demonstrated through many illustrative examples. All the formulae, charts and examples are given in SI units.

Some of the material included in this book has been drawn from the works of other authors. Inspite of sincere efforts, some contributions may not have been acknowledged. The author apologises for such omissions. Thanks are due to all the authors and publishers for their kind permission to reproduce tables, figures and equations.

The author wishes to express his appreciation to Km. Seema Verma, Mr. Dinesh Saini and Sri S.S. Gupta for the typing and drawing work. Thanks are also due to the many colleagues, friends and students of the author who assisted in the writing of this book.

The author would be failing in his duty if he does not acknowledge the support he received from his family members, especially his daughter Vartika Hemant, who encouraged him right through the various stages of study and writing.

This book is dedicated to the author's parents as a token of his gratitude, unbounded affection and blessings which have sustained him all his life.

SWAMI SARAN

Contents

Introduction

1.1 SUBSTRUCTURES—DEFINITION AND PURPOSE

Every structure, whether it be a building, a tower, a bridge, or a dam is founded on soil, and consists of two parts, namely (*i*) superstructure or upper part, and (*ii*) substructure. The foundation is that part of the substructure which provides interface between the substructure and the supporting earth.

In general, a substructure may be defined as that part of a structure which helps in transferring the load of the superstructure and its own load on to the supporting soil. A substructure may be partly or wholly embedded inside the earth. The structures embedded which retain earth, water and similar material, may also be termed as substructures as they help in transferring the load of the structure resting on their backfill. Some of the important substructures are listed below:

(*a*) Pedestals, footings and rafts for supporting walls and columns.
(*b*) Footings and rafts on piles used for supporting tall buildings, towers, etc.
(*c*) Rafts supported on wells used for very heavy structures subjecting to lateral loads.
(*d*) Piers and abutments along with their foundations which may be a footing or raft, footing or raft on piles or a raft on well. These systems are very common for supporting bridges.
(*e*) Retaining walls and their foundations.
(*f*) Sheet piles, anchored bulkheads, breakwaters, wharves, jetties, etc. commonly used as coastal structures.

1.2 ROLE OF FOUNDATION ENGINEER

A foundation engineer is concerned directly with the design of substructures such that the resulting soil stability and estimated deformations are tolerable and within codal provisions. From practical consideration, following points have to be considered carefully:

(*a*) Forces and moments acting on the superstructure and their transfer to the substructure elements.
(*b*) Fixity conditions and various tolerances (*i.e.* settlement, lateral displacement and tilt) keeping in view the behaviour of the superstructure.
(*c*) Identification of soil/rock parameters needed in design, their evaluation from an adequate field exploration and testing programme.

(*d*) Design of substructure elements so that they can be built as economically as possible.

A thorough understanding of the principles of soil mechanics in terms of stability, stress and settlement analysis is a necessary ingredient to the successful practice of foundation engineering. The understanding of the geological processes involved in the formation of soil masses is also important as both soil stability and deformation are dependent on the stress history of the mass.

It may be stressed again that the design of substructures must be adequate from both safety and economic points of view. A foundation engineer must look at the entire system— the purpose, construction methods, and construction costs–to arrive at a design which fulfils the owner's need and does not excessively degrade the environment.

Construction of structures in areas which have been used as sanitary landfills, garbage dumps or even in hazardous waste disposal areas may pose a real challenge to a foundation engineer. Such areas should be well explored and decisions on foundation type be taken very carefully. A one to two per cent overdesign in these areas may prove to be a good investment.

1.3 GENERAL REQUIREMENTS OF SUBSTRUCTURES

A substructure must be capable of satisfying several stability and deformation requirements such as:

1. The substructure should be properly located with respect to any future influence which could adversely affect its performance.
2. The substructure elements must be safe against overturning, rotation, sliding or soil rupture.
3. Total and differential settlements should be tolerable for both the substructure and superstructure elements.

1.4 SCOPE

The primary focus of the book will be on analysis and design of substructures for buildings, bridges, offshore structures, transmission line towers and retaining structures using the fundamental principles of basic soil mechanics and earth pressures which have been discussed in Chapters 2 and 4 respectively. Chapter 3 is devoted to the various methods of geotechnical exploration and evaluation of soil parameters needed in designs. The limit state concept has been utilised for structural design. The designs based on limit state concepts are better than the designs done by ultimate load or working stress method as this concept is more rational and provides economical design. Salient features of limit state design concepts have been included in Chapter 5. In Chapter 6, general principles involved in a foundation design have been discussed.

Both shallow and deep foundations (*i.e.* footings, rafts, piles and wells) have been covered in detail in Chapters 7 to 9. Analysis and design of piers and abutments have been given in Chapter 9. Chapter 10 covers mainly the design of breakwaters and wharves. Retaining structures of concrete (commonly termed as retaining wall) and metal (as sheet piling) are considered in Chapters 11 and 12. Expansive soils are common in many parts of our country. Various types of foundations and methods of their design are given in Chapter 13. Methods of preparing designs of transmission line tower foundation are presented in Chapter 14. Soil stability can significantly be enhanced by reinforcing it with metal strips, geotextiles, geogrid, etc. Principles involved in reinforced earth and design of reinforced earth walls have been given in the last in Chapter 15.

Engineering Properties of Soils

2.1 INTRODUCTION

Most soils have been derived from rock. Rock is defined as hard and compact natural aggregate of mineral grains cemented by strong and permanent bonds. Soil is formed from rock by mechanical disintegration or chemical decomposition, or both. For engineering purposes, soil is defined as a natural aggregate of mineral grains, loose or moderately cohesive, organic or inorganic in nature, that have the capacity of being separated by means of simple mechanical processes.

A thorough understanding of engineering properties of soils is essential to use the current methods in the design of foundations and earth structures. No analysis, regardless of how theoretically exact it may be, can yield useful results if an improper appraisal of soil properties has been made. The properties of the soil that are of primary importance from practical considerations are permeability, compressibility and shear strength. In this chapter, these three properties have been discussed. Before this some important physical properties and methods of soil classification have been described for lucid understanding.

2.2 WEIGHT VOLUME RELATIONSHIPS

The soil mass consists of solids and voids. The voids may be partially or wholly filled with water. Although the solids and voids in a sample of soil do not occupy separate spaces, it is convenient to represent it having separate spaces as shown in Fig. 2.1.

Symbols used in Fig. 2.1 are described below:

W = Total weight of soil mass

W_s = Weight of solid matter

W_w = Weight of water

V = Total volume of soil mass

V_s = Volume of solid matter

V_w = Volume of water in the voids

V_a = Volume of air in the voids

V_v = Volume of total void space

The physical properties of soils are influenced by the relative proportions of water, solids and air that they contain. Following terms are common in use:

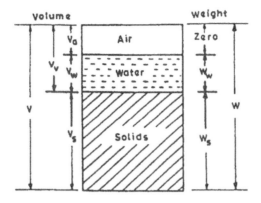

Fig. 2.1. Phase Diagram.

Void Ratio, e : The ratio of the total volume of the voids to the volume of solids, *i.e.*

$$e = \frac{V_v}{V_s} = \frac{V_a + V_w}{V_s} \qquad \text{... (2.1)}$$

Porosity, n : The ratio of the total volume of the voids to the total volume of soil mass, *i.e.*

$$n = \frac{V_v}{V} = \frac{V_a + V_w}{V_a + V_w + V_s} \qquad \text{... (2.2)}$$

Degree of Saturation, S : The ratio of the volume of water to the total volume of voids in the soil mass, *i.e.*

$$S = \frac{V_w}{V_a + V_w} \qquad \text{... (2.3)}$$

Water Content, w : The ratio of total weight of water to the weight of solids, *i.e.*

$$w = \frac{W_w}{W_s} \qquad \text{... (2.4)}$$

Bulk Unit Weight, γ_t : The ratio of total weight of the soil mass to the total volume, *i.e.*

$$\gamma_t = \frac{W}{V} \qquad \text{... (2.5)}$$

Dry Unit Weight, γ_d : The weight of solids in a given soil mass per unit total volume, *i.e.*

$$\gamma_d = \frac{W_s}{V} \qquad \text{... (2.6)}$$

Saturated Unit Weight, γ_sat : The total weight of a completely saturated soil per unit volume, *i.e.*

$$\gamma_{sat} = \frac{W_w + W_s}{V} \qquad \text{... (2.7)}$$

Submerged Unit Weight, γ_b : The unit weight of saturated soil mass minus unit weight of water, *i.e.*

$$\gamma_b = \gamma_{sat} - \gamma_w \qquad \qquad ... (2.8)$$

where γ_w = unit weight of water, 10 kN/m³.

Specific Gravity of Solids, G : The ratio of the weight of the solids to the weight of an equivalent volume of water, *i.e.*

$$G = \frac{W_s}{V_s \cdot \gamma_w} \qquad \qquad ... (2.9)$$

Now

$$e = \frac{V_v}{V_s} = \frac{V_v}{V_s} \cdot \frac{V_w}{V_w} = \frac{V_v}{V_w} \cdot \frac{V_w}{V_s}$$

or

$$e = \frac{1}{S} \cdot \frac{W_w \cdot G \cdot \gamma_w}{W_s \cdot \gamma_w} = \frac{wG}{S} \qquad \qquad ... (2.10)$$

$$n = \frac{V_v}{V_v + V_s} = \frac{V_v / V_s}{1 + V_v / V_s} = \frac{e}{1 + e} \qquad \qquad ... (2.11)$$

$$\gamma_t = \frac{W}{V} = \frac{W_s + W_w}{V_s + V_v} = \frac{W_s (1 + W_w / W_s)}{V_s (1 + V_v / V_s)}$$

or

$$\gamma_t = \frac{G \gamma_w (1 + w)}{1 + e} \qquad \qquad ... (2.12)$$

If the soil mass is dry, $w = 0$, then

$$\gamma_d = \frac{G \gamma_w}{1 + e} \qquad \qquad ... (2.13)$$

and,
$$\gamma_t = \gamma_d (1 + w) \qquad \qquad ... (2.14)$$

For saturated soil mass, $S = 1$, $\quad \therefore \quad e = wG$

$$\gamma_{sat} = \frac{\gamma_w (G + wG)}{1 + e}$$

or,

$$\gamma_{sat} = \frac{\gamma_w (G + e)}{1 + e} \qquad \qquad ... (2.15)$$

$$\gamma_b = \gamma_{sat} - \gamma_w$$

$$= \frac{\gamma_w \, (G + e)}{1 + e} - \gamma_w$$

$$\gamma_b = \frac{\gamma_w \, (G - 1)}{1 + e} \qquad \qquad \dots (2.16)$$

2.3 INDEX AND SIMPLE SOIL PROPERTIES

For the classification of soils, certain tests are carried out in the laboratory. The results obtained from these are termed as index properties of soils. Index properties are of two general types: (a) soil grain properties, and (b) soil aggregate properties.

Soil Grain Properties

The important soil grain properties are grain size and grain shape. Grain size is determined by sieve analysis. A known weight of dry soil is sieved through a set of sieves for about 15 minutes and the result is plotted in the form of a grain size distribution curve on a semi-logarithmic plot (Fig. 2.2). The ordinate is percent finer than the size, D, and abscissa is this grain size in mm.

Uniformity coefficient, C_u and coefficient of curvature, C_c defined below are used for soil classification:

$$C_u = \frac{D_{60}}{D_{10}} \qquad \qquad \dots (2.17)$$

$$C_c = \frac{(D_{30})^2}{D_{10} \times D_{60}} \qquad \qquad \dots (2.18)$$

where
D_{60} = grain size corresponding to 60 per cent finer,
D_{30} = grain size corresponding to 30 per cent finer, and
D_{10} = grain size corresponding to 10 per cent finer.

Low values of C_u imply a uniform close grading. If all particles are of the same size, C_u is unity. A granular soil is termed 'well graded' if C_c lies between 1 and 3 and C_u is greater than 6 in case of sands and greater than 4 in case of gravels. When these requirements are not met, the soil is termed 'poorly graded'.

Soil Aggregate Properties

The soil aggregate properties comprise unconfined compressive strength, sensitivity, thixotropy, relative density, Atterberg limits and activity. The components of aggregate properties are defined below.

Unconfined compressive strength, q_u is defined as the load per unit area at which an unconfined cylindrical specimen of soil will fail in a simple compression test. Unconfined compressive strength is a measure of consistency of clays and is expressed qualitatively by the terms soft, medium, stiff and hard (Table 2.1).

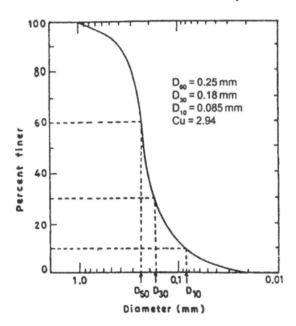

Fig. 2.2. Grain Size Distribution Curve.

Table 2.1. Consistency of Clays.

Consistency	Unconfined Compressive Strength q_u (kN/m²)
Very soft	< 25
Soft	25–50
Medium	50–100
Stiff	100–200
Very stiff	200–400
Hard	> 400

Sensitivity, S_t measures the effect of remoulding of soil on its strength without any change in its moisture content.

Numerically sensitivity, S_t is defined as

$$S_t = \frac{\text{Unconfined compressive strength (undisturbed)}}{\text{Unconfined compressive strength (remoulded)}} \qquad \ldots (2.19)$$

Natural clays are termed normal, sensitive, extra-sensitive and quick, depending upon their sensitivity (Table 2.2).

Some stiff clays having fissures and cracks exhibit sensitivity less than unity.

Thixotropy is the property of a material that enables it to regain the strength lost on remoulding in a relatively short time, without change of moisture content. The increase in strength is attributed to rearrangement of soil and water molecules.

Relative density, D_r of a soil is defined as

Table 2.2. Sensitivity of Clays.

Sensitivitiy	Classification
1-4	Normal
4-8	Sensitive
8-15	Extra-sensitive
>15	Quick

$$D_r = \frac{e_{max} - e_0}{e_{max} - e_{min}} \qquad \ldots (2.20)$$

where e_{max} = void ratio of the soil in the loosest possible state,

e_{min} = void ratio of the soil in the densest possible state, and

e_0 = void ratio in the natural state.

The relative density is commonly used to express the denseness of coarse grained soils and strongly affects its engineering behaviour. It may vary from zero in its loosest possible state ($e = e_{max}$) to 1 (100%) in its densest possible state ($e = e_{min}$). The relative density may be estimated from empirical correlations established with the standard penetration test (SPT) and static cone penetration test (CPT) as given in Table 2.3. The table also gives corresponding values of dry unit weight and angle of internal friction.

Table 2.3. Correlations with Relative Density (Mitchell and Katti, 1981).

State of compaction	Relative density %	SPT blows N	CPT resistance, q_c mN/m^2	Dry Unit weight kN/m^3	Friction angle(ϕ) (degree)
Very loose	0–15	< 4	< 5	< 14	< 30
Loose	15–35	4–10	5–10	14–16	30–32
Medium dense	35–65	10–30	10–15	16–18	32–35
Dense	65–85	30–50	15–20	18–20	35–38
Very dense	85–100	> 50	> 20	> 20	> 38

Atterberg limits

The behaviour of all soils and particularly clay varies considerably with water content. In 1911, Atterberg proposed a series of tests to determine the boundaries between the various physical states of a soil. The different states are characterized by numerical values of constants called liquid limit, plastic limit and shrinkage limit. The limit at which a suspension passes from zero strength to an infinitesimal strength is the true liquid limit, w_L. This point, however, cannot be determined. Therefore, the limit is arbitrarily taken corresponding to the water content at a small measurable strength as determined in the liquid limit test.

At moisture content lower than its liquid limit, the soil gradually loses its plasticity. The moisture content at which the soil has a small plasticity as determined by the standard test is called its plastic limit, w_p. Below plastic limit, the soil displays the properties of a semi-solid. On further reduction in water content, a state is reached when no volume change occurs.

This corresponds to the shrinkage limit which is defined as the maximum water content of a saturated soil at which reduction in its moisture content does not cause a decrease in the volume of the soil.

If the moisture content is reduced beyond shrinkage limit, the soil does not remain fully saturated as the water in the pores is partly replaced by air. A still further reduction of the water content ultimately results in dry soil, which is stiff to very stiff. A diagrammatic representation of different states and limits of soil along with the degree of saturation, change in volume and colour of soil is given in Fig. 2.3.

Plasticity index, I_p is the range of water content over which the soil exhibits plasticity. Numerically it is equal to difference between liquid limit and plastic limit, *i.e.*

$$I_p = w_L - w_P \qquad \qquad \qquad ... (2.21)$$

Liquidity index, I_L is the ratio expressed as percentage between the natural water content, w minus its plastic limit to its plasticity index:

$$I_L = \frac{w - w_P}{I_P} \qquad \qquad \qquad ... (2.22)$$

Relative consistency, C_r is defined as the ratio between liquid limit minus the natural water content of the soil and its plasticity index, *i.e.*

$$C_r = \frac{w_L - w}{I_P} \qquad \qquad \qquad ... (2.23)$$

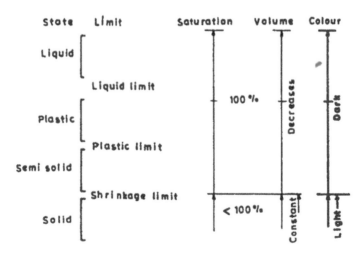

Fig. 2.3. Dependence of Atterberg Limits on Moisture Content Variation.

Activity of clays : If a number of samples are taken from a particular clay stratum and their plasticity index is plotted against clay fraction, usually the plot is a straight line passing through origin. The slope of this line is defined as activity of clays:

$$\text{Activity} = \frac{I_P}{\text{Clay Fraction}} \qquad \qquad ... \ (2.24)$$

Table 2.4 is used for classifying clays on the basis of the activity value.

Table 2.4. Activity of Clays.

Activity	Classification
< 0.75	Inactive
0.75–1.25	Normal
> 1.25	Active

2.4 CLASSIFICATION OF SOILS

Important soil classification systems can be divided into two categories, namely: (*i*) based on the size of the soil, and (*ii*) based on grain size and index properties of the soil. U.S. Bureau of Soil Classification and M.I.T. Soil Classification fall under the first category. AASHO Soil Classification System, Unified Soil Classification System and I.S. Soil Classification belong to the second category. Indian Standard Classification is more commonly used in the country and, therefore, described in the subsequent paragraphs.

The Indian Standard Classification of soils is based both on their grain size and index properties. The description given here is based on IS 1498-1970. The basic soil components are given in Table 2.5.

The criteria for classifying coarse-grained soils are given in Tables 2.6 and 2.7. The coarse-grained soils containing fines between 5 and 12 per cent are classified as border-line cases between the clean and the dirty gravels or sands as for example, *GW – GC*, or *SP – SM*. Similarly, border-line cases might occur in dirty gravels and dirty sands, where plasticity index is between 4 and 7 as, for example *GM – GC*, or *SM – SC*. The rule for correct classification is to favour the non-plastic classification. For example, a gravel with 10 per cent fines, a C_u of 20, a C_c of 2.0 and plasticity index of 6 would be classified as *GW – GM* rather than *GW – GC*.

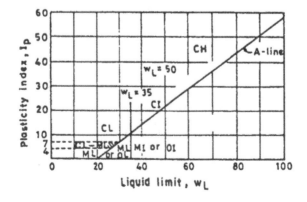

Fig. 2.4. Plasticity Chart.

Table 2.5. Basic Soil Components.

Sl. No.	Soil Component	Soil Component	Symbol	Particle Size, Range and Description
(i)	Coarse-grained	Boulder	None	Rounded to angular, bulky, hard rock paticle; average diameter more than 300 mm.
		Cobble	None	Rounded to angular, bulky, hard, rock particle; average diameter smaller than 300 mm but retained on 80 mm IS Sieve.
		Gravel	G	Rounded to angular, bulky, hard, rock particle passing 80 mm IS Sieve but retained on 4.75 mm IS Sieve.
		Sand	S	Rounded to angular, bulky, hard, rock particle passing 4.75 mm IS Sieve but retained on 75 micron IS Sieve Coarse: 4.75 mm to 2.0 mm IS Sieve Medium: 2.0 mm to 425 micron IS Sieve Fine: 425 micron to 75 micron IS Sieve
(ii)	Fine-grained	Silt	M	Particles smaller than 75 micron IS Sieve; identified by behaviour, that is, slightly plastic or non-plastic regardless of moisture and exhibits little or no strength when air-dried.
		Clay	C	Particles smaller than 75 micron IS Sieve; identified by behaviour, that is, it can be made to exhibit plastic properties within a certain range of moisture and exhibits considerable strength when air-dried.
	Organic matter		O	Organic matter in various sizes and stages of decomposition.

(IS: 1498–1970)

The fine-grained soils are classifed using plasticity chart shown in Fig. 2.4 and Table 2.7. The fine-grained soils whose plot on plasticity chart falls on, or practically on (i) 'A' line, (ii) 'W_L = 50' line shall be assigned proper boundary classification. Soils which plot above the 'A' line, or practically on it, and which have plasticity index between 4 and 7 are classified $ML - CL$.

2.5 PERMEABILITY

A material is said to be permeable if it contains continuous voids. Since such voids are present in all soils including dense sand, stiff clay and sound granite, all these materials are permeable. The soil property which describes quantitatively, the ease with which water flows through that soil is known as permeability. It is an important engineering property of soil since it directly influences the rate of flow of water in a soil. Its knowledge plays a vital role in problems like (i) excavation of open cuts in sands below water table, (ii) seepage through embankments, (iii) stability of foundations, (iv) rate of consolidation through compressible soils, (v) subgrade drainage and many others.

The pores of most of the soils are so small that the flow through them is laminar. As it is not practicable to study the flow through individual soil pores, the combined effect of flow through all the pores of soil element is considered.

In laminar flow, velocity v is proportional to hydraulic gradient i, i.e.

$$v \propto i$$

$$\qquad \qquad \dots (2.25)$$

Table 2.6. Classification of Coarse-grained Soils (Laboratory Classification Criteria).

Group Symbols	Laboratory Classification Criteria	
GW	C_u greater than 4 C_c between 1 and 3	Determine percentages of gravel and sand from grain-size curve. Depending on percentage of fines (fraction smaller than 75-micron IS Sieve) coarse-grained soils are classified as follows:
GP	Not meeting all gradation requirements for GW	Less than 5% GW, GP, SW, SP
GM	Atterberg limits below 'A' line or I_p less than 4	Limits plotting above 'A' line with I_p between 4 and 7 are border-line cases requiring use of dual symbols	More than 12% GM, GC, SM, SC
GC	Atterberg limits above 'A' line with I_p greater than 7		5% to 12% Border-line cases requiring use of dual symbols
SW	C_u greater than 6 C_c between 1 and 3		Uniformity coefficient, $$C_u = \frac{D_{60}}{D_{10}}$$
SP	Not meeting all gradation requirements for SW		Coefficient of curvature, $$C_c = \frac{(D_{30})^2}{D_{10} \times D_{60}}$$
SM	Atterberg limits below 'A' line or I_p less than 4	Limits plotting above 'A' line with I_p between 4 and 7 are border-line cases requiring use of dual symbols	where
SC	Atterberg limits above 'A' line with I_p greater than 7 I_p = plasticity index.		D_{60} = 60 per cent finer than size D_{30} = 30 per cent finer than size D_{10} = 10 per cent finer than size

(IS: 1498–1970)

Since Darcy (1956) demonstrated experimentally that the flow of water through soils is laminar, the law of flow through the soils is known as Darcy's law. It states that the rate of flow is proportional to the gradient, *i.e.*

$$Q = kiA \qquad\qquad ... (2.26)$$

or

$$v = \frac{Q}{A} = ki \qquad\qquad ... (2.27)$$

In the above equation, Q is the discharge passing through the total sectional area of soil, A (including the area of solids and voids) under a gradient i and k is the constant of proportionality known as coefficient of permeability.

Factors Affecting Permeability

Poiseuille gave the following equation for estimating the discharge Q of a fluid of viscosity η through a pipe of cross-sectional area A:

$$Q = Cm^2 \frac{\gamma_w}{\eta} iA \qquad\qquad ... (2.28)$$

where m = hydraulic mean radius,
            γ_w = density of fluid, and
            C = constant.

Now $m = \dfrac{a}{p} = \dfrac{aL}{pL}$

where p is the perimeter of the pipe and L is the length of flow tube.

Therefore $m = \dfrac{\text{Volume of flow channel}}{\text{Surface area of flow channel}}$

For soils, the volume and surface area of flow channel are equal to the volume of pores and the total surface area of soil grains respectively.

Hence $m = \dfrac{eV_s}{A_s} = \dfrac{e\pi d^3/6}{\pi d^2} = \dfrac{1}{6}ed$

where d is the diameter of a hypothetical spherical grain having the same volume to surface area ratio as all the grains in the soil mass collectively. Further, in soils the actual area of flow, a is the porosity times the total area A. Since porosity is $\dfrac{e}{1+e}$, the equation (2.28) may be written as

$$Q = \left[C' d^2 \frac{\gamma_w}{\eta} \frac{e^3}{1+e} \right] iA \qquad\qquad ... (2.29)$$

$$k = C'd^2 \frac{\gamma_w}{\eta} \frac{e^3}{1+e} \qquad \qquad \text{... (2.30)}$$

where C' = constant.

The various terms of the equation (2.30) reveal the effect of different factors on the coefficient of permeability.

(a) Grain Size

Permeability of soil depends on the second power of a representative measure of grain size. Allen Hazen (1911) found that the permeability of clean sands (with less than 5% fines) can be expressed as,

$$k = cD_{10}^2 \qquad \qquad \text{... (2.31)}$$

where k has the unit in m/s, D_{10} is in mm, constant c varies from 0.004 to 0.012 with an average value of 0.01. The equation (2.31) holds good for $k \geqslant 10^{-5}$ m/sec.

(b) Properties of the Pore Fluid

Permeability depends on the term γ_w/η, i.e., directly proportional to the density of fluid and inversely proportional to its viscosity. In engineering problems, the only fluid to be dealt is water. The unit weight of water remains practically constant with temperature while viscosity decreases with the increase in temperature. It is common practice to note the temperature of water during permeability determination and reduce the measured permeability value to the value corresponding to standard room temperature (27°C).

(c) Void Ratio

According to equation (2.30), k is significantly affected by void ratio, e. The following empirical relations can be used for coarse-grained soils:

$$k \propto e^2 \qquad \qquad \text{... (2.32)}$$

$$k \propto e^3/(1 + e) \qquad \qquad \text{... (2.33)}$$

For fine-grained soils, a plot of log k versus e is approximately linear (Taylor, 1948; Whitman, 1973), i.e.

$$\log k \propto e \qquad \qquad \text{... (2.34)}$$

(d) Structural Arrangement of Soil Particles

This is another important factor affecting permeability. For the same soil at the same void ratio, the permeability may vary with different methods of placement or compaction resulting in different arrangement and shape of voids. Thus the soil in situ often has smaller permeability in vertical direction as compared to the horizontal, due to horizontally stratified structure.

Table 2.8. Permeability and Drainage Characteristics of Soils
Coefficient of Permeability k in m per sec (log scale).

	10^0	10^{-1}	10^{-2}	10^{-3}	10^{-4}	10^{-5}	10^{-6}	10^{-7}	10^{-8}	10^{-9}	10^{-10}	10^{-11}
Drainage		Good					Poor			Practically Impervious		
Soil types	Clean gravel	Clean sands, clean sand and gravel mixtures				Very fine sands, organic and inorganic silts, mixtures of sand silt and clay, glacial till, stratified clay deposits, etc.				"Impervious" soils, e.g., homogeneous clays below zone of weathering		
						"Impervious" soils modified by effects of vegetation and weathering						
Direct determination of k		Direct testing of soil in its original position-pumping tests. Reliable if properly conducted. Considerable experience required										
		Constant-head permeameter. Little experience required										
Indirect determination of k		Falling-head permeameter. Reliable. Little experience required			Falling-head permeameter. Unreliable. Much experience required					Falling-head permeameter. Fairly reliable. Considerable experience necessary		
	Computation from grain-size distribution. Applicable only to clean cohesionless sands and gravels									Computation based on results of consolidation tests. Reliable. Considerable experience required.		

Casagrande and Fadum (1940)

Determination of Permeability

There are several laboratory methods for determining permeability, namely (*i*) Constant head permeability method, (*ii*) Falling head permeability methods, and (*iii*) Computation from consolidation test data. The adequacy of a particular test depends on the type of soil. Table 2.8 gives the range of the values of coefficient of permeability of different soils along with drainage conditions and the suitable method of determination.

There are two field methods of determining the coefficient of permeability: (*i*) Pumping tests, and (*ii*) Borehole tests.

Laboratory method of determining *k*

Constant Head Permeameter

The constant head test requires the measurement of the rate of flow through a soil sample under a constant head and, hence, it is suitable for coarse-grained pervious soils ($k > 10^{-5}$ m/sec) where the quantity of flow is measurable in a reasonable test time.

A schematic diagram for a constant head permeameter is shown in Fig. 2.5. The soil sample (which is prepared or undisturbed) is contained in a vertical cylinder between two porous plates. The diameter and height of the cylinder may be of any convenient dimensions. Water is allowed to flow through the sample from a reservoir designed to keep the water level constant by overflow. Once the steady state is attained, the quantity of flow Q during a time interval t is collected and measured. The value of k is computed from the expression

$$k = \frac{QL}{hAt} \qquad\qquad \text{... (2.35)}$$

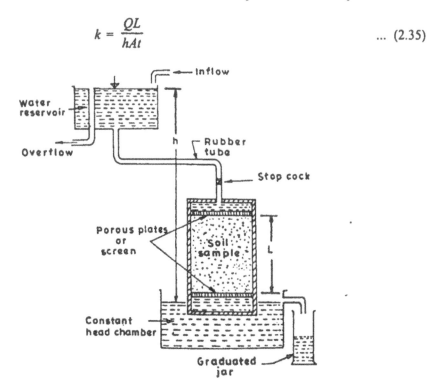

Fig. 2.5. Constant Head Permeameter.

where Q = quantity of flow in time t.

          A = cross-sectional area of the soil sample,

          h = hydraulic head, and

          L = length of sample in the direction of flow.

Variable Head Permeameter

In this test, the decrease in the hydraulic head causing the flow is measured in a given time interval and is suitable for soils having k from 10^{-4} m/s to about 10^{-8} m/s.

A schematic diagram for a variable head permeameter is shown in Fig. 2.6. The soil sample is contained in vertical cylinder between porous plates as in constant head permeameter. A transparent stand pipe of sectional area, a, is attached to the cylinder. After saturating the soil sample, the stand pipe is filled with water up to a height of h_1. The time interval t in which the water level drops from h_1 and h_2 is noted. The value of k is calculated from the following equation:

$$k = 2.3 \frac{aL}{At} \log_{10} \frac{h_1}{h_2} \qquad \qquad \dots (2.36)$$

where a = cross-sectional area of stand pipe,

          L = length of the sample,

          A = cross-sectional area of the sample, and

          t = time for falling of head of water from h_1 to h_2.

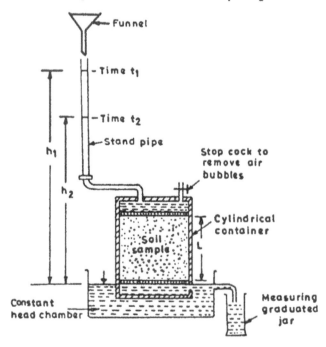

Fig. 2.6. Variable Head Permeameter.

k by Consolidation Test Data

This method is based on the mathematical relationship between k, coefficient of consolidation c_v and coefficient of volume compressibility m_v:

$$k = \gamma_w \cdot m_v \cdot c_v \qquad \qquad \text{... (2.37)}$$

where γ_w = density of water.

Procedure of obtaining m_v and c_v from consolidation test data is given in Sec. 2.9. This test is applicable to clayey soils having permeability less than 10^{-8} m/sec.

Field method of Determining k

As discussed earlier, the permeability of a soil is influenced by factors like soil structure, degree of saturation, heterogeneity stratification, etc., and it is very difficult to stimulate these factors in the laboratory. Therefore, field permeability measurements are desirable to have more reliable information for engineering problems, such as design of cutoff for earth dams, computation of pumping capacity for dewatering excavations and design of drainage systems.

Pumping out Tests

This test consists of pumping out water continuously at a uniform rate from the test well till the water levels in the test and observation wells remain stationary. For performing this test, a tubewell is necessary and it should fully penetrate the test stratum. Observation holes or piezometers should be installed along three radial directions from the well. Firstly, the static water levels in the three observation holes are noted and then pumping at constant rate is started. As the water from the test well is pumped out, a steady state will be attained soon when the water pumped out will be equal to the inflow into the well. At this stage the depth of water in the test well and observation wells will remain constant. The maximum drawdown will occur in the test well. It decreases with the increase in the distance from the test well.

Following two types of situations can occur in the field:

Unconfined acquifer : In this case, the free water surface or the water table is inside the stratum (Fig. 2.7). Observations of drawdown in two wells are sufficient to compute k. However, for better estimate, observations may be taken in more number of boreholes. If Δh_1 and Δh_2 are respectively the values of drawdown in two boreholes located at radial distances r_1 and r_2 then (Fig. 2.7):

$$h_1 = h_w - \Delta h_1$$

$$h_2 = h_w - \Delta h_2 \text{ and}$$

$$k = \frac{Q \cdot \log_e \dfrac{r_2}{r_1}}{\pi \left(h_2^2 - h_1^2 \right)} \qquad \qquad \text{... (2.38)}$$

where Q = steady state pumping rate.

Confined   aquifer : In this case, the pervious stratum is overlain by an impervious stratum, and the water table lies above the test stratum [Figs. 2.8(a) and 2.8(b)]. There may

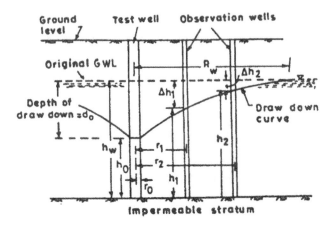

Fig. 2.7. Puming Test in an Unconfined Aquifer.

be two positions of drawdown water table as shown in Figs. 2.8(a) and 2.8(b). If D represents the thickness of aquifer, then for $h_0 > D$ (Fig. 2.8(a))

$$k = \frac{Q}{2\pi D} \cdot \frac{\log_e (r_2 / r_1)}{h_2 - h_1)} \qquad \qquad ... (2.39)$$

For $h_0 < D$ (Fig. 2.8(b))

$$k = \frac{Q}{\pi(2Dh_0 - D^2 - h_0^2)} \log_e (R_w / r_0) \qquad \qquad ... (2.40)$$

The value of R_w can be obtained by joining the position of drawdown water tables by a smooth curve.

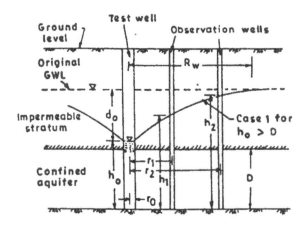

Fig. 2.8(a). Pumping Test in a Confined Aquifer ($h_0 > D$).

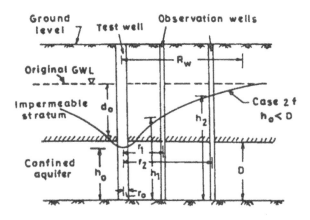

Fig. 2.8(b). Pumping Test in a Confined Aquifer ($h_0 < D$).

Borehole Tests

Two types of borehole tests for determining k are commonly used, namely, (*a*) constant head test, and (*b*) falling head test. The constant head test is applicable for highly permeable strata above water table. In this test, a hole is drilled down to the level at which the test is to be performed. After the required level is reached, the hole is cleaned. The test is then started by allowing water through a metering system to maintain gravity flow at constant head (Fig. 2.9). The permeability is then obtained by:

$$k = \frac{Q}{5.5rH}$$

... (2.41)

where r = internal radius of casing,

Q = constant rate of flow into the hole, and

H = differential head of water.

The falling head test may be conducted both above and below water table. In this test, a hole is drilled up to the bottom of the test horizon and cleaned. A packer is then fixed at the desired depth to enable the testing of the full section of the hole below the packer (Fig. 2.10). The stand pipe is filled up with water and the rate of fall of water inside the pipe is recorded. In case the hole is not stable, casing pipe with perforated section in the strata to be tested is used. The permeability is computed by:

$$k = \frac{d^2}{8L}\left(\log_e \frac{L}{R}\right)\frac{(\log_e h_1 / h_2)}{(t_2 - t_1)}$$

... (2.42)

where d = diameter of stand pipe,

L = length of test section,

R = radius of hole,

h_1 = head of water in the stand pipe at time t_1, and

h_2 = head of water in the stand pipe at time t_2.

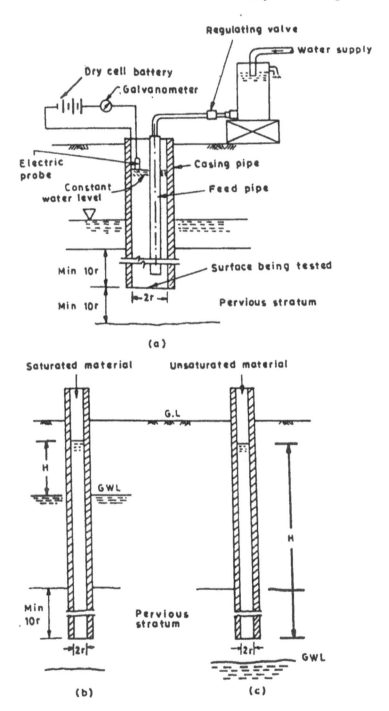

Fig. 2.9. Set Up for Constant Head Method in the Field (Gravity-Open end Type).

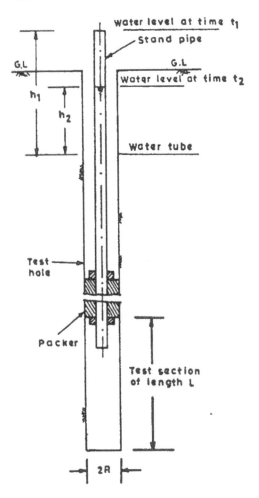

water level at time t_1

Stand pipe

G.L

G.L

Water level at time t_2

h_1

h_2

water tube

Test hole

Packer

Test section of length L

2R

Fig. 2.10. Set Up for Falling Head Method in the Field.

Permeability of Stratified Soils

Soils may be stratified by the deposition of different materials in layers which possess different permeability characteristics. For engineering purposes, it is required to determine the average value of k, applicable for the whole deposit.

The average permeability can be computed if the permeability of each layer is known. Consider the section of a stratified soil mass shown in Fig. 2.11. Let the thickness of the layers be d_1, d_2, ... d_n and k_1, k_2, ..., k_n be their respective coefficients of permeability.

Flow in Horizontal Direction

When the flow is in horizontal direction, the hydraulic gradient i remains the same for all the layers. Let v_1, v_2, v_3, ..., v_n be the discharge velocities in the corresponding strata. Then

$$Q = k_h id = (v_1 d_1 + v_2 d_2 + ... v_n d_n)$$ ... (2.43)

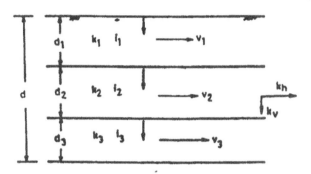

Fig. 2.11. Flow through Stratified Layers of Soils.

$$= (k_1 i d_1 + k_2 i d_2 + \dots k_n i d_n) \qquad \dots (2.44)$$

$$k_h = \frac{k_1 d_1 + k_2 d_2 + \dots k_n d_n}{d} \qquad \dots (2.45)$$

where k_h = average coefficient of permeability parallel to bedding planes (usually horizontal).

Flow in Vertical Direction

In this case, the continuity of flow requires the velocity of flow in each of the layers to be the same. The hydraulic gradients, on the other hand, change from layer to layer. Let h be the total loss of head as water flows from the top layer to the bottom most through a distance of d. Therefore,

$$v = k_v \frac{h}{d} = k_1 i_1 = k_2 i_2 = \dots k_n i_n \qquad \dots (2.46)$$

where k_v = average coefficient of permeability perpendicular to bedding planes (usually vertical).

If $h_1, h_2, \dots h_n$ are the loss of heads in each of the layers, we have

$$h = h_1 + h_2 + \dots h_n \qquad \dots (2.47)$$

$$h = d_1 i_1 + d_2 i_2 + d_n i_n \qquad \dots (2.48)$$

From equations (2.46) and (2.48),

$$k_v = \frac{d}{\dfrac{d_1}{k_1} + \dfrac{d_2}{k_2} + \dots \dfrac{d_n}{k_n}} \qquad \dots (2.49)$$

It should be noted that k_h is generally greater than k_v.

2.6 NEUTRAL AND EFFECTIVE PRESSURES IN SOILS

The total stress σ that acts at any point on a section through a mass of saturated soil may be divided into two parts: (*i*) pore water pressure or neutral stress u, which acts in the water and in the solid in every direction with equal intensity, and (*ii*) effective stress σ', which represents an excess over the neutral stress and acts exclusively between the points of contact of the solid constituents. That is

$$\sigma = \sigma' + u \quad \text{or} \quad \sigma' = \sigma - u \qquad \qquad \text{... (2.50)}$$

The effective stress can induce changes in the volume of a soil mass and can produce frictional resistance in soils. Neutral stress, on the other hand, cannot itself cause volume change or produce frictional resistance. Further the total stress and the neutral stress are measurable, the effective stress is not a physical parameter and cannot be measured. It can only be computed by subtracting the neutral stress from the total stress, both of which are physical parameters.

In Fig. 2.12, the overall regime of ground has been divided into various zones.

When water rises in a soil from below due to capillary action, the lower part becomes fully saturated. In the upper part water fills only the narrowest voids and the uppermost may remain dry. The negative pressure of water (capillary pressure, u_c) held above the water table ($= - h_c \gamma_w$) results in attractive forces between the particles, h_c being the height of capillary water above water table. Therefore, in these zones effective stress will be $\sigma' = \sigma - (- u_c) = \sigma + u_c$. In other words effective stress is increased by u_c.

One approximate relationship suggested by Terzaghi and Peck (1967) for estimating the height of capillary rise, h_c is given below:

$$h_c \, (\text{mm}) = \frac{C}{e D_{10}} \qquad \qquad \text{... (2.51)}$$

where e = void ratio,

D_{10} = effective grain size in mm, and

C = an empirical constant which can have a value between 1.0 and 5.0 sq mm.

Values suggested by Hansbo (1975) for computing h_c are listed in Table 2.9.

The values of h_c given by equation (2.51) and listed in Table 2.9 hold good for zones fully saturated by capillary action.

Referring Fig. 2.12, total and neutral stresses at point A are given below:

$$\sigma = \gamma_d z_1 + \gamma_t z_2 + \gamma_{sat} z_3 + \gamma_{sat} z_4 \qquad \qquad \text{... (2.52)}$$

$$u = - m \, \gamma_w z_2 - \gamma_w z_3 + \gamma_w z_4 \qquad \qquad \text{... (2.53)}$$

In equation (2.53), m is a factor less than unity which takes into account the partial saturation by capillary action. It is not possible to have a good estimate of m. The net neutral stress may be negative or positive depending on the relative magnitudes of capillary pressure and positive pore pressure.

It may be noted that the increase in effective stress due to the pressure of capillary

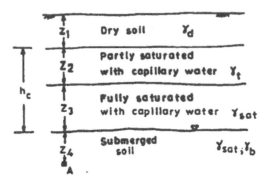

Fig. 2.12. Different Zones of Ground.

moisture increases the shearing resistance of the soil mass. It is due to this fact that excavations can be made in fine sands and silts. Such excavations cannot be made under dry soil conditions or under the water table. However, capillary moisture is easily destroyed by saturation due to rainfall or evaporation. Therefore, stability in fine sands and silts is a temporary phenomenon. Keeping this in view, it is a normal practice to neglect negative pore pressure due to capillary rise in the computation of effective stress to be used in stability analysis. Therefore, in the analysis and design of foundations, neutral stress is taken as below (Fig. 2.12):

$$u = \gamma_w z_4 \qquad\qquad\qquad ... (2.54)$$

Another effect of capillarity is often called capillary siphoning. Water may flow over the crest of an impermeable core in a dam even though the free surface may be lower than the crest of the core. This may be prevented by increasing the core height to the extent height of capillary rise.

Table 2.9. Approximate Height of Capillary Rise in Different Soils (Hansbo, 1975).

Soil Type	Height of Capillary Rise (m)
Coarse sand	0.03–0.15
Medium sand	0.12–1.10
Fine sand	0.30–3.50
Silt	1.50–12.0
Clay	> 10.0

Note : Loose sand/soft clays show smaller rise compared to dense sand/stiff clay.

2.7 QUICK SAND CONDITION

The term 'quick sand' refers to a condition and not to a material. The two factors necessary for the soil to become 'quick' are: the strength of soil must be proportional to effective pressure and the effective pressure must reduce to zero. It is the cohesionless soil whose strength is directly proportional to effective pressure and the strength becomes zero when $\sigma' = 0$. This phenomenon is also termed as 'sand boil'.

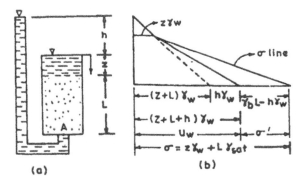

Fig. 2.13. Illustrating the Concept of Quick Sand.

Consider upward flow through a soil sample of length L and cross sectional area A as shown in Fig. 2.13.

Total stress, σ at point $A = \gamma_{sat} L + \gamma_w Z$

Neutral stress, u_w at point $A = \gamma_w (L + z + h)$

For 'quick' condition:

$$\sigma' = \sigma - u_w = 0$$

or

$$u_w = \sigma$$

$$\gamma_w (L + z + h) = \gamma_{sat} L + \gamma_w z$$

$$\gamma_w h = (\gamma_{sat} - \gamma_w) L$$

$$h/L = i_c = \frac{\gamma_{sub}}{\gamma_w} = \frac{G-1}{1+e} \qquad \text{... (2.55)}$$

The gradient, i_c causing quick condition is called the critical hydraulic gradient. Clays may have strength even at zero effective stress, so that quick condition may not necessarily result when the hydraulic gradient equals the critical value. Naturally occurring quick sand conditions are, however, usually confined to fine sands, as being less pervious, they require a smaller discharge to maintain flow at critical gradient.

2.8 SHEAR STRENGTH OF SOILS

The problems involving shear strength of soils are numerous. Stability of (a) foundation, (b) natural slopes, (c) earthen embankments, and (d) cuts and fills behind earth retaining structures, depend upon the shear strength of soils.

The shear strength involves the soil strength parameters of cohesion c and angle of internal friction ϕ. The shear strength S in terms of total stress is

$$S = c + \sigma \tan \phi \qquad \text{... (2.56)}$$

In terms of effective stress the shear strength is

$$S = c' + \sigma' \tan \phi' \qquad \text{... (2.57)}$$

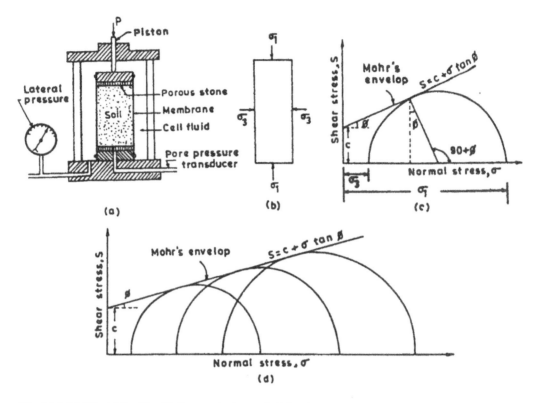

Fig. 2.14. (*a*) Triaxial Set Up, (*b*) Stresses Acting on Soil Specimen, (*c*) Illustrating Relationship between σ_1 and σ_3, and (*d*) Mohr's Circle and Rupture Envelop.

Triaxial compression test

The triaxial compression test is a more adoptable form of shear strength test which can be applied to a wider range of soil types. A diagrammatic sketch of triaxial test set up is shown in Fig. 2.14 (*a*). In this test, the soil sample is first subjected to a cell pressure σ_3 and then the sample is failed by increasing the vertical stress by the amount known as deviator stress, $(\sigma_1 - \sigma_3)$. At failure, the sample is subjected to principal stresses σ_1 and σ_3 as shown in Fig. 2.14 (*b*). From Fig. 2.14 (*c*), one can obtain a relation between σ_1 and σ_3 as:

$$\sigma_1 = \sigma_3 \tan^2(45 + \phi/2) + 2c \tan(45 + \phi/2) \qquad \dots (2.58)$$

In general, since two parameters are involved, two or more tests must be performed to evaluate the envelop equation (Fig. 2.14 (*d*)).

Drainage Conditions : Shear strength of a soil is dependent on the type of test, which may be:

(*a*) Unconsolidated–undrained test or *U* test
(*b*) Consolidated–undrained test or *CU* test
(*c*) Consolidated–drained test or *CD* test.

In the undrained test the soil specimen is not allowed to drain during the application of the all-round pressure or during the application of deviator stress; therefore the pore pressure is not allowed to dissipate at any stage of test. This test gives the total stress parameters. The analysis to determine the ultimate bearing capacity of the foundation soil or initial stability of excavations are carried out in terms of total stresses.

In consolidated-undrained test, drainage is allowed while applying all-round pressure, thus the sample is allowed to consolidate fully during this stage of test. Drainage is not allowed during the application of deviator stress. This test tends to produce small to nearly true ϕ and a measured cohesion intercept depending on the type of soil and degree of saturation. If pore-pressure measurements are taken, the effective parameters ϕ' and c' can be obtained.

In the case of drained test, drainage of pore-water from the sample is allowed both during the stage of consolidation under all-round pressure and during the application of deviator stress. This test gives effective stress parameters. Long-term stability problems are analysed in terms of effective stress.

Triaxial tests are usually carried out on silts, clays and peats. This is due to the fact that samples for the triaxial test can be conveniently obtained from these soils in undisturbed condition. It is generally pointless to test sands and gravels since a test cannot in any case be made directly on an undisturbed sample. Due to their very low permeability value, usually cohesive soils have a degree of saturation close to 100%. However, if the soil is not fully saturated, even then it is advisable to perform the strength test on the sample after complete saturation. It will take into account the possibility of saturation of the soil due to fluctuations of water table and rainfall.

Normally Consolidated and Over Consolidated Clays (S = 100%)

Clay may exist in the field under two conditions: (i) $\sigma_c' \leqslant \sigma_0'$, and (ii) $\sigma_c' > \sigma_0'$, where σ_c' is the effective pressure by which the clay has been subjected in the past while σ_0' represents the effective existing over burden pressure. Clay satisfying former conditions is known as normally consolidated clay and the later one as over consolidated clay. The ratio $\dfrac{\sigma_c'}{\sigma_0'}$ is known as over consolidated ratio (OCR). Strength characteristics of consolidated and over consolidated clays are different and are dependent on OCR.

The U test on a normally consolidated clay gives

$$S = c \quad \text{(Fig. 2.15a)} \qquad \qquad \text{... (2.59)}$$

For CU test

$$S = c + \sigma \tan \phi \quad \text{(Fig. 2.15b)} \qquad \qquad \text{... (2.60)}$$

where ϕ is usually of the order of 3° to 12°.

For CU test with pore pressure measurements or CD test

$$S = \sigma' \tan \phi' \quad \text{(Fig. 2.15c)} \qquad \qquad \text{... (2.61)}$$

i.e., the cohesion intercept is practically zero.

Several relations are available to predict the value of undrained strength of normally consolidated clay.

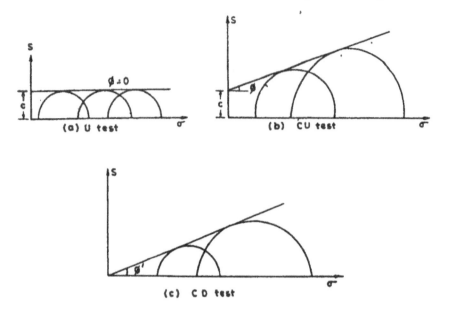

Fig. 2.15. Mohr's Circles and Mohr Envelops under different
Drainage Conditions on *N-C* Soil.

Skempton (1957):

$$\frac{S_u}{\sigma'_0} = 0.11 + 0.0037 \ (I_p\%) \qquad \qquad \dots (2.62)$$

Bjerrum and Simons (1960):

$$\frac{S_u}{\sigma'_0} = 0.45 \ (I_p\%)^{1/2} \text{ for } I_p > 5\% \qquad \qquad \dots (2.63)$$

Karlsson and Viberg (1967):

$$\frac{S_u}{\sigma'_0} = 0.005 \ (w_L\%) \text{ for } w_L > 20\% \qquad \qquad \dots (2.64)$$

The above equations afford a quick but approximate estimation of the increase in the undrained shear strength of normally loaded clay deposit with depth, on the basis of simple Atterberg limit tests. The $\dfrac{S_u}{\sigma'}$ may be computed from the above equations and an average value be used for preliminary design. Preference should be given to the value obtained from strength tests on undisturbed samples.

In over consolidated intact clays, the U test gives a higher strength value than for the same clay in the normally consolidated state. The difference may be attributed to the

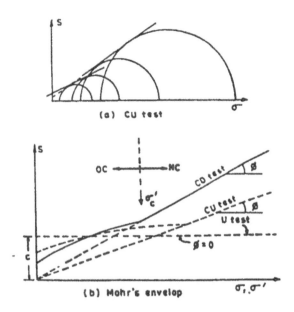

(a) CU test

(b) Mohr's envelop

Fig. 2.16. Mohr's Circle and Mohr's Envelops on *O-C* Soil.

increased density and the negative pore pressures in over consolidated clays. The *CU* test gives higher values if cell pressure, $\sigma_3 < \sigma_c'$ and OCR > 4 due to negative pore pressures (Fig. 2.16a). If cell pressure $\sigma_3 < \sigma_c'$ and OCR < 4, the pore pressures are likely to be positive at failure. It may be noted that the failure envelops for over consolidated clay are non-linear. Typical failure envelops over a wide range of stresses spanning the preconsolidation stress σ_c' are shown in Fig. 2.16b. If the failure envelops of the normally consolidated range are extended backwards they pass through the origin giving $c = 0$.

The value of c' for an over consolidated clay does not usually exceed 30 kPa (Craig, 1978). The failure envelop for the over consolidated range can be approximated by the equations (Fig. 2.16a and b):

$$S = c + \sigma \tan \phi \ (CU \text{ test}) \qquad \qquad \ldots (2.65)$$

$$S = c' + \sigma \tan \phi' \ (CD \text{ test}) \qquad \qquad \ldots (2.66)$$

Direct shear test

The test is performed in a shear box illustrated in Fig. 2.17. The box consists of two parts, the fixed frame and the lower movable frame. The usual, small shear box is 60 mm square; a 100 mm square box has also been introduced (Head, 1982). The large shear box is 300 mm square. A vertical load σ_n per unit area is applied to the upper frame by placing weights on a loading yoke (not shown). The sample is subjected to shear stress at the plane of separation *AA* (Fig. 2.17).

The direct shear test is not used in preference to the triaxial test because of difficulties in controlling drainage condition and the fact that the failure plane is predetermined by the apparatus.

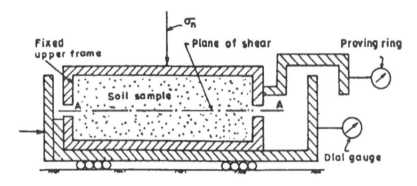

Fig. 2.17. Direct Shear Apparatus.

However, this test is usually performed on dry sand as the same can be placed in box at the required density, further effect of remoulding and saturation is little on the shear strength of sand. This test is performed with normal load held constant. The data obtained from direct shear test is plotted as shown in Fig. 2.18a for a particular value of σ_n. Shear stress corresponding to $(S/\sigma_n)_{peak}$ is a measure of the strength of the soil. In case of cohesionless soil, $\phi = \tan^{-1}$ $(S/\sigma_n)_{peak}$. Usually the peak shear stress values obtained for different values of σ_n are plotted against σ_n as shown in Fig. 2.18b. This type of plot will enable us to get the value of cohesion also if present in the soil.

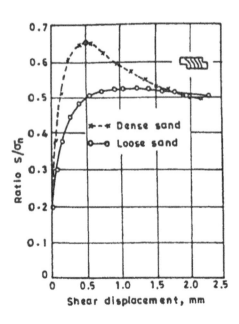

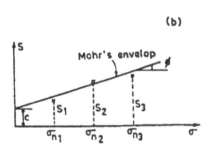

Fig. 2.18. Typical Direct Shear Test Results.

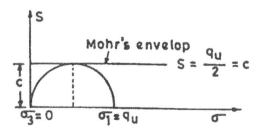

Fig. 2.19. Unconfined Compression Test Results.

Unconfined compression test

In this test, an undisturbed or remoulded sample of cylindrical shape with length diameter ratio equal to 2 is subjected to direct compression, until the failure of sample takes place. Mohr's circle plot for this test is shown in Fig. 2.19 and shear strength of soil is equal to half of the unconfined compressive strength (*i.e.* $S = q_u/2$).

Unconfined compression test is a triaxial test with the confining (cell) pressure equal to zero. This type of test cannot be made on cohesionless soils or on clays and silts which are too soft to stand in the machine without collapsing before the load is applied. Usually the value of shear strength obtained from this test is about 60 to 80 per cent of the true value. Still the current practice tends to use this test for routine recommendations. However, if the value obtained from this test is too low, more elaborate testing procedures (like triaxial tests) may be used.

Vane shear test

The vane shear test is more applicable to field conditions than to laboratory. However, the laboratory vane shear test has a useful application where satisfactory undisturbed tube samples of very soft clays and silts have been obtained, but it becomes impossible to prepare specimens from the tubes, because of their softness for the triaxial or unconfined compression test.

A laboratory vane consists of four blades mounted on a rod as shown in Fig. 2.20. The rod is pushed into a chunk of soil contained in a large mould. The rotating motion is imparted to the vane by means of a disc (not shown) at the top of the rod with the arrangement of measuring the torque. The value of shear strength of soil, S is obtained using the following formula:

$$M_r = \frac{S . \pi . D^2}{2} \left(H + \frac{D}{3} \right) \qquad \qquad \dots (2.67)$$

where M_r = applied torque,
$\quad\quad\quad D$ = width of blade (Fig. 2.20), and
$\quad\quad\quad H$ = height of blade (Fig. 2.20).

In situ vane shear test is often used for determining the sensitivity of clay. Initially, the

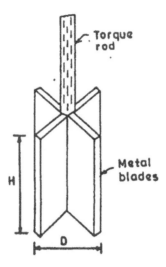

Fig. 2.20. Vane Shear.

soil is undisturbed and the shear strength corresponds to the undisturbed state. When the vane has remoulded the soil, the resistance offered by it corresponds to the remoulded state.

2.9 CONSOLIDATION CHARACTERISTICS OF SOIL

The process of compression by gradual reduction of pore space under steady load is called consolidation. A consolidation test is performed to obtain consolidation parameters to compute the amount and rate of settlement. The test is performed on an undisturbed sample which is placed in a consolidation ring available in diameters ranging from 50 to 100 mm. The relevant Indian Standard, IS:2720 (Part xv—1965) specifies a diameter of 60 mm for normal testing. The thickness of the sample ranged from 20 mm to 30 mm. However, the thickness should not be less than 10 times the maximum diameter of the grain in the specimen. Consolidation tests are restricted to clays and silts, since the theories on which settlement calculations are based are limited to fine-grained soils of these types.

The test is carried out in a special device called an oedometer or a consolidometer, which consists basically of a loading mechanism and a specimen container, known as the consolidation cell. An undisturbed soil specimen, representing the *in situ* soil layer is carefully trimmed and placed in a metallic confining ring which is the main component of the consolidation cell. The ring does not allow any lateral deformation of the soil. Porous stone discs are provided at the top and bottom of the sample to allow drainage in the vertical direction, both ways. Two types of· consolidation cells, the floating-ring cell and the fixed-ring cell, are commonly used. In the floating-ring test, compression occurs from both top and bottom, while in the fixed-ring test, the soil sample moves only downward, relative to the ring. In the floating-ring test, the friction between the ring and the soil is somewhat less than in the fixed-ring test. However, it is only in the fixed-ring test that drainage from the bottom porous stone can be measured or controlled. Hence, measurement of permeability of the soil can be made only in the fixed-ring test. A fixed-ring consolidation cell is shown in Fig. 2.21.

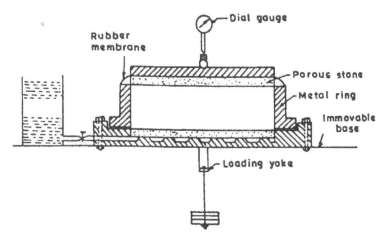

Fig. 2.21. Consolidometer.

The consolidation test proceeds by applying a series of load increments to the soil sample and recording settlements at selected time intervals. For each load interval, observations of compression versus time are taken and the data are plotted on either a semilogarithmic plot (Fig. 2.22) or $\sqrt{t}$ (Fig. 2.23). The purpose of these plots is to obtain the values of t_{50} (time at 50 per cent consolidation) and t_{90} respectively.

From semilogarithmic plot, value of t_{50} is obtained by carrying out following steps (Fig. 2.22):

(*i*) Select a time t_1 in the initial portion of the curve which is parabolic.

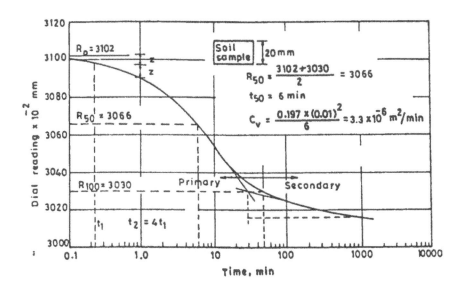

Fig. 2.22. Semilog Plot to Obtain C_v.

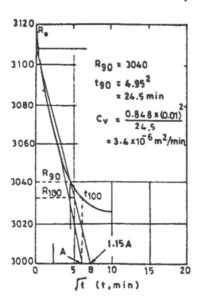

Fig. 2.23. $\sqrt{t}$ Plot to Obtain C_v.

(*ii*) Select a second time $t_2 = 4t_1$ in this parabolic portion of the curve.

(*iii*) Obtain the offset between t_1 and t_2.

(*iv*) Plot this offset distance above t_1 to obtain R_0.

(*v*) R_{100} value is obtained by the intersection of tangents drawn to the mid-curve area and the end portion (Fig. 2.22).

(*vi*) R_{50} is then taken as $\dfrac{R_0 + R_{100}}{2}$ and the time corresponding to this is t_{50}.

From $\sqrt{t}$ plot, value of t_{90} is obtained as follows (Fig. 2.23):

(*i*) The straight line obtained by joining the initial points is extended to abscissa and a point 15 per cent larger is located. This straight line is produced backward to intersect the ordinate at R_0, which is corrected zero reading corresponding to $u = 0$.

(*ii*) Joining points R_0 and B, a second straight line is drawn.

(*iii*) The point of intersection of the second straight line and the original curve locates R_{90} and the corresponding time is t_{90}.

Coefficient of Consolidation

The t_{50} and t_{90} values are used to compute the coefficient of consolidation, C_v which is given by:

$$C_v = \frac{T_v \cdot H^2}{t_i} \qquad\qquad \text{... (2.68)}$$

where T_v = time factor,
         H = length of drainage path; in the laboratory it is the half sample thickness,
              when drainage is from both faces, and
         t_i = time for i per cent consolidation.

The value of time factor depends as degree of consolidation, u and is given by the following equations for linear pore pressure distribution:

(*i*) $u \leq 60\%$

$$T_v = \frac{\pi}{4}u^2$$

... (2.69)

(*ii*) $u > 60\%$

$$T_v = 1.781 - 0.93 \log_{10}(100 - u\%)$$

... (2.70)

For $u = 50\%$, $T_v = 0.197$ and for $u = 90\%$, $T_v = 0.848$.

Therefore $$C_v = \frac{0.197\,H^2}{t_{50}}$$

... (2.71)

or $$C_v = \frac{0.848\,H^2}{t_{90}}$$

... (2.72)

Primary Consolidation Settlements

In a consolidation test, one dimensional compression takes place and, therefore, change in thickness ΔH, per unit original thickness, H_0 of the specimen is equal to change in volume per unit of original volume. The change in volume is a consequence of decrease in void ratio, Δe. If e_0 is the initial void ratio, then (Fig. 2.24)

$$\frac{\Delta H}{H_0} = \frac{\text{Change in volume}}{\text{Original volume}} = \frac{\Delta e}{1 + e_0}$$

... (2.73)

Pressure-deformation data obtained from a consolidation test is plotted as shown in Fig.

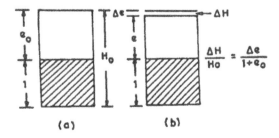

(a)   (b)

Fig. 2.24. (*a*) Initial State of Sample; (*b*) State of Sample after Compression.

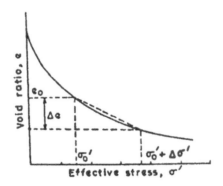

Fig. 2.25. Void Ratio-Effective Stress Plot.

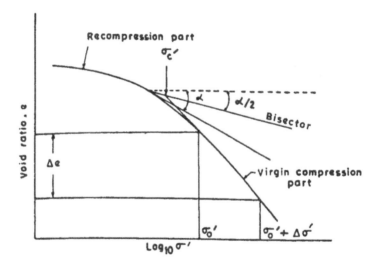

Fig. 2.26. $e \log_{10} \sigma'$ Plot.

2.25 and more commonly as in Fig. 2.26. The slope of the curve for a pressure range obtained from $e - \sigma'$ plot is termed as coefficient of compressibility, a_v. Therefore,

$$a_v = \frac{\Delta e}{\Delta \sigma'} \qquad \qquad ... (2.74)$$

$$\text{Hence} \quad . \quad \Delta H = S_c = \frac{a_v}{1 + e_0} H_0 \cdot \Delta \sigma' \qquad \qquad ... (2.75)$$

where S_c = consolidation settlement.

When the effective stress is plotted against vertical strain or the per cent consolidation, the slope of the curve is called the coefficient of volume compressibility, m_v.

$$m_v = \frac{\Delta H}{H_0 \Delta \sigma'} = \frac{a_v}{1 + e_0} \qquad \text{... (2.76)}$$

The slope of straight line portion of $e - \log_{10} \sigma'$ curve is called compression index, C_c. Therefore,

$$C_c = \frac{\Delta e}{\log_{10} \dfrac{\sigma_0' + \Delta \sigma'}{\sigma_0'}} \qquad \text{... (2.77)}$$

Combining equations (2.73) and (2.77),

$$S_c = \Delta H = \frac{C_c H_0}{1 + e_0} \log_{10} \frac{\sigma_0' + \Delta \sigma_0'}{\sigma_0'} \qquad \text{... (2.78)}$$

The equation holds good for normally consolidated soils. If the soil is preconsolidated, it will not be under virgin compression and C_c cannot be used to compute settlement.

The preconsolidation pressure σ_c' can be obtained by the method proposed by Casagrande (1936). Steps in the method are (Fig. 2.26):

(i) Determine the sharpest curvature and draw a tangent.
(ii) Draw a horizontal line through the tangent point and bisect the angle α.
(iii) Extend the virgin compression portion to intersect α-bisector.
(iv) The intersection point gives the actual preconsolidation pressure, σ_c'.

Once σ_c' is known and it is found to be greater than σ_c', soil is perconsolidated. In this case the settlement is computed by substituting C_r, recompression index in place of C_c in Eq. 2.78. The recompression index is the slope of the recompression part of the $e - \log_{10} \sigma'$ curve (Fig. 2.26).

The compression index, C_c is sometimes approximately computed using one of the following empirical relations:

Terzaghi and Peck (1967)

$$C_c = 0.009 \ (w_L - 10) \qquad \text{... (2.79)}$$

Azzouz et al. (1976)

$$C_c = 0.37 \ (e_0 + 0.003 \ w_L + 0.004 \ w_n - 0.34) \qquad \text{... (2.80)}$$

Koppula (1986)

$$C_c = 0.009 \ w_n + 0.005 \ w_L \qquad \text{... (2.81)}$$

The compression index value obtained from empirical relations should be used only for a preliminary estimate of settlement of normally consolidated soil.

Secondary Consolidation

Secondary consolidation is that settlement continuing beyond primary consolidation. Settlement due to secondary consolidation is usually very small in inorganic soil, while it is significant in organic soils.

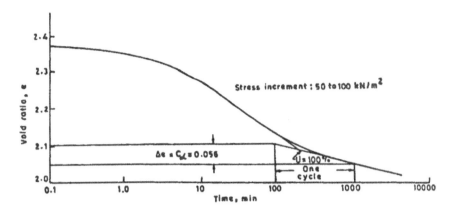

Fig. 2.27. Determination of the Rate of Secondary Compression, C_α.

The slope of the secondary branch of the settlement, *i.e.* after $u = 100\%$, versus log time curve is termed as secondary compression index C_α and is used for computing secondary settlement (Fig. 2.27). Therefore,

$$C_\alpha = \frac{\Delta H_s / H_1}{\log_{10} \dfrac{t_2}{t_1}} = \frac{\Delta e}{\log_{10} \dfrac{t_2}{t_1}} \qquad \qquad \dots (2.82)$$

The compression is noted from plot, usually for one log cycle of time; the corresponding Δe then gives the value of C_α.

The secondary settlement S_s is computed by the following equation:

$$S_s = C_\alpha \cdot H \cdot \log_{10} \frac{t_2}{t_1} \qquad \qquad \dots (2.83)$$

where H_f = thickness of stratum at the end of primary consolidation,

$\qquad H_1$ = laboratory sample thickness at time t_1,

$\qquad \Delta H_s$ = change in laboratory sample height from t_1 to t_2,

$\qquad t_2$ = sometime Δt after t_1, and

$\qquad t_1$ = sometime after primary consolidation.

Mesri and Godlewski (1977) suggested that in normal clays C_α may be taken equal to $0.032\ C_c$.

2.10 ELASTIC PROPERTIES OF SOILS

The elastic modulus E, shear modulus G and Poisson's ratio μ are the important elastic properties of the soil used in computation of settlement of foundations. The shear modulus is related to E and μ as

$$G = \frac{E}{2\,(1 + \mu)} \qquad \qquad \text{... (2.84)}$$

The values of E and μ can be obtained using triaxial compression test. They are dependent on:

(i) Method of performing compression tests
(ii) Confining cell pressure
(iii) Over consolidation ratio
(iv) Density of soil
(v) Water content of soil
(vi) Strain rate
(vii) Type of soil.

A triaxial compression test yields the familiar stress-strain curve illustrated in Fig. 2.28a. Since soils are not perfectly elastic even at very low stress, the value of E may be taken as the slope of the line oa (the secant modulus), if oa is the stress range that we are interested in. Kondner (1963) proposed that the stress-strain curve (Fig. 2.28b) could be represented in hyperbolic form which becomes a straight line on $\varepsilon/\sigma_1 - \sigma_3$ versus ε plot. The value of E is then equal to $1/a$.

Bowles (1996) has listed the values of E and μ as given in Tables 2.10 and 2.11.

Table 2.10. Typical Range of Values for the Static Stress-Strain Modulus
E for Selected Soils (Bowles, 1996).

Soil	E, Mpa
Clay	
Very Soft	2–15
Soft	5–25
Medium	15–50
Hard	50–100
Sandy	25–250
Glacial till	
Loose	10–150
Dense	150–720
Very Dense	500–1440
Loess	15–60
Sand	
Silty	5–20
Loose	10–25
Dense	48–81
Sand and gravel	
Loose	50–150
Dense	100–200
Shale	150–5000
Silt	2–20

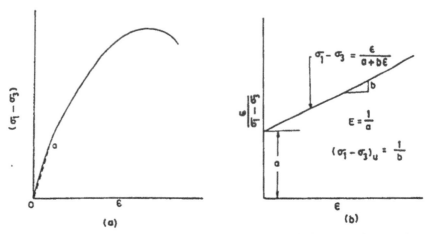

Fig. 2.28. (*a*) Typical Stress-Strain Curve; (*b*) Transformed Stress-Strain Representation (Kondner, 1962).

Table 2.11. Typical Range of Values for Poisson's Ratio (Bowles, 1996).

Type of Soil	μ
Clay, saturated	0.4–0.5
Clay, unsaturated	0.3–0.45
Sandy clay	0.2–0.3
Silt	0.3–0.35
Sand (dense)	0.2–0.4
Coarse (void ratio = 0.4 – 0.7)	0.15
Fine-grained (void ratio = 0.4 – 0.7)	0.25
Rock	0.1–0.4 (depends somewhat on type of rock)
Loess	0.1–0.3
Concrete	0.15

ILLUSTRATIVE EXAMPLES

Example **2.1**

    (*a*) A soil sample of diameter 38.1 mm and length 76.2 mm has wet and dry weight as 0.1843 kg and 0.1647 kg respectively. The specific gravity of the solids is 2.70. Determine dry density, moist density, void ratio, water content, degree of saturation and porosity.

    (*b*) Suppose the diameter and length of the sample were incorrectly measured as 37.6 mm and 75.6 mm respectively. What would be the resulting error in the computed values of water content and degree of saturation.

Solution

    (*a*) Volume of specimen = $\pi/4 \times (38.1)^2 \times (76.2) = 86880$ mm^3

$$\text{Dry denstiy} = \frac{0.1647 \times 10^{-2}}{86880 \times 10^{-9}} = 18.9 \text{ kN/m}^3$$

$$\text{Moist density} = \frac{0.1843 \times 10^{-2}}{86880 \times 10^{-9}} = 21.2 \text{ kN/m}^3$$

$$\text{Water content} = \frac{0.1843 - 0.1647}{0.1647} \times 100$$

$$= 11.9\%$$

$$\gamma_d = \frac{G . \gamma_w}{1 + e}$$

$$1 + e = \frac{2.70 \times 10.0}{18.9}$$

$$e = 0.242, \text{ and}$$

$$e = \frac{WG}{S_r} \times 100 \text{ per cent}$$

$$S_r = \frac{0.119 \times 2.7}{0.424}$$

$$= 75.77\%$$

$$n = \frac{e}{1 + e} = \frac{0.424}{1 + 0.424} = 0.298$$

(b) Volume of sample with incorrect values of diameter and length of sample

$$= \frac{\pi}{4} . (37.6)^2 . (75.6) = 83940 \text{ mm}^3$$

$$\gamma_d = \frac{0.1647 \times 10^{-2}}{83940 \times 10^{-9}} = 19.6 \text{ kN/ m}^3$$

$$\gamma = \frac{0.1843 \times 10^{-2}}{83940 \times 10^{-9}} = 21.9 \text{ kN/ m}^3$$

$$w = \frac{0.1843 - 0.1647}{0.1647} \times 100 = 11.9\%$$

$$e = \frac{G\gamma_w}{\gamma_d} - 1 = \frac{2.7 \times 10}{19.6} - 1 = 0.376$$

$$S = \frac{wG}{e} = \frac{0.119 \times 2.7}{0.376} \times 100 = 85.4\%$$

% error in w = 0%

$$\% \text{ error in } S_r = \frac{85.4 - 75.77}{75.77} \times 100 = 12.73\%$$

Example 2.2

A natural soil deposit has a bulk unit weight of 18.50 kN/m³ and water content of 7%. Calculate the amount of water required to be added to one cubic metre of soil to raise the water content to 15%. Assume the void ratio to remain constant and value of G as 2.70. Determine also degree of saturation.

Solution

$$\gamma_d = \frac{\gamma_t}{1+w} = \frac{18.50}{1+0.07} = 17.29 \text{ kN/ m}^3$$

For one cubic metre of soil,

$$W_d = \gamma_d V = 17.29 \times 1 = 17.29 \text{ kN}$$

Weight of water, $W_w = wW_d = 0.07 \times 17.29 = 1.21$ kN

$$V_w = \frac{W_w}{\gamma_w} = \frac{1.21}{9.81} = 0.123 \text{ m}^3$$

When water content is increased to 15%,

$$W_w = wW_d = 0.15 \times 17.29 = 2.59 \text{ kN}$$

$$V_w = \frac{W_w}{\gamma_w} = \frac{2.59}{9.81} = 0.264 \text{ m}^3$$

Hence additional water required to raise the water content from 7% to 15%

$$= 0.264 - 0.123 = 0.141 \text{ m}^3 = 141 \text{ litres}$$

Void ratio,
$$e = \frac{G\gamma_w}{\gamma_d} - 1 = \frac{2.7 \times 9.81}{17.29} - 1 = 0.53$$

After the water has been added, e remains constant

$$S_r = \frac{wG}{e} = \frac{0.15 \times 2.7}{0.53} = 0.764 = 76.4\%$$

Example 2.3

A 10-m thick layer of saturated clay is underlain by a layer of sand. The density of clay is 19.6 kN/m³. The sand is under an artesian pressure of 6 m. Calculate the depth of the cut that can be made without causing heave.

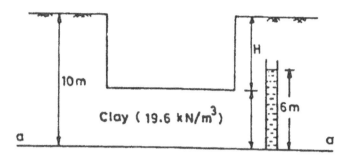

Fig. 2.29. Illustrating the Excavation and Soil Strata.

Solution

Refer Fig. 2.29. Let H be the maximum depth of the cut that can be made without causing heave. Heave will occur when the effective stress at section aa becomes zero.

Total pressure at $aa = (10 - H) \times 19.6$ kN/m²

Neutral pressure at $aa = 6 \times 10$ kN/m²

Effective stress at $aa = [(10 - H) \times 19.6 - 6 \times 10]$ kN/m²

Condition for just avoiding the heave is,

$$(10 - H) \times 19.6 - 6 \times 10 = 0 \text{ or } H = 7 \text{ m}$$

Example 2.4

The water table in a deposit of sand of 7 m thick is 2.5 m below the ground surface. Above the water table the sand is saturated by capillary water. The bulk density of sand is 19.45 kN/m³. Calculate the effective pressure at top and bottom of sand layer and plot the variation of total pressure, effective pressure and neutral pressure with depth.

Solution

Refer Fig. 2.30

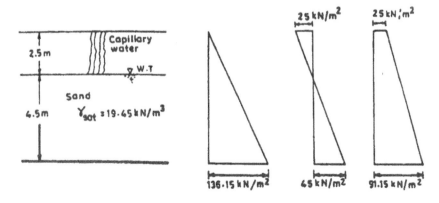

Fig. 2.30 Illustrating the Soil Strata with Capillary Water and Pressure Distributions.

Total pressure at the bottom of the sand layer
$$= 7 \times 19.45 = 136.15 \text{ kN/m}^2$$

Neutral pressure at bottom $= 4.5 \times 10 = 45 \text{ kN/m}^2$

At ground level, neutral pressure (due to capillary rise)
$$= -(2.5 \times 10) = -25 \text{ kN/m}^2$$

Thus, effective stress at the bottom
$$= 136.15 - 45 = 91.15 \text{ kN/m}^2$$

Effective stress at top $= 0 - (-25) = 25 \text{ kN/m}^2$

Example 2.5

If the head lost in soil B is 14 times the head lost in soil A (Fig. 2.31), and the permeability of soil A is 2×10^{-5} m/s, determine (a) quantity of water flowing, (b) permeability of soil B, and (c) rise of water in a stand pipe inserted in soil B at El. 0.15 m.

Solution

(a) Let head loss in soil A is h, then

head loss in soil B will be $14h$.

Total head loss $= h + 14h = 15h$

$15h = 0.75 - 0.60 = 0.15$ or $h = 0.01$ m

Consider the flow of water through soil A

$$Q = K_A \cdot i_A A = 2 \times 10^{-5} \times \frac{0.01}{0.20} \times 10^{-3} = 1 \times 10^{-9} \text{ m}^3/\text{s}$$

(b) For flow of water through soil B

$$Q = K_B \cdot i_B \cdot A$$

$$1 \times 10^{-9} = K_B \frac{0.14}{0.20} \times 10^{-3} \text{ or } K_B = 1.43 \times 10^{-6} \text{ m/s}$$

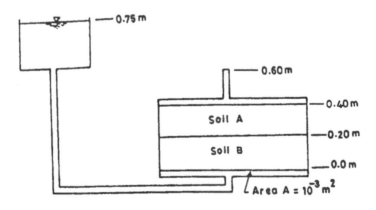

Fig. 2.31. Flow of Water through Soil Sample.

(c) Head loss between El. 0.0 m and El. 0.15 m = $\dfrac{0.14}{0.20} \times 0.15 = 0.105$ m

Therefore if a stand pipe is inserted in soil B at El. 0.15 m, water will rise to a height of $(0.75 - 0.105) = 0.645$ m.

Example 2.6

A 6 m layer of sand is underlain by 3 m thick layer of clay. There is another layer of dense sand below the clay layer. The water table is at the ground surface (Fig. 2.32). The clay layer has a compression index of 0.37 and natural water content is 40%. Compute the ultimate settlement. After two months a settlement 0.044 m is observed. How much time will be required to reach half the settlement.

Solution

Refer Fig. 2.32. Assume $G = 2.7$ and $(\gamma_{sub})_{sand} = 10$ kN/m³

Initial void ratio of clay, $e_0 = \dfrac{wG}{S} = \dfrac{0.4 \times 2.7}{1} = 1.08$

Unit weight of clay, $\gamma_s = \dfrac{(G+e)}{1+e}\gamma_w = \left(\dfrac{2.7 + 1.08}{1 + 1.08}\right) \times 10$

$$= 17.83 \text{ kN/m}^3$$

Initial effective overburden pressure at the centre of the clay,

$$\sigma_0' = 10 \times 6 + 1.5\,(17.83 \times 10) = 72.03 \text{ kN/m}^2$$

After lowering the water table by 6 m, effective pressure

$$\sigma_1' = 20 \times 6 + 1.5\,(17.83 - 10) = 130.89 \text{ kN/m}^2$$

$$S = \dfrac{C_c \cdot H}{1 + e_0}\log_{10}\dfrac{\sigma_1'}{\sigma_0'} = \dfrac{0.37 \times 3.0}{1 + 1.08}\log_{10}\dfrac{130.89}{72.03} = 0.138 \text{ m}$$

$$U = \text{degree of consolidation} = \dfrac{0.044}{0.138} = 0.318 = 31.8\%$$

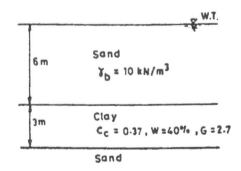

Fig. 2.32. Subsoil Profile.

$$T_v = \frac{\pi}{4}U^2 = \frac{\pi}{4}(0.318)^2 = 0.0794$$

Since $\quad T_v = \dfrac{C_v t}{d^2}$

or $0.0794 = \dfrac{C_v(2)}{(1.5)^2}$ $\quad$ i.e., $C_v = 0.089$ m²/month

For 50% consolidation

$$T_v = \frac{\pi}{4}(0.5)^2 = \frac{0.089 \times t}{1.5^2}$$

or $t = 4.96$ months, say five months.

Example 2.7

A stratum of clay is 3 m thick and has a initial overburden pressure of 50 kN/m² at its middle. Determine the final settlement due to increase in pressure of 40 kN/m² at the centre of clay layer. The clay is over consolidated with a preconsolidation pressure of 80 kN/m². Values of coefficient of recompression and compression index are 0.05 and 0.27 respectively. The initial void ratio is 1.42.

Solution

For over consolidated soils:

$$S = \frac{C_r H}{1+e_0}\log_{10}\frac{\sigma_c{}'}{\sigma_0{}'} + \frac{C_c H}{1+e_0}\log_{10}\left(\frac{\sigma_c{}' + \Delta\sigma'}{\sigma_c}\right)$$

$$= \frac{0.05 \times 3}{1+1.42}\log_{10}\frac{80}{50} + \frac{0.27 \times 3}{1+1.42}\log_{10}\left(\frac{80+10}{80}\right)$$

$$= 0.03 \text{ m} = 30 \text{ mm}$$

Example 2.8

The stresses on a failure plane in a drained test on cohesion soil are as given below:

Normal stress, $\sigma = 200$ kN/m²

Shear stress, $\tau = 80$ kN/m²

Determine the angle of shearing resistance, which the failure plane makes with the major principal plane, and magnitude of principal stresses.

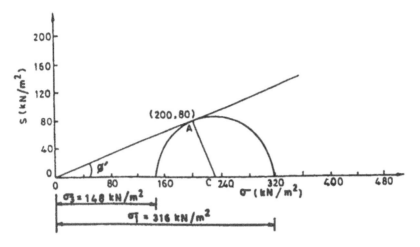

Fig. 2.33. Mohr's Envelops and Mohr's Circle.

Solution

Figure 2.33 shows the Coulomb line passing through origin and the point A with co-ordinates (200, 80)

$$\tan \phi' = \frac{80}{200} = 0.40 \text{ or } \phi' = 21.8°$$

The angle which the failure plane makes with the major principal plane is

$$= 45 + \phi'/2 = 45 + \frac{21.8}{2} = 55.9°$$

A perpendicular to the line OA is drawn at point A, the point at which it intersects the x-axis, locates the position of the centre of Mohr's circle, C. A Mohr's circle is then drawn with centre at point C and radius CA. From Mohr's circle

$$\sigma_1 = 316 \text{ kN/m}^2$$
$$\sigma_3 = 148 \text{ kN/m}^2$$

Example 2.9

A vane 100 mm long and 80 mm diameter was pressed into soft clay at bottom of borehole. Torque was applied and gradually increased to 40 N-m when failure took place. Subsequently the vane was rotated rapidly so as to completely remould the soil. The remoulded soil was sheared at a torque of 15 N-m. Calculate the cohesion of the clay in the natural and remoulded states and also value of sensitivity.

Solution

Natural State:

$$T = \pi d^2 \tau_f \left(\frac{H}{2} + \frac{D}{6} \right)$$

$$40000 = \pi \ (80)^2 \ (\tau_f) \ \left(\frac{100}{2} + \frac{80}{6} \right)$$

$$\tau_f = 0.0314 \ \text{N/m}^2$$

Remoulded State:

$$15000 = \pi \ (80)^2 \ (\tau_f) \ \left(\frac{100}{2} + \frac{80}{6} \right)$$

or $\tau_f = 0.0117 \ \text{N/m}^2$

$$\text{Sensitivity} = \frac{0.0314}{0.0117} = 2.67$$

REFERENCES

1. AASHTO (1978), *Standard Specifications for Transportation Materials and Methods of Sampling and Testing*, 12th Ed., Washington, D.C., Part I, pp. 828-998.
2. Azzouz, A.S. et al. (1976), "Regression Analysis of Soil Compressibility," *Soils and Foundation*, Tokyo, Vol. 16, no. 2, pp. 19-29.
3. Bjerrum, L. and N.E. Simons (1960), "Comparison of Shear Strength Characteristics of Normally Consolidated Clays," 1st PSC, ASCE, pp. 771-726.
4. Bowles, J.E. (1996), *Foundation Analysis and Design*, McGraw-Hill International Book Company.
5. Casagrande, A. (1936), "The Determination of Preconsolidation Load and its Practical Significance," Proc. 1st Intern. Conf. SMFE, Cambridge, Vol. 3, pp. 60-64.
6. Casagrande, A. and R.E. Fadum (1944), "Application of Soil Mechanics in Designing Building Foundations," *Trans. ASCE*, Vol. 109, pp. 383-490.
7. Craig, R.F. (1983), *Soil Mechanics*, Van Nostrand Reinhold, U.K.
8. Darcy, H. (1956), *Les Fontaines Publiques de la Ville de Dijon, Dalmont*, Paris.
9. Hansbo, A. (1975), "Jordmateriallara," Almquist Wiksell Forlag AB, Stockholm.
10. Hazen, A. (1911), Discussion on paper "Dams and Sand Foundation" by A.C. Koenig, *Trans. ASCE*, Vol. 73, pp. 199-203.
11. Head, K.H. (1982), *Manual of Soil Laboratory Testing*, Pentech. Press, London.
12. IS: 1498-1970, "Classification and Identification of Soils for General Engineering Purposes," ISI, New Delhi.
13. IS: 2720 (Part xv)—1965, Consolidation Test, ISI, New Delhi.
14. Karlsson, R. and L. Viberg (1967), "Ratio c/p' in relation to Liquid Limit and Plasticity Index with Special Reference to Swedish Clays," *Proc. Geotechnical Conference*, Oslo, Norway, Vol. 1, pp. 43-47.
15. Kondner, R.L. (1963), "Hyperbolic Stress-Strain Response: Cohesive Soils," JSMFE, *ASCE*, Vol. 89, SMI, pp. 115-143.
16. Koppula, S.D. (1986), "Discussion: Consolidation Parameters Derived from Index Tests", *Geotechnique*, Vol. 36, No. 2, pp. 291-92.
17. Mesri, G. and P.M. Goldewski (1977), "Post Densification Penetration Resistance of Clean Sands," *ASCE*, JGED, Vol. 103, GT5, pp. 417-430.
18. Mitchell, J.K. and R.K. Katti (1981), "Soil Improvement—State of Art Report," 10th Intern. Conf. SMFE, Stockholm, p. 264.
19. Skempton, A.W. (1957), "Discussion on the Planning and Design of the New Hongkong Airport," Proc. Institution of Civil Engineers, London, Vol. 7, pp. 305-307.
20. Terzaghi, K. and R.B. Peck (1967), *Soil Mechanics in Foundation Engineering*, John Wiley and Sons, New York.

PRACTICE PROBLEMS

1. Using phase diagrams, establish the relationship between saturated and dry density (γ_{sat} and γ_d) in terms of specific gravity G, void ratio e and density of water γ_w. Prove that $\gamma = \gamma_d + S(\gamma_{sat} - \gamma_d)$, where S is the degree of saturation.

2. A compacted cylindrical specimen 30 mm diameter and 75 mm length is to be prepared from oven dry soil. If the specimen is required to have a water content of 12% and percentage of air voids as 18%, calculate the weight of soil and water required in the preparation of the sample. Assume $G = 2.70$.

3. (a) Define liquid limit, plastic limit and shrinkage limit and explain their usefulness in understanding engineering behaviour of soils.

   (b) In a Casagrande limit device, specimens of a certain sample of clay at water contents of 15%, 30%, 45% and 55% require 70, 40, 20 and 15 blows respectively. Plastic limit of clay is 22%. Natural water content is 30%. Determine liquidity index and relative consistency of the clay.

4. The dry unit weights of a sand in the loosest and densest states are respectively 13.5 and 22.0 kN/m³. Assuming the specific gravity of solids as 2.65, determine the relative density of sand with porosity of 29%.

5. A sand deposit 10 m thick has a specific gravity as 2.65 and void ratio of 0.70. It overlies a bed of soft clay. The ground water table is proposed to be lowered by another 5 m. Plot the diagrams showing the variation of total pressure, neutral pressure and effective pressure for the following conditions:

   (a) Before lowering of water table.

   (b) After lowering of water table. Assume that the sand between 2.5 m and 7.5 m is fully saturated.

6. (a) Define the coefficient of permeability. Discuss briefly the various factors which influence it.

   (b) A constant head permeability test was run on a sand sample 250 mm in length and 3000 mm² in area. Under a head of 450 mm, the discharge was found 0.25 litre in 120 s. The dry weight of sample was 1.350 kg. Assuming a suitable value of the specific gravity of soil, determine the coefficient of permeability of the soil and the seepage velocity during the test.

7. A soil sample 100 mm long and 2000 mm² in cross sectional area was tested for permeability in a variable head permeater. The stand pipe has a cross sectional area of 100 mm² and head drops from 400 mm to 200 mm in 360 seconds. Determine the coefficient of permeability.

8. A pumping test was carried out in a confined aquifer to determine its 'K'. The thickness of aquifer was 4.5 m. The drawdown in the test well from the original piezometric level at a steady discharge of 0.0267 m³/s was found to be 5.0 m. The radius of influence is about 90 m. The radius of the test well was 100 mm. If the original piezometric level is at a height of 10 m above the bed of aquifer, compute the coefficient of permeability.

9. A drained shear test was performed on a sample of cohesionless soil. If the normal and shear stresses acting on the failure plane are 200 kN/m² and 80 kN/m² respectively, determine

   (a) Angle of shearing resistance,

   (b) Magnitudes of major and minor principal stresses, and

   (c) Angle of failure plane with major principal plane.

10. A consolidated-undrained triaxial test was conducted on normally consolidated clay and the following results were obtained:

σ_3 (kN/m²)	100	300	500
$\sigma_1 - \sigma_3$ (kN/m²)	130	485	780
u (kN/m²)	48	140	260

Determine the total and effective shear strength parameters.

11. When an unconfined compression test was conducted on a specimen of silty clay, it showed a strength of 160 kN/m². Determine the shear parameters of the soil, if the angle made by the failure plane with the axis of specimen was 32°.

12. A saturated clay has a compression index, C_c of 0.18. Its void ratio at a stress of 100 kN/m² is 1.62 and its permeability is 2×10^{-9} m/s. Compute

    (a) Change in void ratio if the stress is increased to 200 kN/m²,

    (b) Settlement of the clay deposit of 1.0 m thickness considering the increase in stress at the centre of clay layer due to superimposed load same as the effective overburden pressure, say 100 kN/m², at that location, and

(c) Time required for 50% consolidation to occur if drainage to one way and time factor is 0.196 for 50% consolidation.

13. Representative samples were obtained from a layer of clay 6.0 m thick, located between the two layers of sand. By means of consolidation tests, it was found that the average value of coefficient of consolidation of these samples was 4×10^{-8} m^2/s. By constructing a building at the site, its gradual settlement was observed. Within how many days did half the settlement occur ?

Soil Exploration

3.1 INTRODUCTION

The field and laboratory investigations required to obtain the physical and engineering properties, and the arrangement of the underlying materials are termed as soil exploration. It is a prerequisite to the economical and safe design of substructure elements. Interpretation of the data obtained through soil exploration generally provides information to (*i*) the type of foundation, (*ii*) safe load capacity of the foundation, (*iii*) settlement of foundation, (*iv*) location of the groundwater level, (*v*) causes of settlements, tilts and cracks in existing structures, and (*vi*) identification and solution of excavation problems.

Steps involved in the planning and execution of a subsurface exploration programme are illustrated in a flowchart (Fig. 3.1). If the site is located in a seismically active region and/or the foundations are subjected to vibratory loads, then liquefaction studies, block vibration and wave propagation tests are carried out in addition to the tests mentioned in the flowchart.

In the following articles, the various procedures of soil exploration are discussed briefly.

3.2 OPEN PITS WITH SAMPLING

Examination of the open pits is the cheapest method of exploration to shallow depths. The pits should be excavated only up to a depth so that the sides of the pit remain stable without the slightest risk of collapse. Although this depth depends on the type of soil, usually it is limited to 3–4 m only. As excavations below the water table are difficult, open pits may not be suitable in such conditions.

Open pits give a clear picture of the soil stratification and enable us to take undisturbed samples of bigger sizes conveniently.

3.3 METHODS OF BORING

Sinking and advancing boreholes is termed as boring. Borings are generally used for investigating the soil strata for a depth greater than 3.0 m or when difficult groundwater conditions are met. The diameter of the borehole usually varies from 50 mm to 250 mm, basically depending on the types of investigation, size of the required samples and the type of available equipment. In case the sides of the hole cannot be left unsupported, the sides are prevented from collapsing by means of casing pipe or by filling the hole with a drilling mud, a viscous suspension of bentonite in water. The common methods of boring are: (*i*) Auger

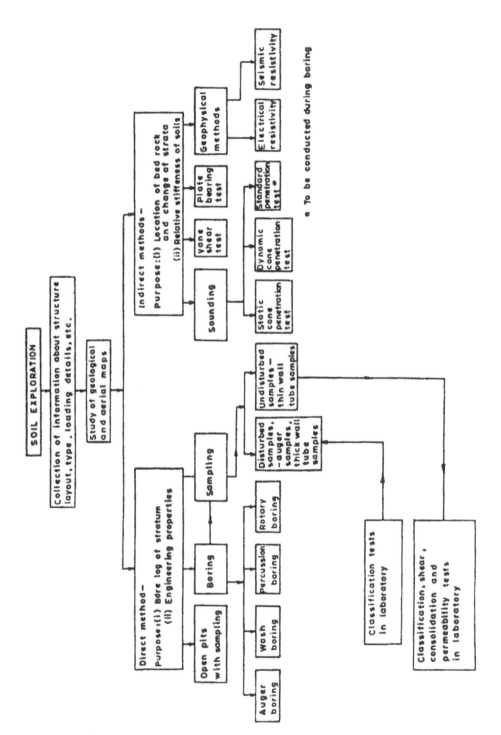

Fig. 3.1. Flowchart Illustrating the Steps Involved in a Soil Exploration Programme.

boring; (*ii*) wash boring; (*iii*) rotary boring; and (*iv*) percussion boring. The suitability of any particular method of boring depends mainly on the nature of soils and the position of the water table (Table 3.1).

Table 3.1. Methods of Boring.

Boring Method	Type of Soil	Observations for Change in Material	Sample Obtained	Utility in Soil Exploration
Auger boring	Partially saturated sands, silts and medium to stiff cohesive soils	Soil shavings removed with the auger	Representative sample, good for identification and classification tests	Reconnaissance or detailed exploration in case of shallow exploration depths
Wash boring	Practically all types of soils except hard and cemented soils or rock	Cutting settled from wash water or drilling fluid	Nonrepresentative sample, inadequate for any useful tests	Reconnaissance and special exploration including ground-water observation
Percurssion boring	All types of soils and rocks. Difficult in loose sands and soft sticky clays	—	Nonrepresentative sample, inadequate for any useful tests	Penetrating gravelly bouldery and rock deposits, useful with auger or wash boring
Rotary boring	All types of soils and rocks except in stony or porous soils and fissured rocks	—	Cores are obtained which are useful for classification and other tests	Detailed and special explorations primarily in rocks

Auger Boring

Boring is usually started with augering. Figure 3.2 shows two types of augers commonly used in practice; size of augers varies from 50 mm to 200 mm. The auger is rotated and pressed down into the soil by means of a T-handle on the upper rod. Once the annular space between the blades is filled up with soil, the auger is withdrawn and cleaned. The cleaned auger is again inserted and the process repeated. As the hole progresses downwards, extension rods are added to the auger.

Hand operated augers are used up to a maximum depth of 10 m. These are generally suitable for all types of soils above water table and below water table in case of clay soils. For deeper exploration, power operated augers can be used. They are generally of the short flight or continuous flight screw type augers (Fig. 3.3) and may be used for boring up to 30 m.

Use of bailer with augering is very common. A bailer is a heavy cutting pipe with a cutting edge (Fig. 3.4). The length and weight of a bailer varies according to requirements. Raising and dropping of bailer in a hole cuts the soil which is pushed into the tube. It is then taken out and emptied. Boring with bailer is proceeded when augering is found difficult.

Soil samples obtained by augers/bailers are severely disturbed and, therefore, used for identification purposes only.

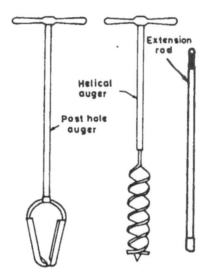

Fig. 3.2. Hand Augers.

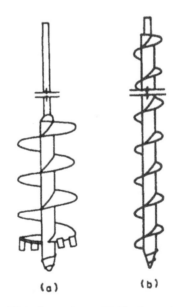

Fig. 3.3 (a). Short Flight Screw Auger; (b) Continuous Flight Screw Auger.

Wash Boring

Wash boring is a commonly used method of boring and can be conveniently used even below water table for all types of soils except hard soils and rock. Firstly, the hole is advanced a short depth by auger and then a casing pipe is pushed to prevent the sides from caving in. In wash boring (Fig. 3.5), water is forced under pressure through an inner tube

Fig. 3.4. Bailer.

which may be rotated or moved up and down inside the casing pipe. The lower end of the tube, fixed with a sharp edge or a tool, cuts the soil which will be floated up through the casing pipe around the tube. The slurry flowing out gives an indication of the soil type. In this method heavier particles of different soil layers remain under suspension in the casing pipe and get mixed up, and hence this method is not suitable for obtaining samples even for soil classification. Therefore, samples of the soil should be obtained through suitable samplers after the borehole has been cleaned.

Percussion Boring

Percussion boring is usually adopted in hard soils or soft rocks where auger boring or wash boring cannot be effectively used. This is the only method suitable for drilling boreholes in bouldery and gravelly strata. In this method, a heavy drilling bit is alternately raised and dropped inside a casing pipe which is driven as for wash boring. Above the water table, the borehole is usually kept dry, except for a limited quantity of water used to form the slurry of pulverized material. The pulverized slurry is bailed out using a bailer or sand pump. A casing pipe may not be necessary if the sides of the hole do not cave in.

Rotary Boring

Rotary boring primarily used for rocks, however, can conveniently be used also in case of sands and clays. It is a fast method of boring and 35 to 60 mm diameter holes can be drilled to great depths.

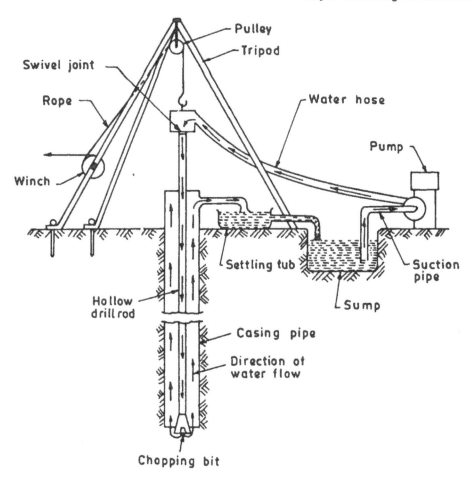

Fig. 3.5. Wash Boring Set Up.

In this method, boring is effected by the cutting action of a rotating bit which should be kept in firm contact with the bottom of the hole. The bit is carried at the end of hollow, jointed drill rods which are rotated by a suitable chuck. A mud-laden fluid or grout is pumped continuously down the hollow drill rods and the side of the hole, and so the protective casing may not be generally necessary. In this method, cores may be obtained by the use of coring tools.

Rotary boring is not suitable in soils containing high percentage of gravel as they tend to rotate under the bit and are not broken up.

3.4 AMOUNT OF BORING

The amount of boring is mainly governed by two quantities, namely, (i) spacing (or number) of boreholes, and (ii) depth of boreholes. They depend on the type of the proposed structure, soil profile, and the properties of the soil that constitutes each individual stratum. In general, the planning of borings should be such that adequate information is available which helps in planning of other *in situ* tests, selection of type and depth of foundation.

Depth of Boreholes

As a general rule, unless bed-rock is encountered, boring should be carried to such a depth that net increase in soil stress, resulting from superimposed loads, is less than 10 per cent of average contact pressure or less than five per cent of the effective stress in the soil at that depth due to over-burden (Task Committee Report, 1972).

IS: 1892–1979 recommended the depth of exploration as shown in Table 3.2.

Tomlinson (1986) has illustrated the probable depth of investigation as shown in Fig. 3.7.

Table 3.2. Depth of Exploration.

S. No.	Type of Foundation	Depth of Exploration
(i)	Isolated spread footing or raft	One and a half times the width of footing, B (Fig. 3.6)
(ii)	Adjacent footings with clear spacing less than twice the width	One and a half times the length of the footing, L (Fig. 3.6)
(iii)	Adjacent rows of footings	Refer Fig. 3.6
(iv)	Pile and well foundations	To a depth of one and a half times the width of structure from the bearing level (toe of pile or bottom of well)
(v)	1. Road cuts	Equal to the bottom width of the cut
	2. Fill	Two metres below ground level or equal to the height of the fill whichever is greater

Spacing and Number of Boreholes

The spacing of borings should be such as to reveal any major changes in thickness, depth or properties of the strata over the base area of the structure and its immediate vicinity.

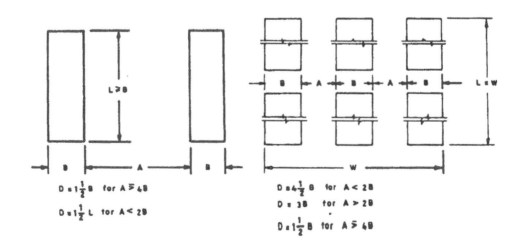

Fig. 3.6. Depth of Exploration.

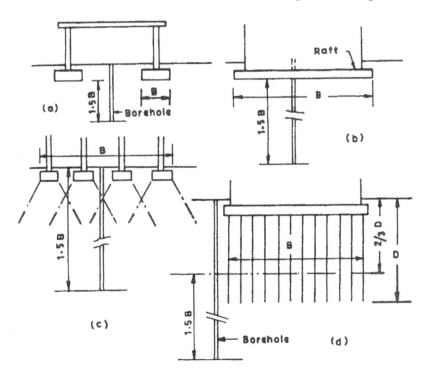

Fig. 3.7. Depths of Boreholes for Various Foundation Conditions (After Tomlinson, 1986).

No clear-cut criteria are available to give directly the number and spacing of borings. At least three borings should be made on a project site where the surface is level, especially if the site is somewhat even. Four to five borings are sufficient to determine if the soil strata is erratic. According to IS: 1892–1972, for a building site covering an area of 10,000 sq metres, one borehole in each corner and one in centre *i.e.* five boreholes in all should be adequate.

Table 3.3 provides some guidelines for the selection of spacing and number of borings for different engineering structures.

Table 3.3. Suggested Number and Spacing of Borings.

Project	Distance between Borings (m) Horizontal Stratification of Soil			Minimum number of borings
	Uniform	Average	Erratic	
High-rise buildings	45	30	15	5
Single or double-storey buildings	60	30	15	3
Bridge abutments, piers, television towers etc.	45	30	7.5	1–2 for each foundation unit
Highways	300	150	50	—
Borrow pits (for compacted fill)	300–150	150–160	45–30	—

3.5 LOCATION OF WATER TABLE

Proper location of groundwater table is essential for carrying out the analysis, design and methodology of construction of foundations. It can be estimated through observations of open wells at the site or in the vicinity. Boreholes can also be used for recording groundwater levels (Hvorslev, 1949). IS: 2132–1972 specifies that in case of cased boreholes, measurement of water level should be taken before and after insertion of casing. In sandy soils, the level should be measured as the casing is pulled and then measured again at least 30 minutes later. In silty and clayey soils, the level should be measured at least 24 hours after the casing is pulled. In case drilling mud is used, casing perforated at the lower end should be lowered into the hole. The hole is then bailed down until all traces of drilling mud are removed from inside the casing. Groundwater levels should then be determined at time intervals of 30 minutes up to 24 hours.

Another method of measuring water table is to fill the hole and bail it out. After bailing a quantity, observe if the water level in the hole is rising or falling. The true water level is between the bailed point where the water was falling and the bailed point where it was rising.

3.6 SOIL SAMPLING

There are two main types of soil samples which can be recovered from boreholes or trial pits:

(a) Disturbed samples are those in which the original natural structure of the soil is completely altered during the sampling operation. If, in addition to the change in the original soil structure, the soils from other layers are mixed up, the samples are called non-representative samples. Such samples are virtually of no use. Soil samples obtained from auger cutting and from sedimentation of water in wash borings belong to this category. On the other hand, if the samples are collected from a particular depth without altering the mineral constituents of strata at that depth, though the original structure has been disturbed, they are known as representative samples. Such samples are suitable for the performance of classification tests, and unsuitable for getting the mechanical properties of the soil. Samples obtained from the split spoon sampler during penetration tests fall under this category.

(b) Undisturbed samples are those in which the material has been subjected to such a small disturbance that it is suitable for all laboratory tests including shear strength and consolidation tests. These are usually obtained by driving a thin walled tube into the soil.

3.7 SAMPLING TOOLS

The fundamental requirement of a sampling tool is that on being forced into the ground it should cause as little displacement, remoulding and disturbance as possible. The extent of disturbance to the samples due to the sampler depends on three features of its design: (i) cutting edge, (ii) inside wall friction, and (iii) non-return valve.

A typical cutting edge is shown in Fig. 3.8. The important factors which affect the disturbance of a sample are:

$$\text{Inside clearance,} \qquad C_i = \frac{D_{si} - D_{ci}}{D_{ci}} 100(\%) \qquad \qquad \dots (3.1)$$

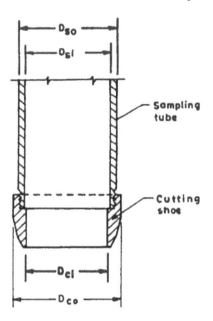

Fig. 3.8. Detail of Cutting Edge.

Outside clearance, $\qquad\qquad C_0 = \dfrac{D_{co} - D_{so}}{D_{so}} 100(\%)$ $\qquad\qquad$ ... (3.2)

Area ratio, $\qquad\qquad A_r = \dfrac{D_{so}^2 - D_{si}^2}{D_{si}^2} 100(\%)$ $\qquad\qquad$ ... (3.3)

The inside clearance allows elastic expansion of the soil as it enters the tube, reduces frictional drag on the sample from the wall of the tube and helps to retain the core. The outside clearance facilitates the withdrawal of the samples from the ground.

As per IS: 1892–1979 :

(i) C_i should be between 1% and 3%.

(ii) C_0 should not be much greater than C_i.

(iii) A_r should be kept as low as possible consistent with strength requirements of the sampling tube.

For stiff formation, $A_r \ngtr 20\%$, and

For soft sensitive clays, $A_r \ngtr 10\%$.

(iv) For a satisfactory undisturbed sample, the recovery ratio calculated as follows should be between 96 and 98 per cent

$$L_r = \frac{L}{H} 100(\%) \qquad\qquad ... (3.4)$$

where L_r = recovery ratio,

L = length of sample within the tube, and

H = depth of penetration of sampling tube.

(*v*) Wall friction can be reduced by: (*a*) suitable inside clearance, (*b*) a smooth finish to the sample tube, and (*c*) oiling the tube properly.

(*vi*) The non-return valve should have a large orifice to allow air and water to escape quickly and easily when driving the sampler.

3.8 TYPES OF SAMPLERS

The more common types of samplers can be divided into three groups: (*i*) open driven samplers, (*ii*) piston samplers, and (*iii*) rotary samplers.

Open Driven Samplers

An open driven sampler is an ordinary seamless steel tube with its lower edge chamfered to make penetration easy. Depending on the requirement of sampling, the sampler may be provided with a separate cutting shoe. The sampler head which connects the tube is provided with vents to permit water to escape when sampling underwater and a check valve to help to retain the sample while withdrawing the sampler. These samplers are either thick wall or thin wall type.

Thick walled samplers are used for obtaining disturbed representative samples and are in the form of a solid tube or a split tube with or without a liner. In Fig. 3.9, assembly of a split spoon sampler is shown which is used in the standard penetration test (IS: 9640–1980).

Thin walled samplers are used for obtaining undisturbed samples. The tubes are cold-drawn seamless tube made out of brass, aluminium or any other suitable material having adequate strength, durability and resistance to corrosion. In Fig. 3.10, a typical thin-walled tube sampler is shown which should satisfy the requirements given in Table 3.4 (IS: 2132–1986).

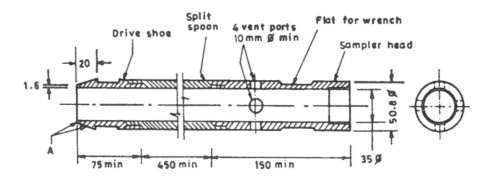

(All dimensions in mm)

Fig. 3.9. Standard Split Spoon Sampler Assembly.

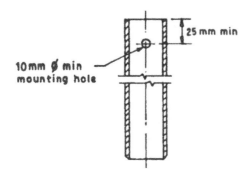

Fig. 3.10. Thin-walled Tube Sampler.

Table 3.4. Requirements of Sampling Tubes.

Inside diameter, mm	38	70	100
Outside diameter, mm	40	74	106
Minimum effective length (that is length available for soil sample), mm	300	450	450

Note 1—The inside and outside diameters specified above give area ratios of 10.9, 11.8 and 12.4 for the 38, 70 and 100 mm sampling tubes respectively.

Note 2—The three diameters recommended in Table 3.4 are indicated for purposes of reducing the number of sizes and fittings to be inventoried. Sampling tubes of intermediate or larger diameters may be used with the permission of the soil engineer-in-charge. Lengths of tubes shown are illustrative. Proper length may be determined to suit field conditions.

Little changes made in the requirements of thin wall tube sampler are given in IS: 2132–1986 and IS: 11594–1985.

Piston Sampler

A piston sampler consists of two separate parts: (*i*) the sampler tube, and (*ii*) the piston system; the latter which is actuated separately, fits tightly in the sampler cylinder (Fig. 3.11). A locking cone provided in the head prevents the piston from moving downwards. At the proposed sampling depth, the piston is fixed in relation to the ground and the sampler cylinder is forced into the soil independently, cutting a sample out of the soil. The piston helps in three ways to retain the sample as it (*a*) prevents the soft soils from squeezing rapidly into the tube, (*b*) increases the recovery length of samples by creating a vacuum that tends to retain the sample, and (*c*) prevents water pressure from acting on the top of the sample.

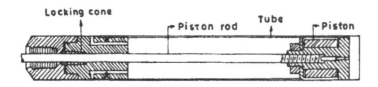

Fig. 3.11. Stationary Piston Sampler.

Piston sampling is relatively a costly procedure and is employed for getting undisturbed samples in saturated sands and other soft and wet soils where open driven thin wall samplers cannot be satisfactorily used.

Rotary Sampler

A rotary sample is a double-walled tube sampler and is commonly used with rotary drilling. The outer tube or the rotating barrel is provided with a cutting bit which cuts an annular ring when the barrel is rotated. The inner tube which is stationary, is provided with a smooth cutting shoe. The inner tube slides over the cylindrical sample cut by the outer rotating barrel. The sample is collected in the inner liner. A spring core catcher is sometimes fitted with the inner tube to retain the sample.

Sampling with a rotary sampler is expensive but a very practical method of continuous core in stiff to hard clays and particularly in rocks. A better estimation of *in situ* rock quality is obtained by a modified core recovery ratio known as rock quality designation, R.Q.D. (Peck et al., 1974). It is defined as

$$R.Q.D. = \frac{L_i}{L_t} \qquad ...(3.5)$$

where L_i = total length of intact hard and sound pieces of core of length greater than 100 mm, and

L_t = total length of drilling.

Breaks caused by drilling should be ignored. The diameter of the core should preferably not be less than 54 mm. Table 3.5 gives the classification of rock quality based on R.Q.D. (Deere, 1964).

Table 3.5. Relation of R.Q.D. and *in situ* Rock Quality (Deere, 1964).

R.Q.D. %	Rock Quality
90–100	Excellent
75–90	Good
50–75	Fair
25–50	Poor
0–25	Very poor

3.9 STANDARD PENETRATION TEST (SPT)

The standard penetration test (SPT) is the most extensively used *in situ* test in India and many other countries. This test is carried in a borehole using a split spoon sampler (Fig. 3.9). As per IS: 2131–1981, steps involved in carrying out this test are as follows:

(i) The borehole is advanced to the depth at which the SPT has to be performed. The bottom of the borehole is cleaned.

(ii) The split spoon sampler attached to standard drill rods of required length is lowered into the borehole and rested at the bottom.

(*iii*) The split spoon sampler is seated 150 mm by blows of a drop hammer of 65 kg falling vertically and freely from a height of 750 mm. Thereafter, the split spoon sampler shall be further driven 300 mm in two steps each of 150 mm. The number of blows required to effect each 150 mm of penetration shall be recorded. The first 150 mm of drive may be considered to be seating drive. The total blows required for the second and third 150 mm of penetration is termed the penetration resistance N.

If the split spoon sampler is driven less than 450 mm (total), then N-value shall be for the last 300 mm penetration. In case, the total penetration is less than 300 mm for 50 blows, it is entered as refusal in the borelog.

(*iv*) The split spoon sampler is then withdrawn and is detached from the drill rods. The split barrel is disconnected from the cutting shoe and the coupling. The soil sample collected inside the barrel is collected carefully and preserved for transporting the same to the laboratory for further tests.

(*v*) Standard penetration tests shall be conducted at every change in stratum or intervals of not more than 1.5 m, whichever is less. Tests may be done at lesser intervals (usually 0.75 m) if specified or considered necessary.

The penetration test in gravelly soils requires careful interpretation since pushing a piece of gravel can greatly change the blowcount.

Corrections to Observed SPT Values (*N*) in Cohesionless Soils

Following two types of corrections are normally applied to the observed SPT values (N) in cohesionless soils:

Correction due to dilatancy

In very fine, or silty, saturated sand, Terzaghi and Peck (1967) recommended that the observed N-values be corrected to N' if N is greater than 15 as

$$N' = 15 + \frac{1}{2}(N - 15) \qquad \text{... (3.6)}$$

Bazaraa (1967) recommended the correction as

$$N' = 0.6N \text{ (for } N > 15) \qquad \text{... (3.7)}$$

This correction is introduced with the view that in saturated dense sand ($N > 15$), the fast rate of application of shear, through the blows of drop hammer, is likely to induce negative pore pressures and thus temporary increase in shear strength will occur. This will lead to an N-value higher than the actual one. Since sufficient experimental evidence is not available to confirm this correction, many engineers are not applying this correction. However, this correction has also been recommended in IS: 2131–1981.

Correction due to overburden pressure

On the basis of field tests, corrections to the N-value for overburden effects were proposed by many investigators (Gibbs and Holtz, 1957; Teng, 1977; Bazaraa, 1967; Peck, Hanson and Thornburn, 1974). The methods normally used are:

Bazaraa (1967)

For $\bar{\sigma}_0 < 75$ kPa

$$N' = \frac{4N}{1 + 0.04\,\bar{\sigma}_0} \qquad \qquad \text{... (3.8)}$$

For $\bar{\sigma}_0 > 75$ kPa

$$N' = \frac{4N}{3.25 + 0.01\,\bar{\sigma}_0} \qquad \qquad \text{... (3.9)}$$

where $\bar{\sigma}_0$ = effective overburden pressure, kPa.

Peck, Hanson and Thornburn (1974)

$$N' = 0.77 \log_{10} \frac{2000}{\bar{\sigma}_0} \cdot N \qquad \qquad \text{... (3.10)}$$

Figure 3.12 gives the correction factor based on Eq. (3.10). Use of this figure has been recommended in IS: 2131–1981. In this figure,

$$C_N = \text{Correction factor} = 0.77 \log_{10} \frac{2000}{\bar{\sigma}_0} \qquad \qquad \text{... (3.11)}$$

There is a controversy whether the correction due to dilatancy should be applied first and then the correction due to overburden pressure or vice versa. However, in IS: 2131–1981, it is recommended that the correction due to overburden should be applied first.

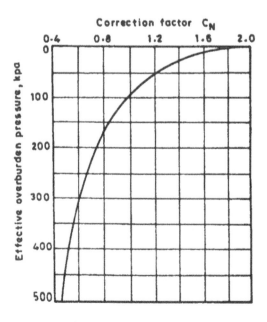

Fig. 3.12. Chart for Correction of N-Values in Sand for Influence of Overburden Pressure (Peck et al., 1967).

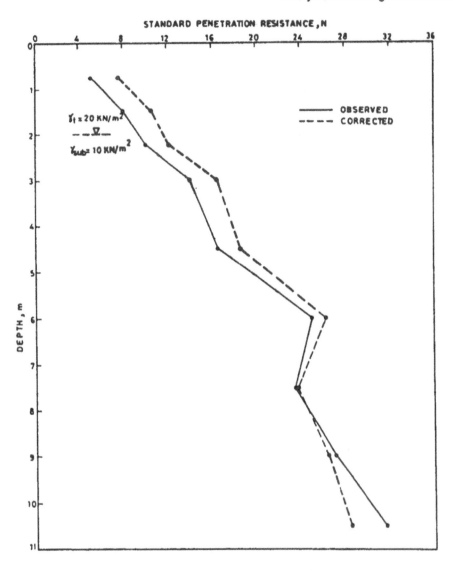

Fig. 3.13. Standard Penetration Resistance 'N' Versus Depth.

A typical set of observed N-values are shown in Fig. 3.13. Corrected N-values as per IS Code recommendations are also shown in the figure.

Empirical Correlations

Empirical correlations between correced value of N and various properties of soils have been obtained in cohesionless soils. Such correlations have already been described in Table 2.3 of Chapter 2.

The SPT was originally developed for cohesionless soils so that undisturbed samples would not have to be taken as it is a difficult and costly procedure in such soils. The test has evolved the current practice of determining N for all soils. Peck, Hanson and Thornburn (1974) have suggested correlation between N-value and consistency of saturated cohesive soils as given in Table 3.6.

Table 3.6. Relation between N and q_u (Peck et al., 1974).

Consistency	N	q_u, kPa
Very soft	0–2	< 25
Soft	2–4	25–50
Medium	4–8	50–100
Stiff	8–15	100–200
Very Stiff	15–30	200–400
Hard	> 30	> 400

q_u = Unconfined compressive strength

Sanglerat (1972) has proposed following correlations:

$$\text{For clay,} \qquad q_u = 25\ N \ (\text{kPa}) \qquad \qquad \dots (3.12a)$$

$$\text{For silty clay,} \quad q_u = 20\ N \ (\text{kPa}) \qquad \qquad \dots (3.12b)$$

$$\text{For silty sand,} \quad q_u = 13\ N \ (\text{kPa}) \qquad \qquad \dots (3.12c)$$

Nixon (1982) suggested

$$\text{For clay,} \qquad q_u = 24\ N \ (\text{kPa}) \qquad \qquad \dots (3.13)$$

and Tomlinson (1986) gave

$$\text{For clay,} \qquad q_u = 25\ N \ (\text{kPa}) \qquad \qquad \dots (3.14)$$

$$\text{For silty clay,} \quad q_u = 20\ N \ (\text{kPa}) \qquad \qquad \dots (3.15)$$

It may be noted that no correction has to be made in N-values taken in cohesive soils.

3.10 DYNAMIC CONE PENETRATION TEST (DCPT)

The dynamic cone penetration is performed in the same way as SPT except that there is no borehole for DCPT. This test is performed in two ways: (i) without bentonite slurry, and (ii) with bentonite slurry.

Without Bentonite Slurry

In this test, a 50-mm diameter 60° cone (Fig. 3.14) fitted to the driving rod (A rod) through an adopter is driven into the soil by blows of 65 kg hammer falling freely from a height of 750 mm (IS: 4968–1980, Part I). Assembly of test equipment for DCPT is shown in Fig. 3.15. The blow count for every 100 mm penetration of the cone is continuously recorded. The cone is driven to the required depth or refusal. The drill rods are withdrawn leaving the cone behind in the ground. The number of blows required for 300 mm penetration is termed as the dynamic cone resistance, N_{cd}. The test gives a continuous record of N_{cd} with depth (Fig. 3.16). In this test no sample can be obtained.

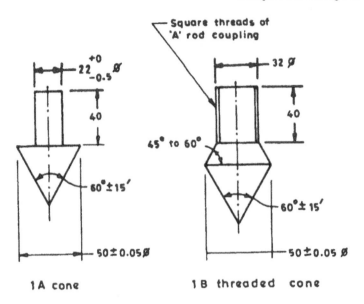

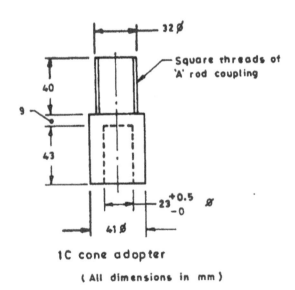

1C cone adopter

(All dimensions in mm)

Fig. 3.14. Cone and Cone Adopter.

Some approximate correlations between N_{cd} and N, applicable for medium to fine sands, are as under:

$$N_{cd} = 1.5 \, N \text{ for depths} < 3 \text{ m} \qquad \qquad ... (3.16a)$$

$$N_{cd} = 1.75 \, N \text{ for depths from 3 m to 6 m} \qquad ... (3.16b)$$

$$N_{cd} = 2.0 \, N \text{ for depths} > 6 \text{ m} \qquad \qquad ... (3.16c)$$

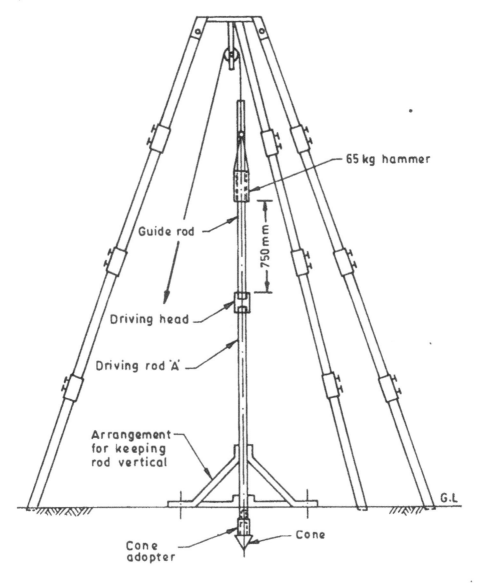

Fig. 3.15. Typical Assembly for Cone Penetration Test.

With Bentonite Slurry

A DCPT with bentonite slurry is conducted to eliminate the surface frictional resistance on the drill rods. In this test, a 62.5 mm diameter 60° cone (Fig. 3.17) is driven in the ground with the arrangement for the drilling mud to flow through the cone (IS: 4968–1980, Part II). The number of blows required for 300 mm penetration of 62.5 mm diameter cone is denoted by N_{cbr}.

The use of bentonite slurry may not be necessary when investigation is up to a depth of 6 m only.

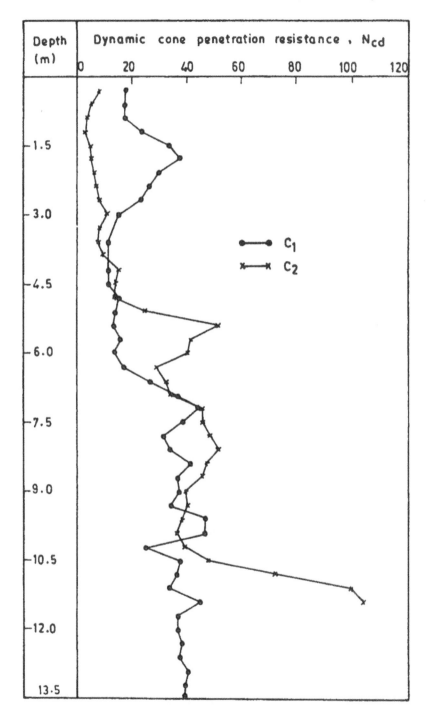

Fig. 3.16. Dynamic Cone Penetration Resistance (N_{cd}) Versus Depth.

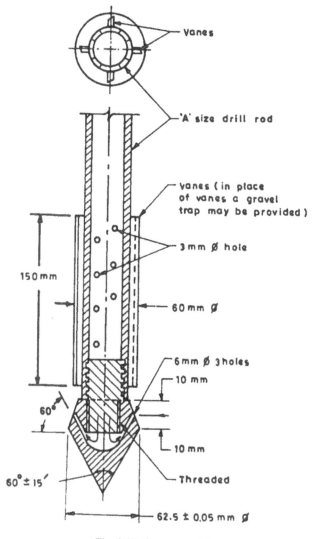

Fig. 3.17. Cone Assembly.

The Central Building Research Institute, Roorkee has suggested the following correlations between N_{cd} and N, and N_{cbr} and N:

(a) Without bentonite slurry

$$N_{cd} = 1.5 \, N \text{ ... upto a depth of 4 m} \qquad \text{... (3.17a)}$$

$$N_{cd} = 1.75 \, N \text{ ... for depths of 4 m to 9 m} \qquad \text{... (3.17b)}$$

$$N_{cd} = 2.0 \, N \text{ ... for depths greater than 9 m} \qquad \text{... (3.17c)}$$

(b) With bentonite slurry

$$N_{cbr} = N \qquad \text{... (3.17d)}$$

Advantages: The DCPT is relatively a quicker and economical test. It may be used to cover a large area under investigation. As it provides a continuous record of the penetration resistance, it is useful in locating the soft pockets and in general the variability of the soil profile. It can also establish the position of rock stratum.

3.11 STATIC CONE PENETRATION TEST (SCPT)

This test consists of pushing a cone into the ground at a steady rate of 20 mm/s either manually or using some power mechanism. The cone used in this test has a base area of 1000 mm^2 and an apex angle of 60°. The cone is screwed to the sounding rod which passes through hollow casing tube of 36 mm external diameter. The resistance, the soil offers to 100 mm penetration of the cone, is termed as static cone penetration resistance, q_c. Firstly, the penetration resistance of cone alone is recorded. Next, the sleeve around the drill rod is pushed down to the level of the cone and both the cone as well as sleeve are then pushed together into the soil to a depth of 100 mm and the combined resistance is recorded. If

Q_c = Total force required to push the cone alone to a distance of 100 mm in the ground

Q_t = Total force required to push the cone and the friction jacket together to a distance of 100 mm in the ground

A_c = Base area of cone

A_f = Surface area of friction jacket

f_c = Local side friction

then

$$q_c = \frac{Q_c}{A_c}$$

... (3.18)

$$f_c = \frac{Q_t - Q_c}{A_f}$$

... (3.19)

In this test also, no sample can be obtained. This test gives a continuous record of variation of both cone resistance and friction resistance (Fig. 3.18).

Empirical Correlations

Empirical correlations between q_c and various soil properties have been made in cohesionless soils. Such correlations have already been described in Table 2.3 of Chapter 2.

Robertson and Campanella (1983) have developed the relationship between q_c and ϕ as a function of effective overburden pressure (Fig. 3.19).

Table 3.7 Approximate Relationships between Cone Point Resistance q_c (kPa) and SPT Value of N and Static Stress-Strain Modulus E_s (kPa) (Bowles, 1982).

Soil Type	q_c/N	E_s, kPa
Silts, fine sands, slightly cohesive soils	150–300	1.5–2 q_c
Fine to medium sands, slightly silty fine to medium sands	300–450	2–4 q_c
Coarse sands	450–700	1.5–3 q_c
Sandy gravel, gravelly sands, stiff clay, sandy clay	700–2000	5–7 q_c

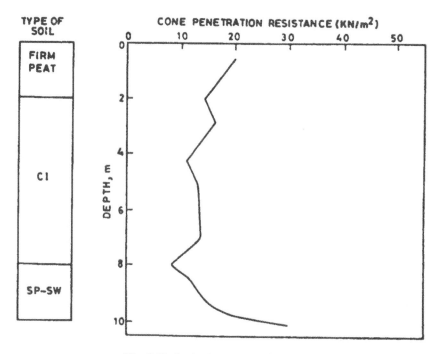

Fig. 3.18. Static Cone Penetration Record.

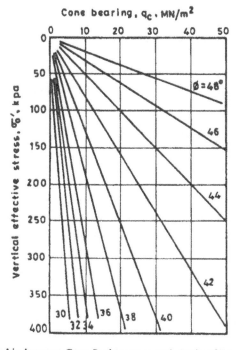

Fig. 3.19. Relationship between Cone Resistance q_c and Angle of Internal Friction ϕ for Uncemented Quartz Sands (Robertson and Campanella, 1983).

Based on the test data of Trofimenkov (1974), Bowles (1982) has suggested the correlations of q_c and N, and q_c and static stress-strain modulus E_s as given in Table 3.7.

Lunne and Kleven (1981) have given following relation between q_c and undrained shear strength of saturated clay, c_u:

$$q_c = N_k c_u + \sigma_0 \qquad \qquad \text{... (3.20)}$$

where N_k = cone factor, and

σ_0 = total overburden pressure.

Values of N_k are recommended as given in Table 3.8.

Table 3.8. Value of Cone Factor, N_k (Lunne and Kleven, 1981).

Type of Clay	Cone factor, N_k
Normally consolidated clay	11 to 19
Over consolidated clay	
at shallow depths	15 to 20
at deeper depths	12 to 18

3.12 PLATE LOAD TEST

Plate load test is a common method of determining bearing capacity and settlement of shallow foundations. The test procedure is explained in IS: 1888–1982. The details of this test with interpretation of data have been given in Sec. 7.8 of Chapter 7 on shallow foundations.

3.13 FIELD VANE SHEAR TEST

Field vane shear test is normally used to determine the undrained shear strength and sensitivity of soft sensitive clays. The details of this test have been described in Sec. 2.20 of Chapter 2.

3.14 LARGE SHEAR BOX TEST

The shear strength characteristics of boulder deposits cannot be investigated in the laboratory in direct shear and triaxial set-ups because of their small sizes and also difficulty of sampling in such soils. Hence a large scale shear box has been designed to suit field conditions (Prakash and Ranjan, 1975). Figure 3.20 shows a typical set-up.

A field shear box with sample size 706 mm × 706 mm (giving an area of 5×10^5 mm^2) was used at the site of a telephone exchange building in Nainital. The normal stresses provided in the form of kentledge were 75 kN/m^2 and 150 kN/m^2 respectively on the two *in situ* samples (Block 1 and Block 2, Fig. 3.20), prepared in the excavation pit of depth 1.0 m. Shearing stress on the two samples were applied through a remote control hydraulic jack and proving ring, with reaction being mutually provided by the two shear boxes. Readings of shear displacement were taken through suitably mounted dial gauges. After the required normal load was placed on the two *in situ* soil samples, the required shear stress was applied by operating the jack and maintained constant till the shear displacement readings got

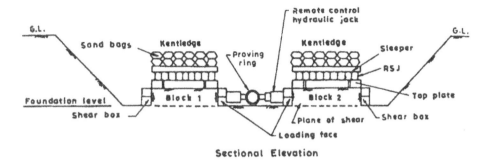

Sectional Elevation

Fig. 3.20. Test Set Up for Large Shear Box Test.

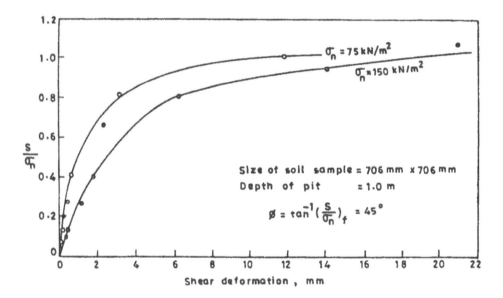

Fig. 3.21. Shear Deformation Versus s/σ_n Plot.

stabilized. The readings of shear displacement dial gauges were taken and then the next increment of shear stress was applied. The process was repeated till the soil sample with smaller normal stress failed in shear along a horizontal shear plane (at the base level of test pit). The space between the failed sample box and the vertical side of the trench was filled by suitable bracing blocks and, thereafter, the test was resumed till the soil sample carrying the large normal stress also failed.

The test data plotted in the form of shear displacement versus s/σ_n is shown in Fig. 3.21.

3.15 FIELD PERMEABILITY TEST

The permeability of soil strata in the field is obtained by pumping out and borehole tests. These tests have been described in Sec. 2.5 of Chapter 2.

3.16 GEOPHYSICAL METHODS

Geophysical methods consist in identifying the stratification in soils and rocks by measuring changes in certain physical characteristics of sub-surface deposits. These methods may be used to explore large areas much more rapidly and economically than is possible by borings. However, the utility of these methods in the design of foundations is very limited since the methods do not quantify the characteristics of the subsoil layers including the position of the water table. The results obtained from these tests must be checked from other methods of exploration.

The two geophysical methods suitable for soil exploration for civil engineering purposes are the seismic and electrical resistivity methods.

Seismic Method

This method is based on the fact that the seismic waves propagate through different types of soils (or rocks) at different velocities and are also refracted when they cross the boundary between two different types of soils. The velocity is faster in denser and more consolidated materials. IS: 1892–1979 specified the probable range of wave velocities in different materials as listed in Table 3.9.

Table 3.9. Wave Velocities in Different Materials.

Material	Wave Velocity (m/s)
Sand and top soil	180–365
Sandy clay	365–580
Gravel	490–790
Glacial till	550–2135
Rock talus	400–760
Water in loose materials	1400–1830
Shale	790–3350
Sandstone	915–2740
Granite	3050–6100
Limestone	1830–6100

The seismic refraction method consists of inducing impact or shock waves by exploding an explosive charge at or near the ground surface. The shock waves are recorded by a device called geophone which records the time of travel of waves. The geophones are installed at suitable known distances on the ground in a line from the source. Out of the three types of the waves (compression, shear and surface waves), usually compression waves are recorded.

Figure 3.22 shows the test results for a three-layer system. G_1, G_2 ..... G_8 represent the locations of the geophones. The last geophone is placed at a distance of about 3 to 5 times of $(h_1 + h_2)$. Following points may be referred from this figure:

(i) Direct wave reaches to the geophones if distance of geophone from the source is less than d_1. Time-distance relationship is represented by the line AB, the slope of which gives $1/v_1$, v_1 being the velocity of compression waves of the first layer.

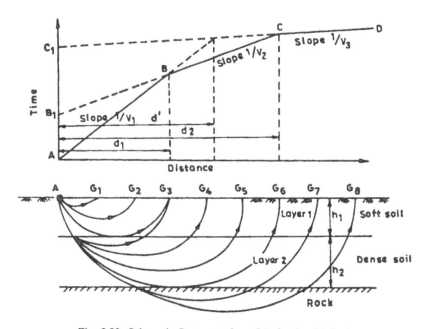

Fig. 3.22. Schematic Representation of Refraction Method.

(*ii*) If the distance of geophone from the source is more than d_1 the waves refracted from the second layer are picked by the distant geophone. If the second layer is denser than the first, time relationship will be as indicated by the line BC. The slope of the line is $1/v_2$, v_2 being the velocity of wave of second layer.

(*iii*) If the distance of geophone from the source is more than d_2, the waves refracted from rock will be picked up by the geophone and the time-distance relationship will be as given by the line CD. The slope of the line is $1/v_3$, v_3 being the velocity of wave in rock.

The thickness of different layers may be obtained using following formulae:

$$h_1 = \frac{d_1}{2}\sqrt{\frac{v_2 - v_1}{v_2 + v_1}} \qquad \text{... (3.21)}$$

and

$$h_1 + h_2 = \frac{d'}{2}\sqrt{\frac{v_3 - v_1}{v_3 + v_1}} \qquad \text{... (3.22)}$$

Seismic refraction method is applicable only when the velocity of wave propagation for each succeeding stratum increases with depth.

Electrical Resistivity Method

This method is based on the principle that the electrical resistance, offered by different types of soils to the flow of current, is different. The resistivity of a soil depends primarily on moisture content and the concentration of dissolved salts. It will decrease with the increase in both moisture and salt contents. Representative values of resistivity are given in Table 3.10 (IS: 1892–1979).

The test is carried out by driving four electrodes into the ground at equal spacing along a straight line. A direct current is imposed between the two electrodes E_1 and E_2 and the same is recorded through a milliammeter. The potential difference between the two inner electrodes, F_1 and F_2 is measured with the help of a potentiometer (Fig. 3.23).

The mean resistivity of the soil is given by

$$\rho = 2\pi d \cdot \frac{E}{I} \qquad \qquad ...(3.23)$$

where ρ = mean resistivity, ohm-m,
         d = distance between electrodes, m (Fig 3.23),
         E = potential difference between electrodes F_1 and F_2, volts, and
         I = current flowing between outer electrodes E_1 and E_2, amperes.

Table 3.10. Mean Resistivity Values of Different Materials.

Material	Mean Resistivity (ohm-m)
Limestone (marble)	10^{12}
Quartz	10^{10}
Rock salt	$10^{6}–10^{7}$
Granite	$5000–10^{6}$
Sandstone	$35–4000$
Moraines	$8–4000$
Limestones	$120–400$
Clays	$1–120$

In sounding by resistivity method, a series of readings are taken, the spacing of the electrodes being increased for each successive reading. However, the centre of the four electrodes remains at a fixed point. As the spacing is increased, the resistivity is influenced by a greater depth of soil. Apparent resistivity is plotted against spacing (Fig. 3.24) and a comparison of such curves with a set of standard curves gives approximate layer thickness.

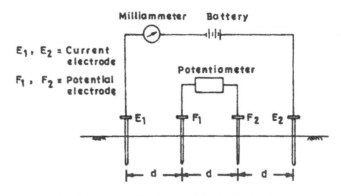

Fig. 3.23. Set up for Electrical Resistivity Method.

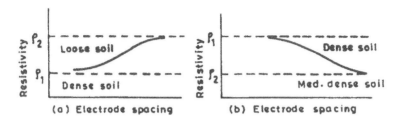

Fig. 3.24. Reistivity Versus Electrode Spacing Plots.

3.17 DYNAMIC PROPERTIES OF SOILS

If the proposed site lies in a seismic zone or the structures are likely to face vibration problems, special dynamic investigation programme must also be planned in addition to the above. Block vibration tests at the location of machinery foundations must be carried out to obtain the dynamic design parameters. Wave propagation tests yield the values of dynamic elastic and shear modulus.

Deposits of fine saturated sands are likely to liquefy under earthquakes. Laboratory and field tests be planned for sand deposits to assess liquefaction potential.

The details of dynamic tests are beyond the scope of this book and the readers are advised to look into the relevant literature on this.

3.18 PLANNING OF EXPLORATION PROGRAMME

Geotechnical exploration programmes vary with the size and nature of projects, the geological conditions of sites, and the type of foundation to be selected. Following steps may help in planning an efficient exploration programme:

(i) Study of the geology of the site through geological maps. This will give an indication of the type of soil, nature and depth of rock.

(ii) Study of layout of proposed structures at the site along with their loading diagrams.

(iii) A visit to the site for visual examination of *in situ* soils should then be made. Few pits at the location of important structures should be made and their logging be prepared during the visit.

(iv) On the layout plan of the site, location of boreholes and dynamic cone penetration tests should be marked to cover the whole site and in particular the location of important structures. Execution of this programme should then be started. If rock is not available at shallow depths, at least one boring should go deeper to ascertain the depth of rock layer. If rock is available at shallow depths, all the borings and DCPT should be carried out up to the rock layer to ascertain its profile.

Samples should be collected carefully for laboratory tests and also examined at the site carefully.

(v) On the basis of the probable soil stratum, planning of tests like plate load tests, vane shear tests, static cone penetration test and *in situ* large shear box tests may be carried out. Usually these tests (if necessary) are conducted at the location of few important structures.

(*vi*) Samples brought from the field are tested in laboratory for identification, classification, strength and consolidation characteristics.

A summary of exploration programme in coarse-grained and fine-grained soils is given in Tables 3.11 and 3.12 respectively (Ranjan et al., 1971).

ILLUSTRATIVE EXAMPLE

A three-storeyed telephone exchange building was proposed to be constructed at Nainital for the Department of Post and Telegraph. The site lies in the north-western side of the Nainital Valley. Site plan with spot levels (R.L.) of different points are given in Fig. 3.25. Loads coming through various columns are shown in Fig. 3.26. The detailed soil exploration programme carried out at the site to give necessary design parameters was planned as given below:

1. *Geology of the area* : The geological study indicated that the test site is covered by slate debris. In the one-third portion close to the existing telephone exchange, it is covered by a slate Boulder Bed extending to a depth of more than 1 m and the rest of two-thirds portion of the site is covered with slate Conglomerate Beds extending up to a depth of 1·m. Below the Boulder Bed/Conglomerate beds lie well bedded and thick-laminated Slate Bed, belonging probably to the lower Krol Formation. It is weathered up to a depth of about 250 mm.

2. *General description of site* : The client was requested to make four pits up to the depth of 2.0 to 3.0 m at different locations at the site. A visit to the site was then made for studying the general conditions, structures around the site and examination of soil stratum through open pits.

On the basis of site plan (Fig. 3.25), it is evident that the ground is not flat, and there is a difference of about 3.5 m in the grounds on the northern and southern sides.

From the visual examination of inspection pits, it was seen that there was rock stratum available at shallow depth. The thickness of soil cover above the rock stratum was found variable from location to location. However, the general trend indicated a thicker soil cover towards the northern side of the site and smaller thickness towards the southern side where the ground elevations were generally lower. Further, the soil cover was heterogeneous mixture of gravel, pebbles and boulders with non-plastic soil as matrix.

3. *About structure* : The proposed structure will transfer its dead and live load through R.C. columns whose positions in plan are as shown in Fig. 3.26. On the basis of preliminary estimates, the maximum load anticipated on each of these columns is shown in this figure.

4. *Planning of field and laboratory testing programme* : Considering the nature of soil/rock strata and the loads anticipated from the structure, it is necessary to adopt the field tests which reflect the special nature of the soil strata as well as the type of foundations likely to be adopted for the structure.

The following tests were planned to be carried out at the site and in the laboratory.

1. Field Tests:
(*a*) Dynamic cone penetration tests (4 Nos)
(*b*) *In situ* large size shear tests (2 Nos)
(*c*) Vertical plate load test (4 Nos)
(*d*) *In situ* unit weight determination (4 Nos)

Table 3.11. Summary of Exploration Programme (Coarse-grained Soils)*.

Type of Foundation	Soil Type	N-Value	Exploration Programme		Method of Estimating Bearing Capacity	Settlement	Remarks
			Field Test	Laboratory Test			
Footings	Well graded gravels, sand gravel mixtures with or without clay binder, clayey sand poorly graded or sandy clay mixtures (*GW, GC, SC, SM, SP*)	15–45	Borings, SPT, plate load test	Classification tests	$D_f/B \leqslant 0.5$ Terzaghi's theory, $D_f/B \geqslant 1.0$ Meyerhof's theory	Plate load test	For soils possessing characteristics of two groups, exploration programme should include shear tests (undrained and consolidated undrained)
Rafts	Well graded gravels, sand gravel mixtures with or without clay binder, clayey sand poorly graded or sandy clay mixtures (*GW, GC, SC, SM, SP*)	10–54	Borings, SPT, plate load test	Classification tests	Using empirical correlations of standard penetration test data		
Piles	Well or poorly graded sands with little or no fines, clayey and silty sands or sandy clay mixtures (*SW, SC, SM, SP*)	5–30	Borings, SPT, pile load test	Classification tests	Meyerhof's theory	Pile load test	

Table 3.12. Summary of Exploration Programme (Fine-grained Soils)*.

Type of Foundation	Soil Type	Consistency		Exploration Programme		Method of Estimating Bearing Capacity	Settlement	Remarks
		L.L.	q_u (kN/m²)	Field Test	Lab. Test			
Footings	Silts and very fine sands, silty to clayey fine sands, sandy clays, silty clays of low plasticity and clays of medium plasticity (ML, CL, CI)	20–50	100–400	Borings, SPT	Classification test, consolidation test, unconfined compression test	Skempton's method	Settlement to be calculated from consolidation test data using Terzaghi's theory. This should be within permissible limits	In case of high silt content, SPT may be carried out. In case of soft clays, field vane shear tests may be carried out
Rafts	Silts and very fine sands, silty to clayey fine sands, sandy clays, gravelly clays, silty clays of low plasticity and clays of medium plasticity. Very compressible micaceous or diatomaceous fine sandy or silty soils, silts and clays of high plasticity (CH, MH, ML, CL, CI)	20–100	25–400	Borings, SPT shelby tube samples, vane shear test	Classification test, consolidation test, unconfined compression test	Skempton's method	Settlement to be calculated from consolidation test data using Terzaghi's theory. This should be within permissible limits	
Piles	Silts and very fine sands, silty to clayey fine sands, sandy clays, gravelly clays, silty clays of low plasticity and clays of medium plasticity. Very compressible micaceous or diatomaceous fine silty soil, silt (CH, MH, MI, CL, CI)	20–100	25–100	Boring, SPT shelby tube samples, vane shear test, pile load test (i) Progressive (ii) Cyclic	Classification test, consolidation test, unconfined compression test	Static formula, correlations of penetration test	From pile load test, theoretical method	

* Ranjan et al. (1971).

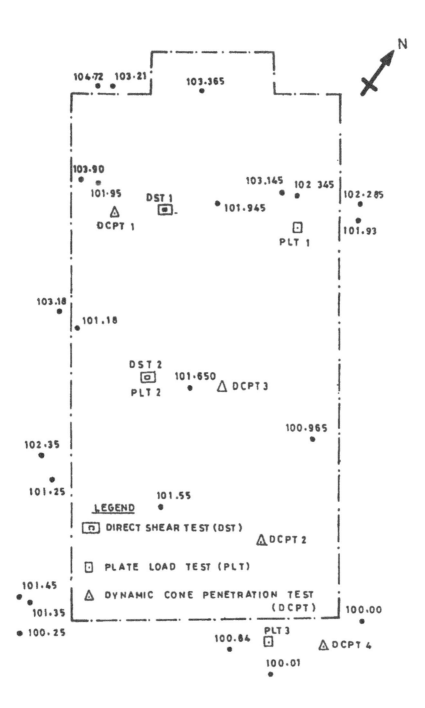

Fig. 3.25. Site Plan Showing the Locations of Field Test.

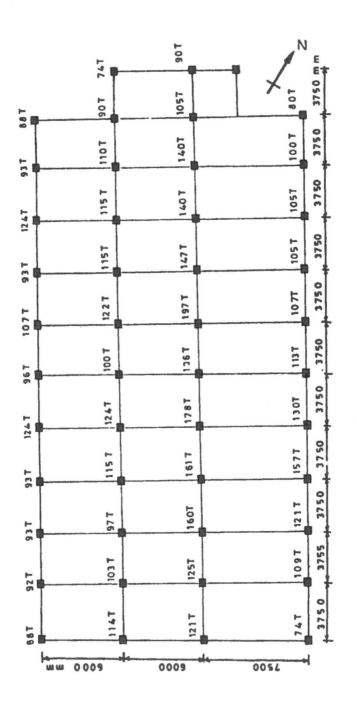

Fig. 3.26. Disposition of R.C. Columns in Plan along with Anticipated Loads.

2. Laboratory Tests:

(a) Natural water content determination

(b) Classification tests, viz., grain size analysis, liquid limit, plastic limit tests

The location of the test points for the field tests and their number are shown in Fig. 3.25. Boring was not included due to the presence of boulders and cobbles at shallow depths.

5. *Results and interpretation:* (a) The classification tests indicated that the soil above the rock stratum is silty gravel/well graded gravel (*GW – GM*). (b) The penetration tests confirmed that the rock stratum is available at about 0.5 m on the southern side of the site. The thickness of soil cover increases towards the northern side of the site. (c) Out of the two *in situ* shear tests, one was performed on the soil and another was performed on rock. Values of angle of internal friction was obtained as 45° for soil (*GM – GP*) and 38° for rock. (d) Out of the four plate load tests, two were performed on soil and two were conducted on the rock stratum. (e) Due to the variable soil cover, it was decided to provide depth of foundation such that all columns rest on rock. Further, it was recommended that the foundations be taken at least 250 mm into the rock stratum. (f) Using the ϕ value of rock as 38°, the safe bearing capacity of the footings of different widths were obtained. Pressures corresponding to 20 mm permissible settlement of footings were read from PLT data on rock after converting it to be equivalent settlement of the plate by extrapolation formula.

An allowable soil pressure of 160 kN/m^2 was then recommended for the foundation design (Saran et al., 1992).

REFERENCES

1. Bazarra, A.R. (1967), "*Use of Standard Penetration Test for Estimating Settlement of Shallow Foundation on Sand*," Ph. D. thesis, Univ. of Illinois, U.S.A.
2. Bowles, J.E. (1982), *Foundation Analysis and Design*, McGraw Hill, U.S.A.
3. Deere, D.U. (1964), "Technical description of cores for engineering purposes," *Rock Mch. Eng. Geol.*, Vol. 1, pp. 16–22.
4. Gibbs, H.J. and W.G. Holtz (1957), "Research on determining the density of sands by spoon penetration testing," *Proc. 4th Intern. Conf. SMFE*, London, Vol. 1, pp. 35–39.
5. Hvorslev, M.J. (1949), "*Surface Exploration and Sampling of Soils for Civil Engineering Purpose*," Waterways Experimental Station, Engineering Foundation, New York.
6. IS: 1892–1979, *Code of Practice for Site Investigations for Foundations*, I.S.I., New Delhi.
7. IS: 2131–1981, *Method for Standard Penetration Test for Soils*, I.S.I., New Delhi.
8. IS: 2132–1972, 1986, *Code of Practice for Thin Walled Tube Sampling of Soils*, I.S.I., New Delhi.
9. IS: 4968 (pt. I)–1976, *Method for Subsurface Sounding for Soils*; Part I, *Dynamic Method Using 50 mm Cone Without Bentonite Slurry*, I.S.I., New Delhi.
10. IS: 4968 (pt. II)–1976, *Method for Subsurface Sounding for Soils*; Part II, *Method Using Cone and Bentonite Slurry*, I.S.I., New Delhi.
11. IS: 4968 (pt. III)–1976, *Method for Subsurface Sounding for Soils*; Part III, *Static Cone Penetration Test* I.S.I., New Delhi.
12. IS: 9640–1980, *Indian Standard Specification for Split Spoon Samplers*, I.S.I., New Delhi.
13. IS: 11594–1985, *Specifications for Mild Steel Thin Walled Sampling Tubes and Sampler Heads*, I.S.I., New Delhi.
14. Lunne, T. and A. Kleven (1981), "Role of CPT in North Sea Foundation Engineering", Proc. Conf. Penetration Testing and Experience, St. Louis, pp. 76–107.
15. Peck, R.B., W.B. Hanson and T.H. Thornburn (1974). *Foundation Engineering*, John Wiley and Sons Inc., New York.

16. Prakash, S. and G. Ranjan (1975), "Large Scale *in situ* direct shear tests on boulder deposits," *Journal of I.E. (India)*, Civil Enggs. Division, Vol. 13, pp. 40–45.

17. Ranjan, G., S. Saran and S. Prakash (1971), *Concept of Standardization in Buildings*, Institution of Engineers (India) Roorkee Local Centre, Roorkee.

18. Roberston, P.K. and R.G. Campanella (1983), Interpretation of Cone Penetration Tests. Part I—Sand, *CGJ*, Ottawa, Vol. 20.

19. Robertson, P.K. and R.G. Campanella (1983), "SPT–CPT correlations," JGED, *ASCE*, Vol. 109.

20. Sanglerat, G. (1972), *The Penetrometer and Soil Exploration*, Elsevier Publishing Co., Amsterdam.

21. Saran, S., A.S.R. Rao and B. Prakash (1992), "Soil Investigations and Recommendations for the Foundation of the Telephone Exchange Building at Nainital," Project report, University of Roorkee, Roorkee.

22. Task Committee Report (1972), "Subsurface investigation for design and construction of foundation of buildings—Part I," *Proc. ASCE*, Vol. 98, No. SM5.

23. Terzaghi, K. and R.B. Peck (1967), *Soil Mechanics in Engineering Practice*, John Wiley and Sons, New York.

24. Teng, W.C. (1977), *Foundation Design*, Prentice-Hall of India Pvt. Ltd., New Delhi.

25. Tomlinson, M.J. (1986), *Foundation Design and Construction*, Pitman Books Ltd., London.

26. Trofimenkov, J.B. (1974), "Penetration testing in USSR, state of the art report," *Proc. European Symposium on Penetration Testing, Stockholm*, Vol. 1, pp. 147–154.

PRACTICE PROBLEMS

1. Describe the general procedure used for making borings and obtaining soil samples for classification and testing.

2. What is a standard penetration test? Explain clearly the procedure of applying correction in observed *N*-values for overburden and position of water table.

3. Give a comparison between standard penetration test, dynamic cone penetration test and static cone penetration test for determining subsurface soil conditions.

4. Describe *in situ* vane shear and *in situ* direct shear tests giving their merits and demerits.

5. Write short notes on:
   (a) Amount of boring;
   (b) Percussion boring;
   (c) Electrical resistivity method; and
   (d) Seismic refraction method.

6. (a) What is area ratio? Describe its significance.
   (b) Compute the area ratio of a sampling tube having outside and inside diameters as 99 mm and 93 mm respectively. Is this tube suitable for undisturbed sampling?
   (c) If this tube is pushed into clay of medium stiffness at the bottom of a borehole, a distance of 500 mm and length of the sample recovered is 482 mm. What is recovery ratio?

7. A ten-storeyed office building complex is planned to cover a plan area of about 300 m × 200 m. Preliminary survey of the site and information on the adjacent area indicated that it is likely to have top 3.0 m silty sand followed by about 7.0 m clay and beyond which it is likely to have medium dense sand extending to great depth. Position of water table is likely to be within 7.0 m from ground surface. Prepare a complete exploration programme that would help in deciding the type of foundation and giving required design parameters.

Lateral Earth Pressure

4.1 INTRODUCTION

The lateral earth pressure is an important design parameter in a number of foundation engineering problems, *e.g.*, retaining and sheet pile walls, braced and unbraced excavations, silo walls and bins, tunnels and other underground structures, etc. It is defined as the force exerted by the retained soil mass on the earth retaining structure.

The magnitude of lateral earth pressure that can exist or develop in a soil mass is related to the strength and stress-strain properties of the material and deformation that occurs within the mass as a result of a lateral movement. The variation in the magnitude of the lateral earth pressure (p) with the movement of the wall (Δ) is illustrated in Fig. 4.1. When the wall is rigidly fixed, the magnlitude of earth pressure is represented by the point 'A' and is termed 'earth pressure at rest'. If the wall yields or moves away from the backfill, pressure starts decreasing from 'at rest' value and reaches a minimum value beyond which there is no decrease in pressure even with the further movement of wall. This is represented by the point 'B' and is termed 'active earth pressure'. If the wall is forced against the backfill, the pressure starts increasing from the 'at-rest' value with the wall movement. A condition will be reached when the pressure reaches a maximum value beyond which increase in pressure does not take place with further movement. This is represented by the point 'C' and is termed as 'passive earth pressure'.

The ratio of the horizontal stress to the effective vertical stress in a soil mass is termed as the coefficient of lateral earth pressure. The order of magnitude of lateral earth pressure coefficients are also given in Fig. 4.1. The order of the amount of the wall movement for creating the active and passive states are given in Table 4.1.

Table 4.1. Amount of Wall Movement.

Soil and Condition	Amount of Wall Movement	
	Active State	*Passive State*
Dense cohesionless soil	0.0005 to 0.001 H	0.0015 to 0.004 H
Loose cohesionless soil	0.002 to 0.004 H	0.006 to 0.008 H
Stiff cohesive soil	0.01 to 0.02 H	0.025 to 0.035 H
Soft cohesive soil	0.02 to 0.05 H	0.04 to 0.08 H

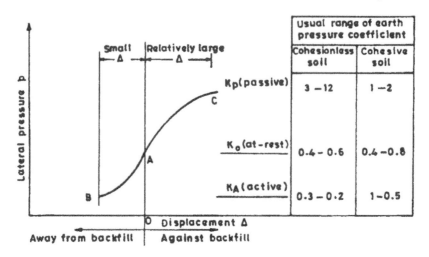

Fig. 4.1. Illustration of Active and Passive Pressures with Respect to Wall Movement.

4.2 EARTH PRESSURE AT REST

In a homogeneous and isotropic soil mass of infinite extent, bounded by a level ground surface, the vertical and horizontal stresses shown on the representative element are principle stresses (Fig. 4.2). The magnitude of the vertical principle stress, σ_v is equal to the weight of the overburden, or

$$\sigma_v = \gamma. Z \qquad \qquad \text{... (4.1)}$$

If the overburden acting as part of the soil mass were to increase, the vertical stress would also increase, and the soil would be subjected to vertical compression. The soil cannot deform laterally because of infinite extent in that direction.

If E and μ are modulus of elasticity and Poisson's ratio respectively of the soil mass, then

$$\text{Vertical strain} = \frac{\sigma_v}{E} \qquad \qquad \text{... (4.2)}$$

Lateral strain in horizontal direction

$$= \frac{\sigma_h}{E} - \mu\left(\frac{\sigma_h}{E} + \frac{\sigma_v}{E}\right) \qquad \qquad \text{... (4.3)}$$

The lateral strain being zero,

$$= \frac{\sigma_h}{E} - \mu\left(\frac{\sigma_h}{E} + \frac{\sigma_v}{E}\right) = 0 \qquad \qquad \text{... (4.4)}$$

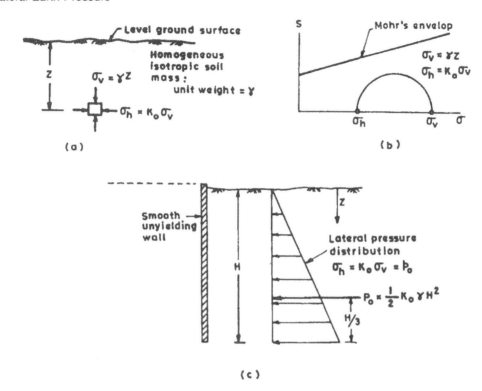

Fig. 4.2. Conditions Related to At-rest Earth Pressure: (*a*) Subsurface in Soil Mass of Infinite Extent; (*b*) Horizontal and Vertical Stresses Related to Failure Envelop; and (*c*) Lateral Earth Pressure Distribution against Smooth, Rigid and Unyielding Wall.

It gives

$$\frac{\sigma_h}{E} = \frac{\mu}{1 - \mu} \qquad\qquad \dots (4.5)$$

Therefore, earth pressure at rest is given by

$$\sigma_h = \sigma_v \left[\mu/(1 - \mu) \right] \qquad\qquad \dots (4.6)$$

or

$$p_0 = \gamma \, Z \, K_0 \qquad\qquad \dots (4.7)$$

where p_0 is designated as 'earth pressure at rest', and K_0 as coefficient of earth pressure at rest. γ represents the effective unit weight of the soil.

Values of K_0 based on experience in the field are given in Table 4.2.

Some of the empirical relationships proposed for getting K_0 are:

$$K_0 = 1 - \sin \phi' \quad \text{(Jaky, 1944)} \qquad\qquad \dots (4.8)$$

$$K_0 = 0.9 \, (1 - \sin \phi') \quad \text{(Fraser, 1957)} \qquad\qquad \dots (4.9)$$

$$K_0 = (1 + 2/3 \sin \phi') \left(\frac{1 - \sin \phi'}{1 + \sin \phi'} \right) \text{ (Kezdi, 1962)} \qquad \qquad ... (4.10)$$

$$K_0 = 0.95 - \sin \phi' \text{ (Brooker and Ireland, 1955)} \qquad \qquad ... (4.11)$$

$$K_0 = 0.19 + 0.233 \log I_p \text{ (Kenny, 1959)} \qquad \qquad ... (4.12)$$

where ϕ' = Effective angle of internal friction
I_p = Plasticity Index.

On the basis of extensive study of the above correlations, Alpan (1959) recommended that Eq. 4.8 and Eq. 4.12 may be used for sands and clays respectively.

Table 4.2. Values of Coefficient of Earth Pressure At-Rest (K_0).

Type of Soil	I_p	K_0
Loose dry sand	—	0.64
Dense dry sand	—	0.49
Loose saturated sand	—	0.46
Dense saturated sand	—	0.36
Compacted clay	9	0.42
Compacted clay	30	0.60
Organic silty clay	45	0.57

4.3 EFFECT OF SUBMERGENCE

Where soil is below water table, the effective stress σ'_v at any depth is reduced by the magnitude of the water pressure at the same depth, *i.e.*

$$\sigma'_v = \gamma_s \, Z - \gamma_w \cdot Z = (\gamma_s - \gamma_w) \, Z$$

or $$\sigma'_v = \gamma_b \cdot Z \qquad \qquad ... (4.13)$$

The submerged soil unit weight is used in Eq. 4.7, the result being that, compared to the nonsubmerged condition, the lateral earth pressure decreases. However, the pressure ($\gamma_w \cdot Z$) acts against the submerged portion of the wall. Hence the presence of water table with the combined effect of lateral earth pressure and hydrostatic pressure causes a greater lateral force to act on the wall (Fig. 4.3).

4.4 RANKINE'S THEORY

Rankine (1857) derived the lateral pressures in a cohesionless soil considering it to be in a state of plastic equilibrium. He considered a semi-infinite mass of soil bound by a horizontal surface and a vertical boundary formed by the vertical face of a smooth wall surface (Fig. 4.4a). If the wall is allowed to move away a slight distance from the retained soil, the soil starts to extend in the lateral direction (Fig. 4.4a). Shearing resistance developed within the

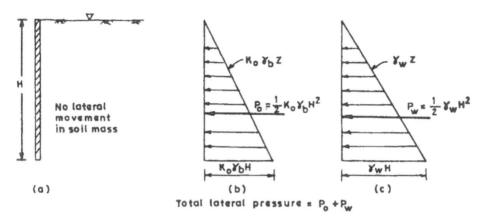

(a)   (b)   (c)

Total lateral pressure $= P_o + P_w$

Fig. 4.3. (*a*) Condition of Submerged Soil, (*b*) Lateral Earth Pressure Diagram and (*c*) Hydrostatic Pressure Diagram.

soil mass acts opposite to the direction of expansion. This makes the lateral earth pressure on the wall decrease and becomes less than at-rest pressure. When the lateral movement is such that soil's maximum shearing resistance occurs, the lateral earth pressure on the wall will be minimum. The soil mass at this condition is said to be in the active Rankine state and the minimum lateral earth pressure is the active earth pressure p_A.

The active earth pressure may be determined by the Mohr's circle plot in reference to the soil's failure envelop (Fig. 4.4b). It gives

$$\sin \phi = \frac{\dfrac{\sigma_v - p_A}{2}}{\dfrac{\sigma_v + p_A}{2}} \qquad \qquad \dots (4.14)$$

or

$$\frac{p_A}{\sigma_v} = \frac{1 - \sin \phi}{1 + \sin \phi} = \tan^2 (45 - \phi/2) = K_A \qquad \dots (4.15)$$

K_A is the coefficient of active earth pressure. When the soil mass is in the active Rankine state, two sets of failure planes develop, each inclined to an angle of $(45 + \phi/2)$ to the horizontal (Fig. 4.4c). The lateral earth pressure diagram is shown in Fig. 4.4d.

If the wall moves into the retained soil mass, the soil compresses in the lateral direction (Fig. 4.5a), with the soil shearing resistance acting to oppose the lateral compression. Therefore, the lateral pressure on the wall increases from the at-rest pressure. For the wall movement at which soil shearing resistance becomes maximum, the lateral earth pressure will also become maximum. The soil mass at this condition is said to be in passive Rankine state and the maximum lateral earth pressure is the passive earth pressure p_p. From Mohr's circle plot (Fig. 4.5b),

$$\frac{p_P}{\sigma_v} = \frac{1 + \sin \phi}{1 - \sin \phi} = \tan^2 (45 + \phi/2) = K_p \qquad \dots (4.16)$$

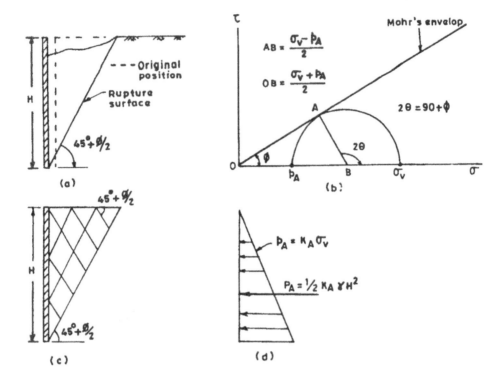

Fig. 4.4. Conditions Related to Active Earth Pressure in Cohesionless Soil: (*a*) Wall and Soil Movement; (*b*) Mohr's Circle Illustrating Relationship between p_A and σ_v; (*c*) Shear Pattern; and (*d*) Earth Pressure Diagram.

K_p is the coefficient of passive earth pressure. When the soil mass is in passive Rankine state, two sets of failure planes develop, each inclined to an angle of $(45 - \phi/2)$ to the horizontal (Fig. 4.5c). The lateral earth pressure diagram is shown in Fig. 4.5d.

For soils possessing both cohesion and angle of internal friction, Mohr's Circle-failure envelop relationships are (Fig. 4.6):

For active pressure condition

$$\sin \phi = \frac{AD}{CD} = \frac{(\sigma_v - p_A)/2}{(\sigma_v + p_A)/2 + c \cot \phi} \qquad \dots (4.17a)$$

$$p_A = \sigma_v \tan^2 (45 - \phi/2) - 2c \tan (45 - \phi/2) \qquad \dots (4.17b)$$

or $$p_A = \gamma Z K_A - 2c \sqrt{K_A} \qquad \dots (4.17c)$$

For passive pressure condition

$$\sin \phi = \frac{BE}{CE} = \frac{(p_P - \sigma_v)/2}{(p_P + \sigma_A)/2 + c \cot \phi} \qquad \dots (4.18a)$$

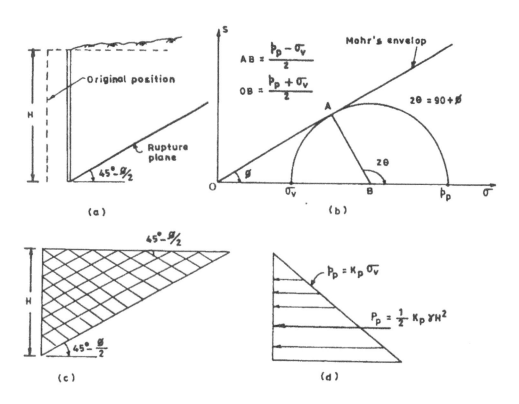

Fig. 4.5. Conditions Related to Passive Earth Pressure in Cohesionless Soil: (*a*) Wall and Soil Movement;
(*b*) Mohr's Circle Illustrating Relationship between p_p and σ_v; (*c*) Shear Pattern; and
(*d*) Earth Pressure Diagram.

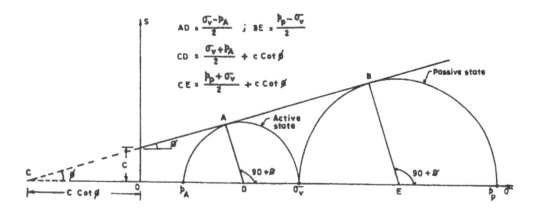

Fig. 4.6. Mohr's Circle Illustrating Relationship between p_A and σ_v, and p_p and σ_v in a $c - \phi$ Soil.

$$p_p = \sigma_v \tan^2 (45 + \phi/2) + 2c \tan (45 + \phi/2) \qquad \qquad \text{... (4.18b)}$$

or $\qquad \qquad p_p = \gamma Z K_p + 2c \sqrt{K_p} \qquad \qquad \qquad \qquad \qquad \text{... (4.18c)}$

The lateral earth pressure diagrams are shown in Figs. 4.7a and 4.7b for active pressure condition and passive pressure condition respectively.

From Eq. 4.17c,

At $p_A = 0$

$$Z = h_0 = \frac{2c}{\gamma \sqrt{K_A}} \qquad \qquad \qquad \text{... (4.19)}$$

It is evident from Fig. 4.7a that negative pressure exists up to h_0, a depth where the active earth pressure becomes equal to zero. Within the zone between the ground surface and depth h_0, the soil is in a state of tension. In practice, this tension cannot be taken to act on the wall since tension cracks tend to develop in the soil within the tension zone and the soil may not remain adhered to the wall. Hence, in computing the total active earth pressure, the tension zone is usually ignored, only the area of the pressure distribution diagram between the depths h_0 and H is considered.

Therefore,

$$P_A = 1/2 \, K_A \gamma \, H^2 - 2c \, H \sqrt{K_A} + \frac{2c^2}{\gamma} \qquad \qquad \text{... (4.20)}$$

It may also be noted from Fig. 4.7a that the net total active earth pressure is zero for a depth equal to $2h_0$. It implies that in a cohesive soil, a vertical cut can be made up to a depth of $2h_0$ without having to provide any lateral support. In a cohesive soil, therefore, the critical depth of a vertical cut, H_c is given by

$$H_c = 2h_0 = \frac{4c}{\gamma \sqrt{K_A}} \qquad \qquad \qquad \text{... (4.21)}$$

Referring Fig. 4.7b, the total passive earth pressure, P_p is given by

$$P_p = 1/2\gamma H^2 K_p + 2cH \sqrt{K_P} \qquad \qquad \qquad \text{... (4.22)}$$

Rankine's theory can also be applied to the case when the backfill surface slopes at an angle i to the horizontal. Figure 4.8a shows an element of soil bound by two vertical planes and two planes parallel to the backfill surface, at depth Z. The vertical stress and the lateral pressure acting on the soil element are conjugate stresses. It is assumed that a similar stress condition would exist against the back of a retaining wall used to support the slope. This assumption requires that the friction between a wall and the retained soil has insignificant

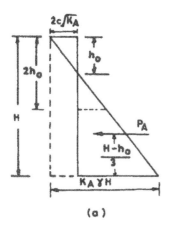

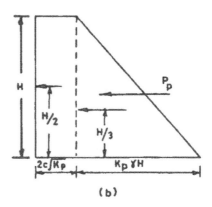

Fig. 4.7. Lateral Earth Pressure Distribution for $c - \phi$ Soil in (*a*) Active State and (*b*) Passive State.

effect on the shearing stresses developed behind the wall. The lateral earth pressure on a vertical plane is thus assumed to act parallel to the backfill surface, and is given by

Active earth pressure:

$$p_A = \gamma Z K_A \qquad \qquad .... \text{(4.23a)}$$

where

$$K_A = \cos i \; \frac{\cos i - \sqrt{\cos^2 i - \cos^2 \phi}}{\cos i + \sqrt{\cos^2 i - \cos^2 \phi}} \qquad ... \text{(4.23b)}$$

Passive earth pressure:

$$p_P = \gamma Z K_P \qquad \qquad ... \text{(4.24a)}$$

where

$$K_P = \cos i \; \frac{\cos i + \sqrt{\cos^2 i - \cos^2 \phi}}{\cos i - \sqrt{\cos^2 i - \cos^2 \phi}} \qquad ... \text{(4.24b)}$$

The lateral active earth pressure diagram is shown in Fig. 4.8b. The total earth pressure is given by

$$P_A = 1/2 \; \gamma \; K_A \; H^2 \qquad \qquad ... \text{(4.25a)}$$

$$P_P = 1/2 \; \gamma \; K_P \; H^2 \qquad \qquad ... \text{(4.25b)}$$

Application of Rankine's theory to some other practical problems is illustrated below.

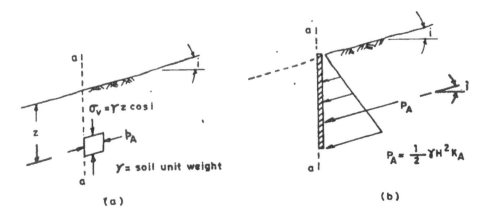

Fig. 4.8. Lateral Earth Pressure in Sloping Backfill; (*a*) Stresses on an Element at Depth Z Below
the Surface of Semi-infinite Sloping Ground, (*b*) Active Earth Pressure Distribution.

(*a*) Backfill carrying uniform surcharge

If a uniformly distributed surcharge pressure of intensity q per unit area acts over the
entire surface of the soil mass (Fig. 4.9a), the vertical stress σ_v at any depth is increased to
$(\gamma z + q)$ which causes an additional lateral pressure of uniform intensity $K_A \cdot q$ behind the
wall.

Typical active and passive earth pressure diagrams are shown in Fig. 4.9b and 4.9c
respectively. The total earth pressures will be as given below:

Active earth pressure (Fig. 4.9b)

(*i*) $qK_A > 2c\sqrt{K_A}$

$$P_A = 1/2\ \gamma H^2 K_A + (qK_A - 2c\sqrt{K_A}) \cdot H \qquad\qquad \dots (4.26a)$$

(*ii*) $qK_A < 2c\sqrt{K_A}$

Pressure at any depth Z is

$$p_A = \gamma Z K_A + qK_A - 2c\sqrt{K_A} \qquad\qquad \dots (4.26b)$$

At $\qquad p_A = 0,\ Z = h_0 = \dfrac{2c\sqrt{K_A} - q \cdot K_A}{\gamma K_A}$

or $\qquad h_0 = \dfrac{2c}{\gamma\sqrt{K_A}} - \dfrac{q}{\gamma} \qquad\qquad \dots (4.27)$

Therefore

$$P_A = 1/2\ \gamma H^2 K_A + qK_A H - 2cH\sqrt{K_A} + 1/2\ (2c\sqrt{K_A} - qK_A) \cdot h_0 \qquad \dots (4.28)$$

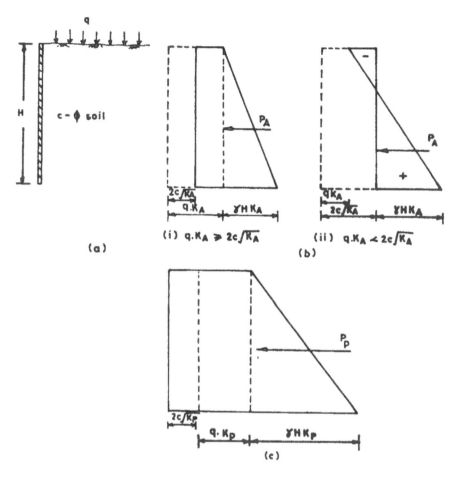

Fig. 4.9. Lateral Earth Pressure in a Wall with Backfill Surface having Uniformly Distributed Surcharge: (*a*) Section of Wall, (*b*) Active Earth Pressure Distribution, and (*c*) Passive Earth Pressure Distribution.

Passive earth pressure (Fig. 4.9c)

$$P_p = 1/2 \ \gamma H^2 K_p + qK_p H + 2cH\sqrt{K_p} \qquad \qquad \dots (4.29)$$

(*b*) Retaining wall with inclined back face and cantilever retaining wall

Figure 4.10 shows as inclined-back wall with a sloping backfill. A vertical line is drawn from the base of the wall '*A*' to intersect the backfill surface at *B*. The total active earth pressure is then computed on the vertical plane *AB* using Eq. 4.25a by substituting *H* as height of vertical plane *AB*, *i.e.*, H_1. The weight of soil included in the triangle *ABC* is shown as *W* and considered as the portion of the wall in stability analysis. Similarly, in a cantilever retaining wall (Fig. 4.11), the total active earth pressure is computed on the vertical plane *AB*, and the weight of soil mass *BEDC* is considered as the portion of the wall.

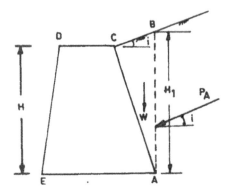

Fig. 4.10. Retaining Wall with Inclined Backface.

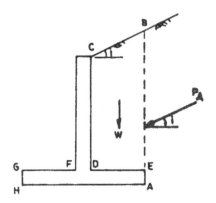

Fig. 4.11. Cantilever Retaining Wall.

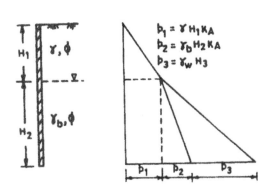

Fig. 4.12. Lateral Pressure on Wall having Partly Submerged Backfill.

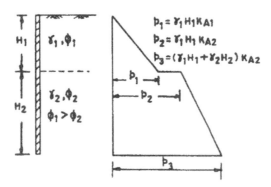

Fig. 4.13. Active Earth Pressure Distribution on Wall with Stratified Backfill.

(c) Partly submerged backfill

For a partly submerged backfill, as shown in Fig. 4.12, K_A is applied to the effective vertical stress. The active earth pressure distribution is computed on the basis of bulk unit weight γ above the water table and submerged unit weight γ_b below the water table. The hydrostatic water pressure below the water table must be added to the active earth pressure to obtain the total lateral pressure.

(d) Stratified backfill

If the backfill is stratified such that K_A and γ are not constant with depth, the pressure will change abruptly at the strata interfaces. The pressure distribution is obtained by using appropriate values of K_A for each strata. For a particular layer the weight of the overlying layers is considered as a surcharge. A typical pressure diagram is shown in Fig. 4.13.

4.5 INFLUENCE OF WALL FRICTION ON THE SHAPE OF THE SURFACE OF SLIDING AND MAGNITUDE OF LATERAL PRESSURE

The basic assumption in Rankine's theory is that the back of the wall is considered smooth. No friction forces are assumed to exist between the soil and the wall and, therefore, the lateral earth pressure is taken to act parallel to the surface of the backfill. The wall face is considered a principle plane. In practice, however, most of the retaining walls are rough. If the wall is considered vertical and backfill soil as cohesionless, the vertical plane in the vicinity of wall is no longer a principle plane. Therefore, stresses in the soil mass will not be the same as in Rankine state.

In active state, the downward movement of the sand with respect to the wall develops frictional force that causes resultant active earth pressure to be inclined at an angle δ to the normal to the wall. This angle is known as angle of wall friction. Advanced theoretical analysis (Ohde, 1938) as well as experiments (Mackey and Kirk, 1967; Narain et al., 1979) have shown that the corresponding surface of sliding bc consists of a curved lower portion and a straight upper part (Fig. 4.14a). The shear pattern in area abd consists of two sets of curved lines, while within section adc of the sliding wedge, the shear pattern is identical with the active Rankine pattern (Fig. 4.4c).

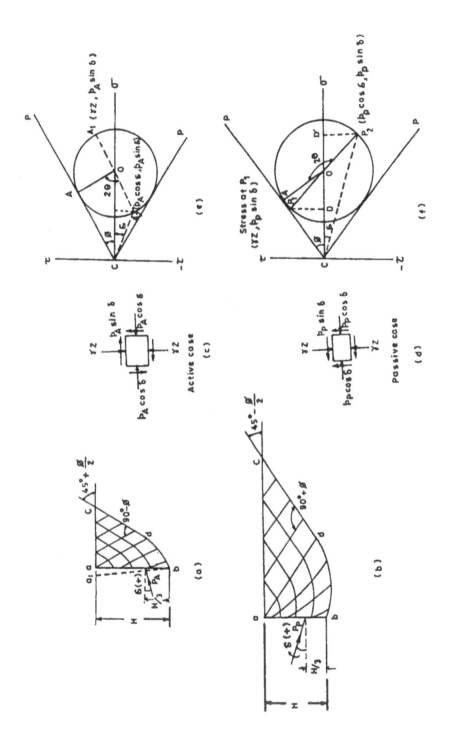

Fig. 4.14. Effect of Wall Friction on Rupture Surface and Earth Pressure. (*a*, *b*) Shear Pattern (*c*, *d*) Stresses on Soil Element (*e*, *f*) Mohr's Circle Representation of Stresses.

In passive state, the sand rises with reference to the wall, and the resultant passive earth pressure acts at an angle δ (Fig. 4.14b) with the normal to the back of the wall. The tangential component of this force tends to prevent the rise of sand. The straight portion of the surface of sliding rises at an angle of $(45 - \phi/2)$ with the horizontal, and the shear pattern is identical to that of passive Rankine state (Fig. 4.5c). Within the area adb both sets of lines which constitute the shear pattern are curved.

If it is assumed that every point on the wall face represents the point of incipient failure, then at any depth Z the stresses on an element near the wall face will be as shown in Figs. 4.14c and d for active and passive states respectively. Stresses on these elements are represented by points A_1 and A_2, and P_1 and P_2 on $\sigma - \tau$ axis (Figs. 4.14e and f). A Mohr's circle can be drawn passing through points A_1 and A_2 and touching the failure envelops (active state, Fig. 4.14e). Similarly, a Mohr's circle can be drawn passing through points P_1 and P_2, and touching the failure envelops (passive state, Fig. 4.14f).

The geometry of the Mohr's circle with respect to failure plane gives:

For active state (Fig. 4.14e)

$$p_A = \frac{1 - \sin\phi \sin(2\theta - \phi)}{[1 + \sin\phi \sin(2\theta + \phi)]\cos\delta}\, \gamma Z \qquad \dots (4.30)$$

$$\theta = 45 - \phi/2 + 1/2 \tan^{-1}\left(\frac{2p_A \sin\delta}{\gamma Z - p_A \cos\delta}\right) \qquad \dots (4.31)$$

The equation (4.31) indicates that the value of θ is $(45 - \phi/2)$ when $\delta = 0$. It will be more than $(45 - \phi/2)$ if friction is considered. Value of active earth pressure (Eq. 4.30) will decrease because the value of $\sin(2\theta + \phi)$ will be less than unity.

For passive state (Fig. 4.14f)

$$p_P = \frac{1 + \sin\phi \sin(2\theta - \phi)}{[1 - \sin\phi \sin(2\theta - \phi)]\cos\delta}\, \gamma . Z \qquad \dots (4.32)$$

$$\theta = 45 + \phi/2 + 1/2 \tan^{-1}\left(\frac{2p_P \sin\delta}{\gamma Z - p_P \cos\delta}\right) \qquad \dots (4.33)$$

The Eq. (4.32) indicates that the effect of friction in passive state will be to increase the wedge angle θ and passive earth pressure p_p.

4.6   COULOMB'S EARTH PRESSURE THEORY

Since in practice the retaining walls are rough, the boundary conditions for the validity of Rankine's theory are not satisfied, and earth pressure computations based on this theory give an appreciable error. Most of this error can be avoided using Coulomb's Theory (Coulomb, 1776). Coulomb's method is based on a simplifying assumption of considering the surface of sliding as planer. However, the error due to this assumption is commonly small compared to that associated with the use of Rankine's theory. The curvature in sliding surface is slight in the active case and the error (under estimation) involved in assuming a plane surface does not exceed 5% in majority of practical problems (Scott, 1974). In passive case the curvature is larger and increases with the increase in δ. The assumptions of planer rupture surface may estimate 50% higher passive earth pressure (Scott, 1974). The main advantage of Coulomb's theory is that it can be applied to any boundary condition. For a smooth vertical wall having horizontal backfill, both the theories give same results.

The basic assumptions in Coulomb's theory are:

(*i*) The backfill soil is dry, homogeneous, isotropic and elastically underformable but breakable granular material.

(*ii*) The wall yields to the extent that the rupture of the backfill soil takes place and a soil wedge is torn off from rest of the soil mass. The wedge itself is considered as a rigid body.

(*iii*) The failure surface is assumed to be a plane passing through the heel of the wall.

(*iv*) Friction between the wall and the soil is considered. Resultant earth pressure acts at an angle δ to the normal of the wall, δ being the angle of wall friction whose magnitude is known.

Active earth pressure

Figure 4.15a shows a section of a rough retaining wall with the trial failure plane BC_1 inclined at an angle θ_1 to the vertical. For the failure condition, the soil wedge ABC_1 is in equilibrium under the following three forces:

(*i*) Weight W_1 of the wedge ABC_1 acting through its centre of gravity.

(*ii*) Active earth pressure P_1 inclined at an angle δ with the normal to the wall in the anticlockwise direction.

(*iii*) Reaction R_1 inclined at angle ϕ to the failure plane BC_1.

The directions of all the three forces W_1, R_1 and P_1 (Fig. 4.15a) are known but the magnitude of only one force W_1 is known. The magnitude of other forces can be obtained by constructing the force polygon as shown in Fig. 4.15b. P_1 is the value of earth pressure corresponding to the trial wedge ABC_1. More trials are made by assuming BC_2, BC_3..... BC_n as failure surfaces inclined at angles θ_2, θ_3,..... θ_n. By constructing force polygons similar to the one in Fig. 4.15b, the values of P_2, P_3, ... P_n can be evaluated. Variation of P with θ is shown in Fig. 4.15c. The maximum value of P is the active earth pressure P_A.

From Fig. 4.15a, considering a general case replacing θ_1 by θ; W_1 by W and P_1 by P:

W = Weight of soil mass of ABD + Weight of soil mass AC_1D

$$= 1/2 \, \gamma H^2 \, (\tan \alpha + \tan \theta) + 1/2 \, \gamma H^2 \, (\tan \alpha + \tan \theta) \times \frac{\sin (\theta + \alpha) \sec \alpha \sin i}{\cos (\theta + i)}$$

or $W = 1/2\gamma H^2 \, (\tan \alpha + \tan \theta) \left(1 + \dfrac{\sin (\theta + \alpha) \sec \alpha \sin i}{\cos (\theta + i)} \right)$ ... (4.34)

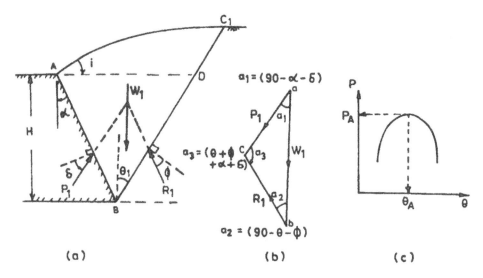

Fig. 4.15. (*a*) Forces Acting on the Failure Wedge in Cohesionless Soil in Active State; (*b*) Force Triangle; and (*c*) Pressure Versus Wedge Angle θ Plot.

From Fig. 4.15b

$$\frac{P}{\cos(\theta + \phi)} = \frac{W}{\sin(\theta + \phi + \alpha + \delta)} \qquad \ldots (4.35)$$

$$P = 1/2\gamma H^2 \frac{\cos(\theta + \phi)(\tan\alpha + \tan\theta)}{\sin(\theta + \phi + \alpha + \delta)} \times \left(1 + \frac{\sin(\theta + \alpha)\sec\alpha\sin i}{\cos(\theta + i)}\right) \qquad \ldots (4.36)$$

Coulomb optimised Eq. 4.36 by solving

$$\frac{\partial P}{\partial \theta} = 0 \qquad \ldots (4.37)$$

with Eq. 4.36 and obtained

$$P_A = 1/2\gamma H^2 K_A \qquad \ldots (4.38a)$$

where

$$K_A = \frac{\cos^2(\phi - \alpha)}{\cos^2\alpha\cos(\alpha + \delta)} \times \frac{1}{\left[1 + \left(\frac{\sin(\phi + \delta)\sin(\phi - i)}{\cos(\alpha - i)\cos(\alpha + \delta)}\right)^{1/2}\right]^2} \qquad \ldots (4.38b)$$

The passive case differs from the active case in the respect that the obliquity angles at the wall and on the failure plane are of opposite signs as shown in Fig. 4.16a. The procedure

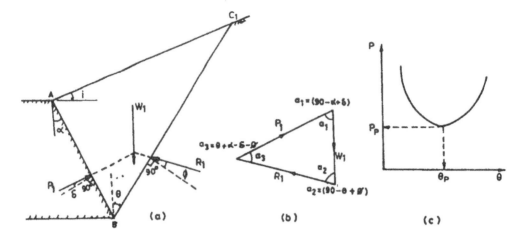

Fig. 4.16. (*a*) Forces Acting on the Failure Wedge in Cohesionless Soil in Passive State; (*b*) Force Triangle; and (*c*) Earth Pressure Versus Wedge Angle θ Plot.

of computation is exactly the same as discussed above for active earth pressure. The force polygon and $P - \theta$ plot are shown in Fig. 4.16b and c respectively. In this case the minimum value of earth pressure is referred as passive earth pressure P_P.

Coulomb gave the following expression for passive earth pressure:

$$P_P = 1/2 \; \gamma H^2 \; K_P \qquad \qquad \ldots (4.39a)$$

where
$$K_P = \frac{\cos^2(\phi + \alpha)}{\cos^2 \alpha \cos(\alpha - \delta)} \times \frac{1}{\left[1 - \left(\dfrac{\sin(\phi + \delta)\sin(\phi + i)}{\cos(\alpha - i)\cos(\alpha - \delta)}\right)^{1/2}\right]^2} \qquad \ldots (4.39b)$$

For $\alpha = \delta = i = 0$, Eqs. (4.38b) and (4.39b) reduce to:

$$K_A = \frac{\cos^2 \phi}{(1 + \sin \phi)^2} = \left(\frac{1 - \sin \phi}{1 + \sin \phi}\right) \qquad \ldots (4.40)$$

$$K_P = \frac{\cos^2 \phi}{(1 - \sin \phi)^2} = \left(\frac{1 + \sin \phi}{1 - \sin \phi}\right) \qquad \ldots (4.41)$$

Therefore, in the case of a smooth vertical wall having horizontal backfill surface, both Rankine's theory and Coulomb's theory give identical values of earth pressure coefficients K_A and K_P.

Coulomb's theory can also be used to determine the active and passive earth pressures for a backfill having both cohesion and angle of internal friction by considering additional forces due to cohesion and adhesion on the failure plane and wall face respectively.

The following forces act on the wedge in the active case (Fig. 4.17a):

(*i*) Weight W_1 of wedge ABC_1.

(*ii*) Reaction R_1 on BC_1 inclined at angle ϕ to the normal to BC_1.

(*iii*) Earth pressure P_1 inclined at angle δ to the normal to the wall face AB.

(*iv*) Cohesion $C_1 = c \times BC_1$ along BC_1.

(*v*) Adhesion $C_a = c_a \times BA$ along BA.

The directions of all the five forces are known, but the magnitude of only three forces W_1, C_1 and C_a are known. The magnitude of the other forces, i.e. P_1 and R_1 can be obtained by constructing the force polygon as shown in Fig. 4.17b. By the method of trials, the slip plane which produces the maximum value of P can be determined (Fig. 4.17c). This value of P is the active earth pressure P_A.

If a uniformly distributed surcharge load of intensity q per unit area is acting over the surface of backfill (Fig. 4.18), additional vertical force of magnitude $q \cdot AC_1$ is to be considered in constructing the force polygon. It can be derived that additional pressure due to surcharge will be:

$$P_{Aq} = \frac{qH \cos \alpha}{\cos (\alpha - i)} \cdot K_A \qquad \qquad \dots (4.42)$$

$$P_{Pq} = \frac{qH \cos \alpha}{\cos (\alpha - i)} \cdot K_P \qquad \qquad \dots (4.43)$$

Values of earth pressure coefficients K_A and K_P are the same as given by Eqs. 4.38b and 4.39b.

The main deficiency in Coulomb's theory is that it does not give the distribution of pressure along the height of wall but gives only the total pressure. Few investigators

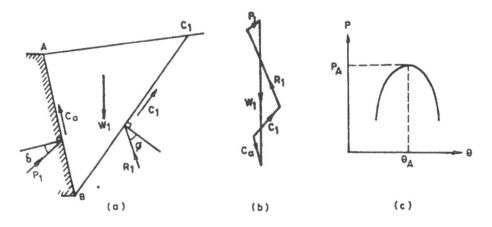

Fig. 4.17. (*a*) Forces Acting on Failure Wedge in $c - \phi$ Soil in Active State:
(*b*) Earth Pressure Versus Wedge Angle θ Plot.

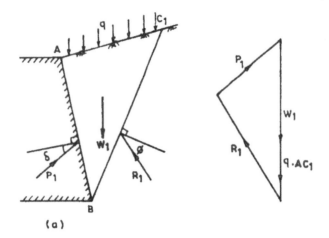

Fig. 4.18. (*a*) Section of Wall with Backfill Surface having Uniformly
Distribution Surcharge; (*b*) Force Triangle.

(Dubrova, 1963; Prakash and Basavanna, 1969; Saran and Prakash, 1970) studied the problem of pressure distribution in retaining walls analytically. Many investigators have studied the problem of distribution of earth pressure by model tests (Mackey and Kirk, 1967; Narain, Saran and Nandkumaran, 1969). The findings of these investigators indicate that the earth pressure distribution is nonlinear both in active and passive states.

For design purposes, the point of application of resultant active earth pressure in an inclined wall with $c - \phi$ backfill material having uniformly distributed load q on it (Fig. 4.19a) may be obtained by the procedures given below:

(*i*) Considering $c = q = 0$, obtain active earth using Eqs. 4.38a and b or by constructing the force polygon as shown in Figs. 4.15b and c. Let this pressure be denoted by $P_{A\gamma}$. The distribution of earth pressure may be assumed linear as shown in Fig. 4.19b.

(*ii*) The earth pressure due to surcharge only (P_{Aq}) is computed using Eq. (4.42) and its distribution along the height of wall may be taken uniform as shown in Fig. 4.19c.

(*iii*) Considering $q = 0$, obtain active earth pressure by constructing the force polygon as shown in Figs. 4.17b and c.

Let this pressure be denoted by $P_{A\gamma c}$. The magnitude of active earth pressure due to cohesion only (P_{Ac}) can be obtained using Eq. 4.44:

$$P_{Ac} = P_{A\gamma c} - P_{A\gamma} \qquad \qquad \dots (4.44)$$

The distribution of earth pressure P_{Ac} may be taken as uniform along the height of the wall as shown in Fig. 4.19d.

The net active earth pressure diagram is then obtained by summing the three diagrams algebraically. Therefore, pressure intensity at any depth z is given by

(*a*) For $p_{Aq} < p_{Ac}$

$$p_A = \frac{2P_{A\gamma}}{H} \cdot Z - \frac{P_{Ac}}{H} + \frac{P_{Aq}}{H} \qquad \qquad \dots (4.45)$$

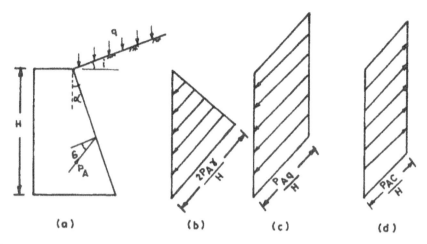

Fig. 4.19. (*a*) Section of Wall; (*b*) Active Earth Pressure Distribution ($c = q = 0$); (*c*) Active Earth Pressure Distribution ($q = \gamma = 0$); (*d*) Active Earth Pressure Distribution ($c = \gamma = 0$).

At $p_A = 0$,

$$Z = h_0 = \frac{P_{Ac} - P_{Aq}}{2 P_{A\gamma}} \qquad \qquad \text{... (4.46)}$$

The total earth pressure is given by

$$P_A = P_{A\gamma} - P_{Ac} + P_{Aq} + 1/2 \, (P_{Ac} - P_{Aq}) \cdot h_0/H \qquad \text{... (4.47)}$$

(*b*) For $P_{Aq} > P_{Ac}$

$$P_A = P_{A\gamma} - P_{Ac} + P_{Aq} \qquad \qquad \text{... (4.48)}$$

4.7 CULMANN'S GRAPHICAL CONSTRUCTION

Culmann (1866) developed a graphical construction for obtaining the earth pressures according to Coulomb's wedge theory for cohesionless soils. The method is applicable for backfill of any shape and different types of surcharge loads. It also permits location of the critical rupture surface.

Graphical procedure

Following steps are involved in the graphical construction to determine the active earth pressure (Fig. 4.20):

(*i*) Draw the wall section along with backfill surface on a suitable scale.

(*ii*) Draw a line *BS* at an angle ϕ with the horizontal through *B*. This line is known as slope line since it represents the natural slope of backfill material.

(*iii*) Draw a line *BL* at an angle $\eta = (90 - \alpha - \delta)$ below the slope line. The line *BL* is known as earth pressure line since it represents the direction of earth pressure with vertical (Fig. 4.15a).

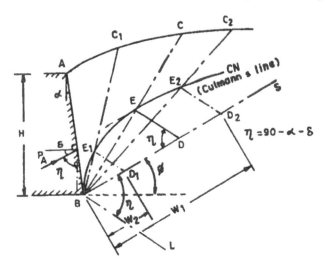

Fig. 4.20. Culmann's Graphical Method for Determining Active Earth Pressure of Cohesionless Soil.

(*iv*) Intercept BD_1, equal to the weight of wedge ABC_1, to any convenient scale along BS.

(*v*) Through D_1, draw a line D_1E_1 parallel to earth pressure line BL intersecting BC_1 at E_1.

(*vi*) Measure D_1E_1 to the same force scale as BD_1. The D_1E_1 is the earth pressure for the trial wedge ABC_1.

(*vii*) A number of trials are made and steps (*iv*) to (*vi*) are repeated for trial wedges BC_2, BC_3, etc. Line joining the points B, E_1, E_2, etc. gives the trace of earth pressures and is known as Culmann's line.

(*viii*) Draw a line parallel to BS and tangential to Culmann's line. The maximum ordinate in the direction of BL is obtained from the point of tangency which represents the active earth pressure P_A. Hence in Fig. 4.20, ED represents the magnitude of P_A. The critical slip plane is BC.

It may be noted that force triangle BD_1E_1 (Fig. 4.20) and force triangle abc (Fig. 4.15b) are identical. The curve shown in Fig. 4.15c for obtaining the maximum earth pressure is same to the Culmann's line with respect to the line BS considering ordinates parallel to earth pressure line BL.

For determination of the passive earth pressure, the slope line is drawn through the point B at an angle ϕ below the horizontal and earth pressure line at an angle of $(90 - \alpha + \delta)$ with the slope line (Fig. 4.21). Steps (*iv*) to (*viii*) described for active earth pressure are exactly same in this case also. By joining the points E_1, E_2 etc., Culmann's line for passive earth pressure is obtained. The minimum value of earth pressure obtained by drawing a tangent on the Culmann's line parallel to BL represents the value of passive earth pressure.

If a uniformly distributed surcharge load of intensity q per unit area is acting over the surface of backfill, the intercepts BD_1, BD_2, etc. will be equal to $(q \cdot AC_1 +$ weight of wedge $ABC_1)$, $(q \cdot AC_2 +$ weight of wedge $ABC_2)$ etc. Rest procedure of graphical construction to determine active and passive earth pressures is same as described above.

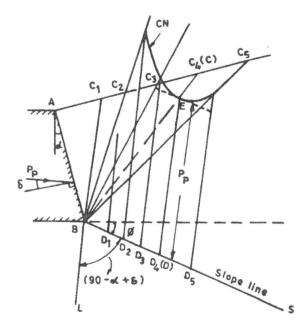

Fig. 4.21. Culmann's Graphical Method for Determining Passive Earth Pressure of Cohesionless Soil.

Earth pressure due to line load

The backfill of a retaining wall may be subjected to a concentrated surcharged load (or line load) acting along a line parallel to the crest of wall due to a railway line, wall of a building, etc. Earth pressure on the wall due to such loads can be determined by Culmann's graphical construction.

Figure 4.22 shows a section of a wall having inclined backfill with a line load of intensity q per unit length of the line at a distance x from the crest of the wall. The procedure against the wall is essentially the same as that illustrated in Fig. 4.20 with the following two differences:

(*i*) For all the sections on the right of the line load, the distance to be laid off on the slope line BS is equal to the weight of the soil wedge plus the line load q.

(*ii*) For the rupture plane through the point of action of line load (i.e. point C', Fig. 4.22), Culmann's line consists of two sections. The sections left of plane BC' is identical with C-line because the wedge bounded by planes to the left of BC' carry no surcharge. On the right side of BC', the Culmann line for the loaded backfill is located above C-line, as indicated by the solid curve CN' (Fig. 4.22), because every wedge bounded by a plane to the right of BC' is acted on by the weight q.

For the line load on the backfill, the active earth pressure is obtained as the maximum ordinate in the direction of BL from the point of tangency on the Culmann's line CN' (*i.e.* $D''E''$). The following points may be noted in Fig. 4.22.

(*a*) If the line load acts at any position on the surface of the fill between A and C'', the greatest distance is $D''E''$, and, therefore, the active earth pressure due to line load only ($D''E'' - DE = \Delta P_A$) will remain same as indicated by curve K of Fig. 4.22.

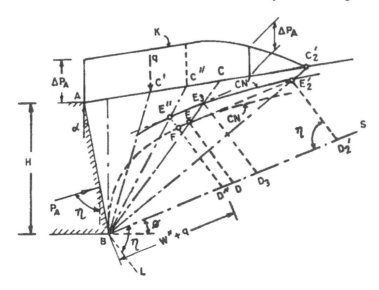

Fig. 4.22. Culmann's Graphical Method for Determining Active Earth Pressure Exerted by
Sand Backfill that Carries a Line Load.

(b) If q moves to the right of C'' to such a position C, the Culmann line consists of the
dash curve CN to the left of BC and the solid curve CN' to the right. The maximum
value of P_A is represented by $E_3 D_3$, and, therefore, ΔP_A decreases.

(c) If the line of action of q is at C_2', the value of the earth pressure $E_2' D_2'$, determined
by means of curve CN', is equal to the value of ED that represents the earth
pressures when there is no surcharge. Therefore, the value of ΔP_A becomes zero.

(d) If q moves to the right of C_2', the earth pressure determined by means of C
becomes smaller than ED. Hence, if the line load acts on the right side of C_2', it no
longer has any effect on the active earth pressure, and the active earth pressure will
be given by ED.

Point of application of earth pressure

The following simplified procedures may be used for getting the position of the point of
application of active earth pressure:

(i) No surcharge on the backfill surface (Fig. 4.23a). The point of location O_1 is located
at the point of intersection of the back of the wall and a line $O_g O_1$, which is parallel
to the surface of sliding BC and passes through the centre of gravity O_g of sliding
wedge ABC.

(ii) Uniformly distributed surcharge load on the backfill surface (Fig. 4.23b).

The point of application O_2 is located at the point of intersection of the back of wall and
a line $O_g O_2$, which is parallel to the surface of sliding BC and passes through the combined
centre of gravity O_g of weight of sliding wedge ABC and total surcharge load $q \cdot AC$.

(iii) Line load surcharge on the backfill surcharge (Figs. 4.23c and d).

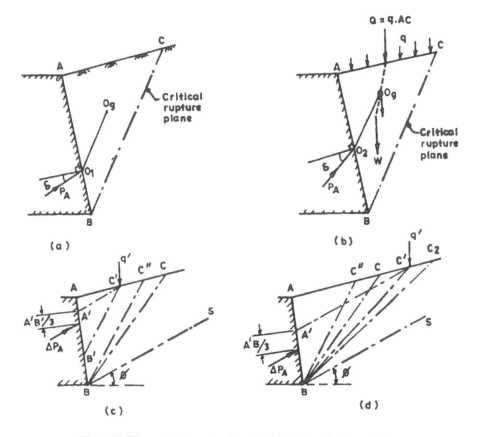

Fig. 4.23. Diagrams Illustrating Simplified Procedure for Determining
Point of Application of Active Earth Pressure.

Figures 4.23c and d illustrate the method for estimating the position of point of application of additional pressure ΔP_A produced by a line load q. The lines BC, BC'', etc. correspond to the lines BC, BC'', etc. in Fig. 4.22. If q acts between A and C'' (Fig. 4.23c), $B'C'$ is traced parallel to the surface of sliding BC'', and $A'C'$ is traced parallel to slope line BS. The force ΔP_A acts at the upper third-point of $A'B'$. If q acts between C'' and C_2, $A'C'$ is traced parallel to BS, and ΔP_A acts at the upper-third point of $A'B$ (Fig. 4.23d).

Stratified backfill

Culmann's graphical construction can also be used for getting earth pressure in stratified backfills with the simplified procedure given below (Fig. 4.24):

(*i*) Firstly, the earth pressure P_1 is obtained by Culmann's construction considering the height of wall as H_1 with backfill (γ_1, ϕ_1) surface having surcharge q per unit area. The point of application of P_1 is obtained by the procedure illustrated in Fig. 4.23b.

(*ii*) Next, the earth pressure P_2 is obtained by considering the wall of height H_2 with backfill (γ_2, ϕ_2) surface having a surcharge q' per unit area,

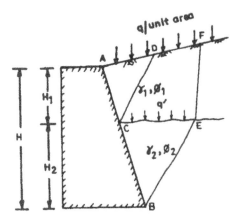

Fig. 4.24. Wall with Stratified Backfill.

where

$$q' = q + \frac{\text{Weight of soil mass } ACEF}{CE} \qquad \ldots (4.49)$$

The point of application of P_2 is obtained by the procedure illustrated in Fig. 4.23b.

(*iii*) The resultant earth pressure will be $(P_1 + P_2)$ and will act at the centre of gravity of the two earth pressures P_1 and P_2.

The procedure explained above is continued when the backfill consists of more than two strata.

4.8 DYNAMIC EARTH PRESSURE

General

In the seismic zones, the retaining walls are subjected to dynamic earth pressure, the magnitude of which is more than the static earth pressure due to ground motion. Since a dynamic load is repetitive in nature, there is a need to determine the displacement of the wall due to earthquakes and their damage potential. This becomes more essential if the frequency of the dynamic load is likely to be close to the natural frequency of the wall-backfill foundation-base soil system. This essentially consists in writing down the equation of motion of the system under free and forced vibrations. This in turn requires the information on the distribution of backfill soil mass and base soil mass participating in the vibrations. It is often difficult to assess these. Therefore, more often, pseudo-static analysis is carried out for getting dynamic earth pressure. In this method, the dynamic force is replaced by an equivalent static force.

In this article, various methods of computing the magnitude and point of application of dynamic earth pressure based on pseudo-static analysis have been discussed.

Coulomb's theory and its modification

Figure 4.25 shows a wall of height H and inclined vertically at an angle α retaining cohesionless soil with unit weight γ and angle of shearing resistance ϕ. BC_1 is the trial failure

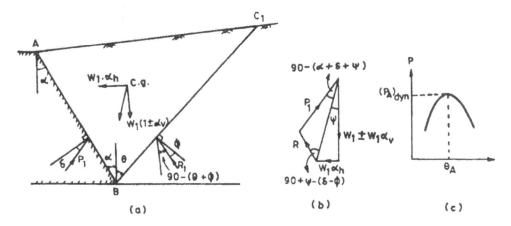

Fig. 4.25. (*a*) Forces Acting on Failure Wedge in Active State; (*b*) Force Polygon;
(*c*) Dynamic Earth Pressure Versus Wedge Angle θ Plot.

plane. During an earthquake the inertia force may act on the assumed failure wedge ABC_1 both horizontally and vertically. If a_h and a_v are the horizontal and vertical accelerations caused by the earthquake on the wedge ABC_1, the corresponding inertial forces are $W_1.a_h/g$ horizontally and $W_1.a_v/g$ vertically, W_1 being the weight of the wedge ABC_1. During the worst condition, $W_1.a_h/g$ acts towards the fill and $W_1.a_v/g$ may act vertically either in the downward or upward direction. Therefore, the direction that gives the maximum increase in earth pressure is adopted in practice.

If α_h and α_v are respectively the horizontal and vertical seismic coefficients, then

$$\alpha_h = \frac{a_h}{g} \qquad \text{... (4.50)}$$

$$\alpha_v = \frac{a_v}{g} \qquad \text{... (4.51)}$$

For the failure condition the soil wedge ABC_1 is in equilibrium under the following forces:

(*i*) W_1 weight of the wedge acting at the centre of gravity of ABC_1.

(*ii*) Earth pressure P_1 inclined at an angle δ to normal to the wall in the anticlockwise direction.

(*iii*) Soil reaction R_1 inclined at an angle φ to the normal on the face BC_1.

(*iv*) Horizontal inertia force ($W_1 \times \alpha_h$) acting at the centre of gravity of the wedge ABC_1.

(*v*) Inertia force $\pm W_1 \alpha_v$ acting at the centre of gravity of the wedge.

Weight W_1 and the inertia forces $W_1\alpha_h$ and $\pm W_1 \alpha_v$ can be combined to give a resultant $\overline{W_1}$, where

$$\overline{W_1} = W_1[(1 \pm \alpha_v)^2 + \alpha_h^2]^{\frac{1}{2}} \qquad \text{... (4.52)}$$

The resultant $\overline{W_1}$ is vertically inclined at angle ψ, such that

$$\psi = \tan^{-1} \left[\frac{\alpha_h}{1 \pm \alpha_v} \right] \qquad \ldots (4.53)$$

The directions of all the three forces $\overline{W_1}$, P_1 and R_1 are known but the magnitude of only one force $\overline{W_1}$ is known. The magnitude of the other forces can be obtained by considering the force polygon as shown in Fig. 4.25b. P_1 is the value of dynamic earth pressure corresponding to the trial wedge ABC_1. More trials are made and the values of P_2, P_3, etc. are obtained. Variation of P and θ is shown in Fig. 4.25c. The maximum value of P is the dynamic active earth pressure $(P_A)_{dyn}$.

Coulomb's modified theory can also be used for determining the dynamic active earth pressure for a backfill having both cohesion and angle of internal friction by considering additional forces due to cohesion and adhesion on the failure plane and wall face respectively. The force polygon may then be drawn as illustrated in Fig. 4.17b for evaluating earth pressure in static condition.

Mononobe and Okabe (1929) were the first to modify Coulomb's expression for incorporating the effect of inertia force. They gave the following relation for the computation of dynamic active earth pressure:

$$(P_A)_{dyn} = \frac{1}{2} \gamma H^2 (K_A)_{dyn} \qquad \ldots (4.54)$$

where

$$(K_A)_{dyn} = \frac{(1 \pm \alpha_v) \cos^2(\phi - \psi - \alpha)}{\cos \psi \cos^2 \alpha \cos(\delta + \alpha + \psi)} \times \left[\frac{1}{1 + \left\{ \frac{\sin(\phi + \delta) \sin(\phi - i - \psi)}{\cos(\alpha - i) \cos(\delta + \alpha + \psi)} \right\}^{\frac{1}{2}}} \right]^2 \qquad \ldots (4.55)$$

The expression of $(K_A)_{dyn}$ gives two values depending on the sign of α_v. For design purposes the maximum of the two should be taken. Mononobe and Okabe also gave the expression for the computation of dynamic passive earth pressure which is:

$$(P_P)_{dyn} = \frac{1}{2} \gamma H^2 (K_P)_{dyn} \qquad \ldots (4.56)$$

where

$$(K_P)_{dyn} = \frac{(1 \pm \alpha_v) \cos^2 (\phi + \alpha - \psi)}{\cos \psi \cos^2 \alpha \cos (\delta - \alpha + \psi)} \times \left[\frac{1}{1 - \left\{ \frac{\sin (\phi + \delta) \sin (\phi + i - \psi)}{\cos (\alpha - i) \cos (\delta - \alpha + \psi)} \right\}^{\frac{1}{2}}} \right]^2 \quad \text{... (4.57)}$$

For design purposes, the minimum value of $(K_P)_{dyn}$ will be taken out of its two values corresponding to $\pm \alpha_v$.

Effect of uniform surcharge

The additional active and passive dynamic earth pressures against the wall due to the uniform surcharge of intensity q per unit area on the inclined earth fill surface shall be:

$$(P_{Aq})_{dyn} = \frac{qH \cos \alpha}{\cos (\alpha - i)} (K_A)_{dyn} \quad \text{... (4.58)}$$

$$(P_{Pq})_{dyn} = \frac{qH \cos \alpha}{\cos (\alpha - i)} (K_P)_{dyn} \quad \text{... (4.59)}$$

Effect of saturation on lateral dynamic earth pressure

For saturated earth fill, the saturated unit weight of soil shall be adopted.

For submerged earth fill, the dynamic increment (or decrement) in active and passive earth pressure during earthquakes shall be found with the following modifications:

(i) The value of δ shall be taken as 1/2 the value of the δ for dry backfill.

(ii) The value of ψ shall be taken as

$$\psi = \tan^{-1} \left[\frac{\gamma_s}{\gamma_s - 1} \right] \frac{\alpha_h}{1 \pm \alpha_v} \quad \text{... (4.60)}$$

where γ_s = saturated unit weight of the soil.

(iii) Submerged unit weight shall be adopted.

(iv) From the value of earth pressure found above, subtract the value of earth pressure by putting $\alpha_h = \alpha_v = \psi = 0$ using submerged unit weight. The remainder shall be the dynamic increment or decrement.

Hydrodynamic pressure on account of water contained in earth fill shall not be considered separately as the effect of acceleration on water has been taken indirectly.

Partially submerged backfill

The ratio of lateral dynamic increment in active pressures to the vertical pressures at various depths along the height of wall may be taken as shown in Fig. 4.26. The pressure

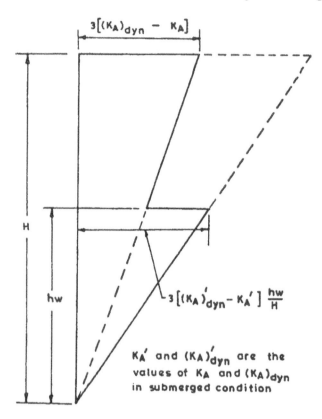

Fig. 4.26. Distribution of the Ratio $\dfrac{\text{Lateral Dynamic Increment}}{\text{Vertical Effective Pressure}}$ with Height of Wall.

distribution of dynamic increment in active pressures may be obtained by multiplying the vertical effective pressures by the coefficients in Fig. 4.26 at corresponding depths.

A similar procedure as described above may be utilized for determining the dynamic decrement in passive pressures.

Modified Culmann construction

Kapila (1962) modified the Culmann's graphical construction for obtaining dynamic active and passive earth pressures.

Different steps in modified construction for determining dynamic active earth pressure are as follows (Fig. 4.27):

(*i*) Draw the wall section along with backfill surface on a suitable scale.

(*ii*) Draw BS at an angle $(\phi - \psi)$ with the horizontal.

(*iii*) Draw BL at an angle of $(90 - \alpha - \delta - \psi)$ below BS.

(*iv*) Intercept BD_1 equal to the resultant of the wedge W_1, and inertial forces ($\pm W_1 \alpha_v$ and

$W_1 \alpha_h$). The magnitude of this resultant is $\overline{W}_1$ and is inclined at an angle ψ to the vertical.

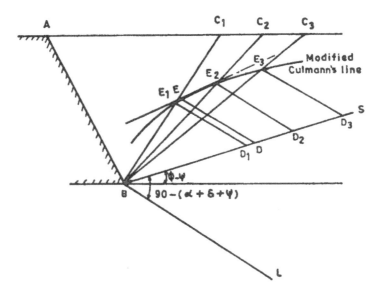

Fig. 4.27. Modified Culmann's Construction for Dynamic Active Earth Pressure.

(*v*) Through D_1 draw D_1E_1 parallel to BL intersecting BC_1 at E_1.

(*vi*) Measure D_1E_1 to the same force scale as BD_1. The D_1E_1 is the dynamic earth pressure for trial wedge.

(*vii*) Repeat steps (*iv*) to (*vi*) with BC_2, BC_3 etc. as trial wedges.

(*viii*) Draw a smooth curve through $BE_1E_2E_3$. This is the modified Culmann's line.

(*ix*) Draw a line parallel to BS and tangential to this curve. The maximum coordinate in the direction of BL is obtained from the point of tangency and is the dynamic active earth pressure, $(P_A)_{dyn}$.

For determining the passive earth pressure draw BS at $(\phi - \psi)$ below horizontal. Next draw BL at $(90 - \alpha - \delta + \psi)$ below BS. The other steps for construction remain unaltered (Fig. 4.28).

Effect of uniformly distributed load and line load on the backfill surface may be handled in a similar way as for the static case.

4.9 DYNAMIC ACTIVE EARTH PRESSURE FOR *c – f* SOILS

The solutions so far discussed consider the soil to be cohesionless. A general solution for the determination of total (static plus dynamic) earth pressures for a $c - \phi$ soil has been developed by Prakash and Saran, 1966 and Saran and Prakash, 1968.

Figure 4.29 shows a section of a wall whose face AB is in contact with soil. The soil retained is horizontal and carries a uniform surcharge. The inclination of the wall AB with vertical is α and inclination of the trial failure surface is θ_1. $AECD$ is the cracked zone in clayey soils, DC being depth h_0 below AD. h_0 is expressed in terms of H as below :

$$h_0 = n (H_1 - h_0) = nH \qquad \qquad \dots (4.61)$$

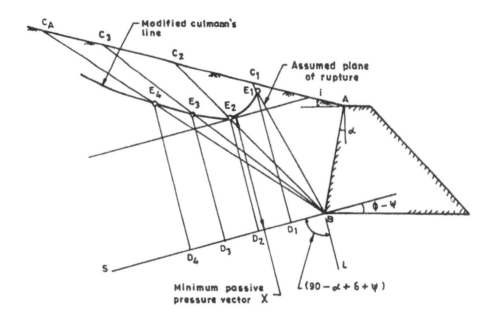

Fig. 4.28. Modified Culmann's Construction for Dynamic Passive Earth Pressure.

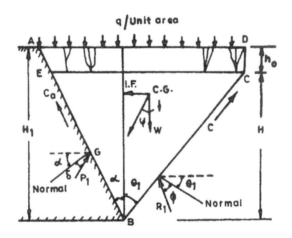

Fig. 4.29. Forces Acting on Failure Wedge in Active State for Seismic Condition in $c - \phi$ Soil.

where H_1 = total height of retaining wall, and
          H = height of retaining wall free from cracks.

In this analysis only horizontal inertia force is considered. All the forces acting on the assumed failure wedge $AEBCD$ are listed in Table 4.3, along with their horizontal and vertical components.

A summation of all the vertical components gives

$$\frac{1}{2}\gamma H^2 (\tan \alpha + \tan \theta_1) + \gamma n H^2 (\tan \alpha + \tan \theta_1) +$$

$$\frac{1}{2}\gamma n^2 H^2 \tan \alpha - cH - c'H + qH (\tan \alpha + \tan \phi + n \tan \alpha) =$$

$$P_1 \sin (\alpha + \delta) + R_1 \sin (\theta_1 + \phi) \qquad \qquad ... (4.62)$$

A summation of all the horizontal components gives

$$- cH \tan \theta_1 + c'H \tan \alpha + (W + Q)\alpha_h = P_1 \cos (\alpha + \delta) - R_1 \cos (\theta_1 + \phi) \quad ... (4.63)$$

Multiply Eq. 4.62 by $\cos (\theta_1 + \phi)$, Eq. 4.63 by $\sin (\theta_1 + \phi)$, substitute for W and Q from Table 4.3 and $c = c'$, and adding, we get

$$P_1 \sin (\beta + \delta) = \gamma H^2 [(n + 1/2) (\tan \alpha + \tan \theta_1) + n^2 \tan \alpha] \times$$

$$[\cos (\theta_1 + \phi) + \alpha_h \sin (\theta_1 + \phi)] + qH [(n + 1) \times$$

$$\tan \alpha + \tan \theta_1] [\cos (\theta_1 + \phi) + \alpha_h \sin (\theta_1 + \phi)] -$$

$$cH [\cos \beta \sec \alpha + \cos \phi \sec \theta_1] \qquad \qquad ... (4.64)$$

where $\beta = \theta_1 + \phi + \alpha$.

Table 4.3. Computation of Force Acting on Wedge *AEBCD* (Fig. 4.28)

Designation	Vertical Component		Horizontal Component	
1. Weight of wedge $ABCD(W)$	$1/2 \gamma H^2(\tan \alpha + \tan \theta_1) +$ $\gamma n H^2(\tan \alpha + \tan \theta_1) +$ $1/2 \gamma n^2 H^2(\tan \alpha)$	↓	—	
2. Cohesion, C $cH \sec \theta_1$	cH	↑	$cH \tan \theta_1$	→
3. Adhesion, C_a $c'H \sec \alpha$	$c'H$	↑	$c'H \tan \alpha$	←
4. Surcharge Q	$qH [(\tan \alpha + \tan \theta_1) +$ $nH \tan \alpha]$	↓	—	
5. Soil reaction R_1	$R_1 \sin (\theta_1 + \phi)$	↑	$R_1 \cos (\theta_1 + \phi)$ ←	
6. Inertia force I_F	—		$(W + Q)\alpha_h$ ←	
7. Earth pressure P_1	$P_1 \sin (\alpha_1 + \delta)$	↑	$P_1 \cos (\alpha_1 + \delta)$ →	

Introducing the following dimensionless parameters:

$$(N_{ac})_{dyn} = \frac{\cos \beta \sec \alpha + \cos \phi \sec \theta_1}{\sin (\beta + \delta)} \qquad \qquad ... (4.65)$$

$$(N_{aq})_{dyn} = \frac{[(n+1)\tan\alpha + \tan\theta_1][\cos(\theta_1+\phi) + \alpha_h \sin(\theta_1+\phi)]}{\sin(\beta+\delta)} \qquad \dots (4.66)$$

$$(N_{a\gamma})_{dyn} = \frac{[(n+1/2)(\tan\alpha + \tan\theta_1) + n^2 \tan\alpha][\cos(\theta_1+\phi) + \alpha_h \sin(\theta_1+\phi)]}{\sin(\beta+\delta)}$$

$$\dots (4.67)$$

we get $\qquad\qquad (P_1)_{dyn} = \gamma H^2 (N_{a\gamma})_{dyn} + qH(N_{aq})_{dyn} - cH(N_{ac})_{dyn} \qquad \dots (4.68)$

where $(N_{ac})_{dyn}$, $(N_{aq})_{dyn}$ and $(N_{a\gamma})_{dyn}$ are earth pressure coefficients which depend on α, n, ϕ, δ and θ_1. For given parameters of wall and soil, the values of $(N_{ac})_{dyn}$, $(N_{aq})_{dyn}$ and $(N_{a\gamma})_{dyn}$ are computed for different wedge angles θ_2, θ_3 etc. The variation of these coefficients with respect to wedge angle θ are shown in Fig. 4.30, and the maximum values of $(N_{aq})_{dyn}$ and $(N_{a\gamma})_{dyn}$ and minimum value of $(N_{ac})_{dyn}$ are obtained.

For such a condition, the earth pressure corresponds to dynamic active earth pressure. Eq. 4.68 can be written as

$$(P_A)_{dyn} = \gamma H^2 (N_{a\gamma m})_{dyn} + qH(N_{aqm})_{dyn} - cH(N_{acm})_{dyn} \qquad \dots (4.69)$$

For static case, $\alpha_h = 0$; earth pressure coefficients become

$$(N_{ac})_{stat} = \frac{\cos\beta \sec\alpha + \cos\phi \sec\theta_1}{\sin(\beta+\delta)} \qquad \dots (4.70)$$

$$(N_{aq})_{stat} = \frac{[(n+1)\tan\alpha + \tan\theta_1]\cos(\theta_1+\phi)}{\sin(\beta+\delta)} \qquad \dots (4.71)$$

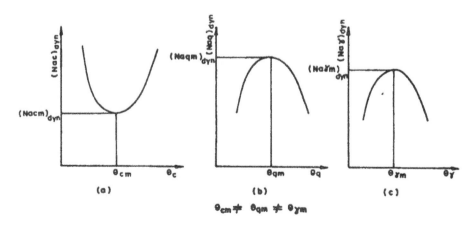

Fig. 4.30. (a) $(N_{ac})_{dyn}$ Versus θ Plot; (b) $(N_{aq})_{dyn}$ Versus θ Plot; (c) $(N_{a\gamma})_{dyn}$ Versus θ Plot.

$$(N_{a\gamma})_{stat} = \frac{[(n+1/2)(\tan\alpha + \tan\theta_1) + n^2 \tan\alpha]\cos(\theta_1 + \phi)}{\sin(\beta + \delta)} \qquad \ldots (4.72)$$

Minimum values of $(N_{ac})_{stat}$, and maximum values of $(N_{aq})_{stat}$ and $(N_{a\gamma})_{stat}$ can be obtained in a similar manner as illustrated above for getting the dynamic earth pressure coefficients. It is found convenient to obtain the dynamic earth pressure coefficients from the following constants:

$$\lambda_1 = \frac{(N_{aqm})_{dyn}}{(N_{aqm})_{stat}} \qquad \ldots (4.73)$$

$$\lambda_2 = \frac{(N_{a\gamma m})_{dyn}}{(N_{a\gamma m})_{stat}} \qquad \ldots (4.74)$$

N_{acm} both for the static and dynamic case is same and has been plotted in Fig. 4.31 for different inclinations of the wall varying from $+20°$ to $-20°$ to the vertical. As is evident from Eq. 4.70, N_{acm} factor is also independent of n.

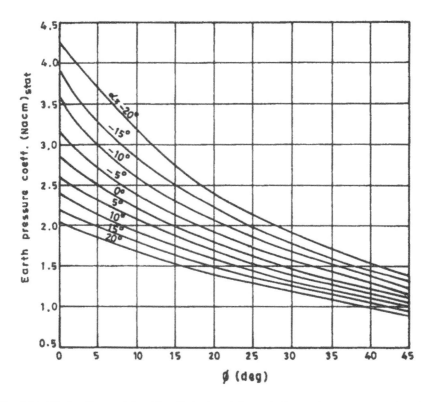

Fig. 4.31. $(N_{acm})_{stat}$ Versus ϕ for All n (Prakash and Saran, 1966, and Saran and Prakash, 1968).

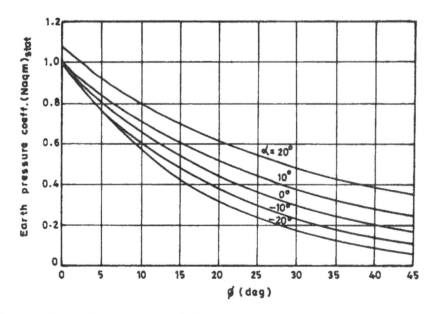

Fig. 4.32. $(N_{aqm})_{stat}$ Versus ϕ for $n = 0$ (Prakash and Saran, 1966, and Saran and Prakash, 1968).

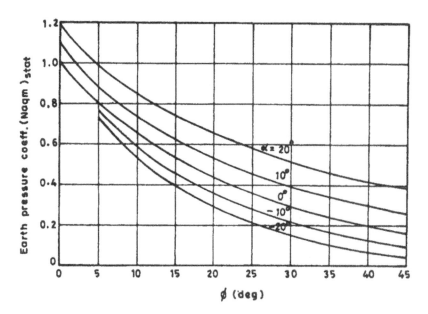

Fig. 4.33. $(N_{aqm})_{stat}$ Versus ϕ for $n = 0.2$ (Prakash and Saran, 1966, and Saran and Prakash, 1968).

Figures 4.32, 4.33 and 4.34 show plots of $(N_{aqm})_{stat}$ versus ϕ for n of 0, 0.2 and 0.4 respectively. These plots consider the inclination of the wall from $+20°$ to $-20°$. Plots of $(N_{a\gamma m})_{stat}$ for the same range of n, ϕ and $\propto$ have been drawn in Figs. 4.35, 4.36 and 4.37.

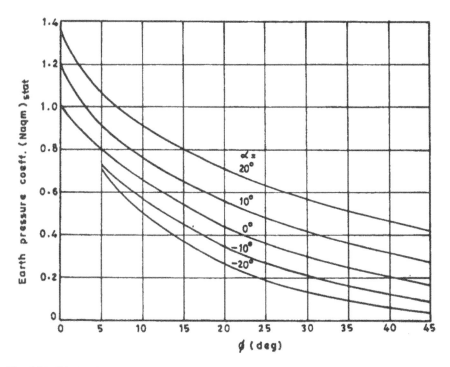

Fig. 4.34. $(N_{aqm})_{stat}$ Versus ϕ for $n = 0.4$ (Prakash and Saran, 1966, and Saran and Prakash, 1968).

It is found that the values of λ_1 and λ_2 alter slightly with increase in n. It is, therefore, recommended that the effect of n on λ_1 and λ_2 may not be considered. Secondly, it is observed that λ_1 and λ_2 are almost same (Prakash and Saran, 1966; Saran and Prakash, 1968). Hence, only one value of $\lambda(= \lambda_1 = \lambda_2)$ is recommended (Fig. 4.38). Since λ is the ratio of earth pressure coefficients in (i) dynamic, and (ii) static case, and both the coefficients decrease with ϕ, the shape of the curves for different α_h values indicate the rate of decrease of one in relation to the other.

4.10 POINT OF APPLICATION

According to Indian Standard (IS: 1893–1984) Specifications, the pressures are located as follows:

From the total pressures computed from Eqs. 4.54 and 4.56 or from graphical construction, subtract the static pressure obtained by putting $\alpha_h = \alpha_v = 0$. The remainder is the dynamic increment (in active case). The static component of the total pressure shall be applied at an elevation $H/3$ above the base of wall. The point of application of the dynamic increment and dynamic decrement shall be assumed to be at an elevation $H/2$ and $2H/3$ respectively above the base of the wall.

The static component of total active and passive earth pressure due to uniformly distributed surcharge on the backfill surface obtained by putting $\alpha_h = \alpha_v = 0$ in Eqs. 4.58 and 4.59 shall be applied at $H/2$ above the base of the wall. The point of application of both the

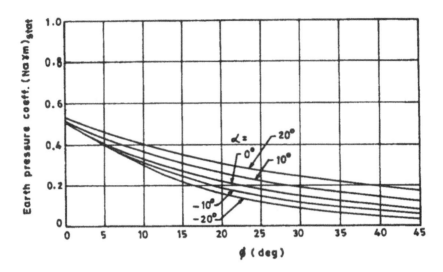

Fig. 4.35. $(N_{a\gamma m})_{stat}$ Versus ϕ for $n = 0.0$ (Prakash and Saran, 1966, and Saran and Prakash, 1968).

dynamic increment and dynamic decrement in this case shall be assumed to be at an elevation $2H/3$ above the base of the wall.

The static and dynamic active earth pressures due to cohesion only ($q = \gamma = 0$) are same. The point of application of this pressure shall be assumed to be at an elevation of $H/2$ above the base of the wall.

ILLUSTRATIVE EXAMPLES

Example 4.1

A smooth vertical retaining wall 7.0 m high retains a cohesionless horizontal backfill. The properties of the fill are $e = 0.49$, $\phi = 30°$ and $G_s = 2.65$. Compute the magnitude of total lateral pressure and its point of application when

(a) The backfill is dry and the wall is restrained against yielding.

(b) Wall free to yield, water table at 2.0 m depth and there is no drainage. Soil above water table is saturated.

Solution

(i)
$$\gamma_d = \frac{G_s \gamma_w}{1+e} = \frac{2.65 \times 10}{1+0.49} = 17.8 \text{ kN/m}^3$$

$$\gamma_s = \frac{G_s + e}{1+e} \gamma_w = \frac{2.65 + 0.49}{1+0.49} \times 10 = 21.1 \text{ kN/m}^3$$

$$\gamma_b = \gamma_s - \gamma_w = 21.1 - 10.0 = 11.1 \text{ kN/m}^3$$

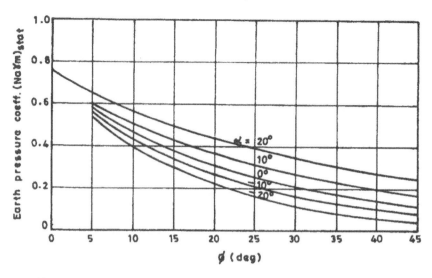

Fig. 4.36. $(N_{a\gamma m})_{stat}$ Versus ϕ for $n = 0.2$ (Prakash and Saran, 1966, and Saran and Prakash, 1968).

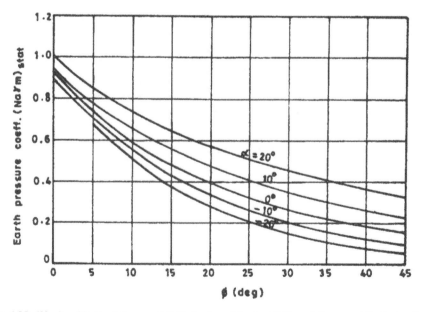

Fig. 4.37. $(N_{a\gamma m})_{stat}$ Versus ϕ for $n = 0.4$ (Prakash and Saran, 1966, and Saran and Prakash, 1968).

(ii) If the wall is restrained against yielding, the lateral pressure that would develop would be 'earth pressure at rest'.

$$K_0 = 1 - \sin \phi = 1 - \sin 30° = 0.5$$

$$p_0 = K_0 \, \gamma_d \, H = 0.5 \times 17.8 \times 7.0 = 62.3 \text{ kN/m}^2$$

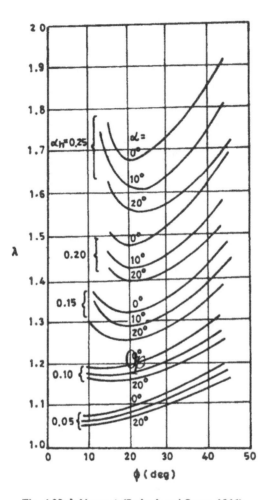

Fig. 4.38. λ Versus ϕ (Prakash and Saran, 1966).

Total pressure per metre length of wall is

$$P_0 = 1/2 \times 62.3 \times 7.0 = 218.0 \text{ kN/m.}$$

Point of application from bottom of wall

$$= \frac{7.0}{3} = 2.34 \text{ m.}$$

(*iii*) If the wall is free to yield, the lateral pressure that would develop would be active earth pressure:

$$K_A = \frac{1 - \sin \phi}{1 + \sin \phi} = \frac{1 - \sin 30}{1 + \sin 30} = \frac{1}{3}$$

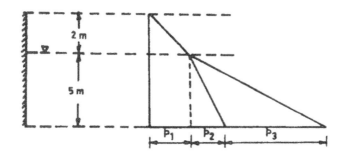

Fig. 4.39. Earth Pressure Distribution Diagram.

Refer Fig. 4.39

$$p_1 = K_A \gamma_s\, 2.0 = \frac{1}{3} \times 21.1 \times 2.0 = 14.06 \text{ kN/m}^2$$

$$p_2 = K_A \gamma_b\, 5.0 = \frac{1}{3} \times 11.1 \times 5.0 = 18.5 \text{ kN/m}^2$$

$$p_3 = \gamma_w\, 5.0 = 10.0 \times 5.0 = 50.0 \text{ kN/m}^2$$

$$P_A = \frac{1}{2} p_1 \times 2.0 + p_1 \times 5.0 + \frac{1}{2} p_2 \times 5.0 + \frac{1}{2} p_3 \times 5.0$$

$$= \frac{1}{2} \times 14.06 \times 2.0 + 14.06 \times 5.0 + \frac{1}{2} \times 18.5 \times 5.0 + \frac{1}{2} \times 50.0 \times 5.0$$

$$= 14.06 + 70.30 + 46.25 + 125.00 = 255.61 \text{ kN/m}$$

Point of application from bottom

$$= \frac{p_1\,(5 + 2/3) + 5p_1\,(5/2) + 2.5p_2\,(5/3) + 2.5p_3\,(5/3)}{p_1 + 5p_1 + 2.5p_2 + 2.5p_3}$$

$$= \frac{14.06 \times 17/3 + 70.30 \times 2.5 + 46.25 \times 5/3 + 125.00 \times 5/3}{255.61}$$

$$= 2.21 \text{ m.}$$

Example 4.2

A smooth vertical retaining wall supports a horizontal backfill with properties $c = 10$ kN/m^2, $\phi = 20°$ and $\gamma_t = 17.5$ kN/m^3. Determine (a) total active earth pressure, (b) position of zero lateral pressure, and (c) distance of centre of pressure from base.

Solution

$$(i)\ \ K_A = \frac{1 - \sin\phi}{1 + \sin\phi} = \frac{1 - \sin 20}{1 + \sin 20} = 0.49.$$

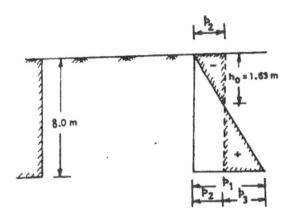

Fig. 4.40. Earth Pressure Distribution Diagram.

Refer Fig. 4.40

$$p_A = \gamma H K_A - 2c \sqrt{K_A}$$
$$p_1 = 17.5 \times 8 \times 0.49 = 68.6 \text{ kN/m}^2$$
$$p_2 = 2 \times 10 \times \sqrt{0.49} = 14.0 \text{ kN/m}^2$$
$$p_3 = 68.6 - 14.0 = 54.6 \text{ kN/m}^2$$

(*ii*) Total pressure taking negative pressure into consideration
$$P_A = 1/2 \, p_1 \times 8.0 - p_2 \times 8.0$$
$$= 1/2 \times 68.6 \times 8.0 - 14.0 \times 8.0 = 274.4 - 112.0 = 162.4 \text{ kN/m}$$

(*iii*) Section of zero pressure intensity
$$h_0 = \frac{2c}{\gamma} \times \frac{1}{\sqrt{K_A}} = \frac{2 \times 10}{17.5} \times \frac{1}{\sqrt{0.49}} = 1.63 \text{ m}$$

(*iv*) Amount of negative pressure
$$= \frac{1}{2} \times p_2 \times h_0 = \frac{1}{2} \times 14.0 \times 1.63 = 11.41 \text{ kN/m}$$

Total pressure neglecting negative pressure
$$P_A = 162.4 + 11.41 = 173.81 \text{ kN/m}$$

(*v*) Point of application of total earth pressure taking negative pressure into consideration

$$= \frac{274.4 \times 8/3 - 112.0 \times 8/2}{162.4} = 1.75 \text{ m}$$

(*vi*) Point of application of total earth pressure neglecting negative pressure

$$= \frac{274.4 \times 8/3 - 112.0 \times 8/2 + 11.41 \{(8 - 1.63)/3\}}{173.81}$$

$= 2.12$ m.

Example 4.3

A 6.0 m high retaining wall with back face inclined 20° to the vertical retains cohesionless backfill ($\phi = 33°$, $\gamma_t = 18$ kN/m³ and $\delta = 20°$). The backfill surface is sloping at an angle 10° to the horizontal.

(*a*) Determine the total active earth pressure using Coulomb's theory and Culmann's graphical construction.

(*b*) If the retaining wall is located in a seismic region ($\alpha_h = 0.1$), determine total active earth pressure using Mononobe's equation and modified Culmann's graphical construction.

(*c*) A vertical line load of 50 kN/m is acting at a horizontal distance of 3.0 m from the wall parallel to the crest of the wall. What is the magnitude of total active thrust. What is the minimum horizontal distance at which the line load can be located from the back of the wall, if there is to be no increase in the total active thrust.

Solution

(*i*) Static active earth pressure

$$P_A = \frac{1}{2} \gamma H^2 \frac{\cos^2(\phi - \alpha)}{\cos^2 \alpha \cos(\delta + \alpha)} \cdot \frac{1}{\left\{ 1 + \left[\frac{\sin(\phi + \delta) \sin(\phi - i)}{\cos(\alpha - i) \cos(\delta + \alpha)} \right]^{\frac{1}{2}} \right\}^2}$$

$$= \frac{1}{2} \times 18 \times 6.0^2 \times \frac{\cos^2(33 - 20)}{\cos^2 20 \cos(20 + 20)} \times \frac{1}{\left\{ 1 + \left[\frac{\sin(33 + 20) \sin(33 - 10)}{\cos(20 - 10) \cos(20 + 20)} \right]^{\frac{1}{2}} \right\}^2}$$

$= 168.42$ kN/m

(*ii*) Refer Fig 4.41 for Culmann's graphical construction for getting static active pressure. $D_s E_s$ gives the total active earth pressure.

$P_A = 1.7 \times 100 = 170$ kN/m

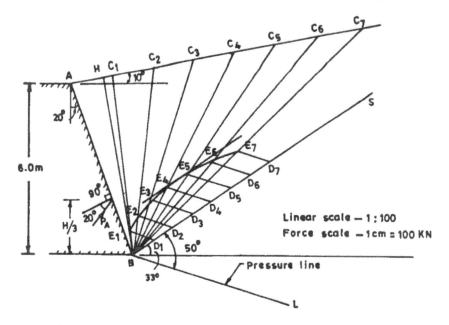

Fig. 4.41 Culmann's Graphical Construction.

(iii) Dynamic active earth pressure

$$(P_A)_{\text{dyn}} = \frac{1}{2}\gamma H^2 \frac{\cos^2(\phi - \psi - \alpha)(1 \pm \alpha_v)}{\cos\psi \cos^2\alpha \cos(\delta + \alpha + \psi)} \times \frac{1}{\left\{1 + \left[\dfrac{\sin(\phi + \delta)\sin(\phi - i - \psi)}{\cos(\alpha - i)\cos(\delta + \alpha + \psi)}\right]^{\frac{1}{2}}\right\}^2}$$

Assuming $\quad \alpha_v = \dfrac{\alpha_h}{2} = \dfrac{0.1}{2} = 0.05$

$$\psi = \tan^{-1}\frac{\alpha_h}{1 \pm \alpha_v} = \tan^{-1}\frac{0.1}{1 \pm .05}$$

$$= 5.44° \text{ with } +\alpha_v \text{ and}$$

$$= 6.0° \text{ with } -\alpha_v.$$

Value of $(P_A)_{\text{dyn}}$ with $(+)\ \alpha_v$

$$= \frac{1}{2} \times 18 \times 6.0^2 \; \frac{\cos^2(33 - 5.44 - 20)\,(1 + 0.05)}{\cos 5.44 \cos^2 20 \cos (20 + 20 + 5.44)} \times$$

$$\cfrac{1}{\left\{ 1 + \left[\cfrac{\sin(33 + 20)\sin(33 - 10 - 5.44)}{\cos(20 - 10)\cos(20 + 20 + 5.44)} \right]^{\frac{1}{2}} \right\}^2}$$

$$= 214.26 \text{ kN/m}$$

Value of $(P_A)_{dyn}$ with $(-)\,\alpha_v$

$$= \frac{1}{2} \times 18 \times 6.0^2 \; \frac{\cos^2(33 - 6 - 20)\,(1 - 0.05)}{\cos 6 \cos^2 20 \cos (20 + 20 + 6)} \times$$

$$\cfrac{1}{\left\{ 1 + \left[\cfrac{\sin(33 + 20)\sin(33 - 10 - 6)}{\cos(20 - 10)\cos(20 + 20 + 6)} \right]^{\frac{1}{2}} \right\}^2}$$

$$= 198.05 \text{ kN/m}.$$

Therefore $(+)\,\alpha_v$ case governs the value of dynamic active earth pressure.

Hence, $(P_A)_{dyn} = 214.26$ kN/m

(*iv*) Refer Fig. 4.42 for modified Culmann's graphical construction for getting dynamic active earth pressure.

D_5E_5 gives the total dynamic active earth pressure.

$(P_A)_{dyn} = 2.1 \times 100 = 210$ kN/m

(*v*) Refer Fig. 4.43 for Culmann's graphical construction for getting the effect of line load.

The magnitude of horizontal thrust when the line load is placed at 3 m distance from the wall is given by E'_2L_2. Hence

$P_A = 1.8 \times 100 = 180$ kN.

For no increase of the total active earth pressure on the wall, the minimum distance of the line load from the wall is AC_5, i.e. 9.0 m.

Example 4.4

A retaining wall 8.0 m high is inclined 20° to the vertical and retains horizontal backfill with following properties:

$$\gamma_t = 18 \text{ kN/m}^3, \; \phi = 30° \text{ and } c = 6.0 \text{ kN/m}^2.$$

There is a superimposed load of intensity 15 kN/m² on the backfill. The wall is located in a seismic region having horizontal seismic coefficient of 0.1. Compute the dynamic active earth pressure and determine the percentage increase in pressure over the static earth pressure.

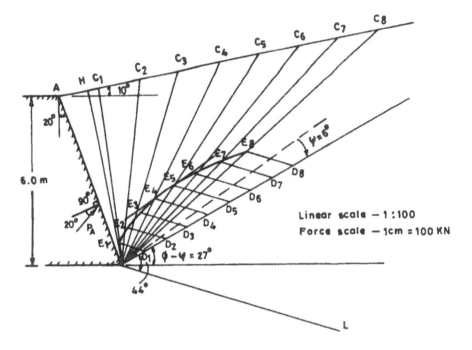

Fig. 4.42. Culmann's Graphical Construction.

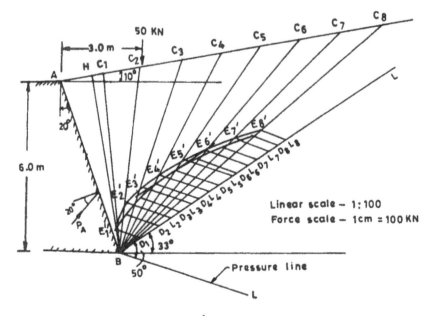

Fig. 4.43. Culmann's Graphical Construction.

Solution

(i) Assumption $\delta = 2/3 \; \phi$

$\qquad\qquad \alpha_v = 0$

(ii) $h_0 = \dfrac{2c}{\gamma} \cdot \dfrac{1}{\sqrt{K_A}}$ where $K_A = \dfrac{1 - \sin 30}{1 + \sin 30} = \dfrac{1}{3}$

$\qquad\quad = \dfrac{2 \times 6.0}{18} \times \sqrt{3} = 1.15$ m

$\qquad H = H_1 - h_0 = 8.0 - 1.15 = 6.85$ m

$\qquad n = \dfrac{h_0}{H} = \dfrac{1.15}{6.85} = 0.167$

(iii) For $\phi = 30°$, $\alpha = +20°$ and $n = 0.167$

Figs. 4.31 to 4.37 give:

$(N_{aym})_{stat} = 0.33$, $(N_{aqm})_{stat} = 0.512$ and $(N_{acm})_{stat} = 1.2$

$(P_A)_{stat} = 18 \times 6.85^2 \times 0.33 + 15 \times 6.85 \times 0.512 - 6 \times 6.85 \times 1.2$

$\qquad\quad = 282.0$ kN/m

(iv) For $\phi = 30°$ $\alpha = +20°$ and $\alpha_h = 0.1$, Fig. 4.38 gives

$\qquad \lambda = 1.19$.

Therefore,

$(N_{aqm})_{dyn} = 1.19 \times 0.512 = 0.609$ and

$(N_{aym})_{dyn} = 1.19 \times 0.33 = 0.393$

$(P_A)_{dyn} = 18 \times 6.85^2 \times 0.393 + 15 \times 6.85 \times 0.609 - 6 \times 6.85 \times 1.2$

$\qquad\quad = 345.18$ kN/m

(v) Percentage increase over static pressure

$$= \frac{345.18 - 282.00}{282.00} \times 100 = 22.4\%.$$

REFERENCES

1. Alpan, I. (1967), "The empirical evaluation of the coefficients K_0 and K_{0r}", *Soils and Foundation*, Vol. VIII, No. 1, pp. 31-40.

2. Brooker, E.W. and H.O. Ireland (1965), "Earth pressure at rest related to stress history," *Canadian Geotechnical Journal*, Vol. II, No. 1, p. 1.

3. Coulomb, C.A. (1776), "Essai sur une application des maximis et minimis a quelques problems des statique relatifs a l' architecture," *Nem. div. Sav. Acad. Sci.*, Vol. 7.

4. Culmann, K. (1866), *Die Graphische Statik*, Mayer and Zeller, Zurich.

5. Dubrova, G.A. (1963), "Interaction of soil and structures," *Izd. Rechnoy Transport*, Moscow, 1963.

6. Fraser, A.M. (1957). "The influence of stress ratio on compressibility and pore pressure coefficients in compacted soils," Ph.D. Thesis, London.

7. IS 1893: (1975). "Earthquake Resistant Design of Structures."

8. Jaky, J. (1944). "Talajmechanika," Budapest.
9. Kapila, I.P. (1962), "Earthquake Resistant Design of Retaining Walls," Proc. Symposium on Earthquake Engineering, University of Roorkee, Roorkee.
10. Kezdi, A. (1962), *Erddruck Theorein*, Springer Verlag, Berlin.
11. Kenny, T.C. (1959), "Discussion on Proc. paper 1732," Proc. ASCE, Vol. 85, No. SM3, pp. 67-69.
12. Mackey, R.D. and D.P. Kirk (1967), "At Rest, Active and Passive Earth Pressures," South East Conference on Soil Engineering, Bangkok.
13. Mononobe, N. and Okabe, S. (1929), "Earth-quake Proof Construction of Masonry Dam," Proc. World Engineering Congress, Vol. 9, p. 275.
14. Nandkumaran, P. (1974), "Behaviour of Retaining Walls under Dynamic Loads," Ph.D. Thesis, University of Roorkee.
15. Narain, J., S. Saran and P. Nandkumaran (1969), "A Passive Pressure in Cohesionless Soils," *Proceedings ASCE*, Vol. 10, No. 1.
16. Ohde, T. (1938), "Zur Theories des Erdrucks under besonderer Beruecksichigung der Erdruck verteilung," *Bautechnik*, Nos. 14, 19, 25, 37, 42, 53, 54.
17. Prakash, S. and B.M. Basavanna (1969), "Earth Pressure Distribution behind Retaining Walls during Earthquakes," 4th World Conference on Earthquake Engineering, Chile.
18. Prakash, S. and S. Saran (1966), "Static and Dynamic Earth Pressures behind Retaining Walls," Symposium on Earthquake Engineering, University of Roorkee.
19. Rankine, W.J.M. (1857), "On the Stability of Loose Earth," Phil. Tras. Royal Soc. (London).
20. Saran, S. and A. Prakash (1970), "Seismic Pressure Distribution in Earth Retaining Walls," European Symposium on Earthquake Engineering, Sofia, Bulgaria.
21. Saran, S. and S. Prakash (1968), "Dimensionless Parameters for Static and Dynamic Earth Pressure for Retaining Walls," *Journal of Indian Geotechnical Society*, Vol. 1, No. 3, July.

PRACTICE PROBLEM

1. Define lateral pressure, and indicate the requirements for the at-rest, active and passive conditions.
2. What are conjugate stresses ? How does Rankine use them for the problem of inclined surface of backfill ?
3. Describe with sketches the effect of wall friction on the shape of rupture surface and lateral earth pressure.
4. How are the point of application of active earth pressures taken in the following cases of an inclined retaining wall with backfill surface (a) having no surcharge, (b) uniformly distributed load, and (c) line load ?
5. Discuss the salient features of the approach suggested by Prakash and Saran (1966) for the computation of dynamic active earth pressure. Give the procedure of locating point of application of the dynamic active earth pressure.
6. Assuming the back of the retaining wall to be smooth and vertical, and the height of wall to be 7.0 m, calculate the magnitude and point of application of total active earth pressure on the wall for the following cases:

    (a) Backfill surface horizontal, two-layer backfill
    $$0 \text{ m to } 4.0 \text{ m}: c = 5 \text{ kN/m}^2, \phi = 35°, \gamma = 19 \text{ kN/m}^3$$
    $$4.0 \text{ m to } 7.0 \text{ m}: c = 8 \text{ kN/m}^2, \phi = 28°, \gamma = 17 \text{ kN/m}^3$$

    (b) Backfill surface horizontal: $c = 0$, $\phi = 38°$, $\gamma = 18 \text{ kN/m}^2$, water table is at 4.0 m above the base of wall.

    (c) Backfill surface sloping away at 15° to the horizontal: $c = 0$, $\phi = 36°$, $\gamma = 16 \text{ kN/m}^3$.

7. A retaining wall 5 m high is inclined at 15° to the vertical. It supports a cohesionless backfill having unit weight of 17 kN/m³. The upper surface of the fill rises from the crest at an angle of 10° to the horizontal. The angle of internal friction of backfill is 30°, and the angle of wall friction is 20°. Compute the total active and passive earth pressures using (i) Coulomb's theory, and (ii) Culmann's graphical construction.
8. A wall with a smooth vertical back 8 m high, supports a cohesive-frictional soil with $c = 10 \text{ kN/m}^2$, $\gamma = 18 \text{ kN/m}^3$ and $\phi = 20°$. Determine (i) total active earth pressure against the wall, (ii) position of zero pressure, and (iii) distance of centre of pressure from the base. If the backfill carries a surcharge of intensity 15 kN/m², how the items (i) to (iii) will get changed.
9. A vertical wall 7 m high supports a cohesionless backfill with γ as 17 kN/m³. The surface of fill is horizontal. The values of ϕ and δ are 34° and 22° respectively. The fill supports two-line loads of 30 kN/m parallel to the crest of wall at distances of 3 m and 4 m respectively. Compute the value of total active earth pressure

against the wall. Determine the horizontal distance from the back of the wall to the point at which the surface of sliding intersects the surface of fill.

10. A retaining wall 7.0 m high inclined at 10° to the vertical is located in a seismic zone ($\alpha_h = 0.1$). The wall carries a horizontal backfill having the following characteristics:

$$\phi = 35°, \delta = 22°, \gamma = 16 \text{ kN/m}^3, c = 8 \text{ kN/m}^2.$$

Determine the total active pressure and its point of application from the base of wall.

11. A vertical retaining wall 10 m high has a backfill the top of which is inclined at 10° to the horizontal. The wall is located in earthquake zone having a value of seismic horizontal coefficient as 0.08. The backfill soil possesses $\phi = 30°, \delta = 20°, \gamma = 17 \text{ kN/m}^3$. Compare the value of total active earth pressure computed by (i) Mononobe-Okabe formula, and (ii) Culmann's graphical construction.

┌─────┐
│ 5 │
└─────┘

Limit State Design—Basic Principles

5.1 INTRODUCTION

In the method of design based on limit state concept, the structure is designed such that it shall (*i*) withstand safely of all loads liable to act on it throughout its life, and (*ii*) satisfy the serviceability requirements, such as limitations on deflection and cracking. The acceptable limit for the safety and serviceability requirements before failure occurs is called a 'Limit State'. The limit states which are usually examined in design are:

(*a*) *Limit state of collapse* : In this limit state, the resistance to bending, compression, shear and torsion at every section shall not be less than the value at that section produced by the probable most unfavourable combination of loads on the structure.

(*b*) *Limit state of serviceability* : In this limit state, the deflection and the width of cracks shall not exceed the permissible values.

The aim of design is to achieve acceptable probabilities that the structure will not become unfit for the use for which it is intended, that is, it will not reach a limit state. This is accomplished by (*i*) decreasing the material strength, and (*ii*) increasing the loads coming on the structure by certain factors known as partial safety factors. Discussion on these is made in subsequent sections.

5.2 PARTIAL SAFETY FACTORS

For materials

The design strength of a material, f_d is given by

$$f_d = \frac{f}{\gamma_m} \qquad \qquad \text{... (5.1)}$$

where f = characteristic strength of material, and

γ_m = partial safety factor appropriate to the material and the limit state being considered.

For assessing the strength of a structure against the limit state of collapse, the value of γ_m is taken 1.5 for concrete and 1.15 for steel. The modulus of elasticity associated with characteristic strength of material is adopted to assess the strength of structure against the limit of serviceability. If f_{ck} and f_{sy} represent respectively the characteristic strengths of concrete and steel, then

$$f_{cd} = \frac{0.67 f_{ck}}{1.5} \quad \text{and} \quad f_{sd} = \frac{f_{sy}}{1.15} \qquad \text{... (5.2)}$$

where f_{cd} and f_{sd} are design values of the strength of concrete and steel in flexure. Usually in foundation M20 or M25 grade of concrete is used for which f_{ck} is 20 N/mm² or 25 N/mm². Tor steel of Fe 415 is used for which f_{sy} is 415 N/mm².

For loads

The design load, F_d is given by

$$F_d = F \cdot \gamma_f \qquad \text{... (5.3)}$$

where F = characteristic load, and

γ_f = partial safety factor appropriate to the nature of loading and the limit state being considered.

As per IS: 456–2000, values of γ_f are given in Table 5.1.

Table 5.1. Values of Partial Safety Factors γ_f for Loads.

Load Combination	Limit State of Collapse			Limit State of Serviceability		
	DL	LL	WL	DL	LL	WL
DL + LL	1.5	1.5	—	1.0	1.0	—
DL + WL	1.5 or 0.9*	—	1.5 1.5	1.0	—	1.0
DL + LL + WL	1.2	1.2	1.2	1.0	0.8	0.8

* This value is to be considered when stability against overturning or safety against stress reversal is critical.

Note 1: While considering earthquake effects, substitute EL for WL.

Note 2: For the limit states of serviceability, the values of γ_f given in this table are applicable for short term effects. While assessing the long term effects due to creep, the dead load and that part of the live load likely to be permanent may only be considered.

5.3 LIMIT STATE OF COLLAPSE

Flexure

Design of a structural component for the limit state of collapse in flexure is based on the following assumptions:

(i) Plane sections normal to the axis remain plane after bending.

(ii) The compressive strength of concrete in the structure is 0.67 times the characteristic strength. The partial safety factor is applied in addition to this.

Therefore, the design value of strength is taken as $\dfrac{0.67 f_{ck}}{1.5} = 0.446 f_{ck}$. The tensile strength of concrete is neglected.

(*iii*) The maximum strain in concrete at the outermost compression fibre is assumed as 0.0035.

(*iv*) The maximum strain in the tension reinforcement in the section at failure shall not be less than:

$$\frac{f_{sy}}{1.15E_s} + 0.002$$

where E_s = modulus of elasticity of steel.

(*v*) Stress block and strain diagrams are taken as shown in Fig. 5.1.

Singly reinforced beam

Refer Fig. 5.1.

$$x_1 = \frac{0.002x}{0.0035} = \frac{4x}{7} \text{ and } x_2 = x - \frac{4x}{7} = \frac{3x}{7}$$

$$C_1 = \frac{2}{3} \cdot \frac{4x}{7} \cdot 0.446f_{ck} \cdot b = 0.167\,f_{ck}\,b \cdot x$$

$$C_2 = \frac{3x}{7} \cdot 0.446f_{ck} \cdot b = 0.191f_{ck}\,b \cdot x$$

$$C = C_1 + C_2 = 0.358f_{ck}\,b \cdot x \qquad \qquad \dots (5.4)$$

$$a = \frac{C_1(3/8x_1 + x_2) + C_2 \cdot x_2/2}{C} = 0.414\,x \qquad \dots (5.5)$$

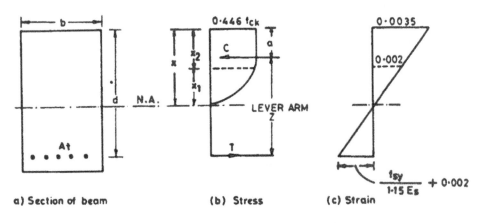

a) Section of beam   (b) Stress   (c) Strain

Fig. 5.1. Stress Block Parameters and Strain Diagram.

For equilibrium,

$$C = T$$

or $$0.358 f_{ck} b.x = 0.87 f_{sy} A_{st}$$

or $$x = \frac{0.87 f_{sy} A_{st}}{0.358 f_{ck} b} \qquad ... (5.6)$$

Lever arm, $$Z = d - 0.414x = d - \frac{0.414.(0.87 f_{sy}.A_{st})}{0.358 f_{ck} b}$$

or $$Z = d - \frac{f_{sy} A_{st}}{f_{ck}.b} \qquad ... (5.7)$$

A compression failure is a brittle failure. The maximum depth of neutral axis (N.A.) is limited to ensure that tensile steel will reach its yield stress before concrete fails in compression, thus brittle failure is avoided. If x_u is the maximum depth of N.A., then

$$\frac{x_u}{d - x_u} = \frac{0.0035}{0.002 + \dfrac{0.87 f_{sy}}{E_s}}$$

or $$\frac{x_u}{d} = \frac{0.0035}{0.0055 + \dfrac{0.87 f_{sy}}{E_s}} \qquad ... (5.8)$$

$$E_s = 2 \times 10^5 \text{ N/mm}^2 \qquad \text{(for all grades of steel)}$$

For Fe 415 steel

$$x_u = 0.48d \qquad ...(5.9)$$

For a balanced section,

$$x = x_u$$

Moment of resistance with respect to concrete,

$$M_{Rc} = 0.358 f_{ck} b x_u [d - 0.414 x_u]$$

$$= 0.358 f_{ck} b . 0.48d [d - 0.414 (0.48d)]$$

or $$M_{Rc} = 0.138 f_{ck} b d^2 \qquad ... (5.10)$$

For M20 concrete, $f_{ck} = 20$ N/mm^2

$$M_{Rc} = 2.76 b d^2 \qquad ... (5.11)$$

Moment of resistance with respect to steel,

$$M_{Rs} = 0.87\ f_{sy}A_{st}\ [d - 0.414\ x]$$

$$= 0.87\ f_{sy}A_{st}\ [d - 0.414.\ (0.48d)]$$

or $M_{Rs} = 0.697\ f_{sy}A_{st}d$ ...(5.12)

The percentage of tensile reinforcement corresponding to limiting moment of resistance is obtained by equating forces of compression and tension, i.e.,

$$0.87f_{sy}A_{st} = 0.358f_{ck}.b.x_u$$

$$A_{st} = p.\ bd$$

$$p(\%) = \frac{0.358.f_{ck}}{0.87f_{sy}}.\frac{x_u}{d}.100$$ ... (5.13 a)

$$= \frac{0.358 \times 20}{0.87 \times 415}.\frac{0.48d}{d}.100$$

$$p = 0.952\% \text{ (for M20 concrete and Fe 415 steel)}$$ ... (5.13 b)

For under, reinforced section

$$x < x_u$$

The moment of resistance (M_R) will be computed with respect to steel and is given by

$$M_R = 0.87f_{sy}A_{st}\left(d - \frac{f_{sy}A_{st}}{f_{ck}b}\right)$$ ... (5.14)

An over-reinforced section $(x > x_u)$ should be redesigned.

Rectangular section with compression reinforcement

Referring to Fig. 5.2, the procedure of designing a rectangular section with compression reinforcement is explained in following steps:

(i) Determine the value of M_u of singly reinforced rectangular beam section using Eq. 5.10 and value of A_{st1} using Eq. 5.13.

(ii) If factored moment M is greater than M_u, then compression reinforcement (A_{sc}) is provided, and computed by

$$A_{sc} = \frac{M - M_u}{f_{sc}(d - d')}$$ ... (5.15)

where f_{sc} = design stress in compression reinforcement corresponding to a strain of $\dfrac{0.0035(x_u - d')}{x_u}$.

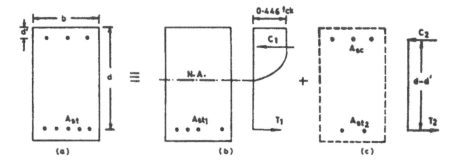

Fig. 5.2. Doubly Reinforced beam.

For this strain the stress can be read from the appropriate stress-strain curve. For Fe 415 steel,

$\dfrac{d'}{d}$	f_{sc} (N/mm^2)
0.05	355
0.10	353
0.15	342
0.20	329

(*iii*) Compute the value of A_{st2} by equating the forces in compression and tension (Fig. 5.2c)

$$f_{sc} A_{sc} = 0.87 f_{sy} \cdot A_{st2}$$

or
$$A_{st2} = \frac{f_{sc} \cdot A_{sc}}{0.87 f_{sy}} \qquad \ldots (5.16)$$

(*iv*) Total tensile reinforcement, A_{st} is

$$A_{st} = A_{st1} + A_{st2} \qquad \ldots (5.17)$$

T-beam

Case 1: $x < d_f$ (Refer Fig. 5.3)

The T-beam will behave as a rectangular beam and the Eq. 5.10 is used for getting the moment of resistance of the section. The value of b will be replaced by b_f.

Case 2: $x > d_f$ (Refer Fig. 5.4)

For limiting case, $x = x_u = 0.48d$ (for Fe 415 steel)

$$x_1 = \frac{4}{7} \cdot (0.48d) = 0.274d$$

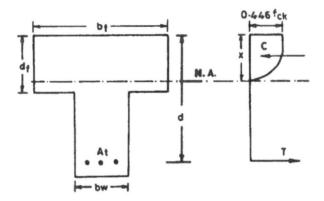

Fig. 5.3. T-Beam Section and Stress Distribution for $x < d_f$.

$$x_2 = \frac{3}{7} \cdot 0.48d = 0.206d \approx 0.2d$$

$$a = 0.42x_u = 0.42 \times 0.48d = 0.198d \approx 0.2d$$

(*i*) $d_f \leqslant 0.2d$ (Fig. 5.4)

$$M_R = 0.446f_{ck} (b_f - b_w) d_f (d - 0.5d_f) + 0.138f_{ck}b_wd^2 \qquad \dots (5.18)$$

(*ii*) $d_f > 0.2d$ (Fig. 5.5)

$$M_R = 0.446f_{ck} (b_f - b_w) y_f (d - y_f/2) + 0.138f_{ck}b_wd^2 \qquad \dots (5.19)$$

where
$$y_f = (0.15x_u + 0.65d_f) \not> d_f \qquad \dots (5.20)$$

Case 3: $x_u > x > d_f$ (*i.e.* under-reinforced section)

(*i*) $d_f \leq \frac{3}{7}x$

$$M_R = 0.446f_{ck} (b_f - b_w) d_f (d - d_f/2) + 0.358 (d - 0.42x) f_{ck}b_w \cdot x \qquad \dots(5.21)$$

(*ii*) $d_f > \frac{3}{7} x$

$$M_R = 0.446f_{ck} (b_f - b_w) y_f (d - y_f/z) + 0.358 (d - 0.42x) f_{ck} \cdot b_w \cdot x \qquad \dots (5.22)$$

where $y_f = 0.15x + 0.65d_f \not> d_f \qquad \dots (5.23)$

Compression

For the limit state of collapse in compression, following assumptions are made in addition to the assumptions made for flexure:

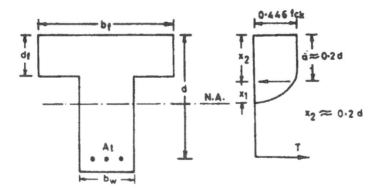

Fig. 5.4. T-beam Section and Stress Distribution for $x = x_u = 0.48d$ and $d_f < 0.2d$.

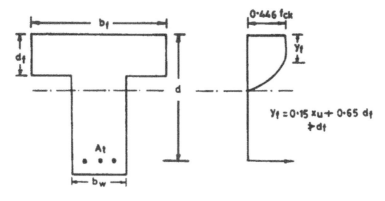

Fig. 5.5. T-Beam Section and Stress Distribution for $x = x_u = 0.48d$ and $d_f \geqslant 0.2d$.

(*i*) The maximum compressive strain in concrete in axial compression is taken as 0.002.

(*ii*) The maximum compressive strain at highly compressed extreme fibre in concrete subjected to axial compression, bending and no tension on the section shall be 0.0035 minus 0.75 times the strain at the least compressed extreme fibre.

Members in compression (e.g. columns) are designed for the actual calculated eccentricity (moment/vertical load) or minimum eccentricity (Eq. 5.24), whichever is larger.

$$e_{min} = \frac{l}{500} + \frac{D}{30}, \text{ subject to a minimum of 20 mm} \qquad \text{... (5.24)}$$

where l = unsupported length of column,

             D = lateral dimension of column in the direction under consideration.

If $e_{min} < 0.05\,D$, an axially loaded short column is designed by the following equation:

$$P_u = 0.4f_{ck}A_c + 0.67f_{sy}A_{sc} \qquad\qquad\qquad\qquad \text{... (5.25)}$$

where P_u = axial load on the member,

       A_{sc} = area of longitudinal reinforcement for column,

$$A_c = \text{area of concrete} = \left(A - \frac{pA_g}{100}\right)$$

       p = percentage of reinforcement,

       A_g = gross area of concrete = bD,

       b = width of column, and

       D = Depth of column.

The equation (5.25) may be rewritten as:

$$\frac{P_u}{f_{ck}bD} = 0.4 + \frac{p}{100 f_{ck}}(0.67 f_{sy} - 0.4 f_{ck}) \qquad\qquad \text{... (5.26)}$$

It may be noted that the Eqs. (5.25) and (5.26) are based on stresses in concrete and steel corresponding to a maximum strain of 0.002.

As per IS: 456 – 2000, the strength of a column with helical reinforcement may be taken as 1.05 times the strength of the column with lateral ties, i.e. equal to 1.05 P_u provided the ratio of the volume of helical reinforcement to the volume of the core shall not be less than

$$0.36(A_g/A_c - 1)f_{ck}/f_{sy}$$

where A_g = gross area of the section,

       A_c = area of the core of the helically reinforced column measured to the outside diameter of the helix,

       f_{ck} = characteristic compression strength of the concrete, and

       f_{sy} = characteristic strength of the helical reinforcement, but not exceeding 415 N/mm^2.

When a column is subjected to combined axial load and uniaxial (or biaxial) bending, the design involves lengthy calculation by trial and error. In order to overcome these difficulties, interaction diagrams have been prepared and published by ISI (SP:16 'Design aids for reinforced concrete, to IS: 456 – 2000). Charts have been developed for different values of f_{ck} (15, 20, 25, 30, 35 and 40 N/mm^2), f_{sy} (250, 415 and 500 N/mm^2) and d'/d (0.05, 0.10, 0.15 and 0.20). Further, charts include both circular and rectangular sections having reinforcement either on two sides or four sides.

Figure 5.6 shows a typical interaction diagram for designing a column section subjected to axial compression only. Figs. 5.7 and 5.8 represent the interaction diagrams for rectangular sections having reinforcement on all the four sides for d'/d equal to 0.05 and 0.10 respectively. Similar charts for circular sections are given in Figs. 5.9 and 5.10.

For a column subjected to combined compression and bending, the condition given by Eq. 5.27 should be satisfied:

$$\left(\frac{M_{ux}}{M_{ux1}}\right)^{\alpha_n} + \left(\frac{M_{uy}}{M_{uy1}}\right)^{\alpha_n} < 1.0 \qquad\qquad \text{... (5.27)}$$

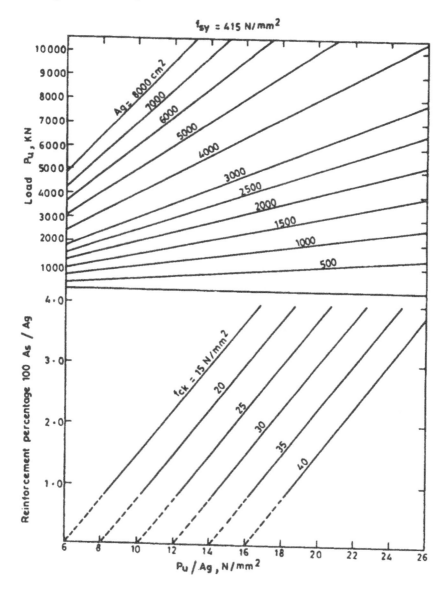

Fig. 5.6. Interaction Diagram for Column under Axial Compression.

where M_{ux}, M_{uy} = moments about x and y axes due to design loads,

M_{ux1}, M_{uy1} = maximum uniaxial moment capacity for an axial load of P_u, bending

about x and y axes respectively, and α_n is related to $\dfrac{P_u}{P_{uz}}$, and

$$P_{uz} = 0.45 f_{ck} A_c + 0.75 f_{sy} A_{sc}$$

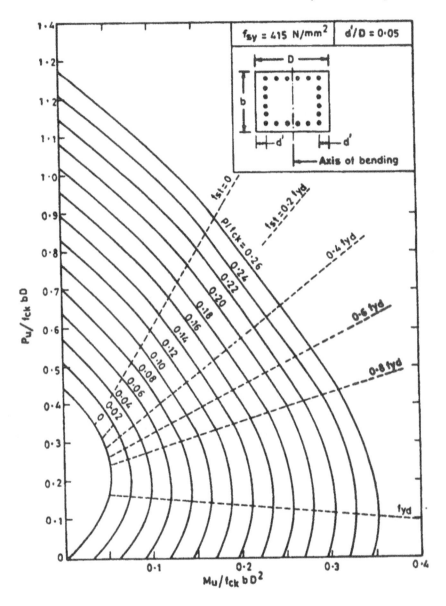

Fig. 5.7. Interaction Diagram for Rectangular Column under Combined
Compression and Bending ($d'/D = 0.05$).

For values of P_u/P_{uz} = 0.2 to 0.8, the values of α_n vary linearly from 1.0 to 2.0. For values less than 0.2, α_n is 1.0; for values greater than 0.8, α_n is 2.0.

The use of design charts has been illustrated in example 5.5.

The slender compression members are designed for additional moments given by Eqs. 5.28.

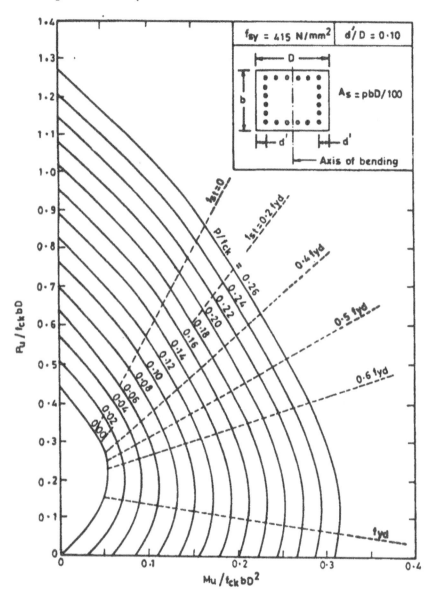

Fig. 5.8. Interaction Diagram for Rectangular Column under Combined Compression and Bending ($d'/D = 0.10$).

$$M_{ax} = \frac{P_u D}{200} \left\{ \frac{l_{ex}}{D} \right\}^2 \qquad \qquad \dots \text{(5.28a)}$$

$$M_{ay} = \frac{P_u b}{2000} \left\{ \frac{l_{ey}}{b} \right\}^2 \qquad \qquad \dots \text{(5.28b)}$$

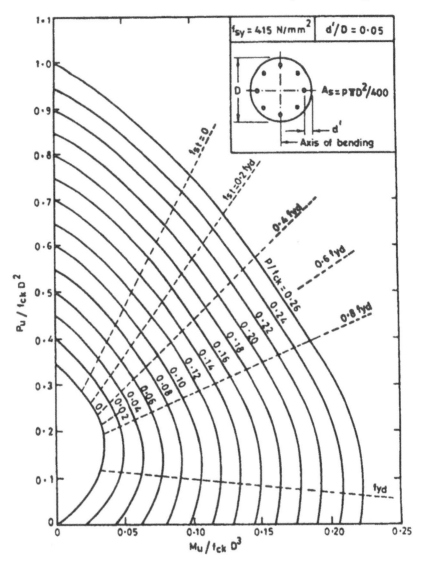

Fig. 5.9. Interaction Diagram for Circular Column under Combined Compression and Bending (d'/D = 0.05).

where P_u = axial load on the member,
          l_{ex} = effective length in respect to major axis,
          l_{ey} = effective length in respect to minor axis,
          D = depth of cross-section at right angles to the major axis, and
          b = width of member.

Shear

Shear in a foundation member is checked according to the bending action namely (a) one-way bending action, or (b) two-way bending action.

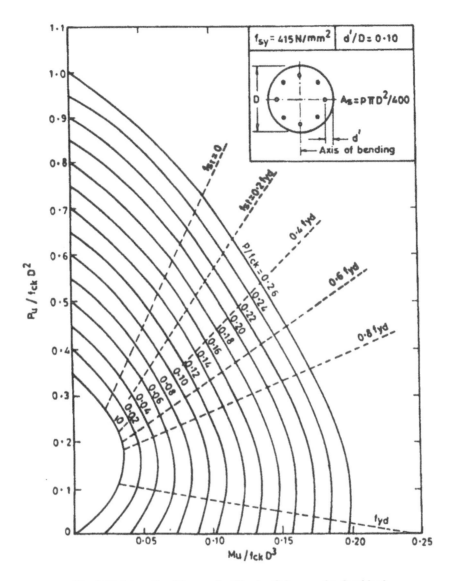

Fig. 5.10. Interaction Diagram for Circular Column under Combined
Compression and Bending (d'/D = 0.10).

The procedure of checking the shear in a foundation member having one-way bending is outlined as below:

(*i*) The nominal shear stress τ_v in beams of uniform depth is obtained by Eq. 5.29

$$\tau_v = \frac{V_u}{bd} \qquad\qquad\qquad ...\ (5.29)$$

where V_u = shear force due to design loads,
         b = width of rectangular beam and width of web (b_w) in T-beam, and
         d = effective depth.
In beams of varying depth, τ_v is given by

$$\tau_v = \frac{V_u \pm M_u / d \tan\beta}{bd} \qquad \qquad \text{... (5.30)}$$

where M_u = bending moment at the section, and

         β = angle between the top and bottom edges of the beam.
    The negative sign in Eq. 5.30 applies when the bending moment M_u increases numerically in the same direction as the effective depth d increases, and the positive sign when the moment decreases numerically in this direction (Fig. 5.11).
    The nominal shear stress is checked at a section at a distance equal to effective depth d from the face of the column when the reaction in the direction of applied shear introduces compression into the end region of the member. Otherwise, the shear is checked at the face of support (Fig. 5.12).

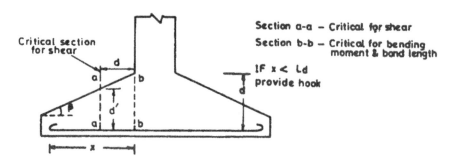

Fig. 5.11. Critical Section for Shear.

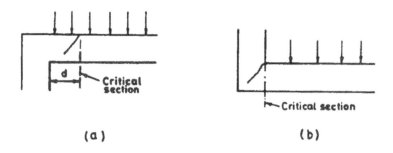

(a)   (b)

Fig. 5.12. Critical Section for Shear.

(ii) The value of τ_v should not exceed the maximum shear stress, $\tau_{c\,max}$. For M20 and M25 concrete, values of $\tau_{c\,max}$ are 2.8 N/mm^2 and 3.1 N/mm^2 respectively.

(iii) The design shear strength of concrete τ_c in beams without shear reinforcement is given in Table 5.2.

Table 5.2. Design Shear Strength of Concrete, τ_c, N/mm^2.

$100\,\dfrac{A_s}{bd}$	Concrete Grade	
	M20	M25
0.25	0.36	0.36
0.50	0.48	0.49
0.75	0.56	0.57
1.00	0.62	0.64
1.25	0.67	0.70
1.50	0.72	0.74
1.75	0.75	0.78
2.00	0.79	0.82
2.25	0.81	0.85
2.50	0.82	0.86

Note : The term A_{st} is the area of longitudinal tension reinforcement which continues at least one effective depth beyond the section being considered, except at supports where the full area of tension reinforcement may be used, provided the detailing conforms to clauses 26.2.2 and 26.2.3 of IS: 456-2000.

(iv) If $\dfrac{\tau_c}{2} \ll \tau_v \gg \tau_c$, minimum shear reinforcement, given by Eq. 5.31, is provided:

$$A_{sv} = \frac{0.4bS_v}{f_{sy}} \qquad \text{... (5.31)}$$

where A_{sv} = total cross-sectional area of stirrup legs effective in shear, and

S_v = stirrup spacing along the length of the member.

No shear reinforcement is needed if $\tau_v < \tau_c$.

(v) If $\tau_v > \tau_c$, shear reinforcement is provided in any of the following forms:

(a) Vertical stirrups

$$A_{sv} = \frac{(V_u - \tau_c bd)S_v}{0.87 f_{sy} d} \qquad \text{... (5.32)}$$

(b) Inclined stirrups

$$A_{sv} = \frac{(V_u - \tau_c bd)S_v}{0.87 f_{sy} d(\sin\alpha + \cos\alpha)} \qquad \text{... (5.33)}$$

(c) Bent-up bars

$$A_{sv} = \frac{(V_u - \tau_c bd)}{0.87 f_{sy} \sin \alpha}$$ ... (5.34)

where α is the angle between the inclined stirrup or bent-up bar with the axis of the member not less than 45°. Terms V_u, τ_c, S_v, b, d, f_{sy} and A_{sv} have the same meaning as explained for Eqs. 5.29, 5.30 and 5.31.

    (vi) When more than one type of shear reinforcement is used to reinforce the same portion of the beam, the total shear resistance shall be computed as the sum of the resistances for various types separately.

    (vii) Spacing of stirrups should not exceed 0.75d for vertical stirrups and d for stirrups inclined at 45°. In no case, the spacing should be more than 450 mm.

When the bending of a foundation slab is primarily two-way, the procedure of checking the shear stress (or commonly known as punching shear stress) is as given below:

    (i) Punching shear stress, τ_v' around the column is checked on a perimeter 0.5 times the effective depth away from the face of column (Fig. 5.13a). τ_v' is given by Eq. 5.35

$$\tau_v' = \frac{\text{column load} - p'.a}{p_0.d'}$$ ... (5.35)

where p_0 = periphery of critical section,
         = 4 (b + d) in a square column of width b (Fig. 5.13b),
         p' = net upward soil reaction,
         a = area of the periphery of critical section = $(b + d)^2$ in a square column, and
         d' = depth of footing at critical section.

    (ii) When shear reinforcement is not provided, τ_v' should not exceed $K_s \tau_c'$

where K_s = (0.5 + β_c) but not greater than 1, β_c being the ratio of short side to long side of the column,

         $\tau_c' = 0.25 \sqrt{f_{ck}}$

            = $0.25 \sqrt{20}$ = 1.12 N/mm^2 (For M20 Concrete).

    (iii) For $\tau_c' < \tau_v' < 1.5 \, \tau_c'$, shear reinforcement is provided.

    (iv) For $\tau' > 1.5\tau_c'$, the foundation slab is redesigned.

Usually the foundation slabs are designed such that no shear reinforcement is required.

5.4 LIMIT STATE OF SERVICEABILITY

In all normal cases, the deflection of foundation will not be excessive and, therefore, there is no need of checking it. For checking cracks, the spacing of the reinforcement bars in beams and slabs should be as per the recommendation of IS: 456–2000.

5.5 DEVELOPMENT LENGTH

    (i) The development length L_d is given by Eq. 5.36.

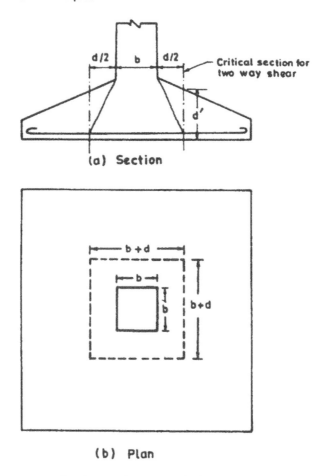

(a) Section

(b) Plan

Fig. 5.13. Critical Section for Two-way Shear.

$$L_d = \frac{0.87\phi f_{sy}}{4\tau_{bd}} \qquad \qquad \ldots (5.36)$$

where ϕ = nominal diameter of the bar,
            f_{sy} = characteristic strength of steel, and
            τ_{bd} = design bond stress.

As per IS: 456–2000, values of τ_{bd} for plain bars in tension are 1.2 N/mm² and 1.4 N/mm² for M20 and M25 concrete respectively. For deformed bars these values shall be increased by 60% and, therefore, are equal to 1.92 N/mm² and 2.24 N/mm² respectively.

For bars in compression, the values of τ_{bd} given for bars in tension shall be increased by 25 per cent.

Therefore, if M20 concrete is used, then:
In tension

$$L_d = \frac{\phi \times 0.87 \times 415}{4 \times 1.2} = 75\phi \qquad \text{(Plain bars)}$$

$$L_d = \frac{\phi \times 0.87 \times 415}{4 \times 1.92} = 47\phi \qquad \text{(Deformed bars of grade Fe 415)}$$

In compression

$$L_d = \frac{75\phi}{1.25} = 60\phi \qquad \text{(Plain bars)}$$

$$L_d = \frac{47\phi}{1.25} = 37\phi \qquad \text{(Deformed bars of grade Fe 415)}$$

(*ii*) When the anchoring bars are used in bundles, the development length of each bundled bar shall be that of the individual bar, increased by 10 per cent for two bars in contact, 20 per cent for three bars in contact and 33 per cent for four bars in contact.

(*iii*) *Anchoring bars in tension:* Deformed bars may be used without end anchorages, provided development length requirement is satisfied. Hooks are normally provided for plain bars. Maximum anchorage length obtained by the provision of hooks and bends are shown in Fig. 5.14.

(*iv*) *Anchoring bars in compression:* The anchorage length of straight bar in compression shall be equal to the development length of bars in compression. The projected length of hooks, bends and straight lengths beyond bends if provided for a bar in compression, shall be considered for development length.

(*v*) The critical sections for checking the bond length in a structural member of a foundation are the sections corresponding to maximum bending moments. In addition, bond length requirement must be satisfied at the location where the reinforcement is curtailed.

(*vi*) As per IS: 456–2000, the positive reinforcement at simple supports shall be limited to a diameter such that

$$L_d < \frac{M_1}{V} + L_0 \qquad\qquad \text{... (5.37)}$$

where M_1 = moment of resistance of the section assuming all reinforcement at the section to be stressed to $0.87 f_y$,

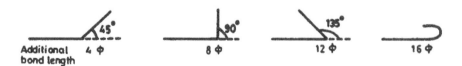

Fig. 5.14. Bends and Hooks with Additional Bond Lengths.

V = shear force at the section due to design load, and

L_0 = sum of anchorage beyond the centre of the support and equivalent anchorage value of any hook or mechanical anchorage at the support.

The value of M_1/V in Eq. 5.37 may be increased by 30 per cent when ends of the reinforcement are confined by a compressive reaction.

5.6 REQUIREMENTS OF REINFORCEMENT FOR STRUCTURAL MEMBERS

Tension reinforcement in beams

(*i*) The area of tension reinforcement shall not be less than given by Eq. 5.38.

$$\frac{A_{st}}{bd} = \frac{0.85}{f_{sy}} \qquad \ldots (5.38)$$

where A_{st} = minimum area of tension reinforcement,
$\quad\quad\quad b$ = breadth of beam or breadth of web of T-beam, and
$\quad\quad\quad d$ = effective depth.

The maximum area of tension reinforcement shall not exceed 0.04 bD.

(*ii*) For curtailment, reinforcement shall extend beyond the point at which resistant moment of the section is equal to the applied moment, considering only the continuing bars for a distance equal to d or 12 ϕ whichever is more.

(*iii*) At least one-third the positive moment reinforcement in simple members and one-fourth the positive moment reinforcement in continuous members shall extend along the same face of the member into the support, to a length equal to $L_d / 3$.

(*iv*) At least one-third of the total reinforcement provided for negative moment at the support shall extend beyond the point of inflexion for a distance not less than d or 12ϕ or one-sixteenth of the clear span whichever is greater.

Compression reinforcement in beams

The maximum area of compression reinforcement shall not exceed 0.04 bD. Compression reinforcement in beams shall be enclosed by stirrups for effective lateral restraint.

Reinforcement in slabs

The reinforcement in either direction in slabs shall not be less than 0.15 per cent of total cross-sectional area. The diameter of reinforcing bars shall not exceed one-eighth of total thickness of slab.

Reinforcement in columns

Longitudinal. Reinforcement

(*i*) The cross-sectional area of longitudinal reinforcement shall not be less than 0.8% nor more than six per cent* of gross cross-sectional area of column.

---

* The use of six per cent reinforcement may involve practical difficulties in placing and compacting concrete; hence lower percentage is recommended. Where bars from the column below have to be lapped with those in the column under consideration, the percentage of steel shall usually not exceed four per cent.

(*ii*) The minimum number of longitudinal bars provided in a column shall be four in rectangular columns and six in circular columns.

(*iii*) The bars shall not be less than 12 mm in diameter.

(*iv*) Spacing of longitudinal bars measured along the periphery of the column shall not exceed 300 mm.

(*v*) In case of pedestals in which longitudinal reinforcement is not taken into account in strength calculations, nominal longitudinal reinforcement not less than 0.15 per cent of the cross-sectional area shall be provided.

Note : Pedestal is a compression member the effective length of which does not exceed three times the least lateral dimension.

Transverse Reinforcement

Transverse reinforcement may be in the form of lateral ties or spirals. The purpose of transverse reinforcement is to hold the vertical bars in position and it does not contribute to the strength of column. Piles are usually designed as columns and transverse reinforcement is provided in the form of ties.

The pitch of lateral ties shall not be more than the least cf the following distances:

(*i*) The least lateral dimension of the compression member;

(*ii*) Sixteen times the smallest diameter of the longitudinal reinforcement bar to be tied; and

(*iii*) Forty-eight times the diameter of the transverse reinforcement.

The diameter of the lateral ties shall not be less than one-fourth of the diameter of the largest longitudinal bar, and in no case less than 5 mm.

5.7 DESIGN FOR SHEAR AND TORSION

Equivalent shear

Equivalent shear, V_e, shall be calculated from the formula:

$$V_e = V_u + 1.6\frac{M_{t}}{b} \qquad \qquad ...(5.39)$$

where V_e = equivalent shear,

V_u = shear,

M_{t} = torsional moment, and

b = breadth of beam.

The equivalent nominal shear stress, τ_{ve} in this case shall be calculated as given by Eq. 5.29, except for substituting V_u by V_e. The values of τ_{ve} shall not exceed the values of $\tau_{c\,max}$ *i.e.* 2.8 N/mm^2 for M-20 grade.

If the equivalent nominal shear stress, τ_{ve} does not exceed τ_c given in Table 5.2, minimum shear reinforcement shall be provided.

If τ_{ve} exceeds τ_c given in Table 5.2, both longitudinal and transverse reinforcement shall be provided in accordance as below.

Reinforcement in members subjected to torsion

Reinforcement for torsion, when required, shall consist of longitudinal and transverse reinforcement.

Longitudinal Reinforcement

The longitudinal reinforcement shall be designed to resist an equivalent bending moment, M_{e1}, given by

$$M_{e1} = M_u + M_t \qquad \text{... (5.40a)}$$

where M_u = bending moment at the cross-section.

$$M_t = M_\psi \left(\frac{1 + \dfrac{D}{b}}{1.7} \right) \qquad \text{... (5.40b)}$$

where M_ψ is the torsional moment, D is the overall depth of the beam and b is the breadth of the beam.

If the numerical value of M_t exceeds the numerical value of the moment M_u, longitudinal reinforcement shall be provided on the flexural compression face, such that the beam can also withstand an equivalent M_{e2} given by $M_{e2} = M_t - M_u$, the moment M_{e2} being taken as acting in the opposite sense to the moment M_u.

Transverse Reinforcement

Two legged closed hoops enclosing the corner longitudinal bars shall have an area of cross-section A_{sv}, given by

$$A_{sv} = \frac{M_\psi S_v}{b_1 d_1 (0.87 f_{sy})} + \frac{V_u S_v}{2.5 d_1 (0.87 f_{sy})} \qquad \text{... (5.41a)}$$

but the total transverse reinforcement shall not be less than

$$\frac{(\tau_{ve} - \tau_c) b.S_v}{0.87 f_{sy}} \qquad \text{... (5.41b)}$$

where
M_ψ = torsional moment,
V_u = shear force,
S_v = spacing of the stirrup reinforcement
b_1 = centre-to-centre distance between corner bars, in the direction of the width
d_1 = centre-10-centre destance between corner bars,
b = breadth of the member,
f_{sy} = characteristic strength of the stirrup reinforcement,
τ_{ve} = equivalent shear stress, and
τ_c = shear strength of the concrete as per Table 5.2.

ILLUSTRATIVE EXAMPLES

Example 5.1

Design a rectangular beam to resist a bending moment equal to 40 kNm.

Solution

Factored B.M. = 40 × 1.5 = 60 kNm
For a balanced section and M20 concrete:

$$M_{Rc} = 2.76 \ bd^2 \qquad \text{(From Eq. 5.11)}$$

Assuming, $d/b = 2.0$

$$\frac{2.76}{2} \ d^3 = 60 \times 10^6$$

or $d = 351.6$

Adopt $d = 360$ mm, $D = 400$ mm and $b = 180$ mm

$$A_{st} = \frac{0.952}{100} bd \text{ (From Eq. 5.13b)}$$

$$= \frac{0.952}{100} \times 180 \times 360 = 617 \text{ mm}^2.$$

Use three bars of 16 mm diameter.

Example 5.2

Design a rectangular beam for an effective span of 5.0 m. The superimposed load is 70 kN/m.

Solution

Assume self-weight of beam = 5 kN/m
Superimposed load   = 70 kN/m
Total load   = 75 kN/m
Factored load   = 1.5 × 7.5 = 112.5 kN/m

Maximum bending moment $= \dfrac{112.5 \times 5^2}{8} = 351.6$ kNm

$= 351.6 \times 10^6$ Nmm

Limiting B.M. $= 2.76 \ bd^2.$

Let the section of beam be 300 mm × 600 mm
If effective cover = 60 mm, $d = 540$ mm
Limiting bending moment = $2.76 \ bd^2$

$$= 2.76 \times 300 \times 540^2 = 241.4 \times 10^6 \text{ Nmm}$$

$$A_{st1} = \frac{0.952}{100} \times 300 \times 540$$

$$= 1542 \text{ mm}^2.$$

From Eq. 5.15,

$$A_{sc} = \frac{M - M_u}{f_{sc}(d - d')}$$

Taking $d'/d = 0.1$, $f_{sc} = 353$ N/mm^2

$$A_{sc} = \frac{(351.6 - 241.4) \times 10^6}{353 \times 0.9 \times 540} = 642 \text{ mm}^2.$$

From Eq. 5.16,

$$A_{st2} = \frac{f_{sc} \cdot A_{sc}}{0.87 f_{sy}} = \frac{353 \times 642}{0.87 \times 415}$$

$$= 628.0 \text{ mm}^2$$

$$A_{st} = A_{st1} + A_{st2} = 1542 + 628 = 2170 \text{ mm}^2.$$

Provide five bars of 25 mm diameter in tension ($A_{st} = 2454$ mm^2) and three bars of 18 mm diameter in compression ($A_{sc} = 763$ mm^2).

Example 5.3

Determine the depth and amount of reinforcement in a singly reinforced T-beam shown in Fig. 5.15 for a bending moment of 145 kN/m.

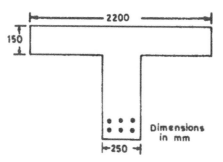

Fig. 5.15. T-Beam Section.

Solution

Let us assume that N.A. lies in the flange, then the T-beam will behave as a rectangular beam of width b_f.

$$M_R = 2.76 \, b_f \, d^2 = 145 \times 10^6 \times 1.5$$

$$d = \sqrt{\frac{145 \times 10^6 \times 1.5}{2.76 \times 2200}} = 189.3$$

Say $d = 200$ mm and $D = 240$ mm

$$A_{st} = \frac{0.952}{100} \times b_f \cdot d$$

$$= \frac{0.952}{100} \times 2200 \times 200 = 4188.9 \text{ mm}^2$$

Check :

$$x = \frac{0.87 f_{sy} \cdot A_{st}}{0.358 f_{ck} \cdot b_f}$$

$$= \frac{0.87 \times 415 \times 4188.9}{0.358 \times 20 \times 2200}$$

$$= 96 \text{ mm} < 150 \text{ mm}.$$

Therefore N.A. lies in the flange.

Use seven bars of 28 mm diameter ($A_{st} = 4308 \text{ mm}^2$).

Example 5.4

A R.C. beam has an effective depth of 500 mm and breadth of 300 mm. It contains four bars of 25 mm diameter. Compute the shear reinforcement needed for a shear force of 200 kN.

Solution

Percentage area of longitudinal steel

$$p = \frac{100 A_{st}}{bd} = \frac{100 \times 4 \times \pi / 4 \times 25^2}{300 \times 500} = 1.306\%$$

$$\tau_c = 0.68 \text{ N/mm}^2 \qquad \text{(Table 5.2)}$$

Nominal shear stress, $\tau_v = \dfrac{V_u}{bd}$

$$= \frac{200 \times 1000 \times 1.5}{300 \times 500} = 2.0 \text{ N/mm}^2$$

$$\tau_{c \, max} = 1.9 \text{ N/mm}^2 \text{ (M 20 mix)}$$

0.68 < 2.0 < 2.8, therefore provide shear reinforcement

$$V_{us} = V_u - \tau_c bd$$
$$= 300 \times 1000 - 0.68 \times 300 \times 500$$
$$= 198000$$

Use two legged vertical stirrups of 10 mm diameter,

Spacing, $\quad S_v = \dfrac{0.87 A_{sv} \cdot f_{sy} \cdot d}{V_{us}}$

$$= \frac{0.87 \times 2 \times 78.5 \times 415 \times 500}{198000}$$

$$= 143.1 \text{ mm (say 140 mm)}.$$

Example 5.5

Design a short column under biaxial bending with the following data:

Size of column	= 500 mm × 500 mm
Concrete grade	= M20
Factored load, P_u	= 1000 kN
Factored moment M_{ux}	= 80 kN
M_{uy}	= 65 kN

Moments due to minimum eccentricity are less than the values given above. Reinforcement is distributed equally on four sides.

Solution

Let us assume percentage of reinforcement, p = 1.0%

$$\frac{p}{f_{ck}} = \frac{1.0}{20} = 0.05$$

Let the effective cover be 50 mm on either side.

$$\frac{d'}{D} = \frac{50}{500} = 0.1$$

$$\frac{P_u}{f_{ck}bD} = \frac{1000 \times 1000}{20 \times 500 \times 500} = 0.2$$

From Fig. 5.8, for $\dfrac{p}{f_{ck}} = 0.05$ and $\dfrac{P_u}{f_{ck}bD} = 0.2$;

$$\frac{M_u}{f_{ck}bD^2} = 0.096$$

Therefore

$$M_{ux1} = M_{uy1} = 0.096 \times 20 \times 500 \times 500^2$$

$$= 240 \times 10^6 \text{ Nmm} = 240 \text{ kNm}$$

$$P_{uz} = 0.45 f_{ck}A_c + 0.75 f_{sy}A_{sc}$$

$$= 0.45 \times 20 \times 500^2 + 0.75 \times 415 \times 0.01 \times 500^2$$

$$= 3028 \times 10^3 \text{ N} = 3028 \text{ kN}$$

$$\frac{P_u}{P_{uz}} = \frac{1000}{3028} = 0.33$$

$$\frac{M_{ux}}{M_{ux1}} = \frac{80}{240} = 0.33$$

$$\frac{M_{uy}}{M_{uy1}} = \frac{65}{240} = 0.27$$

$$\alpha_n = 1.0 + \frac{0.33 - 0.2}{0.8 - 0.2} \times 1.0 = 1.216$$

$$\left(\frac{M_{ux}}{M_{ux1}}\right)^{\alpha_n} + \left(\frac{M_{uy}}{M_{uy1}}\right)^{\alpha_n}$$
$$= (0.33)^{1.216} + (0.27)^{1.216}$$
$$= 0.463 < 1.0$$

Thus the assumed value of reinforcement percentage is O.K.
$$A_{sc} = 0.01 \times 500 \times 500 = 2500 \text{ mm}^2$$
Provide eight bars of 20 mm diameter giving area equal to 2512 mm².

REFERENCES

1. Design Aids for Reinforcement Concrete to IS: 456–1978, special publication 16, Indian Standards Institution, New Delhi, 1980.
2. IS: 456–2000 "Code of Practice for Plain and Reinforced Concrete," Indian Standards Institution, New Delhi.

PRACTICE PROBLEMS

1. Determine the moment of resistance of a beam having rectangular section 300 mm wide, 500 mm deep reinforced with six bars of 18 mm diameter.
2. Design a rectangular section to resist a bending moment of 15 kNm.
3. Design a doubly reinforced section for a rectangular beam of mid span having an effective span of 5 m. The superimposed load is 50 kN/m.
4. Design a T-beam (b_f = 2500, d_f = 150 mm and b_w = 350 mm) to resist a moment of 200 kNm.
5. A R.C. beam has an effective depth of 450 mm and a breadth of 300 mm. It contains six bars of 20 mm as tension reinforcement. Compute the shear reinforcement for a shear force of 250 kN.
6. A R.C. T-beam (b_f= 1300 mm, d_f= 130 mm, b_w = 300 mm and d = 500 mm) is reinforced with four bars of 25 mm diameter. Calculate the shear reinforcement for a shear force of 150 kN.
7. A simply supported beam is 250 mm by 450 mm and is reinforced with three bars of 20 mm diameter. If the shear force at the centre of support is 100 kN, determine the anchorage length.

Foundation Design—
General Principles

6.1 TYPES OF FOUNDATIONS

General

The foundation of a structure is that part of the structure which is in direct contact with the subsoil and transmits the load of the structure to it. The major purpose of a foundation is the proper transmission of the load of the structure to the soil in such a way that the soil is not over-stressed and does not undergo deformations that would cause undesirable settlements.

The various types of foundations can be grouped into two categories: (*i*) shallow foundations, and (*ii*) deep foundations. These are classified on the basis of the depth of foundation D_f which is defined as the vertical distance between the base of the foundation and the ground surface, unless the base of the foundation is located beneath a basement or beneath the bed of a river. In these cases the depth of foundation is referred to the level of the basement floor or the river bed. Terzaghi (1943) suggested that a shallow foundation is one, in which the value of ratio $D_f/B \leqslant 1$, B being the width of the base of the foundation. Foundations with D_f/B ratio greater than 1 fall under the category of deep foundations. For shallow foundations, D_f/B commonly ranges between 0.25 and 1; whereas for deep foundations, it is usually between 5 and 20.

Shallow foundation

Shallow foundations are of five types, namely: (*i*) isolated footings, (*ii*) continuous footings, (*iii*) combined footings, (*iv*) mats or rafts, and (*v*) floating rafts. An isolated or a continuous footing is basically a pad used to spread out a column or wall load over a sufficiently large soil area (Figs. 6.1a and 6.1b). Depending on the shape of column, isolated footing may be square, rectangular or circular in shape. Single footings may be of uniform thickness or either stepped or slopped (Figs. 6.1a, c and d). A footing that supports a group of columns is a combined footing and these may be rectangular or trapezoidal in shape (Figs. 6.2a and 6.2b). Trapezoidal combined footing is provided when the column having more load lies near the property line. When the distance between the columns is large, it is economical to connect the isolated footings by a strap beam (Fig. 6.2c). The strap beam does not

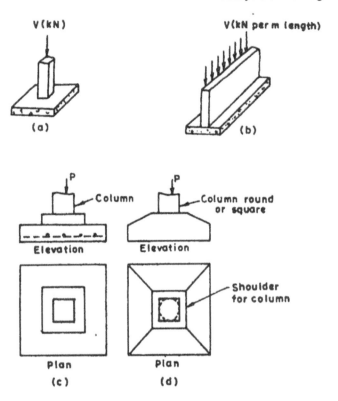

Fig. 6.1. Typical Footings: (*a*) Square Sqread Footing to Support Column Loading; (*b*) Strip Footing to Support Wall Loading; (*c*) Stepped Footing; and (*d*) Sloped Footing.

transfer any pressure to the soil. A raft or mat is a combined footing that covers the entire area beneath a structure and supports all the walls and columns (Fig. 6.3). Raft foundations are required on soils of low bearing capacity, or where structural columns or other loaded areas are so close in both directions that individual pad foundations would nearly touch each other. These are also useful in reducing differential settlement on variable soils or where there is large variations in loading between adjacent columns.

Deep foundations

Piles, piers and wells (or caissons) are the common types of deep foundations used to transmit structural loads through the upper zones of poor soil to a depth where the soil is capable of providing adequate support. Deep foundations are also provided in situations where it is necessary to provide resistance to uplift or where there is a chance of erosion due to flowing water or other reasons. Piles are small diameter shafts, driven or installed into the ground and are constructed in clusters/groups to provide foundations of structures. Basically, piles are of three types, namely (i) driven piles, (ii) cast-in-situ piles, and (iii) driven and cast-in-situ piles. Driven piles may be of timber, steel or concrete. When these piles are of concrete, they are to be precast. Precast piles can be constructed in a variety of forms,

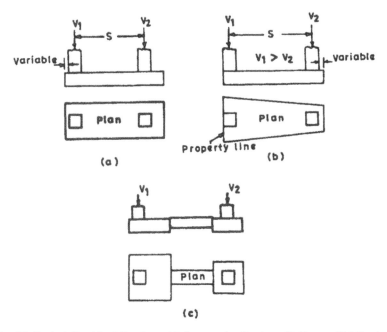

Fig. 6.2. Typical Combined Footings: (*a*) Rectangular Footing; (*b*) Trapezoidal Footing; and (*c*) Strap Beam Footing.

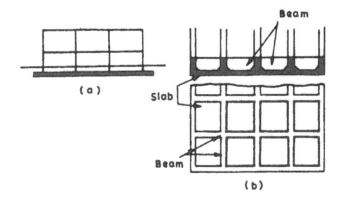

Fig. 6.3. Typical Raft Foundations: (*a*) Plain Slab Type; (*b*) Slab and Beam Type.

two of which are illustrated in Fig. 6.4. Such piles must be reinforced to withstand handling until they are ready to be driven and must also be reinforced to resist stresses caused by driving. They may be driven vertically or at an angle to the vertical using a pile hammer. Cast-in-situ piles are constructed by making holes in the ground to the required depth and then filling them with concrete. Straight bored piles or piles with one or more bulbs at intervals may be cast at site. The latter type is known as under-reamed piles (Fig. 6.5). In driven and cast-in-situ piles, a steel shell is driven into the ground with the aid of a mandrel

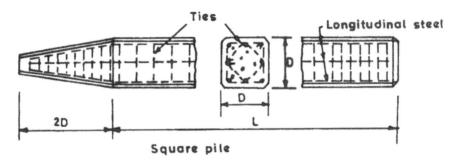

Square pile

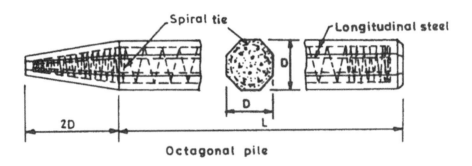

Octagonal pile

Fig. 6.4. Typical Details of Precast Concrete Piles.

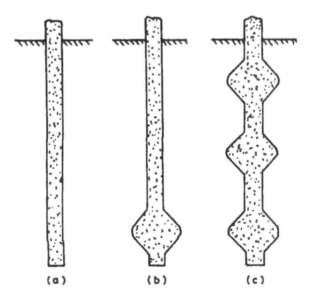

Fig. 6.5. Typical Cast-in-situ Piles: (*a*) Straight Bored Pile; (*b*) Single Bulb Pile; and (*c*) Multi Bulb Pile.

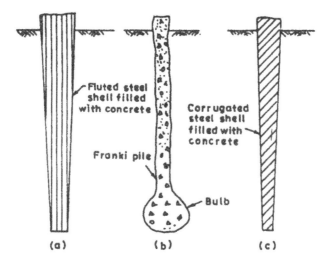

Fig. 6.6. Types of Driven Cast-in-situ Concrete Piles.

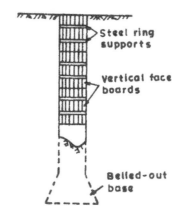

Fig. 6.7. A Typical Deep Pier Foundation.

inserted into the shell. The mandrel is withdrawn and concrete is poured in the shell. The piles of this type are called as shell type (Figs. 6.6a and c). The shell-less type is formed by withdrawing the shell while the concrete is being placed (Fig. 6.6b).

A pier is a heavy structural member acting as a massive strut, for example piers supporting a bridge over a water-way, or the supports to the heavy gate structures of a barrage or spillway (Fig. 6.7).

A well (or a caisson) is the component of a structure which is sunk through ground or water for the purpose of excavating and placing the foundation at the prescribed depth and which subsequently becomes an integral part of the structure. There are three types of caissons, namely (i) Open caisson–a caisson open at the top and bottom (Fig. 6.8a), (ii) Box caisson–a caisson which is closed at the bottom but open at the top (Fig. 6.8b) and (iii)

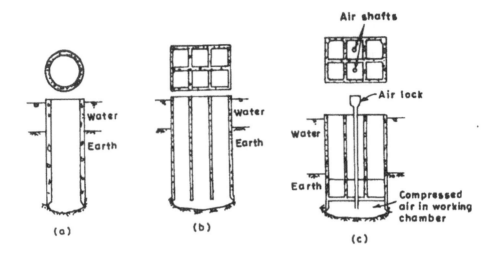

Fig. 6.8. Types of Well (or Caisson): (*a*) Open Caisson; (*b*) Box Caisson; and (*c*) Pneumatic Caisson.

Pneumatic caisson–a caisson with a working chamber in which the air is maintained above atmospheric pressure to prevent the entry of water into the excavations (Fig. 6.8c).

Floating raft

The floating raft is a special type of foundation construction which has application in locations where deep deposits of compressible cohesive soils exist and the use of piles is impracticable. In this, the raft is placed at such a depth that the intensity of superimposed loading due to structure and the foundation does not exceed or nearly equals the vertical pressure existing at that depth prior to excavation (Fig. 6.9). Theoretically, the soil below the foundation is not subjected to any change of loading, and, therefore, there would be no settlement. However, there will be some settlement as the soil at the base of the raft will expand after excavation and then recompress after the construction of the structure.

6.2 SELECTION OF TYPE OF FOUNDATION

Selection of proper type of foundation requires the following information:
   (*i*) soil conditions and soil properties underlying an area,
   (*ii*) position of water table,
   (*iii*) type of the structure along with the loadings, and
   (*iv*) tolerances of the structure, i.e. permissible values of settlement and tilt.

The different types of foundations discussed in Art 6.1 will be taken up in some detail in later chapters. However, the potential application of various types of foundations is approximately listed in Table 6.1.

McCarthy (1988) has also suggested guidelines for selecting an appropriate foundation type based on soil conditions as illustrated in Table 6.2.

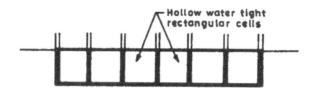

Fig. 6.9. Floating Raft.

6.3 BASIC REQUIREMENTS OF A FOUNDATION

A foundation must be capable of satisfying the following requirements:

1. The foundation must be properly located considering any future influence which could adversely affect its performance, particularly for footings and mats.
2. The soil supporting the foundation must be safe against shear failure.
3. The foundation must not settle or deflect to a degree that can result in a damage to the structure or impair its functioning.
4. The foundation should be safe against sliding and overturning.

These requirements ordinarily should be considered in the order given. The first requirement involves many different factors, most of which cannot be evaluated analytically and have to be answered by engineering judgment. The second is specific. It is analogous to the requirement that a beam in the superstructure must be safe against breaking under its working load. An answer to this requirement can be obtained analytically. Answer to the third requirement can be obtained only partly. Settlement of a structure under the working loads

Table 6.1 Guide to Tentative Selection of Foundation Type.

S. No.	Soil Stratum	Type of Structure	Foundation Type
1.	Medium to dense sand, stiff clay or stiff silt extending up to about 6.0 m to 8.0 m.	Up to three to four storeyed buildings, water tanks of low capacities (< 1000 kL), small towers and chimneys (height < 12 m), retaining walls, etc.	Isolated footings or continuous footings or combined footings.
2.	Loose to medium dense sand to great depth.	Multistoreyed buildings, tall towers and chimneys, overhead tanks (> 1000 kL), structures imposing heavy loads on columns, etc.	Raft foundation: over one half area of building covered by individual footing or driven piles.
3.	Soft clay extending up to large depths (about 10.0 m to 12.0 m) underlain by a firm strata like medium to dense sand.	As mentioned at S. No. 2.	Friction piles if the strength of soft clay increases with depth or bearing piles when very firm strata (e.g. rock) is available beyond soft clay or floating raft if tolerances are very small.
4.	Poor surface and near surface soils followed by soil mixed with boulders firm strata available at deeper depths (10 m–30 m).	Bridges, very heavy column loads, chances of erosion due to flowing water, well foundation structures subjected to large lateral loads.	Well foundation

Table 6.2. Illustrations Relating Soil Conditions and Appropriate Foundation Types (After McCarthy, 1988).

Soil Conditions	Appropriate Foundation Types and Location	Design Comments
El-0.0m ground surface Compact sand, (deposit to great depth)	Installation below frost depth or where erosion might occur	Spread footings most appropriate for conventional foundation needs. A deep foundation such as piles could be required if uplift or other unusual forces were to act.
El-0.0m Firm clay or firm silt and clay (to great depth)	Installation depth below frost depth, or below zone where shrinkage and expansion due to change in water content could occur	Spread footings most appropriate for conventional foundation needs. Also see comment for (1) above.
El-0.0m Firm clay El-3.0 m Soft clay (to great depth)	Comments as for (2) above	Spread footings would be appropriate for low to medium range of loads, if not installed too close to soft clay. If heavy loads are to be carried, deep foundations might be required.
El-0.0m Loose sand (to great depth)	Depth greater than frost or erosion depth	Spread footings may settle excessively or require use of very low bearing pressures. Consider mat foundation, or consider compacting sand by vibroflotation or other method, then use spread footings. Driven piles could be used and would densify the sand. Also consider augered cast-in-place piles.
El-0.0m (soft) Soft clay but firmness increasing with depth El-8.0 (med.firm) (to very great depth) El-16.0 (firmer)	(or)	Spread footings, probably not appropriate. Friction piles or piers would be satisfactory if some settlement could be tolerated. Long piles would reduce settlement problems. Should also consider mat foundation or floating foundation.

... Contd.

Table 6.2. Contd.

Soil Conditions	Appropriate Foundation Types and Location	Design Comments
	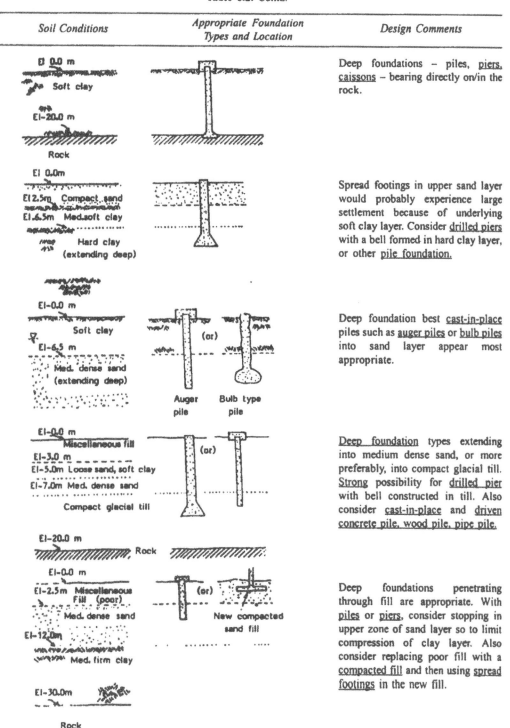	Deep foundations – piles, piers, caissons – bearing directly on/in the rock.
		Spread footings in upper sand layer would probably experience large settlement because of underlying soft clay layer. Consider drilled piers with a bell formed in hard clay layer, or other pile foundation.
		Deep foundation best cast-in-place piles such as auger piles or bulb piles into sand layer appear most appropriate.
		Deep foundation types extending into medium dense sand, or more preferably, into compact glacial till. Strong possibility for drilled pier with bell constructed in till. Also consider cast-in-place and driven concrete pile, wood pile, pipe pile.
		Deep foundations penetrating through fill are appropriate. With piles or piers, consider stopping in upper zone of sand layer so to limit compression of clay layer. Also consider replacing poor fill with a compacted fill and then using spread footings in the new fill.

... Contd.

Table 6.2. Contd.

Soil Conditions	Appropriate Foundation Types and Location	Design Comments
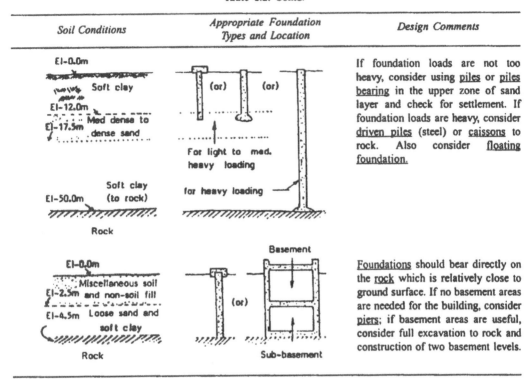		If foundation loads are not too heavy, consider using <u>piles</u> or <u>piles bearing</u> in the upper zone of sand layer and check for settlement. If foundation loads are heavy, consider <u>driven piles</u> (steel) or <u>caissons</u> to rock. Also consider <u>floating foundation.</u>
		Foundations should bear directly on the <u>rock</u> which is relatively close to ground surface. If no basement areas are needed for the building, consider <u>piers</u>; if basement areas are useful, consider full excavation to rock and construction of two basement levels.

depend basically on the type of foundation and soil, and the same can be estimated analytically. However, exact evaluation of the tolerances of different structures with respect to different soils is difficult to estimate and hence one has to depend for this on the engineering judgment keeping in view the functioning of the structure. The fourth requirement is specific and evaluated after obtaining relevant earth pressure against foundation.

The safety criteria against shear failure requires the estimation of the maximum load per unit area of the foundation which the soil can sustain without rupture and is termed as bearing capacity. It also depends on the type of foundation and soil characteristics. Therefore, the methods of determination of bearing capacity and settlement of foundation have been discussed in subsequent chapters on foundations (Chapters 7, 8, 9, etc.). The stability analysis of a foundation has also been covered in the concerned chapter.

Location and depth of foundation

To be safe against adverse environmental influences, the foundation must be carried below:

(i) Zones of volume change due to moisture fluctuations–usually the depth of such zones do not exceed more than 1.5 m except in a few clayey soils, like black cotton soils which exhibit high volume change with variations in moisture content. In case of black cotton soils, strata below 3.5 m depth is normally considered free from seasonal volume changes.

Sowers (1962) classified the soils susceptible to volume change as given in Table 6.3.

(*ii*) Depth of frost penetration-In regions where temperature goes down below the freezing point, foundations of the outside columns and walls should be located below the level at which frost is unlikely to cause a perceptible heave. Fine sands and silts are highly susceptible to frost. In places where average annual temperature is less than 32° F, the deeper soils are permanently frozen and only the surface thaws during the warm weather. This condition is known as perma frost. In such areas, foundations are located at the thaw lines.

Table 6.3. Soils Susceptible to Volume Change.

Likelihood of Volume Change with Changes in Moisture Content	Plasticity Index, PI		Shrinkage Limit
	Arid Region	Humid Region	
Little	0–15	0–30	12 or more
Little to moderate	15–30	30–50	10–20
Moderate to severe	> 30	> 50	< 10

(*iii*) Organic matter, peat, muck, unconsolidated materials such as abandoned garbage dumps and similar fill in areas.

(*iv*) Underground defects—A foundation in any type of soil should go below the zone significantly weakened by root holes or cavities produced by burrowing animals or worms. Foundations should not be located on underground defects such as faults, cavities, mines and man-made discontinuities such as sewers, underground cables and utilities.

(*v*) Scour depth—Foundations for structures in a river have to be protected from the scouring action of the flowing stream. The depth of foundation for a bridge pier or any similar structure must be sufficiently below the deepest scour level.

The property line is considered by law to extend indefinitely into the ground on all sides of the site. Due to this reason the foundation should not cross the property line.

Other conditions, which are relevant to a particular type of foundation, have been discussed in the concerned chapter.

Allowable settlement and tilt

General

For a structure there are certain amounts of tolerances in the form of total settlement, differential settlement and tilt (Fig. 6.10) that can occur without (*i*) over-stressing the structure, and (*ii*) creating unacceptable maintenance or aesthetic problem. If the structure settles uniformly (Figs. 6.10a and 6.11a), it is not likely to suffer structural damage. However, when the settlement becomes excessive, utilities such as water supply, sewage pipes, electric and telephone lines, etc. may get impaired. The difference in total settlement

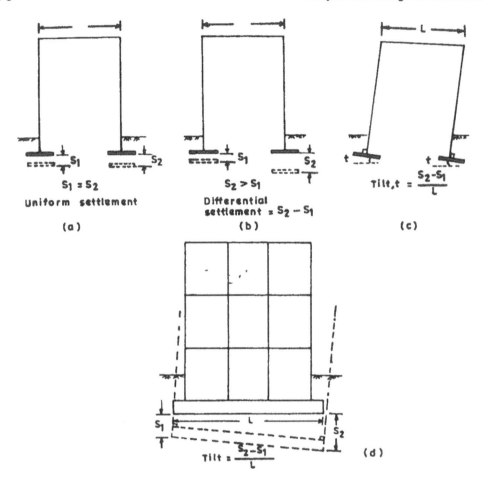

Fig. 6.10. Types of Settlement: (*a*) Uniform Settlement; (*b*) Differential Settlement;
(*c*) Angular distortion; and (*d*) Tilt.

between any two points is the differential settlement (Figs. 6.10b and 6.11b). When the
columns are monolithic with the foundation slab, tilt of the foundations will take place such
that the angles between columns and foundations remain 90° (Fig. 6.10c). Angular distortion
is the ratio of differential settlement between two columns to the spacing between them. If
the entire structure rotates, then tilt of the structure is equal to the ratio of the difference in
settlement of the two ends of the foundation to the width of the foundation (Fig. 6.10d). A
structure may fail if the angular distortion or tilt exceeds the permissible limit.

Differential settlement will occur in all cases because of natural variability of soils even
where total settlements are calculated to be uniform. The magnitude of these differential
settlements may be related to the magnitude of the total settlement (D' Appolonia, 1968). On
the basis of observations taken on several buildings, Bjerrum (1963a, 1963b) developed
charts as shown in Fig. 6.12. The charts give relationships between angular distortion,
differential settlement and total settlement. From these plots, for a permissible value of

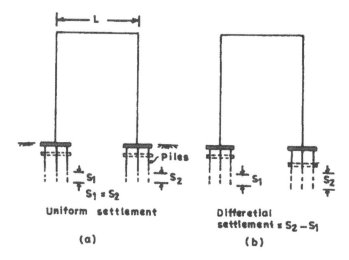

Fig. 6.11. Types of Settlement: (*a*) Uniform Settlement; and (*b*) Differential Settlement.

Table 6.4 Limiting Angular Distortions.

Type of Structural Problem	Angular Distortion δ/L
Difficulties with machinery sensitive to settlement	1/750
Danger for frames with diagonals	1/600
Limit for buildings where cracking is not permissible	1/500
Limit where first cracking in panel walls is to be expected or where difficulties with overhead cranes are to be expected	1/300
Limit where tilting of high, rigid buildings may become noticeable	1/250
Considerable cracking in panel and brick walls. Safe limit for flexible brick walls where $h/L < 1/4$. Limit where structural damage of general buildings may occur	1/150

angular distortion, firstly, maximum differential settlement is obtained and then the permissible total settlement. Bjerrum (1963a) indicated the types of damage that can be expected for various values of the angular distortion (Table 6.4).

Shallow foundation

IS 1904: 1986 gives limits of total settlement, differential settlement and angular distortion for certain type of structures considering two types of foundations namely isolated foundation and raft foundation (Table 6.5). The limits have been specified considering two categories of soils: (*i*) sand and hard clay, and (*ii*) plastic clay. It may be noted from this table that a higher total settlement is permissible in clay than in sand. This is due to the fact that the settlement in sands occurs almost immediately on placement of load while in clays, consolidation settlement occurs over a long period of time. Thus there is time for structures resting on clay to adjust to the differential settlements. In sands, the differential settlement can occur as soon as the total settlement itself has occurred, thus leaving no time for gradual

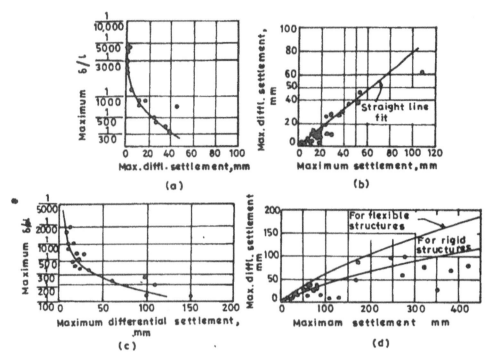

Fig. 6.12. Settlement Characteristics of Structures (*a, b*) on Sand, (*c, d*) on Clay.

adjustment.

For design purposes, the appropriate value of permissible total settlement may be selected with the help of the tolerances given in Tables 6.4 and 6.5 and charts given in Fig. 6.12.

It may be emphasised again that the above-mentioned tables and charts are only the guidelines for getting the value of allowable settlement. Actual tolerances of a given structure should be examined carefully. For example an antenna tower having a raft foundation has a tolerance of total settlement of about 2 mm to 4 mm which is about 1/10 that of the value of allowable total settlement worked out by the procedures given above. Similarly, the tolerances of high rise buildings will be much smaller than the conventional two-three storeyed buildings.

Pile foundation

Tilting of a structure resting on pile foundations due to either differential settlement or by the rotation of whole structure may cause excessive stresses and moments in the piles themselves. Therefore, the permissible limits of distortion settlement and tilt in the case of structures resting on pile foundation should not only be according to the safety and functioning of the superstructure but also on the structural safety of the piles. Therefore, relatively pile foundations are designed for lesser values of permissible settlement and tilt.

IS: 2911 Part iv (1979) recommended permissible value of total settlement for (i) single pile as 8 mm, and (*ii*) group of piles as 25 mm, unless it is specified that a total settlement

Table 6.5 Maximum and Differential Settlements of Buildings.

Sl. No.	Type of Structure	Isolated Foundations Sand and Hard Clay			Isolated Foundations Plastic Clay			Raft Foundations Sand and Hard Clay			Raft Foundations Plastic Clay		
		Maximum Settlement mm	Differential Settlement mm	Angular Distortion	Maximum Settlement mm	Differential Settlement mm	Angular Distortion	Maximum Settlement mm	Differential Settlement mm	Angular Distortion	Maximum Settlement mm	Differential Settlement mm	Angular Distortion
1.	For steel structures	50	.0033L	1/300	50	.0033L	1/300	75	.0033L	1/300	100	.0033L	1/300
2.	For reinforced concrete structure	50	.0015L	1/666	75	.0015L	1/666	75	.002L	1/500	100	.002L	1/500
3.	For load bearing walls in multistoreyed buildings												
	(a) For *L/H† = 2 to 7	60	.002L	1/5000	60	.0002L	1/5000	Not likely to be encountered					
	(b) For L/H = greater than 7	60	.0004L	1/2500	80	.00033L	1/2500						
4.	For framed multistoreyed buildings	60	.002L	1/500	75	.002L	1/500	75	.0025L	1/400	125	.0033L	1/300
5.	For water towers and silos	50	.0015L	1/666	60	.0004L	1/666	100	.0025L	1/400	125	.0025L	1/400

Note: The values given in the table may be taken only as a guide and the permissible settlement and differential settlement in each case should be decided as per requirements of the designer.

* L denotes the length of deflected part of wall/raft of centre-to-centre distance between columns.

† H denotes the height of wall from foundation/footing.

different from these values are permissible on the basis of the nature and type of structure. In the author's opinion, a group of at least nine piles should be considered for the limit of 25 mm. If the number of piles in a group are less than nine, linear interpolation may be done to get the permissible total settlement.

Therefore, in pile foundations, allowable total settlement may be first worked out using the limits of angular distortion given in Table 6.4 and then charts given in Fig. 6.12. The value of total settlement thus obtained is compared with the allowable settlement as per the recommendation of IS: 2911 Part iv (1979). Lesser of the two values is adopted for design.

Well foundation

Well foundations are commonly provided to support bridges. Sometimes these are also used as foundations for heavy machinery and transmission line towers. Tolerance of a well foundation depends upon the tolerance of the superstructure. However, Indian standard (IS: 3955–1967) recommended that tilt should generally be limited to 1 in 60. Also shift should normally be restricted to one per cent of the depth of sunk. As per Indian Roads Congress (IRC: 5–1985), in case of simply supported type of bridges, the differential settlement of the supports should not exceed 1 in 400, unless provision has been made for rectification of this settlement.

6.4 TERMINOLOGY

It will be useful at this stage to define the various terms relating to bearing capacity and bearing pressure. These are as follows:

(*i*) *Total overburden pressure,* σ: It is the intensity of total pressure, due to the weights of both soil and soil water, on any horizontal plane at and below foundation level before construction operations are commenced.

(*ii*) *Effective overburden pressure,* σ': It is the intensity of intergranular pressure on any horizontal plane at and below foundation level before construction operations are commenced. It is, therefore, equal to the total overburden pressure (σ) minus the porewater pressure. Hence

$$\sigma' = \sigma - \gamma_w h \qquad\qquad \ldots (6.1)$$

where h is the height of the water table above the section at which σ' is obtained.

(*iii*) *Total foundation pressure,* q : It is the intensity of total pressure on the ground beneath the foundation after the structure has been erected and fully loaded. It is inclusive of gross load of the foundation substructure, the loading from superstructure, and the gross loading from any backfilled soil and soil water supported by the structure. It is also termed as gross load intensity.

(*iv*) *Net foundation pressure,* q_n: It is the net increase in pressure on the ground beneath the foundation due to the dead load and live load applied by the structure

$$q_n = q - \sigma \qquad\qquad \ldots (6.2)$$

The pressure q_n is used for calculating the distribution of stress at any depth below foundation level.

(v) *Ultimate bearing capacity, q_u:* It is maximum gross intensity of loading that the soil can support before it fails in shear.

(vi) *Net ultimate bearing capacity, q_{nu}:* It is maximum net intensity of loading at the base of the foundation that the soil can support before failing in shear. Therefore,

$$q_{nu} = q_u - \sigma \qquad \qquad \text{... (6.3)}$$

(vii) *Net safe bearing capacity, q_{nF}:* It is the maximum net intensity of loading that the soil can safely support without the risk of shear failure. It is obtained by dividing q_{nu} by a factor of safety, F. Hence

$$q_{nF} = \frac{q_{nu}}{F} \qquad \qquad \text{... (6.4)}$$

(viii) *Safe bearing pressure, q_{ns}:* It is the maximum net intensity of loading that can be allowed on the soil without the settlement exceeding the permissible value.

(ix) *Allowable bearing pressure, q_a:* It is the maximum allowable net loading that can be imposed on the soil with no possibility of shear failure or the possibility of excessive settlement. It is hence the smaller of the net safe bearing capacity (q_{nF}) and the safe bearing pressure (q_{ns}).

6.5 COMPUTATION OF LOADS

A structure may be subjected to a combination of some or all the following loads and forces:

(a) *Dead load :* It includes the weight of the structure and all material permanently attached to it such as fire-proofing and air-conditioning systems. If the weight of the earth is directly supported by elements of the structure, it should be considered as dead load.

(b) *Live load :* It includes all vertical loads that are not a permanent part of the structure but are expected to superimpose on the structure during a part of its useful life. Human occupancy, partition walls, furniture, warehouse goods and mechanical equipment are major live loads.

Foundations of highway and railroad bridges subjected to traffic loadings are designed as per Indian Roads Congress (IRC) and Indian Railways Services (IRS) specifications. These will be discussed in detail in Chapter 9.

(c) *Wind load :* It acts on all the exposed areas of the structure. The relationship between wind pressure and velocity is $p = kV^2$, where p is the pressure, V the velocity and k is coefficient, the value of which depends on a number of factors, such as the wind speed, the type, proportion and shape of structure and the temperature of air. In design aids (part vi, Structural Design–section 1), wind pressure maps and tables as given in Appendix 6.1 have been prepared taking the value of K as 0.006. Using these maps and tables, intensity of wind pressure (in kg/m²) acting on a component of structure can be obtained.

(d) *Seismic Force :* As per IS 1893 (Part 1)-2002, for the purpose of determining seismic forces, the country is classified into four seismic zones as shown in Fig. 6.13a. As this version of code is limited to buildings only, further discussions here have been incorporated on the basis of IS : 1893–1984. According to this the country is divided into five seismic zones (Fig. 6.13b). In author's opinion, seismic zones as shown in Fig. 6.13a should be considered.

Two methods, namely (i) seismic coefficient method, and (ii) response spectrum method are used for computing the seismic force. For pseudo-static design of foundations of buildings, bridges and similar structures, seismic coefficient method is used. For the analysis of earth dams and dynamic designs, response spectrum method is used (IS: 1893–1984).

In seismic coefficient method, the design value of horizontal seismic coefficient α_h is obtained by the following expression:

$$\alpha_h = \beta I \alpha_0$$

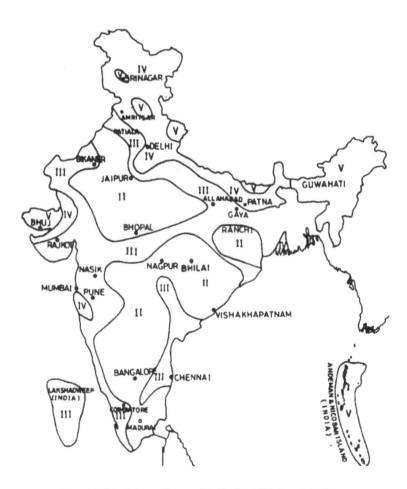

Fig. 6.13(a). Seismic Zones of India (IS 1893 Part 1-2002).

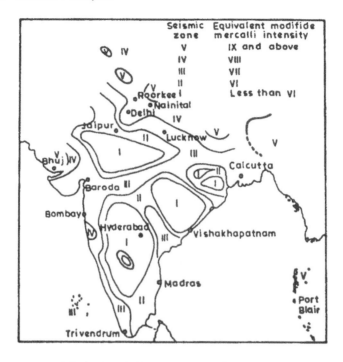

Fig. 6.13(b). Seismic Zones of India. (IS 1893–1984)

where α_0 = basic seismic coefficient, Table 6.6,
            I = coefficient depending upon the importance of structure, Table 6.7, and
            β = coefficient depending upon the soil-foundation system, Table 6.8.

The vertical seismic coefficient, α_v shall be considered in the case of structures in which stability is a criterion of design or for overall stability of structures. It may be taken as half of the horizontal seismic coefficient. Therefore,

$$\alpha_v = \frac{\alpha_h}{2}.$$

In response spectrum method, the response acceleration coefficient is first obtained for the natural period and damping of the structure and the design value of horizontal seismic coefficient is computed using the following expression:

$$\alpha_h = \beta \; I. \; F_0 \frac{S_a}{g}$$

where F_0 = seismic zoning factor for average acceleration spectra (Table 6.9), and

      $\dfrac{S_a}{g}$ = average acceleration coefficient as read from Fig. 6.14 for appropriate natural period and damping of the structure.

Table 6.6. Values of Basic Seismic Coefficients, α_0.

Zone No.	α_0
V	0.08
IV	0.05
III	0.04
II	0.02

Table 6.7. Values of Importance Factor, I.

S. No.	Type of Structure	Value of I
1.	Containment structures of atomic power reactors for preliminary design	3.0
2.	Dams (all types)	2.0
3.	Containers of inflammable or poisonous gases or liquids	2.0
4.	Important service and community structures, such as hospitals, water towers and tanks, schools, important bridges, emergency buildings like telephone exchanges and fire brigades, large assembly structures like cinemas, assembly halls and subway stations.	1.5
5.	All others	1.0

Table 6.8. Values of β for Different Soil-Foundation Systems.

Type of Soil	Values of β for				
	Isolated Footings without Tie Beams	Combined or Isolated Footings with Tie Beams	Raft Foundation	Pile Foundation	Well Foundation
1. Rock or hard soil	1.0	1.0	1.0	1.0	1.0
2. Medium soils	1.2	1.0	1.0	1.0	1.2
3. Soft soils	1.5	1.2	1.0	1.2	1:5

Table 6.9. Values of Seismic Zoning Factor, F_0.

Zone No.	F_0
V	0.40
IV	0.25
III	0.20
II	0.10

Permissible Increase in Allowable Soil Pressure

(*i*) Where the bearing pressure due to wind is less than 25 per cent of that due to dead and live loads, it may be neglected in design. Where this exceeds 25 per cent, foundations may be so proportioned that the pressure due to combined dead, live

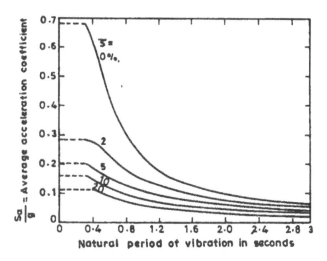

Fig. 6.14. Average Acceleration Spectra.

and wind loads does not exceed the allowable bearing pressure by more than 25 per cent.

(*ii*) When earthquake forces are included, the permissible increase in allowable bearing pressure of pertaining soil shall be as given in Table 6.10, depending upon the type of foundation of the structure.

(*iii*) Wind and seismic forces shall not be considered simultaneously.

(*iv*) If any increase in bearing pressure has already been permitted for forces other than seismic forces, the total increase in allowable bearing pressure when seismic force is also included shall not exceed the limits specified above.

Table 6.10. Permissible Increase in Allowable Soil Pressure for Earthquake Forces.

Type of Soil	Permissible Increase in Allowable Soil Pressure in Per cent				
	Isolated Footings without Tie Beams	Isolated Footings with Tie Beams, Combined Footings and Friction Piles	Raft Foundation & Bearing Piles	Friction Piles	Well Foundation
1. Rock or hard soils ($N^* > 30$)	50	50	50	--	50
2. Medium soils N between 10 and 30	25	25	50	25	25
3. Soft soils $N < 10$	0	25	50	25	25

* N = Standard penetration resistance.

Safety against liquefaction

Desirable field values of N are as follows:

Zone	Depth Below Ground Level in Metres	N Values
III, II and I	up to 5	15
	10	25
I and II (for important structures only)	up to 5	10
	10	20

In submerged loose sands and soils falling under classification SP with standard penetration values less than the values specified above, the vibrations caused by earthquake may cause liquefaction or excessive total and differential settlements. In important projects, this aspect of the problem needs to be investigated and appropriate methods of compaction or stabilization adopted to achieve suitable N. Alternatively deep foundations may be provided and taken to the depths well into the layers which are not likely to liquefy. Marine clays and other sensitive clays are also known to liquefy due to collapse of soil structure and will need special treatment according to site conditions.

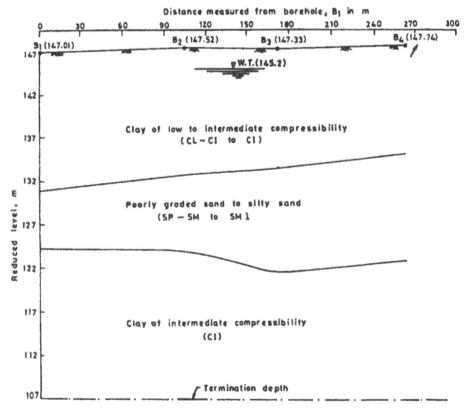

Fig. 6.15. Typical Soil Profile.

6.6 DESIGN STEPS

A foundation may be designed stepwise as given below:

1. Determine loads as explained in Sec. 6.5.
2. Sketch the soil profile showing the stratification at the site along with the position of maximum water level. Note down the physical and engineering properties of each layer on the soil profile itself (e.g. Fig. 6.15).
3. On the basis of the type of structure, amount of loads coming on the foundation and soil profile, decide the type and minimum depth of foundation.
4. Select appropriate dimensions of the foundation. Determine safe bearing capacity of the foundation, taking into consideration the engineering properties of soil, depth and plan dimensions of the foundation. Determine intensity of maximum base pressure, and it should be less than the safe bearing capacity.
5. Determine the settlement, tilt and horizontal displacement of the foundation under the actual forces and moments and compare with the respective permissible values.
6. Check the stability of the foundation against horizontal and uplift forces.
7. Carry out the structural design of the foundation keeping in view the critical sections for bending moment and shear.
8. Check the need of foundation drains, water-profing or damp-proofing.
9. Prepare the complete working drawings for execution of work in the field.

The details of steps 4 to 9 have been given in Chapters 7, 8 and 9 for shallow foundation, pile foundation and well foundation respectively.

REFERENCES

1. Bjerrum, L. (1963a), "Discussion to European Conference on Soil Mechanics and Foundation Engineering," *Wiesbaden*, Vol. II, p. 135.
2. Bjerrum, L. (1963b), "Generalle Krav til fundamentering av forskjellige buggverk; tillatte Setninger;" Den Norske Ingenirforeming, Kurs, fundamentering, Oslo.
3. D' Appolonia, D.J., et al. (1968), "Settlement of Spread Footings on Sand," JSMFD, ASCE, Vol. 94, SM3, pp. 735-760.
4. *Design Aids for Reinforced Concrete to IS: 456–1978*, special publication 16, Indian Standards Institution, New Delhi, 1980.
5. IS: 1904 (1986), "Code of Practice for the Structural Safety of Buildings: Shallow Foundation."
6. IS: 2911 Part IV (1985), " Code of Practice for the Design and Construction of Pile Foundation: Load Test on Piles."
7. IS: 1893 (1984, 2002), "Criteria for Earthquake Resistant Design of Structures."
8. McCarthy, D.F. (1988), *Essentials of Soil Mechanics and Foundations: Basic Geotechnics*, Prentice Hall, Englewood Cliffs, New Jersey.
9. Sowers, G.F. (1962), "Shallow Foundation," *Foundation Engineering*, edited by G.A. Leonards, Chapter 6, McGraw-Hill, New York, U.S.A.
10. Terzaghi, K. (1943), *Theoretical Soil Mechanics*, Ch. 8, John Wiley, New York, U.S.A.

PRACTICE PROBLEMS

1. What factors determine whether a foundation type is in the shallow or deep foundation category?
2. List the steps involved in a foundation design.
3. Describe with suitable examples the principles involved in the design of a floating foundation.
4. What are the basic requirements for the satisfactory performance of a foundation?

APPENDIX 6.1

Fig. 6.16. Basic Maximum Wind Pressure Map of India Including Wind of Short Duration.

Note 1 : For purpose of this map, a short duration wind is that which lasts only for a few minutes, generally less than five minutes.

Note 2 : The relationship between wind pressure and velocity is $p = KV^2$: where p is the pressure, V the velocity and K is a coefficient, the value of which depends on a number of factors such as the wind speed, the type, proportion and shape of structure and the temperature of air. In the preparation of this basic wind pressure map, a value of 0.006 has been assumed for K and p is expressed in kg/m² and V in km/h.

Note 3 : The basic wind pressures for the zones shown in the map shall be as given below :

ZONE	Pressure in kg/m² upto a height of 30 m above the mean retarding surface	Pressure in kg/m² at a height (Expressed in metres) of									
		35	40	45	50	60	70	80	100	120	150
	100	104	105	108	111	115	118	122	127	132	138
	150	156	158	163	167	172	177	183	191	198	207
	200	208	210	217	222	230	236	244	254	264	276

(For intermediate heights interpolated values may be adopted)

Note 4 : The number of severe cyclones which have approached or crossed the coasts during 1891 to 1960 is indicated in circles in 5° latitude zones. (A severe cyclone is one in which the wind speed exceeds 89 km/h corresponding to a wind pressure of 48 kg/m²). The influence of a severe cyclone may be taken to extend from the coast line up to the line demarcating 60 kg/m² zone.

Note 5 : The basic wind pressures indicated in the map are the maximum ever likely to occur in the respective areas, under fully exposed conditions. In case of mountainous areas, the values indicated above should be modified according to the local conditions because the surface wind is known to depend markedly on the local topography, etc.

Fig. 6.17. Basic Maximum Wind Pressure Map of India Excluding Winds of Short Duration.

Note 1 : For purpose of this map, a short duration wind is that which lasts only for a few minutes, generally less than five minutes.

Note 2 : The relationship between wind pressure and velocity is $p = KV^2$; where p is the pressure, V the velocity and K is a coefficient, the value of which depends on a number of factors such as the wind speed, the type, proportion and shape of structure and the temperature of air. In the preparation of this basic wind pressure map, a value of 0.006 has been assumed for K and p is expressed in kg/m^2 and V in km/h.

Note 3 : The basic wind pressures for the zones shown in the map shall be as given below :

ZONE	Pressure in kg/m² upto a height of 30 m above the mean retarding surface	Pressure in kg/m² at a height (Expressed in metres) of									
		35	40	45	50	60	70	80	100	120	150
	60	62	63	65	67	69	71	73	76	79	83
	100	104	105	108	111	115	118	122	127	132	138
	150	156	158	163	167	172	177	183	191	198	207
	200	208	210	217	222	230	236	244	254	264	276

(For intermediate heights interpolated values may be adopted)

Note 4 : The number of severe cyclones which have approached of crossed the coasts during 1891 to 1960 is indicated in circles in 5° latitude zones. (A severe cyclone is one in which the wind speed exceeds 89 km/h corresponding to a wind pressure of 48 kg/m^2). The influence of a severe cyclone may be taken to extend from the coast line up to the line demarcating 60 kg/m^2 zone.

Note 5 : The basic wind pressure indicated in the map are the maximum ever likely to occur in the respective areas, under fully exposed conditions. In case of mountainous areas, the values indicated above should be modified according to the local conditions because the surface wind is known to depend markedly on the local topography, etc.

Shallow Foundation

7.1 INTRODUCTION

Shallow foundation is a very common type of foundation provided for many structures. In general, spread footings, combined footings and raft or mat foundations are the usual types of shallow foundation. Spread footings are undoubtedly the most widely used type because these are usually more economical. Construction of footings requires the least amount of equipment and skill. Furthermore, the conditions of the footing and the supporting soil can be readily examined. Combined footings and rafts are more favourable than footings when the bearing capacity of the soil is so low that disproportionately large footings would be required or when the foundation has to resist large horizontal forces.

In this chapter, firstly, the methods of computing bearing capacity and settlement of shallow foundations have been discussed. This is followed with the procedures of their proportioning and structural design.

7.2 LOCATION AND DEPTH OF FOUNDATION

In general, foundations should rest on good bearing soil strata. However, Indian Standard Code (IS: 1904–1978) makes the recommendation that a foundation should be located at a minimum depth of 500 mm below the zone of volume change. In an expansive clay soil, swelling and shrinkage of the soil mass will occur due to rise or lowering of groundwater table following seasonal weather changes. The zone of seasonal variation in water content varies in thickness from 1.5 m to 3.5 m for the 'black cotton soils' of India. In the case of fine sands and silts, in areas where extremely low temperatures are likely to materialize, the foundation must be placed below the zone of frost heave.

Foundations for structures in a river have to be protected from the scouring action of the flowing stream. The depth of foundation for a bridge pier or any similar structure must be sufficiently below the deepest scour level.

Where footings are adjacent to a sloping ground or where the bases of footings are at different levels or at a level different from those of footings of adjoining structures, Indian Standard Code (IS : 1904–1986) makes the following recommendations:

(a) When the ground surface slopes downward adjacent to a footing, the sloping surface should not encroach upon a frustum of bearing material under the footing having sides which make an angle of 60° with the horizontal for rock and 30° for soil. The horizontal distance from the lower edge of the footing to the sloping surface shall be at least 600 mm for rock and 900 mm for soil.

(b) For footings in granular soils, the line joining the lower adjacent edges of adjacent footings should not have a slope steeper than two horizontal to one vertical (Fig. 7.1a).

(c) In clayey soils, the slope of the line joining the lower adjacent edge of the upper footing and the upper adjacent edge of the lower footing should not be steeper than two horizontal to one vertical (Fig. 7.1b).

To avoid damage to an existing structure, the foundation for a new structure at an adjacent site should be located as illustrated in Fig. 7.2. The adjacent edge of the new footing must be at least at a distance 'S' from the edge of the existing footing where 'S' is equal to the width of the larger footing. The line from the edge of the new footing to the edge of the existing footing should make an angle of 45° or less with the horizontal plane, that is, the distance 'S' should be greater than the difference in elevation between the adjacent footings.

7.3 BEARING CAPACITY OF FOOTINGS

Modes of shear failure

The maximum load per unit area that can be imposed on a footing without causing rupture of soil is its bearing capacity (sometimes termed critical or ultimate bearing capacity). It is usually denoted by q_u. This load may be obtained by carrying out a load test on the footing which will give a curve between average load per unit area and settlement of the footing. Based on pressure-settlement characteristics of a footing and pattern of shearing zones, three modes of shear failure have been identified as (i) general shear failure, (ii) punching shear failure, and (iii) local shear failure (Caquot, 1934; Terzaghi, 1943; DeBeer and Vesic, 1958; Vesic, 1973).

In general shear failure, well defined slip lines extend from the edge of the footing to the adjacent ground. Abrupt failure is indicated by the pressure-settlement curve (Fig. 7.3a). Usually in this type, failure is sudden and catastrophic and bulging of adjacent ground occurs. This type of failure occurs in soils having brittle type stress-strain behaviour (e.g. dense sand and stiff clays).

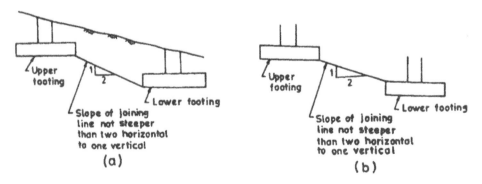

Fig. 7.1. Footing at Different Levels: (a) Granular Soil; (b) Clayey Soil.

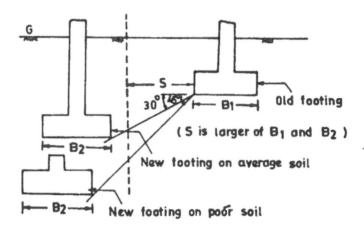

Fig. 7.2. Footings for Old and New Structures.

In punching shear failure, there is vertical shear around the footing perimeter and compression of soil immediately under the footing, with soil on the sides of the footing remaining practically uninvolved. The pressure-settlement curve indicates a continuous increase in settlement with increasing load (Fig. 7.3b).

The local shear failure is an intermediate failure mode and has some of the characteristics of both the general shear and punching shear failure modes. Well defined slip lines immediately below the footing extend only a short distance into the soil mass. The pressure-settlement curve does not indicate the bearing capacity clearly (Fig. 7.3c). This type of failure occurs in soils having plastic stress-strain characteristics (e.g. loose sand and soft clay).

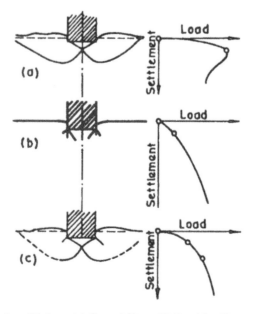

Fig. 7.3 Typical Modes of Failure: (*a*) General Shear; (*b*) Punching Shear; and (*c*) Local Shear.

In Fig. 7.4, types of failure mode that can be expected for a footing in a particular type of sand is illustrated (Vesic, 1973). This figure indicates that the type of failure depends on the relative density and depth-width ratio (D_f/B) of the footing. There is a critical value of D_f/B ratio below which only punching shear failure occurs.

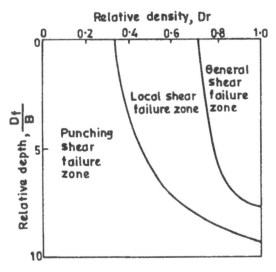

Fig. 7.4. Region for Three Different Modes of Failure.

The criteria given in Table 7.1 may also be followed for identification of type of failure.

Table 7.1. Identification of Type of Failure.

Type of Failure	Relative Density Dr (%)	φ (Deg)	Void Ratio e
1. General shear failure	⩾ 70	⩾ 36°	⩽ 0.55
2. Local shear failure or punching shear failure	< 20	⩽ 29°	⩾ 0.75

Generalized bearing capacity equation

In the design of foundation, usually net bearing capacity is computed and used. It is defined as the maximum net intensity of loading at the base of the foundation that the soil can support before failing in shear. It is denoted by q_{nu}. Therefore,

$$q_{nu} = q_{du} - \gamma_1 D_f \qquad \qquad \text{... (7.1)}$$

where q_{du} = ultimate bearing capacity.

The equation of net bearing capacity developed for strip footing considering general shear failure (Terzaghi, 1943; Meyerhof, 1951) is extended to consider variations from the basic assumptions by applying modification factors that account for the effect of each variation (Hansen, 1970). It may be written as:

$$q_{nu} = cN_c \cdot S_c \cdot d_c \cdot i_c \cdot b_c + \gamma_1 D_f (N_q - 1) \cdot S_q \cdot d_q \cdot i_q b_q r'_w +$$

$$\frac{1}{2} \gamma_2 BN_\gamma \cdot S_\gamma \cdot d_\gamma \cdot i_\gamma \cdot b_\gamma \cdot r'_w \qquad \qquad \ldots (7.2)$$

where q_{nu} = net ultimate bearing capacity,

c = undrained cohesion of soil,

B = width of footing,

D_f = depth of foundation below ground surface,

γ_1 = density of soil above the base level of footing,

γ_2 = density of soil below the base level of footing,

N_c, N_q, N_γ = bearing capacity factors,

S_c, S_q, S_γ = shape factors for square, rectangular and circular foundations,

d_c, d_q, d_γ = depth factors,

i_c, i_q, i_γ = inclination factors,

g_c, g_q, g_γ = ground inclination factors, and

r_w, r_w' = groundwater table factors.

Bearing capacity factors: N_c, N_q and N_γ are non-dimensional factors which depend on angle of shearing resistance of soil (Terzaghi, 1943; Terzaghi and Peck, 1967). Their values may be obtained from Table 7.2.

Table 7.2 Bearing Capacity Factors.

ϕ	N_c	N_q	N_γ
0	5.14	1.00	0.00
5	6.49	1.57	0.45
10	8.35	2.47	1.22
15	10.98	3.94	2.65
20	14.83	6.40	5.39
25	20.72	10.66	10.88
30	30.14	18.40	22.40
35	46.12	33.30	48.03
40	75.31	64.20	109.41
45	138.88	134.88	271.76
50	266.89	319.07	762.89

Shape factors : Approximate values of shape factors which are sufficiently accurate for most practical purposes are given in Table 7.3.

Depth factors : The bearing capacity factors given in Table 7.2 do not consider the shearing resistance of the failure plane passing through the soil zone above the level of the foundation base. If this upper soil zone possesses significant shearing strength, the ultimate value of bearing capacity would be increased (Meyerhof, 1951). For this case, depth factors are applied, whereby

$$d_c = 1 + 0.4 \frac{D_f}{B} \qquad\qquad \text{... (7.3)}$$

$$d_q = 1 + 2 \tan \phi(1-\sin \phi)^2 . \frac{D_f}{B} \qquad\qquad \text{... (7.4)}$$

$$d_\gamma = 1 \qquad\qquad \text{... (7.5)}$$

Table 7.3. Shape Factors.

S. No.	Shape of the Base of Footing	Shape Factor		
		S_c	S_q	S_γ
(i)	Continuous strip	1.00	1.00	1.00
(ii)	Rectangle	1 + 0.2 B/L	1 + 0.2 B/L	1 −0.4 B/L
(iii)	Square	1.3	1.2	0.8
(iv)	Circle	1.3	1.2	0.6
	(B = diameter)			

The use of depth factors is conditional upon the soil above foundation level being not significantly inferior in shear strength characteristics to that below this level.

Factors for eccentric-inclined loads : A foundation engineer frequently comes across the footings subjected to eccentric-inclined loads, e.g. in the case of the foundations of retaining walls, abutments, columns, stanchions, portal framed buildings, etc. (Fig. 7.5). The eccentricity may be one way or two ways as shown in Figs. 7.6a and 7.6b respectively. The effect of eccentricity can be conveniently and conservatively considered as follows:

One-way eccentricity (Fig. 7.6a) : If the load has an eccentricity e, with respect to the centroid of the foundation in only one direction, then the dimension of the footing in the direction of eccentricity shall be reduced by a length equal to $2e$. The modified dimension shall be used in the bearing capacity equation and in determining the effective area of the footing in resisting the load.

Two-way eccentricity (Fig. 7.6b) : If the load has double eccentricity (e_L and e_B) with respect to the centroid of the footing, then the effective dimensions of the footing to be used in determining the bearing capacity as well as in computing the effective area of the footing in resisting the load shall be determined as given below:

$$L' = L - 2e_L \qquad\qquad \text{... (7.6)}$$
$$B' = B - 2e_B \qquad\qquad \text{... (7.7)}$$
$$A' = L' \times B'. \qquad\qquad \text{... (7.8)}$$

In computing the shape and depth factors for eccentrically-obliquely loaded footings, effective width (B') and effective length (L') will be used in place of total width (B) and total length (L).

For a design, eccentricity should be limited to one-sixth of the foundation dimension to prevent the condition of uplift occurring beneath the part of the foundation.

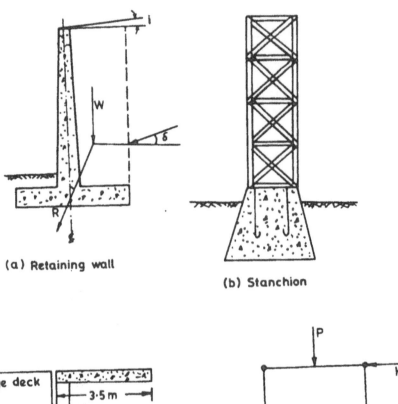

(a) Retaining wall

(b) Stanchion

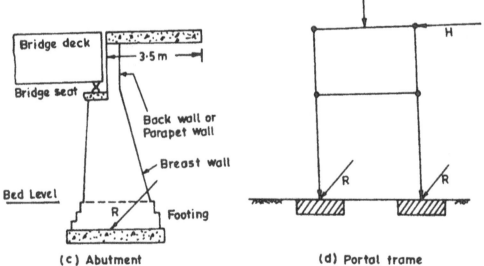

(c) Abutment

(d) Portal frame

Fig. 7.5. Problems of Footings Subjected to Eccentric-Inclined Loads.

Inclination factors: Following inclination factors may be adopted in design:

$$i_\gamma = \left[1 - \frac{Q_h}{Q + BLc\cot\phi}\right]^{m+1} \qquad \dots \ (7.9)$$

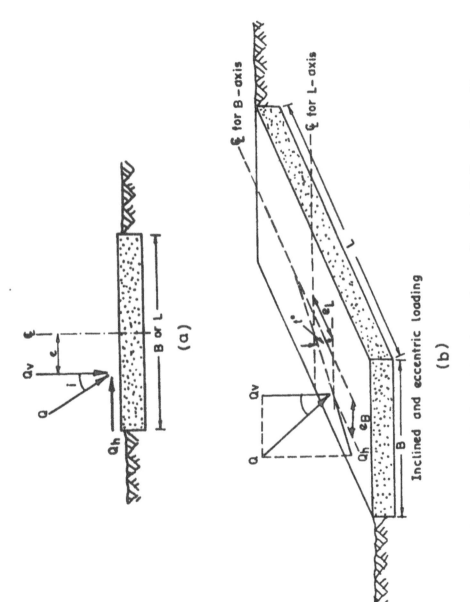

Fig. 7.6. Eccentrically-Obliquely Loaded Footings (*a*) One-way Eccentricity (*b*) Two-way Eccentricity.

$$i_q = \left[1 - \frac{Q_h}{Q + BLc \cot \phi}\right]^m \qquad \dots (7.10)$$

$$i_c = i_q - \left(\frac{1 - i_q}{N_c \tan \phi}\right) \text{ (when } \phi > 0°) \qquad \dots (7.11)$$

$$= 1 - \frac{mQ_h}{cN_c BL} \text{ (when } \phi = 0°) \qquad \dots (7.12)$$

where Q_h is the horizontal component of the load Q acting on the foundation at inclination i with vertical. Values of m are taken as given below:

(i) If the angle of inclination i is in the plane of L-axis

$$m = (2 + L/B) / (1 + L/B) \qquad \dots (7.13)$$

(ii) If the angle of inclination is in the plane of the B-axis

$$m = (2 + B/L) / (1 + B/L) \qquad \dots (7.14)$$

As per IS 6403:1981, the inclination factors are given by

$$i_c = i_q = \left(1 - \frac{i}{90}\right)^2 \qquad \dots (7.15)$$

$$i_\gamma = \left(1 - \frac{\alpha}{\phi}\right)^2 \qquad \dots (7.16)$$

For more accurate estimation of bearing capacity of an one-way eccentrically-obliquely loaded footing, bearing capacity factors as shown in Figs. 7.7 to 7.9 developed by Saran and Agarwal (1991) may be used. These factors have been obtained by carrying out a theoretical analysis based on limit equilibrium and limit analysis approaches.

As evident from these figures, bearing capacity factors ($N_{\gamma ei}$, N_{qei} and N_{cei}) depend on ϕ, i and e/B. Values of these bearing capacity factors are substituted in Eq. 7.2, in place of N_c, N_q and N_r for getting the bearing capacity of eccentrically-obliquely loaded footing. If use of these bearing capacity factors is made, then inclination factors, and reduced dimension B' for accounting the effect of eccentricity and inclination are not used. For one-way eccentricity case, therefore, the modified bearing capacity equation will be as given below:

$$q_{nu} = cN_{cei} \cdot S_c \cdot d_c \cdot b_c + \gamma_1 D_f (N_{qei} - 1) S_q \cdot d_q \cdot b_q \cdot r_w + \frac{1}{2} \cdot \gamma_2 B \cdot N_{\gamma ei} S_\gamma \cdot d_\gamma \cdot b_\gamma \cdot r'_w \qquad \dots (7.17)$$

Use of Eqs. 7.6 to 7.16 make a more conservative estimate of the bearing capacity.

Base inclination factors : If the base of a foundation is inclined from the horizontal and an applied load acts normal to the base (Fig. 7.10), the pattern of rupture surface beneath the foundation will be different from the pattern that develops beneath the level footing carrying

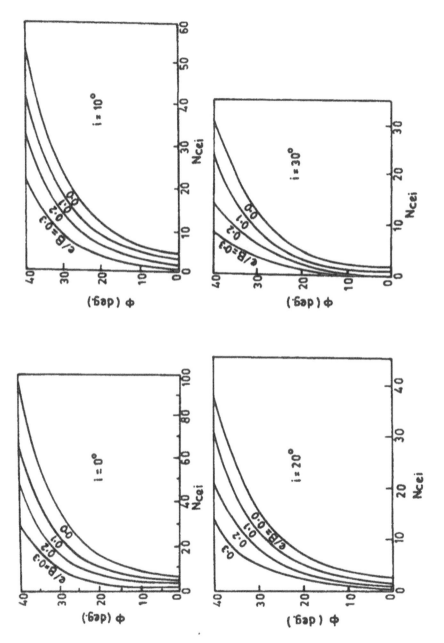

Fig. 7.7. N_{cei} Versus ϕ.

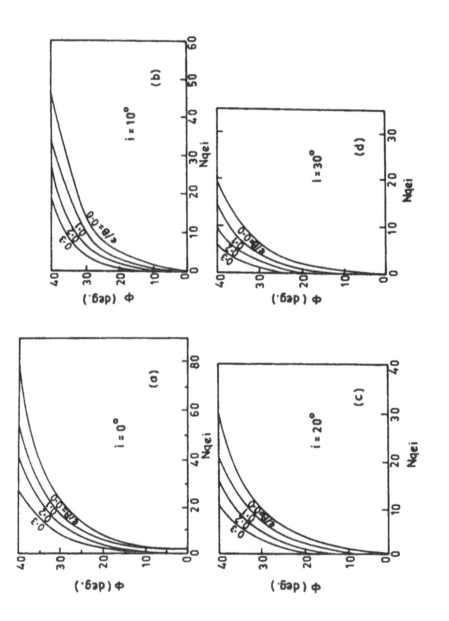

Fig. 7.8. N_{qei} Versus ϕ.

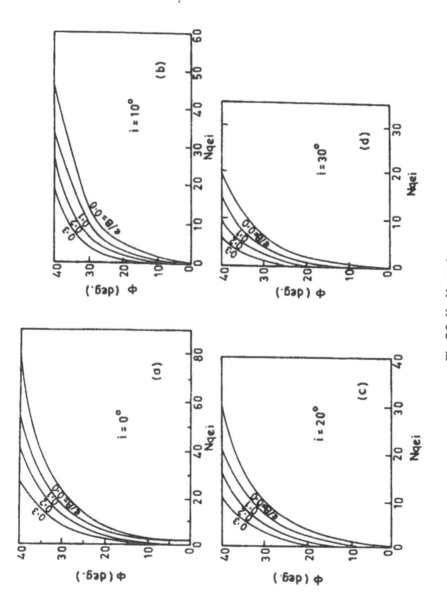

Fig. 7.9. $N_{\gamma ei}$ Versus ϕ.

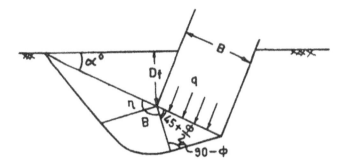

Fig. 7.10. Inclined Footing.

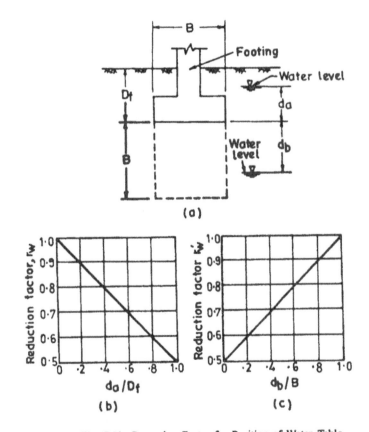

Fig. 7.11. Correction Factor for Position of Water Table.

a vertical load (Meyerhof, 1953). For this condition, base inclination factors as given below may be used:

$$b_c = b_q - \frac{(1 - b_q)}{N_c \tan \phi} \quad \text{for } \phi > 0° \qquad \qquad \text{... (7.18)}$$

$$= 1 - 0.0067\alpha \text{ for } \phi = 0° \qquad \dots (7.19)$$

$$b_q = b_\gamma = \left(1 - \frac{\alpha}{57.3}\tan\phi\right)^2 \qquad \dots (7.20)$$

where α represents the angle of the base inclination in degree with horizontal.

Water table factors : Correction factors r_w and r_w' may be computed using following Eqs. (Fig. 7.11).

$$r_w = 1.0 - 0.5\frac{d_a}{D_f} \qquad \dots (7.21)$$

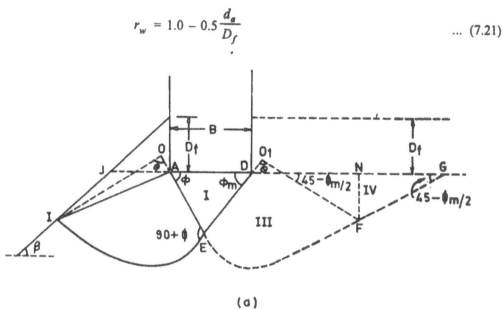

(a)

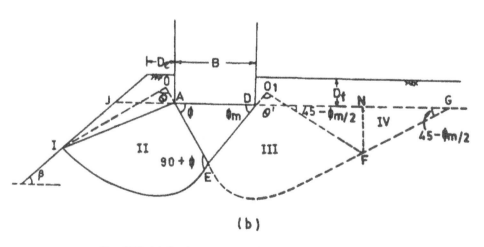

(b)

Fig. 7.12. (a) Footing on Slope; (b) Footing Adjacent to Slope.

$$r'_w = 0.5 + 0.5\frac{d_b}{B} \qquad\qquad \dots (7.22)$$

where d_a and d_b represent the position of water table with respect to the base of the footing as shown in Fig. 7.11. When the water level is below the bottom of footing, $r_w = 1.0$ and when water level is above the bottom of footing, $r'_w = 0.5$.

Footing on a slope: Footings are sometimes placed on or adjacent to a slope (Fig. 7.12). From the figure it can be seen that the lack of soil on the slope side of the footing will tend to reduce the bearing capacity of the footing. Saran and Sud (1989) developed analytical solutions of footings adjacent to slopes using limit equilibrium and limit analysis approaches. Results have been presented in the form of non-dimensional bearing capacity factors $N_{c\beta}$, $N_{q\beta}$ and $N_{\gamma\beta}$. These factors depend on slope angle β, angle of internal friction ϕ, D_f/B ratio and D_e/B ratio, where D_e is the distance of the edge of the footing from the top edge of the slope (Fig. 7.12). Values of bearing capacity factors are shown in Figs 7.13 to 7.15 covering usual

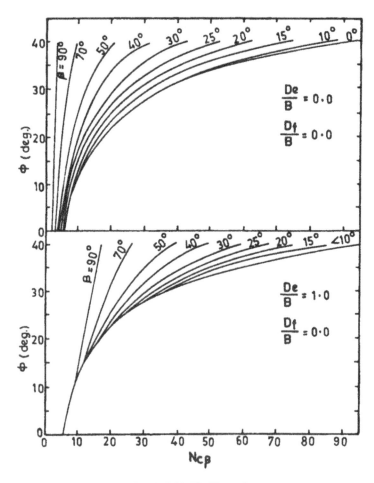

Fig. 7.13(a). $N_{c\beta}$ Versus ϕ.

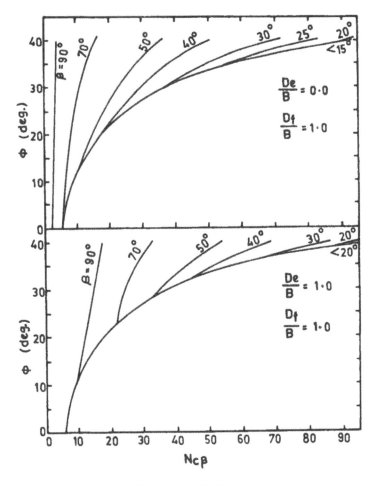

Fig. 7.13(*b*). $N_{c\beta}$ Versus ϕ.

range of footing location and depth of embedment. For footings on slopes, the bearing capacity factors are taken from these figures corresponding to $D_e/B = 0$.

Values of bearing capacity factors ($N_{c\beta}$, $N_{q\beta}$ and $N_{\gamma\beta}$) are substituted in Eq. (7.2) in place of N_c, N_q and N_γ for getting the bearing capacity of a footing placed on or adjacent to a slope. The other factors (i.e. shape factor, depth factor, load inclination factor, base tilt factor) and concept of reduced width for accounting eccentricity has been discussed earlier. The bearing capacity equation for footing on slope can, therefore, be written as given below:

$$q_{nu} = cN_{c\beta}\, S_c d_c i_c + \gamma_1 D_f (N_{q\beta} - 1)\, S_q d_q i_q r_w + \frac{1}{2}\gamma_2 BN_{\gamma\beta}\cdot S_\gamma d_\gamma i_\gamma r'_w \quad \ldots\ (7.23)$$

7.4 LOCAL AND PUNCHING SHEAR FAILURE

The assumption that the soil behaves as a rigid material is satisfied for the case of general

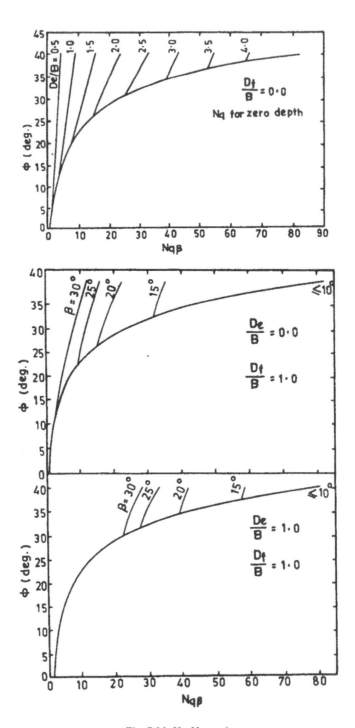

Fig. 7.14. $N_{q\beta}$ Versus ϕ.

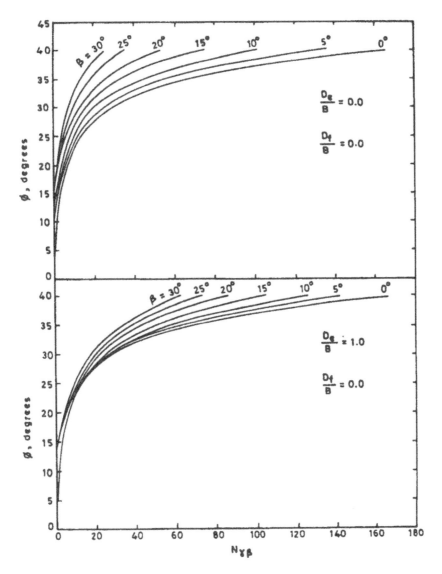

Fig. 7.15(a). $N_{\gamma\beta}$ Versus ϕ.

shear but is not appropriate for punching and local shear. Comparison of the relative pressure-settlement curves (Fig. 7.3) indicates that, for punching and local shear failure cases, the ultimate pressure is less and the settlement is greater than for the condition of general shear failure. For design purposes, the general shear, local and punching shear failures can be identified as per the criterion given in Table 7.1.

Terzaghi (1943) proposed empirical adjustments to shear strength parameters c and ϕ to cover the cases of local and punching shear failures. Shear strength parameters c_m and ϕ_m should be used in the bearing capacity equation and the bearing capacity factors are obtained

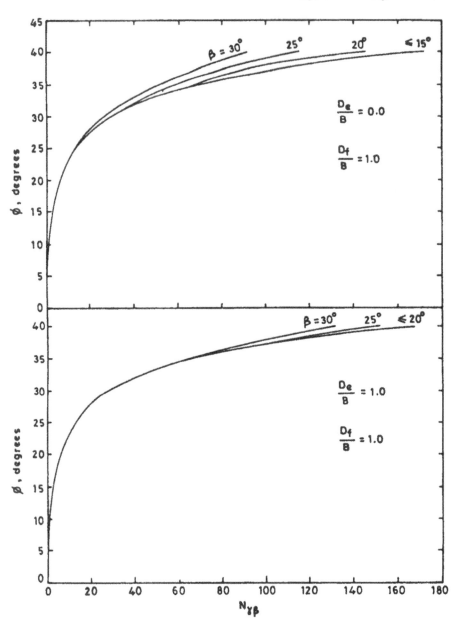

Fig. 7.15(b). $N_{\gamma\beta}$ Versus ϕ.

on the basis of ϕ_m instead of ϕ, where

$$c_m = \frac{2}{3} c$$

... (7.24a)

$$\phi_m = \tan^{-1}\left(\frac{2}{3}\tan\phi\right) \qquad \qquad \dots (7.24b)$$

If the failure lies between general shear and local shear failure, then linear interpolation is done to evaluate the value of bearing capacity factors. For example, value of N_γ for $\phi = 34°$ will be

$$(N_\gamma)_{\phi\,-\,34°} = (N_{\gamma m})_{\phi\,-\,29°} + \frac{(N_\gamma)_{\phi=36°} - (N_{\gamma m})_{\phi=29°}}{36°-29°} \cdot (34° - 29°) \qquad \dots (7.25)$$

where $(N_{\gamma m})_{\phi\,-\,29°}$ = value of N_γ factor for $\phi = 29°$ considering local shear failure condition. Therefore its value will be obtained using Table 7.2 for $\phi_m = \tan^{-1}$ (2/3 tan 29°) = 20.29°.

7.5 SKEMPTON'S BEARING CAPACITY FACTOR, N_C

For saturated clays ($\phi = 0$), Skempton (1951) has proposed the values of N_c factor as shown in Fig. 7.16 for computing the net bearing capacity from following Eq. 7.26a:

$$q_{nu} = c.N_c \qquad \qquad \dots (7.26a)$$

For a factor of safety of 3,

$$q_{nu} = \frac{c.N_c}{3} = \frac{q_u N_c}{6} \qquad \qquad \dots (7.26b)$$

where q_u = unconfined compressive strength of clay.

By making use of Eq. 7.26b and N_C values given in Fig. 7.16a, Peck et al. (1974) developed curves as shown in Fig 7.16b for computing directly the values of q_{na} with the known values of q_u and D_f/B ratio for strip footings. Values of q_{na} for rectangular footings can be obtained by multiplying q_{na} (Strip) by shape factor $(1 + 0.2\ B/L)$.

7.6 FOOTING ON LAYERED SOILS

Footings may be placed on stratified deposits. The usual methods of determining bearing capacity are not applicable. The shear pattern gets distorted and the bearing capacity becomes dependent on the extent of the rupture surface in weaker or stronger material.

A practicable solution which gives reasonable safety is as follows (Fig. 7.17).

(i) Consider the different layers of soil within effective shear depth which is approximately equal to 0.5 B tan (45 + $\phi/2$). If the thickness of the first layer below the base of the footing is more than the significant shear depth, analysis of single layer holds good.

(ii) Average values of c and ϕ are obtained as

$$c_{av} = \frac{c_1 h_1 + c_2 h_2 + \dots c_n h_n}{\sum h_i} \qquad \qquad \dots (7.27)$$

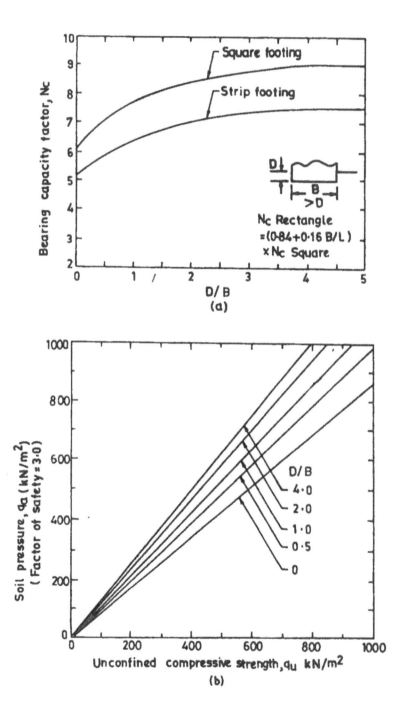

Fig. 7.16. (a) Skempton's Bearing Capacity Factor N_c for Clay Soils; (b) Safe bearing Pressure Versus q_u for Clay Soils (Peck et al., 1974).

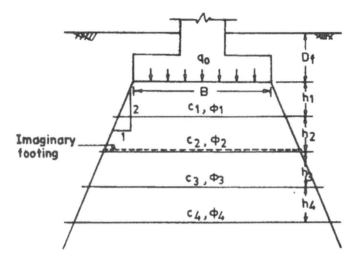

Fig. 7.17. Footing on Layered Soils.

$$\phi_{av} = \tan^{-1} \frac{h_1 \tan\phi_1 + h_2 \tan\phi_2 + ... h_n \tan\phi_n}{\sum h_i} \qquad ... (7.28)$$

Determine the bearing capacity of the footing considering single layer with average shear strength parameter c_{av} and ϕ_{av}.

Alternatively compute the bearing capacity of each layer (considering single layer) using equivalent 'B' based on 2V: 1H slopes and the distance from the base to the top of the layer (Fig. 7.16). For example for layer no. 3, equivalent width = $(B+h_1+h_2)$ and equivalent depth of footing = $D_f + h_1 + h_2$. The pressure on the imaginary footing $[q_0\ BL/(B + h_1 + h_2)\ (L + h_1 + h_2)]$ should not exceed the bearing capacity of this layer, *i.e.*

$$\frac{q_0 BL}{(B + h_1 + h_2)(L + h_1 + h_2)} \not> \text{Safe bearing capacity of layer no. 3} \qquad ... (7.29)$$

This method will give results on the conservative side since the shear strength of the upper layers is neglected.

For two-layers cohesive soil ($\phi = 0$) system, solutions based on limit equilibrium analysis have been developed by Button (1953). The shear strength of the upper layer of thickness d and lower layer of unlimited extent are taken as c_1 and c_2 respectively (Fig. 7.18a). The results have been presented in the form of non-dimensional bearing capacity factor N_c as shown in Fig. 7.18b. The following points may be noted from this chart:

(*i*) When d/b is large, homogenous case is represented and $N_c = 5.51$. Therefore $q_u = 5.51c_1$.

(*ii*) For $d / b = 0$, $N_c = 5.51$ and $q_u = 5.51\ c_2$.

(*iii*) For $c_2/c_1 \leqslant 1$, N_c for a given d/b increases as c_2/c_1 increases unless it reaches a limiting value of 5.51. At this point the rupture surface remains in upper layer.

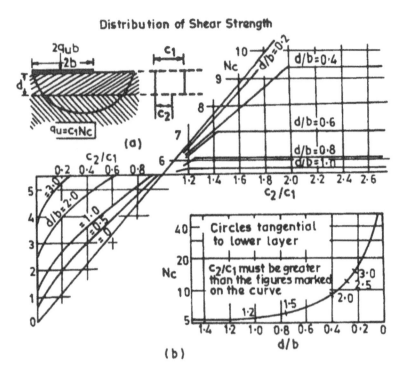

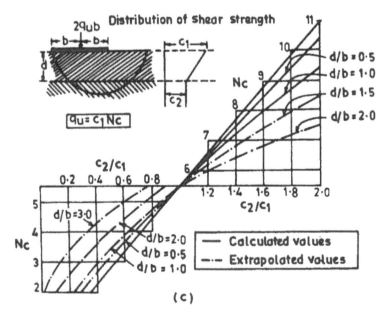

Fig. 7.18. Bearing Capacity Factor N_c for a Footing Located on a Two-layer
Cohesive Soil System (Button, 1953).

(iv) When $c_2/c_1 > 1$, the lower layer has a greater strength than the upper one. For a particular value of d/b, bearing capacity rises, but, at the same time, a small portion of the total length of the slip surfaces passes into the lower stronger layer. At a limiting point the slip circle becomes tangential to the upper surface of the lower layer and after this, any further increase in the strength of the lower layer will not influence the bearing capacity. This is represented by the sudden change in the slope of the lines of values of $c_2/c_1 > 1$. No change in N_c values occurs after this sudden change.

Button also gave the solutions for the case when shear strength of upper clay layer varies linearly with depth. Values of N_c factor for this case are given in Fig. 7.18c.

Reddy and Srinavasan (1967) extended the Button solution to anisotropic soils defined by a coefficient of anisotropy of soil immediately underlying the footing as

$$K = \frac{c_{1V}}{c_{1H}} \qquad \qquad ...\ (7.30)$$

where c_{1V} = vertical shear strength, and

c_{1H} = horizontal shear strength.

A value of $K < 1$ indicates overconsolidation, $K = 1$ is isotropic, and $K > 1$ is normally consolidated. Charts from Reddy and Srinavasan are shown for two K values of 0.8 and 1.2 in Fig. 7.19. Charts for $K = 1.0$ will be the same as given by Button (1953) in Fig. 7.18b.

Meyerhof and Brown (1967) mentioned that the chart developed by Button (Fig. 7.17b, $K = 1$) may also be used for other values of K by adopting c_1 as $(c_{1V} + c_{1H})/2$. Errors associated with this would not be more than 10 per cent.

7.7 BEARING CAPACITY FROM PENETRATION TESTS

There may be certain locations from where it is difficult to recover soil samples for determining shear strength properties, e.g. in cohesionless soils particularly when explorations extend below the groundwater table. In such cases, estimation of bearing capacity is done on the basis of some empirical relations with the stiffness of soil predicted by *in situ* tests. Suitable tests are the standard penetration test or the static or dynamic cone penetration tests (Chapter 3).

From Standard Penetration Test (SPT)

Procedure of using the N-value in the correlations for getting bearing capacity of footing is as under:

(i) Plot the observed N-value with respect to depth. Correct each observed N-value for submergence and overburden pressure using (Eq. 3.6) and Fig. 3.12 respectively. Plot the corrected N-values also as shown in Fig. 7.20.

(ii) Assume a trial width and length of the footing and sketch the layout on the N-plot (Fig. 7.20).

(iii) Determine average of corrected N-values considering these up to significant shear

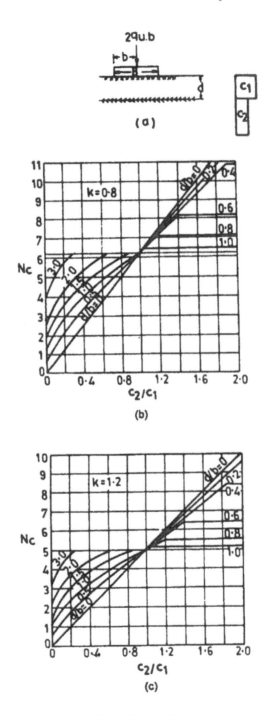

Fig. 7.19. Bearing Capacity Factor N_c for a Footing Located on a Two-layer
Anisotropic Cohesive Soil System (Reddy and Srinavasan, 1967).

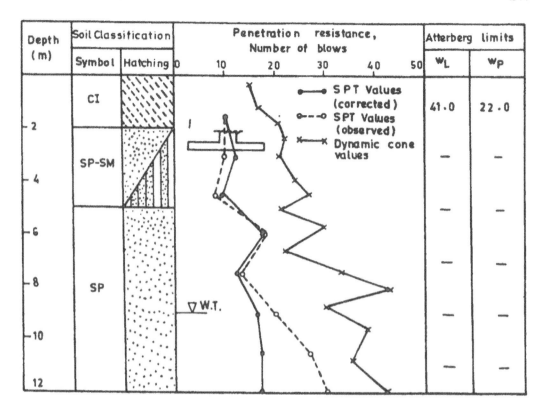

Fig. 7.20. Bore Log Details along with Penetration Test Data.

depth, i.e. $0.5\ B$ tan $(45 + \phi/2)$ from base level of the footing. In computing the average, any individual value more than 50 per cent greater than the average shall be neglected, but the values for all loose seams shall be included. This corrected value of N will be adopted to get the bearing capacity.

(iv) Different correlations for computing bearing capacity are:

(a) Peck, Hanson and Thornburn (1974) have given a relationship to obtain angle of internal friction ϕ of cohesionless soils as shown in Fig. 7.21. The net bearing capacity of a footing shall be calculated using Eq. 7.2 for ϕ obtained from Fig. 7.21.

For cohesionless soils, more correlations are available between N and ϕ as given below:

Meyerhof (1956)

$$\phi = 25 + 0.15\ D_R \qquad\qquad\qquad ...\ (7.31a)$$

$$\phi = 30 + 0.15\ D_R \text{ (for soils having fines more than 5\%)} \qquad ...\ (7.31b)$$

Giuliani and Nicoll (1982)

$$\frac{D_R}{100} = \frac{\sqrt{N}}{4.19 + 0.64(\sigma')^{0.606}} \qquad\qquad ...\ (7.32)$$

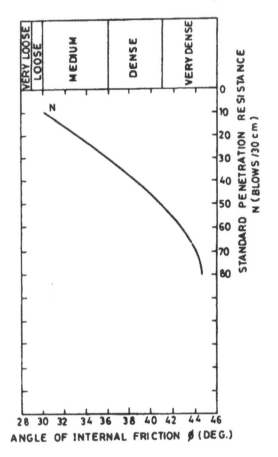

Fig. 7.21. Relationship between ϕ and N.

$$\tan \phi = 0.57 + 0.36 \ (D_R)^{0.87} \qquad \qquad \ldots (7.33)$$

$$\gamma_d = 1.33(1 + 0.23 \ D_R)(C_u)^{0.1} \qquad \qquad \ldots (7.34)$$

where D_R = relative density in per cent,

C_u = uniformity coefficient,

σ' = effective stress at the depth at which N was recorded, and

γ_d = dry density of soil.

(b) Teng (1962) has suggested the following correlations for direct estimation of net bearing capacity from N-values:

(i) Strip footings:

$$q_{nu} = 5[3N^2Br_w + 5(100 + N^2)D_f r_w'] \qquad \qquad \ldots (7.35)$$

(ii) Square and circular footings:

$$q_{nu} = 5[N^2Br_w + 3(100 + N^2)D_f r_w'] \qquad \qquad \ldots (7.36)$$

where q_{nu} = net ultimate bearing capacity in kN/m^2.

B = width of footing in metres. If $D_f > B$, use $D_f = B$, and

r_w and r_w' = correction factors for water table (Eqs. 7.21 and 7.22).

(c) Parry (1977) has proposed the following correlation:

$$q_{nu} = 30N \text{ (kN/m}^2) \qquad \qquad \dots (7.37)$$

(d) Ajayi and Balogun (1988) have recommended following correlation:

$$q_{nu} = 350N \text{ (kN/m}^2). \qquad \qquad \dots (7.38)$$

Bearing capacity of shallow foundations supported on clay are not obtained directly from N-value. A major reason for this situation has been the difficulty in relating N-value to a cohesive soil's stress history. However, N-value is useful to have an approximate estimate of the value of unconfined compressive strength, q_u (Table 3.6, Eqs. 3.7, 3.8 and 3.9 of Chapter 3). The undrained shear strength, c is $q_u/2$.

The value of cohesion, c, can be substituted in bearing capacity equation which may be rewritten as follows for clays ($\phi' = 0$):

$$q_{nu} = cN_c \cdot S_c \cdot d_c \cdot i_c \cdot b_c \qquad \qquad \dots (7.39)$$

The bearing capacity factor N_c for clays ($\phi = 0$) is 7.14.

From Dynamic Cone Penetration Test (DCPT)

As DCPT provides the continuous record of the penetration resistance (N_{cd} or N_{cbr}), it is utilized for locating the soft pockets and in general the variability of the soil profile. However, correlations are available to determine equivalent N-value (Eqs. 3.16 and 3.17 of Chapter 3). N-values thus obtained are corrected for submergence and overburden pressure and then used to get the bearing capacity as illustrated above.

From Static Cone Penetration Test (SCPT)

The static cone point resistance, q_c in kN/m^2 used to determine the bearing capacity of cohesionless soil by obtaining the factor N_γ from the correlation given by Schmertmann (1975):

$$N_\gamma = \frac{q_c}{80} \text{ (kN/m}^2) \qquad \qquad \dots (7.40)$$

From N_γ, ϕ is obtained using the relation between N_γ and ϕ (Table 7.2) by interpolation and thereafter Eq. 7.2 is used for getting bearing capacity of actual footings.

More detailed design information, reflecting the effect of footing width and depth has been developed (Fig. 7.22). This has alos been incorporated in IS 6403:1981.

Nixon (1982) has suggested the relation between q_c and N as given in Table 7.4.

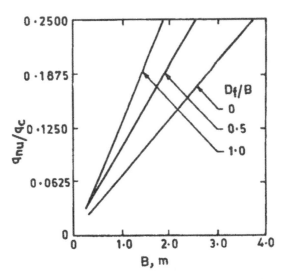

Fig. 7.22. Bearing Capacity Determination from Static Cone Value (IS : 6403–1981).

Table 7.4 Relation between q_c and N (Nixon, 1982).

Type of Soil	$\dfrac{q_c(kN/m^2)}{N}$
Silty or sandy clay	200
Sandy silt	300
Fine sand	400
Fine to medium sand	500
Medium to coarse sand	800
Coarse sand	1000
Sandy gravel	1200 to 1800

7.8 BEARING CAPACITY FROM PLATE LOAD TEST

A semi-direct method of determining bearing capacity in the field is by conducting a plate bearing test according to the procedure laid down by IS : 1888–1982. The method of performing the test is as follows:

(*i*) A suitable size of plate is selected for the test. Normally rough mild steel plates 300 mm, 600 mm or 750 mm size, square in shape, are used. The smaller sizes are used in dense and stiff soils and larger sizes in loose and soft soils.

(*ii*) A pit of dimension not less than five times the width of plate is excavated up to the anticipated depth of foundation. If water table is above the level of foundation, pump out the water carefully and it should be kept just at the level of the foundation. The ground should be levelled and the test plate is seated over the ground at the centre of the pit.

(*iii*) The load on the plate is applied either by gravity loading or reaction loading. The

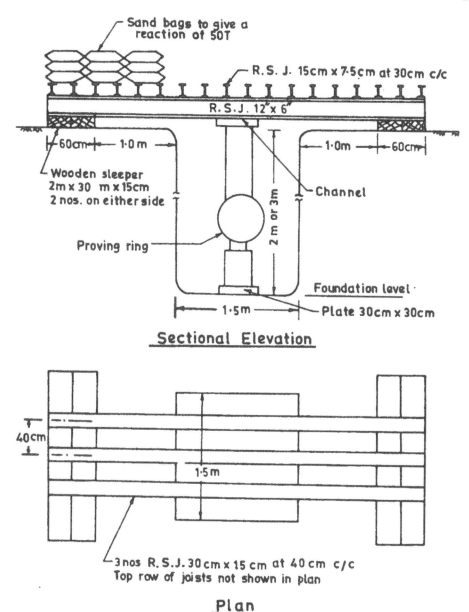

Fig. 7.23. Test Set Up for Vertical Load Test.

settlement of the plate is measured by a set of four dial gauges placed near each corner of plate. The dial gauges are fixed to independent supports which do not get disturbed during the test. A typical set-up of plate load test is shown in Fig. 7.23.

(iv) A seating load of 7 kN/m^2 is first applied and released after some time. Loads are applied on the test plate in increments of one-fifth of the estimated safe load up to failure or at least until a settlement of 25 mm has occurred whichever is earlier. The readings of the settlement dial gauges for each increment of the load are recorded after these become sensibly constant. An average of these four readings is taken as the settlement of plate for the applied load.

Test data of the plate load test is plotted as load intensity versus settlement curve on linear scale or on log-log scale as shown in Fig. 7.24. In the former case, the bearing capacity of the plate is obtained by intersection-tangent method. The log-log plot often indicates more clearly the failure point.

Before proceeding with the further interpretation of plate load test data, following points should be kept in mind:

(i) Plate load test gives reliable results only when the soil condition is uniform from the base of the footing to a depth at least 1.5 times the width of the largest footing. It is due to the fact that the load test reflects the load settlement characteristics of the stratum limited to 1.5 times the width of plate (Fig. 7.25).

(ii) A plate load test is a short duration test. Since settlement in cohesive and partially cohesive soils takes place in a long period of time, load bearing test on such soils are not very practical.

(iii) The plate load test should not be conducted on a soil layer affected by capillary water. A test plate resting on capillary bed undergoes smaller settlement than a plate resting on dry or submerged sand bed (Ramasamy, Rao and Singh, 1981). A plate load test in cohesionless soils should be performed at the water table if it lies within 1.5 m below foundation base.

From the above, it may be concluded that the data of a plate load test requires careful and expert interpretation.

In cohesionless soils, the ultimate bearing capacity of the plate, q_{up} can be expressed as:

$$q_{up} = 0.4 \; \gamma_2 \, B_p \, N_\gamma \, S_\gamma \, r'_w \qquad\qquad \text{... (7.41)}$$

where B_p = width of plate.

The value of density, γ_2 can be easily obtained by taking core-cutter sample from the base of the pit. As q_{up} is known from load intensity-settlement curve (Fig. 7.21), N_γ can be computed using above Eq. 7.41. From N_γ, ϕ is obtained using the relation between N_γ and ϕ (Table 7.2), and thereafter Eq. 7.2 is used for getting the bearing capacity of actual footing.

7.9 FACTOR OF SAFETY

The net bearing capacity of the soil is divided by a safety factor to obtain the net safe bearing capacity. It is denoted by q_{nF}.

A factor of safety is used as a safeguard against (i) natural variations in shear strength of soil; (ii) assumptions made in theoretical methods, (iii) inaccuracies of empirical methods and (iv) excessive settlement of footings near shear failure. A factor of safety of 2.5 to 3.0 is generally used to cover the variations or uncertainties listed above. Therefore

$$q_{nF} = \frac{q_{nu}}{F} \qquad\qquad \text{... (7.42)}$$

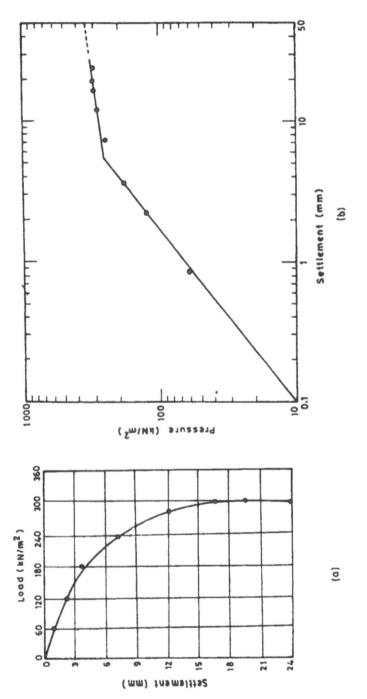

Fig. 7.24. Load Intensity Versus Settlement Curve.

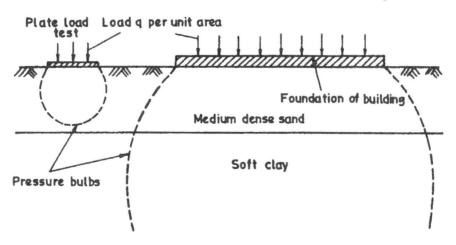

Fig. 7.25. Plate Load Test in Non-homogeneous Soil.

7.10 SETTLEMENT OF FOOTINGS

General

When a load is applied to a footing resting on soil, the soil mass deforms, resulting in the settlement of footing and structure. Settlement produced by loading consists of two components: (*i*) immediate settlement (S_i) which takes place during application of the loading as a result of elastic deformation of the soil without change in water content, and (*ii*) consolidation settlement (S_c) which takes place as a result of volume reduction of the soil caused by extrusion of porewater from the soil. In case of footings on sands, both the immediate and consolidation settlements are of relatively small order. A high proportion of the total settlement is almost completed by the time the full loading comes on the foundation. Settlement of footings on clays are partly immediate and partly consolidation settlements. Usually the consolidation settlement is much higher than the immediate settlement and may take place over a very long period of time.

The differential or relative settlement between one part of a structure and another is much more important to the safety of the super structure than the magnitude of the total settlement. It may occur due to (*i*) variation in strata, (*ii*) variations in foundation loading, (*iii*) difference in time of construction of adjacent parts of a structure, and (*iv*) variation in site conditions. If the whole of the foundation area of a structure settles to the same extent, there is no deterimental effect on the super structure. If, however, there is differential settlement, stresses are set up in the structure which may cause cracking and even collapse of the structure depending on the amount of differential settlement.

There are two procedures of proportioning a foundation against excessive settlement:

(*i*) Settlement of the foundation under actual loads is computed and it is ensured that the same is less than the allowable settlement, and

(*ii*) Maximum net intensity of loading that can be placed on the foundation corresponding to the allowable settlement is computed. This is termed as safe bearing pressure and denoted by q_{ns}. The actual net pressure intensity on the base of the foundation (q_n) must be less than q_{ns}. The value of q_n is given by

$$q_n = q - \gamma_1 D_f \qquad \qquad \dots (7.43)$$

where q = gross loading intensity, *i.e.*, the total pressure at the base of the footing due to the weight of super structure, self-weight of foundation and weight of soil fill above base level of foundation.

Immediate Settlement

The immediate (or elastic) settlement of a footing resting on soil can be computed from an equation from the theory of elasticity as follows:

$$S_i = q_n \cdot B \cdot \frac{1 - \mu^2}{E_s} \cdot I_p \qquad \qquad \dots (7.44)$$

where S_i = immediate settlement,
   q_n = intensity of net footing pressure,
   B = width of footing,
   I_p = influence factor which depends on the shape and rigidity of footing (Table 7.5), and
   E_s, μ = elastic properties of soil.

In a flexible footing, the contact pressure is uniform. A rigid foundation such as a reinforced concrete footing will settle uniformly. Contact pressure beneath a rigid footing depends on the type of soil and is as shown in Fig. 7.26. An approximate method of computing the settlement of rigid foundation is to use Eq. (7.44), taking q_n as mean stress and value of I_p from Table 7.5.

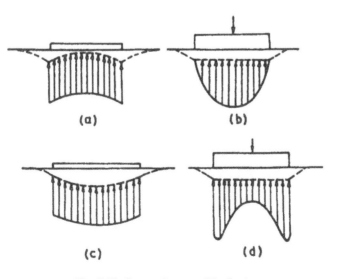

(a)   (b)

(c)   (d)

Fig. 7.26. Contact Pressure Distribution.

Table 7.5 Influence Factor, I_p for Vertical Displacement.

Shape	I_p Flexible Foundation			I_p Rigid Foundation
	Centre	Corner	Average	
Circle*	1.00	0.64	0.85	0.86
Square	1.12	0.56	0.95	0.82
Rectangle				
$L/B = 1.5$	1.36	0.68	1.20	1.06
$L/B = 2.0$	1.52	0.76	1.30	1.20
$L/B = 5.0$	2.10	1.05	1.83	1.70
$L/B = 10$	2.52	1.26	2.25	2.10

* Use diameter for B in Eq. (7.44).

Equation (7.44) is applicable for computing elastic settlement of a footing resting on a soil in which elastic modulus E_s is constant with depth, *i.e.* independent of the confining pressure, *e.g.* saturated clays and rocks. The analysis of immediate settlement of footings on soils such as sands whose elastic modulus increases with the confining pressure is more complex and has not been solved rigorously. Terzaghi (1943) suggested an approximate solution for this problem which is as follows:

The variation of elastic modulus E_s with respect to depth is considered linear, given by

$$E_z = E_0 + mZ \qquad \qquad ... \ (7.45)$$

where E_0 and m are constants. E_z is the value of E_s at depth z.

E_0 represents the value of elastic modulus at zero confining pressure. It is further assumed that pressure on the footing decreases according to the spread as shown in Fig. 7.27. If q represents the stress intensity at depth Z, then

$$q_z(L + Z)(B + Z) = q_n \cdot L \cdot B$$

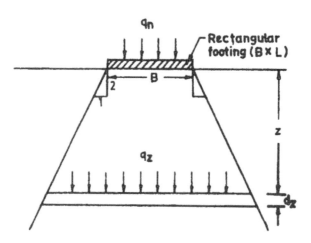

Fig. 7.27. Variation of Pressure with Depth.

$$q_z = \frac{q_n \cdot L \cdot B}{(L+Z)(B+Z)} \qquad \text{... (7.46)}$$

The vertical strain, ε_z of a horizontal strip located at depth Z, below the footing is

$$\varepsilon_z = \frac{q_z}{E_z} = \frac{q_z}{E_0 + mZ} \qquad \text{... (7.47)}$$

Total elastic settlement, S_i will be

$$S_i = \int_0^{d_s} \frac{q_z}{E_0 + mZ} dZ = \int_0^{d_s} \frac{q_n \cdot L \cdot B \cdot dZ}{(L+Z)(B+Z)(E_0 + mZ)}$$

$$= q_n \cdot L \cdot B \left[A' \log_e\left(1 + \frac{d_s}{B}\right) + B' \log_e\left(1 + \frac{d_s}{L}\right) + \frac{C'}{m} \log_e\left(1 + \frac{md_s}{E_0}\right) \right] \qquad \text{... (7.48)}$$

where
$$A' = \frac{-1}{(L-B)(E_0 - mL)} \qquad \text{... (7.49)}$$

$$B' = \frac{1}{(L-B)(E_0 - mB)} \qquad \text{... (7.50)}$$

$$C' = \frac{m^2}{(E_0 - mL)(E_0 - mB)} \qquad \text{... (7.51)}$$

d_s = significant depth below the base of footing.

Figure 7.28 shows the plots of S_i versus B for $m = 0$ and $m > 0$. It may be seen from this figure that an assumption of constant E_s overestimates the settlement.

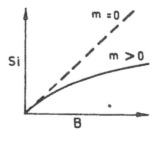

Fig. 7.28. Immediate Settlement Versus Depth.

Consolidation settlement of clay

The actual settlement of a footing, S_{ac} due to consolidation can be calculated by the following equation:

$$S_{ac} = S_c \cdot \mu_1 \cdot \mu_2 \cdot \mu_3 \qquad \qquad \dots (7.52)$$

where μ_1 = the coefficient depending on the geometry and the loading history of clay,
μ_2 = the coefficient which takes into account the rigidity of the footing, and
μ_3 = the depth correction factor which takes into account the effect of embedment of the footing.

S_c is the consolidation settlement obtained using the one-dimensional consolidation theory. The procedure of computing S_c has already been described in Sec. 2.9 of Chapter 2. Skempton and Bjerrum (1957) have recommended the values of coefficient μ_1 as shown in Fig 7.29. IS: 8009 (pt I)–1976 suggested a value of μ_2 as 0.8 for computing the settlement of rigid footing. Fox (1948) has given a chart as shown in Fig. 7.30 to compute the value of μ_3.

7.11 SETTLEMENT FROM PENETRATION TESTS

From standard penetration tests

As mentioned earlier, SPT data is utilized for getting the settlement of footings on

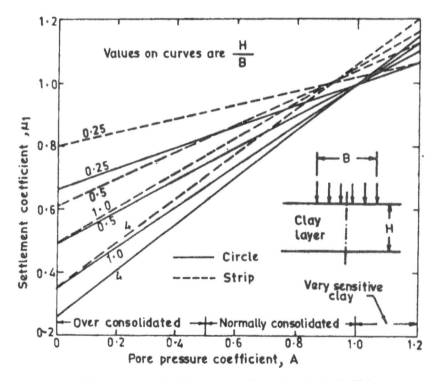

Fig. 7.29. Settlement Coefficient μ_1 for Three-dimensional Coefficient.

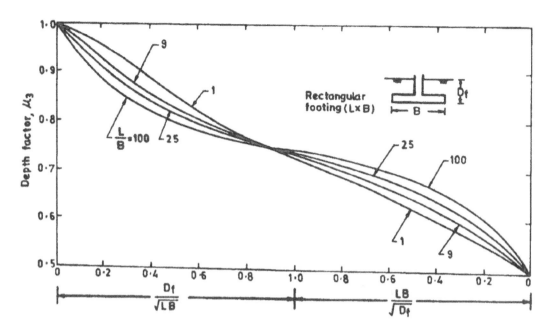

Fig. 7.30. Fox's Depth Correction Factor.

granular soils. Firstly obtain the average of corrected N-values considering these up to significant depth (d_s) from base of the footing. The significant depth is taken as two times the width of footing (2.0 B) in case of square and circular footings and 5.0 B for strip footings ($L/B \geqslant 8$). The significant depth for rectangular footings may be obtained by linear interpolation. This average corrected N-value is used for computing settlement.

Different correlations and charts available for computing net safe bearing pressure from N-value are:

(*i*) Teng (1962)

$$q_{ns} = 1.4 \ (N - 3) \left(\frac{B+0.3}{2B}\right) r'_w \cdot C_D \cdot S_a \qquad \qquad \dots (7.53)$$

where q_{ns} = safe bearing pressure, kN/m^2,
        B = width of footing, m,
        r'_w = water table correction factor (Eq. 7.22),
        C_D = depth correction factor.
          = $(1 + D_f/B) < 2$, and
        S_a = allowable settlement, mm
(*ii*) Peck, Hanson and Thornburn (1974)

They have proposed design charts to estimate net safe bearing pressure on the basis of N-value such that the maximum settlement of footings on sand does not exceed 25 mm. For footings of about 1.0 m or greater in width, the charts can be represented by the following relation:

$$q_{ns} = 0.44 N r_w' \cdot S_a \qquad \qquad ...\ (7.54)$$

where q_{ns} = net safe bearing pressure, kN/m²,
         r_w' = water table correction factor (Eq. 7.22), and
         S_a = allowable settlement, mm.

(iii) Meyerhof (1974)

For $B \leqslant 1.2$ m

$$q_{ns} = 0.49 N \left(\frac{1 + 0.2 D_f}{B} \right) \cdot S_a \qquad \qquad ...\ (7.55)$$

For $B > 1.2$ m

$$q_{ns} = 0.32 N \left(\frac{B + 0.3}{B} \right)^2 \left(\frac{1 + 0.33 D_f}{B} \right) \cdot S_a \qquad \qquad ...\ (7.56)$$

where q_{ns} = net safe bearing pressure, kN/m²,
         B = width of footing, m,
         D_f = depth of footing, m ($D_f < B$), and
         S_a = allowable settlement, mm.

Bowles (1982) has mentioned that Meyerhof's correlations underestimate q_{ns}. He recommended an increase of 50% in the values of q_{ns} obtained from Eqs. 7.55 and 7.56.

(iv) IS : 8009 (pt. I)–1976.

As per IS code recommendations, the net safe bearing pressure can be computed using Fig. 7.31 for any value of permissible settlement. The value   obtained from the chart is multiplied with r'_w (Eq. 7.22) for accounting the effect of water table.

From Static Cone Penetration Tests

Methods available for computing settlement of footings on cohesionless soils from static cone penetration test data are as given below:

(i) DeBeer and Martens (1957)

In this method, the sand stratum is divided into a convenient number of layers such that each layer has an approximately constant value of cone penetration resistance, q_c (Fig. 7.32). The settlement of each sand layer is then obtained using the average value of q_c of that layer from the following equation:

$$S_i = 2.3 \frac{h_i}{C} \log_{10} \frac{\sigma'_0 + \delta \sigma}{\sigma'_0} \qquad \qquad ...(7.57)$$

where S_i = settlement of the ith layer,

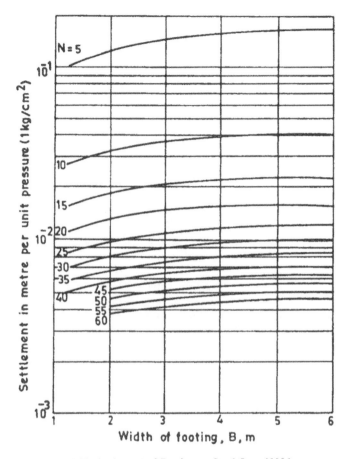

Fig. 7.31. Settlement of Footing on Sand from N-Values.

h_i = thickness of the ith layer,

σ_0' = effective overburden pressure at the middle of the ith layer,

$\delta\sigma$ = increase in vertical stress at the middle of the ith layer,

C = compressibility coefficient

= 1.5 (q_c/σ_0'), and

q_c = average value of q_c of the ith layer.

The total settlement of the footing is the sum of the settlement of all individual layers.

(ii) Meyerhof (1965)

Meyerhof observed that the procedure given by DeBeer and Martens (1957) · overestimates the settlement. He recommended that the value of C to put in Eq. 7.57 may be computed from the following relation:

$$C = 1.9\frac{q_c}{\sigma_0'} \qquad\qquad ... (7.58)$$

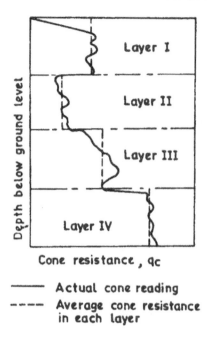

Cone resistance, q_c

――――― Actual cone reading

－－－－ Average cone resistance
in each layer

Fig. 7.32. Static Cone Resistance Diagram.

(*iii*) Schmertmann (1970)

Schmertmann (1970) gave the following equation for computing the settlement:

$$S = C_1 \cdot C_2 \, q_n \sum_{z=0}^{2B} \left(\frac{I_z}{E_s}\right) \delta Z \qquad \qquad ...(7.59)$$

where C_1 = depth correction factor,

$= 1 - 0.5(\sigma_{v1}'/q_n)$

C_2 = creep factor,

$= 1 + 0.2 \log_{10}(t/0.1)$

σ_0' = effective overburden pressure at foundation level,

q_n = net increase in pressure at foundation level,

t = period in years for which settlement is to be calculated,

I_z = vertical strain influence factor (Fig. 7.33), and

E_s = deformation modulus (Fig. 7.33).

The value of I_z is obtained from the curves C_1 and C_2 for square ($L/B = 1$) and strip footing ($L/B > 10$). The deformation modulus for square and strip footings is obtained by multiplying the static cone resistance (q_c) by a factor 2.5 and 3.5 respectively. For rectangular footings value of I_z and E_s are obtained by linear interpolation.

The q_c diagram is divided into a number of layers of thickness δZ and the average q_c of

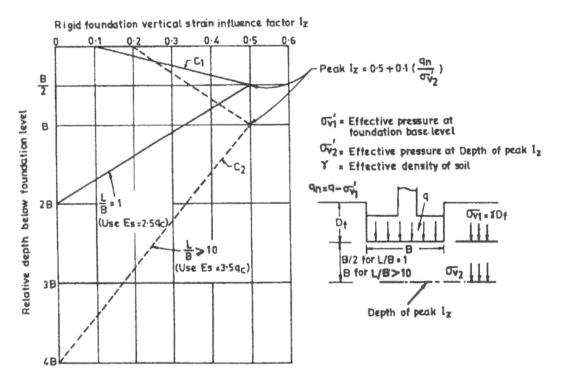

Fig. 7.33. Vertical Strain Influence Factor (Schmertmann et al., 1970).

each layer is obtained. The strain influence factor diagram is placed along side this diagram beneath the footing which is drawn to the same scale. Approximate values of E_s and I_z for each layer are then computed. Sum of the values of I_z / E_s computed for different layers is then obtained upto depth of 2 B from base of footing. This quantity on multiplication with $C_1 C_2 q_n$ gives the total settlement.

7.12 SETTLEMENT, TILT AND HORIZONTAL DISPLACEMENT

An eccentrically-obliquely loaded rigid footing settles as shown in Fig. 7.34 in which S_e and S_m represent respectively the settlement of the point under load and edge of the footing. If 't' is the tilt of the footing, then S_m is given by:

$$S_m = S_e + (B/2 - e) \sin t \qquad \qquad \text{... (7.60)}$$

H_D in Fig. 7.34. represents the horizontal displacement of the footing.

Agarwal (1986) carried out model tests on eccentrically-obliquely loaded footings resting on sand. Footings of different widths and shapes were used. In each test, for a pressure intensity, observations were taken to obtain S_e, t and H_D. Effect of relative density of sand was also studied. In addition to these tests, pressure settlement and pressure tilt characteristics of eccentrically-obliquely loaded footing resting on soil beds were also obtained using non-linear constitutive laws of soils (Saran and Agarwal, 1991). From the

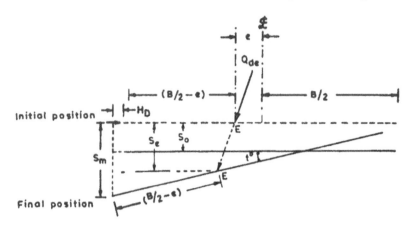

Fig. 7.34. Settlement and Tilt of a Eccentrically-Obliquely Loaded Footing.

model test data and results of analysis based on constitutive laws, plots of S_e/S_0 versus e/B and S_m/S_0 versus e/B were prepared for different load inclinations (Figs. 7.35 and 7.36). S_0 represents the settlement of the footing subjected to central vertical load ($e/B = 0 = i$) and obtained corresponding to the pressure intensity giving the same factor of safety at which S_e and S_m values are taken.

These plots were found independent to the type of soil, factor of safety, size and shape of footing. The average relationships can be represented by the following expressions:

$$\frac{S_e}{S_0} = A_0 + A_1 \left(\frac{e}{B}\right) + A_2 \left(\frac{e}{B}\right)^2 \qquad\qquad \text{... (7.61)}$$

$$\frac{S_m}{S_0} = B_0 + B_1 \left(\frac{e}{B}\right) \qquad\qquad \text{... (7.62)}$$

where $\qquad A_0 = 1 - 0.56 \left(\frac{i}{\phi}\right) - 0.82 \left(\frac{i}{\phi}\right)^2 \qquad\qquad \text{... (7.63)}$

$$A_1 = -3.51 + 1.47 \left(\frac{i}{\phi}\right) + 5.67 \left(\frac{i}{\phi}\right)^2 \qquad\qquad \text{... (7.64)}$$

$$A_2 = 4.74 - 1.33 \left(\frac{i}{\phi}\right) - 12.45 \left(\frac{i}{\phi}\right)^2 \qquad\qquad \text{... (7.65)}$$

$$B_0 = 1 - 0.48 \left(\frac{i}{\phi}\right) - 0.82 \left(\frac{i}{\phi}\right)^2 \qquad\qquad \text{... (7.66)}$$

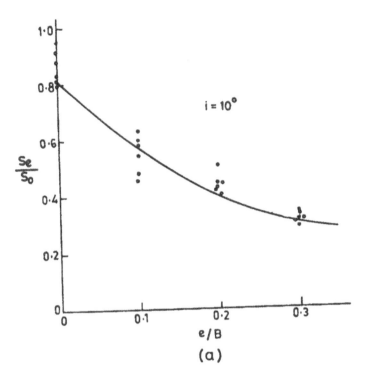

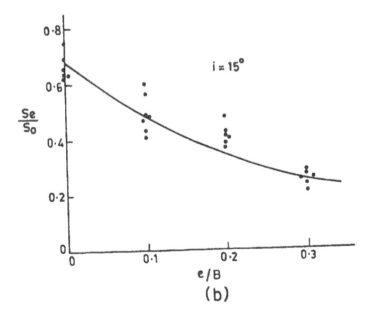

Fig. 7.35. S_e/S_0 Versus e/B for $i = 15°$.

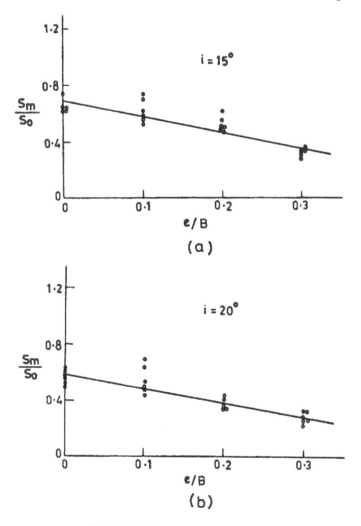

Fig. 7.36. S_m/S_0 Versus e/B for $i = 20°$.

$$B_1 = -1.80 + 0.94 \left(\frac{i}{\phi} \right) + 1.63 \left(\frac{i}{\phi} \right)^2 \qquad \qquad \text{... (7.67)}$$

Value of S_0 can be obtained using the data of plate load test or standard penetration test in cohesionless soils, and consolidation test data in clays in conventional manner.

Similarly, a unique correlation was obtained between H_D/B and i/ϕ.

$$\frac{H_D}{B} = 0.121 \left(\frac{i}{\phi} \right) - 0.682 \left(\frac{i}{\phi} \right)^2 + 1.99 \left(\frac{i}{\phi} \right)^3 - 2.01 \left(\frac{i}{\phi} \right)^4 \qquad \text{... (7.68)}$$

The above correlation is found independent to the type of soil, factor of safety, size and shape of the footing. The effect of e/B was found to be small and the displacement value decreased little with the increase in eccentricity. This effect is neglected considering the results slightly on the safe side.

7.13 SETTLEMENT OF FOOTINGS ON SLOPES

Sud (1985) have conducted large number of model tests on footings on and adjacent to cohesionless slopes. He has also analyzed the footings on slopes of $c - \phi$ materials using constitutive relations. The model test data and results of the analysis based on constitutive relations have been utilized in making plots between S_β/S_0 versus D_e/B for different values of β, and a unique correlation is obtained as given below:

$$\frac{S_\beta}{S_0} = 0.00385\beta° + (1 - 0.0125\beta°) \frac{D_e}{B} \qquad \qquad ... \ (7.69)$$

where S_β = settlement of the footing on or adjacent to a slope,

   S_0 = settlement of the footing on flat ground,

   $\beta°$ = angle of slope with horizontal,

   B = width of footing, and

   D_e = distance of the edge of footing from the top edge of slope.

S_β and S_0 correspond to the same factor of safety.

7.14 ALLOWABLE BEARING PRESSURE (q_{na})

Allowable bearing pressure is the maximum net intensity of loading that can be imposed on the soil with no possibility of shear failure or possibility of excessive settlement. It is denoted by q_{na}. It is hence smaller of the net safe bearing capacity (q_{nF}) and the safe bearing pressure (q_{ns}), i.e.

$$q_{na} = q_{nF} \quad \text{if} \quad q_{nF} < q_{ns} \qquad \qquad ... \ (7.70)$$

and $q_{na} = q_{ns} \quad \text{if} \quad q_{ns} < q_{nF} \qquad \qquad ... \ (7.71)$

7.15 ALLOWABLE BEARING PRESSURE FOR RAFT FOUNDATION

The raft (or mat) foundation is a large footing extending over a great area, frequently covering the entire area of the structure. It is used where base soil has a low bearing capacity and/or the column loads are so large that more than 50 per cent of area is covered by conventional spread footings. Raft foundations are commonly provided beneath multistoreyed buildings, overhead tanks, siloclusters, chimneys, and various other tower structures. A footing foundation usually having a width more than 6.0 m is considered as a raft.

A raft tends to bridge over local soft pockets or any other heterogeneity of the strata. Due to this, chances of differential settlement become less as compared to spread or combined footings. Therefore, the rafts are usually proportioned for higher values of permissible differential and total settlement.

Granular soils

The allowable bearing pressure of a raft is always governed by settlement consideration, since the large size of the raft gives a very high safe bearing capacity. Teng (1962) has suggested the following equation for getting net allowable soil pressure of a raft:

$$q_{na} = 0.7(N - 3) r'_w C_D S_a \qquad \qquad ... (7.72)$$

where q_{na} = net allowable soil pressure, kN/m^2,

r'_w = water table correction factor (Eq. 7.22),

C_D = depth correction factor

= $(1 + D_f/B) < 2$, and

S_a = permissible value of settlement, mm.

Peck, Hansen and Thornburn (1974) recommended the following equation

$$q_{na} = 0.88 N r'_w S_a \qquad \qquad ... (7.73)$$

q_{na}, r'_w, N and S_a have the same meaning as explained above. This equation underestimates q_{na} for $N > 50$. If N-value is less than 5, either the sand should be compacted before constructing a raft or a deep foundation should be used.

The value of net allowable soil pressure of a raft in cohesionless soil may also be obtained by replacing N with $q_c/2$ in the Meyerhof's correlations (Eqs. 7.55 and 7.56). q_c is static cone resistance in kN/m^2.

Cohesive soils

The net bearing capacity (q_{nu}) and settlement of a raft foundation can be computed using the same methods as employed for footings. However, by increasing the depth of the raft, the pressure at its base can be reduced. Thus in very soft soils, the factor of safety can be increased by increasing the depth of foundation. This is the principle of floating foundation and discussed subsequently.

7.16 FLOATING RAFT

General

As mentioned in Chapter 6, the function of a raft foundation is to spread the load over a wider area, and to give a measure of rigidity to the substructure to enable it to bridge over local areas of weaker or more compressible soil. The degree of rigidity given to the raft also reduces the differential settlement. Floating rafts are designed on the same principles with an additional function in which they utilize the principle of buoyancy to reduce the net load on

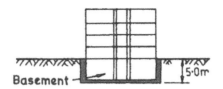

Fig. 7.37. Floating Raft.

the soil. This is achieved by providing a hollow substructure of such a depth that the weight of the soil removed in excavating for it almost balances the weight of the superstructure and substructure. For example, excavation to a depth of 5.0 m from the ground having density 18 kN/m^2, relieves the soil at the foundation level (*i.e.* at 5.0 m depth) of a pressure of 90 kN/m^2 (Fig. 7.37). If the weight of the substructure itself imposes a pressure of 30kN/m^2, a loading of 60 kN/m^2 can be placed on the foundation without causing any additional pressure at the foundation level. A bearing pressure of 60 kN/m^2 is roughly equivalent to overall loading of a four-storeyed building, and therefore, a structure of this magnitude may be placed without having any appreciable settlement.

In practice, it is rarely possible to balance the loading so that no additional pressure comes on the soil because factors like fluctuations in the water table and distribution of live loads cannot be predicted with accuracy. Therefore, floating rafts also experience some settlements. Another factor causing settlement of floating raft is the reconsolidation of soil which swells as a result of the removal of overburden pressure in excavating the soil for the substructure.

Floating foundations are used in very weak soils where there is no possibility of using any other type of foundation. The main purpose of a floating foundation is to reduce the settlement which otherwise would occur. From economy considerations, it is the usual practice to allow some net additional loading to come on the soil. The allowable intensity of pressure of this loading is determined by the permissible settlement and angular distortion of the structure. Such foundations are sometimes called partly compensated foundations or partly floating foundations.

Basement and floating raft

Basements are a form of floating rafts. The main function of a basement is to provide additional space in the building. However, it also reduces the net bearing pressure by weight of displaced soil and, thus, employs the same principle as of floating raft.

Design and construction

Floating rafts are constructed either in the form of caissons or they are built *in situ* in an open excavation. The caisson ·method is suitable for soft clays where the soil can be removed by grabbing as the raft sinks down under its own weight. Construction in open excavations is suitable for site conditions where the groundwater level can be kept down by pumping without the risk of 'boiling', and where heave of soil at the base is not excessive.

The following points are to be considered during the design and construction stage of a floating foundation in open excavations:

(*i*) The excavation for the foundation has to be done with care. The sides of the excavation should be suitably supported with proper sheeting and bracing systems.

(*ii*) Pumping out of the water is necessary when excavation has to be done below the water table. Effect of lowering of water table on adjoining structures should be carefully examined.

(*iii*) In clayey soils ($\phi = 0$), the depth to which excavation can be made without having shear failure is given by:

$$d_c = \frac{5.7c}{\gamma - (c/B)\sqrt{2}} \quad \text{(Terzaghi, 1943)} \qquad \qquad \text{... (7.74)}$$

Skempton (1951) suggested the following equations for the computation of d_c and factor of safety (F_s) against bottom failure for an excavation:

$$d_c = \frac{cN_c}{\gamma} \qquad \qquad \text{... (7.75)}$$

$$F_s = \frac{cN_c}{\gamma d + q} \qquad \qquad \text{... (7.76)}$$

where N_c = bearing capacity factor (Fig. 7.16a),

q = surcharge load, and

d = depth of excavation.

(*iv*) Excavation for foundation reduces the pressure on the soil below the foundation depth which may result in the heaving of the bottom of excavation. Heaving can be minimised by phasing out excavation in narrow trenches and placing the foundation soon after excavation. It can also be minimised by lowering the water table during the excavation process. Friction piles are also sometimes used to minimise the heave (Fig. 7.38).

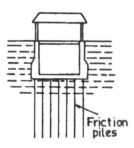

Fig. 7.38. Anchorge of Floating Structures in Soil.

7.17 UPLIFT CAPACITY OF FOOTINGS

Footings are sometimes subjected to uplift or tension forces, *e.g.* in the case of footings of legs of elevated water tanks, anchorages for the anchor cables of transmission towers, and in number of industrial equipment installations. Balla (1961) considered this problem and idealised the footings subjected to develop tension resistance as shown in Fig. 7.39. He considered the failure surface as circular shown by line ab and developed highly complicated mathematical expressions for circular footings. Meyerhof and Adams (1968) also analysed the problem and proposed that the footings may be considered as shallow or deep depending on the depth of base of footing from ground surface D_f (Fig. 7.39). They have considered circular and rectangular footings on both cohesive and cohesionless soils. The expressions developed by them are as given below along with the procedure of their use.

   (*i*) A footing may be considered as shallow or deep if D_f/B ratio is smaller or greater than limiting H/B ratio which depends on angle of internal friction (Table 7.6).

Table 7.6 Limiting H/B Values.

ϕ	H/B	ϕ	H/B
20°	2.5	35°	5.0
25°	3.0	40°	7.0
30°	4.0	45°	9.0

  (*ii*) For shallow footing $(D_f/B \leqslant H/B)$
    (*a*) Circular footing

$$P_u = \pi B c D_f + S_f (\pi/2) B_\gamma D_f^2 K \tan \phi + W \qquad \ldots (7.77)$$

    (*b*) Rectangular footing

$$P_u = 2cD_f (B + L) + \gamma D_f^2 (2S_f B + L - B)K \tan \phi + W \qquad \ldots (7.78)$$

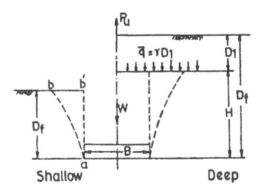

Fig. 7.39. Footings Subjected to Uplift or Tension Forces.

(*iii*) For deep footing $(D_f/B > H/B)$
    (*a*) Circular footing

$$P_u = \pi c B H + S_f(1.57)\gamma B(2D_f - H) \cdot H \cdot K \tan \phi + W \qquad \ldots \text{(7.79)}$$

    (*b*) Rectangular footing

$$P_u = 2cH (B + L) + \gamma(2D_f - H) (2S_f B + L - B)HK \tan \phi + W \quad \ldots \text{(7.80)}$$

where
        P_u = ultimate uplift resistance,
        B = diameter or width of footing,
        L = length of footing,
        c = cohesion,
        D_f = depth of footing,
        γ = density of soil,
        K = coefficient of earth pressure and may be taken a value between K_0 and 1.0, and
        S_f = shape factor = $1 + mD_f/B$ for shallow footings
                  = $1 + mH/B$ for deep footings.

Values of m for various ϕ angles are as follows:

ϕ	20°	25°	30°	35°	40°	45°
m	0.05	0.10	0.15	0.25	0.35	0.50

(*iv*) A factor of safety of about 2 to 2.5 is recommended to get the allowable pull-out capacity of footing.
(*v*) In very poor soils, side soil resistance will be very small and, therefore,

$$P_u = W \text{ (Robinson and Taylor, 1969)} \qquad \ldots \text{(7.81)}$$

In this case, a lower factor of safety of about 1.5 may be used to obtain the allowable pull-out capacity.

7.18 STRUCTURAL DESIGN

Continuous footing
Under masonry wall: Refer Fig. 7.40
        b_w = width of wall, m
        q_a = allowable soil pressure, kN/m²
        V = load from wall per metre run, kN/m

Width of footing,
$$B = \frac{V + W_f}{q_a}$$

where
                W_f = weight of foundation

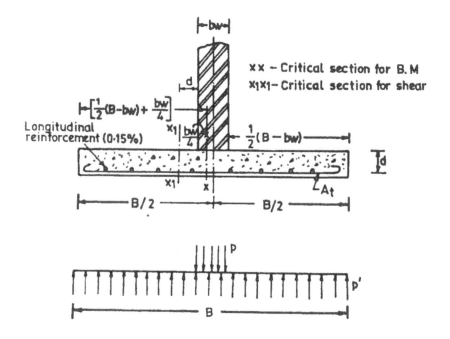

Fig. 7.40. Continuous Footing under Masonry Wall.

Net upward soil reaction,
$$p' = \frac{V}{B}.$$

Pressure intensity at the contact of wall and footing, $p = \dfrac{V}{b_w} = \dfrac{p'B}{b_w}$.

The maximum *B.M.* is considered to act at a section *xx* which is midway between the edge of the wall and centre of the wall, *i.e.* at a distance $b_w/4$ from the edge of wall. Its magnitude per metre length is given by

$$M_{xx} = p'\frac{1}{2}\left[\frac{B-b_w}{2}+\frac{b_w}{4}\right]^2 - \frac{p}{2}\left[\frac{b_w}{4}\right]^2$$

or
$$M_{xx} = \frac{p'}{8}(B-b_w)\left(B-\frac{b_w}{4}\right)\times 10^6 \text{ N mm/m} \qquad \ldots (7.82)$$

Determine depth by:

$$M_R = 0.138 f_{ck}b\cdot d^2 = M_{xx} \text{ [put } b = 1000 \text{ mm]} \qquad \ldots (7.83)$$

Reinforcement by:

$$0.87 f_y A_{st}\left(d - \frac{f_y A_{st}}{f_{ck} b}\right) = M_{xx} \text{ [put } b = 1000 \text{ mm]} \qquad \qquad \text{... (7.84)}$$

Equation 7.84 is quadratic in A_{st} and, therefore, two values of A_{st} will be obtained. Use the lesser value of A_{st}.

In addition to reinforcement (A_{st}) computed as above, provide longitudinal reinforcement @ 0.15 per cent of cross-sectional area of footing.

Check for nominal shear at section $x_1 x_1$, which is at a distance of effective depth from the face of the wall. Redesign if necessary for no shear reinforcement.

Check for development length with respect to the section xx. Provide hooks in the reinforcement if necessary.

Under concrete wall: Refer Fig. 7.41.

The design procedure will be the same as discussed above except that the critical section xx for B.M. will be at the face of wall as shown in Fig. 7.41. Therefore,

$$M_{xx} = p' \frac{1}{2}\left[\frac{B - b_w}{2}\right]^2$$

or $M_{xx} = \dfrac{p'}{8}(B - b_w)^2 \times 10^6 \ N \ mm/m \qquad \qquad \text{... (7.85)}$

Provide dowels to ensure monolithic construction.

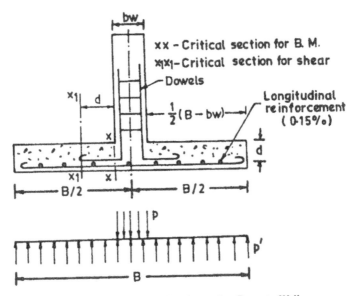

Fig. 7.41. Continuous Footing under Concrete Wall.

Isolated Footing of Uniform Depth
Square footing (Fig. 7.42)

$$b_c \times b_c = \text{column dimensions, m} \times \text{m}$$
$$q_a = \text{allowable soil pressure, kN/m}^2, \text{ and}$$
$$V = \text{column load, kN}$$

Area of footing, $A = \dfrac{V + W_t}{q_a}$, $A = B \times B$, where B is width of footing

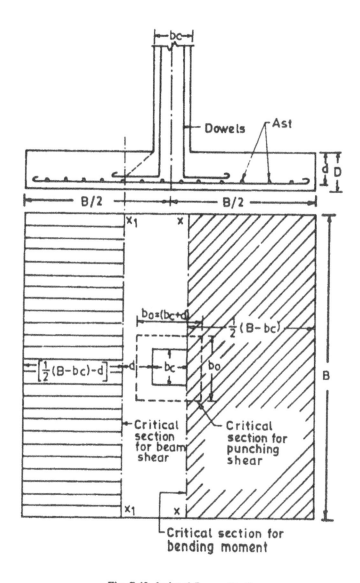

Fig. 7.42. Isolated Square Footing.

Net upward soil reaction, $p' = \dfrac{V}{A}$

$$M_{xx} = \frac{p'}{8} B (B - b_c)^2 \times 10^6 \; N \; mm \qquad \qquad \text{... (7.86)}$$

Compute the depth of slab using Eq. 7.83. Width of the section (b) will be used as B.

Check for punching shear.

Compute reinforcement using Eq. 7.84.

Reinforcement is provided uniformly along the total width of the footing in both directions.

Check for nominal shear at section $x_1 x_1$, redesign if necessary for no shear reinforcement.

Check for development length.

Examine the necessity of dowels and pedestal.

Rectangular footing (Fig. 7.43)

$b_c \times l_c$ = column dimensions, m × m

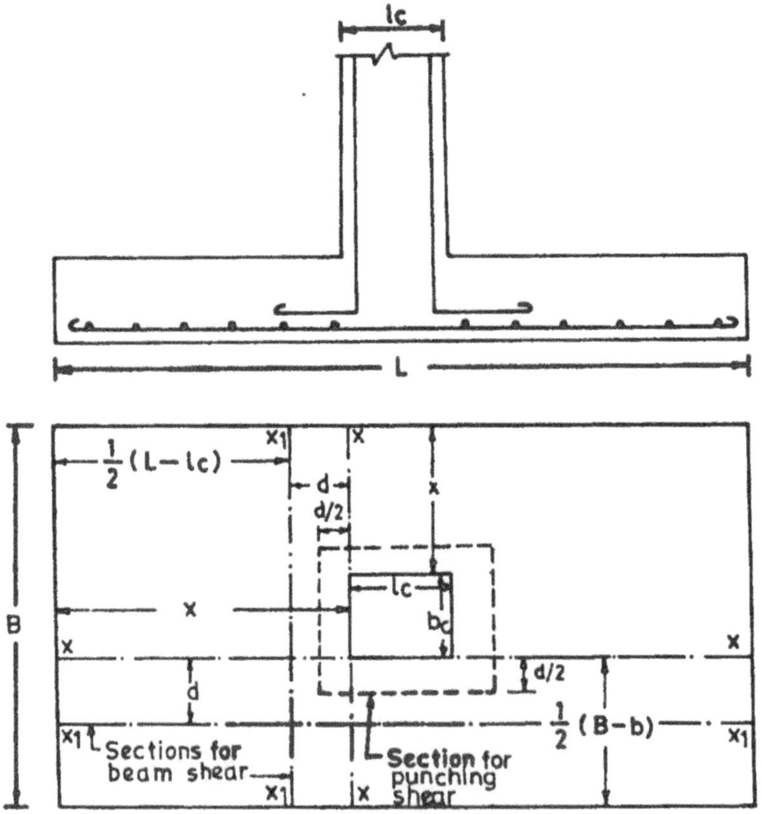

Fig. 7.43. Isolated Rectangular Footing.

From economic considerations, the footing is proportioned by leaving equal projections on all the sides of column as shown in Fig. 7.43.

Hence, area of footing, $\quad A = \dfrac{V + W_f}{q_a} = (2x + l_c)(2x + b_c)$ $\qquad$ ... (7.87)

From Eq. 7.87, value of x can be obtained.

$$M_{xxl} = p'x(2x + l_c) \cdot \frac{x}{2} \qquad ... (7.88a)$$

$$M_{xxb} = p'x(2x + b_c) \cdot \frac{x}{2} \qquad ... (7.88b)$$

Therefore $\qquad M_{xxl} > M_{xxb}.$

Compute the depth of the beam using Eq. 7.83. For moment M_{xxl} width of section (b) will be $(2x + l_c)$ while for moment M_{xxb}, width of section will be $(2x + b_c)$. It may be noted that the depth of section will work out to be same in both the cases. Alternatively, the length and width of the footing may be taken in the same proportion as of the column, then

$$A = \frac{V + W_f}{q_a} = L \times B \qquad ... (7.89a)$$

$$\frac{L}{B} = \frac{l_c}{b_c} \qquad ... (7.89b)$$

Solving the above two equations, L and B may be obtained. Now if x_b and x_l are the projections of footing from the longer and shorter sides of the column respectively. Then

$$M_{xxl} = P'x_b(2x_l + l_c)x_b/2 \qquad ... (7.90a)$$
$$M_{xxb} = p'x_l(2x_b + b_c)x_l/2 \qquad ... (7.90b)$$

Check for punching shear.

Using Eq. 7.83 compute reinforcement A_{tl} and A_{tb} for moments M_{xxl} and M_{xxb} respectively. The reinforcement A_{tb} will be provided parallel to the longer side and it is distributed equally along the whole width.

The reinforcement A_{tl} is divided in two parts A'_{tl} and A''_{tl},

where $\qquad A'_{tl} = A_{tl}\dfrac{2}{L/B + 1}$ $\qquad$ ... (7.91a)

$\qquad A'_{tl} = A_{tl} - A'_{tl}$ $\qquad$ ... (7.91b)

where $\qquad L = 2x + l_c$ $\qquad$ ... (7.92a)

$\qquad B = 2x + b_c$ $\qquad$ ... (7.92b)

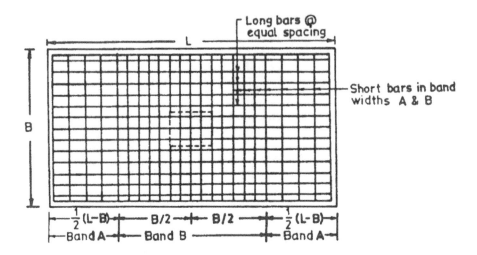

Fig. 7.44. Reinforcement Plan in a Rectangular Footing.

The reinforcement A'_u is spread over the band of dimension equal to the width of footing. A''_u is placed in the outer portion of the footing (Fig. 7.44).

Check for nominal shear.

Check for development length.

Examine the necessity of pedestal and dowels.

Footing for a circular column

r = radius of column, m.

The footing for a circular column may be either square or circular at the base.

For the purpose of design of a square footing, the circular column may be replaced by an equivalent square column of the same area as the circular column (Fig. 7.45).

For circular footing (Fig. 7.46)

Area of footing,
$$A = \frac{V + W_f}{q_a} = \pi R^2.$$

The maximum bending moment in the footing is calculated at the edge of the column for one quadrant as shown by shaded portion.

Net upward reaction, $p' = V/\pi R^2$.

Distance of C.G. of shaded area from the face of column

$$= \frac{0.2}{R + r}(3R^2 - 2Rr - 2r^2). \qquad \qquad \dots (7.93)$$

Bending moment at the face of column

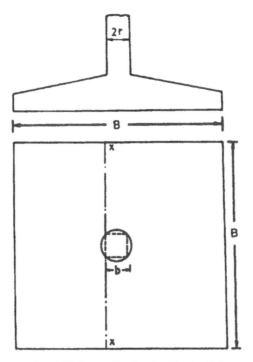

Fig. 7.45. Square Footing for Circular Column.

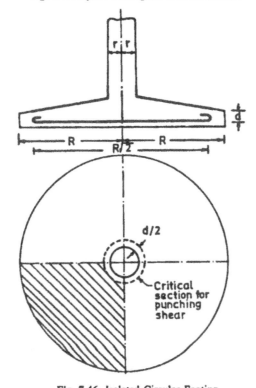

Fig. 7.46. Isolated Circular Footing.

$$M_{xx} = p' \frac{\pi}{4} (R^2 - r^2) \frac{0.2}{R+r} (3R^2 - 2Rr - 2r^2) \qquad \text{... (7.94a)}$$

Breadth of section, $b = \dfrac{2\pi r}{4}.$ ... (7.94b)

Compute the depth of footing using Eq. 7.83.

    Check for punching shear.

    Compute reinforcement using Eq. 7.84.

    The reinforcement is provided in the form of a square mesh centrally located under the column, the width of the square mesh being equal to $R\sqrt{2}$.

    Check for nominal shear at section x_1x_1 (Fig. 7.47).

    Check for development length.

    Examine the necessity of pedestals and dowels.

Isolated sloped footing

    The depth of the footing near the face of the column is governed by the bending moment, punching shear and nominal shear. This depth can be reduced towards the edges of the footing where the bending moment and shear force decreases rapidly. If this decrease is achieved linearly, a sloped footing (Fig. 7.48) is obtained. A minimum thickness of 150 mm is kept at the edges.

Square footing (Fig. 7.48)

    The moment of resistance of the trapezoidal section consists of the moment resistance of the central rectangular portion plus the moment of resistance of two triangular portions (Fig. 7.48). It is given by

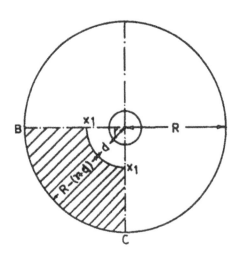

Fig. 7.47. Section for Nominal Shear in Circular Footing.

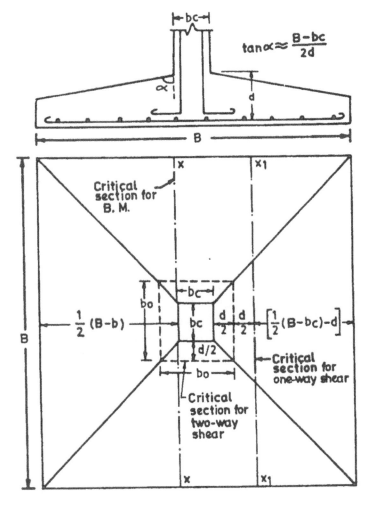

Fig. 7.48. Isolated Sloped Square Footing.

$$M_R = 0.138 f_{ck} b_c d^2 + 0.0483 \tan \alpha f_{ck} d^3 \qquad \text{... (7.95)}$$

where $\qquad \tan \alpha = \dfrac{B - b_c}{2d}$ $\qquad\qquad\qquad\qquad$ ... (7.96)

The depth, d of the footing can be calculated by equating the moment at the face of column (M_{xx}, Eq. 7.86) to the moment of resistance, M_R given by Eq. 7.95. Alternatively, the trapezoidal section can be replaced by an equivalent rectangular section of width $b_c + 1/8$ ($B - b_c$) and of depth d to calculate the moment of resistance.

Check for punching shear.

Determine reinforcement using Eq. 7.84 by putting width of section as $\left[b_c + \dfrac{B - b_c}{8} \right]$.

Check for shear, nominal shear stress $= \dfrac{V}{b'd'}$,

where $b' = $ minimum width in tensile zone or width of section at N.A., and
                         $d' = $ depth of section at which nominal shear stress is computed.

Check for development length.
Examine the necessity of pedestal and dowels.

Rectangular footing

A rectangular sloped footing can be designed by considering equivalent rectangular section of width $l_c + (L - l_c/8)$ and depth d for computing moment of resistance along a section passing through the longer side of column, and of width $[b_c + (B - b_c)/8]$ for computing moment of resistance along shorter side of column.

Combined footing
General

A footing that supports the load of two or more adjacent columns is known as combined footing. Such footings are provided when (*i*) the columns are very near to each other so that their footings overlap, (*ii*) the allowable soil pressure is low, requiring more area below the individual footing, and (*iii*) when the end column is near a property line, so that its footing cannot spread in that direction. The performance of a combined footing against lateral load is better than an islated footing.

A combined footing may be rectangular or trapezoidal in plan. These footings are designed with the principle that the centre of gravity of the footing area coincides with the centre of gravity of the external loads. It will give uniform base pressure.

Proportioning

Rectangular footing: If a rectangular footing is to be designed for two columns having equal loads then the length of the footing is determined keeping hogging and sagging moments equal. If x and l are respectively the projections and centre-to-centre spacing of columns (Fig. 7.49a), we have

$$P'\frac{x^2}{2} = V_1 \cdot \frac{l}{2} - P'\frac{(x+l/2)^2}{2} \qquad \qquad \text{... (7.97)}$$

Solving Eq. (7.97), x can be obtained. Then

Length of footing $= 2x + l$ ... (7.98a)
Width of footing $= A/(2x + l)$ ... (7.98b)
The area of the footing A is given by

$$A = \frac{V_1 + V_2 + W_f}{q_a} = \frac{2V_1 + W_f}{q_a} \quad \text{(when } V_1 = V_2\text{)} \qquad \qquad \text{... (7.99)}$$

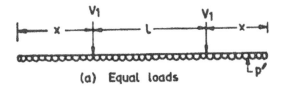

(a) Equal loads

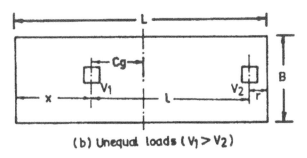

(b) Unequal loads ($V_1 > V_2$)

Fig. 7.49. Proportioning of a Combined Rectangular Footing.

where q_a = allowable soil pressure, kN/m², and
W_f = weight of foundation,

and
$$p' = \frac{2V_1}{l+2x}.$$
... (7.100)

The Eq. (7.97) may therefore be written as :

$$\frac{x^2}{(l+2x)} = \frac{l}{2} - \frac{l+2x}{4}$$

$$4x^2 = 2l\,(l + 2x) - l^2 - 4x^2 - 4lx$$

$$8x^2 = l^2$$

$$x = \frac{l}{2\sqrt{2}}.$$
... (7.101)

If the two columns have unequal loads, length of the rectangular footing is determined by fixing the projection on the lighter column side (Fig. 7.49b). If C_g is the distance of the centre of gravity of loads from the heavier column, then from Fig. 7.49b,

$$x + C_g = \frac{x+l+r}{2}$$
... (7.102)

From Eq. (7.102), the value of x can be determined.

Trapezoidal footing : If the outer column, near the property line, carries heavier load, provision of trapezoidal footing becomes essential to bring the centre of gravity of the two column loads.

For proportioning trapezoidal footing, projections x and x' (Fig. 7.50) are first fixed. The values of a and b are then obtained by solving the following two equations:

$$\frac{a+b}{2}(l + x + x') = A \qquad\qquad ... (7.103)$$

$$\frac{(2b+a)}{a+b} \cdot \frac{l+x+x'}{3} = x + C_g. \qquad\qquad ... (7.104)$$

Bending pattern

Rectangular footing: A combined footing will bend both in longitudinal as well in transverse direction as shown in Figs. 7.51a and b. In longitudinal direction, the footing will have sagging bending moments in the cantilever portions, and hogging moment in some length of middle portion. In the transverse direction the footing will have sagging bending moment. The transverse bending moment will decrease at a distance away from the column, where the bending will be primarily longitudinal. Therefore, the sections near and around the column will be subjected to heavier bending stresses.

A precise analysis of a combined footing having no transverse or longitudinal beam is not possible since the exact bending pattern is not known. The design is based primarily on empirical practices. Usually the combined footing is considered as longitudinal beam of width B. Therefore, the footing will be a beam loaded with uniformly distributed load $w = p'B$ per unit length having two columns with loads V_1 and V_2 (Fig. 7.51c). The bending moment and shear force diagrams are respectively shown in Figs. 7.51d and 7.51e. The footing will be designed for sagging moment at the outer face of each column and for maximum hogging-bending moment. The reinforcement will be placed on the bottom face for sagging bending-moment and on the top face for the hogging moment.

For transverse bending, width of an imaginary transverse beam is considered as $B_1 = b_c + 2d$ and $B_1' = b_c' + 2d$ below the columns V_1 and V_2 respectively with the condition that B_1 and B_1' individually are not greater than B. It is based on the concept that failure due to shear occurs approximately along $45°$ lines (Figs. 7.52a and 7.52b). If the clear projections on the left side of column V_1 and right side of the column V_2 are less than d, then

$\qquad\qquad B_1$ = actual projection on left of column $V_1 + b_c + d$ $\qquad$ ... (7.105a)

and $\qquad B_1'$ = actual projection on right of column $V_2 + b_c' + d$ $\qquad$ ... (7.105b)

Compute $\qquad p_1' = \dfrac{V_1}{B_1 \times B}$ and $\qquad\qquad$ ... (7.106a)

$\qquad\qquad p_2' = \dfrac{V_2}{B_1' \times B}$. $\qquad\qquad$ ... (7.106b)

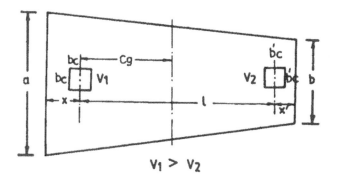

Fig. 7.50. Proportioning of a Combined Trapezoidal Footing.

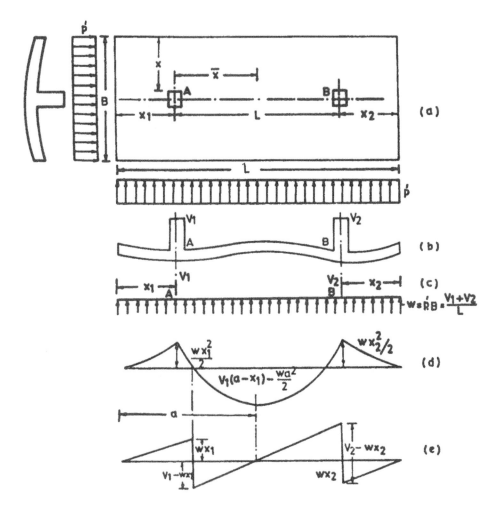

Fig. 7.51. Bending Pattern of a Combined Rectangular Footing.

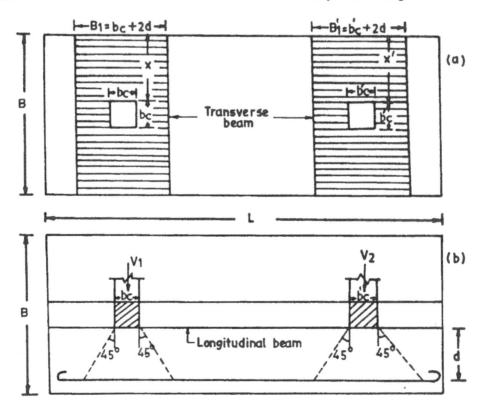

Fig. 7.52. Design Principle of Combined Rectangular Footing in Transverse Direction.

B.M. in transverse beam $\quad B_1 = \dfrac{p_1' \, x^2}{2}$ $\qquad$ ... (7.107a)

B.M. in transverse beam $\quad B_1' = \dfrac{p_2' \, x'^2}{2}$ $\qquad$ ... (7.107b)

The thickness of the footing will then be obtained for a higher value of *B.M.* obtained from Eqs. 7.107. The reinforcement in the transverse direction will be obtained separately on the basis of moments obtained by Eqs. 7.107 and will respectively be provided in width B_1 and B_1'. For the rest of the portion of the footing, minimum transverse reinforcement (0.15% on the area of cross-section of the footing) should be provided.

A more common method of designing the combined footing is by providing a longitudinal beam of width equal to width of wider column as shown in Fig. 7.53a. This beam is monolithic with footing slab. In such a case, the slab on both sides of beam will act as a cantilever against the upward soil reaction and will transmit the total soil pressure to the beam through shear. The beam then transmits this pressure to the columns and balances their loads.

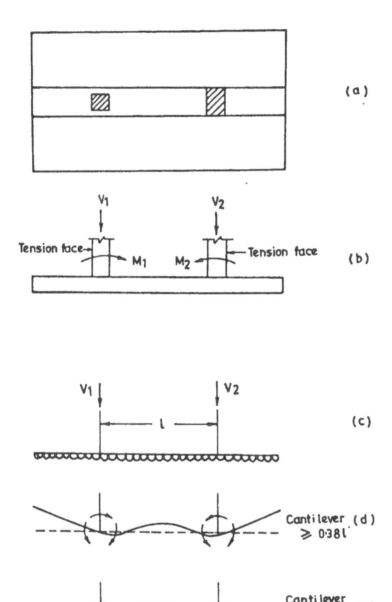

Fig. 7.53. Rectangular Combined Footing with a Longitudinal Beam.

The bending moments and shear forces in the longitudinal beam will be as shown in Figs. 7.51d and 7.51e. This beam will behave as a T-beam at places where the slab lies on the compression side. Therefore, its web should project below the slab if sagging moments in the beam are more. If the hogging moments are excessive, the web should project at the top of slab.

In drawing the *B.M.* and *S.F.* diagrams, it was assumed that the columns are pin jointed to the footing and are incapable of transmitting any moment to it. In practice, columns are always made monolithic with the footing, and they will always transmit some bending moments to the footing as shown in Fig. 7.53b. This may further be explained with the deflection pattern of the footing. A combined footing under column loads deflects as shown in Figs. 7.53d and 7.53e. In the first case, the angle between the columns and the internal span of the footing tends to increase, while in the second case this angle tends to decrease. Due to monolithic construction, change of angle is not possible. Therefore, there will be additionnal joint moments as shown by arrows (Figs. 7.53d and e). When cantilevers are long (> 0.38*l*) (Fig. 7.53d), the full cantilever moment is not transferred to the central span and is shared by the columns. This increases the hogging moment in the footing between the columns. In other cases (Cantilever < 0.38 *l*), the sagging moment near the columns in the central span is increased. As the amount of fixity is unknown, it is difficult to obtain the actual values of these moments. This effect is taken care by the following provisions:

(*i*) Hogging moment $\not< \dfrac{wl^2}{20}$

(*ii*) Sagging moment $\not< \dfrac{wl^2}{16}$ (Cantilever > 0.38 *l*)

$\not< \dfrac{wl^2}{24}$ (Cantilever < 0.38 *l*).

Trapezoidal footing: The bending moment and shear force diagrams in longitudinal direction will be the same as shown in Figs 7.51d and e and, therefore, will be designed exactly in the same manner as discussed for rectangular footing.

The footing will bend transversely near each column face. The width of the imaginary transverse beam near the column V_1 will be $[x + b_c + d]$ (Fig. 7.54). The net upward soil reaction, p' will be $V_1/[(x + b_c + d) . a']$, where a' is average width of the footing (Fig. 7.54). Clear projection beyond column $V_1 = (a - b_c)/2$. Maximum *B.M.* in the transverse direction $= p'/2[(a - b_c)/2]^2$. Similarly, the bending in the transverse direction near column V_2 can be obtained. Rest design will be exactly the same as for rectangular beams.

If a longitudinal beam is provided connecting the columns, design of beam and slab remain same as illustrated for rectangular footing.

The depth determined on the basis of bending is to be checked for punching shear.

The footing is tested for one way shear at a distance *d* from the column faces (Fig. 7.55). For the central portion, four locations a_1, a_2, a_3 and a_4 should be tried for diagonal tension. At points a_1 and a_4, tension cracks will occur only if sagging bending moments occurs. Another possibility of tension cracks are at points a_2 and a_3 just to the right

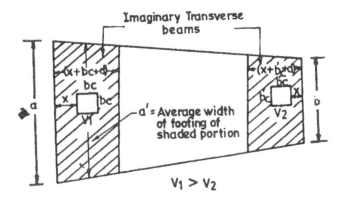

Fig. 7.54. Transverse Beams in a Trapezodial Footing Checking for Diagonal Tension in a Combined Footing.

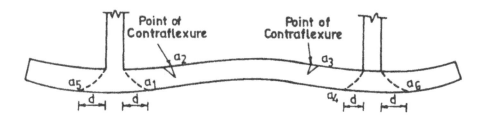

Fig. 7.55. Checking for Diagonal Tension in a Combined Footing.

and left of the point of inflexion, so that there are hogging bending moments giving rise to cracks on the top face. In cantilever portions, diagonal tensions will be checked at points a_5 and a_6.

If the shear stress comes out to be of more than permissible value, shear stirrups should be provided for the required portion. However, nominal shear stirrups should always be provided throughout the length of the beam.

The reinforcement should be checked for development length in usual way.

Foundations with strap beams

The depth of a combined footing is dependent on the value of maximum bending moment. If the area of the footing is so arranged that the footing has large areas under the columns and these are connected by a narrow beam, the upward reaction will be very near the column loads resulting in reduction in maximum bending momnet. The connecting beam is called the strap beam (Fig. 7.56a). This arrangement is specially suited in case of footing for a wall column where it is difficult to make the footing concentric with the column. The two areas under the columns are arranged in such a way that the c.g. of the areas coincides with the c.g. of the column loads. It will give uniform base pressure on the footing.

$$\text{Area of footing, } A = \frac{V_1 + V_2 + W_f}{q_a}. \qquad \qquad \text{... (7.108)}$$

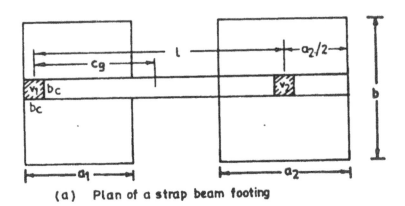

(a)   Plan of a strap beam footing

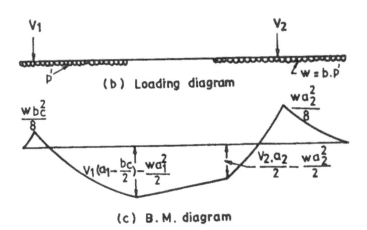

(b) Loading diagram

(c) B.M. diagram

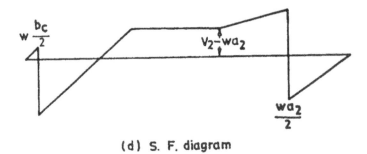

(d) S. F. diagram

Fig. 7.56. Footing with Strap Beam.

If a_1, a_2 and b are taken as the dimensions of the footing as shown in Fig. 7.56a

$$b(a_1 + a_2) = A \qquad \qquad \text{... (7.109a)}$$

$$\frac{ba_1 \cdot a_1/2 + b \cdot a_2(l + b_c/2)}{b(a_1 + a_2)} = C_g + b_c/2. \qquad \qquad \text{... (7.109b)}$$

Assuming a suitable value of b, values of a_1 and a_2 may be obtained by solving Eqs. 7.109a and b.

$$p' = \frac{V_1 + V_2}{b(a_1 + a_2)}. \qquad \qquad \text{... (7.110)}$$

The loading diagram, *B.M.* and *S.F.* diagrams are shown in Fig. 7.56. The rest of the procedure of design is exactly the same as for rectangular combined footing.

Raft foundation

A raft foundation consists of an R.C. slab acted upon by upward soil reaction at the bottom and supported by beams on its top which ultimately neutralise their upward reaction with downward column loads (Fig. 7.57). It is like an inverted roof slab supported on beams and columns. If the centre of gravity of the loads coincide with the centroid of the raft, the upward soil reaction will be uniform. When the column loads are nearly equal, flat slab construction may be adopted.

When the raft foundation is subjected to wind moments also, the upward soil reaction will be linear. In the design of slab, if the wind load reactions are less than 33% of the reactions due to other loads, wind effect is not considered. The beams are designed with or without wind effect, whichever is more critical.

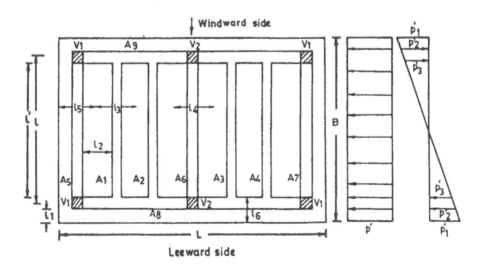

Fig. 7.57. Layout of a Raft with Base Pressures.

Salient features of the design are briefly described in the following steps:
(*i*) Determine the outer dimensions of the raft leaving clear projection of about 0.5 m. If

ΣV_i = total vertical column loads,
ΣM_i = total wind moment acting on columns,
q_a = allowable soil pressure, and
W_f = weight of foundation, then

$$e = \frac{\Sigma M_i}{\Sigma V_i + W_f} \qquad \qquad \text{... (7.111)}$$

Check,
$$\frac{\Sigma V_i + W_f}{L \times B}\left[1 + \frac{6e}{B}\right] \ngtr q_a \qquad \qquad \text{... (7.112)}$$

Upward soil reactions p_1' and p_2' are :

$$p_1', p_2' = \frac{\Sigma V_i + W_f}{L \times B}\left[1 \pm \frac{6e}{B}\right] - \frac{W_f}{L \times B}. \qquad \qquad \text{... (7.113)}$$

Prepare a layout of the raft with the provisions of the beams connecting the columns. From economic considerations, maximum cantilever moment in the slab should be approximately equal to the maximum moment in the interior panel of the slab. To satisfy this condition, provide intermediate beams if necessary.
(*ii*) Design of slab (Fig. 7.57) : Maximum cantilever moment,

$$M_1 = \frac{p_1' + p_2'}{2} \cdot l_1 \cdot \frac{2p_1' + p_2'}{p_1' + p_2'} \cdot \frac{l_1}{3} + p' \frac{l_1^2}{2} \qquad \qquad \text{... (7.114)}$$

Maximum moment in the interior panels, $M_2 = \dfrac{(p' + p_3')l_2^2}{12} \qquad \qquad \text{... (7.115)}$

Maximum S.F. in cantilever portion, $Q_1 = \dfrac{p_1' + p_2'}{2} \cdot l_1 + p'l_1 \qquad \qquad \text{... (7.116)}$

Maximum S.F. in the interior panels, $Q_2 = \dfrac{(p' + p_3')l_2}{2}. \qquad \qquad \text{... (7.117)}$

The thickness of the slab will be obtained considering the higher of the two moments M_1 and M_2. Reinforcement computed corresponding to moment M_1 will be provided at the bottom of slab under the beams, and for moment M_2 at the top of the slab in the central portion between beams.

Check for one-way shear and bond length.
(*iii*) Intermediate secondary beams (Beams A_1, A_2, A_3 and A_4, Fig. 7.57).

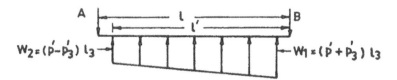

Fig. 7.58. Loading on the Secondary Beam A_1.

These beams are supported on main beams and their span is taken equal to centre to centre distance of main beams, but the loading may be considered only on clear span (Fig. 7.58). The load acting directly under the main beams will be borne by the main beams themselves. The width of slab giving upward reaction will be l_3. This beam will act as T-beam because the slab is at the bottom which is the compression side.

From (Fig. 7.58),

Reaction at A,
$$R_A = \frac{w_2 . l'}{2} + (w_1 - w_2)\left(\frac{l'}{4} - \frac{l'^2}{6l}\right). \qquad \ldots (7.118)$$

Reaction at B,
$$R_B = \frac{w_2 l'}{2} + (w_1 + w_2)\left(\frac{l'}{4} + \frac{l'}{6l}\right). \qquad \ldots (7.119)$$

If x is the section of zero S.F. from point A, then

$$w_2\left[x - \frac{l-l'}{2}\right] + \frac{(w_1 - w_2)\left[x - \frac{l-l'}{2}\right]^2}{2l'} = R_A. \qquad \ldots (7.120)$$

Maximum hogging moment,

$$M_3 = R_A . x - \frac{w_2\left(x - \frac{l-l'}{2}\right)^2}{2} - \frac{(w_1 - w_2)}{l'} . \frac{\left(x - \frac{l-l'}{2}\right)^3}{6} \qquad \ldots (7.121)$$

The depth of beam and amount of reinforcement are computed on the basis of moment M_3. The reinforcement is provided at the top of the beam. Nominal reinforcement is provided at the bottom of the beam to take up positive moment due to any fixity.

The beam should be checked for one-way shear, and reinforcement for bond length. The beam may need web reinforcement which will consist of vertical stirrups and can be designed in usual manner.

Secondary beams joining columns: (Beams A_5, A_6 and A_7, Fig. 7.57).

The width of slab giving upward reaction to beams A_5 and A_7 is l_5 (Fig. 7.57) and for beam A_6 is l_4. The loading diagrams on these beams will be as shown in Fig. 7.59 for beam A_6, and Fig. 7.60 for beams A_5 and A_7. From Fig. 7.59,

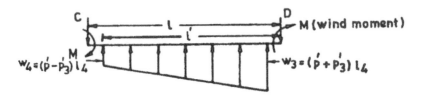

Fig. 7.59. Loading on the Central Beam A_6.

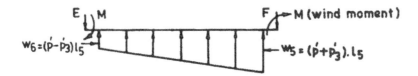

Fig. 7.60. Loading on the Beam A_5 or A_7.

Reaction at C, $R_c = \dfrac{w_4 \cdot l'}{2} + (w_3 - w_4)\left(\dfrac{l'}{4} - \dfrac{l'^2}{12l}\right) + \dfrac{2M}{l}$ ... (7.122)

Reaction at D, $R_D = \dfrac{w_4 \cdot l'}{2} + (w_3 - w_4)\left(\dfrac{l'}{4} + \dfrac{l'^2}{12l}\right) - \dfrac{2M}{l}$ ... (7.123)

If x is the distance of section of zero S.F. from point C, then

$$w_4 \cdot \left(x - \dfrac{l-l'}{2}\right) + \dfrac{(w_3 - w_4)\left(x - \dfrac{l-l'}{2}\right)^2}{2l'} = R_A.$$ ... (7.124)

Maximum B.M.

$$M_4 = R_c \cdot x - M - w_4 \dfrac{[x - (l-l')/2]^2}{2} - \dfrac{(w_3 - w_4)}{l'} \cdot \dfrac{[x - (l-l')/2]^3}{6}.$$ ... (7.125)

Sagging B.M. near column, $M_5 = \dfrac{(R_c + R_D) \cdot l}{24} + M.$ ... (7.126)

For hogging moment, the beam will act as T-beam. Reinforcement will be provided at the top of the beam for hogging moment, and at the bottom of the beam for sagging moment.

The beam should be checked for two-way shear (punching shear) and one-way shear.

Vertical stirrups be provided as the web reinforcement. Reinforcement is checked for bond length.

Main beams (A_8 and A_9, Fig. 7.57): The main beams A_8 and A_9 will receive reactions from intermediate secondary beams A_1, A_2, A_3, A_4, A_5, A_6 and A_7, and reaction from the cantilever portion of slab and portion of slab directly under the beam. Therefore, the loading diagrams of beams A_8 and A_9 will be as shown in Figs. 7.61 and 7.62 respectively. Bending moment and shear force diagrams of beams A_8 and A_9 may be drawn in usual way. The thickness of the main beam is obtained using the values of maximum hogging and sagging moments. Thickness of the beam from hogging moment is obtained considering it as T-beam. This thickness is then checked for sagging moment considering it as rectangular beam. Reinforcement computed on the basis of hogging moment is provided at the top of beam, and at bottom for sagging moment. These reinforcements are checked for bond length. The beam should be provided vertical stirrups in the usual manner.

7.19 MODULUS OF SUBGRADE REACTION

The modulus of subgrade reaction is a conceptual relationship between soil pressure and deflection. It is defined as the ratio between the soil pressure and the corresponding settlement, and mathematically expressed as (Fig. 7.63a):

$$K_s = \frac{q_n}{S_v}. \qquad \qquad \text{... (7.127)}$$

The value of K_s may be determined by performing a plate load test and plotting a curve between q_n and S_v (Fig. 7.63b). It is evident that the curve is nonlinear and, therefore, the value of K_s is dependent on the strain level. In the opinion of author, the value of K_s may be taken as modulus corresponding to 4% to 6% strain, i.e. for settlement equal to .04 B_p to .06 B_p, B_p being the width of plate. The lower value may be used for footings on dense/stiff soils

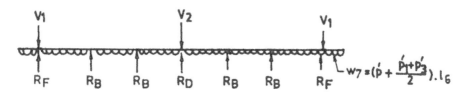

Fig. 7.61. Loading on the Main Beam A_8.

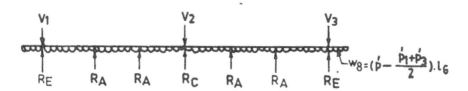

Fig. 7.62. Loading on the Main Beam A_9.

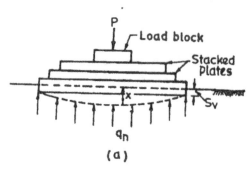

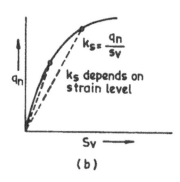

Fig. 7.63. Determination of Modulus of Subgrade Reaction.

and higher value for loose/soft soils.

The value of K_s is significantly effected by size and shape of the footing and depth of foundation. Terzaghi (1955) has suggested the following correlations for determining the value of modulus of subgrade reaction of the actual footing.

1. *Effect of size* : If K_f and K_p are respectively the values of K_s for footings of widths B_f and B_p expressed in centimetres then

For clayey soils

$$\frac{K_f}{K_p} = \frac{B_p}{B_f}.$$

... (7.128)

For sandy soils

$$\frac{K_f}{K_p} = \left(\frac{B_p(B_f + 30)}{B_f(B_p + 30)}\right)^2$$

... (7.129)

2. *Effect of shape* : If K_{fl} and K_{fB} respectively are the values of K_s for rectangular $(L \times B)$ and square $(B \times B)$ footings, then

$$\frac{K_{fl}}{K_{fB}} = \frac{1 + 0.5B/L}{1.5}$$

... (7.130)

3. *Effect of depth of footing* : If K_{fD} and K_{f0} are the values of K_s for footings located at depth D_f and at surface, then

$$\frac{K_{fD}}{K_{f0}} = 1 + \frac{2D_f}{B} \qquad \qquad \text{... (7.131)}$$

Value of modulus of subgrade reaction of the actual footing considering it as the surface footing may also be obtained using the values of elastic modulus (E_s) and Poisson ratio (μ) of the soil from the following equation:

$$K_{fl} = \frac{E_s}{B_f(1-\mu^2)I_p} \qquad \qquad \text{... (7.132)}$$

where I_p = influence factor depending on the shape of footing (Table 7.5).

Bowles (1982) suggested an indirect method of approximate estimation of the value of modulus of subgrade reaction. According to him, it may be assumed that net ultimate bearing capacity of a footing occurs at a settlement of 25 mm (*i.e.* 0.025 m). The value of net ultimate bearing capacity may be obtained using following equation:

$$q_{nu} = c \cdot N_c \cdot S_c + \gamma_1 D_f N_q S_q \cdot r_w + 1/2 \, \gamma_2 B N_\gamma S_\gamma r_w'. \qquad \qquad \text{... (7.133)}$$

Therefore

$$K_{fD} = \frac{q_{nu}}{0.025} = 40 q_{nu} \text{ kN/m}^3. \qquad \qquad \text{... (7.134)}$$

Indian Standard (IS : 2950, part 1) recommended the values of K_s obtained from plate load test conducted on 300 mm × 300 mm plate. These values are listed in Tables 7.6 and 7.7 for granular and cohesive soils respectively.

Bowles (1982) has suggested the values of K_s as given in Table 7.8.

Table 7.6. Values of K_s (kN/m³) for Granular Soils.

Relative Density N-value	Loose < 10	Medium 10-30	Dense 30 and above
K_s for dry or moist soils	15000	47000	180000
K_s for submerged soil	9000	29000	108000

Table 7.7. Values of K_s (kN/m³) for Cohesive Soils.

Consistency	Stiff	Very Stiff	Hard
Unconfined strength (kN/m²)	10-20	20-40	40 and above
K_s	27000	54000	108000

Table 7.8. Range of Values of Modulus of Subgrade Reaction K_s.

Soil	K_s, kN/m³
Loose sand	4800–16000
Medium dense sand	9600–80000
Dense sand	64000–128000
Clayey medium dense sand	32000–80000
Silty medium dense sand	24000–48000
Clayey soil:	
q_u < 200 kPa	12000–24000
200 < q_u < 400 kPa	24000–48000
q_u < 800 kPa	> 48000

7.20 BEAMS ON ELASTIC FOUNDATION

When flexure rigidity of the footing is taken into account, a solution based on the concept of beams on elastic foundation is used. In this a foundation is considered a bed of springs (Winkler, 1867). The basic differential equation is (Fig. 7.64)

$$EI\frac{d^4y}{dx^4} = q_n b = K_f B.y \qquad \qquad \text{... (7.135)}$$

where E = modulus of elasticity of beam,

  I = moment of inertia of the beam section, and

  K_f = modulus of subgrade reaction of footing.

In solving Eq. (7.135), a variable is introduced:

$$\lambda L = \left(\frac{K_f BL^4}{4EI}\right)^{\frac{1}{4}} \qquad \qquad \text{... (7.136)}$$

For infinite beam, solution exists for the cases listed in Table 7.9.

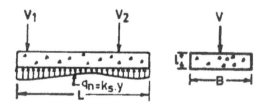

Fig. 7.64. Beam on Elastic Foundation.

Table 7.9. Infinite Beam of Elastic Foundations.

Case	Solution for		
	Deflection (y)	Moment (M)	Shear Force (Q)
(i) Concentrated load, V at centre	$\dfrac{V\lambda}{2K_f B} A_{\lambda x}$	$\dfrac{V}{4\lambda} C_{\lambda x}$	$-\dfrac{V}{2} D_{\lambda x}$
(ii) Moment M_0, at centre	$\dfrac{M_0 \lambda^2}{K_f B} B_{\lambda x}$	$\dfrac{M_0}{2} D_{\lambda x}$	$-\dfrac{M_0}{2} A_{\lambda x}$
(iii) Concentrated load, V at free end	$\dfrac{2V\lambda}{K_f B} D_{\lambda x}$	$-\dfrac{V}{2} B_{\lambda x}$	$-VC_{\lambda x}$
(iv) Moment, M_0 at free end	$\dfrac{-2M_0 \lambda^2}{K_f B} C_{\lambda x}$	$M_0 A_{\lambda x}$	$-2M_0 B_{\lambda x}$

where
$$A_{\lambda x} = e^{-\lambda x}(\cos \lambda x + \sin \lambda x) \qquad \dots (7.137)$$
$$B_{\lambda x} = e^{-\lambda x}\sin \lambda x \qquad \dots (7.138)$$
$$C_{\lambda x} = e^{-\lambda x}(\cos \lambda x - \sin \lambda x) \qquad \dots (7.139)$$
$$D_{\lambda x} = e^{-\lambda x}\cos \lambda x \qquad \dots (7.140)$$

For the convenience of designers, values of $A_{\lambda x}$, $B_{\lambda x}$, $C_{\lambda x}$, and $D_{\lambda x}$ for different values of λ_x are given in Table 7.10.

Table 7.10. Coefficient for the Solution of an Infinite Beam on an Elastic Foundation.

λx	$A_{\lambda x}$	$B_{\lambda x}$	$C_{\lambda x}$	$D_{\lambda x}$
0.0	1.0000	0.0000	1.0000	1.0000
0.1	0.9907	0.0903	0.8100	0.9003
0.5	0.8231	0.2908	0.2415	0.5323
1.0	0.5083	0.3096	−0.1108	0.1988
1.5	0.2384	0.2226	−0.2068	0.0158
2.0	0.0667	0.1231	−0.1794	−0.0563
3.0	−0.0423	0.0070	−0.0563	−0.0493
4.0	−0.0258	−0.0139	0.0019	−0.0120
5.0	−0.0045	−0.0065	0.0084	0.0019
6.0	−0.0017	−0.0007	0.0031	0.0024
7.0	0.0013	0.0006	0.0001	0.0007
8.0	0.0003	0.0003	−0.0004	0.0000
9.0	0.0000	0.0000	−0.0001	−0.0001

Hetenyi (1946) developed equations of deflection, moment and shear force for a load at any point along a beam (Fig. 7.65) measured from the left end as follows:

$$y = \frac{V\lambda}{K_f B} \cdot A' \qquad \dots (7.141)$$

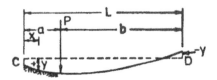

Fig. 7.65. Beam of Finite Length on an Elastic Foundation.

$$M = \frac{V}{2\lambda} \cdot B' \qquad \qquad \text{... (7.142)}$$

$$Q = V \cdot C' \qquad \qquad \text{... (7.143)}$$

where $\quad A' = \dfrac{1}{(\sinh^2 \lambda L - \sin^2 \lambda L)}$ [2 cosh λx cos λx(sinh λL cos λa cosh $\lambda b-$

sin λL cosh λa cos λb) + (cosh λx sin λx +
sinh λx cos λx) [sinhλL(sin λa cosh λb −
cos λa sinh λb) + sin λL(sinh λa cos λb − cosh λa sin λb)]]. ... (7.144a)

$B' = \dfrac{1}{(\sinh^2 \lambda L - \sin^2 \lambda L)}$ [2 sinh λx sin λx(sinh λL cos λa cosh $\lambda b-$

sin λL cosh λa cos λb) + (cosh λx sin λx − sinh λx cos λx) ×
[sinh λL(sin λa cosh λb − cos λa sinh λb) +
sin λL(sinh λa cos λb − cosh λa sin λb)]].   ... (7.144b)

$C' = \dfrac{1}{\sinh^2 \lambda L - \sin^2 \lambda L}$ [(cosh λx sin λx + sinh λx cos λx) ×

(sinh λL cos λa cosh λb − sin λL cosh λa cos λb) +
sinh λx sin λx [sinh λL(sin λa cosh λb − cos λa sinh λb) +
sin λL(sinh λa cos λb − cosh λa sin λb)]].   ... (7.144c)

For the convenience of designers, values of A', B' and C' are given in Appendix 7.1 for different load locations and values of λL.

Whether a foundation should be analysed on the basis of the conventional method considering linear pressure distribution or as a beam on an elastic foundation, is decided on the value of λL (Eq. 7.136) as follows:

1. $\lambda L < \pi/4$: The foundation is treated as rigid member and its design may be carried out by conventional method.

2. $\pi/4 \leqslant \lambda L \leqslant \pi$: The foundation is treated as a flexible beam of finite length on elastic foundation.

3. $\lambda L > \pi$: The foundation is treated as a flexible beam of infinite length on elastic foundation. Solutions of flexible beam of finite length are also applicable in this case.

7.21 ANALYSIS OF FOOTING BY FINITE DIFFERENCE

In Sec. 7.20, closed form solutions for getting deflections, moments and shear forces in a footing based on the concept of beams on an elastic foundation have been presented. These solutions have the limitations of their applicability to the uniform section of the beam with constant modulus of subgrade reaction. Finite difference method is a technique to give the solution of differential equation of bending moment. This method requires writing and solving a large number of simultaneous equations. The solution can be easily obtained on a computer. The main advantage of finite difference method is that the modulus of subgrade reaction and cross-section of the beam can vary in any manner.

Central differences, involving pivotal points symmetrically located with respect to i (Fig. 7.66), are more accurate than forward and backward difference and are particularly useful in boundary value problems.

According to the definition of central differences
First central difference of y at i

$$\ddot{\cdot} \; \delta y_i = y_{i+1/2} - y_{i-1/2} \qquad \qquad \text{... (7.145)}$$

Second central difference of y at i

$$\delta^2 y_i = y_{i+1} - 2y_i + y_{i-1}. \qquad \qquad \text{... (7.146)}$$

Using Tayler's series expansion, it can be proved that second derivative of y at $i = 1/h^2$ (second central difference of y at i) approximately

or
$$D^2 y_i = \frac{1}{h^2} \delta^2 y_i \qquad \qquad \text{... (7.147a)}$$

$$D^2 y_i = \frac{1}{h^2} (y_{i+1} - 2y_i + y_{i-1}). \qquad \qquad \text{... (7.147b)}$$

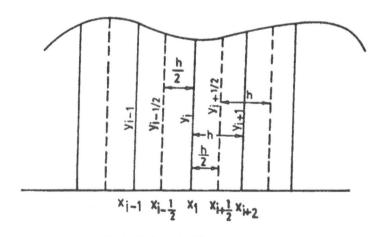

Fig. 7.66. Central Difference Illustration.

From the solution of Eq. (7.135):

$$\frac{d^2y}{dx^2} = \frac{M}{EI} \qquad \qquad \text{... (7.148)}$$

The above equation may be written as follows in terms of central difference

$$M = \frac{EI}{h^2}(y_{i+1} - 2y_i + y_{i-1}). \qquad \qquad \text{... (7.149)}$$

The following procedure can be adopted to solve a problem:
1. Divide the beam into n portions (Fig. 7.67) such that

$$h = \frac{L}{n}. \qquad \qquad \text{... (7.150)}$$

2. Compute equivalent concentrated loads at points 1, 2, 3, ...n along the beam using the relevant values modulus of subgrade reaction. Values of R_1, R_2...R_n will be as follows for a constant value of modulus of subgrade reaction:

$$R_1 = \frac{h}{2}K_f y_1$$

$$R_2 = hK_f y_2$$

$$R_3 = hK_f y_3$$

$$\cdots\cdots\cdots\cdots\cdots$$

$$R_n = \frac{h}{2}K_f y_n \qquad \qquad \text{... (7.151)}$$

3. Apply finite difference equation at points 2, 3, 4, ...$n - 1$:

$$\frac{EI}{h^2}(y_3 - 2y_2 + y_1) = -R_1 h + V_1 x$$

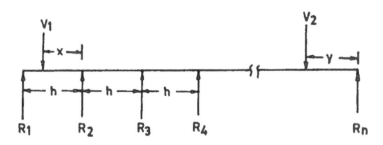

Fig. 7.67. Computation of Bending Moment in a Beam by Finite Difference Method.

$$= -\frac{h}{2}K_f y_1 h + V_1 x$$

$$\frac{EI}{h^2}(y_4 - 2y_3 + y_2) = -R_1 2h - R_2 h + V_1 (x + h)$$

$$= -\frac{h}{2}K_f y_1 2h - hK_f y_2 h + V_1 (x + h)$$

....

....

$$\frac{EI}{h^2}(y_n - 2y_{n-1} + y_{n-2}) = R_n y = \frac{h}{2}K_f y_n y. \qquad\qquad ...\ (7.152)$$

Here we get $(n - 2)$ simultaneous equations.

4. Two more equations can be obtained by writing:
   (*i*) Summation of all vertical forces equal to zero, i.e. $\Sigma V = 0$

$$R_1 + R_2 + R_3 \ ..... \ R_n = V_1 + V_2 \qquad\qquad ...\ (7.153)$$

   (*ii*) Moment of all the forces about the point (1) or (*n*) are equal to zero, i.e. $\Sigma M = 0$, or

$$R_2 h + R_3 2h + ... \ R_n L = V_1 (h - x) + V_2 (L - y) \qquad\qquad ...\ (7.154)$$

   Values R_1, R_2, ...R_n in Eqs. (7.153) and (7.154) can be expressed in terms of y_1, y_2, ...y_n.

5. Solving the *n*-simultaneous equations, we can obtain the values of y_1, y_2,....y_n.

6. Moments at different points can then be determined using Eq. (7.149).

7.22 DESIGN OF CIRCULAR AND ANNULAR RAFTS

Circular and annular rafts are commonly used for the foundations of axisymmetric tower-shaped structures such as water tanks, chimneys, silos, telecommunication towers etc. If the superstructure consists of a cylindrical shaft such as chimney, the shaft can be directly rested on the raft which will then behave as a flat slab raft. In the case when the superstructure is supported on a circular row of columns e.g. in the case of an overhead tank, the columns are usually supported on a ring beam, and the foundation becomes beam and slab type. Beam and slab type raft is more economical in comparison to flat slab construction and, therefore, commonly adopted in practice.

Circular and annular rafts are usually subjected to a vertical load (DL + LL), and horizontal load and moment due to wind or seismic forces. Figure 7.68 shows an annular raft alongwith base pressure distribution. As per IS:11089-1984, curves A and B shown in Fig. 7.69 give the most economical locations of the ring beam on the raft when the soil pressures are uniform and when they vary linearly.

Few investigators have studied the behaviour of annular footings using model tests

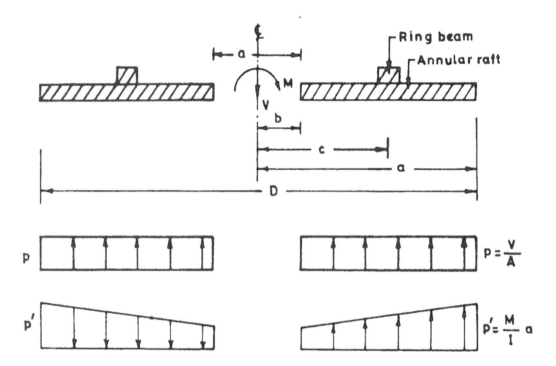

Fig. 7.68. Annular Raft with Ringh Beam Showing Base Pressures.

(Gupta, 1985; Galav, 1997; Al-Smadi, 1998). One of the main observations from these studies was that upto annular ratio (b/a) of about 0.4, the ultimate bearing capacity of the footing increases and the settlement for the same pressure intensity decreases. Keeping this in view, it is recommended that the annular ratio be kept within 0 and 0.4 The allowable soil pressure, therefore, of an annular raft may be obtained conservatively considering it as circular raft of diameter equal to the external diameter of the annular raft (a, Fig. 7.68). Hence the centre of the ring beam may be located such that c/a lies between 0.7 and 0.85 (Figs. 7.68 and 7.69).

Design of Slab

Chu and Afandi (1966) have given following expressions for radial and tangential moments, M_r and M_θ, and shear force Q_r.

(i) For Vertical load

Zone 1 ($r < c$, Fig. 7.68)

$$M_{r1} = \frac{pa^2}{64} [4(3 + \mu)\rho^2 - 16\alpha^2 (1 + \mu) \log_e \rho - 8\alpha^2 (3 + \mu) +$$

$$2 (1 + \mu) K_1 - (1 - \mu) K_2\rho^{-2}] \qquad \qquad \text{... (7.155)}$$

$$M_{\theta 1} = \frac{pa^2}{64} \, [4(1 + 3\mu)\rho^2 - 16\alpha^2 \, (1 + \mu) \, \log_e \rho - 8 \, \alpha^2 \, (1 + 3\mu) +$$

$$2 \, (1 + \mu) \, K_1 + (1 - \mu) \, K_2 \rho^{-2}] \qquad \dots (7.156)$$

$$\dot{Q}_{r1} = \frac{pa}{2} \, (\rho - \alpha^2 \, \rho^{-1}) \qquad \dots (7.157)$$

where $\quad K_1 = 2 \, (1 + \alpha^2) \left[\dfrac{3 + \mu}{1 + \mu} \right] - \dfrac{8\alpha^4 \, \log_e \alpha}{1 - \alpha^2} - 4\beta^{\,2} \left[\dfrac{1 - \mu}{1 + \mu} \right] - 8 \, (1 + \log_e \beta) \quad \dots (7.158)$

$$K_2 = 4\alpha^2 \left[\frac{3 + \mu}{1 - \mu} \right] - \frac{16\alpha^4 \, \log_e \alpha}{1 - \alpha^2} \left[\frac{1 + \mu}{1 - \mu} \right] - 16\alpha^2 \left[\frac{1 + \mu}{1 - \mu} \right] (1 + \log_e \beta) - 8\alpha^2 \beta^2$$

$$\dots (7.159)$$

$$\alpha = \frac{b}{a}; \qquad \beta = \frac{c}{a}; \qquad\qquad \rho = \frac{r}{a}$$

Zone 2 ($r > c$, Fig. 7.68)

$$M_{r2} = \frac{pa^2}{64} \, [4(3 + \mu)\rho^2 - 16 \, (1 + \mu) \, \log_e \rho - 8 \, (3 + \mu) + 2 \, (1 + \mu) \, K'_1 - (1 - \mu) \, K'_2 \rho^{-2}]$$

$$\dots (7.160)$$

$$M_{\theta 2} = \frac{pa^2}{64} \, [4(1 + 3\mu)\rho^2 - 16 \, (1 + \mu) \, \log_e \rho - 8 \, (1 + 3\mu) + 2 \, (1 + \mu) \, K'_1 - (1 - \mu) \, K'_2 \, \rho^{-2}]$$

$$\dots (7.161)$$

$$Q_{r2} = \frac{pa}{2}(\rho - \rho^{-1}) \qquad \dots (7.162)$$

where

$$K'_1 = K_1 + 8 \, (1 + \log_e \beta) \, (1 - \alpha^2) \qquad \dots (7.163)$$

$$K'_2 = K_2 - 8\beta^2(1 - \alpha^2) \qquad \dots (7.164)$$

(ii) For Moment
Zone 1 ($r < c$, Fig. 7.68)

$$M'_{r1} = \frac{p'a^2}{192} \left[4(5 + \mu)\rho^3 + 2(3 + \mu)K_3\rho + 2(1 - \mu)K_4\rho^{-3} + (1 + \mu)K_5\rho^{-1} \right] \qquad \dots (7.165)$$

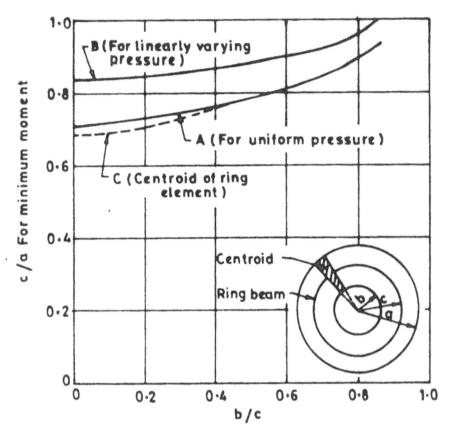

Fig. 7.69. Position of Ring Beam for Minimum Moments.

$$M_{\theta 1} = \frac{p'a^2}{192}\left[4(1+5\mu)\rho^3 + 2(1+3\mu)K_3\rho - 2(1-\mu)K_4\rho^{-3} + (1+\mu)K_5\rho^{-1}\right]$$

$$\dots (7.166)$$

$$Q'_{r1} = \frac{p'a}{192}\left[72\rho^2 + 8K_3 - 2K_5\rho^{-2}\right]$$

$$\dots (7.167)$$

where

$$K_3 = \frac{3}{\beta^2} - 3\beta^2\left[\frac{1-\mu}{3+\mu}\right] - 8\left[1 + \frac{\alpha^4}{1+\alpha^2}\right]\left[\frac{2+\mu}{3+\mu}\right]$$

$$\dots (7.168)$$

$$K_4 = 3\beta^2\alpha^4 - \frac{3\alpha^4}{\beta^2}\left[\frac{3+\mu}{1-\mu}\right] + 8\left[\frac{\alpha^4}{1+\alpha^2}\right]\left[\frac{2+\mu}{1-\mu}\right]$$

$$\dots (7.169)$$

$$K_5 = 12\,\alpha^4$$

$$\dots (7.170)$$

Zone 2 ($r > c$, Fig. 7.68)

The values of M_{r2}, $M_{\theta2}$ and Q'_{r2} can be obtained from Eqs. (7.165), (7.166) and (7.167) respectively by replacing K_3, K_4 and K_5 by K'_3, K'_4, and K'_5, where

$$K'_3 = K_3 - \frac{3}{\beta^2}(1 - \alpha^4) \qquad \qquad \text{... (7.171)}$$

$$K'_4 = K_4 + 3\beta^2 (1 - \alpha^4) \qquad \qquad \text{... (7.172)}$$

$$K'_5 = 12.0$$

Design of Ring Beam

Krishna and Jain (1980) gave a method for analyzing a circular beam loaded uniformly. The beam is considered simply supported on columns placed equidistant along its circumference as shown in Fig. 7.70.

The bending moment at any point P on the ring beam located at angle ϕ with respect to OA is given by,

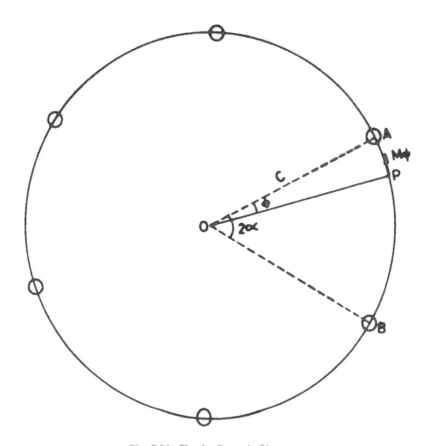

Fig. 7.70. Circular Beam in Plan.

$$M_\phi = wc^2 (- 1 + \alpha \sin \phi + \alpha \cos \phi \cdot \cot \alpha) \qquad \dots (7.173)$$

Twisting moment at point P

$$M_{t\phi} = wc^2(\phi - \alpha + \alpha \cos \phi - \alpha \cot \alpha \cdot \sin \phi) \qquad \dots (7.174)$$

Shear force at any point P

$$V_{u\phi} = wc \, (\alpha - \phi) \cdot \qquad \dots (7.175)$$

For obtaining the point of maximum torsional moment, Eq. (7.174) may be differentiated with respect to ϕ and equated to zero, i.e.,

$$\frac{d \, M_{t\phi}}{d\phi} = 0 \qquad \dots (7.176)$$

It gives

$$\sin \phi = \left(\frac{1}{\alpha}\right) [\sin^2 \alpha \pm \cos \alpha \, (\alpha^2 - \sin^2 \alpha)^{1/2}] \qquad \dots (7.177)$$

Based on Eqs. (7.173) to (7.177), the values of support moment (M_0), mid span moment (M_α) and maximum twisting moment ($M_{t\phi m}$) are given by

$$M_0 = \beta \, w \, c^2 \, (2\alpha) \qquad \dots (7.178)$$
$$M_\alpha = \beta' \, w \, c^2 \, (2\alpha) \qquad \dots (7.179)$$
$$M_{t\phi m} = \beta'' \, w \, c^2 \, (2\alpha) \qquad \dots (7.180)$$

The values of β' β'' and β'' are given in Table 7.11.

Table 7.11 Coefficients of Bending and Twisting Moments in Circular Beams.

No. of supports	2α (Deg.)	β	β′	β″	Value of φ for $M_{t\phi m}$ (Deg.)
4	90	0.137	0.070	0.021	19.25
5	72	0.108	0.054	0.014	15.25
6	60	0.089	0.045	0.009	12.75
7	51.43	0.077	0.037	0.007	10.75
8	45	0.066	0.033	0.005	9.5
9	40	0.060	0.030	0.004	8.5
10	36	0.054	0.027	0.003	7.5

In general, the diagrams of shear force, bending moment and twisting moment in one span of the beam are as shown in Fig. 7.71.

ILLUSTRATIVE EXAMPLES

Note : In all illustrative examples concrete of grade M20 and steel of grade Fe 415 have been used.

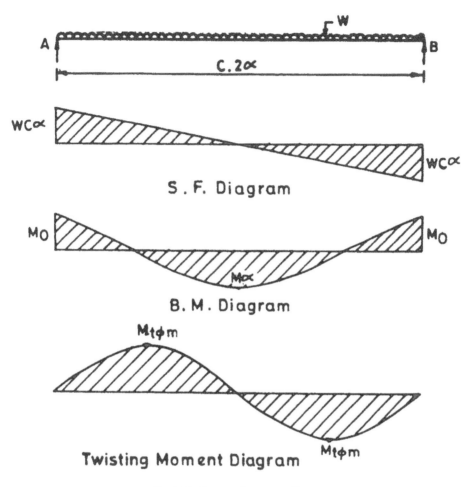

Fig. 7.71. Force and moment diagram.

Example 7.1

Design an isolated footing for a column of 500 mm × 500 mm size subjected to a vertical load of 2400 kN, moment of 400 kN-m and shear load of 360 kN. The soil properties are as follows:

$$c = 6 \text{ kN/m}^2, \ \phi = 39° \text{ and } \gamma_t = 18 \text{ kN/m}^3.$$

A plate load test was performed at the anticipated depth of foundation on a plate of size 600 mm × 600 mm and a pressure settlement record as given below was obtained. Design the footing completely giving reinforcement drawing assuming that permissible values of settlement, tilt and lateral displacement are 50 mm, one degree and 25 mm respectively.

Pressure (kN/m²)	0.0	240	480	720	960	1200	1440	1680
Settlement (mm)	0.0	2.0	5.0	7.5	12.0	16.0	23.0	28.0

Solution

(i) *Safe bearing capacity*

$$\text{Eccentricity of load} = e = \frac{M}{V} = \frac{400}{2400} = 0.1667 \text{ m}$$

$$\text{Inclination of load} = i = \tan^{-1}\left(\frac{Q_h}{V}\right) = \tan^{-1}\left(\frac{360}{2400}\right)$$

$$= 8.53°.$$

Assuming the size of footing as 2.0 m × 2.0 m provided at a depth of 1.0 m below ground surface, and say i is in the plane B-axis,

hence, $B' = B - 2e = 2 - 2(0.1667) = 1.667$ m

and $L' = L = 2$ m

$$q_{nu} = cN_cS_cd_ci_c + \gamma_1D_f(N_q - 1) S_qd_qi_q\gamma_w + \frac{1}{2}\gamma_2BN_\gamma S_\gamma d_\gamma i_\gamma \gamma_w'.$$

Since $\phi = 39°$, it is the case of general shear failure. From Table 7.2 for $\phi = 39°$, $N_c = 70.79$, $N_q = 59.62$ and $N_\gamma = 100.71$.

Depth factors are given as

$$d_c = 1 + 0.4\frac{D_f}{B'} = 1 + 0.4\frac{1}{1.667} = 1.24$$

$$d_q = 1 + 2 \tan \phi (1 - \sin \phi)^2\frac{D_f}{B'} = 1 + 2 \tan 39(1 - \sin 39)^2\frac{1}{1.667} = 1.13$$

and $d_\gamma = 1$.

Shape factors are calculated as

$$S_c = 1 + 0.2\frac{B'}{L} = 1 + 0.2\frac{1.667}{2} = 1.167$$

$$S_q = 1 + 0.2\frac{B'}{L} = 1 + 0.2\frac{1.667}{2} = 1.167$$

$$S_\gamma = 1 - 0.4\frac{B'}{L} = 1 - 0.4\frac{1.667}{2} = 0.667.$$

Inclination factors

$$Q = \sqrt{Q_h^2 + Q_v^2} = \sqrt{360^2 + 2400^2} = 2426.85 \text{ kN},$$

and $$m = \left(2+\frac{B'}{L}\right)\left(1+\frac{B'}{L}\right) = \left(2+\frac{1.667}{2}\right)\left(1+\frac{1.667}{2}\right) = 5.195.$$

Therefore

$$i_\gamma = \left[1 - \frac{Q_h}{Q + B'\,Lc\cot\phi}\right]^m = \left[1 - \frac{360}{2426.85 + 1.667(2)(6)\cot 39}\right]^{5.195} = 0.428$$

and $\qquad i_c = i_q - \dfrac{(1 - i_q)}{N_c \tan\phi} = 0.438 - \dfrac{(1 - 0.438)}{70.79 \tan 39} = 0.428$

$$i_\gamma = \left[1 - \frac{Q_h}{Q + BLc\cot\phi}\right]^{m+1}$$

$$= \left[1 - \frac{360}{2426.85 + 1.667(2)(6)\cot 39}\right]^{6.195} = 0.374.$$

Assuming water table to be below the ground surface at a depth greater than $(D_f + B)$, hence $r_w = r'_w = 1$, and also say $\gamma_1 = \gamma_2 = \gamma_r$

Hence $\qquad q_{nu} = cN_c S_c d_c i_c + \gamma_1 D_f(N_q - 1)S_q d_q i_q r_w + \dfrac{1}{2}\gamma_2\, BN_\gamma S_\gamma d_\gamma i_\gamma r'_w$

$\qquad q_{nu} = 6(70.79)\,(1.167)\,(1.24)\,(0.428) +$
$\qquad\qquad 18(1)\,(59.62 - 1)\,(1.167)\,(1.13)\,(0.438)\,(1) +$

$$\frac{1}{2}(18)\,(1.667)\,(100.71)\,(0.6677)\,(1)\,(0.374)\,(1)$$

$$= 263.06 + 609.45 + 376.92 = 1249.43 \text{ kN/m}^2.$$

According to Meyerhof

$$i_c = i_q = \left(1 - \frac{i}{90}\right)^2 = \left(1 - \frac{8.53}{90}\right)^2 = 0.82$$

$$i_\gamma = \left(1 - \frac{i}{\phi}\right)^2 = \left(1 - \frac{8.53}{39}\right)^2 = 0.61.$$

Therefore,
$\qquad q_{nu} = 6(70.79)\,(1.167)\,(1.24)\,(0.82) + 18(1)\,(59.62 - 1)\,(1.167)\,(1.13)\,(0.82) +$
$\qquad\qquad 1/2(18)\,(1.667)\,(100.71)\,(0.667)\,(1)\,(0.61)\,(1)$
$\qquad\qquad = 503.99 + 1140.98 + 614.76 = 2259.74 \text{ kN/m}^2.$

From charts (According to Saran and Agarwal, 1991),

$$\phi = 39°, \quad \frac{e}{B} = \frac{0.1667}{2} = 0.083 \text{ and } i = 8.53°$$

	For $\dfrac{e}{B} = 0.0$		(Figs. 7.7 to 7.9)	For $\dfrac{e}{B} = 0.10$		(Figs. 7.7 to 7.9)
	$i = 0°$	$i = 10°$	$i = 8.53°$	$i = 0°$	$i = 10°$	$i = 8.53°$
N_γ	144	58	70.64	75	42	46.85
N_q	68	41	44.97	47	30	32.50
N_c	88	48	53.88	58	37	40.09

Hence for $\dfrac{e}{B} = 0.083$ and $i = 8.53°$,

$N_c = 50.89$, $N_q = 34.62$ and $N_\gamma = 42.43$.
In this case
$i_c = i_q = i_\gamma = 1$[As effect of eccentricity and inclination has been considered already]
and $B = 2$ m
Depth factors will be

$$d_c = 1 + 0.4\frac{D_f}{B} = 1 + 0.4\frac{1}{2} = 1.2,$$

$$d_q = 1 + 2 \tan 39 (1 - \sin 39)^2 \frac{1}{2} = 1.11$$

and $d_\gamma = 1$
and shape factors will be (Saran and Agarwal, 1991) :
$S_c = 1.3$, $S_q = 1.2$ and $S_\gamma = 0.8$ (As footing is square).

Thus $q_{nu} = 6(50.89)\ (1.3)\ (1.2)\ (1) + 18(1)\ (34.62 - 1)\ (1.2)\ (1.11)\ (1)\ (1)$

$$+ \frac{1}{2}\ (18)\ (2)\ (42.43)\ (0.8)\ (1)\ (1)\ (1)$$

$$= 476.33 + 806.07 + 610.99 = 1893.39 \text{ kN/m}^2.$$

Therefore, value of q_{nu} from charts lies between values of q_{nu} obtained by Eq. 7.2 and Meyerhof's method.
Hence $Q_{nu} = q_{nu} \times$ Area of footing
$Q_{nu} = 1893.39 \times 2 \times 2$ kN $= 7573.56$ kN.

Factor of safety $= 7573.56/2400 = 3.16 > 3$. Therefore, foundation is safe against shear.

(ii) Settlement computation
When footing is subjected to a central vertical load only, in that case $e/B = 0$ and $i = 0$.
From charts of Saran and Agarwal, 1991:
For $\phi = 39°$, $N_c = 88$, $N_\gamma = 144$ and $N_q = 68$

thus $q_{nu} = 6 \times 88 \times 1.3\ (1.2)\ (1) + 18\ (1)\ (68 - 1)\ (1.2)\ (1.11)\ (1)\ (1) +$

$$\frac{1}{2}\ (18)\ (2)\ (144)\ (0.8)\ (1)\ (1)\ (1)$$

$$= 823.68 + 1606.39 + 2703.6 = 4503.67 \text{ kN/m}^2.$$

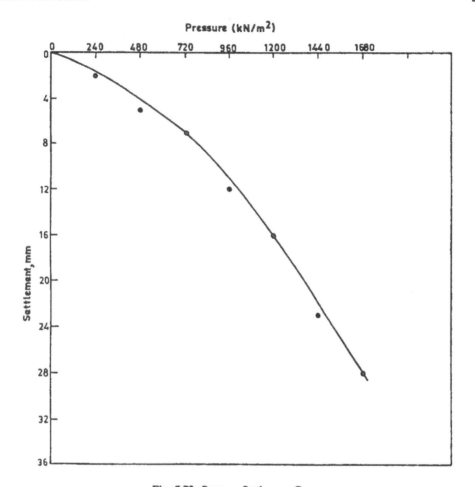

Fig. 7.72. Pressure-Settlement Curve.

Pressure on footing corresponding to a F.O.S. = 3.16, will be

$$= \frac{4503.67}{3.16} = 1425.22 \text{ kN/m}^2.$$

From plate load test data (Fig. 7.72) corresponding to pressure 1425.22 kN/m²,

$$S_p = 22.2 \text{ mm.}$$

Now since $\dfrac{S_f}{S_p} = \left[\dfrac{B_f(B_p + 30)}{B_p(B_f + 30)}\right]^2.$

Thus settlement of footing when it is subjected to central vertical load will be

$$S_0 = S_f = \left[\frac{200(60+30)}{60(200+30)}\right]^2 \times 22.2 = 37.77 \text{ mm.}$$

Now
$$\frac{i}{\phi} = \frac{8.53}{39} = 0.2187$$

Hence
$$A_0 = 1 - 0.56\left(\frac{i}{\phi}\right) - 0.82\left(\frac{i}{\phi}\right)^2 = 1 - 0.56(0.2187) - 0.82(0.2187)^2 = 0.838$$

$$A_1 = -3.51 + 1.47\left(\frac{i}{\phi}\right) + 5.67\left(\frac{i}{\phi}\right)^2$$

$$= -3.51 + 1.47\,(0.2187) + 5.67\,(0.2187)^2 = -2.917$$

$$A_2 = 4.74 - 1.38\left(\frac{i}{\phi}\right) - 12.45\left(\frac{i}{\phi}\right)^2 = 4.74 - 1.38\,(0.2187) - 12.45\,(0.2187)^2 = 3.843$$

$$\frac{S_e}{S_0} = A_0 + A_1\left(\frac{e}{B}\right) + A_2\left(\frac{e}{B}\right)^2 = 0.838 - 2.917\left(\frac{0.1667}{2}\right) + 3.843\left(\frac{0.1667}{2}\right)^2 = 0.6216$$

$$S_e = 0.6216 S_0 = 0.6216 \times 37.77 = 23.48 \text{ mm}$$

and
$$B_0 = 1 - 0.48\left(\frac{i}{\phi}\right) - 0.82\left(\frac{i}{\phi}\right)^2 = 1 - 0.48(0.2187) - 0.82(0.2187)^2 = 0.856$$

$$B_1 = -1.80 + 0.94\left(\frac{i}{\phi}\right) + 1.63\left(\frac{i}{\phi}\right)^2$$

$$= -1.80 + 0.94(0.2187) + 1.63(0.2187)^2 = -1.516$$

$$\frac{S_m}{S_0} = B_0 + B_1\left(\frac{e}{B}\right) = 0.856 + (-1.516)\left(\frac{0.1667}{2}\right) = 0.7296$$

$$S_m = 0.7296\, S_0 = 0.7296(37.77) = 27.56 \text{ mm} < 50 \text{ mm (safe)}$$

$$\sin t = \frac{S_m - S_e}{B/2 - e} = \frac{27.56 - 23.48}{2000/2 - 166.7} = 0.0049$$

$$t = 0.28° < 1° \text{ (safe)}$$

and $\quad \dfrac{H_D}{B} = 0.12 \ (0.2187) - 0.682 \ (0.2187)^2 + 1.99 \ (0.2187)^3 - 2.01 \ (0.2187)^4$

$\qquad\qquad = 0.0101$

$\qquad H_D = 0.0101 \ B = 0.0101 \ (2000) = 20.21$ mm < 25 mm (safe)

(iii) *Proportioning of foundation*

$$\text{Column load} = 2400 \text{ kN}$$
$$\text{Weight of foundation} = 240 \text{ kN (10\% of column load)}$$
$$\text{Total } V = 2400 + 240 = 2640 \text{ kN.}$$

Taking allowable soil pressure $q_a = 1425.22$ kN/m², as settlements are under permissible limit corresponding to this soil pressure, and considering the square foundation i.e. $L = B$

$$\frac{V}{A}\left(1 + \frac{6e}{L}\right) = q_a$$

$$\frac{2640}{L \times L}\left(1 + \frac{6 \times 0.1667}{L}\right) = 1425.22$$

$$L^3 - 1.852L - 1.856 = 0 \text{ or } L = 1.713 \text{ m.}$$

The size of foundation 1.75 m × 1.75 m is adopted for design. This may be shown that this size is also adequate from shear failure and settlement considerations.

(iv) *Maximum bending moment*

$$p'_{1,2} = \frac{V}{A}\left(1 \pm \frac{6e}{L}\right) = \frac{2400 \times 1.5}{(1.75)^2}\left(1 \pm \frac{6 \times 0.1667}{1.75}\right)$$

$$p'_1 = 1848.57 \text{ kN/m}^2, \quad p'_2 = 502.45 \text{ kN/m}^2.$$

Pressure distribution is shown in Fig. 7.73a.

$$M_{xx} = 1367.81 \times 1.75 \times 0.625 \times \frac{0.625}{2} + \frac{1}{2} \times 480.76 \times 0.625 \times 1.75 \times \frac{2}{3} \times 0.625$$

or $\qquad M_{xx} = 467.51 + 109.55 = 577.06$ kNm.

(v) *Effective depth*

(a) Resisting width $= b' = b_c + \dfrac{B - b_c}{8} = 500 + \dfrac{1750 - 500}{8}$

$$= 656.25 \text{ mm.}$$

Since for Fe 415 steel, $M_R = 0.138 f_{ck} \ b' \ d^2$.

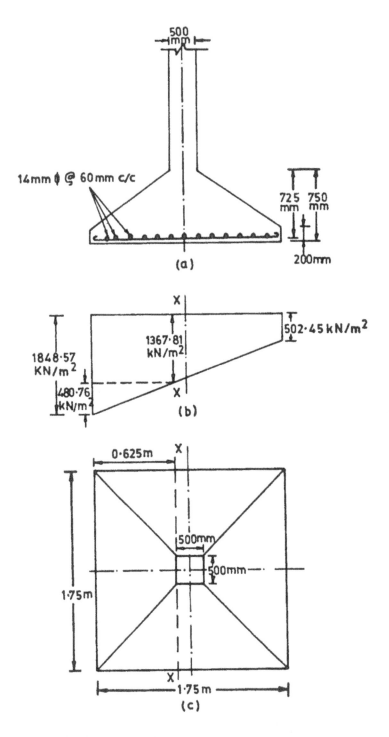

Fig. 7.73. Details of Design of Isolated Square Footing.

Hence, $577.06 \times 10^6 = 0.138 \times 20 \times 656.25 \times d^2$

                                       $d = 564.44$ mm

(b) $M_R = 0.138 f_{ck} b_c d^2 + 0.0483 \tan \alpha . f_{ck} d^3$

where $\tan \alpha = \dfrac{B - b_c}{2d}$

          $577.06 \times 10^6 = 0.138 \times 20 \times 500 \times d^2 + 0.0483 \dfrac{(1750 - 500)20}{2d} d^3$

                     $d = 539.34$ mm.

Adopt $d = 550$ mm and $D = 575$ mm

(vi) *Check for punching shear*

$$\text{Punching shear stress} = \frac{1.5 \times 2400 \times 10^3}{4(500 + 550)(550)}$$

$$= 1.558 \text{ N/mm}^2$$
$$> 1.12 \text{ N/mm}^2 \text{ (unsafe)}$$

Hence increase $d = 725$ mm and $D = 750$ mm

$$\text{Now punching shear stress} = \frac{1.5 \times 2400 \times 10^3}{4(500 + 725)(725)}$$

$$= 1.013 \text{ N/mm}^2 \ (< 1.12 \text{ N/mm}^2, \text{ hence safe})$$

(vii) *Check for one-way shear*

Critical section for one-way shear is at a distance 'd' from face of column. In this case it lies outside the foundation section. Therefore, it will be safe in one-way shear.

(viii) *Reinforcement*

$$M_{xx} = 0.87 f_y A_{st} \left(d - \frac{f_y A_{st}}{f_{ck} b'} \right)$$

$$577.06 \times 10^6 = 0.87 \times 415 A_{st} \left(725 - \frac{415 A_{st}}{20 \times 656.25} \right)$$

$$A_{st} = 2470.77 \text{ mm}^2.$$

Use 14 mm ϕ bars at a spacing $s = \dfrac{1000 \times \pi / 4(14)^2}{2470.77} = 62.27$ mm

Provide 14 mm ϕ bars at a spacing 60 mm c/c.

(ix) Development length

Development length required.

$$L_d = \frac{\sigma_{st}\phi}{4\tau_{bd}} = \frac{0.87 \times 415 \times 14}{4 \times 1.92} = 658 \text{ mm}$$

and available length = (600 + 16 × 14) = 824 mm > L_d (safe)

The details of reinforcement as shown in Fig. 7.73.

Example 7.2

Design a strip footing which is placed at a distance of 1.5 m from the edge of a slope making 30° with horizontal. The footing is subjected to a load intensity of 300 kN/m². The slope consists of silty soils with following properties.

$$c = 0, \phi = 35° \text{ and } \gamma_t = 17 \text{ kN/m}^3$$

Standard penetration tests were performed at the site. An average corrected value of standard penetration resistance was found as 17. Adopt the permissible value of settlement as 60 mm.

Solution

(i) Safe bearing capacity: Assuming 1.5 m wide foundation located at 0.75 m depth,

$$\frac{D_e}{B} = \frac{1.5}{1.5} = 1 \text{ and } \frac{D_f}{B} = \frac{0.75}{1.5} = 0.5, \beta = 30° \text{ (given).}$$

Since $\phi = 35°$, it is neither a case of general shear failure nor of local shear failure. Hence we will find out bearing capacity factors for $\phi = 39°$ and for $\phi = 28°$ and then by interpolation we may find N_c, N_q and N_γ corresponding to $\phi = 35°$.

For $\qquad \phi = 39°, \beta = 30°, \dfrac{D_e}{B} = 1 \text{ and } \dfrac{D_f}{B} = 0.5,$

$\qquad\qquad N_\gamma = 90, N_q = 13 \text{ and } N_c = 69.5.$

For $\qquad \phi = 28°, \phi' = \tan^{-1} (2/3 \tan \phi) = 19.52°, \text{ and } \beta = 30°, \dfrac{D_e}{B} = 1$

and $\qquad \dfrac{D_f}{B} = 0.5;$

$\qquad\qquad N_\gamma' = 5.5, N_q' = 6.5 \text{ and } N_c' = 17.$

Thus for $\qquad \phi = 35°, \dfrac{D_e}{B} = 1, \dfrac{D_f}{B} = 0.5 \text{ and } \beta = 30°,$

$\qquad\qquad N_\gamma = 59.27, N_q = 10.64 \text{ and } N_c = 50.41$

$\qquad\qquad q_u = cN_c + \gamma D_f N_q + 0.5\gamma_2 BN_\gamma$

$$q_u = 0(50.41) + 17(0.75)(10.64) + 0.5(17)(1.5)(59.27)$$

or $\qquad q_u = 0 + 135.66 + 755.69 = 891.35 \text{ kN/m}^2.$

Hence, $\qquad \text{F.O.S.} = \dfrac{891.35}{300} = 2.97 \approx 3 \text{ (o.k.)}$

(ii) Settlement computation

For $\qquad\qquad \beta = 0, \dfrac{D_f}{B} = 0.5, \text{ and}$

$\qquad\qquad\qquad \phi = 39°, N_\gamma = 150 \text{ and } N_q = 40$

For $\qquad\qquad \phi = 28° \ (\phi' = 19.52°), \ N_\gamma' = 6 \text{ and } N_q' = 6.5. \text{ Interpolation gives}$

For $\qquad\qquad \phi = 35°, N_\gamma = 97.64, N_q = 27.82$

$\qquad\qquad q_{u0} = \gamma D_f N_q + 0.5\gamma B N_\gamma = 17 \times 0.75 \times 27.82 + 0.5 \ (17) \ (97.64) \ (1.5)$

$\qquad\qquad q_{u0} = 354.70 + 1244.91 = 1599.61 \text{ kN/m}^2.$

Pressure intensity corresponding to FOS of 2.97 will be

$$= \dfrac{1598.61}{2.97} = 538.25 \text{ kN/m}^2.$$

The value of corrected N found by $SPT = 17$

Settlement caused by pressure intensity of 538.25 kN/m^2 is

$$S_0 = \dfrac{538.25}{0.44 \times 17} = 71.96 \text{ mm}.$$

Settlement of footing corresponding to FOS = 2.97 and when $\beta = 0$ is

$$S_0 = 71.96 \text{ mm}.$$

Now $\qquad\qquad \dfrac{S_\beta}{S_0} = \alpha_1 \dfrac{D_e}{B} + \alpha_2.$

where $\qquad\qquad \alpha_1 = 1 - 0.0125\beta° = 1 - 0.0125 \times 30 = 0.625, \text{ and}$

$\qquad\qquad\qquad \alpha_2 = 0.00385 \times 30 = 0.1155$

Hence $\qquad\qquad \dfrac{S_\beta}{S_0} = 0.625 \times 1 + 0.1155; \ S_\beta = 0.7405 \times 71.96$

or $\qquad\qquad S_\beta = 53.28 \text{ mm} < 60 \text{ mm}.$

(iii) Pressure distribution

Considering that strip footing is provided for concrete wall of 450 mm thickness,

$$p' = 1.5 \times 300 = 450 \text{ kN/m}^2$$

Pressure distribution is shown in Fig. 7.74b.

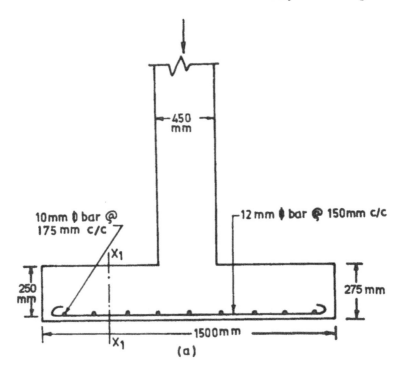

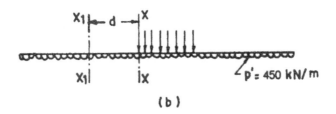

Fig. 7.74. Details of Design of a Wall Footing.

(iv) Maximum bending moment

$$M_{xx} = 450 \times 0.525 \times \frac{0.525}{2}$$

or $\qquad M_{xx} = 62.016$ kNm.

(v) Effective depth

$$M_{xx} = 0.138 f_{ck} bd^2$$
$$62.016 \times 10^6 = 0.138 \times 20 \times 1000 \times d^2$$
$$d = 149.89 \text{ mm}$$

Adopt $\qquad d = 160$ mm and $D = 185$ mm.

(vi) Check for nominal shear

S.F. at critical section x_1x_1, $V_{us} = 450 \times 0.325 = 146.25$ kN

Hence shear stress $\tau_v = 146.25 \times 10^3/(160 \times 1000) = 0.914$ N/mm^2.

Now for balance section, percentage area of steel $= A_{st}/bd \times 100$

$$p\ (\%) = \frac{A_{st}}{bd} \times 100 = \frac{x_u}{d} \times \frac{0.36f_{ck}}{0.87f_y} \times 100 = \frac{0.48 \times 0.36 \times 20}{0.87 \times 415} \times 100 = 0.96\%$$

Corresponding strength of concrete in shear $\tau_c = 0.61$ N/mm$^2 < \tau_v$.

Hence, revise the depth of section to

$$0.61 = \frac{450(525 - d)}{d \times 1000} \quad \text{or } d = 222.8 \text{ mm.}$$

Adopt $\qquad\qquad d = 250$ mm and $D = 275$ mm.

(vii) Reinforcement

$$M_{xx} = 0.87 f_y A_{st}\left(d - \frac{f_y A_{st}}{f_{ck}b}\right)$$

$$62.016 \times 10^6 = 0.87 \times 415 A_{st}\left(250 - \frac{415 A_{st}}{20 \times 1000}\right)$$

$$A_{st}^2 - 12048.2\ A_{st} + 8277864.2 = 0$$

$$A_{st} = 731.50 \text{ mm}^2.$$

Using 12 mm ϕ bars, spacing $s = \dfrac{1000 \times \pi/4(12)^2}{731.50} = 154.5$ mm.

Provide 12 mm ϕ bars at a spacing of 150 mm c/c.

Distribution reinforcement (in transverse direction)

$$A_d = \frac{0.15}{100} \times bD = \frac{0.15}{100} \times 1000 \times 275 = 412.5 \text{ mm}^2.$$

Use 10 mm ϕ bars, spacing $s = \dfrac{1000 \times \pi/4(10)^2}{412.5} = 190.4$ mm

$$s = 190.4 \text{ mm. c/c.}$$

Provide 10 mm ϕ bars @ 175 mm c/c in transverse direction.

(viii) Development length

$$L_d = \frac{\sigma_{st}\phi}{4\tau_{bd}} = \frac{0.87 \times 415 \times 12}{4 \times 1.92} = 564.2 \text{ mm}$$

We provide a standard U-type hook whose anchorage value is equal to 16 ϕ at each end. So net development length available $= 500 + 16 \times 12$

$$= 692 \text{ mm } (> L_d, \text{ safe}).$$

The details of reinforcement is shown in Fig. 7.74.

Example 7.3

A rectangular column 450 mm $\times$ 600 mm transfers a vertical load of 1000 kN, without any moment. The allowable soil pressure is 120 kN/m^2. Design a rectangular footing to support the column.

Solution

(i) *Size of footing*

$$\text{Column load} = 1000 \text{ kN}$$
$$\text{Self weight of the footing} = 100 \text{ kN (10\% of column load)}$$
$$\text{Total load} = 1100 \text{ kN}$$

$$A = \frac{1100}{120} = 9.17 \text{ m}^2.$$

Taking the ratio of width and length of the footing same as that of column dimensions, Width of footing $= (450/600 \text{ or } 3/4)$ of the length of footing.

or $\qquad (3/4 \ L) \ (L) = 9.17$

or $\qquad L = 3.49$ m and $B = 2.62$ m.

Width and length of footing are adopted as 2.70 m and 3.60 m respectively as shown in Fig. (7.75).

Thus actual area of footing $= 2.7 \times 3.6 = 9.72 \text{ m}^2$

and $\qquad\qquad p' = \dfrac{1000 \times 1.5}{9.72} \quad 154.32 \text{ kN/m}^2.$

. (ii) *Maximum bending moment*

Bending moment at section $x_1 x_1$, for 1 m strip along the length

$$M_1 = 1/2.p'[(L-l_c)/2]^2 = 1/2 \times 154.32 \times [(3.6 - 0.6)/2]^2$$
$$M_1 = 173.61 \text{ kNm.}$$

Bending moment at the section $x_2 x_2$ for 1 m strip along the width

$$M_2 = 1/2 \times p' \times [(B - b_c)/2]^2 = 1/2 \times 154.32 \times [(2.7 - 0.45)/2]^2$$
$$M_2 = 97.66 \text{ kNm.}$$

Hence, design moment $= M_{Fd} = 173.61$ kNm.

(iii) *Depth of footing*

For Fe415 steel and M20 concrete

$$M_R = 0.138 f_{ck} bd^2$$
$$173.61 \times 10^6 = 0.138 \times 20 \times 1000 \times d^2$$
$$d = 250.8 \text{ mm}$$

Adopt $\qquad\qquad d = 300$ mm and $D = 350$ mm.

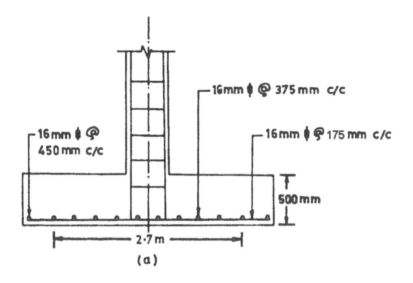

16mm ϕ @ 375mm c/c

16mm ϕ @ 175mm c/c

16mm ϕ @ 450mm c/c

500mm

2·7 m

(a)

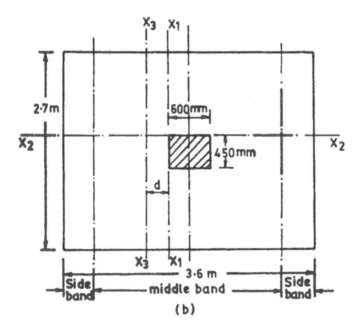

X_3 X_1

2·7m

600mm

X_2 X_2

450mm

d

X_3 X_1

3·6 m

Side band — middle band — Side band

(b)

Fig. 7.75. Details of Design of an Isolated Rectangular Footing.

(iv) Check for punching shear

$$\tau_v = \frac{1.5 \times 1000 \times 10^3}{2[(600+300)+(450+300)] \times 300}$$

$$= 1.52 \text{ N/mm}^2 > 1.12 \text{ N/mm}^2 \text{ (unsafe)}$$

Increase, $d = 450$ mm and $D = 500$ mm

$$\tau_v = \frac{1.5 \times 1000 \times 10^3}{2[(600+450)+(450+450)] \times 450}$$

$$= 0.86 \text{ N/mm}^2 < 1.12 \text{ N/mm}^2 \text{ (safe)}.$$

(v) Check for one-way shear

Shear force at a section $x_3 x_3$ at a distance d from column face,

$$V = 154.32 \left(\frac{3.6 - 0.6}{2} - 0.45 \right)$$

$$= 162.04 \text{ kN}$$

Hence $$\tau_v = \frac{162.04 \times 10^3}{450 \times 1000} = 0.36 \text{ N/mm}^2.$$

Corresponding 100 $A_{st}/bd = 0.25\%$

Hence $A_{st} = 0.25/100 \times 1000 \times 450 = 1125$ mm^2.

Thus minimum $A_{st} = 1125$ mm^2 should be provided to resist one-way shear.

(iv) Reinforcement for bending moment

$$M_R = 0.87 f_y A_{st} \left(d - \frac{A_{st} f_y}{b f_{ck}} \right)$$

$$173.61 \times 10^6 = 0.87 \times 415 A_{st} \left(450 - \frac{A_{st} 415}{1000 \times 20} \right)$$

$$A_{st} = A_{tb} = 1126.78 \text{ mm}^2 \text{ (> 1125 mm}^2).$$

Provide 16 mm ϕ bars, along the length, at a spacing

$$S_v = \frac{1000 \times \pi/4 \times (16)^2}{1126.78} = 178.43 \text{ mm} = 175 \text{ mm c/c}.$$

The cross-sectional area of steel bars required along the width for 1 m strip

$$97.66 \times 10^6 = 0.87 \times 415 A_{st} \left(450 - \frac{A_{st} \times 415}{1000 \times 20} \right)$$

$$A_{st} = A_{tl} = 618.39 \text{ mm}^2.$$

Reinforcement in middle band

$$A'_{tl} = \frac{2}{L/B + 1} \times A_{tl} = \frac{2}{3.6/2.7 + 1} \times 618.39$$

$$A'_{tl} = 530.05 \text{ mm}^2.$$

Reinforcement in two side bands

$$A''_{tl} = 618.39 - 530.05 = 88.34 \text{ mm}^2.$$

Sapcing of bars of 16 mm ϕ in central band

$$S_v = \frac{1000 \times (\pi/4) \times (16)^2}{530.05} = 379 \text{ mm c/c. Say 375 mm c/c.}$$

Sapcing of bars in each edge band = $S_v = \dfrac{1000 \times (\pi/4) \times (16)^2}{44.17}$

$$= 455.2 \text{ mm} > 450 \text{ mm}$$

$$\text{Hence } S_v = 450 \text{ mm c/c.}$$

(vii) *Check for development length*
    Required $L_d = 47 \phi = 47 \times 16 = 752 \text{ mm.}$

L_d available along the length = $\left(\dfrac{L - l_w}{2} - \text{end distance} \right)$

$$= \left(\frac{3600 - 600}{2} - 25 \right)$$

$$= 1475 \text{ mm} > 752 \text{ mm (safe).}$$

L_d available along the width $= \left(\dfrac{B - b_w}{2} - \text{end distance} \right)$

$$= \left(\frac{2700 - 450}{2} - 25 \right)$$

$$= 1100 \text{ mm} > 752 \text{ mm (safe).}$$

The details of reinforcement is shown in Fig. 7.75.

Example 7.4

   Design a combined footing, having longitudinal beam joining two columns 400 mm ×
400 mm, and 300 mm × 300 mm in sections carrying loads of 750 kN and 300 kN
respectively spaced at 4.5 m c/c. Interpretation of some of the *in situ* tests suggested that
the footing may be designed for an allowable soil pressure of 130 kN/m².

Solution

(*i*) Proportioning of foundation

$$\text{Total column load} = 750 + 300 = 1050 \text{ kN}$$
$$\text{Weight of footing plus soil} = 105 \text{ kN}$$
$$\text{Total} = 1155 \text{ kN}$$

$$A = \frac{1155}{130} = 8.885 \text{ m}^2.$$

   As shown in Fig. 7.76a, if c.g. of the loads is taken at a distance x from the centre of
lighter column, then

   $300 x = 750 (4.5 - x)$ or $x = 3.214$ m, leaving a clear projection of 0.386 m from the
lighter column.

Therefore, $L = (3.214 + 0.15 + 0.386) \times 2 = 7.5$ m

and $B = \dfrac{8.885}{7.5} = 1.18$ m, Adopt $B = 1.30$ m

         A (actual) $= 7.5 \times 1.3 = 9.75$ m²

         $$p' = \frac{1050 \times 1.5}{9.75} = 161.54 \text{ kN/m}^2.$$

Provide a beam of width 400 mm connecting the two columns.

(*ii*) *Design of slab*

                Span as cantilever $= 0.65 - 0.2 = 0.45$ m.

   Maximum cantilever moment $= \dfrac{161.54 \times (0.45)^2}{2} = 16.35$ kNm.

   Maximum shear force $= 161.54 \times 0.45 = 72.69$ kN.

   Hence thickness of slab $= \sqrt{\dfrac{16.35 \times 10^6}{0.138 \times 20 \times 10^3}}.$

                $d = 76.97$
            Adopt $d = 100$ mm.

Shear force at a distance 'd' from column face $= 161.54(0.45 - 0.10)$
                                        $= 56.539$ kN.

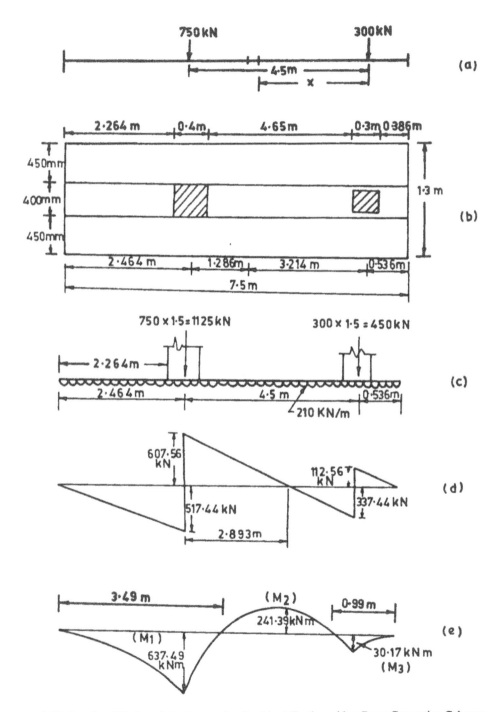

Fig. 7.76. Details of Design of the Rectangular Combined Footing with a Beam Connecting Columns.

$$\tau_v = \frac{56.539 \times 10^3}{100 \times 1000} = 0.565 \text{ N/mm}^2,$$

Corresponding to $\tau_v = 0.565$ N/mm², $A_t/bd \times 100 = 0.80\%$
Provide a reinforcement of 0.96%, $\tau_c = 0.61$ N/mm²

$$d = \frac{161.54(450 - d)}{100 \times 0.61} \text{ or } d = 107.81 \text{ mm.}$$

Provide $d = 110$ mm and $D = 150$ mm.

$$A_t = \frac{0.96}{100} \times 1000 \times 110 = 1056 \text{ mm}^2.$$

Provide 10 mm ϕ bars at a spacing $= \dfrac{1000 \times (\pi/4) \times (10)^2}{1056} = 74.38$ mm

$$= 70 \text{ mm c/c (say)}$$

$$A_{st} \text{ actual} = \frac{1000 \times (\pi/4) \times (10)^2}{70} = 1121.98 \text{ mm}^2$$

$$16.35 \times 10^6 = 0.87 \times 415 A_{st} \left[110 - \frac{415 A_{st}}{20 \times 1000} \right]$$

$$A_{st} = 450 \text{ mm}^2 \ (< 1121.98 \text{ mm}^2, \text{ safe})$$
Bond length required $= 47 \times 10 = 470$ mm
Bond length available $= 450 - 15 = 435$ mm.

Hence provide standard U-hook to increase bond length. Net bond length available = 435 + 16 × 10 = 595 mm (> L_d, safe).

(iii) Design of beam
Shear force, bending moment diagrams are shown in Fig. 7.76d and Fig. 7.76e respectively.

$$\text{Maximum cantilever moment} = 210 \times \frac{(2.264)^2}{2}$$

$$= 538.20 \text{ kNm.}$$

$$\text{Moment at the inside face of column} = \frac{210(2.464 + 0.2)^2}{2} - 1125 \times 0.20$$

$$= 520.17 \text{ kNm (Sagging).}$$
Distance of section of zero S.F. from centre of heavier column

$$= \frac{1125}{210} - 2.464 = 2.89 \text{ m.}$$

Maximum hogging moment at the section of zero S.F.

$$= 1125 \times 2.89 - \frac{210 \times 5.354^2}{2} = 241.39 \text{ kNm.}$$

From empirical relations

$$\text{Hogging moment} = \frac{wl^2}{20} = \frac{210 \times (4.5)^2}{20} = 212.62 \text{ kNm.}$$

$$\text{Sagging moment} = \frac{wl^2}{16} = \frac{210 \times (4.5)^2}{16} = 265.78 \text{ kNm.}$$

Therefore design moments are:

        Hogging moment = 241.39 kNm,

        Sagging moment = 538.20 kNm.

For sagging moment the beam will be designed as T-Beam and for hogging moment it will act as rectangular beam.

T-beam

For sagging moment = 538.20 kNm.

(a) Assume $D_f/d \leq 0.2$

$$538.20 \times 10^6 = 0.138 \, b_w d^2 f_{ck} + 0.45 f_{ck}(b_f - b_w)D_f\left(d - \frac{d_f}{2}\right)$$

$$538.20 \times 10^6 = 0.138 \times 400 \times d^2 \times 20 + 0.45 \times 20 \times (1300 - 400) \times 150(d - 150/2)$$

        $d = 384$ mm.

Now $\dfrac{D_f}{d} = \dfrac{150}{384} = 0.390 > 0.2$; hence it needs revision.

$$y_f = 0.15 x_u + 0.65 D_f = 0.15 \times 0.48d + 0.65 \times 150$$
$$= 0.072d + 97.5$$

$$538.20 \times 10^6 = 0.446 f_{ck}(b_f - b_w)y_f\left(d - \frac{y_f}{2}\right) + 0138 f_{ck} b_w d^2$$

$$538.20 \times 10^6 = 0.446 \times 20 \times (1300 - 400)(0.072d + 97.5)$$

$$\times \left(d - \frac{0.072d + 97.5}{2}\right) + 0.138 \times 20 \times 400 \times d^2$$

        $d = 366$ mm.

Depth considering rectangular beam for hogging moment = (241.39 kNm)

$$d = \sqrt{\frac{241.31 \times 10^6}{0.138 \times 400 \times 20}} = 467 \text{ mm.}$$

Depth for safe punching shear,

$$\tau_v' = 1.12 = \frac{1125 \times 10^3}{2(400 + d) \times d + 2 \times (400 + 150)150}$$

$d = 478$ mm

Hence provide $d = 500$ mm and $D = 540$ mm.

Reinforcenment:

For hogging moment

$$0.87 f_y A_{st} \left(d - \frac{f_y A_{st}}{f_{ck} b} \right) = 241.39 \times 10^6$$

$$0.87 \times 415 \times A_{st} \left(500 - \frac{415 \times A_{st}}{20 \times 400} \right) = 241.39 \times 10^6$$

$$A_{st} = 1603.85 \text{ mm}^2.$$

Provide 15 bars of 12 mm ϕ.

Extend these bars beyond point of inflexion by d or 12 ϕ whichever is greater, i.e. 500 mm.

(a) S.F. at point of inflexion (at a distance 3.49 m from left end)

$$= 1125 - 210(3.49) = 392.1 \text{ kN}$$

and $$M_R = 0.87 \times 415 \times 15 \times \pi/4 \times 12^2 \left(500 - \frac{415 \times 15 \times \pi/4 \times 12^2}{20 \times 400} \right)$$

$$= 252.25 \text{ kNm.}$$

Development length $L_d = 47 \phi = 564$ mm,

$$\frac{M_1}{V} + L_0 = \frac{252.25 \times 10^6}{392.1 \times 10^3} + 500$$

$$= 1143.33 \text{ mm} > L_d$$

Extend five bars in the support by a distance of $L_d/3 = 188$ mm say 200 mm.

(b) S.F. at point of inflexion (at a distance 0.99 m from right support) = 450 − 210 (0.99) = 242.01 kN.

Hence $M_1/V + L_0 = 252.25 \times 10^6/(242.1 \times 10^3) + 500 = 1541.92$ mm $> L_d$

Therefore, here also extend five bars in the support by a distance of 200 mm.

For sagging moment

Force in compression = $0.36 f_{ck} b_w x_u + 0.446 f_{ck} (b_f - b_w) y_f$

(since $D_f/d > 0.2$)

$$= 0.36 \times 20 \times 400 \times 0.48 \times 500 + 0.446 \times 20 \times (1300 - 400) \times$$
$$(0.072 \times 500 + 97.5)$$
$$= 1762.93$$

Moment of compressive force about extreme fibre

$$= 0.36 f_{ck} b_w x_u (0.42 x_u) + 0.446 f_{ck} (b - b_w) \cdot y_f \frac{y_f}{2}$$

$$= 0.36 \times 20 \times 400 \times (0.48 \times 500)(0.42 \times 0.48 \times 500) +$$

$$0.446 \times 20 (1300 - 400)(0.072 \times 500 + 97.5)^2 \times \frac{1}{2}$$

$$= 141.21 \text{ kNm,}$$

Hence $$a = \frac{141.21 \times 10^6}{1762.93 \times 10^3} = 80.10 \text{ mm,}$$

Therefore, lever arm $= z = 540 - 80.10 = 459.90$ mm
so $0.87 f_y A_{st} z = 538.20 \times 10^6$
          $0.87 \times 415 \times A_{st} \times 459.9 = 538.20 \times 10^6$
                          $A_{st} = 3241.24 \text{ mm}^2.$

Provide 22 mm ϕ bars, number of bars $= 3241.24 \{\pi/4(22)^2\} = 8.51 = 9$.
Extend the 1/3 (9) = 3 bars beyond point of inflexion by 500 mm distance.

Check for shear
    Critical section for checking shear stress will be at a distance 'd' from the face of columns as point of inflexion is away from this section.
          S.F. $= 1125 - 210(2.464 + 0.70) = 460.5$ kN

$$\tau_v = \frac{460.5 \times 10^3}{500 \times 400} = 2.30 \text{ N/mm}^2$$

and $$\frac{100 A_{st}}{bd} = \frac{100 \times \pi / 4(22)^2 \times 9}{500 \times 400} = 1.71\%$$

          $$\tau_c = 0.745 \text{ N/mm}^2 < 2.30 \text{ N/mm}^2$$

Hence shear reinforcement is required.
          $V_c = \tau_c bd = 0.745 \times 500 \times 400 = 149.0$ kN
          $V_{us} = 460.5 - 149.0 = 311.5$ kN

Use 12 mm ϕ bar four legged stirrups as shear reinforcement
          $A_{sv} = 4 \times \pi/4 \times (12)^2 = 452.39 \text{ mm}^2.$
    Spacing between vertical stirrups $= 0.87 f_y d A_{sv}/V_{us}$

$$S_v = \frac{0.87 \times 415 \times 500 \times 452.39}{311.5 \times 10^3}$$

$$= 262.1 \text{ mm}$$

$$S_v = 250 \text{ mm (say)}$$

and from S.F.D. (Fig. 7.72(d))

$$\frac{311.5 \times 10^3}{x} = \frac{607.56 \times 10^3}{2.893} \text{ or } x = 1.483 \text{ m}$$

Distance from face of columns = 2.893 − 0.20 − 1.483 = 1.21 m say 1.25 m.

Hence provide 12 ϕ bar four legged stirrups @ 250 mm c/c up to 1.25 m and after that provide normal shear reinforcement that is these stirrups @ 450 mm c/c.

The details of reinforcement is shown in Fig. 7.77.

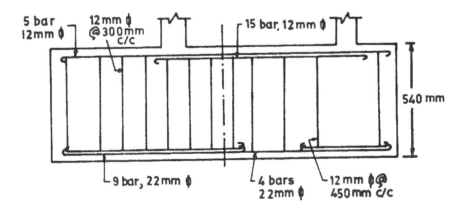

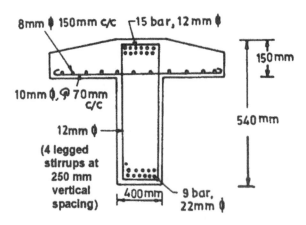

Fig. 7.77 Details of Reinforcement in the Rectangular Combined Footing with a Beam Connecting Columns.

Example 7.5

Design combined rectangular footing for two columns, having the crosssection and loads same as is Example 7.4. Allowable soil pressure may be taken as 130 kN/m³ and value of coefficient of subgrade reaction as 5×10^4 kN/m³. Also determine bending moments by

(*i*) Subgrade reaction method, and

(*ii*) Finite difference method.

Solution

Design of footing without considering longitudinal beam.

Considering the size of footing be the same as in case of Example 7.4. Hence bending moment and shear force diagrams for footing in longitudinal direction will also remain the same (Fig. 7.76).

(*i*) Effective depth for maximum bending moment

$$d = \sqrt{\frac{538.20 \times 10^6}{0.138 \times 20 \times 1300}} = 387.3 \text{ mm.}$$

(*ii*) Depth from punching shear criterion

$$\tau_v = \frac{1125 \times 10^3}{4(400 + d) \times d} = 1.12 \text{N/mm}^2$$

or $d = 368.2$

Thus adopt $d = 400$ mm and $D = 440$ mm.

(*iii*) Reinforcement

For sagging moment (M_a)

$$538.20 \times 10^6 = 0.87 f_y A_{st}\left(d - \frac{f_y A_{st}}{f_{ck} b}\right)$$

$$538.20 \times 10^6 = 0.87 \times 415 A_{st}\left(400 - \frac{415 A_{st}}{20 \times 1300}\right)$$

$$A_{st} = 4546 \text{ mm}^2.$$

Use 16 mm ϕ bars. No. of bars = $4546/\{\pi/4(16)^2\}$ = 22.6 = 24 (say)
At the point of contraflexure near the heavy column, S.F. = 392.1 kN

and $$M_R = 0.87 \times 415 \times 24 \times \pi/4(16)^2\left(400 - \frac{415 \times 24 \times \pi/4(16)^2}{20 \times 1300}\right)$$

$$M_R = 562.5 \text{ kNm}$$

$$\frac{M_1}{V} + L_0 = \frac{562.5 \times 10^6}{392.1 \times 10^3} + 400 = 1834 \text{ mm.} > L_d (47\phi).$$

Hence extend 1/3 (24) = 8 bars for a distance 400 mm from point of contraflexure and remaining bars should be curtailed at this point.

Reinforcement for hogging moment (M_b)

$$241.39 \times 10^6 = 0.87 \times 415 \times A_{st} \left(400 - \frac{415 A_{st}}{20 \times 1300} \right)$$

$$A_{st} = 1796 \text{ mm}^2.$$

Provide 16 bars of 12 mm ϕ and out of these five bars should be extended beyond the point of inflexion by a distance of 400 mm and rest of the bars may be curtailed at this point.

Reinforcement for sagging moment (M_c)

$$30.17 \times 10^6 = 0.87 \times 415 \times A_{st} \left(400 - \frac{415 A_{st}}{20 \times 1300} \right)$$

$$A_{st} = 212.5 \text{ mm}^2.$$

Minimum reinforcement $\qquad = \dfrac{0.15 \times 400 \times 1300}{100} = 780 \text{ mm}^2.$

No. of 16 mm ϕ bars $\qquad = \dfrac{780}{\left(\dfrac{\pi}{4} \times 16^2 \right)} = 3.9 = 4$

and reinforcement from development side

$$\frac{M_1}{V} + L_0 > L_d \ (47\phi = 752 \text{ mm})$$

V = Shear force at point of contraflexure near support = 242.01 kN
L_d = 12 ϕ or d whichever is more = 400 mm

$$\frac{M_1}{242.01 \times 10^3} + 400 > 752$$

$$M_1 = 85.18 \times 10^6 \text{ Nmm} = 85.18 \text{ kNm.}$$

Thus $\qquad\qquad A_s = 604.5 \text{ mm}^2.$

Number of 16 mm ϕ bars = 3.

Hence eight bars of 16 mm ϕ, which are provided for M_a moment are able to resist M_c sagging moment.

(*iv*) Check for one-way shear

The critical section for one-way shear will be at a distance d from the face of column. Hence S.F. at this section will be (considering heavier column),

$$V = 517.44 - 210(0.2 + 0.5)$$
$$= 370.4 \text{ kN}$$

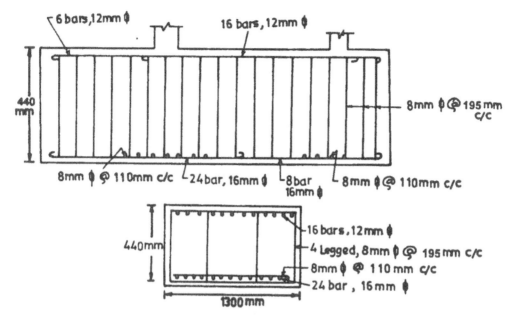

Fig 7.78. Details of Reinforcement in the Rectangular Combined Footing.

$$\tau_v = \frac{370.4 \times 10^3}{1300 \times 400} = 0.71 \text{ N/mm}^2.$$

For balanced section $\tau_c = 0.60$ N/mm$^2 < \tau_v$. S.F. at a distance d from the inner face of heavy column, is

$$V = \frac{607.56}{2.89}(2.893 - 0.7) = 460.6 \text{ kN.}$$

Thus
$$\tau_v = \frac{460.6 \times 10^3}{1300 \times 400} = 0.88 \text{ N/mm}^2 > 0.60 \text{ N/mm}^2.$$

Hence shear reinforcement is required.
Use 8 mm ϕ four legged vertical stirrups at a spacing

$$S_v = \frac{0.87 \times 415 \times 400 \times 4 \times \pi/4 \times (8)^2}{460.6 \times 10^3 - 0.60 \times 1300 \times 400} = 195 \text{ mm}$$

$$S_v = 195 \text{ mm c/c.}$$

Transverse reinforcement
Transverse reinforcement is provided to resist the bending in this direction.

Projection beyond the face of heavy column = (1300 − 400)/2 = 450 mm and width of bending strip = $b + 2d$ = 400 + 2 (500) = 1400 mm.

Net upward pressure p_0' = 1125/(1.3 × 1.4) = 618.13 kN/m².

Maximum *BM* (at the face of column) = 618.13 × $\dfrac{(0.45)^2}{2}$ = 62.59 kNm

$$d = \sqrt{\dfrac{62.59 \times 10^6}{0.138 \times 20 \times 1000}} = 150.6 \text{ mm} < 400 \text{ mm (safe)}.$$

Since transverse reinforcement is provided above longitudinal one, hence d = 400 − 16 = 394 mm.

Transverse reinforcement A_{st} is given by

$$62.59 \times 10^6 = 0.87 \times 415 \times A_{st}\left(394 - \dfrac{415 A_{st}}{20 \times 1000}\right)$$

$$A_{st} = 450 \text{ mm}^2.$$

Use 8 mm ϕ at a spacing $s = \dfrac{1000 \times (\pi/4)(8)^2}{450} = 111$ mm c/c.

Hence provide 8 mm ϕ 110 mm c/c with a standard U hook for adequate bond length.

Provide similar reinforcement under lighter column for a width bending strip = $b + 2d$ = 300 + 1000 = 1300 mm.

The details of reinforcement are shown in Fig. 7.78.

Subgrade Reaction Method

Coefficient of subgrade reaction K_p = 5 × 10⁴ kN/m³.

Since D = 440 mm,

$$I = \dfrac{1.3 \times (0.44)^3}{12} = 9.228 \times 10^{-3} \text{ m}^4$$

$$\lambda L = \sqrt[4]{\dfrac{K_f B L^4}{4EI}},$$

where $K_f = K_p\left[\dfrac{B_p(B_f + 30)}{B_f(B_p + 30)}\right]^2 \times \left(\dfrac{m + 0.5}{1.5m}\right) \times \left(1 + \dfrac{2D_f}{B}\right)$

$$m = \dfrac{L}{B} = \dfrac{7.5}{1.3} = 5.77 \text{ and assuming } D_f = 1.5 \text{ m}$$

$$K_f = 5 \times 10^4 \left(\dfrac{30(130 + 30)}{130(30 + 30)}\right)^2 \times \left(\dfrac{5.77 + 0.5}{1.5 \times 5.77}\right) \times \left(1 + \dfrac{2 \times 1.5}{1.3}\right)$$

$K_f = 3.56 \times 10^4$ kN/m^3 and taking $E_C = 5000 \sqrt{20} = 2.24 \times 10^4$ N/mm^2

$$\lambda L = \sqrt[4]{\frac{3.56 \times 10^4 \times 1.3 \times (7.5)^4}{4 \times 2.24 \times 10^7 \times 9.228 \times 10^{-3}}} = 3.65$$

$\lambda L = 3.65 > \pi$, hence foundation may be treated as flexible beam of infinite length.
1125 kN load is acting at $= 2.464/7.5L = 0.33L$
450 kN load is acting at $= 0.536/7.5L = 0.07L$

$$M = \frac{P_1}{2\lambda} B_1' + \frac{P_2}{2\lambda} B_2'$$

$$= \frac{1125 \times 7.5}{2 \times 3.65} B_1' + \frac{450 \times 7.5}{2 \times 3.65} B_2'$$

$$M = 1155.82 \; B_1' + 462.33 \; B_2'$$

From tables of Appendix 7.1

X/L	B_1'	$1155.82\,B_1'$	B_2'	$462.33\,B_2'$	M(kNm)
0.0	0	0	0	0	0
0.1	0.0483	55.82	−0.0095	−4.392	51.42
0.2	0.1612	186.31	−0.0392	−18.123	168.18
0.3	0.1996	230.7	−0.1004	46.417	184.28
0.4	0.2892	334.26	−0.1615	−74.66	259.6
0.5	0.1963	226.88	−0.2416	−111.69	115.19
0.6	−0.0202	−23.34	−0.3216	−148.68	−171.94
0.7	−0.0205	−23.69	−0.3053	−141.14	−164.83
0.8	−0.0210	−24.27	−0.2893	−133.75	−158.02
0.9	−0.0133	−15.37	−0.0353	−16.32	−31.69
1.0	0	0	0	0	0

Finite Difference Method

For illustration the soil reaction is considered equivalent to point load reactions R_1, R_2, R_3 and R_4 as shown in Fig. 7.79. For a precise analysis, more number of point load reactions should be considered.

$$\frac{EI}{h^2} = \frac{2.24 \times 10^7 \times 9.228 \times 10^{-3}}{2.5^2} = 0.529 \times 10^4$$

$$K_f = 3.56 \times 10^4 \text{ kN/m}^3$$

$$R_1 = \frac{h}{2} K_f B y_1 = \frac{2.5}{2} \times 3.56 \times 10^4 \times 1.3 \times y_1 = 5.78 \times 10^4 y_1$$

$$R_2 = 11.56 \times 10^4 y_2$$

$$R_3 = 11.56 \times 10^4 y_3$$

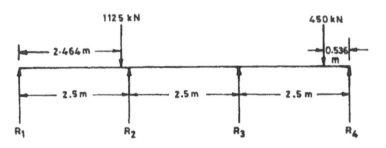

Fig. 7.79. Showing the Point Load Reactions.

$$R_4 = 5.78 \times 10^4 y_4$$

At point R_2, $\Sigma M = 0$

$$\frac{EI}{h^2}(y_3 - 2y_2 + y_1) = V_1 x - R_1 h$$

$$0.529 \times 10^4 (y_3 - 2y_2 + y_1) = 1125(2.5 - 2.464) - 5.78 \times 10^4 \times 2.5 y_1$$

$$y_3 - 2y_2 + 28.31 y_1 = 76.56 \times 10^{-4} \qquad \qquad \text{... (i)}$$

At point R_3, $\Sigma M = 0$

$$0.529 \times 10^4 (y_4 - 2y_3 + y_2) = 1125(2.5 + 0.036) - 5.78 \times 10^4 \times 2.5 \times 2 \times y_1$$
$$- 11.56 \times 10^4 \times y_2 \times 2.5$$

$$y_4 - 2y_3 + 55.63 y_2 + 5.91 y_1 = 53.88 \times 10^{-2} \qquad \qquad \text{... (ii)}$$

Now $\Sigma V = 0$

$$5.78 \times 10^4 (y_1 + 2y_2 + 2y_3 + y_4) = 1125 + 450$$

or $\qquad\qquad y_1 + 2y_2 + 2y_3 + y_4 = 2.72 \times 10^{-2} \qquad \qquad \text{... (iii)}$

$\Sigma M_1 = 0$

$$R_2 h + R_3(2h) + R_4(3h) = 1125 \times 2.464 + 450 \times 6.964$$

$$y_2 + 2y_3 + 1.5 y_4 = 5.11 \times 10^{-2} \qquad \qquad \text{... (iv)}$$

Solving eqns. (i), (ii), (iii), and (iv), we get

$$y_1 = 1.8 \times 10^{-3} \text{ m}$$

$$y_2 = 7.3 \times 10^{-3} \text{ m}$$

$$y_3 = - 27.5 \times 10^{-3} \text{ m}$$

$$y_4 = 65.9 \times 10^{-3} \text{ m}$$

Hence $\qquad M_2 = \dfrac{EI}{h^2}(y_3 - 2y_2 + y_1) = 0.529 \times 10^4(-27.5 - 2 \times 7.3 + 1.8) \times 10^{-3}$

$$= -213 \text{ kNm}$$

$$M_3 = 0.529 \times 10^4(65.9 + 2 \times 27.5 + 7.3) \times 10^{-3}$$
$$= 678 \text{ kNm.}$$

Example 7.6

Design a combined trapezoidal footing for two columns 400 mm × 400 mm and 300 mm × 300 mm in section carrying loads of 750 kN and 450 kN respectively spaced at 3.5 m c/c. There is a restriction on extending the footing on the heavier column side by a distance not more than 100 mm. Adopt allowable soil pressure of 130 kN/m² for design purposes.

Solution

(i) Proportioning of foundation

$$\text{Column load} = 750 + 450 = 1200 \text{ kN}$$
$$\text{Weight of foundation (10\%)} = 120 \text{ kN}$$
$$\text{Total} = 1320 \text{ kN}$$

$$\text{Area required} = \frac{1320}{130} = 10.15 \text{ m}^2.$$

If x is the distance of centre of gravity of the loads from the centre of column carrying 750 kN load, then

$$x = \frac{450 \times 3.5}{1200} = 1.3125 \text{ m (Fig. 7.80)}$$

Now, $\quad L = 3.5 + 0.2 + 0.15 + 2(0.1) = 4.05 \text{ m}$

$$\frac{a+b}{2} \cdot L = A \text{ or } \frac{a+b}{2} \cdot 4.05 = 10.15 \text{ or } a + b = 5.02$$

$$\frac{a+2b}{a+b} \cdot \frac{L}{3} = (1.3125 + 0.3) \text{ or } a + 2b = 6.0.$$

Solving the above equations:

$$a = 4.04 \text{ m and } b = 0.98 \text{ m.}$$

Upward reaction $= p' = 1200 \times 1.5/[1/2(4.04 + 0.98) \times 4.05] = 177.06 \text{ kN/m}^2.$

Let two columns be connected by means of a beam of size 400 mm.

(ii) Design of slab

$$\text{Maximum projection of slab from beam} = \frac{4.04 - 0.40}{2} = 1.82 \text{ m}$$

$$\text{Maximum moment} = 177.06 \times \frac{(1.82)^2}{2} = 293.3 \text{ kNm.}$$

Maximum S.F. in slab $= 177.06 \times 1.82 = 322.18 \text{ kN.}$

$$\text{Depth of slab, } d = \sqrt{\frac{253.3 \times 10^6}{0.138 \times 20 \times 1000}}$$

$$= 302.94 \text{ mm; say 310 mm}$$

$$\tau_v = \frac{322.18 \times 10^3}{310 \times 1000} = 1.04 \text{ N/mm}^2.$$

Restricting $\tau_c = 0.67$ N/mm^2, corresponding to $100 A_{st}/bd = 1.25$

$$d = \frac{322.18 \times 10^3}{0.67 \times 1000} = 480 \text{ mm}.$$

Provide $d = 480$ mm and $D_f = 500$ mm.

$$A_{st} = \frac{1}{100} \times 540 \times 1000 = 5400 \text{ mm}^2.$$

Use 24 mm ϕ at spacing $= \dfrac{1000 \times \pi/4(24)^2}{5400} = 93.23$ mm; say 90 mm.

Bond length required $= 47\phi = 1128$ mm.
Available bond length $= 1820$ mm $> L_d$ (safe).
    Also provide 6 mm bars @ 50 mm c/c as distribution reinforcement.

(*iii*) Design of beam
    Let the section of zero S.F. be at a distance x_1 from left face of footing.

Width at section $xx = 0.98 + \dfrac{4.04 - 0.98}{4.05}(4.05 - x_1)$

$$= 4.04 - 0.756x$$

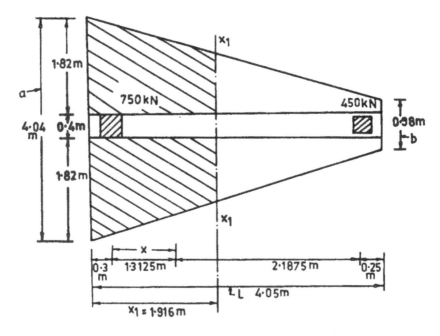

Fig. 7.80. Proportioning of the Trapezoidal Combined Footing.

Hence, $\frac{1}{2}[4.04 + (4.04 - 0.756x_1)]x_1 \times 177.06 = 750 \times 1.5$

or $x_1 = 1.916$ m

and width of footing at zero S.F. section = 2.59 m

c.g. of shaded area (Fig. 7.80) from zero shear force line (x_1x_1)

$$= \frac{2 \times 4.04 + 2.59}{4.04 + 2.59} \times \frac{1.916}{3} = 1.42 \text{ m.}$$

Distance of c.g. of shaded area from column A

$$= 1.916 - 1.42 - 0.1 - 0.2 = 0.196 \text{ m.}$$

Therefore BM at the section of zero S.F. = $1.5 \times 750 \times 0.196$

$$= 220.5 \text{ kNm (Hogging).}$$

Average value of variation of w along with span

$$= \frac{4.04 + 0.98}{2} \times 177.06 = 444.42 \text{ kNm.}$$

Maximum sagging moment = $697.01 \times \dfrac{(0.3)^2}{2} = 31.39$ kNm.

From empirical relations

$$\text{Hogging moment} = \frac{wl^2}{20} = \frac{444.42 \times (3.5)^2}{20} = 272.21 \text{ kNm.}$$

$$\text{Sagging moment} = \frac{wl^2}{24} = \frac{444.42 \times (3.5)^2}{24} = 226.84 \text{ kNm.}$$

Therefore, the design moments will be

$$\text{Hogging moment} = 272.21 \text{ kNm.}$$
$$\text{Sagging moment} = 226.84 \text{ kNm.}$$

Designing beam as rectangular beam for both hogging as well for sagging moment

Depth of beam, $d = \sqrt{\dfrac{272.21 \times 10^6}{0.138 \times 20 \times 400}} = 496.55$ mm.

Adopt $d = 500$ mm and $D_f = 530$ mm.

Reinforcement
   For hogging moment

$$272.21 \times 10^6 = 0.87 f_y A_{st}\left(d - \frac{f_y A_{st}}{bf_{ck}}\right)$$

$$= 0.87 \times 415 A_{st}\left(500 - \frac{415 A_{st}}{400 \times 20}\right)$$

$A_{st} = 1871$ mm^2.

Provide 12 mm ϕ bars. Number of bars $= 1871/[\pi/4(12)^2] = 16.57 = 17$ (say)

For sagging moment

$$31.39 \times 10^6 = 0.87 \times 415 A_{st}\left(500 - \frac{415 A_{st}}{400 \times 20}\right)$$

$A_{st} = 177$ mm^2.

Provide 8 mm ϕ bars. Number of bars $= \dfrac{177}{\pi/4(8)^2} = 3.52 = 4$ (say)

Shear reinforcement

Width of footing at the inside face of 750 kN column,

$$= 4.04 - \left(\frac{4.04 - 0.98}{4.05}\right) \times 0.5 = 3.66 \text{ m.}$$

Shear force at the inside face of column of 750 kN load

$$V = 750 \times 1.5 - \frac{1}{2}(4.04 + 3.66) \times 177.06 \times 0.5$$

$$V = 784.06 \text{ kN.}$$

Shear stress $\tau_v = \dfrac{784.06 \times 10^3}{400 \times 500} = 3.92$ N/mm^2

$$\frac{A_{st}}{bd} \times 100 = \frac{(1243.4 + 177)}{400 \times 500} \times 100 = 0.71\%$$

It gives $\tau_c = 0.55$ N/mm^2.

Thus $V_{us} = 784.06 - \dfrac{0.55 \times 400 \times 500}{1000} = 674.06$ kN

Provide 14 mm ϕ vertical four legged stirrups at a spacing

$$= \frac{0.87 \times 415 \times 4 \times \pi/4(14)^2 \times 500}{674.06 \times 10^3} = 164.82 \text{ mm } c/c; \text{ say 160 mm } c/c$$

The details of reinforcement is shown in Fig. 7.81.

Example 7.7

Design the footings for the columns given in Example 7.6 with a strap beam.

Solution

(*i*) Proportioning of foundation

Column load $= 1200$ kN

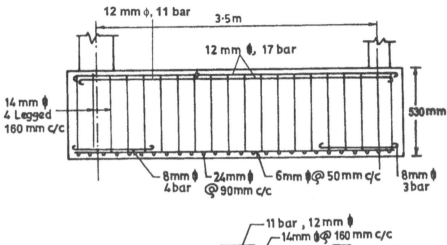

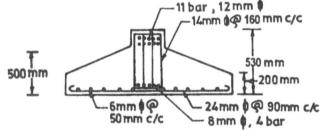

Fig. 7.81. Details of Reinforcement of the Trapezoidal Combined Footing.

Weight of foundation = 120 kN
Total = 1320 kN

$$\text{Area of foundation} = \frac{1320}{130} = 10.15 \text{ m}^2.$$

Centre of gravity of column loads from centre of column of 750 kN

$$= \frac{450 \times 3.5}{1200} = 1.3125 \text{ m.}$$

From Fig. 7.82a we get

$$B(L_1 + L_2) = 10.15.$$

Say we adopt
$$B = 2 \text{ m,}$$
$$L_1 + L_2 = 5.08 \text{ m} \qquad \qquad \dots \text{(i)}$$

and
$$\frac{BL_1 \times L_1 / 2 + BL_2 (3.5 + L_1 / 2)}{B(L_1 + L_2)} = 1.3125 + \frac{L_1}{2}$$

$$L_1^2 + 2L_2 (3.5 + L_1/2) = 13.32 + 5.075 L_1 \qquad \qquad \dots \text{(ii)}$$

Solving (i) and (ii),
$$L_1 = 3.17 \text{ m}$$
$$L_2 = 1.91 \text{ m.}$$

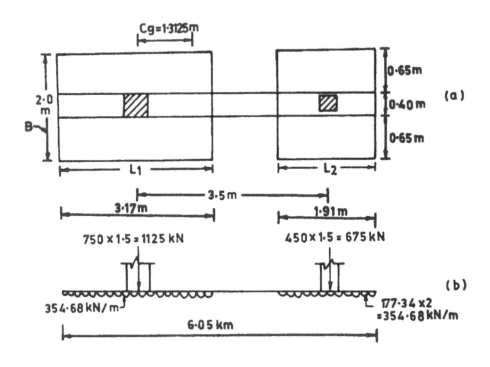

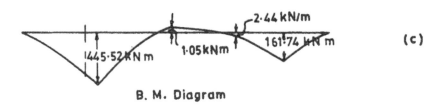

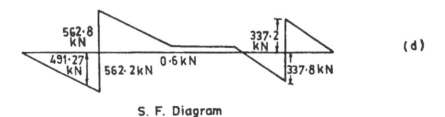

Fig. 7.82. Details of Design of Footing with Strap Beam.

(*ii*) Design of slab

$$\text{Upward reaction} = \frac{1200 \times 1.5}{10.15} = 177.34 \text{ kN/m}^2.$$

$$\text{Maximum bending moment} = 177.34 \frac{(1-0.2)^2}{2} = 56.75 \text{ kNm.}$$

$$\text{Maximum S.F.} = 177.34 \ (1 - 0.2) = 141.87 \text{ kN.}$$

$$\text{Depth of slab} = d = \sqrt{\frac{56.75 \times 10^6}{0.138 \times 20 \times 1000}}$$

$$d = 143.4 \text{ mm.}$$

For balanced section $A_{st}/bd \times 100 = 0.96\%$ and $\tau_c = 0.61 \text{ N/mm}^2$.

$$\text{Depth required for safe shear stress} = \frac{141.87 \times 10^3}{0.61 \times 1000}$$

$$= 232.57 \text{ mm}$$

Provide 250 mm effective and 280 mm overall depth of slab,

$$A_{st} = \frac{0.96}{100} \times 1000 \times 250 = 2400 \text{ mm}^2.$$

$$\text{Use 14 mm } \phi \text{ bars at a spacing} = \frac{\pi/4(14)^2 1000}{2400} = 64.10 \text{ mm}$$

$$= 60 \text{ mm c/c (say).}$$

Also provide 6 mm ϕ bars @ 50 mm c/c as distribution reinforcement.

(*iii*) Design of beam

As shown in Fig. 7.82b, the intensity of base pressure acting on beam will be 354.68 kN/m. Bending moment and shear force diagrams are shown in Figs. 7.82c and 7.82d respectively.

From empirical relations

$$\text{Hogging moment} = \frac{wl^2}{20} = \frac{354.69 \times (3.5)^2}{20} = 217.24 \text{ kNm}$$

$$\text{Sagging moment} = \frac{wl^2}{24} = \frac{354.69 \times (3.5)^2}{24} = 181.03 \text{ kNm.}$$

Thus   design hogging moment = 217.24 kN-m
                 design sagging moment = 445.52 kN-m.

Design of beam will be done considering it as T-beam for sagging moment and for hogging moment its depth will be checked considering it as a rectangular beam.

$$b_f = \frac{l_0}{6} + b_w + 6D_f = \frac{3170}{6} + 400 + 6 \times 300 = 2728.3 \text{ mm.}$$

But available $b_f = 2000$ mm.

(a) Assume that NA lies in flange

$$d = \sqrt{\frac{445.52 \times 10^6}{0.138 \times 20 \times 2000}} = 284.1 \text{ mm} > D_f$$

(b) Assume $D_f > 0.2d$, $y_f = 0.15x_u + 0.65D_f = 0.072d + 195$

$$M_R = 0.138 f_{ck} b_w d^2 + 0.446 f_{ck} (b_f - b_w) y_f \left(d - \frac{y_f}{2}\right)$$

$$445.52 \times 10^6 = 0.138 \times 20 \times 400 d^2 + 0.446 \times 20(2000 - 400) \times$$

$$(0.072d + 195)\left(d - \frac{0.072d + 195}{2}\right)$$

$$d^2 + 1233d - 342242 = 0$$
$$d = 233.39 \text{ mm}$$
$$0.2\, d = 46.67 \text{ mm} < D_f.$$

Depth for hogging moment

$$d = \sqrt{\frac{217.24 \times 10^6}{0.138 \times 20 \times 400}} = 443.59 \text{ mm.}$$

Adopt $d = 450$ mm and $D_f = 500$ mm.

Reinforcement

For sagging moment

Force in compression $= 0.36\, f_{ck} b_w x_u + 0.446\, f_{ck} (b_f - b_w) y_f$

$$= 0.336 \times 20 \times 400 \times 0.48 \times 450 + 0.446 \times 20 \times (2000 - 400)$$
$$(0.072 \times 450 + 195)$$
$$= 3826.06 \text{ kN.}$$

Moment of compressive force about extreme fibre

$$= 0.36 f_{ck} b_w x_u (0.42x_u) + 0.446 f_{ck} (b - b_w) y_f \frac{y_f}{2}$$

$$= 0.36 \times 20 \times 400 \times (0.48 \times 450)^2 \times 0.42 + 0.446 \times 20 \times$$

$$(2000 - 400)\frac{(0.072 \times 450 + 195)^2}{2}$$

$$= 4.26 \times 10^8 \text{ Nmm.}$$

Hence $$a = \frac{4.26 \times 10^8}{3826.06 \times 10^3} = 111.3 \text{ mm.}$$

Therefore lever arm $= z = 450 - 111.3 = 338.7$ mm.

So
$$0.87 f_y A_{st} \, z = 4.26 \times 10^8$$
$$A_{st} = 3483.6 \text{ mm}^2$$

Provide 20 mm ϕ bars. Number of bars $= \dfrac{3483.6}{\pi / 4(20)^2} = 11.09 = 11$ say.

Shear reinforcement

Maximum shear stress at the face of heavy column

$$\tau_v = \frac{491.27 \times 10^3}{400 \times 450} = 2.73 \text{ N/mm}^2$$

$$\frac{100 A_{st}}{bd} = \frac{3454}{400 \times 450} \times 100 = 1.92\%. \text{ Thus } \tau_c = 0.78 \text{ N/mm}^2$$

$$V_{us} = 491.27 \times 10^3 - 0.78 \times 400 \times 450 = 351.37 \text{ kN.}$$

Provide 12 mm ϕ four legged vertical stirrups at a spacing

$$= \frac{0.87 \times 415 \times 4 \times \pi / 4(12)^2 \times 450}{351.37 \times 10^3} = 209 \text{ mm c/c.}$$

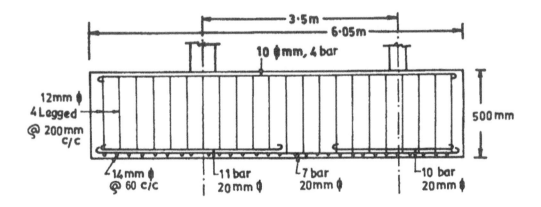

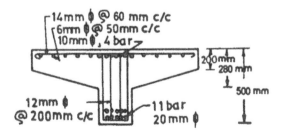

Fig. 7.83. Details of Reinforcement in the Footing with Strap Beam.

For hogging moment

As actual hogging moment is of negligible quantity

$$\text{Minimum reinforcement} = \frac{0.15 \times 400 \times 450}{100} = 270 \text{ mm}^2.$$

$$\text{Provide 10 mm } \phi \text{ bars. Number of bars} = \frac{270}{\pi/4(10)^2} = 3.43; \text{ say 4}$$

The details of reinforcement are given in Fig. 7.83.

Example 7.8

A building consists of 12 columns 400 mm × 400 mm sizes arranged in three rows of four each. The distance between the columns is 5.0 m each. The load carried by four corner columns is 500 kN each, that carried by exterior column is 550 kN each and that carried by interior column is 900 kN each. The allowable soil pressure is 50 kN/m². Design the raft foundation.

Solution

(*i*) Proportioning of raft: Let the clear projection of the raft be 600 mm and columns are

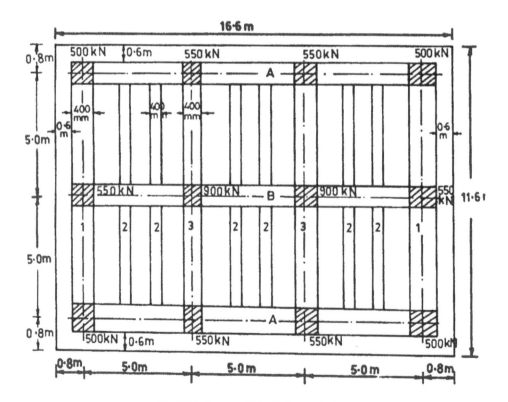

Fig. 7.84. Layout of the Raft Foundation.

connected with beams known as primary beams. Two intermediate beams of 400 mm width in each panel are also provided as shown in Fig. 7.84. These are termed as secondary beams. The width of primary beams is kept equal to width of column, *i.e.* 400 mm.

$$\text{Length of raft} = 15 + 1.2 + 0.4 = 16.6 \text{ m}$$
$$\text{Width of raft} = 10 + 1.2 + 0.4 = 11.6 \text{ m}$$
$$\text{Area of raft} = 16.6 \times 11.6 = 192.50 \text{ m}^2.$$

(*ii*) Design of slab

$$\text{Total vertical load } V \text{ of 12 columns} = 4 \times 500 + 6 \times 550 + 2 \times 900$$
$$V = 7100 \text{ kN}$$
$$\text{Weight of raft} = 710 \text{ kN}$$
$$\text{Total} = 7810 \text{ kN}$$

$$\text{Area required} = \frac{7810}{50} = 156.2 \text{ m}^2 < 192.50 \text{ (safe)}.$$

$$\text{Upward reaction} = p' = \frac{7100 \times 1.5}{192.50} = 56.0 \text{ kN/m}^2.$$

$$\text{Cantilever moment in the slab} = 56.0 \times \frac{0.6^2}{2} = 10.08 \text{ kNm.}$$

$$\text{Moment in central portion of slab} = \frac{56.0 \times (1.27)^2}{12} = 7.53 \text{ kNm.}$$

$$\text{Depth } d = \sqrt{\frac{10.08 \times 10^6}{0.138 \times 1000 \times 20}} = 60.43 \text{ mm.}$$

$$\text{Maximum S.F. in cantilever portion} = 56 \times 0.6 = 33.6 \text{ kN.}$$

$$\text{Maximum S.F. in central portion of slab} = \frac{56 \times 1.27}{2} = 35.56 \text{ kN.}$$

$$\text{Depth for safe shear stress} = \frac{35.56 \times 10^3}{1000 \times 0.61} = 58.29 \text{ mm.}$$

Take $d = 70$ mm and $D = 100$ mm

and provide $A_{st} = \dfrac{0.96}{100} \times 70 \times 1000 = 672 \text{ mm}^2.$

Hence $M_R = 0.87 \times 415 \times 672\left(70 - \dfrac{672 \times 415}{20 \times 1000}\right) = 16.02 \times 10^6 \text{ Nmm}$

$$[> 10.08 \times 10^6 \text{ Nmm, safe}].$$

Use 10 mm ϕ bars at a spacing $= \dfrac{\pi/4(10)^2 \times 1000}{672} = 116.8 \text{ mm}$

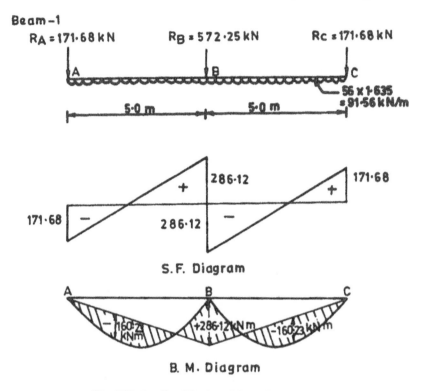

Fig. 7.85. Details of Design of Secondary Beam 1.

$$\text{say} = 110 \text{ mm.}$$

Available bond length = 600 – 40 = 560 mm > L_d = 470 mm (safe).

Transverse reinforcement = $\dfrac{0.15}{100} \times 1000 \times 100 = 150 \text{ mm}^2$.

Use 10 mm ϕ bars at 400 mm c/c spacing.

Secondary beams
Beam (1) Fig. 7.85
   Maximum sagging moment (below support B)
$$= \text{Coeff.} \times w \times L^2 = 0.125 \times 91.56 \times 5 \times 5$$
$$= 286.12 \text{ kNm.}$$
   Maximum hogging moment = 0.07 × (91.56 × 5) × 5 = 160.23 kNm.
$$R_A = R_C = 0.375 \times (91.56 \times 5) = 171.68 \text{ kN}$$
$$R_B = 1.25 \times (91.56 \times 5) = 572.25 \text{ kN.}$$
   Hence, Maximum B.M. = 286.12 kN-m (sagging), and
       maximum S.F. = 286.12 kN.
Beams 2 and 3 [since beams 2 and 3 are identical, Fig. 7.86]
Maximum sagging moment (below support B)

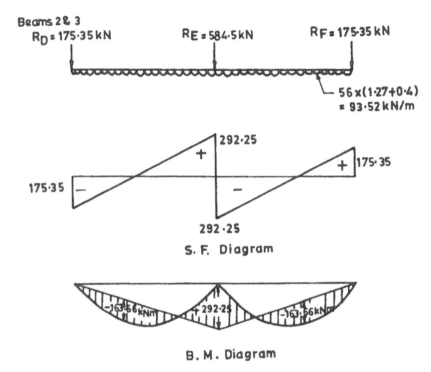

Fig. 7.86. Details of Design of Secondary Beams 2 and 3.

$$= 0.125 \times (93.52 \times 5) \times 5 = 292.25 \text{ kNm.}$$

Maximum hogging moment $= 0.07 \times (93.52 \times 5) \times 5 = 163.66 \text{ kNm.}$

$$R_D = R_F = 0.375 \times (93.52 \times 5) = 175.35 \text{ kN}$$

$$R_E = 1.25 \times (93.52 \times 5) = 584.5 \text{ kN}$$

Maximum B.M. $= 292.25$ kNm (Sagging), and

Maximum S.F. $= 292.25$ kN.

Summary of computations

Beam	S.F.	B.M.
1	286.12 kN	286.12 kNm (sagging)
2	292.25 kN	292.25 kNm (sagging)
3	292.25 kN	292.25 kNm (sagging)

Hence design moment $= 292.25$ kNm (sagging)

Design shear force $= 292.25$ kN.

Since sagging moments are maximum, all secondary beams will be designed as T-beam.

Design

$$b_f = \frac{L_0}{6} + b_w + 6D_f = \frac{0.7 \times 10 \times 10^3}{6} + 400 + 6 \times 100$$

$$= 2166.7 \text{ mm}$$

Available $b_f = 1.27 + 0.4 = 1.67 \text{ m} = 1670 \text{ mm}$.

Adopt $b_f = 1670 \text{ mm}$.

(*a*) Assume that N.A lies in flange

$$d = \sqrt{\frac{292.25 \times 10^6}{0.138 \times 20 \times 1670}} = 251.80 \text{ mm}.$$

Now $x_u = 0.48 \ d = 120.86 \text{ mm} > D_f$

Hence N.A. lies outside of flange.

(*b*) Assume $D_f < 0.2d$,

$$M_R = 0.138 f_{ck} b_w d^2 + 0.446 f_{ck}(b_f - b_w)D_f \times \left(d - \frac{D_f}{2}\right)$$

$$292.25 \times 10^6 = 0.138 \times 20 \times 400 d^2 + 0.446 \times 20 \times (1670 - 400) \, 100\left(d - \frac{100}{2}\right)$$

or $d^2 + 1026d - 316006 = 0$.

It gives $d = 248.03 \text{ mm}$.

Now $0.2 \ d = 49.60 \text{ mm} < D_f$

(*c*) Assume $D_f > 0.2d$

$$M_R = 0.138 f_{ck} b_w d^2 + 0.446 f_{ck}(b_f - b_w)y_f\left(d - \frac{y_f}{2}\right)$$

$$y_f = 0.15 x_u + 0.65 D_f = 0.15 \times 0.48d + 0.65 \times 100$$

$$= (0.072d + 65) \text{ mm}.$$

$$292.25 \times 10^6 = 0.138 \times 20 \times 400 d^2 + 0.446 \times 20(1670 - 400)$$

$$\times (0.072d + 65) \left(d - \frac{0.072d + 65}{2}\right)$$

or $d^2 + 361.41 \ d - 167229 = 0$.

It gives $d = 266 = 270 \text{ mm (say)}, \ D = 300 \text{ mm}$.

Reinforcement

For sagging moment

For balanced section $C = T$

$$0.36 f_{ck} b_w x_u + 0.446 f_{ck}(b_f - b_w)y_f = 0.87 f_y \cdot A_{st}$$

$$y_f = 0.072(270) + 65 = 84.44 \text{ mm}$$

$$x_u = 0.48 \, d = 0.48 \times 270 = 129.6 \text{ mm}$$

$$0.36 \times 20 \times 400 \times 129.6 + 0.446 \times 20 \times (1670 - 400) \times 84.44 = 0.87 \times 415 \times A_{st}$$

$$A_{st} = 3683 \text{ mm}^2.$$

Provide 20 mm ϕ bars. Hence number of bars required $= \dfrac{3683}{\pi / 4(20)^2}$

$$= 11.7; \text{ say } 12$$

Therefore, provide 12 bars of 20 mm dia.
For hogging moment

$$163.66 \times 10^6 = 0.87 \times 415 A_{st}(270)\left(1 - \frac{A_{st}415}{400 \times 270 \times 20}\right)$$

It gives $\qquad A_{st} = 1229.5 \text{ mm}^2.$
Provide eight bars of 14 mm ϕ.
Check in shear

$$\tau_v = \frac{V_u}{b_w d} = \frac{292.25 \times 1000}{400 \times 270} = 2.70 \text{ N mm}^2$$

$$p = \frac{100 A_{st}}{b_w d} = \frac{100 \times 3683}{400 \times 270} = 3.41\%$$

$\tau_c = 0.82 \text{ N/mm}^2 < \tau_v,$ shear stirrups are required.

$$V_{us} = V_u - \tau_c bd = 292.25 \times 10^3 - 0.82 \times 400 \times 270 = 203.7 \text{ kN}.$$

Provide 12 mm ϕ four legged stirrups (Vertical)

$$A_{sv} = 4 \times \pi/4 \times (12)^2 = 452.39 \text{ mm}^2.$$

Spacing, $\qquad S_v = \dfrac{0.87 \times 415 \times 452.39 \times 270}{203.7 \times 10^3} = 216.5 \text{ mm}.$

$$> 0.75 \, d \ (202.5 \text{ mm})$$
$$< 450 \text{ mm}$$

Hence adopt $\qquad S_v = 200 \text{ mm}$

Minimum reinforcement provided at a spacing

$$s = \frac{415 \times 452.39}{0.4 \times 400} = 1173 \text{ mm} > 200 \text{ mm (safe)}$$

Therefore, provide four legged vertical stirrups of 12 mm ϕ at a spacing of 200 mm c/c.
The details of reinforcement are shown in Fig. 7.87.

Secondary Beam

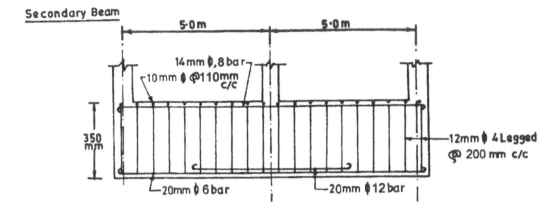

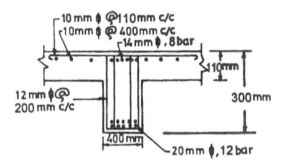

Fig. 7.87. Details of Reinforcement in the Secondary Beam.

Design of primary beams
Beam (A)

Total downward force = 3150 kN
Total upward force = 171.68 × 2 + 175.35 × 8 + 56.0 × 16.6 × 1
= 2675.76 kN.

The unbalance in downward and upward vertical load is due to the unknown fixity between columns and foundation raft. The reactions from secondary beams have been obtained considering full fixity. For design purposes, the system may be made balanced by adjusting the balance amount of load in the upward soil reaction or by changing the extreme column loads.

An upward uniformly distributed load of 84.57 kN/m make the system balanced (Fig. 7.88). The shear force and bending moment obtained by conventional method are shown in Fig. 7.88.

Design moment = 914.0 kNm (hogging)
Maximum S.F. = 510.66 kN.

The beam will be designed as inverted T-beam.

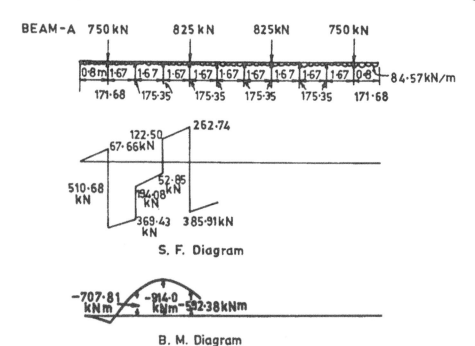

Fig. 7.88. Details of Design of Primary Beam A.

Design

$$b_f = \frac{l_0}{6} + b_w + 6D_f = \frac{0.7 \times 15 \times 10^3}{6} + 400 + 6 \times 100 = 2750 \text{ mm.}$$

(a) Assume that $D_f < 0.2d$

$$M_R = 0.138 f_{ck} b_w d^2 + 0.446 f_{ck}(b_f - b_w) D_f \times \left(d - \frac{D_f}{2}\right)$$

$$914.0 \times 10^6 = 0.138 \times 20 \times 400d^2 + 0.446 \times 20 \times (2750 - 400) \, 100 \left(d - \frac{100}{2}\right)$$

or $\qquad d^2 + 1899d - 922950 = 0.$

It gives $\qquad\qquad d = 401 \text{ mm}$

$$0.2d = 80.3 \text{ mm} < D_f.$$

(b) Assume $D_f > 0.2d$

$$M_R = 0.138 f_{ck} b_w d^2 + 0.446 \times f_{ck}(b_f - b_w) \, y_f \times \left(d - \frac{y_f}{2}\right)$$

$$y_f = 0.15 \, X_u + 0.65 D_f$$
$$= 0.15 \times 0.48d + 0.65 \times 100$$

$$y_f = (0.072d + 65) \text{ mm}$$

$$914.0 \times 10^6 = 0.138 \times 20 \times 400d^2 + 0.446 \times 20 \times$$

$$(2750 - 400) \times (0.072d + 65) \times \left(d - \frac{0.072d + 65}{2}\right)$$

or, $d^2 + 494.15d - 374559.6 = 0$

$$d = 412.9 \text{ mm.}$$

Adopt $d = 415 \text{ mm and } D = 455 \text{ mm.}$

Reinforcement

For balanced section $C = T$

$$0.36f_{ck}b_w x_u + 0.446f_{ck}(b_f - b_w)D_f = 0.87f_y A_{st}$$

$$X_u = 0.48 \ d = 0.48 \times 425 = 204 \text{ mm}$$

$$0.36 \times 20 \times 400 \times 199.2 + 0.446 \times 20 \ (2750 - 400) \ 100$$

$$= 0.87 \times 415 \times A_{st}$$

or $A_{st} = 7394.8 \text{ mm}^2.$

Provide 28 mm ϕ bars. Number of bars $= \dfrac{7394.8}{\pi/4(28)^2} = 12$

Check in shear

$$\tau_v = \frac{510.66 \times 10^3}{400 \times 415} = 3.07 \text{ N/mm}^2 < \tau_{c \ max} \text{ (safe)}$$

$$p = \frac{100 \times \pi/4(28)^2 \times 12}{400 \times 415} = 4.45\%, \ \tau_c = 0.82 \text{ N/mm}^2$$

Hence $V_{us} = 510.66 - \dfrac{1}{1000} \times (0.82 \times 400 \times 415) = 374.54 \text{ kN.}$

Provide 12 mm ϕ four legged stirrups at the spacing

$$= \frac{0.87 \times 415 \times 4 \times \pi/4(12)^2 \times 415}{374.54 \times 10^3}$$

$$= 180.88 \text{ mm c/c; say 175 mm c/c}$$

Beam (B) (Fig. 7.89)

Total downward force $= 825 \times 2 + 1350 \times 2 = 4350 \text{ kN.}$

Total upward force $= 572.25 \times 2 + 584.5 \times 8 + 56 \times 0.4 \times 16.67$

$$= 6192.84 \text{ kN.}$$

To make upward and downward forces equal we increase the end column loads to 1746.42 kN.

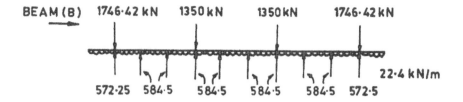

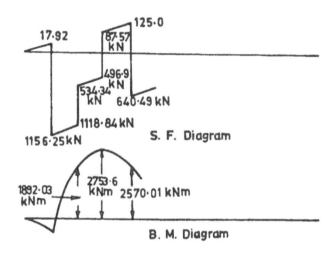

Fig. 7.89. Details of Design of Primary Beam B.

Design moment = 2753.6 kNm (hogging)
Maximum S.F. = 1156.25 kN.

The beam will be designed as inverted T-beam.
Design

$$b_f = \frac{l_0}{6} + b_w + 6D_f = \frac{0.7 \times 15 \times 10^3}{6} + 400 + 6 \times 100$$

$$= 2750 \text{ mm.}$$

(1) Assume that $D_f < 0.2d$,

$$M_R = 0.138 f_{ck} b_w d^2 + 0.446 f_{ck}(b_f - b_w)D_f \times \left(d - \frac{D_f}{2}\right)$$

$$2753.6 \times 10^6 = 0.138 \times 20 \times 400 \times d^2 + 0.446 \times 20 \times (2750 - 400) \, 100 \left(d - \frac{100}{2}\right)$$

or, $d^2 + 1898.7d - 2589138 = 0.$
It gives $d = 919.26$ mm
Adopt $d = 1000$ mm and $D = 1040$ mm.

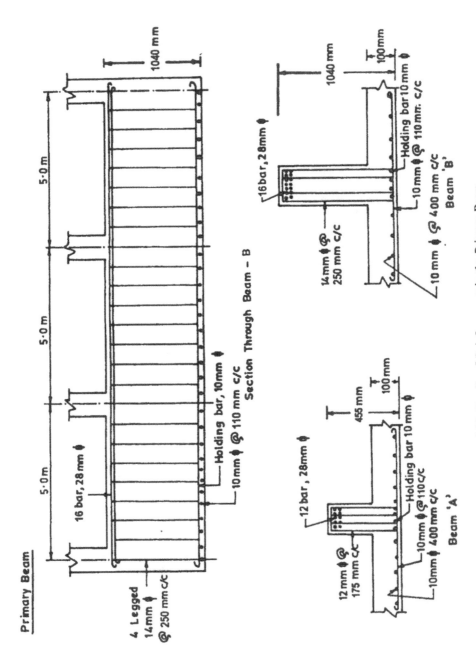

Fig. 7.90. Details of Reinforcement in the Primary Beam.

Reinforcement

For balanced section $\qquad C = T$

$$0.36 f_{ck} b_w x_u + 0.446 f_{ck} (b_f - b_w) D_f = 0.87 f_y A_{st}$$

$$X_u = 0.48 \ d = 516 \text{ mm}$$

$$0.36 \times 20(400) \times 480 + 0.446 \times 20(2750 - 400)100 = 0.87 \times 415 A_{st}$$

$$A_{st} = 9634.6 \text{ mm}^2$$

Provide 28 mm ϕ bars. Number of bars $= \dfrac{9634.6}{\pi / 4(28)^2} = 15.65$; say 16.

Check in shear

$$\tau_v = \frac{1156.25 \times 10^3}{400 \times 1000} = 2.89 \text{ N/mm}^2 < \tau_{c \ max}$$

$$p = \frac{100 \times \pi / 4(28)^2 \times 16}{400 \times 1000} = 2.46\%, \ \tau_c = 0.82 \text{ N/mm}^2.$$

Hence $\qquad V_{us} = 1156.25 \times 10^3 - 0.82 \times 400 \times 1000 = 872 \text{ kN}.$

Provide 14 mm ϕ four legged stirrups at spacing

$$= \frac{0.87 \times 415 \times 4 \times \pi / 4(14)^2 \times 1000}{872 \times 10^3}$$

$$= 254.8 \text{ mm c/c; say 250 mm}$$

The details of reinforcement is shown in Fig. 7.90.

Example 7.9

Design an annular raft for an overhead tank of capacity 2000 kL having staging height of 22.0 m. The depth of raft below ground surface is 2.5 m. The tank is situated in seismic zone III. The earthquake force results a moment of 8670 kN-m at the base of footing. The soil is cohesionless and the allowable soil pressure works out as 85 kN/m² limiting total settlement to 50 mm.

Solution

(*i*) $\qquad$ Load of water = 20000 kN

Load of superstructure and foundation = 10000 kN (assumed)

Weight of soil above foundation = 1000 kN (assumed)

Total load at the base of foundation = 31000 kN

Bearing area required = 31000/8.5 = 364.7 m²

Use annular raft with outer and inner diameters as 22.6 m and 6.4 m respectively.

Bearing area provided = (π/4) (22.6² – 6.4²) = 368.98 m²

Moment of inertia of raft = (π/64) (22.6⁴ – 6.4⁴) = 12723.4 m²

$$p_{max} = \frac{31000}{368.98} + \frac{8670 \times 11.3}{12723.4} = 91.6 \text{ kN/m}^2 \ [< 1.5 \times 85 = 127.5 \text{ kN/m}^2, \text{ safe}]$$

$$p_{min} = \frac{31000}{368.98} - \frac{8670 \times 11.3}{12723.4} = 76.3 \text{ kN/m}^2 \ [> 0.0, \text{ safe}]$$

(*ii*) p = net uniform pressure intensity on raft slab

$$= \frac{30000}{368.98} = 8.12 \text{ kN/m}^2$$

p' = pressure intensity due to moment

$$= \frac{8670 \times 11.3}{12723.4} = 7.7 \text{ kN/m}^2$$

Provide a ring beam of width 1000 mm (1.0 m) with its centre at 8.0 m from the centre of raft.

(*iii*) Design of annular raft

Refer Fig. 7.68

$\qquad$ $a = 11.3$ m, $\quad b = 3.2$ m, $\quad c = 8.0$ m. Therefore,

$\qquad$ $\alpha = b/a = 3.2/11.3 = 0.283$

$\qquad$ $\beta = c/a = 8.0/11.3 = 0.708$

$\qquad$ $\rho = r/a = r/11.3$

Using Eqs. 7.155 to 7.164 for p and Eqs. 7.165 to 7.172 for p', the total values of radial moment (M_r), circumferential moment (M_θ) and shear force Q_r works out as given below in Table 7.12.

Table 7.12 Values of M_r, M_θ and Q_r.

r (m)	M_r (kN-m)	M_θ (kN-m)	Q_r (kN)
3.2	25.0	−323.0	132.0
4.0	73.0	−234.0	166.0
6.0	358.0	−38.0	252.0
7.5	684.0	129.0	319.0
8.5	387.0	120.0	−286.0
10.0	69.0	90.0	−121.0
11.0	1.6	73.0	−26.0
11.3	−0.2	70.0	0.5

Therefore,

Maximum factored M_r = 684 × 1.5 $\quad$ = 1026 kN-m

Maximum factored M_θ = − 323 × 1.5 = − 485 kN-m

Maximum factored Q_r = 319 × 1.5 $\quad$ = 479 kN

$$\text{Depth of raft slab} = d = \sqrt{\frac{1026 \times 10^6}{0.138 \times 20 \times 1000}} = 610 \text{ mm}$$

Provide 680 mm (effective) and 720 mm (overall) thickness.

Radial Steel

$$0.87 \times 415 \times A_{st} \times 680 \left(1 - \frac{A_{st} \times 415}{1000 \times 680 \times 20}\right) = 1026 \times 10^6$$

or $\qquad 7.492 \, A^2_{st} - 245514 \, A_{st} + 1026 \times 10^6 = 0$

It gives, $\quad A_{st} = 4916 \;\; \text{mm}^2$

Use seven bars of 32 mm ϕ per metre. Provide these bars at spacing 150 mm c/c. Curtail the length of every three bars towards the edges due to variation of M_r.

Circumferential Steel

Maximum sagging moment $= 129 \times 1.5 = 194$ kN-m
It gives $A_{st} = 811 \;\; \text{mm}^2$.
Use 5 bars of 16 mm. *i.e.* at 200 mm c/c.
Maximum hogging moment $= 485$ kN-m
It gives $A_{st} = 2116 \;\; \text{mm}^2$.
Use 10 bars of 16 mm ϕ i.e. at 100 mm c/c.

Spacing of circumferential steel may be varied as per the variation of M_r given in Table 7.12.

Checking for Shear

Shear force at the location of maximum $M_r = 479$ kN

$$\frac{100 \, A_{st}}{bd} = \frac{100 \times 7 \times \dfrac{\pi}{4} \times 32^2}{1000 \times 680} = 0.827\%$$

From Table 5.2, $\qquad\qquad \tau_c = 0.58$ N/mm^2

Depth for safe shear stress $= 850$ mm (effective) and 900 mm (overall)

$$A_{st} = \frac{0.827}{100} \times 1000 \times 850 = 7029.5 \;\; \text{mm}^2$$

Use nine bars of 32 mm ϕ *i.e.* at spacing 110 mm c/c.

Revised circumferential steel

For sagging moment $= 642 \;\; \text{mm}^2$
Use four bars of 16 mm ϕ *i.e.* at 200 mm c/c.
For hogging moment $= 1646 \;\; \text{mm}^2$
Use nine bars of 16 mm ϕ at 120 mm c/c.

(*iv*) Design of ring beam

Assuming that the load on the ring beam is transferred through six equidistant columns, then from Table 7.11,

$$2\,\alpha = 60°$$

Maximum twisting moment will occur at $\phi = 12.75°$.

$$w = \text{load per unit length}$$

$$= \frac{30000}{\pi \times 16.0} = 597 \text{ kN/m}$$

Using Eqs. 7.173, 7.174 and 7.175, values of M_ϕ, $M_{t\phi}$ and $V_{u\phi}$ are obtained respectively. Knowing these values, equivalent longitudinal reinforcement M_{e1}, equivalent shear V_e and transverse reinforcement A_{sv} are obtained using Eqs. 5.40, 5.39 and 5.41 respectively.

The values so obtained are given in Table 7.13 for different values of ϕ.

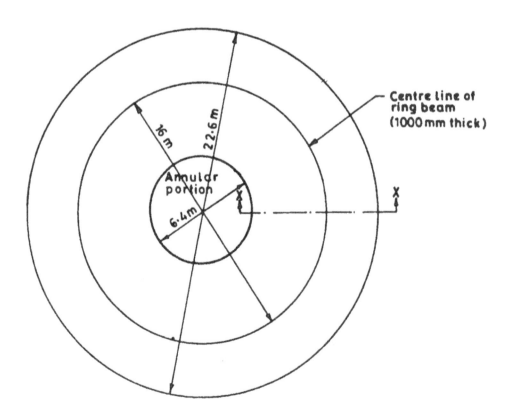

Fig. 7.91. Plan of Annular Ring Raft.

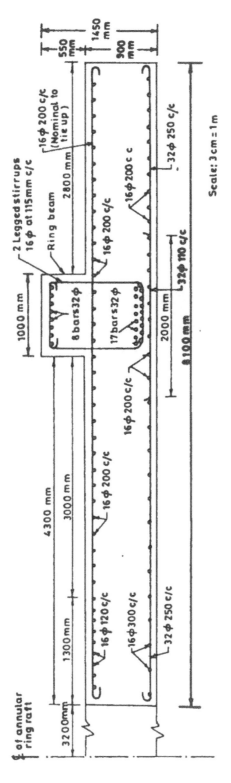

Fig. 7.92. Section at XX through Ring Beam.

Table 7.13 Values of Moment, Shear Force and Reinforcements.

ϕ	M_ϕ	M_{ψ}	M_e	Reinforcement		V_u	V_e	A_{sv} for
(Deg.)	(kN-m)	(kN-m)	(kN-m)	A_{st} (mm²)	No. 32 of mm ϕ bars	(kN)	(kN)	$S_v = 150$ mm (mm²)
0	−3559	0	−3559	13108	17	2500	2500	145
6.5	−1559	283	−1035	3226	4	1958	2410	160
12.75	0.0	360	667	2041	3	1437	2013	143
20	1185	282	1708	5520	7	833	1286	95
30	1800	0	1800	5849	8	0	0	0

From Table 7.13,

Maximum factored moment = 3559 × 1.5 = 5339 kN-m

Maximum factored shear force = 2500 × 1.5 = 3750 kN

$$\text{Thickness of ring beam} = \sqrt{\frac{5839 \times 10^6}{0.138 \times 20 \times 1000}} = 1390 \text{ mm}$$

Provide 1400 mm effective and 1450 mm overall

$$\frac{100 \, A_{st}}{bd} = \frac{100 \times 13108}{1000 \times 1400} = 0.93\%$$

From Table 5.2, $\tau_c = 0.60$ N/mm²

$$\tau_{ve} = \frac{3750 \times 1000}{1000 \times 1400} = 2.7 \text{ N/mm}^2 \; [< \tau_{c \, max} = 2.8 \text{ N/mm}^2, \text{ safe}].$$

As τ_{ve} exceeds τ_c, both longitudinal and transverse reinforcements will be provided.

$$A_{sv} = \left(\frac{\tau_{ve} - \tau_c}{0.87 \, f_{sy}}\right) b \cdot S_v = \frac{(2.7 - 0.60)}{0.87 \times 1400} \times 1000 \times 150$$

= 258 mm² (Governs as more from A_{sv} values given in Table 7.13).

If $S_v = 115$ mm, then $A_{sv} = 198$ mm². Therefore, use two legged stirrups of 16 mm ϕ at 115 mm c/c. Provided $A_{sv} = 201$ mm².

The plan of annular ring raft is shown in Fig. 7.91. The details of reinforcement on a section x–x is shown in Fig. 7.92.

REFERENCES

1. Agarwal, R.K. (1986), Behaviour of Shallow Foundations Subjected to Eccentric-inclined Loads, Ph.D. Thesis, University of Roorkee, Roorkee.
2. Ajayi, L.A. and Balogun, L.A. (1988), Penetration Testing in Tropical Lateritic and Residual Soils— Nigerian Experience, Proc. First Intern. Symp. on Penetration Testing, Orlando, pp. 315-328.
3. Balla, A. (1961), The Resistance to Breaking out of Mushroom Foundations for Pylons, 5th ICSMFE, Vol. 1, pp. 569-576.

4. Bjerrum, L. (1963a), Discussion to European Conference on Soil Mechanics and Foundation Engineering, *Wiesbadan*, Vol. II, p. 135.

5. Bjerrum, L. (1963b), Generally Krav till fundamentering av forskjelligebuggverk, tillate setninque, Den Norske Ingenirforening, Kurs, Fudamentering, Oslo.

6. Bowles, J.E. (1982), *Foundation Analysis and Design*, McGraw Hill, New York.

7. Button, S.J. (1953), *The Bearing Capacity of Footings on a Two-layer Cohesive Sub-soil*, 3rd, ICSMFE, Vol. 1, pp. 332-335.

8. Caquot, A. (1934), *Equilibre des Massifs a Frottement Interene*, Ganthiev-Villars, Paris.

9. Chu, K.H. and Afandi, O.F. (1966), Analysis of Circular and Annular Slabs for Chimney Foundations, Journal American Concrete Institute, Vol. 63, No. 12, pp. 1425-1447.

10. De Beer, E. and A. Martens (1957), Method of Computation of an Upper Limit for the Influence of Heterogenity of Sand Layers in Settlement of Bridges, *4th ICSMFE*, London, Vol. 1, pp. 275-281.

11. De Beer, E. and Vesic, A. (1958), Etude experimentale de la capacite portante du sable sons des fondations Directes etablies en surface, *Annales des Travaux Public de Delgigue*, 59 (33) : 5-58.

12. Fox, L. (1948), "The Mean Elastic Settlement of a Uniformly Loaded Arwa at a Depth Below the Ground Surface," *2nd ICSMFE*, Rotterdam, Vol. 2, pp. 236–246.

13. Giuliani, F. and F.L.G. Nicoll (1982), "New Analytical Correlations between SPT, Overburden Pressure and Relative Density," Proc. Second European Symp. on Penetration Testing, Amsterdam, Vol. 1, pp. 47–50.

14. Hansen, J.B. (1970), "A Revised and Extended Formula for Bearing Capacity," *Bull. No. 28*, Danish Geotechnical Institute, Copenhagen.

15. Hetenyi, M. (1946), "Beams on Elastic Foundation," Ann Arbor, Univ. of Michigan Press, p. 255.

16. IS : 1888 (1982), *Method of Load Test*.

17. IS : 1904 (1986), *Code of Practice for the Structural Safety of Buildings : Shallow Foundations*.

18. IS : 6403 (1981), *Code of Practice for the Determination of Bearing Capacity of Shallow Foundations*.

19. IS : 8009, Part I (1976), *Code of Practice for Calculations of Settlement of Foundations Subjected to Symmetrical Vertical Loads–Shallow Foundations*.

20. Krishna, J. and Jain, O.P. (1980), "Plain and Reinforced Concrete", Nemchand & Bros, Roorkee.

21. Meyerhof, G.G. (1951), "The Ultimate Bearing Capacity of Foundations," *Geotechnique*, Vol. 2, No. 4, pp. 301–331.

22. Meyerhof, G.G. (1953), "The Bearing Capacity of Foundations under Eccentric and Inclined Loads," *3rd ICSMFE*, Zurich, Vol. 1, pp. 669.

23. Meyerhof, G.G. (1956), "Penetration Tests and Bearing Capacity of Cohesionless Soils," *JSMFE*, ASCE, Vol. 82, SM1, pp. 1–19.

24. Meyerhof, G.G. (1963), "Some Recent Research on Bearing Capacity of Foundations," *Canadian Geotechnical Journal*, Vol. 1, No. 1, pp. 16–26.

25. Meyerhof, G.G. (1965), "Shallow Foundations," *J. SMFE*, ASCE, Vol. 91, SM2, pp. 21–31.

26. Meyerhof, G.G. (1974), "General report: Outside Europe," *Proc. Conf. on Penetration Testing*, Stockholm, Vol. 2, pp. 40–46.

27. Meyerhof, G.G and J.D. Brown (1967), "Discussion: Bearing Capacity of Footings on Layered Clays," *J. SMFD*, ASCE, Vol. 93, SM5, Part 1, pp. 361–363.

28. Meyerhof, G.G. and J.I. Adams (1968), "The ultimate uplift Capacity of Foundations", CGJ, Otawa, Vol. 5, No. 4, pp. 225–244.

29. Nixon, I.K. (1982), "Standard Penetration Test-State-of-the Art Report", Proc. secod Europeansymp. on Penetration Testing, Amsterdam, Vol. 1, pp. 3–24.

30. Parry, R.H.G. (1977), "Estimating Bearing Capacity of Sand from SPT Values," *J GED*, ASCE, Vol. 103, GT9, pp. 1014–1019.

31. Peck, R.B., W.E. Hanson and W.T. Thornburn (1974), *Foundation Engineering*, 2nd Edition, John Wiley and Sons, New York.

32. Ramasamy, G., A.S.R. Rao and P.S. Singh (1986), "Influence of Capillary Zone on Settlement of Footings on Sand," *Indian Geotechnical Journal*, Vol. 16, No. 4, pp. 383–389.

33. Reddy, A.S. and R.J. Srinavasan (1967), "Bearing Capacity of Footings on Layered Clays," *J. SMFD*, ASCE, Vol. 93, SM2, March, pp. 83–99.

34. Saran, S. and R.K. Agarwal (1991), "Bearing Capacity of Eccentrically-Obliquely Loaded Footings," *Journal of Geot. Engg.*, ASCE, Vol. 117, No. 11, Nov. 1991.

35. Saran, S., V.K. Sud and S.C. Handa (1989), "Bearing Capacity of Footings Adjacent to Slopes," *Journal of Geot. Engg.*, ASCE, Vol. 115, No. 4, April, pp. 553–573.

36. Schmertmann, J.H. (1970), "Static Cone to Compute Static Settlement Over Sand," *J. SMFD*, ASCE, Vol. 96, SM3, pp. 1011–1043.

37. Schmertmann, J.H. (1975), "The Measurement of In-situ Shear Strength," *7th PSC*, ASCE, Vol 2, pp. 57–138.

38. Schmermann, J.H., J.R. Hartman and P.R. Brown (1978), "Improved Strain Influence Factor Diagrams," *J. GED*, ASCE, 104 (GT8), pp. 1131–1135.

39. Skempton, A.W. (1951), "The Bearing Capacity of Clays," Building Research Congress, London, Institution of Civil Engineers, Division I: 180.

40. Skempton, A.W. and Bjerrum, L. (1957), "A Contribution to the Settlement Analysis of Foundations on Clay," *Geotechnique*, Vol. 7, p. 168.

41. Sud, V.K. (1985), "Behaviour of Footings on Cohesionless Soils," Ph.D. Thesis, University of Roorkee, Roorkee, India.

42. Teng, W.C. (1962), *Foundation Design*, Prentice-Hall Inc., Englewood Cliffs, New Jersey.

43. Terzaghi, K. (1943), *Theoretical Soil Mechanics*, John Wiley and Sons, New York.

44. Terzaghi, K. and Peck, R.B. (1948), Soil Mechanics in *Engineering Practice*, Ist Ed., John Wiley and Sons, New York.

45. Vesic, A.S. (1973), "Analysis of Ultimate Loads of Shallow Foundation," *J. SMFD*, ASCE, Vol. 95, SMI.

46. Winkler, E. (1967), "Die Lehre von Elastizitat und Festigkeit", Publisher Prague.

PRACTICE PROBLEMS

1. Give the various types of shallow foundation. Explain the situations in which each is preferred.

2. Give the general formula of obtaining bearing capacity of shallow foundation. Explain clearly each term.

3. Starting from fundamentals, derive the expressions for getting the bearing capacity of footings on layered clays. Describe the salient features of the results.

4. Give the procedure of computing bearing capacity, settlement and tilt of eccentrically-obliquely loaded footings.

5. How is bearing capacity, settlement and tilt of a footing situated on slope estimated?

6. Define coefficient of vertical subgrade reaction. How is it determined? What are the factors on which it depends? Discuss stepwise the procedure of designing combined footings using finite difference technique.

7. Design an isolated footing for a column of 500 mm × 500 mm size subjected to a vertical load of 1200 kN, moment of 200 kNm and shear load of 180 kN. The soil properties are as follows:

$$\phi = 36° \text{ and } \lambda_t = \text{kN/m}^3.$$

A plate load test was performed at the anticipated depth of foundation on a plate of size 600 mm × 600 mm and a pressure-settlement curve shown in Fig. 7.68 was obtained. Design the footing completely giving reinforcement drawings assuming that permissible values of settlement, tilt and lateral displacement are 50 mm, 1 degree and 25 mm respectively.

8. Design a strip footing, which is placed at a distance of 1.0 m from the edge of a slope making 25° with the horizontal. The footing is subjected to a load intensity of 250 kN/m². The slope consists of silty soils with following properties:

$$C = 30 \text{ kN/m}^2, \phi = 37° \text{ and } \lambda_t = 17 \text{ kN/m}^3.$$

Standard penetration tests were performed at the site. An average corrected value of standard penetration resistance was found as 20. Take the permissible value of settlement as 40 mm.

9. Design a combined footing for two columns 500 mm × 500 mm and 400 mm × 400 mm in sections carrying loads of 900 kN and 400 kN respectively spaced at 5.0 m c/c. Interpretation of some of the *in situ* tests suggested that the footing may be designed for an allowable soil pressure of 130 kN/m² and coefficient of subgrade reaction of 7.5×10^4 kN/m³. Illustrate the designs as follows:

   (i) Design of slab

   (ii) Computation of S.F. and B.M in beam by

(*a*) Conventional method

(*b*) Subgrade reaction method

(*c*) Finite difference method.

(*iii*) Discuss the reasons of differences in the values of shear forces and bending moments obtained by the three methods. Carry out the structural design adopting (*ii b*).

10. If in example No. 3 there is a restriction on extending the footing on the heavier column side by a distance not more than 250 mm, design the combined footing by conventional method as (*a*) trapezoidal footing, and (*b*) strap beam footing.

11. A building consists of 12 columns 400 mm × 400 mm sizes arranged in three rows of four each. The distance between the columns is 4.5 m each. The load carried by interior column is 500 kN each, that carried by exterior column is 600 kN each and that carried by interior column is 800 kN each. The allowable soil pressure is 60 kN/m². Design the raft foundation.

APPENDIX 7.1

Coefficients for obtaining deflections, moments, and shear for beam on an elastic foundation with finite length

$$\lambda L = 1.0$$
LOAD AT 0.0L

x/L	A'	B'	C'
0.0	4.0378	0.0000	0.0000
0.0	4.0378	0.0000	−1.0000
0.1	3.4176	−0.1616	−0.6272
0.2	2.8004	−0.2550	−0.3163
0.3	2.1883	−0.2923	−0.0669
0.4	1.5820	−0.2858	0.1214
0.6	0.3856	−0.1900	0.3179
0.8	−0.7957	−0.0631	0.2767
1.0	−1.9716	0.0000	0.0000

LOAD AT 0.1L

x/L	A'	B'	C'
0.0	3.4176	0.0000	0.0000
0.1	2.9295	0.0325	0.3173
0.1	2.9295	0.0325	−0.6826
0.2	2.4413	−0.0763	−0.4141
0.4	1.4706	−0.1572	−0.0230
0.6	0.5116	−0.1205	0.1749
0.8	−0.4379	−0.0427	0.1822
1.0	−1.3839	0.0000	0.0000

LOAD AT 0.2L

x/L	A'	B'	C'
0.0	2.8004	0.0000	0.0000
0.2	2.0816	0.1024	0.4882
0.2	2.0816	0.1024	−0.5117
0.4	1.3587	−0.0286	−0.1676
0.5	0.9976	−0.0497	0.0498
0.6	0.6375	−0.0509	0.0318
0.8	−0.0800	−0.0222	0.0875
1.0	−0.7957	0.0000	0.0000

LOAD AT 0.3L

x/L	A'	B'	C'
0.0	2.1883	0.0000	0.0000
0.2	1.7204	0.0813	0.3909
0.3	1.4845	0.1759	0.5511
0.3	1.4845	0.1759	−0.4488
0.4	1.2457	0.1001	−0.3122
0.6	0.7629	0.0186	−0.1113
0.8	0.2783	−0.0017	−0.0072
1.0	−0.2062	0.0000	0.0000

LOAD AT 0.4L

x/L	A'	B'	C'
0.0	1.5820	0.0000	0.0000
0.2	1.3587	0.0603	0.2941
0.4	1.1299	0.2292	0.5431
0.4	1.1299	0.2292	−0.4568
0.6	0.8875	0.0885	−0.2548
0.8	0.6375	0.0187	−0.1023
1.0	0.3856	0.0000	0.0000

LOAD AT 0.5L

x/L	A'	B'	C'
0.0	0.9814	0.0000	0.0000
0.2	0.9976	0.0394	0.1979
0.4	1.0102	0.1587	0.3988
0.5	1.0124	0.2486	0.4999
0.5	1.0124	0.2486	−0.4999
0.6	1.0102	0.1587	−0.3988
0.8	0.9976	0.0394	−0.1979
1.0	0.9814	0.0000	0.0000

λL = 2.0
LOAD AT 0.0L

x/L	A'	B'	C'
0.0	2.2751	0.0000	0.0000
0.0	2.2751	0.0000	−0.9999
0.1	1.8262	−0.3150	−0.5900
0.2	1.4015	−0.4837	−0.2678
0.3	1.0148	−0.5401	−0.0269
0.4	0.6708	−0.5150	0.1409
0.6	0.0990	−0.3273	0.2903
0.7	−0.1433	−0.2105	0.2855
0.8	−0.3688	−0.1051	0.2340
1.0	−0.7998	0.0000	0.0000

LOAD AT 0.1L

x/L	A'	B'	C'
0.0	1.8262	0.0000	0.0000
0.1	1.5284	0.0690	0.3354
0.1	1.5284	0.0690	-0.6645
0.2	1.2296	-0.1395	-0.3887
0.4	0.6724	-0.2845	-0.0109
0.6	0.2005	-0.2120	0.1608
0.7	-0.0080	-0.1424	0.1798
0.8	-0.2051	-0.0735 ·	0.1584
1.0	-0.5856	0.0000	0.0000

LOAD AT 0.2L

x/L	A'	B'	C'
0.0	1.4015	0.0000	0.0000
0.2	1.0526	0.2058	0.4915
0.2	1.0526	0.2058	-0.5084
0.4	0.6715	-0.0525	-0.1631
0.6	0.3015	-0.0956	0.0305
0.8	-0.0404	-0.0415	0.0820
1.0	-0.3688	0.0000	0.0000

LOAD AT 0.3L

x/L	A'	B'	C'
0.0	1.0148	0.0000	0.0000
0.2	0.8640	0.1545	0.3762
0.3	0.7739	0.3384	0.5403
0.3	0.7739	0.3384	-0.4596
0.4	0.6617	0.1841	-0.3156
0.6	0.4002	0.0239	-0.1023
0.7	0.2638	0.0028	-0.0359
0.8	0.1276	0.0084	-0.0031
0.9	0.0080	0.0039	-0.0151
1.0	0.1433	0.0000	0.0000

LOAD AT 0.4L

x/L	A'	B'	C'
0.0	0.6708	0.0000	0.0000
0.2	0.6715	0.1074	0.2688
0.4	0.6322	0.4289	0.5322
0.4	0.6322	0.4289	-0.4677
0.6	0.4928	0.1495	-0.2398
0.8	0.3015	0.0266	-0.0801
1.0	0.0990	0.0000	0.0000

LOAD AT 0.5L

x/L	A'	B'	C'
0.0	0.3675	0.0000	0.0000
0.2	0.4829	0.0650	0.1703
0.4	0.5733	0.2840	0.3832
0.5	0.5892	0.4605	0.4999
0.5	0.5892	0.4605	-0.4999
0.6	0.5733	0.2840	-0.3832
0.8	0.4829	0.0650	-0.1703
1.0	0.3675	0.0000	0.0000

$$\lambda L = 3.0$$

LOAD AT 0.0L

x/L	A'	B'	C'
0.0	2.0131	0.0000	0.0000
0.0	2.0131	0.0000	-0.9999
0.1	1.4283	-0.4366	-0.4849
0.2	0.9182	-0.6151	-0.1353
0.4	0.2260	-0.5434	0.1856
0.6	-0.0853	-0.2858	0.2128
0.8	-0.1874	-0.0771	0.1248
0.9	-0.2087	-0.0198	0.0652
1.0	-0.2260	0.0000	0.0000

LOAD AT 0.1L

x/L	A'	B'	C'
0.0	1.4283	0.0000	0.0000
0.1	1.1343	0.1197	0.3845
0.1	1.1343	0.1197	-0.6154
0.2	0.8336	-0.1564	-0.3203
0.3	0.5589	-0.2822	-0.1125
0.4	0.3335	-0.3066	0.0199
0.6	0.0393	-0.2037	0.1222
0.7	-0.0470	-0.1296	0.1203
0.8	-0.1099	-0.0636	0.0963
0.9	-1611	-0.0173	0.0555
1.0	-0.2087	0.0000	0.0000

LOAD AT 0.2L

x/L	A'	B'	C'
0.0	0.9182	0.0000	· 0.0000
0.2	0.7324	0.3101	0.4986
0.2	0.7324	0.3101	−0.5013
0.4	0.4359	−0.0614	−0.1479
0.6	0.1647	−0.1168	0.0283
0.8	−0.0295	−0.0488	0.0657
0.9	−0.1099	−0.0145	0.0446
1.0	−0.1874	0.0000	0.0000

LOAD AT 0.3L

x/L	A'	B'	C'
0.0	0.5164	0.0000	0.0000
0.2	0.5913	0.1960	0.3344
0.3	0.5869	0.4501	0.5125
0.3	0.5869	0.4501	−0.4874
0.4	0.5178	0.2089	−0.3202
0.6	0.2897	−0.0147	0.0755
0.8	0.0596	−0.0294	0.0280
0.9	−0.0470	−0.0105	0.0298
1.0	−0.1517	0.0000	0.0000

LOAD AT 0.4L

x/L	A'	B'	C'
0.0	0.2260	0.0000	0.0000
0.2	0.4359	0.1071	0.1996
0.4	0.5499	0.5230	0.5054
0.4	0.5499	0.5230	−0.4945
0.6	0.4081	0.1169	−0.1969
0.8	0.1647	−0.0007	−0.0237
0.9	0.0393	−0.0039	0.0068
1.0	−0.0853	0.0000	0.0000

LOAD AT 0.5L

x/L	A'	B'	C'
0.0	0.0327	0.0000	0.0000
0.2	0.2903	0.0428	0.0972
0.4	0.5045	0.2926	0.3407
0.5	0.5452	0.5442	0.4999
0.5	0.5452	0.5442	−0.4999
0.6	0.5045	0.2926	−0.3407
0.8	0.2903	0.0428	−0.0972
0.9	0.1623	0.0068	−0.0292
1.0	0.0327	0.0000	0.0000

$$\lambda L = 4.0$$
LOAD AT 0.0L

x/L	A'	B'	C'
0.0	2.0015	0.0000	0.0000
0.0	2.0015	0.0000	−1.0000
0.1	1.2351	−0.5218	−0.3559
0.2	0.6251	−0.6441	0.0095
0.4	−0.0157	−0.4040	0.2059
0.6	−0.1397	−0.1288	0.1222
0.8	−0.0805	−0.0140	0.0293
0.9	−0.0367	−0.0011	0.0058
1.0	0.0076	0.0000	0.0000

LOAD AT 0.1L

x/L	A'	B'	C'
0.0	1.2351	0.0000	0.0000
0.1	0.9451	0.1823	0.4366
0.1	0.9451	0.1823	−0.5633
0.2	0.6313	−0.1337	−0.2480
0.4	0.1546	−0.2441	0.0462
0.6	−0.0321	−0.1241	0.0791
0.8	−0.0562	−0.0281	0.0378
0.9	−0.0477	−0.0064	0.0169
1.0	−0.0367	0.0000	0.0000

LOAD AT 0.2L

x/L	A'	B'	C'
0.0	0.6251	0.0000	0.0000
0.2	0.6001	0.4020	0.5003
0.2	0.6001	0.4020	−0.4996
0.4	0.3216	−0.0642	−0.1216
0.6	0.0811	−0.1119	0.0298
0.8	−0.0277	−0.0410	0.0444
0.9	−0.0562	−0.0116	0.0273
1.0	−0.0805	0.0000	0.0000

LOAD AT 0.3L

x/L	A'	B'	C'
0.0	0.2158	0.0000	0.0000
0.2	0.4777	0.1972	0.2820
0.3	0.5333	0.5036	0.4877
0.3	0.5333	0.5036	−0.5122
0.4	0.4658	0.1769	−0.3089
0.5	0.3377	−0.0021	−0.1474
0.6	0.2075	−0.0731	0.0388
0.8	0.0140	−0.0488	0.0434
0.9	−0.0552	−0.0156	0.0349
1.0	−0.1190	0.0000	0.0000

LOAD AT 0.4L

x/L	A'	B'	C'
0.0	−0.0157	0.0000	0.0000
0.2	0.3216	0.0625	0.1233
0.4	0.5354	0.5263	0.4862
0.4	0.5354	0.5263	−0.5137
0.6	0.3461	0.0224	−0.1427
0.7	0.2075	−0.0438	−0.0322
0.8	0.0811	−0.0434	0.0249
0.9	−0.0321	−0.0166	0.0344
1.0	−0.1397	0.0000	0.0000

LOAD AT 0.5L

x/L	A'	B'	C'
0.0	−0.1180	0.0000	0.0000
0.1	0.0310	−0.0109	−0.0174
0.2	0.1832	−0.0118	0.0252
0.4	0.4747	0.2113	0.2932
0.5	0.5399	0.5269	0.4999
0.5	0.5399	0.5269	−0.4999
0.6	0.4747	0.2113	−0.2932
0.8	0.1832	−0.0118	−0.0252
1.0	−0.1180	0.0000	0.0000

Pile Foundation

8.1 INTRODUCTION

Piles are relatively long and slender members used to transfer loads through weak soil or water to deeper soil or rock strata having a high bearing capacity. They are also used in normal ground conditions to resist heavy uplift forces and horizontal forces as in foundations of multistoreyed buildings, transmission line towers, retaining walls, bridge abutments and dolphins (Fig. 8.1). Piles also provide a convenient method of construction of works over water, such as jetties or bidge piers. Piles are sometimes used to control earth movements.

When a pile passes through poor material and its tip penetrates a small distance into a stratum of good bearing capacity, it is called a bearing pile (Fig. 8.2a). When a pile is installed in a deep stratum of limited supporting ability and the pile develops its carrying capacity by friction on its sides, it is called friction pile (Fig. 8.2b). Many times, the load carrying capacity of piles results from the combination of point resistance and skin friction (Fig. 8.2c). Piles used to resist upward forces (Fig. 8.1b) or anchor down the structures subject to uplift, such as building with basements below groundwater table, or buried tanks are known as tension piles (Fig. 8.2d). In case of large lateral loads piles driven at an angle to the vertical are useful (Fig. 8.1c). These piles are known as batter piles. Short piles driven in loose granular soil to compact these deposits are called compaction piles (Fig. 8.2e). A pile passing through a swelling or consolidating layer may be subjected to skin friction in the upward direction as shown in Fig. 8.2f. Such piles are termed as negative skin friction piles.

A pile may be installed by any of the following methods:

(i) *Driven precast*—Precast unit, usually of timber, concrete, or steel is driven into the soil by the blows of a hammer.

(ii) *Driven cast-in-situ*—It is fromed by driving a tube with a closed end into the soil, and filling the tube with concrete. The tube may or may not be withdrawn.

(iii) *Bored cast-in-situ*—This is formed by boring a hole into the soil and filling it with concrete.

(iv) *Jacking*—Steel or concrete units are jacked into the soil.

Piles are usually installed in clusters/groups to provide foundations for structures. A pile foundation may have vertical piles (Fig. 8.1a) or batter piles or a combination of vertical and batter piles (Fig. 8.1c and d). A pile foundation may be subjected to a vertical load or a

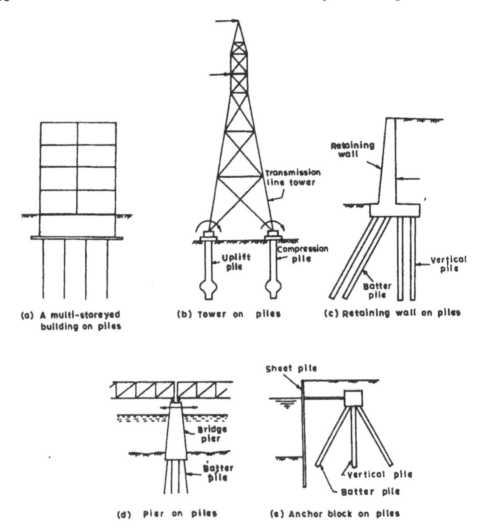

Fig. 8.1. Uses of Piles.

lateral load or a combination of vertical and lateral loads. Dynamic loads may act on piles during earthquake and under machine foundation.

8.2 CLASSIFICATION OF PILES BASED ON MATERIALS

Piles may be classifed according to the materials of which they are made. The main materials used in making piles are timber, concrete and steel.

Timber piles

Timber piles fall under the category of driven piles. These are made of tree trunks with branches carefully trimmed off. Timber should be of good quality and free of defects. The

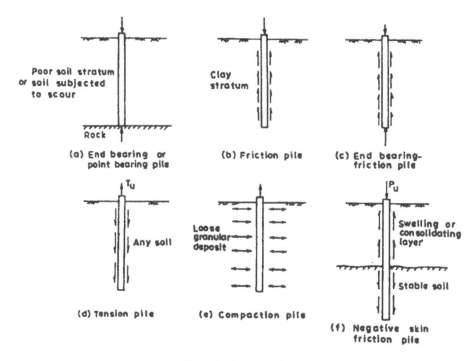

Fig. 8.2. Types of Piles.

length of the pile usually varies from 6.0 to 15.0 m. If greater lengths are required, they are to be spliced. The tip of the pile may be pointed or provided with a metal shoe for hard driving. An iron cap may also be provided to prevent damage during driving.

The design load of a timber pile usually does not exceed 300 kN. Piles above water table may be treated with creosote under pressure. Timber piles work out cheaper at places where timber is available in plenty.

Concrete piles

These are prismatic or circular in section either with taper or without taper. Concrete piles are either precast or cast-in-situ.

Precast concrete piles are constructed in a central casting yard to the specified length, cured, and then transported to the construction site. If space is available and a sufficient quantity of piles are needed a casting yard may be provided at the site to reduce transportation costs. Precast concrete piles are designed for bending stresses during pickup and transport to the site, for bending moment from lateral loads, and to provide sufficient resistance to vertical loads, and any tension forces that may develop during driving. Figure 8.3 illustrates typical bending moments developed during pick up, depending on the location of pick up point. In this figure w represents the weight per metre of pile.

Precast concrete piles are used in lengths up to about 20 m. The optimum load range is 300 kN to 600 kN. Piles may also be prestressed. Maximum load on a prestressed concrete pile may go up to 2000 kN.

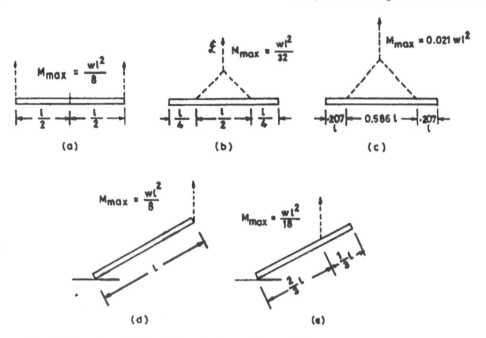

Fig. 8.3. Location of Pick-up Points for Precast Piles with the Resulting Bending Moments.

Cast-in-situ piles are concrete piles. As mentioned earlier cast-in-situ piles are of two types: (*i*) cased piles, and (*ii*) uncased piles. Different commercial firms have developed different techniques of installation of both the types of piles. Raymond pile is a patented cased pile of Raymond concrete pile company of USA. Franki pile is an uncased pile and patented by Braithwaite and Jessop Construction Co. Ltd., Kolkata. Under-reamed pile is also an example of uncased cast-in-situ pile.

Steel piles

A steel pile may be a pipe, a rolled section or a fabricated shape. The most commonly used steel piles are H-piles and pipe-piles. Pipe piles are made of seamless or welded pipes and are usually filled with concrete. These pipe piles may be driven close-ended or open-ended. The open-end piles may be driven to the required depth and the close-ended piles are formed by fixing a drive point to the pile tip.

Steel piles required heavy equipment for driving. H-piles are specially proportioned to withstand heavy impact stresses during hard driving.

Steel piles can be used to carry heavy loads. Splicing of steel piles is easy. The steel piles are likely to be affected by corrosive action and so will require painting or encasement in concrete to resist corrosion.

8.3 BEARING CAPACITY OF PILES

The bearing capacity of a single pile depends upon (*a*) type, size and length of pile, (*b*) type of soil, and (*c*) the method of installation.

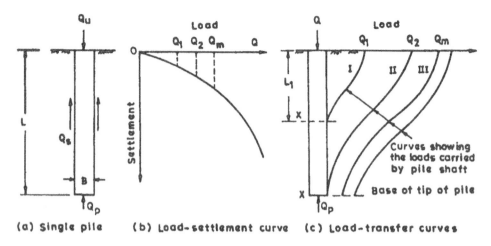

Fig. 8.4. Load Transfer Mechanism.

Consider a pile (Fig. 8.4a) loaded to failure by gradually increasing the load on the top. A load-settlement curve shown in Fig. 8.4b can be obtained by measuring the settlement of the pile top at every stage of loading. The transfer of the load takes place as explained below:

(*i*) When a load Q_1 acts on the pile head, the axial load at the ground level is also Q_1 but at level L_1, the axial load is zero. The total load Q_1 is distributed as friction load within the length of the pile L_1. The lower section of the pile is not affected by this load (curve I of Fig. 8.4c).

(*ii*) As the load is increased to Q_2, the axial load at the base of tip of the pile is just zero. The total load Q_2 is distributed as friction load along the whole length of pile L (curve II of Fig. 8.4c).

(*iii*) If the load put on the pile is greater than Q_2, a part of this load is transferred to the soil at the base as point load and the rest is transferred to the soil surrounding the pile. With the increase of the load Q on the top, both the friction and point loads go on increasing. The friction load attains an ultimate value Q_s at a particular load level, say Q_m, at the top, and any further increment of load added to Q_m will not increase the values of Q_s.

(*iv*) On further increasing the load on the pile top, the point load still goes on increasing and attains a value Q_p at which the soil fails by punching shear.

The load placed at the pile top which mobilises full friction load Q_s and full point bearing load Q_p is the ultimate bearing capacity of the pile, Q_u. Thus,

$$Q_u = Q_p + Q_s \qquad\qquad ...\ (8.1)$$

Methods available to estimate the ultimate capacity of a single pile in compression can be grouped in following categories:

(*i*) Static analysis,
(*ii*) Dynamic analysis,
(*iii*) Static-in-situ test.

The ultimate pile capacity can be determined by any of the above methods. Using an adequate factor of safety, the allowable load on the piles from shear considerations is obtained.

Static-in-situ test, popularly known as pile load test, is the only direct method for determining the allowable load on piles. It is considered to be the most reliable of all the approaches, primarily due to the fact that it is an in-situ test performed on pile of prototype pile dimension. Pile load tests are very useful in cohesionless soils. However, in case of cohesive soils, the data from pile load test should be used with caution on account of disturbance due to pile driving, development of pore pressure and the inadequate time allowed for consolidation settlement.

Pile load test is a costly test and is used to confirm whether the actual pile installed in the field can take the load predicted by static or dynamic analysis.

Of the static and dynamic analysis, static methods are more common in use for determining ultimate bearing capacity of piles. Dynamic formulae are sometimes used for predicting bearing capacity of driven piles in cohesionless soils.

In the following sections, approaches common in use to determine the bearing capacity of pile have been discussed, keeping in view the method of installation and type of soil.

8.4 ULTIMATE BEARING CAPACITY OF A DRIVEN PILE IN COHESIONLESS SOILS

In the case of a driven pile in cohesionless soil, the soil gets compacted up to a distance of 3.5 times the diameter of the pile (Fig. 8.5). As a result of this, there will be an increase in the value of ϕ i.e. angle of internal friction within this zone. According to Kishida (1967),

$$\phi_2 = \frac{\phi_1 + 40}{2} \qquad \qquad \dots (8.2)$$

where ϕ_2 = maximum value of ϕ at the tip of pile, and

   ϕ_1 = value of ϕ before installation of pile.

However, it is recommended that the value of ϕ to be used in the analysis should be that which existed before pile driving (Tomlinson, 1986).

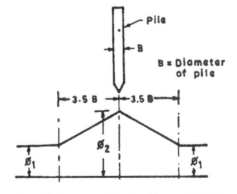

Fig. 8.5. Influence Zone of Changes of Angle of Internal Friction with Pile Driving.

A pile should be driven at least five times the diameter into the bearing stratum.

Standard penetration test and static cone penetration test are the two important tests by using which the bearing capacity of driven pile in cohesionless soil can be predicted with reasonable accuracy.

Using standard penetration test data

The value of ultimate capacity of a driven pile in cohesionless soil can be obtained by the procedure described below:

(i) Perform standard penetration tests up to the expected depth of penetration of pile. Let N_1 represent the average of observed N-values along the length of pile and N_2 is the observed N-value near the pile tip. It may be noted that the N-values are not corrected for effective overburden pressure.

(ii) The point bearing load (or base resistance), Q_p is given by:

$$Q_p = A_b \cdot \sigma'_v \, (N_q - 1) \qquad \qquad \text{... (8.3)}$$

where A_b = area of cross-section of pile tip, m²

σ'_v = effective overburden pressure at the pile tip, kN/m², and

N_q = bearing capacity factor.

For determining N_q, firstly using Fig. 7.21, the value of ϕ is obtained corresponding to SPT value N_2. The value of N_q is then read from Fig. 8.6 in case the pile is penetrated into the bearing stratum greater than five times the pile diameter. For lesser penetration, it is recommended to get the value of N_q-factor using Fig. 8.7.

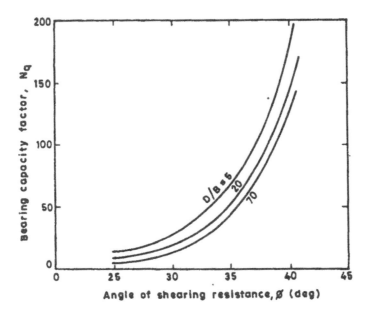

Fig. 8.6. Bearing Capacity Factor N_q (Berezantesv, 1961).

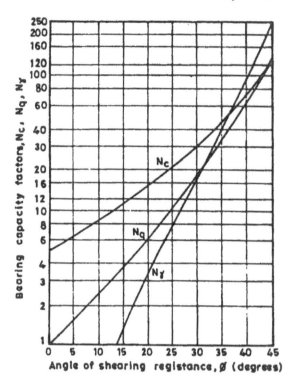

Fig. 8.7. Bearing Capacity Factor N_q (Hansen, 1961).

As per Tomlinson (1986), the maximum unit base resistance ($q_p = Q_p/A_b$) is normally limited to 11000 kN/m^2 whatever might be the penetration depth.

(*iii*) The side friction resistance, Q_s is given by:

$$Q_s = f \cdot A_s$$

$$= K \cdot \sigma'_v \tan \delta \cdot A_s \qquad \qquad \dots (8.4)$$

where f = unit skin friction, kN/m^2,

K = earth pressure coefficient,

σ'_v = average effective overburden pressure over embedded depth of pile, kN/m^2,

δ = angle of friction between pile material and soil, and

A_s = embedded surface area of pile, m^2.

For determining K and δ, firstly using Fig. 7.21, the value of ϕ is obtained for SPT value N_1. Broms (1966) has suggested the values of K and δ as shown in Table 8.1.

Equation (8.4) implies that the unit skin-friction (f) increases continuously with increasing depth of embedment of pile. Vesic (1970) showed that at some penetration depth between 10 and 20 pile diameters, a peak value of unit skin friction is reached which is not exceeded at greater penetration depths. Therefore, the use of Eq. (8.4) is recommended if the depth of pile penetration is limited to 20 diameter. For piles driven deeper than 20 pile diameters, Tomlinson (1986) suggested the value of unit skin friction as given in Table 8.2.

Table 8.1. Values of K and δ.

Pile Material	δ	Values of K	
	(Deg)	Low Relative Density $\phi < 29°$	High Relative Density $\phi > 36°$
Steel	20	0.5	1.0
Concrete	3/4 ϕ	1.0	2.0
Wood	2/3 ϕ	1.5	4.0

Using static cone penetration test data

The procedure of obtaining bearing capacity of pile using static cone penetration test is described in the following steps:

(i) Make the plot of cone resistance (q_c) versus depth.

Table 8.2. Values of Unit Skin Friction ($L_f/B > 20$).*

Relative Density %	Unit Skin Friction, f (kN/m²)
Less than 35 (loose)	10
35-65 (medium dense)	10-25
65-85 (dense)	25-70

*L_f = Embedded length of pile, B = Diameter of pile

(ii) Determine the average cone resistance over the range of depth as shown in Fig. 8.8. Let it be q_{c1}, kN/m².

The ultimate point resistance is given by

$$Q_p = q_{c1} \cdot A_b \text{ (kN)} \qquad \qquad ... (8.5)$$

(iii) Determine the average cone resistance along the pile shaft. Let it be q_{c2}, kN/m².

The ultimate side resistance of a driven concrete pile is given by

$$Q_s = \frac{q_{c2}}{200} \cdot A_s \text{ (kN)} \qquad \qquad ... (8.6)$$

For driven steel H-piles,

$$Q_s = \frac{q_{c2}}{400} \cdot A_s \text{ (kN)} \qquad \qquad ... (8.7)$$

Working load of a pile can be used using a factor of safety of 2.5. Therefore

$$\text{Working load} = \frac{Q_p + Q_s}{2.5} \qquad \qquad ... (8.8)$$

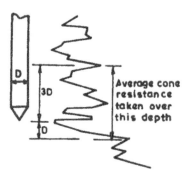

Fig. 8.8. Zone Considered for Computing end Resistance of Pile from SCPT.

8.5 ULTIMATE BEARING CAPACITY OF A BORED AND CAST-IN-SITU PILE IN COHESIONLESS SOILS

In all cases of bored piles formed in cohesionless soils, the soil will get loosened due to boring even though it may be initially in dense or medium dense state. Equations (8.3) and (8.4) may be used for computing point bearing and skin friction resistance adopting the value of ϕ corresponding to loose condition ($\phi < 29°$). However, if the piles are installed by rotary drilling under a bentonite slurry, computations of both the skin friction and point bearing resistance may be done for the ϕ corresponding to undisturbed conditions (Fleming and Sliwinski, 1977).

8.6 ULTIMATE BEARING CAPACITY OF DRIVEN AND CAST-IN-SITU PILES IN COHESIONLESS SOILS

Driven and cast-in-situ piles are formed by driving a tube into the ground. The bearing capacity of pile in which the tube is left into the ground can be obtained by the procedure given in Sec. 8.4 for driven piles. In the case, the tube is withdrawn with the placing of concrete, the procedure given in Sec. 8.5 for bored and cast-in-situ piles may be adopted for computing bearing capacity.

8.7 ULTIMATE BEARING CAPACITY OF DRIVEN PILES IN COHESIVE SOILS

The ultimate bearing capacity of a driven pile in cohesive soil is given by:

$$Q_u = N_c \cdot c_b \cdot A_b + \alpha \, \bar{c}_u \, A_s \qquad \qquad \dots (8.9)$$

where
N_c = bearing capacity factor = 9.0,
c_b = undisturbed shear strength of clay at the pile tip,
A_b = area of cross-section of pile tip,
α = adhesion factor,
$\bar{c}_u$ = average undrained shear strength of clay adjacent to shaft, and
A_s = embedded surface area of pile.

Driven piles cause displacement of soil. Due to this pore pressures get increased which reduces the effective stress and, therefore, skin friction capacity will be smaller. Keeping this in view, Tomlinson (1986) recommneded the charts shown in Fig. 8.9 to get the value of adhesion factor α. These charts are essentially applicable to piles carrying light to moderate loading driven to a relatively shallow penetration into the bearing stratum. Where heavy loads are carried, the piles may be driven very deeply into the bearing stratum. For such cases the ultimate bearing capacity of pile is given by:

$$Q_u = N_c \cdot c_b \cdot A_b + F\alpha_p \, \bar{c}_u \, A_s \qquad\qquad \dots (8.10)$$

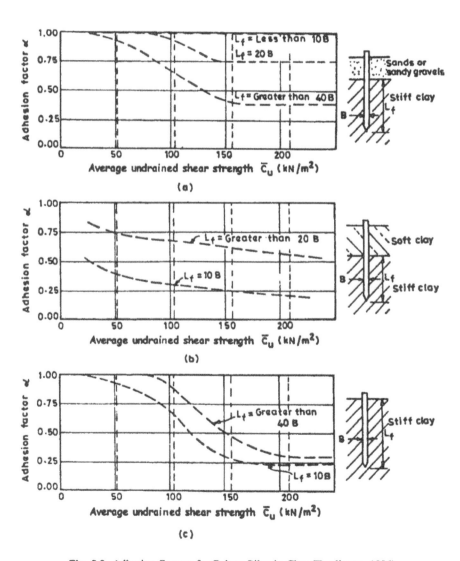

Fig. 8.9. Adhesion Factors for Driven Piles in Clay (Tomlinson, 1986).

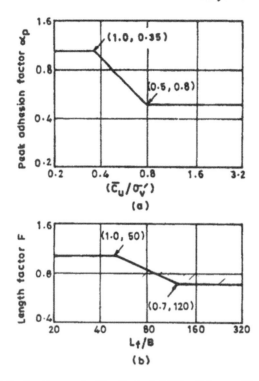

Fig. 8.10. Adhesion Factors for Heavily-loaded Piles Driven to Deep Penetration.

where N_c, c_b, A_b, $\bar{c}_u$ and A_s have the same meaning as in Eq. (8.9). α_p is peak adhesion factor depending on the ratio of average undrained shear strength and average effective overburden pressure $(\bar{c}_u/\sigma_v')$. This is shown in Fig. 8.10a. F is a factor depending on the slenderness ratio of pile (L/B). This is shown in Fig. 8.10b.

8.8   ULTIMATE BEARING CAPACITY OF BORED AND CAST-IN-SITU PILES IN COHESIVE SOILS

Ultimate bearing capacity of bored and cast-in-situ piles can be obtained using Eq. (8.9) by taking the value of adhesion factor α as 0.45 in medium clays and 0.3 in highly fissured clays.

8.9   ULTIMATE BEARING CAPACITY OF PILES IN c-ϕ SOILS

The point bearing capacity of a pile in $c - \phi$ soil is given by:

$$Q_p = A_b \left[1.2\ cN_c + \gamma L_f\ (N_q - 1) + 0.4\ \gamma\ BN_\gamma \right] \qquad \ldots (8.11)$$

where A_b = area of cross-section of pile tip,
            c = unit cohesion,
            γ = unit weight of soil,
            B = diameter of pile,

N_c, N_q, N_γ = bearing capacity factors (Table 7.2), and

L_f = embedded length of pile.

The side soil resistance can be computed as the sum of adhesion and skin friction given by the following equation:

$$Q_s = A_s[K\bar{\sigma}_v \tan \delta + \alpha \bar{c}_u] \qquad \qquad ... (8.12)$$

Values of K and α should be appropriately selected keeping in view the method of installation of pile and density/stiffness of the soil.

If the pile toe is terminated in a layer of stiff clay or dense sand underlain by soft clay or loose sand there is a risk of pile punching through the weak layer. Meyerhof (1976) suggested the following equation to compute the base resistance of the pile in the strong layer where the thickness H between the pile tip, and the top of the weak layer is less than the critical thickness of about 10 B (Fig. 8.11):

$$Q_p = Q_0 + \left(\frac{Q_1 - Q_0}{10B}\right) H < Q_1 \qquad \qquad ... (8.13)$$

where Q_0 and Q_1 are the ultimate base resistances of the pile in lower weak and upper strong layers respectively.

8.10 BEARING CAPACITY OF PILE FROM DYNAMIC ANALYSIS

The resistance offered by the soil to penetration of pile during driving gives an indication of its bearing capacity. A pile which meets greater resistance during driving is capable of taking more load. Dynamic pile formulae have been derived on the principle that the energy supplied to the pile is utilized in overcoming the resistance of the ground to the penetration of the pile. Allowance is made for losses of energy due to elastic compression of pile cap, pile and

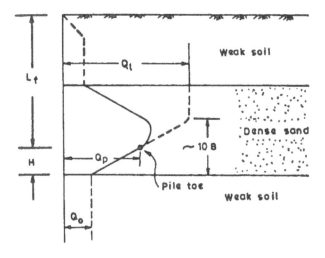

Fig. 8.11. End Bearing Resistance of Piles in Layered Soils (Meyerhof, 1976)

subsoil and also the losses due to the inertia of the pile. Some of these formulae are discussed in the following sections. Figure 8.12 shows a schematic diagram of pile driving.

Engineering News formula

The Engineering News formula is the simplest and most commonly used dynamic pile formula. Using a factor of safety of 6, the allowable pile capacity Q_a is given by Eq. (8.14):

$$Q_a = \frac{166.7\,WH}{S+C} \qquad\qquad ...\ (8.14)$$

where Q_a = allowable pile load in kN,

   W = weight of hammer in kN,

   H = drop of hammer in m,

   S = average penetration of pile per blow for the final 150 mm of driving in mm, and

   C = additional penetration of the pile tip that would have occurred if there were no energy losses

   = 25.4 mm for piles driven with drop hammer

   = 2.54 mm for piles driven with steam hammer.

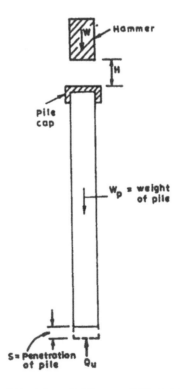

Fig. 8.12. Schematic Diagram of Pile Driving.

Modified Hiley formula

Modified Hiley formula is considered to be superior to the Engineering News formula, as it takes into account various energy losses during driving in a more realistic manner. According to this formula, the ultimate capacity of pile is given by

$$Q_u = \frac{W.H.\eta}{\left[S + \frac{1}{2}(C_1 + C_2 + C_3)\right]} \qquad \dots (8.15)$$

where Q_u = ultimate pile capacity in kN. The safe load is estimated by dividing it by a factor of safety of 2.5.

W = weight of hammer in kN,

H = height of free fall of hammer in mm taken at its full value for trigger operated drop hammers, 80 per cent of the fall of normally proportioned winch-operated drop hammers, and 90 per cent of the stroke for single acting hammers. In case of McKiernan-Terry type of double acting hammers, 90 per cent of maker's rated energy in kN mm per blow be used for WH. The set should be taken corresponding to the maximum speed of the hammer.

η = efficiency of the blow that represents the ratio of energy after impact of the striking energy of the ram.

$$= \frac{W + Pe^2}{W + P} \quad \text{(for } W > P \cdot e)$$

$$= \frac{W + Pe^2}{W + P} - \left[\frac{W - Pe}{W + P}\right]^2 \quad \text{(for } W < P \cdot e)$$

P = weight of pile, anvil, helmet and follower if any in kN.

e = coefficient of restitution of the materials under impact. IS : 2911 (Part 1) – 1979 recommended the value of e as given in Table 8.3.

Table 8.3. Values of e.

S. No.	Description	e
1.	For steel ram of double-acting hammer striking on steel anvil and driving reinforced concrete pile	0.50
2.	For cast-iron ram of single acting or drop hammer strking on head of reinforced concrete pile	0.40
3.	For single-acting or drop hammer striking a well-conditioned driving cap and helmet with hard wood dolly while driving reinforced concrete piles or directly on head of timber pile	0.25
4.	For a deteriorated condition of the head of pile or of dolly	0.00

The values of η in relation to e and to the ratio of P/W as recommended by IS : 2911 (Part I) – 1979 are tabulated in Table 8.4.

Table 8.4. Values of η in relation to e and P/W.

P/W	$e = 0.5$	$e = 0.4$	$e = 0.32$	$e = 0.25$	$e = 0$
0.5	0.75	0.72	0.70	0.69	0.67
1.0	0.63	0.58	0.55	0.53	0.50
1.5	0.55	0.50	0.47	0.44	0.44
2.0	0.50	0.44	0.40	0.37	0.33
2.5	0.45	0.40	0.36	0.33	0.28
3.0	0.42	0.36	0.33	0.30	0.25
3.5	0.39	0.33	0.30	0.27	0.22
4.0	0.36	0.31	0.28	0.25	0.20
5.0	0.31	0.27	0.24	0.21	0.16
6.0	0.27	0.24	0.21	0.19	0.14
7.0	0.24	0.21	0.19	0.17	0.12
8.0	0.22	0.20	0.17	0.15	0.11

Where the pile finds refusal in rock, $0.5\,P$ should be substituted for P in the above expressions for η.

S = final set or penetration per blow in mm,
C_1 = temporary compression of dolly and packing,
$\quad$ = $0.177\,Q_u/A$, where the driving is without dolly or helmet and cushion about 25 mm thick,
$\quad$ = $0.905\,Q_u/A$, where the driving is with short dolly upto 600 mm long, helmet and cushion about 75 mm thick,
C_2 = temporary compression of pile,
$\quad$ = $0.0657\,Q_u L/A$,
C_3 = temporary compression of ground
$\quad$ = $0.355\,Q_u/A$,
L = length of pile in metres, and
A = area of cross-section of pile in mm^2.

Comments on the use of dynamic formulae

Pile driving formulae in general have limited value in piling work mainly as the static resistance is not necessarily equal to the dynamic resistance. Also, the assumption of free impact between two bodies is not justified since the pile is far from being a free body. Also, the pile is not a massive body but is a slender body.

Pile driving formulae are based on the premise that all blows of a given hammer deliver the same energy. However, recent investigations (Housel, 1965, Tavenas and Audy, 1972) showed that the energy per blow delivered to a pile by the same driving equipment varies by ± 70 per cent of the average energy, and that for steam or diesel hammers, the average energy is generally 30 to 60 per cent lower than the rated energy.

These formulae also assume that the bearing capactiy of a driven pile is a direct function of the energy delivered to it during the last blow of driving. However, it has been reported

(Canadian Manual, 1975) that the bearing capacity of a pile is related more to the distribution of the energy with time at and after the impact and also by the magnitude and duration of the peak impact force.

Dynamic formulae could be used with more confidence in free draining materials such as coarse sand. If the pile is driven to saturated loose fine sand and silt, there is every possibility of development of liquefaction which reduces the bearing capacity of the pile.

Dynamic formulae are not recommended for computing allowable loads of piles driven into cohesive soils. In cohesive soils, the resistance to driving increases through the sudden increase in stress in porewater and decreases because of the decreased value of the internal friction between soil and pile because of porewater. These two oppositely directed forces do not lend themselves analytical treatment and as such the dynamic penetration resistance to pile driving has no relationship to static bearing capacity. There is another effect of pile driving in cohesive soils. During driving, the soil gets remoulded and the shear strength of soil gets reduced considerably. Though there will be a regaining of shear strength after a lapse of some days after the driving operation, this will not be reflected in the resistance value obtained from the dynamic formulae.

Therefore, the dynamic formulae for estimating the pile capacity should be used with caution.

8.11 PILE LOAD TEST

General

Pile load test is the most reliable and acceptable method of determining pile load capacity. It may be performed either on a driven pile or a cast-in-situ pile. The purpose of a load test is to establish the load-settlement relationship of the pile under compression load. Load test may be made either on single pile or a group of piles. Load test on a pile group is very costly and may be undertaken only in very important projects.

Two categories of tests on piles, namely, initial test and routine test are usually carried out. Initial test should be carried out on test piles to estimate the allowable load, or to predict the settlement at a working load. Routine test is carried out as a check on working piles and to assess the displacement corresponding to the working load. A test pile is one pile which is used in load test only and does not carry the load of superstructure. The minimum test load on such piles should be twice the safe load (safe load calculated for this purpose using static formula) or the load at which the total settlement attains a value of 10 per cent of pile diameter in case of a single pile and 40 mm in case of a pile group. A working pile is a pile which is driven or cast-in-situ along with other piles to carry load from the superstructure. The test load on such piles should be up to one-and-a-half times the safe load or up to the load at which total settlement attains a value of 12 mm for a single pile and 40 mm for a group of piles, whichever is earlier.

Where specific information about the subsoil strata and past experience is lacking for a sizable work involving more than 200 piles, there should be a minimum of two initial tests. For routine test, the minimum number of tests should be half per cent of the piles used. The number of tests to be conducted may vary up to two per cent or more, depending upon the nature of soil strata and importance of the structure.

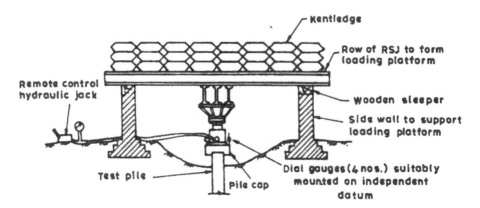

Fig. 8.13. Typical Pile Load Test Set-up.

Load tests may be of two types, namely (*i*) Progressive loading test, and (*ii*) cyclic loading test.

Figure 8.13 shows a typical test set up for a pile load test.

Progressive loading test

IS : 2911 Part IV-(1985) details the procedure for carrying out the load tests and assessing the allowable load. According to the code, the test shall be carried out by applying a series of vertical downward loads on a R.C.C. cap over the pile. The load shall preferably be applied by means of a remote controlled hydraulic jack taking reaction against a loaded platform. The test load shall be applied in increments of about 20 per cent of the assumed safe load. Settlement shall be recorded with at least three dial gauges of sensitivity 0.02 mm, suitably mounted on independent datum bars at least five times the pile diameter away with a

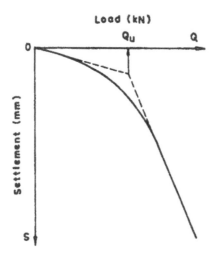

Fig. 8.14. Typical Load-settlement Curve Obtained from Pile Load Test.

minimum distance of 1.5 m. Each stage of loading shall be maintained till the rate of movement of pile top is not more than 0.1 mm per hour in a sandy soil and 0.02 mm per hour in a clayey soil or for a maximum of two hours, whichever is later. Figure 8.14 shows a typical load-settlement curve obtained from a pile load test. The ultimate capacity of the pile, Q_u may be obtained by intersection-tangent method as illustrated in Fig. 8.14.

The allowable load on a single pile shall be the lesser of the following:

(a) Two-third of the final load at which the total settlement attains a value of 12 mm, unless it is specified that a total settlement different from 12 mm is permissible in a given case on the basis of the nature and type of structure. In the latter case, the allowable load shall correspond to the actual permissible total settlement.

(b) Fifty per cent of the final load at which the total settlement equals 10 per cent of the pile diameter in case of uniform diameter piles and 7.5 per cent of bulb diameter in case of underreamed piles.

The allowable load on a group of piles shall be the lesser of the following:

(a) Final load at which the total settlement attains a value of 25 mm, unless specified in a given case on the basis of the nature and type of structure.

(b) Two-third of the final load at which the total settlement attains a value of 40 mm.

Cyclic load test on pile

The vertical cyclic pile load test is specially carried out when it is required to separate the pile load into skin friction and point bearing. The load shall be applied in increments of about 20 per cent of the estimated safe load. Loading and unloading should be carried out alternately at each stage and the elastic rebound in the pile is measured.

Van Veele (1957) postulated that the point load Q_p increases linearly with the elastic compression of the soil at the base. It is this principle that is used in separating the frictional load from the point load.

The total settlement S_t of pile obtained from a pile load test comprises of two components, namely, elastic settlement, S_e and plastic settlement S_p.

$$S_t = S_e + S_p \qquad \dots (8.16)$$

The elastic settlement, S_e is due to the elastic recovery of the pile material and the elastic recovery of soil at the base of the pile, S_{es}.

In the cyclic loading procedure of the pile load test, it is easy to obtain the elastic settlement and the plastic settlement at every stage of loading (Fig. 8.15).

If the load on the pile is greater than a particular value Q_m, a part of the load will be transferred to the soil at the base of the pile and the rest to the soil around the pile. The load transferred to the base will compress the soil at the base of the pile. Hence for $Q > Q_m$

$$Q = Q_p + Q_s$$

The total settlement of the pile, S_t at any load level Q can be written as

$$S_t = \Delta L + S_b \qquad \dots (8.17)$$

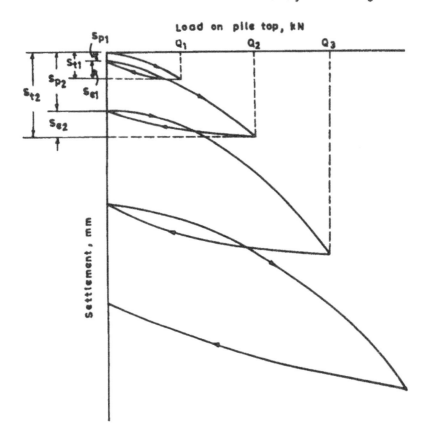

Fig. 8.15. Typical Load-settlement Curve Obtained from a Cyclic Pile Load Test.

where ΔL = compression of the pile, and

                S_b = compression of the soil at the base.

The compression of the soil at the base consists of elastic compression of the soil at base (S_{eb}) and plastic compression of the soil at base (S_{pb}).

Therefore $S_b = s_{eb} + S_{pb}$ ... (8.18)

Equations (8.16), (8.17) and (8.18) give

$$S_e + S_p = \Delta L + S_{eb} + S_{pb} \qquad\qquad\qquad ... (8.19)$$

or $$S_{eb} = (S_p - S_{pb}) + (S_e - \Delta L) \qquad\qquad ... (8.20)$$

The cyclic plate load test data (Fig. 8.15) gives S_p and S_e. Value of S_p consists of plastic settlement of pile material and plastic settlement of soil at the base (S_{pb}). The plastic settlement of pile material is usually very small and, therefore, can be neglected. Therefore, S_p is approximately equal to S_{pb}. Hence Eq. (8.20) can be written as

$$S_{eb} = (S_e - \Delta L) \qquad\qquad\qquad\qquad ... (8.21)$$

Value of ΔL is given by

$$\Delta L = \frac{(Q - Q_s/2)L}{AE} \qquad \ldots (8.22)$$

where Q = total laod on pile,
Q_s = side resistance load,
L = length of pile,
A = average cross-sectional area of the pile, and
E = modulus of elasticity of the pile material.

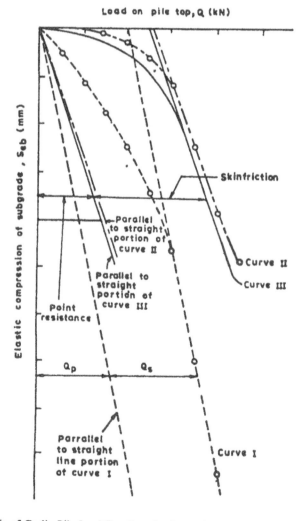

Fig. 8.16. Analysis of Cyclic Pile Load Test Data for Separation of Skin Friction and Point Bearing.

S_{eb} is to be computed from Eq. (8.20) but ΔL is not known. ΔL cannot be determined without a knowledge of Q_s (Eq. 8.21). Since the objective of the exercise is to determine Q_p and Q_s, the problem does not have a direct solution. The following indirect procedure is adopted:

(i) Assuming that there is no compression in the pile, (i.e., $\Delta L = 0$) plot a graph between total elastic recovery at pile head, S_{eb} and the load on the pile top Q, as shown by curve I of Fig. 8.16.

(ii) Draw a straight line parallel to the straight portion of the curve I (Fig. 8.16) to divide the load into two parts, and obtain the approximate values of point resistance, Q_p and skin friction resistance, Q_s.

(iii) Using the approximate values of skin friction Q_s for different load levels Q, compute the corresponding elastic compressions ΔL of the pile, using Eq. (8.22).

(iv) Obtain the corresponding values of elastic compressions of the soil at the base of the pile, from Eq. (8.21).

(v) Plot a new curve II using these new values of S_{eb} against corresponding loads Q.

(vi) Draw a new straight line parallel to the straight portion of curve II.

(vii) Repeat the steps (iii) and (iv) to obtain new values of S_{eb} and then step (v) to obtain a new curve III.

The process may be repeated till reasonably accurate values of Q_p and Q_s are obtained. However, it is found that the third trial would give sufficiently accurate values, for all practical purposes.

To obtain the safe pile capacity, a factor of safety of 2 on the ultimate skin friction resistance and 2.5 on the ultimate point bearing resistance can be applied.

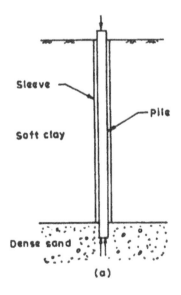

Fig. 8.17(a). Elimination of Skin Friction Through Sleeve.

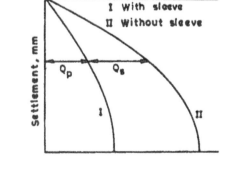

Fig. 8.17(b). Load-settlement Curve in Two Cases.

Another way of obtaining point bearing resistance and skin friction resistance separately is by performing two pile load tests. In one test, a sleeve is put around the pile such that friction is not acting on the pile shaft (Fig. 8.17a). Another test is performed without putting sleeve around the pile. The difference on the load from the two tests will give the shaft friction (Fig. 8.17b).

8.12 COYLE AND REESE (1966) METHOD OF ESTIMATING LOAD-SETTLEMENT BEHAVIOUR OF PILES

The pile capacity estimated from load tests may be valid only for the particular pile and soil conditions under test. In a real situation, the soil conditions at different points of the site may vary, the ground conditions which prevail at the time of load test such as change in moisture content in the top layer, thixotropic hardening in clays, removal of soil due to scouring during floods in the case of bridge foundations, etc. Under these circumstances, the load-settlement behaviour observed from the pile load test cannot be used directly to estimate the pile capacity. In such cases, the procedure proposed by Coyle and Reese (1966) to analytically estimate the load-settlement behaviour of the pile can be adopted to estimate its capacity. The method is based on the premise that as the pile undergoes vertical displacement under load (the displacement may be either elastic compression, point movement or both), the soil provides a shear resistance on the pile, the magnitude of which is dependent upon the stress deformation resistance characteristics of the soil.

The procedure of constructing the load-settlement curve is presented stepwise as below (Bowles, 1982):

1. Divide the pile into a number of segments (say four as shown in Fig. 8.18a).
2. Assume a small tip movement, Y_{T1}.
3. Compute the tip load (T), corresponding to the movement Y_{T1} using the following equation:

$$T = A_b \cdot K_s \cdot Y_{T1} \qquad \qquad \text{... (8.23)}$$

where A_b = area of pile at tip,

K_s = vertical subgrade modulus of the soil (Table 8.5), and

Y_T = tip movement (assumed).

4. Estimate mid point movement Y_4 of the bottom segment. For the first trial, the mid-point movement is assumed equal to the tip movement (Y_{T1}).
5. Using the estimated mid-point movement, obtain the ratio of load transfer to soil shear strength from Fig. 8.18b.
6. From a curve of soil shear strength versus depth (Fig. 8.18c) obtain the shear strength of the soil at the depth of segment. Curve shown in Fig. 8.18c is obtained from soil borings where shear strength is determined by field and laboratory testing.
7. Using the ratio determined in step 5 and shear strength at the depth of segment in step 6, compute the load transfer per unit area τ_4 on the segment.

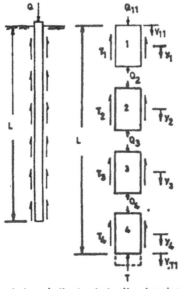

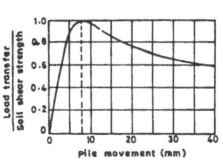

(b) Ratio of load transfer to soil shear
      strength versus pile movement

(a) Axially loaded pile showing
      forces acting on segments

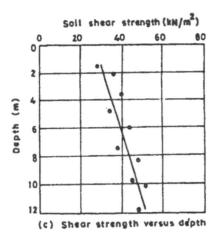

(c) Shear strength versus depth

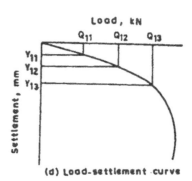

(d) Load-settlement curve

Fig. 8.18. Procedure of Computing Load-settlement Relationship (Coyle and Reese, 1966).

8. Using the value of load transfer computed in step 7, compute the load Q_4 at the top of
   the bottom segment (Fig. 8.18b)

$$Q_4 = T + \tau_4 \cdot L_4 \cdot p \qquad \qquad \dots (8.24)$$

where Q_4 = load at top of segment,
           T = load at the tip from step 3,
           τ_4 = load transfer per unit area in bottom segment from step 7,
           p = perimeter of pile, and

L_4 = length of the bottom segment considered (say $L/4$ in four segments of equal length).

9. Compute the elastic deformation of the bottom half segment by

$$\Delta y_4 = \frac{\left(\dfrac{Q_{mid} + T}{2}\right)\left(\dfrac{L_4}{2}\right)}{AE} \qquad \text{... (8.25)}$$

Assuming a linear variation of load distribution within a segment, Q_{mid} is computed as

$$Q_{mid} = \frac{Q_4 + T}{2} \qquad \text{... (8.26)}$$

A and E are cross-sectional areas of pile and elastic modulus of pile material respectively.

10. Compute the new mid-point movement of the bottom segment by

$$Y_4 = \Delta y_4 + Y_{T1} \qquad \text{... (8.27)}$$

Table 8.5. Values of K_s [IS : 2950 (Part I)-1981].

(A) For Granular Soils

Relative Density N-value/30 cm	Loose <10	Medium 10-30	Dense above 30
K_s for dry or moist soils ($\times 10^4$ kN/m³)	1.5	1.5–4.7	4.7–18
K_s for submerged soil ($\times 10^4$ kN/m³)	0.9	0.9–2.9	2.9–10.8

(B) For Cohesive Soils

Consistency	Stiff	Very Stiff	Hard
Unconfined strength ($\times 10^2$ kN/m³)	1–2	2–4	above 4
K_s value ($\times 10^4$ kN/m³)	2.7	2.7–5.4	5.4–10.8

11. Compare the computed new mid-point movement from step 10 with the estimated mid-point movement from step 4.
12. If the computed mid-point movement does not agree with the assumed mid-point movement within a specified tolerance, repeat steps 3 through 11 and compute a new mid-point movement.
13. When convergence is achieved, go to the next segments above the bottom segment and work up the pile to compute a value of 'Q_{11}' and 'Y_{11}' at the top.
14. The same procedure is followed for different tip movement Y_{T2}, Y_{T3} .... (Y_{Tn}) values and a set of (Q_{12}, Y_{12}), (Q_{13}, Y_{13}) values are obtained. The same are plotted to give the necessary load settlement curve (Fig. 8.18d).

The method outlined above provides adequate flexibility for taking into account the actual field pile-soil situation. The method can be easily programmed and various factors affecting pile capacity such as layered soils, variation in shear strength with depth, the thixotropic strength gain with time and the effects of pile installation can be accounted for (Ramaswamy, 1990). Such a computer programme can also be used to moderate the observed load-settlement behaviour of a test pile to account for any deviation in the ground conditions that actually exist at the time of the load test and that is likely to prevail at the time of actual loading of the foundation. Such a moderate load-settlement curve may be used to predict the safe pile capacity.

Vijayavergiya (1969) suggested a method for the determination of load transfer curve to describe the pile-soil interface behaviour along the pile shaft and the pile tip, i.e., mobilized shaft resistance 'f' versus shaft displacement 'z' and base resistance 'q' versus tip displacement 'z' curves. He expressed the f-z and q-z curves by the following relations:

$$f = f_{max} \left[2\sqrt{\frac{z}{z_s}} - \frac{z}{z_s} \right] \qquad \dots (8.28)$$

$$q = q_{max} \left(\frac{z}{z_t} \right)^{1/3} \qquad \dots (8.29)$$

where $\quad f$ = unit shaft resistance mobilized at any shaft movement, z

z_s = critical movement of pile segment at which f_{max} is mobilized, and $z < z_s$

f_{max} = maximum unit shaft resistance mobilized at pile shaft movement z_s

$\quad = K\sigma'_v \tan \delta \qquad$ ..... for sands

$\quad = \alpha \bar{c}_u \qquad$ ..... for clays

K = lateral earth pressure coefficient (Table 8.1)

σ'_v = effective overburden pressure over the pile segment under consideration

δ = angle of friction between pile segment material and soil (Table 8.1)

α = adhesion factor (Fig. 8.9)

$\bar{c}_u$ = average undrained shear strength of clay adjacent to shaft

q = unit base resistance for a particular tip movement z

q_{max} = maximum unit base resistance mobilized at critical pile tip movement z_t

$\quad = \sigma'_v N_q \qquad$ ..... for sands

$\quad = c_u N_c \qquad$ ..... for clays

N_q = bearing capacity factor (Fig. 8.6)

N_c = bearing capacity factor, 9.0

c_u = undrained shear strength of clay at pile tip. It is usually taken as the average value over a distance of twice the pile diameter below the pile tip.

Vijayavergiya (1977) stated that the value of Z_s may be appropriately chosen within the range 5 mm to 7.5 mm. The value of Z_t is found to be a function of pile tip dimension, B and has an average range from 0.04B to 0.06B for both clay and sand.

The procedure of constructing the load-settlement curve is almost similar as given by Coyle and Reese (1966). However the steps involved in this procedure are given below:

(1)   Divide the pile into a number of segments (say four as shown in Fig. 8.18a).
(2)   Assume a small pile tip movement, Y_{T1}.
(3)   Using q-z relation (Eq. 8.29), compute q.
        The pile tip load T is then given by

$$T = q \cdot A_b \qquad\qquad ... (8.30)$$

(4)   Estimate the mid-point movement Y_4 of the bottom segment. For the first trial the mid-point movement is assumed equal to the tip movement Y_{T1}.
(5)   Using f-z relation (Eq. 8.28), determine f. Then the value of frictional resistance acting on the bottom segment will be $f . L_4 . p$, p being the perimeter of the pile.
(6)   Compute the load Q_4 at the top of the bottom segment as below :

$$Q_4 = T + f . L_4 . p \qquad\qquad ... (8.31)$$

After estimating Q_4, the procedure is exactly same as described in steps 9 to 14 of Coyle and Reese (1966) method.

Hazarika and Ramasamy (2000) implemented the above steps in a computer programme. To verify the analytical procedure described above, the computer programme was used for computing load-settlement curves for four full scale pile load tests reported in literature.

The cases were so selected as to represent a wide range of pile-soil situations, *viz.,* layered soil deposits, piles of non-uniform cross-sections, different types of installation procedures, etc. The soil and pile conditions for each of these tests are described briefly. The input parameters are selected suitably from the field test results. In all the cases, z_s is taken as 8 mm and z_t equal to six per cent of tip dimension. The measured and predicted load-settlement curves for each of the cases are also presented.

Case I : Driven Precast Piles at Kakinada (India)

These tests were conducted on precast concrete piles installed for two major fertiliser plants at Kakinada on the South-Eastern coast of India. Load tests on 22 precast driven piles of size 400 mm × 400 mm have been reported (Raju and Gandhi, 1989). The subsoil consists of three layers, viz. dense fine sand (top), soft marine clay (middle) and very stiff clay (bottom). Being very near to the coastline, the water table was assumed to be at the ground surface level. The soil details of the load test selected for this study and the input parameters used in the computations are shown in Fig. 8.19 along with the computed and field load-settlement curves. The estimated load-settlement curve is in good agreement with the observed one.

Case II : Cast, in-situ Pile reported by Tomlinson (1987)

The results of a load test on a bored cast-in-situ concrete pile installed through loose fine sand and peat into dense sand has been reported by Tomlinson (1987). The hole for installing the pile was constructed by conventional drilling techniques. The pile and soil details along with the input parameters are shown in Fig. 8.20. Shaft resistance in the sandy clay fill and peat layer are neglected in the analysis; however the weight of these layers are taken into

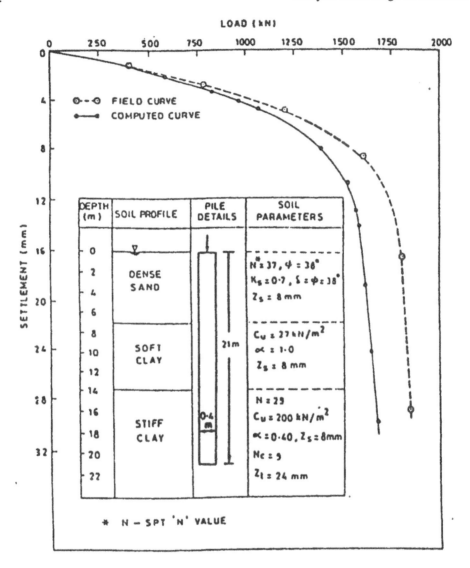

Fig. 8.19. Pile-Soil Details and Load-Settlement Curve for Case I.

account in calculating the overburden pressures in the dense sand. The load settlement curve predicted for this example is compared in Fig. 8.20 along with the field curve. A reasonable match is observed.

Case III : Step Tapered Piles in Mississippi Valley (USA)

Raymond step-taper piles were required to be constructed for the development of a chemical plant of the Gulf Oil Corporation in the west bank of the Mississippi River. Load tests at this site were reported by Darragh and Bell (1969). The pile-soil details along with the input parameters are shown in Fig. 8.21. The tip portion of the step-tapered pile consists of

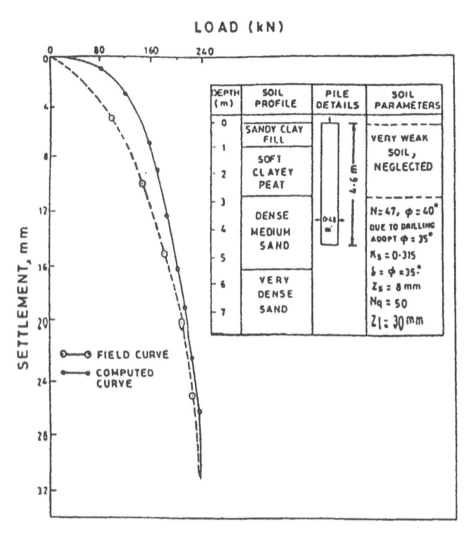

Fig. 8.20. Pile-Soil Details and Load-Settlement Curve for Case II.

a pipe section with closed end. The undrained shear strength values for the cohesive soil layers have been adopted from the design shear strength curve provided in the report. The resulting load-settlement curve obtained from the programme is shown along with the field curve in Fig. 8.21. Both the curves are found to be in good agreement.

Case IV : Pile Prediction Event of 1989 Foundation Engineering Congress, USA

A series of axial load tests were performed for conducting a pile prediction event associated with the 1989 Foundation Engineering Congress North-Western University, USA (Finn, 1989). The load test reported on a steel pile (section depth = 346 mm, flange width = 371 mm) is taken for study. The pile and soil details are shown in Fig. 8.22 along with the

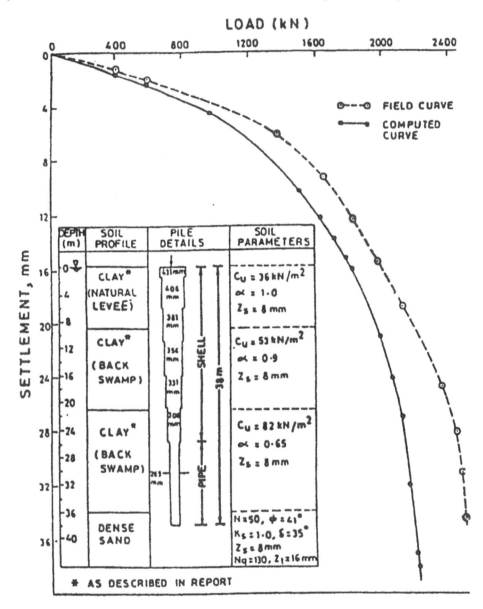

Fig. 8.21. Pile-Soil Details and Load-Settlement Curve for Case III.

input parameters. The formation of soil plugs between the flanges of the pile has been taken into account, as suggested by Bowles (1996). The load settlement curve reported from the field test is plotted along with the predicted curve in Fig. 8.22. The computed curve is found to match the field curve quite closely.

It is evident from the above case studies that the procedure of obtaining load-settlement curve of a pile using *f*-*z* and *q*-*z* relations is satisfactory.

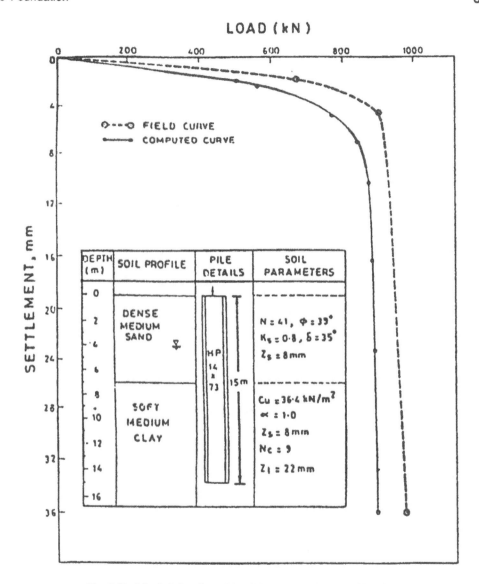

Fig. 8.22. Pile-Soil Details and Load-Settlement Curve for Case IV.

8.13 NEGATIVE SKIN FRICTION

In the case of driven or bored piles installed in compressible fill or any soil showing appreciable consolidation under its own weight, the resulting downward movement of soil around the pile induces down drag on the pile (Fig. 8.23). This drag on the pile surface is called negative skin friction and it reduces the allowable working load on the pile by adding to the structural loads. A small relative movement between the soil and the pile of the order of 10 mm may be sufficient for full negative skin friction to occur. Negative skin friction may

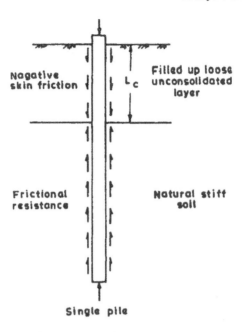

Fig. 8.23. Negative Skin Friction Pile.

also develop in loose sand deposits. Piles in clay deposits subjected to general subsidence resulting from lowering of groundwater table or other causes may also be subjected to this downdrag. Reconsolidation of the remoulded clay layer around any driven pile may also result in negative skin friction.

The magnitude of negative skin friction, F_n for a single pile in a filled up soil deposit may be estimated as given below:

In cohesive soils

$$F_n = \alpha \cdot c_u \cdot A_s \qquad \qquad \ldots (8.32)$$

In cohesionless soils

$$F_n = \frac{1}{2} K \gamma L_c \cdot \tan \delta \cdot A_s \qquad \qquad \ldots (8.33)$$

where α = adhesion factor,

c_u = unit undrained cohesion, and

A_s = surface area of pile in compressible stratum = $p \cdot L_c$

K = coefficient of earth pressure,

γ = density of soil,

L_c = length of pile in compressible layer,

δ = angle of wall friction, and

p = perimeter of the pile.

Values of α, K and S may be taken as per Sec. 8.7.

As the load on the pile having negative skin friction is equal to the sum of working load plus negative skin friction, safety factor is given by

$$\text{Safety factor} = \frac{\text{ultimate carrying capacity}}{\text{working load + negative skin friction}} \qquad \text{... (8.34)}$$

Negative skin friction can be reduced in precast piles by painting the pile surface with bitumen.

8.14 VERTICAL PILE SUBJECTED TO LATERAL LOAD

General

Lateral load and moments may act on piles in addition to axial loads. Two types of piles are encountered in practice, namely (*i*) long piles, and (*ii*) short piles. When a pile length is greater than a particular length, the length loses its significance. The behaviour of the pile will not be affected if the length is greater than this particular length. Such piles are called as long flexible piles. In case of short piles, the flexural stiffness (EI) of the material of pile loses its significance. The pile behaves as a rigid member and rotates as a unit.

Three types of boundary conditions occur in practice, namely (*i*) free-head pile, (*ii*) fixed-head pile, and (*iii*) partially-restrained head pile. In the case of free-head pile, the lateral load may act at or above ground level and the pile head is free to rotate without any restraint. A fixed-head pile is free to move only laterally but rotation is prevented completely, whereas a pile with partially restrained head moves and rotates under restraint.

Methods of calculating lateral resistance of vertical piles can broadly be divided into two categories:

(*i*) Methods of calculating ultimate lateral resistance
   (*a*) Hansen's method (1961): This method is based on earth pressure theory and is applicable only to short piles.
   (*b*) Brom's method (1964a, b): This method is also based on earth pressure theory with the simplifying assumption for distribution of ultimate soil resistance along the pile length. This method is applicable for both short and long piles.
(*ii*) Methods of calculating acceptable deflection at working load
   (*a*) Modulus of subgrade reaction approach (Reese and Matlock, 1956): In this method it is assumed that the soil acts as a series of independent linearly elastic springs.
   (*b*) Elastic approach (Poulos, 1971a and b): In this method, the soil is assumed as an ideal elastic continuum.

Modulus of subgrade reaction approach is relatively simple and has been used in practice for a long time. This method can incorporate factors such as nonlinearity, variation of subgrade reaction with depth and layer system. In this chapter, only this method has been dealt with.

Subgrade reaction approach

In this approach a laterally loaded pile is treated as a beam on elastic foundation. It is assumed that the pile is supported by a series of infinitely closed spaced independent and elastic springs as shown in Fig. 8.24b. The stiffness of these springs K_h (also called the

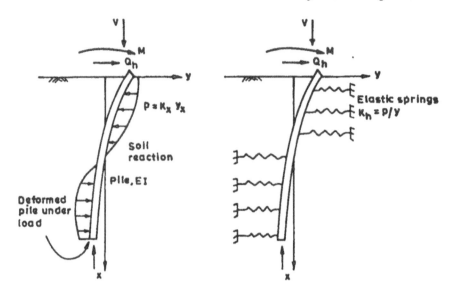

Fig. 8.24. Behaviour of Laterally Loaded Pile as Beam on Elastic Foundation.

modulus of horizontal subgrade reaction) is expressed as below:

$$K_h = \frac{p}{y} \qquad\qquad\qquad ...\ (8.35)$$

where p = soil reaction per unit length of pile, and
         y = pile deformation.

Palmer and Thompson (1948) employed the following form to express the modulus of subgrade reaction:

$$K_x = K_h \left(\frac{x}{L}\right)^n \qquad\qquad\qquad ...\ (8.36)$$

where K_h = value of K_x at tip of pile, i.e. at x equal to L,
         x = any point along pile length measured from ground surface, and
         n = a coefficient equal to or greater than zero.

The most commonly used value of n for sands and normally consolidated clays is unity. For over consolidation clays, n is taken as zero. For the value of $n = 1$, the variation of K_h with depth is expressed by the following relationship:

$$K_x = \eta_h \cdot x \qquad\qquad\qquad ...\ (8.37)$$

where η_h is the constant of modulus of subgrade reaction expressed in kN/m³. For the value of $n = 0$, the modulus will be constant with depth. Value of η_h and K_h depend on soil properties (density for granular soils and shear strength for clayey soils), the width or diameter of pile, the flexural stiffness of the pile material and the deflection of the pile. Values of η_h and K_h can be obtained from Tables 8.6 and 8.7 respectively (IS : 2911-Part I).

The behaviour of a pile can be analysed by using the equation of an elastic beam supported on elastic foundation and is given by the following equation:

$$EI \frac{d^4y}{dx^4} + p = 0 \qquad \qquad \cdots (8.38)$$

where E = modulus of elasticity of pile,
         I = moment of inertia of pile section, and
         p = soil reaction = $K_h \cdot y$.

Table 8.6. Typical values of η_h (IS : 2911-Part I, Sec. 1).

Soil Type	η_h in kN/m^3	
	Dry	Submerged
Loose sand	2.6×10^3	1.5×10^3
Medium sand	7.7×10^3	5.2×10^3
Dense sand	20×10^3	12.5×10^3
Very loose sand under repeated loading	—	0.41×10^3

Table 8.7. Typical Values of K_h for Preloaded Clays (IS : 2911-Part I, Sec. 1).

Unconfined Compressive strength, kN/m^2	Range of Values of K_h, kN/m^2	Probable Value of K_h, kN/m^2
20 to 40	700 to 4200	773
100 to 200	3200 to 6500	4879
200 to 400	6500 to 13000	9773
More than 400	—	19546

Equation (8.38) can be written as

$$\frac{d^4y}{dx^4} + \frac{K_h y}{EI} = 0 \qquad \qquad \cdots (8.39)$$

Solution of Eq. (8.39) is obtained to get the values of deflection, shear force and moment for the pile in cohesionless soil or in cohesive soil.

Laterally loaded pile in cohesionless soil

Free-head pile

Figure 8.25 shows the distribution of pile deflection y, pile slope variation, moment, shear, and soil reaction along the pile length due to a lateral load Q_{hg} and a moment M_g applied at the pile head. The behaviour of this pile can be expressed by Eq. (8.39). In general, solution for this equation can be expressed by the following formulation:

$$y = f(x, T, L, K_h, EI, Q_{hg}, M_g) \qquad \qquad \cdots (8.40)$$

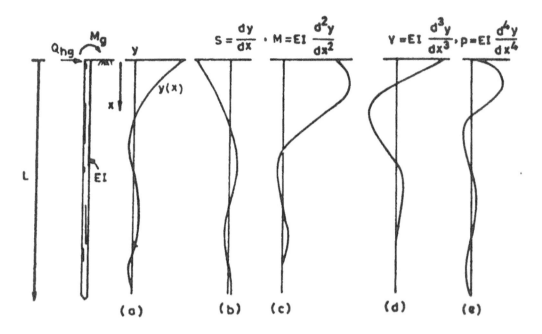

Fig. 8.25. Pile Subjected to Lateral Load and Moment.

where x = depth below ground level,

T = relative stiffness factor,

L = pile length,

$K_h = \eta_h x$ = modulus of horizontal subgrade reaction,

η_h = constant of subgrade reaction,

EI = pile Stiffness,

Q_{hg} = lateral load applied at the pile head, and

M_g = moment applied at the pile head.

Elastic behaviour can be assumed for small deflections relative to the pile dimensions. For such a behaviour, the principle of superposition may be applied. By utilizing the principle of superposition, the effects of lateral load Q_{hg} on deformation y_A and the effect of moment M_g on deformation y_B can be considered separately. Then the total deflection y_x at the depth x can be given by the following:

$$Y_x = Y_A + Y_B \qquad\qquad\qquad \text{... (8.41)}$$

where $$\frac{Y_A}{Q_{hg}} = f_1(x,\ T,\ L,\ k_h,\ EI) \qquad\qquad\qquad \text{... (8.42a)}$$

and $$\frac{Y_B}{M_g} = f_2\ (x,\ T,\ L,\ k_h,\ EI) \qquad\qquad\qquad \text{... (8.42b)}$$

f_1 and f_2 are two different functions of the same terms. In Eqs. (8.42a) and (8.42b) there are six terms and two dimensions; force and length are involved. Therefore, following four independent nondimensional terms can be determined (Matlock and Reese, 1962):

$$\frac{Y_A EI}{Q_{hg} T^3}, \frac{x}{T}, \frac{L}{T}, \frac{k_h T^4}{EI} \qquad \text{... (8.43a)}$$

$$\frac{y_B EI}{M_g T^2}, \frac{x}{T}, \frac{L}{T}, \frac{k_h T^4}{EI} \qquad \text{... (8.43b)}$$

Furthermore, the following symbols can be assigned to these nondimensional terms:

$$\frac{Y_A EI}{Q_{hg} T^3} = A_y \text{ (deflection coefficient for lateral load)} \qquad \text{... (8.44a)}$$

$$\frac{Y_B EI}{M_g T^2} = B_y \text{ (deflection coefficient for moment)} \qquad \text{... (8.44b)}$$

$$\frac{x}{T} = Z \text{ (depth coefficient)} \qquad \text{... (8.45a)}$$

$$\frac{L}{T} = Z_{max} \text{ (maximum depth coefficient)} \qquad \text{... (8.45b)}$$

$$\frac{k_h T^4}{EI} = \phi(x) \text{ (soil modulus function)} \qquad \text{... (8.46)}$$

From equations (8.44a) and (8.44b), one can obtain:

$$y_x = y_A + Y_B = A_y \frac{Q_{hg} T^3}{EI} + B_y \frac{M_g T^2}{EI}. \qquad \text{... (8.47)}$$

Similarly, one can obtain expressions for moment M_x, slope S_x, shear V_x, and soil reaction p_x as follows :

$$M_x = M_A + M_B = A_m Q_{hg} T + B_m M_g \qquad \text{... (8.48)}$$

$$S_x = S_A + S_B = A_s \frac{Q_{hg} T^2}{EI} + B_s \frac{M_g T}{EI} \qquad \text{... (8.49)}$$

$$V_x = V_A + V_B = A_v Q_{hg} + B_v \frac{M_g}{T} \qquad \text{... (8.50)}$$

$$P_x = P_A + P_B = A_p \frac{Q_{hg}}{T} + B_p \frac{M_g}{T^2} \qquad \text{... (8.51)}$$

Referring to the basic differential equation (8.39) of beam on elastic foundation and utilizing the principle of superposition, we get:

$$\frac{d^4 y_A}{dx^4} + \frac{k_h Y_A}{EI} = 0 \qquad \text{... (8.52)}$$

$$\frac{d^4 y_B}{dx^4} + \frac{k_h}{EI} y_B = 0 \qquad \text{... (8.53)}$$

Substituting for y_A and y_B from equations (8.44a) and (8.44b), K_h/EI from equation (8.46) and x/T from equation (8.45a), we get:

$$\frac{d^4 A_y}{dz^4} + \phi(x) A_y = 0 \qquad \text{... (8.54)}$$

$$\frac{d^4 B_y}{dz^4} + \phi(x) B_y = 0 \qquad \text{... (8.55)}$$

For cohesionless soil where soil modulus is assumed to increase with depth $K_h = \eta_h x$, $\phi(x)$ may be equated to $Z = x/T$. Therefore, equation (8.46) becomes

$$\frac{\eta_h x T^4}{EI} = \frac{x}{T} \qquad \text{... (8.56)}$$

This gives

$$T = \left(\frac{EI}{\eta_h} \right)^{1/5} \qquad \text{... (8.57)}$$

Solutions for equations (8.54) and (8.55), by using finite-difference methods, were obtained by Reese and Matlock (1956) for values of A_y, A_s, A_m, A_v, A_p, B_y, B_s, B_m, B_v and B_p for various $Z = x/T$.

It has been found that pile deformation is like a rigid body (small curvature) for $Z_{max} < 2$. Therefore, piles with $Z_{max} < 2$ will behave as rigid piles or poles. Also, deflection coefficients are same for $Z_{max} = 5$ to 10. Therefore, pile length beyond $Z_{max} = 5$ does not change the deflection. In practice, in most cases pile length is greater than $5T$; therefore, coefficients given in Tables 8.8 and 8.9 can be used. Figure 8.26 provides values of A_y, A_m, B_y and B_m for different values of Z_{max}.

Fixed-head pile

For a fixed-head pile, the slope (S) at the ground surface is zero. Therefore, from equation (8.49)

$$S_x = S_A + S_B = A_s \frac{Q_{hg} T^2}{EI} + B_s \frac{M_g T}{EI} = 0. \qquad \text{... (8.58)}$$

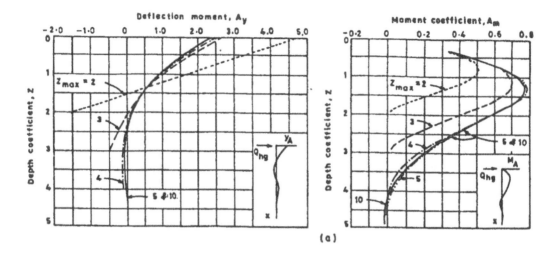

(a)

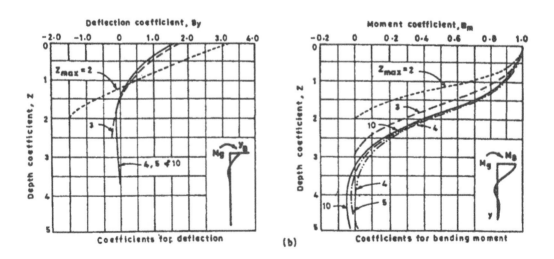

(b)

Fig. 8.26. Coefficients for Free-headed Piles in Cohesionless Soils (Reese and Matlock, 1956).

Table 8.8. Coefficients A for Long Piles ($Z_{max} \geqslant 5$): Free Head (Matlock and Reese, 1961, 1962).

Z	A_y	A_s	A_m	A_v	A_p
0.0	2.435	−1.623	0.000	1.000	0.000
0.1	2.273	−1.618	0.100	0.989	−0.227
0.2	2.112	−1.603	0.198	0.956	−0.422
0.3	1.952	−1.578	0.291	0.906	−0.586
0.4	1.796	−1.545	0.379	0.840	−0.718
0.5	1.644	−1.503	0.459	0.764	−0.822
0.6	1.496	−1.454	0.532	0.677	−0.897
0.7	1.353	−1.397	0.595	0.585	−0.947
0.8	1.216	−1.335	0.649	0.489	−0.973
0.9	1.086	−1.268	0.693	0.392	−0.977
1.0	0.962	−1.197	0.727	0.295	−0.962
1.2	0.738	−1.047	0.767	0.109	−0.885
1.4	0.544	−0.893	0.772	−0.056	−0.761
1.6	0.381	−0.741	0.746	−0.193	−0.609
1.8	0.247	−0.596	0.696	−0.298	−0.445
2.0	0.142	−0.464	0.628	−0.371	−0.283
3.0	−0.075	−0.040	0.225	−0.349	0.226
4.0	−0.050	0.052	0.000	−0.106	0.201
5.0	−0.009	0.025	−0.033	0.013	0.046

Table 8.9. Coefficients B for Long Piles ($Z_{max} > 5$): Free Head (Matlock and Reese, 1961, 1962).

Z	B_y	B_s	B_m	B_v	B_p
0.0	1.623	−1.750	1.000	0.000	0.000
0.1	1.453	−1.650	1.000	−0.007	−0.145
0.2	1.293	−1.550	0.999	−0.028	−0.259
0.3	1.143	−1.450	0.994	−0.058	−0.343
0.4	1.003	−1.351	0.987	−0.095	−0.401
0.5	0.873	−1.253	0.976	−0.137	−0.436
0.6	0.752	−1.156	0.960	−0.181	−0.451
0.7	0.642	−1.061	0.939	−0.226	−0.449
0.8	0.540	−0.968	0.914	−0.270	−0.432
0.9	0.448	−0.878	0.885	−0.312	−0.403
1.0	0.364	−0.792	0.852	−0.350	−0.364
1.2	0.223	−0.629	0.775	−0.414	−0.268
1.4	0.112	−0.482	0.688	−0.456	−0.157
1.6	0.029	−0.354	0.594	−0.477	−0.047
1.8	−0.030	−0.245	0.498	−0.476	−0.054
2.0	−0.070	−0.155	0.404	−0.456	0.140
3.0	−0.089	0.057	0.059	−0.213	0.268
4.0	−0.028	0.049	−0.042	0.017	0.112
5.0	0.000	0.011	−0.026	0.029	−0.002

Therefore

$$\frac{M_g}{Q_{hg}T} = -\frac{A_s}{B_s} \text{ at } x = 0.$$

From Tables 8.8 and 8.9 for $Z = x/T = 0$

$$A_s/B_s = -\frac{-1.623}{-1.75} = -0.93$$

Therefore, $M_g/Q_{hg}T = -0.93$. The term $M_g/Q_{hg}T$ has been defined as the nondimensional fixity factor by Prakash (1962). Then the equations for deflection and moment for fixed head can be modified as follows:

From equation (8.47),

$$y_x = A_y \frac{Q_{hg}T^3}{EI} + B_y \frac{M_g T^2}{EI}$$

Substituting $M_g = -0.93 Q_{hg}T$ for fixed head, we get

$$y_x = (A_y - 0.93\ B_y) \frac{Q_{hg}T^3}{EI}$$

or
$$y_x = C_y \frac{Q_{hg}T^3}{EI}. \qquad \qquad \dots (8.59)$$

Similarly, $\qquad M_x = C_m Q_{hg}T. \qquad \qquad \dots (8.60)$

Values of C_y and C_m can be obtained from Fig. 8.27.

Partially fixed head pile

In cases where the piles undergo some rotation at the joints of their head and the cap, these are called partially fixed piles. In such a situation, the coefficient C needs modification as follows:

$$C_y = (A_y - 0.93\ \lambda B_y) \qquad \qquad \dots (8.61)$$

$$C_m = (A_m - 0.93\ \lambda B_m), \qquad \qquad \dots (8.62)$$

where λ is per cent fixity (i.e. $\lambda = 1$ for 100 per cent fixity for fully restrained pile head and $\lambda = 0$ for fully free pile head). At intermediate fixity levels, proper λ can be taken.

Laterally loaded pile in cohesive soil

Free-head pile

For normally consolidated clays, the modulus of subgrade reaction increases linearly with depth. Therefore, for such clays the above analysis for cohesionless soil shall apply.

For over consolidated clays, subgrade modulus is constant with depth. For such clays, deflection coefficients are defined as

$$\frac{y_A EI}{Q_{hg}R^3} = A_{yc} \qquad \qquad \dots (8.63)$$

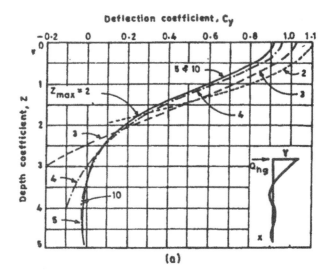

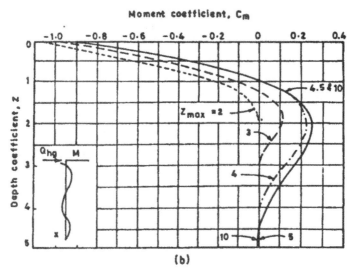

Fig. 8.27. Coefficient for Fixed-head Pile in Cohesionless Soils (Reese and Matlock, 1956).

$$\frac{y_B EI}{M_g R^2} = B_{yc} \qquad \qquad \dots (8.64)$$

where A_{yc}, B_{yc} = deflection coefficients in clay for Q_{hg} and M_g. Letting $y = y_A + y_B$ as in equation (8.47), we get deflection y at any depth.

$$y = A_{yc}\frac{Q_{hg}R^3}{EI} + B_{yc}\frac{M_g R^2}{EI} \qquad \qquad \dots (8.65)$$

Similarly, moment M at any depth is

$$M = A_{mc}\, Q_{hg}\, R + B_{mc}\, M_g \qquad\qquad ... (8.66)$$

Solutions for A and B coefficients similar to those presented for cohesionless soil had been obtained by Davisson and Gill (1963). In equation (8.54), by replacing A_y with A_{yc}, we get

$$\frac{d^4 A_{yc}}{dz^2} + \phi(x) A_{yc} = 0 \qquad\qquad ... (8.67)$$

Now putting $\phi(x) = 1$, and replacing T with R, equation (8.46) becomes:

$$\frac{K_h R^4}{EI} = 1 \qquad\qquad ... (8.68)$$

$$R = \left(\frac{EI}{K_h}\right)^{1/4} \qquad\qquad ... (8.69)$$

and

$$Z = \frac{x}{R}. \qquad\qquad ... (8.70)$$

Substituting the above equations in equation (8.67), the solutions for A and B coefficients can be obtained in a similar manner as for cohesionless soils.

The solutions for A_{yc} and A_{mc} have been plotted with nondimensional depth coefficient Z in Fig. 8.28a and B_{yc} and B_{mc} in Fig. 8.28b. It will be seen in Fig. 8.28 that if Z_{max} ($= L/R$) < 2, the pile behaves as a rigid pile or a pole. And for Z_{max} ($= L/R$) > 4, the pile behaves as an infinitely long pile.

8.15 LATERAL LOAD CAPACITY OF SINGLE PILE

Broms (1964) developed an analysis for getting the ultimate lateral capacity of a single pile. He made certain simplifying assumptions regarding distribution of shear resistance with depth, considered short rigid and long flexible piles separately, and also dealt with free-head and fixed-head cases separately. He developed the analysis both for pile in cohesionless soil and cohesive soil. The expressions developed by Broms (1964) are given below:

In cohesionless soils

Short piles
(a) Free-head (unrestrained) piles

$$Q_{hu} = \frac{0.5\gamma\, L^3 B K_p}{(h + L)} \qquad\qquad ... (8.71)$$

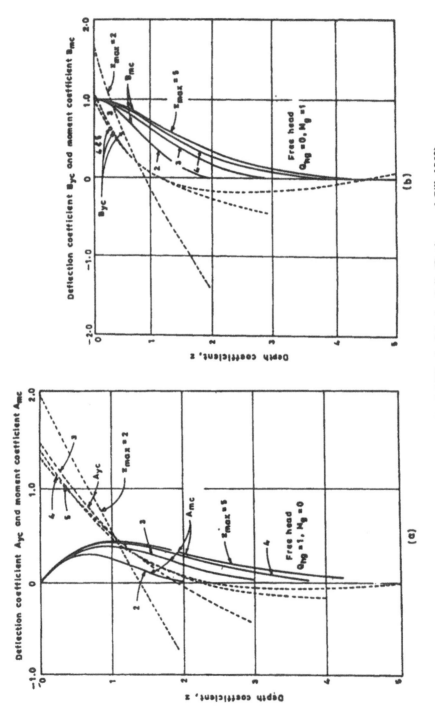

Fig. 8.28. Coefficients for Free-headed Piles in Cohesive Soils (Davisson and Gill, 1963).

(b) Fixed-head (restrained) piles

$$Q_{hu} = 1.5 \; \gamma L^2 B K_p \qquad \qquad \cdots \text{ (8.72)}$$

where Q_{hu} = ultimate lateral capacity of pile,
γ = density of soil around the pile,
L = embedded length of pile,
h = unsupported length of pile,
B = diameter of pile or width of pile,
K_p = coefficient of passive earth pressure

$$= \frac{1+\sin\phi}{1-\sin\phi} \text{, and} \qquad \qquad \cdots \text{ (8.73)}$$

ϕ = angle of internal friction.

Long piles
Fixed-head (restrained) piles

$$Q_{hu} = \frac{2M_u}{h+0.55\left(\dfrac{Q_{hu}}{\gamma BK_p}\right)^{0.5}} \qquad \qquad \cdots \text{ (8.74)}$$

where M_u = ultimate moment capacity of the pile shaft.

Knowing the cross-sectional area and amount of the reinforcement, value of M_u can be obtained in conventional manner.

The ultimate lateral resistance of a restrained long pile is considerably higher than the ultimate lateral resistance of a corresponding free-headed pile.

Figure 8.29 can be used to determine Q_{hu} of long piles by using $Q_{hu}/(K_p B^3 \gamma)$ versus $M_u/(B^4 \gamma \; K_p)$ plot.

In cohesive soils

Short piles
(a) Free-head (unrestrained) piles

$$2.25Bc_u\left(L-\frac{Q_{hu}}{9c_u B}\right)^2 = Q_{hu}\left(h+1.5B+\frac{Q_{hu}}{18c_u B}\right) \qquad \qquad \cdots \text{ (8.75)}$$

The Eq. (8.75) can be solved by trial for Q_{hu}. The solution is provided in Fig. 8.30 where if L/B and h/B ratios are known then $Q_{hu}/(c_u B^2)$ can be obtained. Thus the Q_{hu} value can be calculated.

(b) Fixed-head (restrained) piles

$$Q_{hu} = 9c_u B(L - 1.5B) \qquad \qquad \cdots \text{ (8.76)}$$

where c_u = undrained cohesion.

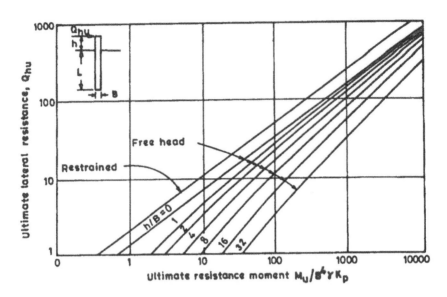

Fig. 8.29. Ultimate Lateral Load Capacity of Long Piles in Cohesionless Soils (Broms, 1964b).

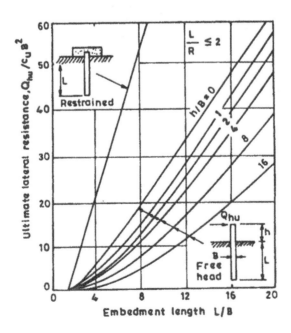

Fig. 8.30. Ultimate Lateral Load Capacity of Short Piles in Cohesive Soils (Broms, 1964b).

Long piles
Fixed-head (restrained piles)

$$Q_{hu} = \frac{2M_u}{\left(1.5B + \dfrac{Q_{hu}}{18c_u B}\right)} \qquad \text{... (8.77)}$$

The solution of Eq. 8.77 is given in Fig. 8.31 both for unrestrained and restrained piles. Thus for a known $M_u/(c_u B^3)$, one can obtain $Q_{hu}/(c_u B^2)$ and finally Q_{hu} can be calculated.

A factor of safety of 2.5 is recommended for obtaining safe lateral capacity of the pile.

Jain (1983) developed a method of computing lateral load capacity of a pile with a known value of pile head deflection. The method is based on analysing the pile as an equivalent cantilever (Fig. 8.32). The length of the equivalent cantilver can be obtained using chart given in Fig. 8.33. The results presented in Fig. 8.33 are obtained from an analysis based on modulus of subgrade reaction approach. Values of relative stiffness (T or R) may be obtained using Eq. (8.57) or Eq. (8.69). For a given limiting pile head deflection, y_0, the lateral load capacity can be obtained using following expressions (Ramasamy et al., 1987).
(*a*) For free-head pile

$$Q_h = \frac{3y_0 EI}{\left(h + L_f\right)^3} \qquad \text{... (8.78)}$$

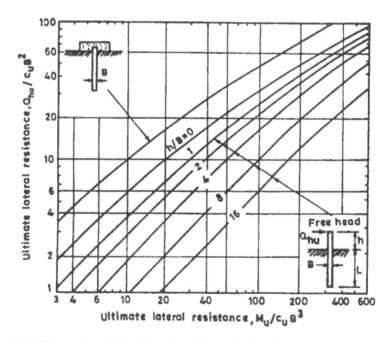

Fig. 8.31. Ultimate Lateral Load Capacity of Long Piles in Cohesive Soils (Broms, 1964).

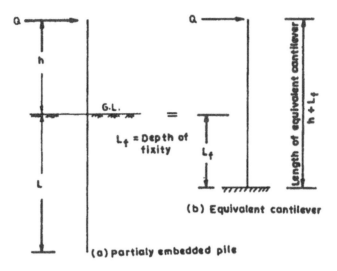

Fig. 8.32. Concept of Equivalent Cantilever.

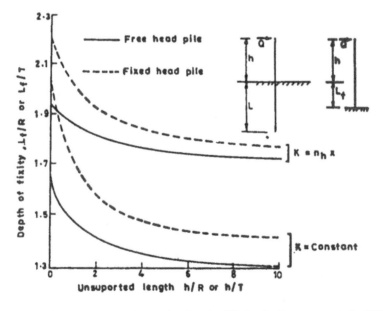

Fig. 8.33. Charts for Getting Depth of Fixity, L_f (Ramasamy et al., 1987).

(b) For fixed-head pile

$$Q_h = \frac{12y_0 EI}{\left(h + L_f\right)^3}.$$

... (8.79)

8.16 BATTER PILES UNDER LATERAL LOADS

In the case of large lateral loads, it is desirable to provide batter piles. These are also called inclined piles or raker piles. If the lateral load acts on the pile in the direction of batter, it is called as 'in' batter or negative batter pile (Fig. 8.34a). If the lateral load acts in the direction opposite to that of the batter, it is called an 'out' batter or positive batter pile (Fig. 34c). The degree of batter, that is the angle made by the pile with the vertical may go up to 30°.

Murthy (1965) conducted extensive model tests on 19 mm outside diameter and 762 mm long instrumented piles in cohesionless soils. The piles were tested at batters of 0°, ±15°, ±30° and ±45°. The sand was placed at a relative density of 67%. Lateral loads normal to the pile's axis were applied in the direction of batter and against batter also. The flexural strains were measured during the test. The data were processed by fitting orthogonal polynomials. The first and second integration of fitted moments yielded curves for rotation and deflection respectively whereas the first and second differentiation gave the shear and soil reaction curves. The findings of his model test data can be summarised as below:

(*i*) The ratio of ultimate lateral capacity of batter pile to the ultimate lateral capacity of vertical pile $\dfrac{P_u^\beta}{P_u^0}$ varies with batter angle β as shown in Table 8.10.

(*ii*) The following relationships between the relative stiffness factor of a batter pile with that of vertical pile have been established.

For piles of batter −22.5° to + 45°

$$T_b = T_0 (1 + 7.5 \times 10^{-3}\beta). \qquad\qquad \text{... (8.80)}$$

For piles of batter −22.5° to − 45°

$$T_b = 0.86 T_0 \qquad\qquad \text{... (8.81)}$$

where T_b = relative stiffness factor of a pile of batter β, and

T_0 = relative stiffness factor of a vertical pile.

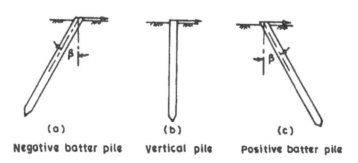

(a) \
Negative batter pile

(b) \
vertical pile

(c) \
Positive batter pile

Fig. 8.34. Batter Piles.

Table 8.10. Ratio $\dfrac{P_u^\beta}{P_u^0}$ for Piles of Different Batter.

Batter Angle	−45°	−37.5°	−30°	−15°	0°	+15°	+22.5°
$\dfrac{P_u^\beta}{P_u^0}$	0.94	1.16	1.20	1.10	1.0	0.90	0.70

Table 8.11. Non-dimensional A Factors for Deflection, Moment and Soil Reaction (Murthy, 1965).

Batter in Pile	0°	+15°	+30°	+45°	−15°	−30°	−45°
A_y	2.27	2.40	2.52	2.76	2.20	2.36	1.76
A_m	0.75	0.78	0.88	0.98	0.74	0.74	0.58
A_p	0.90	1.02	1.14	1.43	0.89	0.98	0.78

In Eqs. 8.80 and 8.81 the pile batter angle β expressed in degree is negative for 'in' batter and positive for 'out' batter piles.

(iii) Non-dimensional parameters A_y, A_m and A_p were obtained as given in Table 8.11.

8.17 UPLIFT CAPACITY OF PILES

Piles are subjected to uplift forces, for example, in the case of foundations of tall chimneys, transmission towers, jetty structures, etc. Resistance to uplift is given by the friction between the pile and the surrounding soil. It may be increased in the case of bored piles by under-reaming or by belling-out the bottom of piles, or by the bulb end of a driven and cast-in-situ pile.

The resistance of circular and straight sided piles to uplift is calculated in exactly the same way as the skin friction on compression piles and is given by,

$$Q_{ut} = A_s\left(\alpha c_d + K\bar{\sigma}_v \tan\delta\right) \qquad \dots (8.82)$$

where Q_{ut} = ultimate pull-out resistance of pile,

   c_d = average undisturbed shear strength adjacent to pile shaft,

   K = coefficient of lateral earth pressure,

   $\bar{\sigma}_v$ = average of effective vertical stress along the length of pile,

   α = adhesion factor, and

   A_s = surface area of pile.

The values of K, δ and α depend on many factors such as type of soil, material of pile, type of construction, etc. and approximate values for the same can be chosen from Table 8.1 and Fig. 8.9.

Meyerhof and Adams (1968) have developed the following expressions for computing the uplift capacity of bulb pile (Fig. 8.35).

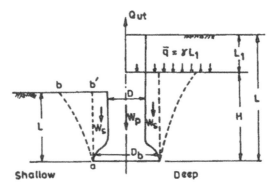

Fig. 8.35. Piles Subjected to Uplift or Tension Forces.

In cohesive soils
(a) Based on mobilization of shear strength along the cylindrical surface above base diameter

$$Q_{ut} = m_f c_u \cdot A_s' + W_s + W_p \qquad \qquad \text{... (8.83)}$$

(b) Based on bearing capacity failure

$$Q_{ut} = 9 c_u \cdot \frac{\pi}{4}\left(D_b^2 - D^2\right) + W_p \qquad \qquad \text{... (8.84)}$$

where m_f = mobilization factor as given in Table 8.12,

A_s' = surface area of vertical cylinder above base

   $= \pi \cdot D_b \cdot L$

D_b = diameter of enlarged base,

L = embedded length of pile,

W_s = weight of soil included in the annulars between the pile shaft and the vertical cylinder above the base,

W_p = weight of the pile,

c_u = unit undrained cohesion, and

D = diameter of pile shaft.

The ultimate uplift capacity of the pile is the smaller of the values given by Eqs. (8.83) and (8.84).

In c-ϕ soils
(a) Based on modification of shear strength along the cylindrical surface above the base diameter
   (i) For shallow depths when embedded length of pile L is less than a certain value H, referred to as the limiting height of failure surface above the base (Fig. 8.35).

Table 8.12. Values of Mobilization Factor m_f.

Type of Soil	m_f
Soft clay	1.0
Medium clay	0.7
Stiff clay	0.5
Stiff fissured clay	0.25

$$Q_{ut} = \pi c_u D_b L + \pi/2 S \gamma_b D_b L^2 K_u \tan\phi + W_p \qquad \dots (8.85)$$

(ii) For larger depths (i.e. $L > H$)

$$Q_{ut} = \pi c_u D_b H + S \cdot \gamma_b D_b (2L - H) H K_u \tan\phi + W_p \qquad \dots (8.86)$$

where S = shape factor,

$= 1 + mL/D_b$ with a maximum value of $(1 + mH/D_b)$,

m = a coefficient depending on ϕ,

γ_b = effective unit weight of soil, and

K_u = earth pressure coefficient varying from 0.9 to 0.95 for values of ϕ from 25° to 40°.

Values of H, m and maximum value of shape factor, i.e. S_{max} can be taken from Table 8.13.

(b) Based on bearing capacity failure

$$Q_{ut} = \pi/4 \left(D_b^2 - D^2 \right) \left(c_u N_c + \gamma_b L N_q \right) + W_p \qquad \dots (8.87)$$

where N_c, N_q = bearing capacity factors (Table 7.2),

f_s = unit skin friction. It may be taken as equal to half of the value taken for compression loads.

The values of Q_{ut} is taken as the smaller of the values obtained from Eqs. (8.85), (8.86) and (8.87). For design purposes, it is common to apply a factor of safety of 2 to 3 depending on the type of soil stratum.

Table 8.13. Values of H, m and S_{max}.

ϕ	20°	25°	30°	35°	40°	45°
H/D_b	2.5	3	4	5	7	9
m	0.05	0.10	0.15	0.25	0.35	0.50
S_{max}	1.12	1.30	1.60	2.25	3.45	5.50

8.18 PILE GROUPS

General

Piles are usually installed in groups. The codes do not permit the use of less than three piles to support a major column and less than two piles to support a foundation wall. The bearing capacity and settlement of pile groups are, therefore, needed for the design of foundation.

According to theory of elasticity the stress in soil at the level of tip of a bearing pile under an axial load Q is as shown in Fig. 8.36 a-i. If a group of bearing piles are arranged at ordinary pile spacing and each is acted upon by an axial load Q, the stress in soil overlaps. The total stress at any point is equal to sum of stresses induced by each of these piles (Fig. 8.36 a-ii). Thus, total stress may be several times greater than that under a single pile.

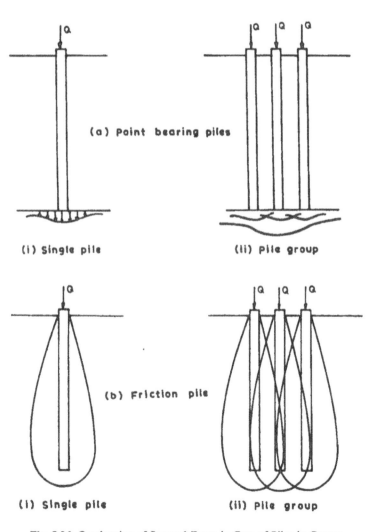

Fig. 8.36. Overlapping of Stressed Zones in Case of Piles in Groups.

Similarly, the overlapping of stresses takes place in case of group of friction piles (Fig. 8.36b). The sufficient overlapping of stresses may either cause failure of the soil or excessive settlement of the pile group.

The pile tops are connected together with a reinforced slab or beam called the pile cap. The pile cap helps the pile group to act as an integral unit. A pile group having pile cap standing clearly above the ground is known as free standing pile group and this type of construction is used when it is required to keep the pile cap away from direct contact with soils which are expansive in nature. A pile group in which pile cap rests on the soil, partially or fully buried below ground level is known as piled foundation. In a piled foundation, the pile cap may, under certain soil conditions, help in transmitting a part of the load to the soil on which it rests.

Spacing of piles

The spacing of piles in a group depends on many factors such as length, size and shape of piles, soil characteristics and magnitude and type of loads. When the piles are spaced closely, the soil is highly stressed and it may cause the shear failure of soil or excessive settlement of pile group. Besides this, following are the disadvantages of closer spacing:

(i) When the piles are penetrated through incompressible strata, great horizontal forces are set up. If the piles are closely spaced then due to these horizontal forces, damage of adjacent piles may take place.

(ii) Due to lifting force, necking of green concrete and separation of unreinforced set concrete may occur.

(iii) The carrying capacity of the soil upon which the group acts may be less than the capacities of the soils surrounding the piles.

(iv) Number of piles driven in clays may be reduced.

The spacing of the piles could be increased, but large spacings will require massive and heavy pile caps. This will further increase the load on piles.

The minimum spacing of piles is recommended by different codes of practice. The British Code of Practice (2004: 1972) suggests a minimum spacing equal to the perimeter of the pile for a friction pile and twice the least width for an end bearing pile. The Norwegian Code of Practice on Piling, Den Norske Pelekomite (1973) gives the recommendations for minimum pile spacing as given in Table 8.14.

Table 8.14. Spacing of Piles.

Length of Pile	Friction Piles in Sand	Friction Piles in Clay	Point Bearing Piles
Less than 12 m	3 B	4 B	3 B
12 to 24 m	4 B	5 B	4 B
Greater than 24 m	5 B	6 B	5 B

B is the pile diameter or largest side.

IS: 2911 (Part I)-1979 recommends a minimum spacing of 2.5 times the shaft diameter for point bearing piles, three times the shaft diameter for friction piles, whereas in loose sands or fill deposits, a minimum spacing of two times the diameter of the shaft is suggested. In case of piles of non-circular cross-section the diameter of the circumscribed circle shall be adopted.

In case, driven piles are used, the spacing of piles must be greater and the rate of driving in clays must be such as not to allow the development of porewater pressure. A good practice is to drive piles from the centre of the group towards the edges.

Efficiency of a pile group

The efficiency of a pile group (E_g) is the ratio of the actual group capacity to the sum of the individual pile capacities. The group efficiency, E_g depends mainly on the spacing between piles, type of soil in which the piles are installed and the manner of pile installation, that is, driven, bored, cast-in-situ, etc.

Model pile-group tests in remoulded clay by Sowers et al. (1961) and Whitaker (1957) for pile caps not in contact with the soil indicated group efficiencies varying from about 0.7 for 16-pile groups at S / B ratios of 1.75 to about 0.9 for larger spacings and smaller pile groups (two to four piles).

Model pile-group tests by Hanna (1963) indicated efficiencies varying from about 0.8 to greater than 1.0 for piles in sands. The lower efficiencies are obtained in dense sand which may be loosened somewhat by pile driving and smaller S / B ratios and the higher values in loose sand which is densified during driving and/or larger S / B ratios.

Vesic (1967) carried out tests on pile groups having 4 and 9 piles driven into sand under controlled conditions with pile spacings 2, 3, 4 and 6 times pile diameter. His tests were conducted in homogeneous and medium dense sand. His findings are given in Fig. 8.37. As per his findings the point efficiency of all the tests is unity. The skin efficiency increases with the increase in the pile spacing. The efficiency increases with the cap resting on the soil.

For a group configuration shown in Fig. 8.38, efficiency of the pile group consisting of friction piles may be obtained as described below.

Let, m = number of rows, and
             n = number of columns.

Then total number of piles = mn.
The perimeter of the group is given by

$$p = 2[(m - 1) S + S(n - 1)] + 8 \times B / 2$$

or $$p = 2(m + n - 2) S + 4B \qquad \text{... (8.88)}$$

where S = centre to centre spacing between piles, and
             B = diameter of piles.

The group efficiency E_g is the ratio of the skin resistance on the group perimeter p.f.L to the sum of the skin resistances of individual piles $\pi\, BmnfL$, or

$$E_g = \frac{p.f.L}{\pi BmnfL} = \frac{p}{\pi Bmn}$$

or $$E_g = \frac{2(m+n-2)S + 4B}{\pi Bmn} \qquad \text{... (8.89)}$$

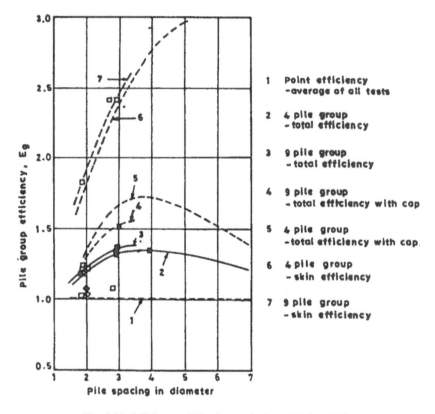

Fig. 8.37. Efficiency of Pile Groups in Sand (Vesic, 1967).

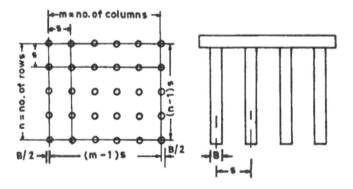

Fig. 8.38. Computation of Pile Group Efficiency.

For 100% efficiency

$$S = \frac{1.57\,Bmn + 2B}{m+n-2}$$

... (8.90)

With square piles, width of pile may be substituted for B and $4 \times$ width for πB.

Feld (1943) proposed a simple thumb rule of computing group efficiency by simply reducing pile capacity by 1/16 for each adjacent pile.

The converse-Labour equation had been widely used to compute pile group efficiencies and is given by

$$E_g = 1 - \frac{(n-1)m+(m-1)n}{90mn}\theta \qquad (8.91)$$

where $\theta = \tan^{-1}\dfrac{B}{S}$.

It may be noted that Eq. (8.91) is also based on the skin resistances of a single pile and block skin resistance.

Arrangement of piles

As mentioned earlier, the pile tops are connected together with a R.C.C. pile cap. Pile cap is designed as an individual footing subjected to column load plus weight of the pile cap and the soil above the cap. Wherever the conditions permit, the piles should be arranged in the most compact geometric form in order to keep the stresses in the pile cap to a minimum. These geometric forms are shown in Fig. 8.39.

8.19 BEARING CAPACITY OF A PILE GROUP

Terzaghi and Peck (1967) suggested that the pile groups may fail as a unit (Fig. 8.40) and recommended Eq. (8.92) for getting the ultimate bearing capacity of the pile group,

$$Q_{ug} = fLp + q_{ult}A - \gamma LA , \qquad \ldots (8.92)$$

where Q_{ug} = upper limit of pile group capacity, not exceeding the ultimate capacity of a single pile times the number of piles in the group,

f = shear resistance of soil along the vertical surface of the block

= 1/2 × unconfined compression strength for cohesive soils = earth pressure at rest × tan δ, for granular soil with angle of wall friction δ,

L = length of pile embedment in soil,

p = perimeter of area enclosing all the piles in the group,

q_{ult} = ultimate bearing capacity of soil at the level of pile tip,

A = area enclosing all the piles in the group, and

γ = unit weight of soil within the block $L \times A$.

The Eq. (8.92) is based on the assumptions that (i) the pile cap is perfectly rigid, and (ii) the soil contained within the periphery circumscribing all the piles behaves as a solid block. The following points are important to note:

(a) If $Q_{ug} < nQ_u$, the bearing capacity of the pile group is governed by Q_{ug}. n represents the

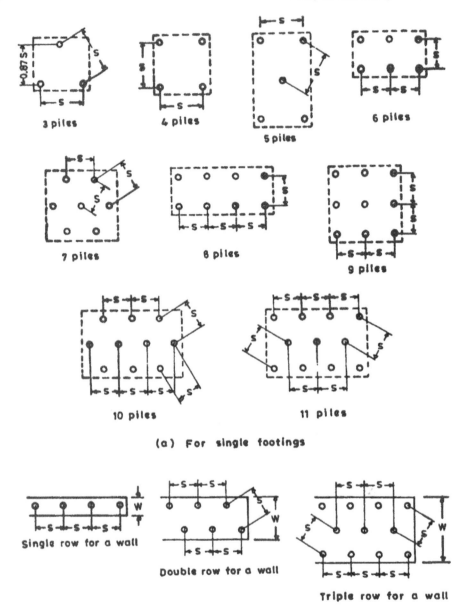

(a) For single footings

(b) For foundation walls

Fig. 8.39. Typical Arrangement of Pile Group.

number of piles and Q_u is the ultimate bearing capacity of the individual pile. This case indicates that the pile spacing is too small to utilise the full capacity of individual piles.

(b) If $Q_{ug} > nQ_u$, then nQ_u will represent the capacity of the pile group. This indicates that the spacing is large and can be reduced to the limit where $Q_{ug} = nQ_u$.

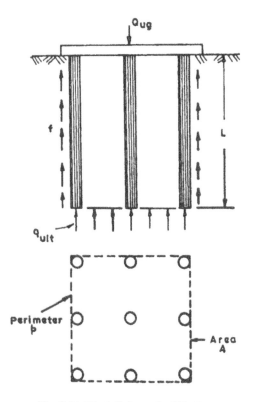

Fig. 8.40. Block Failure of a Pile Group.

8.20 SETTLEMENT OF PILE GROUP

General

The installation piles usually alters the deformation and compressibility characteristics of the soil mass which controls the behaviour of single pile under the load, yet this influence extends only to a few pile diameters below the pile points. Accordingly, as suggested by Terzaghi and Peck (1967), the total settlement of a group of driven or bored piles under the safe design load not exceeding one-third to one half of the ultimate group capacity can generally be estimated roughly as for an equivalent pier foundation. The deformation and compressibility properties of the soil beneath the pile group can be estimated from emprical correlations with the results of field tests, e.g. penetration tests, plate load tests, etc. or from laboratory tests on undisturbed samples of cohesive soil. The estimates should also be checked by pile load tests for possible extrapolation to group behaviour.

Stresses on underlying strata

Since the use of elastic theory for determination of vertical stresses in the soil surrounding and below the pile tips is extremely laborious for practial cases, approximate methods are proposed. These methods are discussed below:

(i) For friction piles two cases may be considered. In case I, the load is assumed to spread

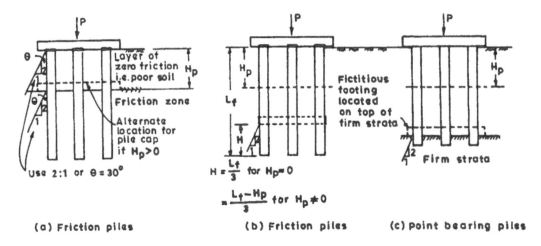

Fig. 8.41. Simplified Computation of Soil Stresses Beneath a Pile Group.

from a fictitious rigid footing located at the top of the layer, providing friction resistance at a 2:1 slope (or 30°) (Fig. 8.41a). For a homogeneous stratum this is at the ground surface. In case II, the load is placed on a fictitious rigid footing located at $L_f/3$ from the bottom of the piles (average depth), with L_f as in Fig. 8.41b. The spread-out of load is also taken at either 2:1 or 30°. Case I or II should be used, whichever gives the largest computed stresses on underlying strata.

(ii) For point-bearing piles in dense sand-gravel deposits, the fictitious footing is placed on the deposit in which the piles penetrate. Again, the load is spread at a 2:1 or 30° slope, as shown in Fig. 8.41c.

The soil at or below the imaginary raft must carry the load without excessive deformation.

The above methods are approximate and must be used with caution. In some cases, the settlement calculated by these methods are smaller than the measured values. However, these simple approximations give sufficient information for determining the supporting strength of the lower strata.

Settlement of pile group in sand

On the basis of imaginary raft (Fig. 8.41), the preliminary estimate of settlement of a pile group in sand can be made from standard penetration test or static cone penetration test data as for shallow foundations. These methods have already been described in Chapter 7.

The common practice of estimating the settlement of pile group in sand is to extrapolate this from the settlement of an individual pile measured in a load test. On the basis of few available observations, Skempton (1953) obtained the variation of settlement ratio S_g / S_i with B_g as shown in Fig. 8.42, where S_g and S_i are respectively the settlement of pile group of width B_g and individual pile, for the same average load Q per pile in the group. It is evident from this figure that the settlement ratio can have a value up to 16 which means that the settlement of a pile group in sand may be up to 16 times that of individual pile depending on the width of pile group. Regression analysis yields:

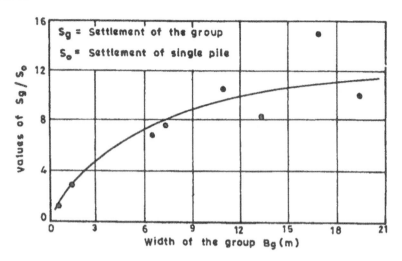

Fig. 8.42. S_g/S_i Versus B_g Plot.

$$\frac{S_g}{S_i}=\left(\frac{4B_g+2.7}{B_g+3.6}\right)^2 \qquad \text{... (8.93)}$$

where B_g is width of pile group in metres.

Meyerhof (1959) expressed the settlement ratio for square pile groups driven in sand as

$$\frac{S_g}{S_i}=\frac{S(5-S/(3B))}{B(1+1/m)} \qquad \text{... (8.94)}$$

where S = spacing between piles,

B_g = width of pile group,

m = number of rows in the pile group, and

B = pile diameter.

As a preliminary guide, for normal loads level (safety factor > 3), the settlement S_i of displacement piles can be estimated from the following empirical formula (Vesic, 1970, 1977):

$$S_i=\frac{B}{1000}+S_e \qquad \text{... (8.95)}$$

where B = pile diameter, mm,

S_e = elastic compression of pile, mm, and

S_i = settlement of single pile, mm.

The elastic compression of the pile Eq. (8.95) is commonly obtained as:

$$S_e = \frac{1000QL}{AE}$$ ... (8.96)

where Q = applied load, kN,
             A = average cross-sectional area, m^2,
             L = length of pile, m, and
             E = modulus of elasticity of pile (kPa).

Settlement of pile group in clay

The settlement of pile group in clay is due to consolidation settlements. It cannot be estimated from the data of a load test on a single pile because the time effect, effect of remoulding of soil due to pile driving and scale effect are quite different for the single test pile and the group of piles. The widely used approach to calculate the settlement of a pile group in clay is the equivalent raft approach. The equivalent raft may be considered at a position as illustrated in Fig. 8.41 depending on the type of pile.

8.21 NEGATIVE SKIN FRICTION IN A PILE GROUP

Negative skin friction will act on the piles of a group where the fill and any underlying compressible clay moves downwards relative to the shaft (Fig. 8.43). The magnitude of negative skin friction in a pile group may be taken as the higher value given by Eqs. (8.97) and (8.98),

$$F_{ng} = F_n \cdot n$$ ... (8.97)

$$F_{ng} = c_u \cdot L_c \cdot p_g + \gamma \cdot L_c \cdot A_g$$ ... (8.98)

where F_{ng} = negative skin friction of pile group,
             F_n = negative skin friction of a single pile Eqs. (8.32) and (8.33)
             n = number of piles in the group,
             p_g = perimeter of the group,
             γ = unit weight of the soil within the pile group up to a depth L_c,
             A_g = area of the pile group within the perimeter p_g, and
             L_c = length of pile on which negative skin friction acts.

The factor of safety is defined as

$$\text{Factor safety} = \frac{\text{ultimate load capacity of the pile group}}{\text{working load} + \text{negative skin friction load}}$$

8.22 UPLIFT CAPACITY OF A PILE GROUP

The uplift resistance of a pile group is taken as the lesser of the following two values:

(1) Sum of uplift resistance of piles in the group, and

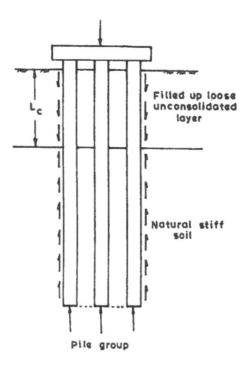

Fig. 8.43. Negative Skin Friction in a Pile Group.

(2) Sum of shear resistance mobilized on the surface perimeter of the group plus total weight of soil and piles enclosed in the perimeter.

8.23 ULTIMATE LATERAL LOAD RESISTANCE OF A PILE GROUP

The ultimate load resistance of a pile group may be obtained on the concept of group efficiency.

$$(Q_{hu})_G = n \cdot Q_{hu} \cdot \eta_1$$

where $(Q_{hu})_G$ = ultimate lateral load capacity of a group,

n = number of piles in a group,

Q_{hu} = ultimate lateral load capacity of a single pile, and

η_1 = group efficiency under lateral loads.

Cohesionless soils

A series of model pile groups in cohesionless soils were tested for lateral loads by Oteo (1972). The model test results indicated the values of η_1 as given in Table 8.15.

Cohesive soils

A series of model pile groups had been tested for lateral loads in clay by Prakash and Saran (1967). The model test results indicated the values of η_1 as given in Table 8.16.

It is evident from Tables 8.15 and 8.16, that η_1 value increases as S / B increases, S and B are respectively the centre to centre spacing between the piles and width/diameter of pile. Further the η_1 value in cohesive soil is smaller than in cohesionless soil. However, to have more definite information, there is a need to carry out further laboratory and full-scale tests on pile groups.

For $S / B > 8$, η_1 value may be taken as unity.

Alternatively the lateral load capacity of a single pile using subgrade reaction method for a given horizontal displacement may be obtained for the reduced value of coefficient of subgrade reaction which is equal to the η_h or K_h multiplied by group reduction factor. Davisson (1970) has suggested the values of group reduction factor as given in Table 8.17.

Pile group capacity under lateral load will be then sum of the individual pile capacities calculated on the basis of reduced η_h or K_h value.

The permissible value of lateral deflection of a pile is recommended as 12 mm (McNulty, 1957).

Table 8.15. Group Efficiency η_1 for Cohesionless Soils (Oteo, 1972).

S/B	η_1
3	0.50
4	0.60
5	0.68
6	0.70

Table 8.16. Group Efficiency η_1 for Cohesive Soils (Prakash and Saran, 1967).

S/B	η_1
3.0	0.40
3.5	0.45
4.0	0.50
4.5	0.55
5.0	0.60

Table 8.17. Group Reduction Factor for Coefficient of Subgrade Reaction (Davisson, 1970).

S/B	Group Reduction Factor for η_h or K_h
3	0.25
4	0.40
6	0.70
8	1.00

8.24 DISTRIBUTION OF LOAD BETWEEN VERTICAL PILES OF PILE GROUP SUBJECTED TO ECCENTRIC LOADING

Eccentric vertical loadings may result from space limitations or variable loads. Individual pile loads may be obtained using elastic theory by the following methods.

(a) Eccentricity about one axis

When eccentricity is about one axis only (Fig. 8.44a):

$$V_{pi} = \frac{V}{n} \pm \frac{V \cdot e \cdot x_i}{I_g} \cdot A \qquad \qquad \text{... (8.99)}$$

where V_{pi} = load on ith pile,

V = total vertical load acting on the pile group,

n = total number of piles,

e = amount of eccentricity with respect to the centre of the pile group,

x_i = distance of the centre of the ith pile from the centre of the pile group, measured parallel to e,

I_g = moment of inertia of the piles about the axis normal to the direction of eccentricity

$$= Ax_1^2 + Ax_2^2 + \dots Ax_n^2 \,,$$

A = area of pile cross-section, and

$x_1, x_2 \dots x_n$ = distances from centre of gravity of pile group to the line of each pile, measured parallel to e.

Since all the piles in the group are assumed to be identical, Eq. (8.99) may be written as:

$$V_{pi} = \frac{V}{n} + \frac{V \cdot e \cdot x_i}{\Sigma x_i^2}. \qquad \qquad \text{... (8.100)}$$

(b) Eccentricity about two axes

When there is eccentricity about both axes, individual pile loads may be determined by the method of superposition (Fig. 8.44b). The formula in this case is

$$V_{pi} = \frac{V}{n} \pm \frac{V \cdot e_y \cdot y_i}{I_x} \cdot A \pm \frac{V \cdot e_x \cdot x_i}{I_y} \cdot A \qquad \qquad \text{... (8.101)}$$

where V, n and A have same meaning as in Eq. (8.99).

e_x = amount of eccentricity with respect to centre of pile group measured along X-axis,

e_y = amount of eccentricity with respect to centre of pile group measured along Y-axis,

I_x = moment of inertia of the piles about X-axis

$$= A \cdot y_1^2 + A \cdot y_2^2 + \dots + A \cdot y_n^2 = A \Sigma y_i^2 \,,$$

I_y = moment of inertia of piles about Y-axis

$$= A \cdot x_1^2 + A \cdot x_2^2 + \dots + A \cdot x_n^2 = A \Sigma x_i^2$$

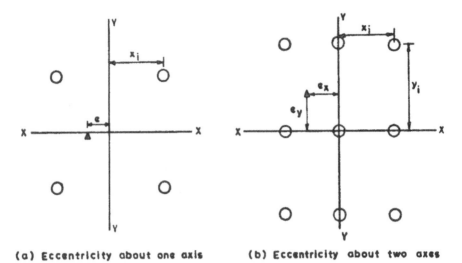

(a) Eccentricity about one axis   (b) Eccentricity about two axes

Fig. 8.44. Eccentric Loadings on Vertical Piles from Vertical Loads.

x_i = distance from the centre of gravity of the pile group to the line of each pile, measured parallel to X-axis, and

y_i = distance from the centre of gravity of the pile group to the line of each pile, measured parallel to Y-axis.

Equation (8.101) may be written as

$$V_{pi} = \frac{V}{n} \pm \frac{V \cdot e_y \cdot y_i}{\Sigma y^2} \pm \frac{V \cdot e_x \cdot x_i}{\Sigma x^2}$$

... (8.102)

8.25 DISTRIBUTION OF LOAD BETWEEN VERTICAL AND BATTER PILES OF A PILE GROUP

There are several methods for analysis of the pile foundations involving vertical and batter piles. Some of the methods commonly used are discussed below.

(a) Culmann's graphical method

In this method piles are grouped according to their slopes. It is assumed that all piles are subjected to axial load only and that all piles in each group are subjected to equal axial load. Based on these assumptions the centre of reactions can be located. The method may be described stepwise as follows (Fig. 8.45):

 (i) Sketch a profile of the pile foundation and locate the centre line of each group of parallel piles.
 (ii) Draw the resultant R of all external forces applied on the pile foundation. R intersects the centre line of the pile group 1 (vertical piles) at point a.
 (iii) Intersect centre line of group 2 and centre line of group 3 at b. Connect ab.

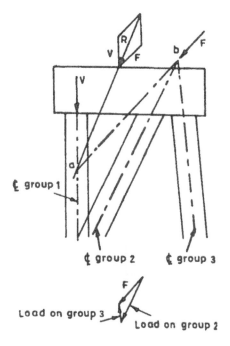

Fig. 8.45. Culmann's Graphical Method.

(iv) Resolve R into components V and F. V is vertical and F is parallel to line ab.
(v) Group 1 is subjected to total axial force V. Group 2 and group 3 are subjected to force F.
(vi) Resolve force F into axial loads along centre line of group 2 and centre line of group 3.

In the cases where piles are arranged in not more than three directions, the solution by this method is statically determinate. For cases where piles are driven in more than three directions, this method may be used if further compounding of pile reactions is made.

(b) Analytical method

In this method, the forces in each pile are determined. These forces are not equal because all piles in a group are not at the same distance from the resultant. The method is described below (Fig. 8.46).

(i) Resolve resultant force R into a vertical component V and a horizontal component H.
(ii) Ignore the horizontal component and treat the pile group as if all piles were vertical. Therefore,

Pile load $$V_i = \frac{V}{n} \pm \frac{V \cdot e_x \cdot x_i}{\Sigma x_i^2} \qquad \qquad \dots (8.103)$$

(iii) Each pile is assumed to be subjected to an axial load, R_1, R_2 ..., whose vertical component is equal to the corresponding vertical reaction determined in step ii. This

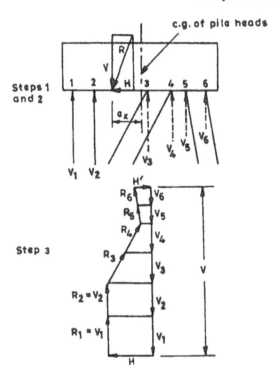

Fig. 8.46. Analytical Method.

can be done analytically:

$$R_i = \frac{V_i}{\cos\theta}$$

where θ = the angle between the pile and the vertical.

Hypothetically, the batter piles should have slopes such that the force polygon should close up (i.e. $H' = 0$), Fig. (8.46). However, it is considered acceptable if H' is less than 4 kN per pile.

8.26 DISTRIBUTION OF LATERAL LOAD IN A PILE GROUP

The response of group of piles under lateral loads differs from that of a single pile because of the following reasons:

(i) The behaviour of one pile in a group is quite different from that of others due to interference of zone of influence of one pile with that of the adjacent and other piles. For example, in group of pile shown in Fig. 8.47, pile 'a' stresses the soil outside the pile group, whereas piles 'b' and 'c' stress the soil immediately in front of their locations inside the pile group itself (Prakash, 1962).

Wen (1955) and Prakash (1962) carried out model tests on the pile groups containing four piles and nine piles. The loads taken by the piles were found as given in Table 8.18.

(ii) Depending upon the rigidity of the connection of pile cap with the superstructure, the pile

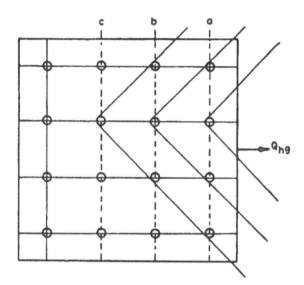

Fig. 8.47. Pile Group Subjected to a Lateral Load.

heads in a group may be fully or partially fixed. The fixity of the pile head reduces the deflection by a large amount.

Keeping the above facts in view, the maximum lateral load taken by a pile in a pile group may be taken as

$$Q_h = 1.2 \frac{Q_{hg}}{n} \qquad \dots (8.104)$$

where Q_h = lateral load per pile in a group,
$\quad n$ = total number of piles, and
$\quad Q_{hg}$ = lateral load acting on the pile group.

Table 8.18. Distribution of Lateral Load in Piles of a Pile Group.

	Front Pile	*Central Pile*	*Rear Pile*
Wen (1955)	57%	—	43%
Prakash (1962)	55 to 57%	—	43 to 45%
Wen (1955)	43%	34%	23%
Prakash (1962)	40%	32%	28%

8.27 HRENNIKOFF'S METHOD

Hrennikoff (1949) suggested a method for analysis of pile foundations with vertical and batter piles subjected to external forces such as vertical load V, lateral load H and moment M. Figure 8.48 represents a foundation comprising vertical and batter piles. Group 1 comprises n_1 piles making an angle θ_1 with the foundation base; group 2 comprises n_2 piles at an angle

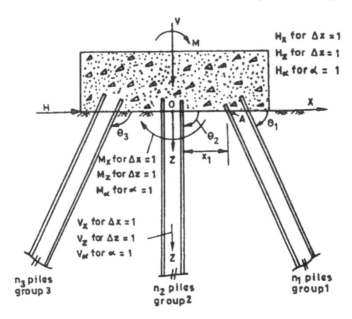

Fig. 8.48. Pile Group Subjected to Combined Vertical and Horizontal Loads, and Moment with Definition of Foundation Constants.

θ_2; group 3 comprises n_3 piles at angle θ_3 and so on. Let the total number of pile groups be N. The angle θ is measured from the positive direction of the X-axis clockwise to the given pile. The analysis gives (i) the lateral, vertical and rotational deflections of the pile cap, and (ii) the axial loads, shear loads and moments at each of the pile heads.

Assumptions
1. The pile cap in which the pile heads are embedded is rigid.
2. All piles behave alike with regards to the load deformation relation. The load carried by each pile is proportional to the displacement of pile head which consists of three components, namely, (a) the axial displacement; S_a, (b) the transverse displacement, y_t, and (c) the rotational displacement, α.
3. The pile cap movements are small.
4. The problem is two-dimensional, that is, the piles, as well as the external forces, are arranged in planes transverse to the length of the foundation and they are symmetrical with regard to the transverse middle plane.

Pile constants
The pile constants are defined as the forces with which the pile acts on the foundation when pile head is given a unit displacement (Fig. 8.49). There are five pile constants (K_a, m_t, K_t, m_α, K_α) associated with the displacements S_a, y_t and α. By Betti's reciprocal theorem, $K_\alpha = m_t$; there are only four independent constants characterising the load deformation relation of a pile when the pile is embedded in the footing.

Foundation constants

The foundation constants are defined as the resultant force with which all piles act together on the footing, when the footing is given a unit translation displacement in the positive direction of one of the axes or a unit rotation about the origin in the clockwise direction. Each unit displacement in general brings into play a resisting force, which may be represented by its two components acting along the coordinate axes and the moment about the origin as shown in Fig. 8.48.

The constants (H_x, V_x, M_x), (H_z, V_z, M_z), and $(H_\alpha, V_\alpha, M_\alpha)$ are obtained by giving the foundation (i.e. pile cap) displacement $\Delta x = 1$, $\Delta z = 1$ and $\alpha = 1$ as shown in Fig. 8.48. Thus, there are nine constants. According to Betti's reciprocal theorem

$$V_x = H_z, \ M_x = H_\alpha \text{ and } M_z = V_\alpha.$$

There are, therefore, only six independent constants which may be taken as H_x, H_z, H_α, V_z, V_α and M_x. The positive signs of these functions correspond to the positive directions of the axes and to the clockwise direction of moments.

These foundation constants can be evaluated in terms of the pile constants K_a, K_t, m_t and m_α, and the geometry of the foundation.

Equations for foundation constants

Firstly consider the movement of the footing by a unit distance in x-direction (i.e. $\Delta x = 1$, Fig. 8.50a). Consider the head of one typical pile A. The forces that would be developed due to the movement of this pile from A to A_1 are shown in the figure.

Resolving the forces $K_a \cos \theta_1$ and $K_t \sin \theta_1$ in x and z directions, we get

$$H_{x1} = -\left(K_a \cos^2 \theta_1 + K_t \sin^2 \theta_1\right) \qquad \text{... (8.105)}$$

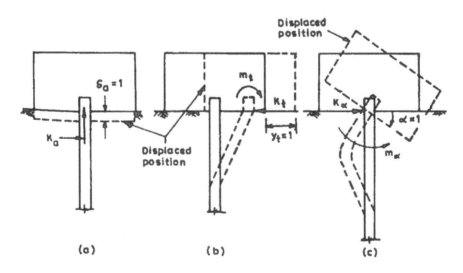

Fig. 8.49. Definition of Pile Constants.

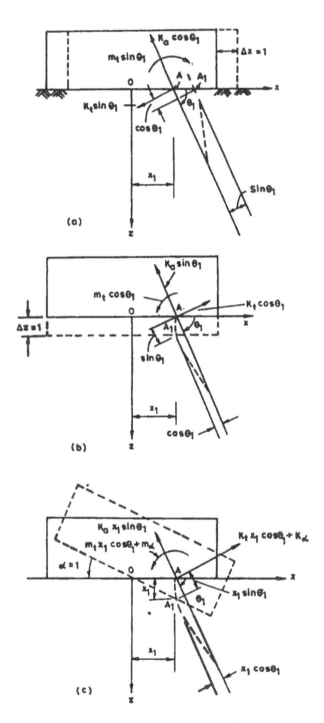

Fig. 8.50. Development of Foundation Constants.

$$V_{x1} = -\frac{1}{2}(K_a - K_t)(\sin 2\theta_1) \qquad \dots (8.106)$$

The sums of the components of the induced forces in all the N groups of piles in the directions of axes represent H_x and V_x.
Therefore,

$$H_x = -\Sigma n\left(K_a \cos^2\theta + K_t \sin^2\theta\right) \qquad \dots (8.107)$$

$$V_x = -\frac{1}{2}(K_a - K_t)\Sigma\left(n\sin 2\theta\right). \qquad \dots (8.108)$$

The symbol Σ covers N terms of the form expressed in the bracket.

The third foundation constant M_x due to $\Delta x = 1$ may be obtained by taking the moments of the resistance of the piles about the origin. Summing up the moments of all N groups, we get

$$M_x = -\frac{1}{2}(K_a - K_t)\Sigma\left(n\bar{x}\sin 2\theta\right) + m_t\Sigma\left(n\sin\theta\right) \qquad \dots (8.109)$$

where $\bar{x}$ = coordinate of the centre of the group of n piles having the angle θ.

The forces generated by giving the displacement of the footing $\Delta z = 1$ and $\alpha = 1$ are shown in Fig. 8.50b and 8.50c respectively.

Proceeding exactly in a similar manner as illustrated above for $\Delta x = 1$, we may obtain

$$V_z = -\Sigma\left[n\left(K_a \sin^2\theta + K_t \cos^2\theta\right)\right] \qquad \dots (8.110)$$

$$M_z = -\Sigma\left(K_a \sin^2\theta + K_t \cos^2\theta\right)n\bar{x} - m_t\Sigma\left(n\cos\theta\right) \qquad \dots (8.111)$$

$$M_\alpha = -\Sigma\left[\left(K_a \sin^2\theta + K_t \cos^2\theta\right)\Sigma(x)^2\right] - 2m_t\Sigma n\bar{x}\cos\theta - m_\alpha\Sigma n. \qquad \dots (8.112)$$

The foundation constants determined by Eqs. (8.107) through (8.112) with their proper signs represent forces and moments exerted by the piles on the footing and not by the footing on piles. Positive signs of these functions indicate forces acting in the positive directions of the coordinate axes or, in the case of moments, in the clockwise direction. The X-coordinates and the trigonometric functions in their expressions must be taken with proper algebraic signs.

The equations (8.107) through (8.112) assume full fixity of piles in the footing. If pin-ended conditions exist, the corresponding formula for pile constants may be obtained by substituting $m_t = m_\alpha = 0$ in the equations.

Equations in equilibrium

The equations of equilibrium of the footing may be written as:

$$H_x\Delta x + H_z\Delta z + H_\alpha\alpha + H = 0 \qquad \dots (8.113)$$

$$H_z \Delta x + V_z \Delta z + V_\alpha \alpha + V = 0 \qquad \qquad \text{... (8.114)}$$

$$H_\alpha \Delta x + V_\alpha \Delta z + M_\alpha \alpha + M = 0 \qquad \qquad \text{... (8.115)}$$

The component displacements of the footing ΔX, ΔZ and α may be determined from Eqs. (8.113) to (8.115), provided values of the foundation constants are known. Once these component displacements are known, the displacements of the individual pile heads found by geometry and the axial, lateral and rotational pile loads may then be computed.

Pile displacements and loads

Expressions for pile head displacements may be written as follows with reference to an arbitrary pile A (shown in Fig. 8.48) situated at a distance of x_1 from the origin O.

The longitudinal displacement δ_a downward,

$$\delta_a = \Delta x \cos\theta_1 + \Delta z \sin\theta_1 + \alpha x_1 \sin\theta_1. \qquad \qquad \text{... (8.116)}$$

The transverse displacement y_t to the right,

$$y_t = \Delta x \sin\theta_1 - \Delta z \cos\theta_1 - \alpha x_1 \cos\theta_1. \qquad \qquad \text{... (8.117)}$$

The rotation of the pile head is α, clockwise. The expressions for the pile loads are as given below. The axial load of pile (compression)

$$V_g = K_a \delta_a \qquad \qquad \text{... (8.118)}$$

The transverse load, acting on the foundation to the right

$$H_g = -K_t \cdot y_t + m_t \cdot \alpha \qquad \qquad \text{... (8.119)}$$

The moment, acting on a foundation clockwise,

$$M_g = m_t y_t - m_\alpha \alpha. \qquad \qquad \text{... (8.120)}$$

Values of pile constants

Values of pile constant may be obtained using following equations:

$$K_a = \frac{AE}{L} \quad \text{for point bearing piles} \qquad \qquad \text{... (8.121a)}$$

$$= \frac{2AE}{L} \quad \text{for friction piles} \qquad \qquad \text{... (8.121b)}$$

$$K_t = \frac{2.56EI}{A_y T^3} \qquad \qquad \text{... (8.122)}$$

$$m_t = \frac{2.35}{A_y} \cdot \frac{EI}{T^2} \qquad \qquad \text{... (8.123)}$$

$$m_\alpha = \frac{3.54}{A_y} \cdot \frac{EI}{T}. \qquad \qquad \text{... (8.124)}$$

The values of A_y may be taken from Table 8.11 for different batters. The relative stiffness factor T is obtained using Eqs. 8.57 and 8.80.

8.28 LATERAL PILE LOAD TEST

Lateral load capacity of a single pile can be obtained by performing a lateral pile load test. In this, the lateral load is applied to the test pile by using a hydraulic jack and a suitable reaction system (Fig. 8.51). The amount of the lateral load applied is measured either by a calibrated load cell or a pressure gauge. The pile is tested by applying the lateral load in 10 steps with a maximum of twice the design lateral load. For each increment of lateral load, the lateral movement of the pile head is measured by dial gauges.

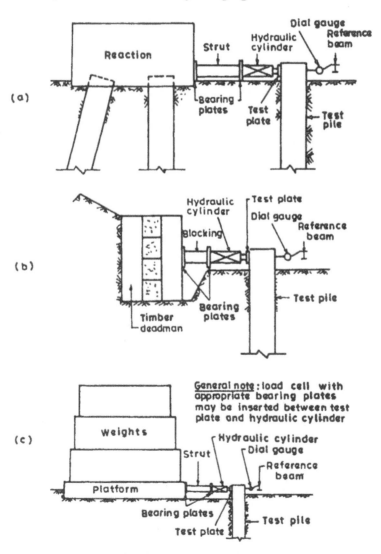

Fig. 8.51. Typical Set-ups for Applying Lateral Loads.

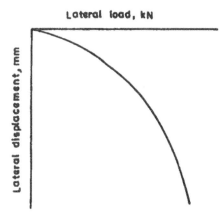

Fig. 8.52. Lateral Load Versus Lateral Displacement of Curve.

Lateral load versus deflection curve is then plotted (Fig. 8.52). Allowable lateral load capacity will be the least from the following criteria:

  (i) Half the final lateral load at which lateral movement of pile is 12 mm.
 (ii) Lateral load corresponding to 5 mm lateral movement.
(iii) Lateral load corresponding to allowable lateral movement.

8.29 PROPORTIONING AND DESIGN OF PILE FOUNDATIONS

The selection of the type of pile, shape, size and length, number of piles required to carry the design load, spacing between the piles and arrangement of piles in the group—all these put together constitute proportioning. To proportion a pile foundation, with the knowledge of the various aspects of pile foundations discussed in the previous sections, one may have to use judgement based on his own experience and that of others reported in the literature.

There may be three cases that can occur in practice, namely (a) group of vertical piles subjected to a central vertical load, (b) group of vertical piles subjected to an eccentric vertical load, and (c) group of piles subjected to vertical load, lateral load and moment. The method of proportioning the pile foundation has been discussed below for the three cases separately.

(a) Group of vertical piles subjected to a central vertical load

The various steps involved in proportioning a pile foundation can be outlined as follows:

(i) *Selection of type of pile*—First a soil profile representing the results of exploratory boring should be prepared. Usually, the soil profile provides all the information required to decide appropriate type of pile for the given situation. The selection of the type of pile (i.e. whether driven or cast-in-situ) is governed at least partly by practical considerations.

Driven piles can be precast and installed according to specifications. The pile work can be programmed in advance. But they cause vibrations which may affect the neighbouring structures. They are not suitable in soils of poor drainage qualities and this would lead to development of pore pressures or lifting of adjacent piles.

Cast-in-situ piles are suitable where the length can be adjusted according to pile conditions. They are ideal in places where vibrations are required to be avoided or where the soil has poor drainage qualities. But installation of these piles require great care and are not suitable in underground flowing water. The cast-in-situ piles take very much less load as compared to driven piles. The advantages and disadvantages of driven and cast-in-situ piles are midway between the two types of piles mentioned above.

(*ii*) *Length of pile*—If point-bearing piles are appropriate, it may be possible to judge the required length with reasonable accuracy on the basis of the soil profile. A reliable estimate of the length of friction piles in sand can be made only by carrying out load tests on piles (initial test). The length of friction piles in soft clay can be determined by making an estimate of the factor of safety of the pile group against complete failure.

(*iii*) *Size of pile*—Factors affecting the selection of size of pile are as follows:
   (a) The permissible compressive stress on each pile determines the area required or diameter of pile.
   (b) Pile size should be minimised in order to reduce downdrag force if applicable, minimise weight and reduce cost.
   (c) It may in some cases be economical to increase pile size if by so doing the number of piles can be reduced to a practical minimum.
   (d) The question of drive ability of a chosen pile size in a given stratum also needs to be examined.

(*iv*) *Estimation of pile capacity*—The ultimate bearing capacity of a single pile is either computed using methods described in Sec. 8.3 to 8.10 or else determined by means of load tests. This value is divided by appropriate factor of safety to obtain the 'safe design load load per pile'.

(*v*) *Spacing between piles in a group*—The distance S between the centres of piles depends mainly on the type of the pile and soil. It should be carefully chosen as detailed in Sec. 8.18.

(*vi*) *Geometry of the pile lay-out*—The piles are arranged in most compact form (Fig. 8.39). Usually, the piles are arranged in either a square or triangular pattern. By multiplying the number of piles by S^2 (square pattern) or by $1/2S^2$ $\sqrt{3}$ (triangular pattern) the total area required for the pile supported parts of the foundation is obtained. If this area is considerably smaller than half the total area covered by the structure, the structure is established on pile supported footings.

If it is considerably greater, the structure is founded on a pile supported raft, and the spacing on the piles is increased so that the pile layout forms a continuous pattern. If the intensity of loading on different parts of the raft is very different, the spacing between piles is adopted to the intensity on each of the parts. Finally, if doubtful whether the structure should be established on footings or on a raft, the decision is made after a comparison of the costs of the two alternatives.

(*vii*) *Stability of the pile group*—If the foundation is supported by friction piles in soft clay or plastic silt, the ultimate bearing capacity of the pile group should be estimated and the design load must not be allowed to exceed one third of the ultimate value. If this condition is not satisfied, the piles and the soil located between the piles may sink into the ground, although the load per pile may be within the "safe design load".

(*viii*) *Estimation of settlement*—Adequate bearing capacity does not ensure that the pile will not settle excessively, because the settlement of a pile group may have no relation whatsoever to the settlement of a single pile under the load per pile assigned to the foundation. The settlement of a pile group may range from a few millimetres to several centimetres, depending on the soil conditions, the number of piles and the area covered by the structure. Settlements of less than about 50 mm are commonly not harmful, but settlements of 150 mm or more may have very undesirable effects on the superstructure. Hence if a foundation rests on friction piles driven into soft clay or if the points of point bearing piles are located above soft strata, settlement computations should be carried out.

(*ix*) *Design of pile*—Load shared by different piles: Firstly the load taken by each pile is obtained as given below:

$$V_p = \frac{V}{n} \qquad\qquad \ldots (8.125)$$

where V_p = load taken by each pile,

V = total load acting on the pile, and

n = number of piles.

Effective length of pile—A pile is usually designed as a column for the load coming on it.

When a finished pile projects above ground level and is not secured against buckling by adequate bracing, its effecitve length will be governed by the end conditions imposed on it by the structure it supports and by the nature of the ground into which it is embedded (IS : 2911-pt. I-1979).

In good ground, the lower point of contraflexure can be taken to be at a depth below ground level of about one-tenth of the exposed length. If the top stratum is soft clay or silt, the lower point of contraflexure is taken at about half the depth of penetration into this stratum, but not less than one-tenth of the exposed length of pile. In case the ratio of effective length of pile to the least lateral dimension exceeds 12, the pile should be designed as a long column. If the pile is fixed in position and direction at the top end, the upper point of contraflexure may be taken at one-fourth of the exposed length below the top of pile. If the pile is not so fixed the end conditions should be taken into account in determining the effective length.

Krishna and Jain (1966) suggest that when piles are designed as columns, the length of pile may be considered as 2/3 of the length when embedded in soft strata or 1/3 of the length when embedded in firm stratum and the rest of the length of pile projecting above ground. The end conditions are nearly one end fixed and the other end hinged.

Reinforcement in pile—The reinforcement in the pile is obtained using Eq. (5.25) or (Eq. 5.26). The procedure has already been described in Chapter 5.

IS : 2911 (Part I–1979) recommends the following reinforcement for precast concrete piles.

(a) *Longitudinal reinforcement*—The area of main longitudinal reinforcement should not be less than the following percentages of the cross-sectional area of the pile:

(1) For piles with length less than 30 times the least width $1\frac{1}{4}$ per cent

(2) For piles with length 30 to 40 times the least width $1\frac{1}{2}$ per cent

(3) For piles with length greater than 40 times the least width two per cent

Permissible increase in stress for reinforcement should be as per IS: 456-2000.

(b) *Lateral reinforcement*—These should be in the form of hoops or links and should not be less than 5 mm in diameter. The volume of reinforcement should be as follows:

(1) In the body of the pile — Not less than 0.2 per cent of the gross volume of piles

(2) At each end of pile for a length of about three times the least width — Not less than 0.6 per cent of the gross volume of piles

The transition between closer spacing and the maximum should be gradual over a length of three times the least width.

The cover, clear from the main or longitudinal reinforcement including binding wire should not be less than 50 mm and generally, where piles are exposed to sea water or other corrosive content, the cover should not be less than 60 mm.

Stresses during driving—Precast concrete piles should also be checked for stresses developed in the pile during driving. Driving stress is given by

$$\text{Driving stress} = \frac{\text{Driving resistance}}{A}\left(\frac{2}{\sqrt{\eta}}-1\right) \qquad \dots (8.126)$$

where A = cross-sectional area of pile.

= area of concrete + equivalent area of reinforcement, and

η = efficiency of blow.

Driving resistance may be obtained either using Engineering News formula (Eq. 8.14) or Hiley's formula (Eq. 8.15). Driving stress should not exceed the permissible stress of concrete in compression.

Stresses during handling—Precast concrete piles should be checked for the bending moments developed during lifting and erection of piles (Fig. 8.3).

(x) *Design of pile cap*—Pile caps are structural elements that tie a group of piles together. Pile caps may support bearing walls, isolated columns or groups of several columns. Pile caps are used to transmit the forces from the columns or walls to the piles. Plan dimensions of a pile cap depend on the spacing between the piles and their arrangement. The design of a pile cap is based on the same principles as involved in the design of footings and rafts. Its depth is based on the shear and/or development length for the column bars. Small pile caps may be designed as a reinforced concrete truss as shown in Fig. 8.53. Steel acts as tension

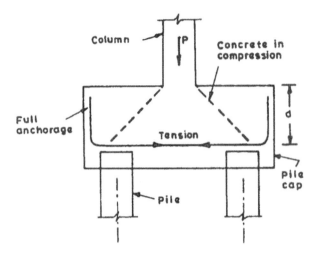

Fig. 8.53. Truss Action in Pile Cap.

chord and concrete as diagonal struts. General guidelines for the design of pile caps are as follows:

(1) Piles should be arranged so that centroid of the group coincides with the line of action of load.

(2) A clear overhang of pile cap beyond the outermost pile should be kept greater than 100 mm. This takes into account the nonverticality of the piles.

(3) (a) The pile cap should be deep enought to allow for necessary overlap of the column and pile reinforcement. (b) The pile cap should be deep enough to ensure full transfer of load from the column to the cap in punching shear.

(4) Computations of moments and shears may be based on the assumption that the reaction from any pile is concentrated at the centre of the pile. Section for determining the maximum bending moment will be the face of column (Fig. 8.54). Shear will be checked at the section located at effective depth 'd' of pile cap from the face of the column.

(5) In computing the external shear on any section a-a through a footing supported on piles as shown in Fig. 8.55, the entire reaction from any pile of diameter B whose centre x_1 is located at $0.5\ B$ or more outside the section a-a will be assumed as producing shear on the section. The shear force will be zero at a section a'-a', due to reaction from a pile whose centre x_2 is at $0.5\ B$ or more inside the section. For intermediate positions of the pile centre, the portion of the pile reaction to be assumed as producing shear on the section is based on straight line interpolation between full value when the pile centre is at $x_1 = 0.5\ B$ outside the section a-a and zero value when the pile centre is at $x_2 = 0.5\ B$ inside the section a'-a'.

(6) Minimum thickness of pile caps at edges should not be less than 300 mm.

Design of capping beams—Piles supporting a load bearing wall or closed spaced columns may be tied together by a continuous capping beam. The piles can be staggered along the line of the beam to take care of small eccentricity of loading. Where piles are used

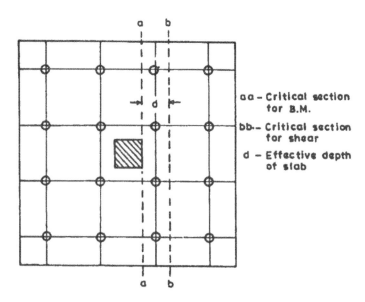

Fig. 8.54. Critical Sections for Bending Moment and Shear.

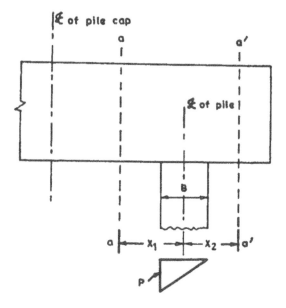

Fig. 8.55. Critical Section in Shear in Pile Cap.

to support light structures with load bearing walls, the piles may be placed at wide spacing beneath the centre of wall.

The beam is designed for a bending moment given by Eq. 8.127.

$$\text{Bending moment} = \frac{WL}{12} \qquad \qquad \text{... (8.127)}$$

where W = total weight of equilateral triangle of brick work above the beam plus self-
           weight of beam, and

           L = spacing between piles.

Design of grade beams—The beam interconnecting all the pile caps of a system of piles is referred to as a grade beam. The minimum depth of grade beam should be 150 mm and may not exceed 200 mm for usual spans of about 3 m. If grade beams are supported during construction, the maximum bending moment is taken as

$$M = wl^2 / 50 \qquad \qquad \text{... (8.128)}$$

where w = uniformly distributed load taken upto a maximum of two-storey, and

           l = effective span.

If grade beams are not supported during construction, the maximum bending moment is taken as $wl^2 / 30$. The panel action or arch action is considered in the grade beam if the ratio of depth to span is more than 0.6.

(b) Group of vertical piles subjected to eccentric vertical load

Steps (i) to (vii) explained above for vertical pile group subjected to central vertical load are applicable in this case also. Other steps may be carried out as described below:

(viii) Settlement analysis of the pile group may be carried on the concept of imaginary footing (Fig. 8.41) subjected to eccentric vertical load. The procedure of computation of settlement has already been described in Chapter 7.

(ix) Load taken by the individual pile of the group is obtained using Eq. (8.100) or Eq. (8.102). After determining the effective length of the pile, the pile carrying the maximum load is designed as column in compression. The reinforcement in the pile subjected to upward load (tension load) is obtained by.

$$\text{Reinforcement} = \frac{\text{upward load}}{0.87 f_y} \qquad \qquad \text{... (8.129)}$$

where f_y = permissible stress in steel.

(x) The pile cap is designed computing the shear and moments considering the piles on the eccentricity side.

(c) Group of vertical piles or vertical and batter piles subjected to vertical load, lateral load and moment

Group of vertical piles

Steps (i) to (viii) explained above for vertical pile group subjected to eccentric vertical load are applicable in this case also. Amount of the vertical load taken by a pile of the group is obtained using Eq. (8.100) or Eq. (8.102) neglecting the lateral load. The lateral load carried by a pile of the group is computed using Eq. (8.104). The pile having the maximum compression load and lateral load is then designed.

For the lateral load, the maximum moment in the pile is obtained by using subgrade reaction approach (Sec. 8.14). The section of the pile is then designed using reaction diagrams (Figs. 5.6 to 5.10).

The lateral deflection of the pile is then obtained using subgrade reaction approach. This deflection may be considered approximately equal to the lateral deflection of the pile group and should not exceed the permissible value of the lateral deflection (4 mm to 12 mm).

Group containing vertical and batter piles

If the lateral loads and moments acting on the pile group are substantial, it is economical to have a pile group with both vertical and batter piles. The inclination of batter piles is usually selected in such a way that all the piles are subjected to axial loads. The load in a pile may be obtained using either Culmann's graphical method or analytical method (Sec. 8.25). The pile can then be designed as an axially loaded column.

(x) The pile cap is designed structurally by computing shear and moments neglecting the lateral load and considering all the piles as vertical piles.

ILLUSTRATIVE EXAMPLES

Example 8.1

Determine the capacity of a 7.0 m long fully embedded precast concrete pile in cohesionless soil. Both standard penetration and static cone penetration tests were performed at the site. Values of standard penetration resistance and cone resistance were obtained as given below:

Depth (m)	N-Value	q_c kN/m²
1.0	5	2200
2.0	7	2400
3.0	8	2800
4.0	6	2700
5.0	10	3400
6.0	15	3100
7.0	16	3300
8.0	13	3200
9.0	20	3500
10.0	22	4000

The moist density of soil is 19 kN/m³ and water table is 5.0 m below the ground surface.

Solution

From N-values

(i) Let the diameter of pile be 0.40 m

(ii) Average value of N near the pile tip

$$= \frac{16 + 13}{2} = 14.5$$

For $N = 14.5$, $\phi = 32°$ (Fig. 7.2)

$$\frac{L}{B} = \frac{7.0}{0.4} = 17.5$$

For $L/B = 17.5$ and $\phi = 32°$, $N_q = 32$ (Fig. 8.6).
Effective overburden pressure at the pile tip,

$\qquad \sigma_v' = 19 \times 5 + 10 \times 2 = 115 \ kN/m^2$

$$A_b = \frac{\pi}{4}(0.4)^2 = 0.1256 \ m^2$$

$$Q_p = A_b \sigma_v' (N_q - 1)$$

$\qquad = 0.1256 \times 115(32 - 1) = 447.7 \ kN.$

Sand having $\phi = 32°$ may be considered medium dense sand.

From Table 8.1, $\delta = \frac{3}{4}\phi = \frac{3}{4} \times 32 = 24°$

$$K = \frac{1.0 + 2.0}{2} = 1.5$$

$$A_s = \pi \times 0.4 \times 7 = 8.796 \ m^2$$

$$\overline{\sigma}_v' = \frac{\frac{1}{2} \times 19 \times 5 \times 5 + 19 \times 5 \times 2 + \frac{1}{2} \times 10 \times 2 \times 2}{7} = 63.92 \ kN/m^2$$

$Q_s = K\overline{\sigma}_v' \tan\delta A_s$

$\qquad = 1.5 \times 63.92 \times \tan 24 \times 8.796$

$\qquad = 375.5 \ kN$

Ultimate capacity of pile = 447.7 + 375.5 = 823.2 kN

Safe capacity of pile $\quad = \dfrac{823.2}{3} = 274.4 \ kN$.

From SCPT data

(i) For point bearing resistance, take the average q_c values, as illustrated in Fig. 8.8.

$$\text{Average } q_c = \frac{3100 + 3300 + 3200}{3} = 3200 \ kN/m^2 .$$

$$\text{There} \quad Q_p = q_{c1} \cdot A_b = 3200 \times 0.1256$$

$$= 402 \ kN$$

(ii) Average of q_c values along the pile shaft

$$= \frac{2200 + 2400 + 2800 + 2700 + 3400 + 3100 + 3300}{7}$$

$$= 2842.85 \ kN/m^2$$

$$Q_s = \frac{q_{c2}}{200} \cdot A_s$$

$$= \frac{2842.85}{200} \times 8.796 = 125 \text{ kN}.$$

(iii) Safe capacity of pile $= \dfrac{402 + 125}{3} = 175.67 \text{ kN}.$

Example 8.2

Determine the capacity of a 6.0 m long bored cast-in-situ pile in medium stiff clay having variation of undisturbed strength as given below:

Depth (m)	c_u (kN/m²)
1.0	50
2.0	65
3.0	55
4.0	75
5.0	80
6.0	70

The density of soil is 18 kN/m³.

Solution

(i) Assume the pile diameter as 0.5 m

(ii)
$$Q_u = N_c \cdot c_b \cdot A_b + \alpha \, \bar{c}_u \, A_s$$

$N_c = 9.0$

$c_b = 70 \text{ kN/m}^2$

$$A_b = \frac{\pi}{4} \times 0.5^2 = 0.19625 \text{ m}^2$$

$\alpha = 0.45$ (Sec. 8.8)

$$\bar{c}_u = \frac{50 + 65 + 55 + 75 + 80 + 70}{6} = 65.8 \text{ kN/m}^2$$

$A_s = \pi \times 0.5 \times 6.0 = 9.42 \text{ m}^2$

$Q_u = 9.0 \times 70 \times 0.19625 + 0.45 \times 65.8 \times 9.42$

$\quad = 402.56 \text{ kN}$

Safe capacity $= \dfrac{402.56}{2.5} = 161.0 \text{ kN}.$

Exampsle 8.3
    An isolated 400 mm diameter reinforced concrete pile in a jetty structure is required to carry a maximum compression load of 510 kN and a net uplift load of 250 kN. The soil consists of a loose to medium-dense saturated sand (average $N = 14$) extending to a depth of 10 m below sea bed followed by dense sand and gravel (average $N = 40$). Determine required depth of penetration of the pile. Submerged density of dense sand and gravel is 12 kN/m³. Also design the pile section.

Solution
    (i) Frictional resistance of pile in the upper 10 m strata
        For $N = 14$, $\phi = 31.5°$ (Fig. 7.2)

        From Table 8.1, $\delta = \dfrac{3}{4}\phi = \dfrac{3}{4} \times 31.5 = 23.6°$ and $K = 1.5$

$$\sigma'_v = \frac{9 \times 10}{2} = 45\,\text{kN/m}^2 \ (\gamma_b = 9 \ \text{kN/m}^3)$$

$$A_s = \pi \times 0.4 \times 10 = 12.56\,\text{m}^2$$

$$Q_{s1} = K.\sigma'_v \cdot \tan\delta \cdot A_s = 1.5 \times 45 \times \tan 23.6 \times 12.56$$
$$= 370 \text{ kN.}$$

    (ii) Let us try 3.5 m penetration of pile into dense sand and gravel
        For $N = 40$, $\phi = 38.5°$ (Fig. 7.2)
        From Table 8.1, $\delta = 3/4 \times 38.5 = 28.9°$ and $K = 2.0$
        Unit skin friction in dense sand and gravel at 10.0 m depth
                = 2.0 × 90 tan 28.9 = 99.4 kN/m²
        Effective stress at 13.5 m depth
                = 90 + 12 × 3.5
                = 132 kN/m²
        Unit skin friction at 13.5 m depth
                = 2.0 × 132 tan 28.9 = 146 kN/m²
        Unit skin friction in dense sand and gravel

$$= \frac{99.4 + 146}{2} = 122\,\text{kN/m}^2 \ [> 110 \text{ kN/m}^2]$$

        Therefore, skin friction resistance in dense sand and gravel
                $Q_s = 110 \times \pi \times 0.4 \times 3.5 = 483$ kN
        Total skin friction resistance = 370 + 483 = 853 kN.

    (iii) Point bearing resistance
        For $\phi = 38.5°$ and $L/B = 13.5 / 0.4 = 33.75$, $N_q = 115$ (Fig. 8.6)

        Unit base resistance $= \overline{\sigma}'_v \left(N_q - 1\right)$

$$= 132 \times (115 - 1)$$

$$= 15048 \text{ kN/m}^2 \ [>11000 \text{ kN/m}^2]$$

Therefore, $Q_p = 11000 \times \pi/4 \times (0.4)^2$
$$= 1382.6 \text{ kN}.$$

(iv) Total capacity of pile $= 853 + 1382.6 = 2234.3 \text{ kN}$

Safe capacity of pile in compression

$$= \frac{2234.3}{3} = 744 \text{ kN} \ (>510 \text{ kN, safe})$$

Ultimate uplift capacity of pile = Ultimate skin friction resistance
$$= 853 \text{ kN}$$

Safe uplift capacity of pile $= \dfrac{853}{3} = 284 \text{ kN} \ (>250 \text{ kN, safe})$

It may be noted that the required penetration depth of the pile is governed by considerations of uplift resistance.

(v) (a) For compression load of 650 kN

The pile is designed as a column. The effective length is taken as $0.33 \times 10 = 3.3$ m considering the pile length in medium dense sand only.

Slenderness ratio $= \dfrac{3.3}{0.4} = 8.25 \ [< 12]$.

Therefore, column may be designed considering it as short column.

$$e_{min} = \frac{L_e}{500} + \frac{B}{30} = \frac{3300}{500} + \frac{400}{30} = 19.9 \text{ mm } [< 0.05 \times 400 = 20 \text{ mm}].$$

Hence the eccentricity may be neglected.

$$P_u = 0.4 f_{ck} \left(A_g - \frac{pA_g}{100} \right) + 0.67 f_y \frac{pA_g}{100}; \ A_g = \frac{\pi}{4} \times 400^2 = 125600 \text{ mm}^2$$

$$510 \times 1.5 \times 1000 = 0.4 \times 20 \left(125600 - \frac{125600p}{100} \right) + 0.67 \times 415 \times \frac{p \times 125600}{100}$$

or $6.097 = 8 - 0.08p + 2.78p$

or $p = -0.706\%$

(b) For uplift load of 250 kN

Area of steel required $= \dfrac{1.5 \times 250 \times 1000}{0.87 \times 415} = 1035 \text{ mm}^2$

$$p = \frac{1035}{125600} \times 100 = 0.824\%$$

As per IS : 2911-Pt 1

Minimum value of reinforcement in driven piles = 1.25% > 0.824%.

Therefore provide 1.25% reinforcement.

$$A_{st} = \frac{1.25}{100} \times 125600$$
$$= 1570 \text{ mm}^2.$$

Use 11 bars of 14 mm diameter.

Total area of longitudinal reinforcement $= 11 \times \frac{\pi}{4} \times 14^2 = 1692 \text{ mm}^2.$

Use 10 mm diameter lateral ties

Pitch < 400 mm

$< 16\phi_L$ or 224 mm

$< 48\phi_T$ or 480 mm.

Provide pitch as 250 mm.
Percentage of lateral reinforcement

$$= \frac{4(370 \times \pi + 100) \times \frac{\pi}{4} \times 10^2}{125600 \times 1000} \times 100$$

$$= 0.31\% \ [>0.2\%, \text{ o.k.}].$$

In top and bottom 1.2 m length of pile, decrease the spacing of lateral ties to 80 mm to give the minimum requirement of 0.6% of gross volume as required.

Example 8.4

It is required to support a tower on bored piles on a site where stiff fissured clay is affected by seasonal swelling and shrinkage movements to a depth of 1.0 m. The unconfined compressive strength of stiff clay increases linearly from 40 kN/m² at 1.0 to 160 kN/m² at 8.0 m. Design the pile assuming a total load = 2500 kN, and a F.O.S. equal to 3.

Solution

(i) The pile group is designed for 100% efficiency. Assuming the following dimensions of a single pile:

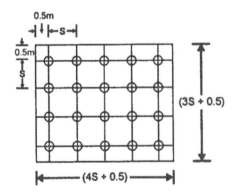

Fig. 8.56. Arrangement of Piles.

Diameter of pile = 0.50 m (circular).

Length of pile = 8.0 m.

Ultimate load capacity of pile, $Q_u = N_c \cdot c_b \cdot A_b + \alpha \bar{c}_u \cdot A_s$

$\alpha = 0.3$ (adhesion factor for fissured clays)

$$c_b = \frac{160}{2} = 80 \text{ kN/m}^2, \text{ and } \bar{c}_u = \frac{40+160}{2 \times 2} = 50 \text{ kN/m}^2, N_c = 9.0$$

$$Q_u = 80 \times 9.0 \times \pi/4(0.5)^2 + 0.3 \times 50 \times \pi \times 0.5 \times 7.0 = 306.30 \text{ kN}$$

Number of piles required $= \dfrac{2500 \times 2.5}{306.30} = 20$

Arrange piles in a manner as shown in Fig. 8.56.

(ii) Ultimate load capacity of the pile group considering block failure is given by

$$Q_{ug} = N_c \cdot c_b \cdot A_{gb} + p_g \cdot \bar{c}_u$$

where A_{gb} = cross-section area of the block,

p_g = perimeter of the block, and

L = embedded length of the pile.

Thus $Q_{ug} = 9.0 \times 80 \times (4S + 0.5)(3S + 0.5) + [2(4S + 0.5) + 2(3S + 0.5)] \times 7.0 \times 50$

$2500 \times 3 = 720 (12S^2 + 0.25 + 3.5S) + 700 (7S + 1)$

or $7500 = 8640S^2 + 7220S + 880$

or $S^2 + 0.836S - 0.893 = 0$; it gives $S = 0.70$ m = 700 mm.

Hence provide spacing $S = 700$ mm c/c for 100% pile group efficiency.

(iii) Structural design

Effective length $L_e = 1 + \dfrac{7.0}{3} = 3.33$ mm.

Hence $\dfrac{L_e}{B} = \dfrac{3.33}{0.5} = 6.66 < 12$.

Pile will act as a short column.

$P_u = 0.4 \times 20(196250 - A_{st}) + 0.67 \times 415 \times A_{st} = 2500 \times 1000 \times 1.5$

or $A_{st} = 8072$ mm^2.

Provide 25 mm diameter Tor bars

$$\text{Number of bars} = \frac{8072}{\dfrac{\pi}{4} \times (25)^2} = 16.45 \text{ say } 18$$

Distribute the bars in the pile section uniformly.

Lateral ties—

Minimum volume = 0.2% of volume of pile

$$= \frac{0.2}{100} \times \frac{\pi}{4} \times (500)^2 \times 10^3 = 392700 \text{ mm}^3.$$

Assume 8 mm diameter *lateral ties*

$$\text{Volume} = \frac{\pi}{4} \times 8^2 \times \pi \times 400 = 63169.46 \text{ mm}^3.$$

Total bars required per metre = 6.14, hence provide at a spacing of 160 mm c/c.
Pitch of ties < 500 mm
$$< 16 \times 25 = 400 \text{ mm}$$
$$< 48 \times 8 = 384 \text{ mm}.$$
Therefore, spacing on 160 mm c/c is adequate.

Example 8.5

The following data refer to a cyclic pile load test on a circular pile of diameter 250 mm and length of 10 m.

Load on pile top (kN)	150	200	250	300	350	400	600	650
Total settlement of pile top (mm)	1.35	2.15	2.65	3.55	4.65	5.55	28.90	64
Net settlement of pile top (mm)	0.36	0.55	0.72	0.95	1.28	1.56	21.60	55

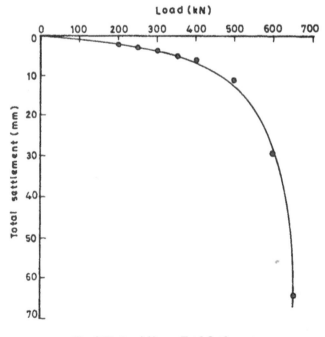

Fig. 8.57. Load Versus Total Settlement.

(a) Plot the load settlement curve and estimate the allowable load of the pile according to Indian Standard Code.

(b) Assuming a value of E as 2.8×10^7 kN/m² for the composite section of the pile, separate the load taken by point bearing and skin friction.

Solution

(a) Plot the load versus total settlement curve as shown in Fig. 8.57. The allowable capacity of pile will be lesser of the following:

(i) 2/3 of load corresponding to 12 mm settlement 2/3(510) = 340 kN.

(ii) 1/2 of load at which the total settlement equals to 10% of pile dia (25 mm) = 1/2 (592) = 296 kN.

(b) Determine the elastic recovery of the pile as given below in column number 4 of Table 8.19.

$$E = 2.8 \times 10^7 \text{ kN/m}^2, \ A = 0.0490625 \text{ m}^2, \ L = 10 \text{ m}.$$

The elastic recovery consists of two parts: (i) elastic compression of subgrade, and (ii) elastic compression of pile. In the first trial, assume the later equal to zero. Plot the load versus elastic recovery curve (Fig. 8.58a). Draw the line parallel to the later portion of the curve as shown in Fig. 8.58a. Obtain elastic compression of pile using Eq. 8.21, i.e.

$$\Delta L = \frac{(Q - Q_s/2)L}{AE}.$$

Values of the elastic compression of pile thus obtained are given in column No. (5a) of Table 8.19. Determine elastic compression of subgrade by subtracting column (5a) from column (4), and plot the same with respect to the load as shown in Fig. 8.58b. Similarly, carry out the second and third trials and plot the curves as shown in Fig. 8.58c and 8.58d. The values of point bearing resistance and frictional resistance corresponding to the Fig. 8.58d may be used in design. Therefore, the capacity of pile 296 kN consists of (i) point resistance, Q_p = 95 kN, and (ii) frictional resistance, Q_s = 291 kN.

Table 8.19. Details of Computations of Elastic Settlement of Subgrade.

S. No.	Load (kN)	Total Sett. (mm)	Net Sett. (mm)	Elastic Recov. (mm)	S_{ep}^* (mm)			S_{eb}^* (mm)		
1	2	3	4	5(a)	5(b)	5(c)	6(a)	6(b)	6(c)
1.	150	1.35	0.36	0.99	0.69	0.58	0.60	0.30	0.41	0.39
2.	200	2.15	0.55	1.60	0.94	0.82	0.86	0.66	0.78	0.74
3.	250	2.65	0.72	1.93	1.20	1.06	1.10	0.73	0.87	0.83
4.	300	3.55	0.95	2.60	1.48	1.32	1.37	1.12	1.28	1.23
5.	350	4.65	1.28	3.37	1.77	1.61	1.68	1.60	1.76	1.69
6.	400	5.55	1.56	3.99	2.06	1.89	1.94	1.93	2.10	2.05
7.	500	10.65	4.80	5.85	2.70	2.46	2.61	3.15	3.39	3.24
8.	600	28.90	21.60	7.30	3.40	3.13	3.28	3.90	4.17	4.02
9.	650	64.00	55.20	8.80	3.78	3.47	3.62	5.02	5.33	5.18

$^*S_{ep}$ = Elastic recovery of pile; S_{eb} = Elastic compression of subgrade.

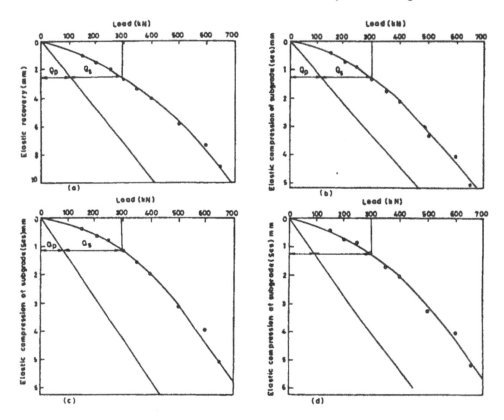

Fig. 8.58. Load versus Elastic Settlement of Subgrade.

Examples 8.6

Design a pile group consisting of RCC piles for a column of size 650 mm × 650 mm carrying load of 5000 kN. The soil exploration data reveal that the sub-soil consists of deposit of soft clay extending to a great depth.

The other data of the deposit are
Compression index, $C_c = 0.10$
Initial void ratio, $e_0 = 0.9$
Saturated unit weight = 19 kN/m³
Unconfined compression strength = 40 kN/m².

Proportion the pile group for the permissible settlement of 50 mm. Design the piles and pile cap.

Solution

(i) Let bored piles of 600 mm diameter and 18 m length are used.

Now
$$Q_u = N_c \cdot c_b \cdot A_b + \alpha \bar{c}_u A_s$$
$$N_c = 9.0, \; c_b = \bar{c}_u = 20 \text{ kN/m}^2, \; \alpha = 0.45$$

$$A_b = \frac{\pi}{4}(0.6)^2 = 0.2826 \text{ m}^2, \quad A_s = \pi \times 0.6 \times 18 = 33.9 \text{ m}^2$$

$$Q_u = 9.0 \times 20 \times 0.2826 + 0.45 \times 20 \times 33.9$$

$$Q_u = 356 \text{ kN.}$$

$$\text{Allowable pile capacity} = \frac{356}{2.5} = 142 \text{ kN}.$$

$$\text{Number of piles required} = \frac{5000}{142} = 35.2.$$

Provide 36 piles in a square pattern as shown in Fig. 8.59.
Now for 100% efficiency of pile group, spacing of piles

$$= \frac{1.57\,B\,mn - 2B}{m + n - 2}$$

where m = number of rows in the pile group = 6,
n = number of columns in the pile group = 6, and
B = pile diameter = 600 mm

$$\text{Spacing} = \frac{1.57 \times 600 \times 6 \times 6 - 2 \times 600}{6 + 6 - 2} = 3271 \text{ mm}$$

Provide spacing 3250 mm c/c.

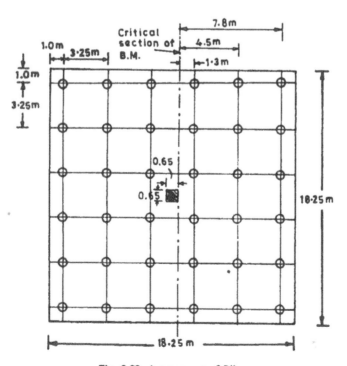

Fig. 8.59. Arrangement of Piles.

(ii) Settlement of pile group

Settlement of the pile group is obtained in two cases, firstly considering an imaginary footing at the ground surface, and secondly at a depth of 2/3L from the ground surface, L being the embedded length of the pile. Settlement should be worked out in both cases and a higher value should be compared with the permissible settlement. In this particular example, an imaginary footing located at the ground surface will give higher value of settlement and, therefore, computations of settlement have been illustrated only in this case.

Plan dimension of imaginary footing = 17.25 m × 17.25 m (Fig. 8.59).

Settlement of pile group is obtained as the settlement of imaginary footing placed at the ground surface considering the soil stratum up to 17.25 m depth.

Following equation is used for getting the settlement.

$$S = \frac{C_c H}{1 + e_0} \log_{10} \frac{p_0 + \Delta p}{p_0}$$

where $C_c = 0.10$, $e_0 = 0.9$,

H = thickness of clay layer,

p_0 = effective overburden pressure at the centre of clay layer

= 19 × d,

Δp = pressure at the centre of clay layer due to superimposed load

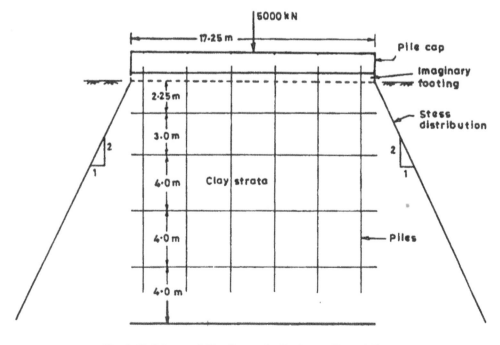

Fig. 8.60. Scheme of Clay Layers for Settlement Computations.

$$= \frac{5000}{(17.25 + d)^2}, \text{ and}$$

d = distance of the centre of clay layer from ground surface.

Scheme of clay layers for settlement computations is taken as shown in Fig. 8.60. Details of computations are given in Table 8.20.

Table 8.20. Details of Settlement of Computations

Layer No.	Thickness of layer (mm)	$\frac{C_c H}{1+e_0}$ (mm)	p_0 (kN/m²)	Δp (kN/m²)	$\log_{10} \frac{p_0+\Delta p}{p_0}$	S (mm)
1	2250	118.4	21.40	14.8	0.228	26.99
2	3000	157.9	71.25	11.3	0.064	10.10
3	4000	210.5	137.75	8.3	0.0254	5.35
4	4000	210.5	213.75	6.1	0.0122	2.57
5	4000	210.5	251.75	4.7	0.008	1.68

Total settlement = 46.69 mm

As permissible settlement is 50 mm, the proposed scheme of pile foundation is adequate.

(iii) Design of pile

$$L_e = \frac{2}{3} \times 18 = 12 \text{ m (Assumed)}$$

$$\frac{L_e}{B} = \frac{12}{0.6} = 20 \text{ As it is greater than 12, pile will be considered as a long column.}$$

$$e_{min} = \frac{18000}{500} + \frac{600}{30} = 56 \text{ mm} > 20 \text{ mm}$$

$$M_u = \frac{1.5 \times 5000}{36} \times 0.056 = 11.67 \text{ kNm}$$

Since bored piles are provided, there will be no handling stresses.

Correction factor for long columns $= 1.25 - \dfrac{L_e}{48B}$

$$= 1.25 - \frac{12 \times 10^3}{48 \times 600} = 0.833$$

Design moment $= \dfrac{11.67}{0.833} = 14 \text{ kNm}$

Design load $= \dfrac{1.5 \times 5000}{36} \times \dfrac{1}{0.833} = 251 \text{ kN.}$

Refer interaction diagrams corresponding to $d'/D = 0.1$ (Figs. 5.7 and 5.8)

Thus $\quad\quad \dfrac{P_u}{f_{ck}bD} = \dfrac{251 \times 10^3}{20 \times 600 \times 600} = 0.03486$

$$\dfrac{M_u}{f_{ck}bD^2} = \dfrac{14 \times 10^6}{20 \times 600 \times 600^2} = 0.0032407$$

As per interaction diagrams, amount of reinforcement works out very little. Therefore, provide nominal reinforcement.

$$A_{st} = \dfrac{0.8}{100} \times \dfrac{\pi}{4} \times 600^2 = 2262 \text{ mm}^2.$$

Provide eight bars of 20 mm diameter.

Lateral ties

Minimum volume of lateral reinforcement per metre length of pile

$$= \dfrac{0.2}{1000} \times \dfrac{\pi}{4} \times (600)^2 \times 1000 = 565486.7 \text{ mm}^2$$

Volume of a tie of 8 mm $\phi = \dfrac{\pi}{4} \times 8^2 \times \pi \times 500 = 78956 \text{ mm}^3$.

Number of ties/metre of pile $= \dfrac{565486.7}{78956} = 7.2$

Hence spacing for lateral reinforcement $= \dfrac{1000}{72} = 140$ mm c/c

$$< 600 \text{ mm}$$
$$< 48 \times 8 = 384 \text{ mm}$$
$$< 16 \times 20 = 320 \text{ mm (o.k.)}$$

(iv) Design of pile cap

Pile cap is 18.25 m × 18.25 m square (Fig. 8.59)

Bending moment at the face of column $= \dfrac{5000}{36} \left(\dfrac{1.3 + 4.55 + 7.8}{3.25} \right)$

$$= 583.3 \text{ kNm per metre width of pile cap}$$

Factored moment $\quad\quad = 1.5 \times 583.3 = 758.3 \text{ kNm}$

Effective depth, $d \quad\quad = \sqrt{\dfrac{758.3 \times 10^6}{0.138 \times 20 \times 1000}} = 524.3 \text{ mm}$

Adopt effective thickness of pile cap as 550 mm. Total thickness of pile cap = 600 mm. Punching shear force = 1.5 × 5000 = 7500 kN

$$\text{Punching shear stress} = \frac{7500 \times 10^3}{4 \times (600 + 650) 600}$$

$$= 2.5 \text{ N/mm}^2 > 1.1 \text{ N/mm}^2 \text{ (unsafe)}.$$

Depth required from punching shear consideration

$$1.1 = \frac{7500 \times 10^3}{4 \times (d + 650) d} \text{ ; it gives } d = 1098 \text{ mm}$$

Provide $d = 1125$ mm and $D = 1175$ mm.

Reinforcement

$$758.3 \times 10^6 = 0.87 f_y A_{st} \left(d - \frac{f_y A_{st}}{f_{ck} b} \right)$$

or
$$758.3 \times 10^6 = 0.87 \times 415 A_{st} \left(1125 - \frac{415 \times A_{st}}{20 \times 1000} \right)$$

It gives, $A_{st} = 1936 \text{ mm}^2$.

$$\text{Provide 16 mm } \phi \text{ bar, spacing of bars} = \frac{1000 \times \frac{\pi}{4} \times (16)^2}{1936}$$

$$= 103.8 \text{ mm c/c}$$

$$= 100 \text{ mm c/c (say)}.$$

Thus provide 16 mm φ bars @ 100 mm c/c in both directions of pile cap at bottom.
Check for one-way shear

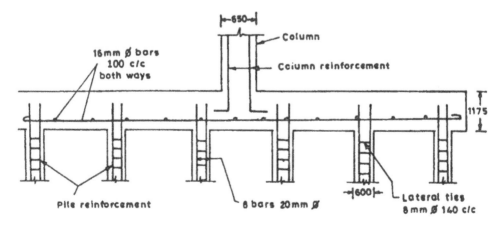

Fig. 8.61. Details of Reinforcement.

S.F. per metre width at section $xx = \dfrac{5000 \times 1.5}{36} \times \dfrac{3}{3.25}$

$$= 192 \text{ kN}$$

Shear stress $\qquad = \tau_v = \dfrac{192 \times 10^3}{1125 \times 1000} = 0.17 \text{ N/mm}^2$

$\dfrac{100 A_{st}}{bd} = \dfrac{100 \times 1936}{1000 \times 1125} = 0.172$, Corresponding $\tau_c = 0.30 \text{ N/mm}^2$

Therefore $\tau_v < \tau_c$ [safe]

Details of reinforcement are shown in Fig. 8.61.

Example 8.7

Design a 8.0-m long precast driven pile in cohesionless soil to a vertical load of 800 kN and a lateral load of 160 kN at the top of the pile which is 0.2 m above the ground surface (Fig. 8.62). The value of constant of subgrade reaction is $5.24 \times 10^4 \text{ kN/m}^3$.

Solution

(i) Let the pile be a square pile of 500 mm × 500 mm cross-section.

$$I = \frac{0.5 \times (0.5)^3}{12} = 0.00521 \text{ m}^4$$

$$E = 2.24 \times 10^7 \text{ kN/m}^2 \text{ (assumed)}$$

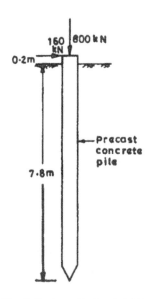

Fig. 8.62. Laterally Loaded Pile.

$$T = \left(\frac{EI}{\eta_h}\right)^{\frac{1}{5}} = \left(\frac{2.24 \times 10^7 \times 0.00521}{5.24 \times 10^4}\right)^{\frac{1}{5}} = 1.173 \, \text{m}.$$

$$Z_{max} = \frac{L_s}{T} = \frac{7.8}{1.173} = 6.64 \quad [> 5, \text{ long pile}].$$

(iii) Variation of bending moment along the length of pile may be obtained using the following equation:

$$M_x = A_m \cdot Q_{hg} \cdot T + B_m \cdot M_g$$

$$Q_{hg} = 160 \times 1.5 = 240 \quad \text{kN}; \quad M_g = 240 \times 0.2 = 48 \, \text{kNm}.$$

Values of A_m and B_m are obtained from Table 8.8 for different values of depth coefficients x/T.

x/T	0	0.5	1.0	1.35	2.0
A_m	0	0.46	0.73	0.77	0.63
B_m	1.0	0.98	0.85	0.70	0.40
M_x (kNm)	48	176.5	246.3	250.4	196.5

Maximum moment (250.4 kNm) occurs at the depth coefficient $x/T = 1.35$, i.e., $x = 1.35 \times 1.173 = 1.583$ m below the ground surface.

(iii) Slenderness ratio $= \frac{8.0}{0.5} = 16$ (effective length is assumed equal to the full length of pile)

Therefore, $\quad e_{min} = \frac{8000}{500} + \frac{50}{30} = 16.0 + 1.67 = 17.67 \, \text{mm}$

Actual eccentricity $= \frac{250.4 \times 1000}{1200} = 208.67 \, \text{mm}.$

Therefore $e > e_{min}$.

(iv) Moment due to handling

From Fig. 8.3

$$M_{max} = \frac{WL}{23.3} = \frac{25 \times 8 \times 0.5 \times 0.5 \times 8}{23.3} \times 1.5 = 25.75 \, \text{kNm}.$$

Hence design moment $= 250.4$ kNm.

(v) Let the pile be reinforced with 20 mm diameter bars provided with 40 mm clear cover. Refer interaction diagram (Figs. 5.7 and 5.8)

$$d' = 40 + \frac{20}{2} = 50 \, \text{mm}$$

$$\frac{d'}{D} = \frac{50}{500} = 0.1$$

$$\frac{P_u}{f_{ck}bD} = \frac{800 \times 1.5 \times 10^3}{20 \times 50 \times 50 \times 10^2} = 0.24$$

$$\frac{M_u}{f_{ck}bD^2} = \frac{250.4 \times 10^6}{20 \times 50 \times 50^2 \times 10^3} = 0.100$$

From charts (Fig. 5.8), for above values, $p/f_{ck} = 0.053$
$p = 20 \times 0.053 = 1.06\%$
Therefore provide minimum reinforcement, i.e., 1.25%

$$A_{st} = \frac{1.25}{100} \times 500 \times 500 = 3125 \text{ mm}^2.$$

Use 10 bars equally distributed in the section.
Amount of lateral reinforcement
$$= 0.2\% \text{ of the volume of pile}$$

$$= \frac{0.2}{100} \times 500 \times 500 \times 1000$$

$$= 500,000 \text{ mm}^3 \text{ in one metre of pile.}$$

Let us use 10 mm diameter lateral ties
Pitch < 500 mm
$< 16\phi_L = 16 \times 20 = 320$ mm
$< 48\phi_T = 48 \times 10 = 480$ mm
Adopt pitch of 250 mm c/c.
Total amount of lateral reinforcement actually provided

$$= 4 \times [4 \times 420 \times \frac{\pi}{4} \times 10^2]$$

$$= 527787.57 \text{ in one metre of pile (Hence o.k.)}$$

Reduce pitch to 80 mm c/c in the top and bottom 1.5 m pile.
(vi) Driving stress
Assuming that the piles will be driven in the field with a hammer of 10 kN falling freely from a height of 3.0 m. If the average penetration of pile per blow for the final 150 mm of driving is 25 mm, then

Driving resistance $\quad = \dfrac{166.7 \, WH}{S + C}$

$$= \frac{166.7 \times 10 \times 3}{25 + 25.4}$$

$$= 99.2 \text{ kN}$$

Driving stress $\quad = \dfrac{\text{Driving resistance}}{\text{Cross-sectional area}} \left(\dfrac{2}{\sqrt{\eta}} - 1 \right).$

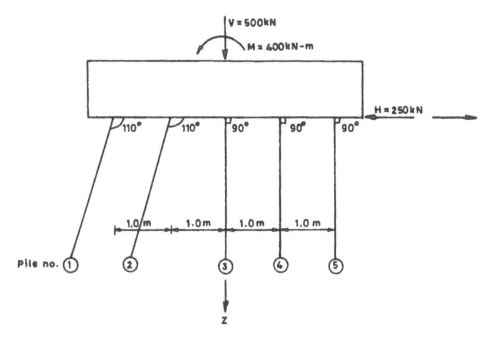

Fig. 8.63. Pile Foundation.

$$\text{Cross-sectional area} = 500 \times 500 + (19 - 1) \times 10 \times \frac{\pi}{4} \times 20^2$$

$$= 250000 + 56520 = 306520 \text{ mm}^2$$

$$\eta = 0.9 \text{ (assumed).}$$

$$\text{Driving stress} = \frac{99.2 \times 1000}{306520}\left(\frac{2}{\sqrt{0.9}} - 1\right)$$

$$= 0.36 \text{ N/mm}^2 \text{ [<15 N/mm}^2\text{, o.k.].}$$

Example 8.8

A retaining wall is supported on piles comprising vertical and batter piles as shown in Fig. 8.63. The piles are rigidly fixed to the foundation which is subjected to:

Vertical load, $V = 500$ kN
Horizontal load, $H = 250$ kN, and
Moment, $M = 400$ kNm per unit length of wall.

The piles are spaced 1.0 m apart parallel to the face of wall. The number of piles per metre length of wall is five as shown in the Fig. 8.63. The piles are R.C.C. piles of 400 mm diameter of length 10 m driven in medium dense sand. The constant of modulus of subgrade reaction η_h is 3×10^4 kN/m^3. Determine (a) the vertical, lateral and rotational deflections of the foundation, and (b) the axial loads, shear loads and moments at each of the pile heads.

Solution

(i)

$$A = \frac{\pi \times 0.4^2}{4} = 0.1256 \text{ m}^2$$

$$I = \frac{\pi \times 0.4^4}{64} = 1.256 \times 10^{-3}$$

$L = 10$ m

$E = 1.4 \times 10^7$ kN/m^2 (assumed)

$\eta_h = 3 \times 10^4$ kN/m^3

$$T = \left(\frac{EI}{\eta_h}\right)^{\frac{1}{5}} = \left(\frac{1.4 \times 10^7 \times 1.256 \times 10^{-3}}{3 \times 10^4}\right)^{\frac{1}{5}} = 0.9 \text{ m}$$

A_y = Deflection coefficient at ground level
= 2.435 (Table 8.8).

(ii) Pile constants

The pile constans computed for vertical piles are also used for batter piles.

$$K_a = \frac{AE}{L} = \frac{0.1256 \times 1.4 \times 10^7}{10} = 17.58 \times 10^4 \text{ kN/m}$$

$$K_t = \frac{2.56}{A_y} \times \frac{EI}{T^3} = \frac{2.56 \times 1.4 \times 10^7 \times 1.256 \times 10^{-3}}{2.435 \times 0.9^3} = 2.54 \times 10^4 \text{ kN/m}$$

$$m_t = \frac{2.35}{A_y} \times \frac{EI}{T^2} = \frac{2.35 \times 1.4 \times 10^7 \times 1.256 \times 10^{-3}}{2.435 \times 0.9^2} = 2.09 \times 10^4 \text{ kNm/m.}$$

$$m_\alpha = \frac{3.54}{A_y} \times \frac{EI}{T} = \frac{3.54 \times 1.4 \times 10^7 \times 1.256 \times 10^{-3}}{2.435 \times 0.9} = 2.84 \times 10^4 \text{ kNm/rad.}$$

(iii) Foundation constants

For $\theta = 110°$

$\sin \theta = 0.94$, $\sin^2 \theta = 0.883$
$\cos \theta = -0.342$, $\cos^2 \theta = 0.117$
$\sin 2\theta = -0.644$

For $\theta = 90°$

$\sin \theta = 1.0 = \sin^2 \theta$
$\cos \theta = 0.0 = \cos^2 \theta$
$\sin 2\theta = 0$

$$H_x = -\sum n \left(K_a \cos^2 \theta + K_t \sin^2 \theta\right)$$

$$= -\left[2(17.58 \times 0.117 + 2.54 \times 0.883) + 3(17.58 \times 0 + 2.54 \times 1)\right] \times 10^4$$

$$= -16.22 \times 10^4$$

$$H_z = -\frac{1}{2}(K_a - K_t) \sum n \sin 2\theta$$

$$= -\frac{1}{2}(17.58 - 2.54)\ [2 \times (-0.644) + 3 \times 0] \times 10^4$$

$$= 9.68 \times 10^4 \text{ kN} = V_x$$

$$H_\alpha = \frac{1}{2}(K_a - K_t)\sum(n\bar{x}\sin 2\theta) + m_t \sum(n\sin\theta)$$

$$= -(17.58 - 2.54)\ [2\ (-1.5)\ (-0.644) + 3(+1.0)\ (0)] \times 10^4 + 2.09\ [2 \times 0.94 + 3 \times 1.0]$$
$$\times 10^4$$

$$= -14.53 \times 10^4 + 9.95 \times 10^4 = -4.58 \times 10^4 \text{ kN m/m} = M_x$$

$$V_z = -\sum n\left(K_a \sin^2\theta + K_t \cos^2\theta\right)$$

$$= -[2(17.58 \times 0.883 + 2.54 \times 0.117) + 3(17.58 \times 1.0 + 2.54 \times 0] \times 10^4$$
$$= -84.38 \times 10^4 \text{ kN}$$

$$M_z = -\sum\left(K_a \sin^2\theta + K_t \cos^2\theta\right)n\bar{x} - m_t \sum(n\cos\theta)$$

$$= -[2(17.58 \times 0.883 + 2.54 \times 0.117) \times (-1.5) + 3(17.58 \times 1.0 +$$
$$2.54 \times 0)(+1.0)] \times 10^4 - 2.09[2\times (-0.342) + 3 \times 0.0] \times 10^4$$
$$= -5.28 \times 10^4 + 1.43 \times 10^4 = -3.85 \times 10^4 \text{ kNm} = V_\alpha$$

$$M_\alpha = -\sum[(K_a \sin^2\theta + K_t \cos^2\theta)\sum x^2] - 2m_t \sum n\bar{x}\cos\theta - m_\alpha \sum n$$

$$= -[(17.58 \times 0.883 + 2.54 \times 0.117)[(-2)^2 + (-1)^2 + (0)^2 + (1)^2 + (2)^2]] \times 10^4$$
$$- [2 \times 2.09[2 \times (-1.5) \times (-0.342) + 3 \times (1.0)(0.0)]] \times 10^4 - 2.84 \times 5 \times 10^4$$
$$= -158.2 \times 10^4 - 4.29 \times 10^4 - 14.2 \times 10^4$$
$$= -176.69 \times 10^4 \text{ kNm/rad.}$$

Equilibrium equations

$$H_x\Delta x + H_z\Delta z + H_\alpha\alpha + H = 0$$
$$H_z\Delta x + V_z\Delta z + V_\alpha\alpha + V = 0$$
$$H_\alpha\Delta x + V_\alpha\Delta z + M_\alpha\alpha + M = 0.$$

Substituting

$$-16.22 \times 10^4\Delta x + 9.68 \times 10^4\Delta z - 4.58 \times 10^4\alpha - 250 = 0$$

or $\quad \Delta x - 0.596\Delta z + 0.283\alpha = -15.41 \times 10^{-4}$

$$9.68 \times 10^4\Delta x - 84.38 \times 10^4\Delta z - 3.85 \times 10^4\alpha + 500 = 0$$

or $\quad \Delta x - 8.72\Delta z - 0.398\alpha = -51.65 \times 10^{-4}$

$$-4.58 \times 10^4\Delta x - 3.85 \times 10^4\Delta z - 176.69 \times 10^4\alpha - 400 = 0$$

or $\quad \Delta x + 0.84\Delta z + 34.43\alpha = -87.34 \times 10^{-4}.$

Solving the above equations, we get

$$\Delta x = -11.98 \times 10^{-4} \text{ m}$$

$$\Delta z = 4.656 \times 10^{-4} \text{ m}$$

$$\alpha = -2.311 \times 10^{-4} \text{ radians.}$$

(v) Pile displacements

The longitudinal and transverse displacements of the pile heads can be obtained using Eqs. (8.116) and (8.117) respectively. The details of computations are given in Tables 8.21 and 8.22.

Each pile will have a rotation of $\alpha = -2.311 \times 10^{-4}$ radians, i.e., the piles will be rotated in anticlockwise direction from their original position.

Table 8.21. Computations of Longitudinal Displacement $\delta_a = (\Delta x \cos \theta + \Delta z \sin \theta + \alpha x \sin \theta)$

Pile No.	$\Delta x \cos \theta$	$\Delta z \sin \theta$	$\alpha x \sin \theta$	δ_a (m)
1.	$-11.98 \times 10^{-4} \times$ (-0.342) $= -4.097 \times 10^{-4}$	$4.656 \times 10^{-4} \times (0.94)$ $= 4.374 \times 10^{-4}$	$-2.311 \times 10^{-4} \times$ $(-2.0) \times 0.94$ $= 4.344 \times 10^{-4}$	4.621×10^{-4}
2.	-4.097×10^{-4}	4.374×10^{-4}	$+2.172 \times 10^{-4}$	2.449×10^{-4}
3.	$-11.98 \times 10^{-4} \times 0$ $= 0$	$4.656 \times 10^{-4} \times 1.0$ $= 4.656 \times 10^{-4}$	$-2.311 \times 10^{-4} \times 0$ $= 0$	4.656×10^{-4}
4.	0	4.656×10^{-4}	-2.311×10^{-4}	2.345×10^{-4}
5.	0	4.656×10^{-4}	-4.622×10^{-4}	0.034×10^{-4}

Table 8.22. Computations of Transverse Displacement $y_t = (\Delta x \sin \theta - \Delta z \cos \theta - \alpha x \cos \theta)$

Pile No.	$\Delta x \sin \theta$	$\Delta z \cos \theta$	$\alpha x \cos \theta$	y_t (m)
1.	$-11.98 \times 10^{-4} \times 0.94$ $= -11.26 \times 10^{-4}$	$4.656 \times 10^{-4} \times (-0.342)$ $= -1.59 \times 10^{-4}$	$-2.311 \times 10^{-4} \times$ $(-2.0) \times (-0.342)$ $= -1.58 \times 10^{-4}$	-14.43×10^{-4}
2.	-11.26×10^{-4}	-1.59×10^{-4}	-0.79×10^{-4}	-13.64×10^{-4}
3.	$-11.98 \times 10^{-4} \times 1.0$ $= -11.98 \times 10^{-4}$	$4.656 \times 10^{-4} \times 0$ $= 0.0$	$-2.311 \times 10^{-4} \times 0$ $= 0.0$	-11.98×10^{-4}
4.	-11.98×10^{-4}	0.0	0.0	-11.98×10^{-4}
5.	-11.98×10^{-4}	0.0	0.0	-11.98×10^{-4}

(vi) Pile loads

The axial load, transverse load and moment acting on a pile can be respectively obtained using Eqs. (8.118), (8.119) and (8.120).

Axial load $(K_a \delta_a)$ in

Pile no. 1 $= 17.58 \times 10^4 \times 4.621 \times 10^{-4} = 81.24$ kN

Pile no. 2 $= 17.58 \times 10^4 \times 2.449 \times 10^{-4} = 43.05$ kN

Pile no. 3 $= 17.58 \times 10^4 \times 4.656 \times 10^{-4} = 81.85$ kN

Pile no. 4 $= 17.58 \times 10^4 \times 2.345 \times 10^{-4} = 41.22$ kN

Pile no. 5 $= 17.58 \times 10^4 \times 0.034 \times 10^{-4} = 0.60$ kN

Transverse load $(-K_t y_t + m_t \alpha)$

Pile no. 1 $= -2.54 \times 10^4 \times (-14.43 \times 10^{-4}) + 2.09 \times 10^4 \times (-2.311 \times 10^{-4})$
$\qquad = 36.65 - 4.83 = 31.82$ kN

Pile no. 2 $= -2.54 \times 10^4 \times (-13.64 \times 10^{-4}) - 4.83$
$\qquad = 34.64 - 4.83 = 25.6$ kN

Pile no. 3 $= -2.54 \times 10^4 \times (-11.98 \times 10^{-4}) - 4.83$
$\qquad = 30.43 - 4.83 = 25.6$ kN

Pile no. 4 $= 25.6$ kN

Pile no. 5 $= 25.6$ kN

Moment $(m_t y_t - m_\alpha \alpha)$ acting on

Pile no. 1 $= 2.09 \times 10^4 \times (-14.43 \times 10^{-4}) - 2.84 \times 10^{-4} \times (-2.311 \times 10^{-4})$
$\qquad = -30.16 + 6.56 = -23.6$ kNm

Pile no. 2 $= 2.09 \times 10^4 \times (-13.64 \times 10^{-4}) + 6.56$
$\qquad = -28.51 + 6.56 = -21.95$ kNm

Pile no. 3 $= 2.09 \times 10^4 \times (-11.98 \times 10^{-4}) + 6.56$
$\qquad = -25.04 + 6.56 = -18.48$ kNm

Pile no. 4 $= -18.48$ kNm

Pile no. 5 $= -18.48$ kNm

REFERENCES

1. Bowles, J.E. (1996), *Foundation Analysis and Design*, 5th Edition, McGraw Hill Book Co., New York.
2. Braithwaite Burn Jessop Construction Co. Ltd., Kolkatta, pamphlet: *The Franki System of Compressed Piling*.
3. Broms, B. (1964a), "The Lateral Resistance of Piles in Cohesive Soils," *J. Soil Mech. Found. Div., ASCE*, Vol. 90, No. SM 2, pp. 27–63.
4. Broms, B. (1964b), "The Lateral Resistance of Piles in Cohesive Soils," *J. Soil Mech. Found. Div., ASCE*, Vol. 90, No. SM 3, pp. 123–156.
5. Broms, B. (1966), "Methods of Calculating the Ultimate Bearing Capacity of Piles—A Summary," *Sols-Soils*, Vol. 5, No. 18-19, pp. 21–31.
6. Berezantsev, V.G., V. Khristoforov and Golubkov, V. (1961), "Load Bearing Capacity and Deformation of Pile Foundations," *Proc. 5th Int. Conf. SM & FE*, Vol. 2.
7. Coyle, H.M. and Reese, L.C. (1966), "Load Transfer for Axially Loaded Piles in Clay," *Proceedings of ASCE*, Vol. 92, SM 2.
8. Coyle, H.M. and Castello, R.R. (1981), "New Design Correlations for Piles in Sand," *Journal of Geotechnical Engineering Division, ASCE*, 107, No. GT 7.
9. Crowther, C.L. (1988), "Load Testing of Deep Foundation—The Planning, Design and Conduct of Pile Load Tests," John Wiley and Sons, New York.
10. Culmann, C. (1866), "Die Graphische Statik," Section 8, *Theoric der Statz and Futtermauern*, Meyer und Zeller, Zurich.
11. Darragh, R.D. and Bell, R.A. (1969), "Load Tests on Long Bearing Piles", ASTM, STP 444, pp. 41–46.
12. Davisson, M.T. and Robinson, K.E. (1955), "Bending and Buckling of Partially Embedded Piles," *6th Int. Conf. on SMFE*, Montreal, Vol. 89, No. SM3, pp. 63–64.
13. Davisson, M.T. and Gill, H.L. (1963), "Laterally Loaded Piles in a Layered Soil System," *Journal of Soil Mech. Found. Div.*, ASCE, Vol. 89, SM-3, p. 63.
14. Davisson, M.T. (1970), "Evaluation of Coefficient of Subgrade Reaction," *Geotechnique*, Vol. 5, pp. 297–326.

15. Davisson, M.T. (1971), "Lateral Load Capacity of Piles," Highway Research Record, Washington, DC, pp. 104–112.

16. Den Norske Pelekomite (1973), "Veiledning ved pelefundamentering", Norwegian Geotechnical Institute, Veiledning, No. 1.

17. Feld, J. (1943), "Discussion: Timber Pile Foundations," *Transactions ASCE*, Vol. 1, pp. 517–522.

18. Finn, R.J. (1989), "Predicted and Observed Axial Behaviour of Piles", Proceedings of a Symposium 1989, Foundation Engineering Congress, North-Western University, USA.

19. Fleming, W.G.K. and Sliwinski, Z.J. (1977), "The use and influence of bentonite in bored pile construction,"*Construction Industry Research and Information Association Report*, No. PDP2.

20. Hanna, T.H. (1963), "Model Studies of Foundation Groups in Sand," *Geotechnique*, Vol. 13, No. 4, December, pp. 334–351.

21. Hanson, J.B. (1961), "A general formula for bearing capacity," *Danish Geotechnical Institute Bulletin*, No. 11.

22. Hazarika, P.J. and Ramasamy, G. (2000), "Response of Piles under Vertical Loadings", *Indian Geotechnical Journal*, Vol. 30, No. 2, pp. 73–91.

23. Hindustan Construction Const. Company Ltd., Bombay, Personal Communications with authors, August 1972, May 1977.

24. Housel, W.S. (1965), "Michigan Study of Pile Driving Hammers," *Proc. Am. Soc. of Civil Engrs.*, Vol. 92, No. SM4, 1965, p. 37.

25. Hrennikoff, A. (1950), "Analysis of Pile Foundations with Battle Piles," *Transactions of American Society of Civil Engineers*, Vol. 115, Paper No. 2401, p. 351.

26. IS 456-2000, "Code of Practice for Plain and Reinforced concrete", I.S.I., New Delhi.

27. IS 2911: Part 1 : Sec. 1 – 1979, "Code of Practice for Design and Construction of Pile Foundation–Driven Cast-in-situ Concrete Piles", I.S.I., New Delhi.

28. IS 2911: Part 1 : Sec. 2 – 1979, "Code of Practice for Design and Construction of Pile Foundation–Bored Cast-in-situ Piles", I.S.I., New Delhi.

29. IS 2911: Part 1 : Sec. 3 – 1979, "Code of Practice for Design and Construction of Pile Foundation–Driven Precast Concrete Piles", I.S.I., New Delhi.

30. IS 2911: Part 1 : Sec. 4 – 1979, "Code of Practice for Design and Construction of Pile Foundation–Bored Precast Concrete Piles", I.S.I., New Delhi.

31. IS 2911: Part 3 – 1980, "Code of Practice for Design and Construction of Pile Foundation–Under-Reamed Piles", I.S.I., New Delhi.

32. IS 2911: Part 4 – 1985, "Code of Practice for Design and Construction of Pile Foundation–Load Test on Piles", I.S.I., New Delhi.

33. IS 2950: Part 1 – 1981, "Code of Practice for Design and Construction of Raft Foundations–Design", I.S.I., New Delhi.

34. Jain, N.K. (1983), "Flexural Behaviour of Partially Embedded Pile Foundations," Ph.D. Thesis, Deptt. of Civil Engg., University of Roorkee.

35. Jain, V. and Ramasamy, G. (1985), "Computer Aided Analysis of Flexible Piles," Second International Conference on Computer Aided Design in Civil Engineering, University of Roorkee, Roorkee, pp. VI, 60–64.

36. Jai Krishna and Jain, O.P. (1966), "Plain and Reinforced Concrete," Vol. I, Nem Chand, Roorkee.

37. Kishida, H. (1964), "The Bearing Capacity of Pile Groups in Sand under Central and Eccentric Loads," Report submitted to National Research Council, Canada, March 1964.

38. Kishida, H. (1967), "Ultimate Bearing Capacity of Piles Driven into Loose Sand," *Soil and Found*, Vol. 7.

39. Kishida, H. and Meyerhof, G.G. (1965), "Bearing Capacity of Pile Groups under Eccentric Loads in Sand," Proc. VI International Conference on Soil Mechanics and Foundation Engineering (Montreal), Canada, Vol. 2, p. 270.

40. Kulwahy, F.H. (1984), "Limiting Tip and Side Resistance, Fact or Fallacy," Symposium on Analysis and Design of Pile Foundation, *ASCE* Proceedings, pp. 80–98.

41. Matlock, H. and Reese, L.C. (1961), "Foundation Analysis of Offshore Pile Supported Structures," Proceedings Fifth International Conference on Soil Mechanics and Foundation Engineering, Paris, Vol. 2, pp. 91–97.

42. Matlock, H. and Reese, L.C. (1962), "Generalised Solutions for Laterally Loaded Piles," *Transactions of the American Society of Civil Engineers*, Vol. 127, Part I, pp. 1220–1247.

43. McCarthy (1988), "Essentials of Soil Mechanics and Foundations," *Canadian Geotechnical Journal*, Vol. V, No. 4.

44. McNulty, J.F. (1956), "Thrust Loading on Piles," *Journal of the Soil Mechanics and Foundations Division*, ASCE, Vol. 82, No. SM2, Proc. Paper 940, pp. 1–25.

45. Meyerhof, G.G. (1959), "Compaction of Sands and Bearing Capacity of Piles," *JSMFD*, ASCE, Vol. 85, SM6, pp. 1–29.

46. Meyerhof, G.G. and Adams, J.I. (1968), "The Uplift Capacity of Foundations," *Canadian Geotechnical Journal*, Vol. V, No. 4, pp. 225–244.

47. Meyerhof, G.G. (1976), "Bearing Capacity and Settlement of Pile Foundations," *J. Geotech. Div. ASCE*, Vol. 102, No. GT 3, pp. 197–228.

48. Murthy, V.N.S. (1965), "Behaviour of Batter Piles Subjected to Lateral Loads," Ph.D. Thesis, I.I.T., Kharagpur, India.

49. Nair, K., Gray, H. and Donovan, N.C. (1969), "Analysis of the Pile Group Behaviour," *Performance of Deep Foundation*, Special Technical Publication, 444, ASTM, pp. 118–159.

50. Oteo, C.S. (1972), "Displacement of a Vertical Pile Group Subjected to Lateral Loads," *Proceedings 5th European Conference on Soil Mechanics and Foundation Engineering*, Madrid, Vol. 1, pp. 397–405.

51. Palmer, L.A. and Thompson, J.B. (1948), "The Earth Pressure and Deflection along the Embedded Lengths of Piles Subjected to Lateral Thrust," *Proceedings Second International Conference on Soil Mechanics and Foundation Engineering*, Rotterdam, Holland, Vo. V, pp. 397–405.

52. Poulos, H.G. (1971a), "Behaviour of Laterally Loaded Piles: Pile Group," *Journal of the Soil Mechanics and Foundation Division*, ASCE, Vol. 97, No. SM5, pp. 711–731.

53. Poulos, H.G. (1971b), "Behaviour of Laterally Loaded Piles: I—Single Piles," *J. Soil Mech. Found. Div.*, ASCE, Vol. 97, No. SM5, pp. 733–751.

54. Poulos, H.G. (1974), "Analysis of Pile Groups Subjected to Vertical and Horizontal Loads," *Aust. Geomechanics J.*, Vol. G4, No. 1, pp. 26–32.

55. Prakash, S. (1962), "Behaviour of Pile Groups Subjected to Lateral Loads," Ph.D. Thesis, University of Illinois, *Urbana*, p. 397.

56. Prakash, S. and Saran, D. (1967), "Behaviour of Laterally Loaded Piles in Cohesive Soils," *Proceedings 3rd Asian Conference on Soil Mechanics and Foundation Engineering*, Halifa (Israel), pp. 235–238.

57. Raju, V.S. and Gandhi, S.R. (1989), "Ultimate Capacity of Precast Driven Piles in Stiff Clay", *Indian Geotechnical Journal*, Vol. 19, No. 4, pp. 273–289.

58. Ramasamy, G. (1974), "Flexural Behaviour of Axially and Laterally Loaded Individual Piles and Group of Piles," Ph. D. Thesis submitted to Indian Institute of Science, Bangalore, India.

59. Ramasamy, G., Ranjan, G. and Jain, N.K. (1987), "Modification to Indian Standard Code Procedure on Lateral Capacity of Piles," *Indian Geotechnical Journal*, Vol. 17, No. 3, pp. 249–258.

60. Ramasamy, G. (1989), "Estimation of Lateral Capacity of Pile—A Note of Caution," *Pile Talk International* 89, Kuala Lumpur, Malaysia, pp. 261–272.

61. Ramasamy, G. (1990), "Estimation of Pile Capacity—Pitfalls and Remedies," *Proceedings of the National Workshop on Geotechnical Engg.*, I.I.Sc. Bangalore.

62. Reese, L.C. and Matlock, H. (1956), "Non-dimensional Solutions for Laterally-loaded Piles with Soil Modulus Assumed Proportional to Depth," *Proceedings of the 8th Texas Conference on Soil Mechanics and Foundation Engineering*, Austin, Texas, pp. 1–41.

63. Skempton, A.W. (1953), "Discussion on Settlement of Pile Groups in Sand," *Proc. 3rd ICSMFE* (Zurich), Vol. 3, p. 172.

64. Skempton, A.W. (1953), "Discussions: Piles and Pile Foundations," *Proc. 3rd ICSMFE*, Zurich, Vol. 3, p. 172.

65. Sowers, G.F. et al. (1961), "The Bearing Capacity of Friction Pile Groups in Homogeneous Clay from Model Studies," *5th ICSMFE*, Vol. 2, pp. 155–159.

66. Tavenas, F.A. and Audy, R. (1972), "Limitations of the Driving Formulas for Predicting the Bearing Capacity of Piles in Sand," *Canadian Geotechincal Journal*, No. 9, p. 47.

67. Terzaghi, K. (1955), "Evaluation of Coefficient of Subgrade Reaction," *Geotechnique*, Vol. 5, pp. 297–326.

68. Terzaghi, K. and Peck, R.B. (1967), *Soil Mechanics in Engineering Practice*, John Wiley & Sons, NY.

69. Tomlinson, M.J. (1986), *Foundation Design and Construction*, Longman Singapore Publishers, Singapore.

70. Tomlinson, M.J. (1987), *Pile Design and Construction Practice*, A View Point Publication.

71. Tomlinson, M.J. (1971), "Some Effects of Pile Driving on Shaft Resistance", Proc. conf. on Behaviour of Piles, Institution of Civil Engineers, London, pp. 107–114.

72. Van Veele, A.F. (1964), "Negative Skin Friction of Pile Foundation in Holland," *Proc. of Symposium on Bearing Capacity of Piles*, Roorkee, Vol. 1, pp. 1–10.

73. Van der Veen, C. (1957), "The Bearing Capacity of a Pile Predetermined by a Cone Penetration Test," *Proc. 4th ICSMF*, London.

74. Vesic, A.S. (1967), "A Study of Bearing Capacity of Deep Foundations Final Report," School of Civil Eng., Georgia Inst. Tech. Atlanta, U.S.A.

75. Vesic, A.S. (1970), "Tests on Instrumented Piles—Ogeechee River Site," *Journal of Soil Mechanics and Foundation Division*, ASCE, Vol. 96, No. SM 2.

76. Vesic, A.S. (1977), "Design of Pile Foundations," Transportation Research Board, National Research Council, Washington, D.C.

77. Vijayavergiya, V.N. (1969), "Load Distribution for a Drilled Shaft in Clay Shale", Ph.D. Dissertation, The University of Texas at Austin, Texas.

78. Whitaker, T. (1957), "Experiments with Model Piles in Groups," *Geotechnique*, Vol. 7, No. 4, December, p. 147.

PRACTICE PROBLEMS

8.1 What are the situations suitable for providing pile foundation ? How are piles classifed?

8.2 Describe the various methods of determining the safe capacity of single pile.

8.3 Describe stepwise the procedure of separately point bearing resistance and skin friction resistance in a cyclic plate load test.

8.4 What do you understand by the term 'Efficiency of a pile group'? Derive the expression of spacing between friction piles for hundred per cent efficiency of the pile group.

8.5 Give the methods of determining the settlement of a pile group in (i) cohesionless soil, and (ii) cohesive soil.

8.6 Describe the subgrade reaction method of obtaining the variation of deflection and moment along the length of pile subjected to lateral load. Give the analysis considering that the pile is embedded in (i) sandy strata, and (ii) clayey strata. Discuss the effect of fixity of pile head on deflection and bending moment.

8.7 Give the methods of determining the loads in the various piles of a pile group containing both vertical and batter piles. The pile group is subjected to a vertical load, a lateral load and a moment.

8.8 A 400 mm × 400 mm size concrete pile is driven through a deposit of loose sand ($\phi = 25°$) 8.0 m thick into a deposit of dense sand ($\phi = 41°$). The pile penetrates to a depth of 2.5 m in the dense sand deposit which extends to a great depth. If the unit weights of loose sand and dense sand deposits are 16 kN/m^3 and 19 kN/m^3 respectively, estimate the ultimate pile capacity.

8.9 Estimate the required length of 500 mm diameter concrete pile driven in a deposit of clay having an unconfined compressive strength of 120 kN/m^2. The pile is required to carry a load of 280 kN. If the cast-in-situ bored pile of the same length is used, what will be its capacity ?

8.10 A 20 mm thick, 400 mm outer diameter steel casing was driven for installation of piles at a site where the soil encountered is silty sand up to great depth. The water table is close to the ground surface. The rated energy of the steam hammer, used for installation of piles is 50 kN/m per blow. The length of casing is 10 m. The average penetration per blow in the last blow is 2.5 mm. Determine the ultimate resistance of pile using Hiley's formula when:

(a) Driving is without dolly or helmet or cushion about 25 mm thick.

(b) Driving is with short dolly 500 mm long, helmet and cushion 60 mm thick.

Assume suitable data not given.

8.11 The following data refer to a cylic plate load test on a pile having a length of 10 m and size 25 cm × 25 cm.

Load on pile top (kN)	100	150	200	250	300	400	500
Total settlement of pile (mm)	1.2	1.9	2.6	3.8	5.6	10.5	23.0
Net settlement of pile (mm)	0.3	0.5	0.7	1.3	1.7	3.2	8.4

(a) Plot the load-settlement curve and estimate the allowable load.

(b) Assuming the value of E as 2.8×10^7 kN/m^2, separate the load taken up in point bearing and skin friction.

8.12 Proportion a friction pile group to carry a load of 3000 kN including the weight of pile cap of a site where the subsoil consists of uniform clay to a depth of 20 m underlain by rock. The undrained cohesion of clay is 50 kN/m^2. The clay is normally loaded and has a void ratio of 0.95 and the liquid limit of 55%. Compute the settlement of the pile group. Design the piles and pile cap.

8.13 A 15 m long pile is embedded 14 m into cohesionless soil whose soil modulus increases at the rate of 10^5 kN/m^2/m. The lateral load acting at its top is 90 kN. EI of the piles is 160×10^{12} kNm^2. Determine the maximum bending moment in the pile if

(a) The pile head is completely free, and

(b) The pile head is completely restrained against rotation.

If the pile is subjected to an axial load of 2000 kN in addition to moment caused by the lateral load, design the pile section.

8.14 A 300 mm diameter 10 m long RCC pile is embedded in medium stiff clay having unconfined compressive strength of 50 kN/m^2. The pile head is fixed at the ground surface and is subjected to an axial load of 450 kN and a lateral load of 95 kN. Selecting an approximate value of modulus of subgrade reaction, compute the maximum bending moment and the deflection of the pile.

8.15 Four precast concrete piles, 500 mm diameter each are driven in a deposit of medium dense sand ($\phi = 32°$) to form a square group. This strata extends upto 7.5 m depth. The piles are 10 m long and 2.5 m length of pile is embedded in dense sand strata ($\phi = 41°$). The spacing between the piles in both directions is 2.0 m. The water table is close to the ground surface. Estimate the pile group capacity. Design the piles and the pile cap. Sketch neatly the details of reinforcement.

8.16 It is proposed to provide pile foundation for the retaining wall shown in Fig. 8.64. The soil strata is also shown in the figure. Draw a scheme of vertical and batter piles so that each pile is subjected to axial load. Design the piles and pile group by (i) Conventional method, and (ii) Hrennikoff's method.

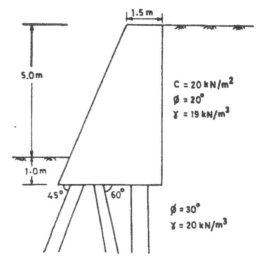

Fig. 8.64. Retaining Wall with Soil Profile.

Bridge Substructures

9.1 INTRODUCTION

A bridge is a structure providing passage over an obstacle without closing the way beneath. The required passage may be for a road, a railway, pedestrians, a canal or a pipeline. The obstacle to be crossed may be a river, a road, railways or a valley. Bridges are very important engineering structures as they enhance the vitalities of the cities and aid the social, cultural and economic improvements of the locations around them. The mobility of an army at war is often affected by the availability or otherwise of bridges to cross rivers.

It may not be possible always to have a wide choice of sites for a bridge. This is particularly so in case of bridges in urban areas and flyovers. For river bridges, the characteristics of an ideal site are: (i) a straight reach of the river; (ii) a narrow channel with firm banks; (iii) rock or other inerodable strata close to the river bed level; (iv) proximity to a direct alignment of the road to be connected; (v) suitability of bridge to the surrounding area; and (vi) absence of expensive river training works.

Bridges are mainly of four types: (a) Arch bridges, (b) Suspension bridges, (c) Truss and girder bridges, and (d) Simply supported bridges. The main parts of a bridge structure are: (i) Decking, consisting of deck slab, girders, trusses, etc., (ii) Bearings for the decking, (iii) Abutments and the piers, (iv) Foundations for the abutments and the piers, (v) River training works, (vi) Approaches to the bridges to connect the bridge proper to the roads on either side, and (vii) Handrails, guard stones, etc.

The following information is required for designing the various component parts of a bridge:
(a) Present and anticipated future volume and nature of traffic on the road at the bridge site.
(b) Hydraulic data pertaining to the river, including the highest flood level (HFL), low water level (LWL), size, shape, slope and nature of the catchment, intensity and frequency of rainfall in the catchment.
(c) Soil profile along the probable bridge sites over the length of the bridge and approaches.
(d) Liability of site to earthquake disturbances.

In this chapter, design of piers and abutments, and their foundations have been covered.

9.2 DEFINITIONS

Bridge: It is a structure for carrying the road traffic or other moving loads over a depression or obstruction such as channel, road or railway.

Culvert: It is a bridge having a gross length of 6 m or less between the faces of abutments.

Foot bridge: It is a bridge exclusively used for carrying pedestrians, cycles and animals.

High flood level: It is the level of the highest flood ever recorded or calculated level for the highest possible flood. It is denoted by HFL.

Low water level: It is the level of the water surface obtaining generally in dry season and it shall be specified in case of each bridge. It is denoted by LWL.

Channel: It is a natural or artificial water course.

Catchment area: The area by which a channel is fed or gets water.

Highlevel bridge: It is a bridge which carries the roadway above the HFL or the channel.

Submersible bridge: It is a bridge designed to be overtopped in floods.

Freeboard: At any point, it is the difference between the HFL after allowing for afflux, if any, and the formation level of road embankment on the approaches or top level of guide bunds at that point.

Abutment: The end support of a bridge superstructure is known as an abutment.

Length of bridge: It is the overall length measured along the centre line of the bridge from end of the bridge deck.

Linear waterway: It is the length available in the bridge between the extreme edges of water at the HFL, measured at right angles to the abutment faces.

Effective linear waterway: It is the total width of the waterway of the bridge minus the effective width of obstruction.

Piers: The intermediate supports of a bridge superstructure are known as piers.

Wing walls: The walls constructed on either side of an abutment to support and protect the embankment are known as wingwalls.

9.3 WIDTH OF ROADWAY AND FOOTWAY

For high level bridges constructed for the use of road traffic only, the width of roadway shall not be less than 4.25 m for a single-lane bridge and 7.5 m for a two-lane bridge and shall be increased by 3.5 m for every additional lane of traffic for a multiple of two lanes.

Causeways and submersible bridges shall provide for at least two lanes of traffic (i.e. 7.5 m width) unless one lane of traffic is specially permitted for design.

The width of footway or safety kerb shall be taken as the minimum clean width anywhere within a height of 2.25 m above the surface of the footway or safety kerb, such width being measured at right angles to the centre line of the bridge and shall not be less than 1.5 m.

9.4 ELEMENTS OF A BRIDGE SUBSTRUCTURE

The bridge substructure consists of piers, abutments and their foundations. In Fig. 9.1, typical cross-sections of piers supported by different types of foundation are shown. The main parts of the substructure are explained below.

(i) Pier cap: It is a reinforced concrete raft slab which distributes the load of the superstructure onto the pier. The bearings of the bridge rest on the cap and are held to it by means of steel bolts. A minimum thickness of about 250 mm is adopted. The pier cap also serves as a coping and is kept projecting beyond the pier by 150 to 300 mm.

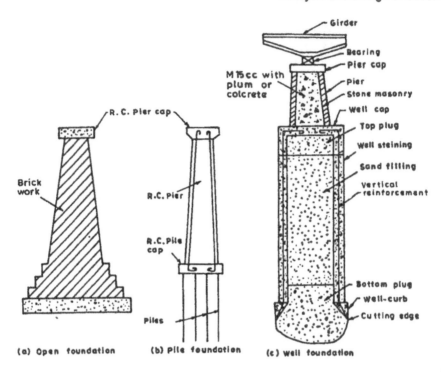

Fig. 9.1. Section of Piers with Different Types of Foundations.

(*ii*) *Pier:* The pier may be made in brick or stone masonry, plain or reinforced concrete, solid or cellular with filling material inside. The top width of the pier should be sufficient to accommodate the bearings over it with about 150 mm projection beyond the outer edges of bearing so that the bearing reaction does not fall just on the edge of the pier. The minimum length of the pier is equal to the distance between the outer edges of bearings. The sides of the pier are usually kept sloping at 1/24 to 1/12 so that pier dimensions increase towards the base.

Piers are generally provided with noses to reduce the water pressures. The noses are provided starting at the base of the pier either up to top or little above HFL. When the noses are taken to the top of the pier, they are extra to the length of the pier. In any case the bearings should not be located on the noses.

(*iii*) *Foundation:* Depending on type of base soil and scour depth, a pier may have either of the following types of foundation.

(a) Open foundation: It is the simplest and most economical type of foundation. Minor bridges are founded at shallow depth and open foundations are adopted for them when the base soil is of reasonable strength. Open foundations are also used for important bridge when rock is available at the surface or at shallow depth.

(b) Pile foundation: The load of the bridge can be transferred to medium depths through piles in case adequate bearing strata is not available at shallow depths. The piles may be bearing piles or friction piles depending upon the soil strata.

Timber piles have been used by Indian Railways for the construction of bridges for restoration of traffic in the event of breaches. In the Assam region they have been used even for the construction of permanent bridges where the' intensity of traffic is light. A few road-over bridges have been constructed under the main broad-gauge lines also recently using R.C.C. piles, but for bridges across rivers having scourable beds, well foundations are used. Usually pile foundations are not used in soils having boulders due to construction difficulties.

(c) Well foundation: It is the more common type of foundation in India for both road and railway bridges. Such foundations can be sunk to great depths and can carry very heavy vertical and lateral loads. Well foundations can be conveniently installed in a boulder stratum. It is a massive structure and relatively rigid in its structural behaviour. Figure 9.1c shows a typical section of a well foundation.

The various elements of a well foundation are as follows:

Well cap: It is an R.C.C. slab of sufficient strength to transmit the load of the superstructure to the body of the well. Its top level is generally kept at LWL or river bed level. The length of the well cap should be sufficient to accommodate the pier and its noses. It is monolithically constructed with the well steining.

Main body of well: It is cellular in cross-section, consisting of hollow cylinder of masonry or concrete of sufficient thickness so as to carry all the forces transmitted to it by the well cap. The minimum inside diameter of about 2 m may be used so that a bucket could work in it freely for taking out the excavated material. The minimum thickness of steining is about 700 to 900 mm since it provides weight during sinking.

The inside open space in a well in which excavation is done during sinking is known as dredge hole. It is later filled with granular material to distribute the load of the superstructure on the bottom plug.

The steining is reinforced with vertical steel bars placed at regular intervals along the periphery of the well at mid-thickness of steining. Such bars are kept in vertical position by means of flat iron strips (Fig. 9.2). The vertical bars are properly embedded in the well cap and well curb so that the three component parts of the well become monolithic.

Well curb: It is R.C.C. ring beam with a steel cutting edge at its bottom. The curb supports the well steining and resist hoop forces to which it is subjected during sinking and afterwards. The well curb is slightly projected beyond the steining by about 25 mm to reduce the skin friction during the sinking of well.

Bottom plug: After the well is sunk to the required depth, the base of the well is plugged with concrete. This is called the bottom plug. It transmits the load to the subsoil and acts as a raft against soil pressures from below. As mentioned above, the hollow space above the bottom plug is filled with granular material.

Top plug: After filling the dredge hole with granular material, a concrete plug covering the filling is usually provided. It provides a contact between the well cap and sand filling, thus helps in transferring some load through the sand filling. It, therefore, relieves the steining to some extent.

In Fig. 9.3, a typical cross-section of an abutment supported by an open foundation is shown. The main parts of an abutment are explained below:

Back wall or parapet: It is that portion of the wall which retains the earth in the height of bridge

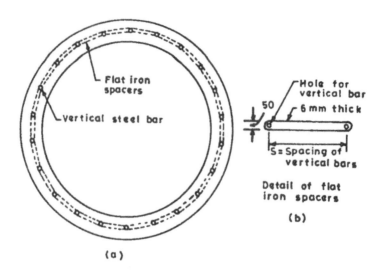

Fig. 9.2. Sectional Plan of Wall Steining.

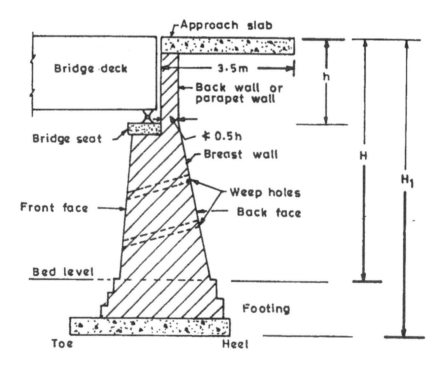

Fig. 9.3. Section of an Abutment with Open Foundation.

deck. Its top thickness is kept about 600 mm which is increased towards the bottom so that at the bridge seat level, the thickness is not less than half of its height.

Approach slab: It is an R.C.C. slab of about 150 mm thickness reinforced with 12 mm diameter bars 150 mm apart bothways at bottom. This slab is provided to cover the full formation width of road up to a length of about 3.5 m and is supported on the backwall. It is not necessary to provide the approach slab, but if it is provided, the surcharge due to live load at the back of the abutment may be neglected.

Bridge seat: It is a plain or R.C.C. bearing plate used to support the bridge bearings and distribute their loads to the breast wall uniformly. The width of the bridge seat may be taken about 300 mm plus the width of the bearing.

Breast wall: It is the main load carrying member. The front face may be vertical or battered. Table 9.1 gives recommended front face batter for different heights.

Table 9.1. Front Face Batter.

S. No.	Height of Breast Wall	Front Face Batter
1	upto 5.0 m	Vertical
2	5 m to 10 m	1 in 24
3	10 m to 15 m	1 in 12
4	> 15 m	1 in 6

The back face of the breast wall has to be sloped appreciably so that the required thickness for stability at bed level (approximately equal to 0.5 H) becomes available. Weep holes are provided to drain off water which gets access to the earth filling. The length of the abutment may be taken equal to 600 mm plus the distance between the outer edges of bearings or sole plates of the bridge.

Foundation: Depending on the type of the base soil and scour depth, an abutment may also have either open foundation, or pile foundation or well foundation. In the case of open foundation, the projection of the base slab on the toe side is kept larger than on the heel side to accommodate the resultant thrust which slopes towards the stream. In the case of pile foundation, batter piles may advantageously be employed. In the case of well foundation the abutment may be placed on the well cap eccentrically so that the resultant thrust intersects the well cap, giving rise to the minimum eccentricity effects on the well.

Wing walls: These walls may be of three types, namely, (i) Straight wing wall—When wing walls are constructed in line with the abutment, they are known as straight wing walls (Fig. 9.4a). Such type of wing walls are economical when there is no danger of material washing down from the banks of the river. (ii) Splayed wing wall—when wing walls are given inclination in plan, they are known as splayed wing walls (Fig. 9.4b). The wing walls may also be curved, instead of being splayed. The splay or inclination is usually 45°. (iii) Return wing walls—When angle of splay becomes 90°, as shown in Fig. 9.4c, wing walls are known as return wing walls. Such wing walls are preferred to splayed wing walls in case of very high embankments (Fig. 9.5).

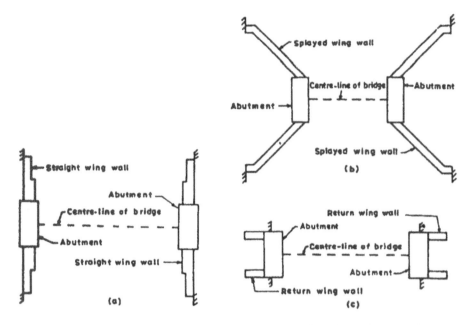

Fig. 9.4. Wing Walls: (*a*) Straight Wing Wall, (*b*) Splayed Wing Wall, (*c*) Return Wall.

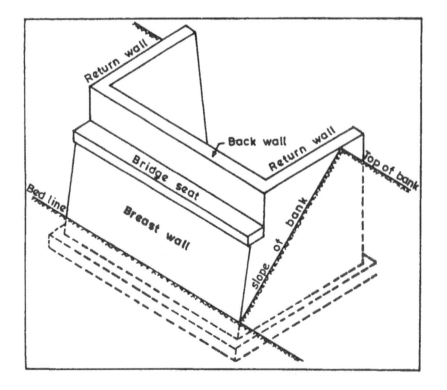

Fig. 9.5. View of an Abutment with Return Wall and Embankment.

9.5 DETERMINATION OF MAXIMUM FLOOD DISCHARGE

The maximum flood discharge for which the waterway of the bridge is to be designed, shall be the maximum flood discharge on record for a period of not less than 50 years. In case where the requisite information is not available, the maximum flood discharge can be estimated by the following methods:

(a) By using one of the empirical formulae applicable to the region;
(b) By using a rational method involving the rainfall and other characteristics for the area;
(c) By the area-velocity method, using the hydraulic characteristics of the stream such as cross-sectional area, and the slope of stream; and
(d) By unit hydrograph method.

Wherever possible, the maximum flood discharge is obtained by at least two different methods, and the higher value is adopted as the discharge for designing the bridge.

Empirical formulae

Empirical formulae for the flood discharge from a catchment area are of the following form:

$$Q = CM^n \qquad \text{... (9.1)}$$

where
Q = maximum flood discharge in m³/sec,
M = catchment area in kilometre square, and
C = constant depending upon the intensity of rainfall in the catchment area and the characteristics of the catchment such as slope of the ground, the nature of soil and type of vegetation, etc.

Some of the empirical formulae are given below:

(i) Dicken's formula

$$Q = CM^{\frac{3}{4}} \qquad \text{... (9.2)}$$

The value of constant is more for rivers with residual soil formed by decomposition of basalts and it is less for rivers with alluvial drift soils. The value of constant C varies from 11 to 22. This formula is used in northern and central part of India.

(ii) Ryve's formula

$$Q = CM^{\frac{2}{3}} \qquad \text{... (9.3)}$$

The value of C is taken as 6.8 for flat tracts within 25 km of the coast, 8.5 for areas between 25 km and 160 km of the coast and 10 for limited areas near the hills. This formula is used in southern parts of India.

(iii) Inglis's formula

$$Q = \frac{125M}{\sqrt{M+10}} \qquad \text{... (9.4)}$$

This formula is used in the Bombay region.

The above empirical formulae are oversimplified and much depends on the value of C. A reliable value of C for any particular region can only be derived by careful statistical analysis of a large volume of observed flood and catchment data. Therefore, reliability of the empirical formulae for estimating the maximum flood discharge is extremely limited in practical field.

Rational formula

The flood discharge depends on (i) intensity, distribution and duration of rainfall, and (ii) area, shape, slope, permeability and initial wetness of the catchment. A typical formula which takes the above factors into consideration is:

$$Q = MI_0 \lambda_0 \qquad \qquad \text{... (9.5)}$$

where Q = maximum flood discharge, m³/s,

M = catchment area, km²,

I_0 = peak intensity of rainfall, mm/hour,

λ_0 = A function depending on the characteristics of the catchment in providing the peak run off

$$= \frac{0.56 P_r f_l}{t_c + 1} \qquad \qquad \text{... (9.6)}$$

t_c = concentration time, hours

$$= \left(0.88 \frac{d_s}{d_e} \right)^{0.385} \qquad \qquad \text{... (9.7)}$$

d_s = distance from the critical point to the bridge site, km

d_e = difference in elevation between the critical point and bridge site, m

P_r = Percentage coefficient of run-off for the catchment characteristics, from Table 9.2

f_l = a factor to correct for the variation of intensity of rainfall I_0 over the area of the catchment, from Table 9.3

Table 9.2. Value of P_r in Rational Formula.

Surface	P_r
Steep bare rock, and also city pavements	0.90
Rock, steep but with thick vegetation	0.80
Plateaus, lightly covered	0.70
Clayey soils, stiff and bare	0.60
Clayey soils, lightly covered	0.50
Loam, lightly cultivated	0.40
Loam, largely cultivated	0.30
Sandy soil, light growth	0.20
Sandy soil, heavy brush	0.10

Table 9.3. Value of Factor f_l in Rational Formula.

Area (km²)	f_l	Area (km²)	f_l
0	1.000	80	0.760
10	0.950	90	0.745
20	0.900	100	0.730
30	0.875	150	0.675
40	0.845	200	0.645
50	0.820	300	0.625
60	0.800	400	0.620
70	0.775	2000	0.600

Area-velocity method

The area-velocity method based on the hydraulic characteristics of the stream is probably the most reliable among the methods of determining the flood discharge. The discharge Q is given by:

$$Q = A \cdot V \qquad \ldots (9.8)$$

where Q = discharge, m³/s,

       A = wetted area, m²

       V = velocity of stream, m/s

$$= C\sqrt{m_d i_h} \text{ (Chezy's formula)} \qquad \ldots (9.9)$$

      m_d = hydraulic mean depth,

      i_h = slope of the hydraulic grade line, and

      C = Chezy's constant.

The value of C can be obtained either by Kutter's formula (Eq. 9.10) or Bazin's formula (Eq. 9.11).

Kutter's formula:

$$C = \frac{23 + \dfrac{0.00155}{i_h} + \dfrac{1}{r_f}}{1 + \dfrac{r_f}{\sqrt{m_d}}\left[23 + \dfrac{0.00155}{i_h}\right]} \qquad \ldots (9.10)$$

where r_f is called rugosity coefficient or roughness factor and its value varies from 0.02 to 0.03, depending on the condition of bed and sides of river.

Bazin's formula:

$$C = 1.81 + \frac{157.6}{\dfrac{K}{\sqrt{m_d}}} \qquad \ldots (9.11)$$

where K is a constant and its value varies from 1.54 to 3.17 depending on the condition of the bed and sides of river.

Unit hydrograph method

A unit hydrograph is defined as the storm run off hydrograph representing a unit depth (1 cm) of direct run off as a result of rainfall excess occurring uniformly over the basin and at a uniform rate for a specified duration (e.g. 6 hours or 12 hours). The area under a unit hydrograph represents the volume of rainfall excess due to a rain of 1cm over the entire basin.

In unit hydrograph method, it is assumed that the storm occurs uniformly over the entire basin and that the intensity of rainfall is constant for the duration of the storm. Direct runoff for any given storm can be calculated by multiplying the maximum ordinate of the unit hydrograph by the depth of runoff over the area. The maximum runoff is then obtained by adding the base flow to the maximum direct runoff rate.

Where possible, more than one method shall be adopted, results compared, and the maximum discharge is fixed by judgement by the engineer responsible for design.

9.6 DETERMINATION OF LINEAR WATERWAY AND EFFECTIVE LINEAR WATERWAY

In the case when the water course to be crossed is an artificial channel for irrigation or navigation, or when the banks are well defined for natural streams, the linear waterway shall be the distance between banks at the water surface elevation corresponding to maximum design discharge.

For natural channels in alluvial beds and having undefined banks, the required effective linear waterway is obtained from the designed discharge using Lacey's formula given in Eq. (9.12).

$$W = C\sqrt{Q} \qquad\qquad\qquad\qquad\qquad ... (9.12)$$

where W = effective linear waterway or regime width, m
          Q = design maximum discharge, m³/s, and
          C = constant, usually taken as 4.8 for regime channels but it may vary from 4.5 to
               6.3 according to local conditions.

It is not desirable to reduce the linear waterway below that for regime condition. If reduction is affected, velocity would increase and greater scour depths would be involved, requiring deeper foundations.

9.7 NUMBER OF SPANS

The number of spans for a bridge should be carefully decided by considering all the aspects of the bridge such as cost of construction, nature of flow, importance of bridge, materials available, etc. Following points are important in deciding the number of spans for a bridge:

(i) Foundations for piers and abutments: Piers and abutments should be so located as to make the best use of the soil conditions available.

(ii) Aesthetic consideration: It is desirable to adopt odd number of spans for a bridge keeping central spans larger than end spans. The velocity of flow is the highest at centre of the length of a bridge. Larger length of the central span will also provide an opening in the centre length of a bridge.

(iii) The number of supports and their locations are so fixed as to provide the most economical design of the bridge.

Considering only the variable items, for a given linear waterway, the total cost of the superstructure increases and the total cost of substructure decreases with increase in span length. The most economical span is that for which the cost of superstructure equals the cost of substructure.

It is possible to derive the expression for economical span of a bridge. Following assumptions are made in the derivation given below:

(a) The bridge consists of equal spans.
(b) The bridge is an arch bridge or a girder bridge.
(c) The cost of superstructure per span varies directly as the square of span.
(d) The cost of railings, flooring, etc. is proportional to the total length of the bridge.

Let A = cost of approaches,

          B = cost of two abutments, including foundations,

          L = total linear waterway,

          l = length of one span,

          n = number of spans,

          P = cost of one pier, including foundation, and

          C = total cost of the bridge.

Therefore, $C = A + B + (n-1)P + nKl^2 + K'L$ ... (9.13)

where K and K' are constants.
For minimum cost, dC/dl should be zero.

Substituting $n = L/l$ and differentiating, and equating the results of differentiation to zero, we get,

$$P = Kl^2$$ ... (9.14)

Therefore, for an economical span, the cost of superstructure of one span is equal to the cost of substructure of same span.

The economical span (l_e) can then be computed as

$$l_e = \sqrt{\frac{P}{K}}$$ ... (9.15)

Values of P and K have to be evaluated as average over a range of possible span lengths.

9.8 DISCHARGE FOR DESIGN OF FOUNDATIONS

To provide for an adequate margin of safety the foundations are designed for a larger discharge which is over the maximum flood discharge obtained by the methods described in Sec. 9.5. This percentage may be 30 per cent for small catchments up to 500 sq kilometres, 25 to 20 per cent for medium catchments of 500 to 5000 sq kilometres, 20 to 10 per cent for large catchments of 5000 to 25000 sq kilometres and less than 10 per cent for larger catchments above 25000 sq kilometres at the discretion of the engineer to cover the possibility of floods of longer return period during the life of the structure.

9.9 DETERMINATION OF THE MAXIMUM DEPTH OF SCOUR

The maximum scour depth should be measured by soundings in the vicinity of the bridge site. Such soundings are best done during or immediately after a flood. Due allowance should be made in the observed values for additional scour that may occur due to the design discharge being greater than the flood discharge for which the scour was observed, and also due to increased velocity due to obstruction to flow caused by the construction of bridges.

Table 9.4. Silt Factor.

Type of Bed Material	d_m mm	Silt Factor
Very fine silt	0.08	0.50
Fine silt	0.12	0.60
Medium silt	0.16	0.70
Coarse silt	0.23	0.85
Fine sand	0.32	1.00
Medium sand	0.50	1.25
Coarse sand	0.73	1.50
Fine bajri & sand	1.0	1.75
Heavy sand	1.29	2.00

The mean depth of scour, d_{sm} below HFL may be calculated from the following equation:

$$d_{sm} = 1.34 \left(\frac{D_b^2}{K_{sf}} \right)^{1/3} \qquad \qquad \dots (9.16)$$

where d_{sm} = mean depth of scour below HFL, m,
D_b = discharge in m³/s per m width, obtained as the total design discharge divided by the effective linear waterway,
K_{sf} = silt factor for a representative sample of bed material (Table 9.4)
$= 1.76 \sqrt{d_m}$
d_m = weighted mean diameter, mm.

When the constricted linear waterway L is less than the regime width W, the value of d_{sm} computed from Eq. (9.16) is to be increased by multiplying the same by the factor $(W/L)^{0.67}$

The maximum depth of scour D is to be taken as below:

(i) in a straight reach   1.27 d_{sm}
(ii) at a moderate bend   1.50 d_{sm}

(iii) at a severe bend	1.75 d_{sm}
(iv) at a right angled bend	2.00 d_{sm}
(v) at noses of piers	2.00 d_{sm}.

It may be noted that Eq. (9.16) gives the scour depth in alluvial river bed condition. Melville (1997) presented an integrated approach for computing the scour depth around piers and abutments. According to him local scour is a time dependent process. The temporal maximum scour depth (for a given set of conditions) is termed the equilibrium depth of scour. At equilibrium, the local scour depth of bridge foundation is described by:

$$d_s = K_{yw} \cdot K_I \cdot K_d \cdot K_s \cdot K_\theta \cdot K_G \qquad \qquad ... (9.17)$$

where, K^s are the factors accounting for the various influences on scour depth. In Eq. (9.17), various terms are as given below:

d_s = maximum scour depth below river bed level,

K_{yw} = depth size, K_{yb} for piers and K_{yl} for abutments,

K_I = factor accounting flow intensity,

K_d = factor accounting sediment size,

K_s = factor accounting shape of pier/abutment,

K_θ = factor accounting pier/abutment alignment, and

K_G = channel geometry.

K_{yw} and d_s have the dimension of length, while the other K^s are dimensionless. Melville (1997) gave the procedure to get the values of K^s. For more details his paper may be referred. In general, this equation gives the less value of maximum scour depth obtained by Eq. (9.16), but it is more realistic.

9.10 DEPTH OF FOUNDATION

The depth of piers and abutments shall be taken down to a level sufficient to secure firm a foundation from consideration of scour, settlement and overall stability.

Shallow foundations in erodible strata: Foundations should be taken down to a comparatively shallow depth bed surface provided the bearing stratum is practically incompressible (e.g. sand), is prevented from lateral movement and is also protected against scour. In any case, the depth of foundation below the deepest scour line should not be less than 2.0 m for pier and abutments with arches and 1.2 m for piers and abutments supporting other type of superstructure.

Deep foundations in erodible strata: The grip below the deepest scour line should not be less than 1/3 the maximum scour depth (D). The grip length should be adequate against lateral stability.

Shallow foundations in inerodible strata: For open foundations resting on inerodible strata like rocks, the minimum embedment of foundations in the rock below shall be as given in Table 9.5.

Deep foundations in inerodible strata: Wells shall be taken to the foundation level and shall be evenly seated all around the periphery on sound rock (i.e. devoid of fissures, cavities, weathered zone, likely extent of erosion, etc.) by providing adequate embedment. In some cases, small patches of soil and weak material continues for a considerable depth while 80% to 90% of the well is resting on good hard rock. In such case it may not be necessary to insist on sinking the well further to make it rest on rock on its entire area.

Table 9.5. Minimum Depth of Embedments.

S. No.	Type of Rock	Ultimate Crushing Strength, kN/m²	Minimum Depth of Embedment (m)
1	Hard rocks e.g. igneous, gneissic, granite	> 10000	0.6
2	Other rocks not covered at S. No. 1	2000	1.5
3	Rocks having fissures, cavities and unfavourable bedding planes	Relatively low	More than 1.5 m after treatment of foundation

9.11 ALLOWABLE BEARING PRESSURE

The allowable bearing pressure of an open foundation and a pile foundation of a pier of an abutment can be obtained by the procedures described in Chapters 7 and 8 respectively.

In the case of well foundations, the allowable bearing pressure can also be estimated by satisfying the two basic design criteria of safety against shear failure and permissible settlement and tilt.

For cohesionless (sandy) soils, the Indian standard (IS: 3955–1967) recommends the following equation to estimate the allowable bearing pressure of a well foundation:

$$q_a = 5.4N^2 B + 16(100 + N^2)D_f \qquad \qquad \text{... (9.18)}$$

where q_a = allowable bearing pressure, kg/m²,

N = corrected standard penetration resistance value,

B = smaller dimension of well cross-section, m, and

D_f = depth of foundation below scour level, m.

In case of cohesive soils, undisturbed samples need be obtained to ascertain the shear and consolidation characteristics of the deposit. The ultimate bearing capacity of a well foundation can be determined using Skempton's method (Sec. 7.5 of Chapter 7). The settlement of the well foundation can be obtained using Terzaghi's equation of one dimensional theory of consolidation (Eq. 2.78 of Chapter 2).

The allowable bearing pressure of a well resting on rock strata can be obtained by evaluating the crushing strength of rock cores collected from the field. However, there is a possibility of fissures, faults and joints in the rock bed which cannot be easily detected from the core samples. Teng (1962) has suggested that the allowable bearing pressure of a well on bed rock should not exceed that of concrete seal. As per Indian Road Congress specifications (IRC: 78–1983), normally allowable bearing pressure exceeding 2MPa shall not be adopted. Further, following values of allowable bearing pressures have been suggested (Table 9.6):

Table 9.6. Allowable Bearing Pressure in Rock.

Types of Rocks	Suggested Allowable Bearing Pressure Values for Average Condition, MPa
Hard igneous and gneissic rocks in sound condition	10
Hard limestones and hard sand stones	4
Schists and slates	3
Hard shales, hard mud stones and soft sand stones	2
Soft shales and soft mud stones	0.6 to 1
Soft limestone	0.6
Heavily shattered rocks, conglomerates and laterites	to be assessed by *in-situ* tests

9.12 LOADS TO BE CONSIDERED

While designing the foundation of a bridge, the following loads and forces should be considered:

   (i)   Dead load
   (ii) Live load
  (iii) Impact load
  (iv) Wind load
   (v) Horizontal forces due to water currents
  (vi) Longitudinal forces caused by tractive efforts of vehicles or by braking of vehicles and/or those caused by restraint to movement of free bearings
 (vii) Buoyancy
(viii) Centrifugal force
  (ix) Seismic force
   (x) Earth pressure
  (xi) Temperature effects

These loads and forces are considered below:

(i) Dead load

The dead load carried by a foundation member consists of its own weight and the portions of the weight of superstructure and substructure supported by the member.

(ii) Live load

Live loads are transient in nature and caused by vehicles which pass over the bridge. These loads cannot be estimated precisely. However, hypothetical loadings which are reasonably realistic have been recommended for design (IRC: 6–2000).

As per I.R.C. recommendations (IRC: 6–2000), the loadings are divided into four categories: (a) class AA loading, (b) class A loading, (c) class B loading and (d) class 70 R loading.

(a) Class AA loading: The I.R.C. class AA loading is based on heavy military vehicles likely to run on certain routes. It is to be adopted for bridges within municipal limits in certain existing industrial areas, certain specified highways, etc. It is the usual practice to design the structures on National and State Highways for class AA loading. It is also desirable that the structures

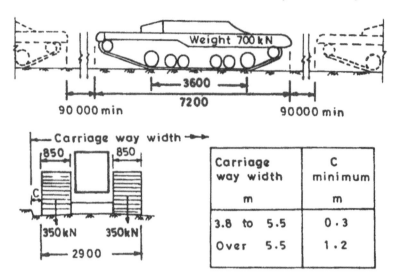

Fig. 9.6. IRC Class AA-Loading (Tracked Vehicle).

designed for class AA loadings should be checked for class A loading because under certain conditions, it is likely to get heavier stresses under class A loading.

In class AA loading, the following two types of vehicles are specified:

(a) Tracked vehicle.

(b) Wheeled vehicle.

Figure 9.6 shows the tracked vehicle. It consists of a packed load of 700 kN which is equally distributed over two tracks of 0.85 m width. The length of vehicle is 7.20 m and out to out distance between the tracks is 2.90 m.

Figure 9.7 shows the wheeled vehicle. The maximum load for single axle is 200 kN and for double axles at 1.2 m centre, it is 400 kN. The maximum wheel load is 62.5 kN.

After considering the vehicle likely to cross the bridge, the choice is made either for tracked vehicle or wheeled vehicle. It is to be assumed for the design purpose that no other live load covers any part of the carriageway of bridge, when a train of tracked or wheeled vehicle is passing over it. The nose to tail spacing between two successive vehicles should not be less than 90 m and the minimum clearance between the road surface of the kerb and the outer edge of wheel or track should be as shown in Fig. 9.6.

(b) Class A loading: The I.R.C. class A loading is based on the heaviest type of commercial vehicle which is considered likely to run on Indian roads. Hence all important permanent road bridges and culverts, which are not covered up by class AA loading, are to be designed for class A loading.

Figure 9.8 shows the train for class A loading. It consists of an engine and two bogies. The specified axle loads with specified distances are also shown in Fig. 9.8. The axle loads are assumed to act simultaneously so as to cause maximum stresses. The train is assumed to move parallel to the length of bridge. No other live load is to occupy any part of the carriageway of bridge when the standard train is crossing the bridge.

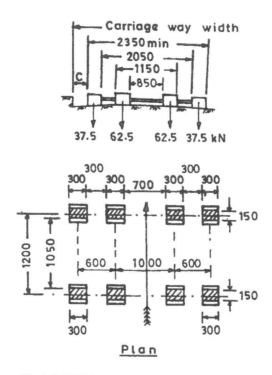

Fig. 9.7. IRC Class AA-Loading (Wheeled Vehicle).

The nose to tail spacing between two successive trains shall not be less than 18.5 m. No other live load shall cover any part of the carriageway when a train of vehicles (or train of vehicles in multilane bridge) is crossing the bridge. The ground contact area for the different wheels and the minimum specified clearanes are also shown in Fig. 9.8.

(c) Class B loading: It comprises a wheel load train similar to that of class A loading but with smaller axle loads as shown in Fig. 9.8. This loading is intended to be adopted for temporary structures, timber bridges and bridges in specified areas.

(d) Class 70 R loading: This is an additional loading which is sometimes specified for use in place of class AA loading. This loading consists of a tracked vehicle of 700 kN or a wheeled vehicle of total load of 1000 kN. The tracked vehicle is similar to that of class AA except that the contact length of the tracks is 4.57 m, the nose to tail length of the vehicle is 7.92 m and the specified minimum spacing between successive vehicles is 30 m. The wheeled vehicle is 15.22 m long and has seven axles with loads totalling to 1000 kN. In addition, the effects on the bridge components due to a bogie loading of 400 kN are also to be checked. The dimensions of the class 70 R loading vehicles are shown in Fig. 9.9. The specified spacing between vehicles is measured from the rearmost point of ground contact of the leading vehicle to the forwardmost point of ground contact of the following vehicle in case of tracked vehicles; for wheeled vehicles, it is measured from the centre of the rearmost wheel of the leading vehicle to the centre of the first axle of the following vehicle.

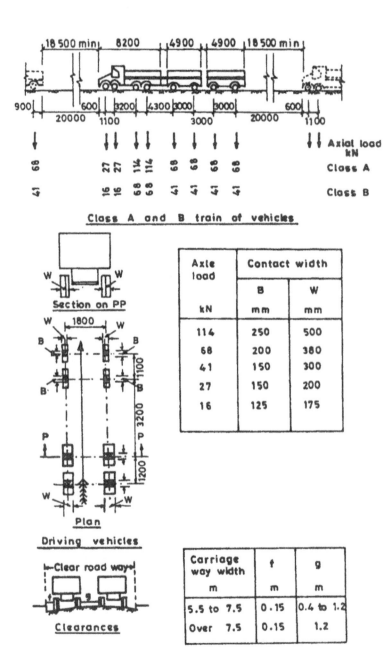

Fig. 9.8. IRC Class A and B Loadings.

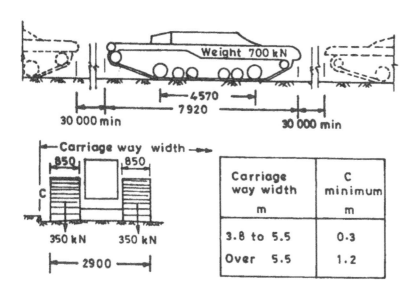

(a) **Tracked vehicle**

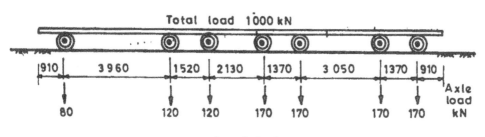

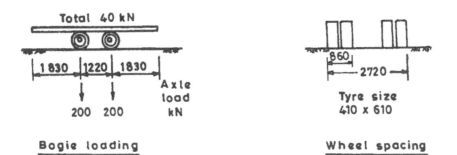

(b) **Wheeled vehicle**

Fig. 9.9. IRC Class 70 R Loading.

The carriageway live load combination shall be considered for the design as given in Table 9.7.

The footways and floors of bridge, which are accessible only to pedestrians and animals, are to be designed for a live load of 4 kN/m². Where crowd loads are likely to occur, such as on

Table 9.7. Live Load Combination.

Sl. No.	Carriageway Width	Number of National Lanes for Design Purpose	Load Combination
1.	Less than 5.3 m	1	One lane of Class A considered to occupy 2.3 m. The remaining width of carriageway shall be loaded with 500 Kg/m²
2.	5.3 m and above but less than 9.6 m	2	One lane of Class 70R OR two lanes of Class A
3.	9.6 m and above but less than 13.1 m	3	One lane of Class 70R with one lane of Class A OR three lanes of Class A
4.	13.1 m and above but less than 16.6 m	4	One lane of Class 70R for every two lanes with one lane of Class A for the remaining lanes, if any, or one lane of Class A for each lane.
5.	16.6 m and above but less than 20.1 m	5	
6.	20.1 m and above but less than 23.6 m	6	

bridges located near pilgrimage towns, the intensity of footway loading shall be taken as 5 kN/m².

Kerbs, 0.6 m or more in width, shall be designed for the above loads, and for a local lateral force of 7.5 kN per metre, applied horizontally at the top of the kerb. If kerb is less than 0.6 m, no live load shall be applied in addition to the lateral load specified above.

The standard loads are to be arranged in such a manner as to produce the maximum reaction on the abutment or pier under consideration for design.

(iii) Impact load

The stresses developed due to fast moving heavy vehicles over uneven surfaces are known as stresses due to impact. The provision made in the design of bridge for impact is expressed as a fraction of the live load. Such a fraction is termed as impact factor, I.

For class A or class B loading for spans between 3 m and 45 m

$$I = \frac{45}{6 + L} \text{ (for R.C.C. bridges)} \qquad \text{... (9.19a)}$$

$$I = \frac{9}{13.5 + L} \text{ (for steel bridges)} \qquad \text{... (9.19b)}$$

For spans less than 3 m, the impact factor is 0.5 for R.C.C. bridges and 0.545 for steel bridges. When the span exceeds 45 m, the impact factor is taken as 0.088 for R.C.C. bridges and 0.154 for steel bridges.

For IRC class AA or 70 R loading

(i) For spans less than 9 m

    (a)   For tracked vehicle   25% for spans up to 5 m linearly reducing to 10% for spans of 9 m

    (b)   For wheeled vehicle   25%

(ii) For spans of 9 m and more

    (a)   For tracked vehicle   For R.C. bridges, 10% up to span of 40 m and in accordance with Fig. 9.10 for spans exceeding 40 m. For steel bridges, 10% for all spans.

    (b)   For wheeled vehicle   For R.C. bridges, 25% for spans up to 12 m and in accordance with Fig. 9.10 for spans exceeding 12 m. For steel bridges, 25% for spans up to 23 m and as in Fig. 9.10 for spans exceeding 23 m.

The span length to be considered in the above computations is determined as below:

(i)   Simply supported, continuous or arch spans—the effective span on which the load is placed.

(ii) Bridges having cantilever arm without suspended span—0.75 of effective cantilever arm for loads on the cantilever arm and effective span between supports for loads on the main span.

When there is a filling of not less than 0.6 m including the road crust as in spandrel filled arches, the impact allowance may be taken as half that computed by the above procedure.

Full impact allowance should be made for design of bearings. But for computing the pressure at different levels of the substructure, a reduced impact allowance is made by multiplying the appropriate impact fraction by a factor as below:

    (i) At the bottom of bed block   0.5

    (ii) For the top 3 m of the   0.5 decreasing uniformly
         substructure below the bed block   to zero

    (iii) For portion of substructure more
         than 3 m below the bed block   0.0

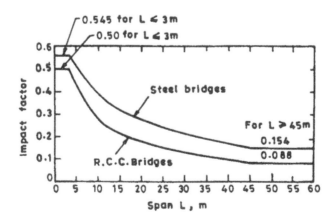

Fig. 9.10. Impact Factor.

(iv) Wind load

Wind forces act horizontally on the exposed area of bridge structure. The intensity of wind force is taken from Table 9.8. The pressures given in this table are to be doubled for bridges situated in areas such as the Kathiawar Peninsula and the coastal regions of West Bengal and Orissa.

Table 9.8. Wind Velocities and Wind Pressures.

H m	V km/hr	P kN/m²	H m	V km/hr	P kN/m²
0	80	0.40	30	147	1.41
2	91	0.52	40	155	1.57
4	100	0.63	50	162	1.71
6	107	0.73	60	168	1.83
8	113	0.82	70	173	1.93
10	118	0.91	80	177	2.02
15	128	1.07	90	180	2.10
20	136	1.19	100	183	2.17
25	142	1.30	110	186	2.24

Note: H = Average height in metre of the exposed surface above the mean retarding surface (ground or bed level or water level).

V = Horizontal velocity of wind in kilometres per hour at height H.

P = Horizontal wind pressure in kN/m² at height H.

The area on which the wind force is assumed to act is determined as below:

(i) For a deck structure:

The area of the structure as seen in elevation including the floor system and railing, less area of perforations in the railings or parapets.

(ii) For a through or half-through structure:

The area of the elevation of the windward truss, plus half the area of elevation above the deck level of all other trusses or girders.

The lateral wind force against any exposed moving live load is taken to act at 1.5 m above the roadway with the following values:

Highway bridges, ordinary	3 kN/linear m
Highway bridges, carrying tramway	4.5 kN/linear m.

For calculating wind force on live load, the clear distance between the trailers of a train should not be omitted.

The bridges will not be considered to be carrying any live load when the wind velocity at the deck level exceeds 130 km/hour.

The total assumed wind force shall be not less than 4.5 kN per metre in the plane of the loaded chord and 2.25 kN per linear metre in the plane of the unloaded chord on through or half-through truss, latticed or similar spans and not less than 4.5 kN per linear metre on deck spans.

A wind pressure 2.4 kN/m² on the unloaded structure shall be used if it produces greater stresses than those produced by the combined wind forces as stated above.

(v) Horizontal forces due to water currents

Any part of a road bridge which may be submerged in running water should be designed to sustain safely the horizontal pressure due to the force of the water current.

On piers parallel to the direction of the water current, the intensity of pressure shall be calculated from the following equation:

$$p = 0.52 \, KV^2 \qquad\qquad\qquad ... (9.20)$$

where p = intensity of pressure due to the water current, in kN/m²,

V = the velocity of the current at the point where the pressure intensity is being calculated, in metres per second, and

K = a constant having different values for different shapes, Table 9.9.

In case of the water current striking the piers at an angle, for calculating the pressure due to the components of velocity perpendicular to the pier, constant K should be taken as 1.5 in all cases except in the case of circular piers where the constant should be taken as 0.66.

For calculating the pressure on the pier, angle θ that the current makes with the axis of the pier should be taken into account. The pier should then be designed for variation of current angle between $20° \pm θ°$. Thus the pressure along the axis of the pier and transverse to it will respectively be given by

$$p_1 = 0.52KV^2 \cos^2(20° \pm θ°) \qquad\qquad ... (9.21a)$$

and
$$p_2 = 0.52KV^2 \sin^2(20° \pm θ°). \qquad\qquad ... (9.21b)$$

The value of V^2 in Eqs. (9.20) and (9.21) shall be assumed to vary linearly from zero at the scour depth to the square of maximum velocity at the free surface of water. The maximum velocity shall be assumed to be $\sqrt{2}$ times the maximum mean velocity of the current.

(vi) Longitudinal forces

Longitudinal forces are caused in road bridges due to any one or more of the following:

1.   Tractive effort caused through acceleration of the driving wheels;
2.   Braking effect due to application of brakes to the wheels;
3.   Frictional resistance offered to the movement of free bearings due to change of temperature or any other cause. Braking effect is invariably greater than the tractive effort. It is computed as follows:

(a) Single lane or two-lane bridge: 20% of the first train load plus 10% of the loads in succeeding trains or parts thereof on any single lane only. If the entire first train is not on the full span, the braking force is taken as 20% of the loads actually on the span. No impact allowance is included for this computation.

(b) Multi-lane bridge: As in (a) above for the first two lanes plus 5% of the loads on the lanes in excess of two.

The force due to braking effect shall be assumed to act along a line parallel to the roadway and 1.2 m above it. While transferring the force to the bearings, the change in the vertical reaction at the bearings should be taken into account.

Table 9.9. Value of Constant K for Pressure Intensity due to Water Currents.

Shape	K values
Square ended piers	1.5
Circular piers or piers with semicircular ends	0.66
Piers with triangular cut and ease waters, the angle included between the faces being 30° or less	0.50
Piers with triangular cut and ease waters, the angle included between the faces being more than 30° but less than 60°	0.50 to 0.70
Piers with triangular cut and ease waters, the angle included between the faces more than 60° but less than 90°	0.70 to 0.90
Piers with cut and ease waters of equilateral arc of circles	0.45
Piers with arcs of the cut and ease waters intersection at 90°	0.45

For a simply supported span with fixed and free bearings (other than elastomeric type) on stiff supports, horizontal forces at the bearing level in the longitudinal direction is greater of the two values given below:

	Fixed bearing	Free bearing
(i)	$F_h - \mu (R_g + R_q)$	$\mu (R_g + R_q)$
or (ii)	$\dfrac{F_h}{2} + \mu (R_g + R_q)$	$\mu (R_g + R_q)$

where F_h = applied horizontal force,

R_g = reaction at the free end due to dead load,

R_q = reaction at free end due to live load, and

μ = coefficient of friction at the movable bearing

which shall be assumed to have the following values:

(i)	For steel roller bearings	0.03
(ii)	For concrete roller bearings	0.05
(iii)	For sliding bearings	
(a)	Steel on cast iron or steel on steel	0.50
(b)	Grey cast iron on grey cast iron (Mechanite)	0.40
(c)	Concrete over concrete with bitumen layer in between	0.60
(d)	Teflon on stainless steel	0.05

In case of simply supported small spans up to 10 metres resting on unyielding supports and where no bearings are provided, horizontal force in the longitudinal direction at the bearing level shall be

$$\frac{F_h}{2} \text{ or } \mu R_g \text{ whichever is greater.}$$

For a simply supported span siting on identical elastomeric bearings at each end resting on unyielding supports. Forces at each end

$$= \frac{F_h}{2} + V_r\, l_{tc}$$

where V_r = shear rating of the elastomer bearings,

l_{tc} = movement of deck above bearing, other than that due to applied forces.

The substructure and foundation shall also be designed for 10 per cent variation in movement of the span on either side.

For continuous bridge with one fixed bearing and other free bearings:

Fixed bearing	*Free bearing*

Case I

$(\mu R - \mu L)$ +ve and F_h acting in +ve direction

(a) If $F_h > 2\ \mu R$ μRx

$F_h - (\mu R + \mu L)$

(b) If $F_h < 2\ \mu R$

$$\frac{F_h}{1 + \sum n_R} + (\mu R - \mu L)$$

Case II

$(\mu R - \mu L)$ +ve and F_h acting in –ve direction

(a) If $F_h > 2\ \mu L$ μRx

$F_h - (\mu R + \mu L)$

(b) If $F_h < 2\ \mu L$

$$\frac{F_h}{1 + \sum n_L} + (\mu R - \mu L)$$

whichever is greater.

where

n_L or n_R = number of free bearings to the left or right of fixed bearings, respectively,

μL or μR = the total horizontal force developed at the free bearings to the left or right of the fixed bearing respectively, and

μRx = the net horizontal force developed at any one of the free bearings considered to the left or right of the fixed bearings.

In seismic areas, the fixed bearing shall also be checked for full seismic force and braking/tractive force.

(vii) Buoyancy

The effect of buoyancy is considered in the design of a bridge, only if the strata of soil are permeable or in other words, if bridge foundations are resting on homogeneous and impermeable strata of soil, no provision is made for buoyancy in the design of bridge. Following points are taken into consideration for the computation of the buoyancy force.

(a) The effects of buoyancy are to be considered in the design of an abutment in case of a submersible bridge. In such a case, it is assumed that the filling behind the abutment is washed away or removed by scouring action.

(b) For the design of submerged masonry or concrete structure, the buoyancy effect through pores is limited to the extent of 15 per cent of full buoyancy effect.

(c) If the member under consideration displaces water only, the reduction in weight due to buoyancy for that member is taken equal to the volume of the displaced water.

(viii) Centrifugal force

For a road bridge located on a curve, effects of centrifugal forces due to movement of vehicles should be taken into account.

The centrifugal force is given by Equation (9.22)

$$C = \frac{WV^2}{127R} \qquad\qquad ... (9.22)$$

where C = centrifugal force in kN acting normal to the traffic (a) at the point of action of the wheel loads or (b) uniformly distributed over every metre length on which the uniformly distributed loads act,

W = live load (a) in kN for wheel loads and (b) in kN/m for uniformly distributed live load,

V = design speed in km/hour, and

R = radius of curvature in metres.

The centrifugal force is assumed to act at a height of 1.2 m above the level of the carriageway. The force is not increased for impact effect.

(ix) Seismic force

For a bridge located in seismic region, allowance should be made in the design for the seismic force. The following seismic forces on the substructure above the scour depth are to be considered:

(a) Horizontal and vertical forces due to dead, live and seismic loads transferred from superstructure to the substructure through the bearings.

(b) Horizontal and vertical seismic forces due to self weight applied at the centre of mass ignoring reduction due to buoyancy or uplift.

(c) Hydrodynamic forces acting on piers and modified earth pressure on abutments due to earthquake.

Piers should be checked for the above seismic forces assumed to act parallel to the current and traffic directions taken separately.

For submerged portions of the pier, the hydrodynamic force F_{Hy}, assumed to act in a horizontal direction corresponding to that of earthquake motion, is given by Equation (9.23).

$$F_{Hy} = C.\,\alpha_h W \qquad\qquad ... (9.23)$$

where C = a coefficient from Table 9.10 depending on the ratio of height of submerged portion of pier (H) to the radius (a) of the enveloping cylinder as in Fig. 9.11,

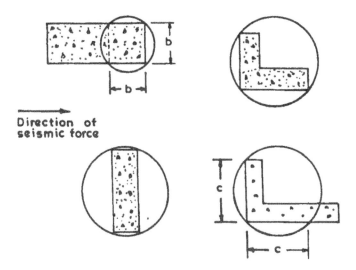

Fig. 9.11. Enveloping Cylinders for Some Typical Cases of Piers.

α_h = horizontal seismic coefficient, and

W = the weight of water of the enveloping cylinder (Fig. 9.11).

Table 9.10. Values of C in Equation (9.22).

H/a	C
1.0	0.390
2.0	0.575
3.0	0.675
4.0	0.730

For submerged superstructure of submersible bridges, the hydrodynamic pressure is determined from Equation (9.24)

$$p = 8.75\alpha_h \sqrt{Hy} \qquad \qquad ... (9.24)$$

where
p = hydrodynamic pressure in kN/m^2,

α_h = horizontal seismic coefficient,

H = height of water surface from the level of deepest scour in m, and

y = depth of the section below the water surface in m.

The details of procedure of obtaining seismic coefficient α_h and α_v have already been described in Sec. 6.5 of Chapter 6.

(x) Earth pressure

Structures designed to retain earth fills are propotioned to withstand lateral pressure calculated in accordance with any rational theory.

Coulomb's theory is acceptable subject to the modification that the centre of pressure

exerted by the backfill, when considered dry, is located at an elevation of 0.42 of the height of
the wall above the base, instead of 0.33 of that height. The minimum pressure for a retaining
structure is that from a fluid weighing 4.8 kN/m³.

In case of bridge abutments, the concentrated surface loads due to any of the I.R.C.
standard vehicles or trains placed on the backfill may be considered to have the same effect as
the equivalent heights of surcharge of earth shown in Table 9.11 which are based on Spangler's
equation (Leonard, 1962). These heights of surcharge are assumed to act over the entire length
of the abutment.

The figures given in Table 9.11 are based on the following values for the constants for the
abutments and the backfill:
1. Length of abutments (L) = 4.5 m for single-lane bridges and 7.6 m for multilane bridges.
2. Angle of internal friction of the backfill (ϕ) = 30°.
3. Density of backfill (γ) = 16 kN/m³.
4. The resultant earth pressure acts in a horizontal direction.

For different values, say L_1, ϕ_1 and γ_1 for the constants, the figures given in Table 9.11
should be multiplied by the following factors:

$$\frac{L(4.5\,\text{or}\,7.6\,\text{m})}{L_1} \;,\; \frac{(1+\sin\phi_1)}{3(1-\sin\phi_1)} \;\text{and}\; \frac{16}{\gamma_1} \;\text{respectively.}$$

Table 9.11. Equivalent Heights of Surcharge of Earth in Metres for IRC Standard Loadings.

Depth of Abutments Below Road Level	Equivalent Heights of Surcharge of Earth in Metres					
	Class AA and 70 R		Class A		Class B	
	Single-lane Bridges	Multi-lane Bridges	Single-lane Bridges	Multi-lane Bridges	Single-lane Bridges	Multi-lane Bridges
0.2	26.0	15.4	14.3	17.2	8.3	10.0
1.0	15.0	9.1	8.5	10.0	5.1	5.8
2.0	8.0	5.5	5.1	6.1	3.0	3.7
3.0	6.8	4.1	3.8	4.6	2.3	2.7
4.0	5.5	3.3	3.0	3.5	1.8	2.1
6.0	3.8	2.3	2.2	2.6	1.3	1.5
8.0	3.0	1.8	1.7	2.0	1.0	1.2
10.0 and above	2.6	1.5	1.4	1.7	0.9	1.0

No live load surcharge need to be considered for the design of an abutment, if an adequately
designed reinforced concrete approach slab covering the entire width of roadway with one end
resting on the abutment and extending for a length of not less than 3.5 m into the approach is
provided. It is usually desirable to provide such approach slabs for most major bridges.

Thorough drainage of backfilling material should be ensured by means of weep holes and
gravel drains, pipe drains, etc. If drainage arrangement is not available, the additional effects
due to submerged soil should be carefully taken into account.

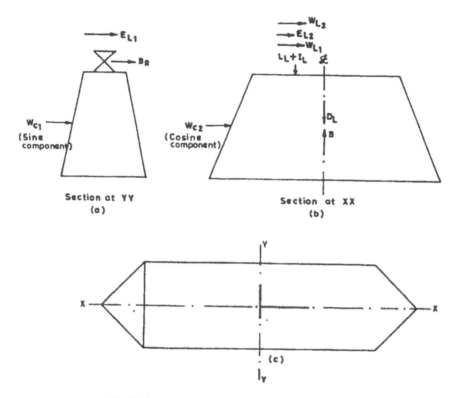

Fig. 9.12. Forces Acting on a Pier and its Foundation.

In the design of return walls, live load surcharge should be considered for loads placed beyond the length of the approach slab.

Temperature movements are partially restrained in girder bridges because of friction at the movable end. A horizontal force equal to the frictional force (i.e. equal to μR, where μ is the coefficient of friction and R the reaction) is applied at the bearing level at the top of pier or abutment and considered in the design of the portions below.

The loads and forces discussed above are shown in Fig. 9.12. All the horizontal forces, which act in the direction of the traffic above the bearing level, will be first transferred at the bearing level. These forces have an overturning effect and produce vertical reactions also at the bearing in addition to the horizontal forces. The horizontal forces acting at the bearing level will give moments about x-x axis of the section of pier or abutment under consideration. The horizontal forces acting across the traffic give moments about y-y axis (Fig. 9.12). As indicated above, temperature effects are accounted for by considering a horizontal force (μR) at the movable end acting at the bearing level.

As per IRC: 21–1972, a section of a pier or an abutment and their foundation should be checked for three combinations of forces:

(i) N case

$$D_L + L_L + I_L + W_C + B_R + C_F + B + E_P + W_L$$

(ii) $N + T$ case

Forces mentioned in (i) + temperature effects

(iii) $N + T + S$ case

Forces mentioned in (ii) + E_L

where D_L = dead load,

L_L = live load,

I_L = impact load,

W_C = water current forces,

B_R = braking effects,

C_F = centrifugal forces,

B = buoyancy,

E_P = earth pressure,

W_L = wind load, and

E_L = earthquake load.

Permissible stresses in masonry, concrete and steel are increased in the above cases as given below:

N case	No increase
$N + T$ case	Increase by 15%
$N + T + S$ case	Increase by 50%.

9.13 LATERAL STABILITY OF WELL FOUNDATION

General

The stability of well under the action of lateral loads depends on the resistance of the soil on its sides and base. For a given vertical load, the deformation of soil increases with the increase in the lateral load and, therefore, the resistances offered by the sides and base of the well also change. The behaviour of the well at ultimate failure is different than at the elastic stage.

Indian Road Congress on the basis of observed behaviour of models of well foundations and related research work formulated the recommendations (IRC: 45–1972) for estimating the resistance of soil below the maximum scour level for design of well foundation in cohesionless soils. The standard suggests the elastic theory to estimate the pressures at the side and the base under design loads, but to determine the actual factor of safety against shear failure, it will be necessary to calculate the ultimate soil resistance. Therefore, the design of well foundation should be checked by both (i) elastic theory, and (ii) ultimate resistance approach.

Elastic theory method

Assumptions

The following assumptions are made in deriving the equations based on elastic theory:

 (i) The soil surrounding the well and below the base is perfectly elastic, homogeneous and follows Hooke's Law.
 (ii) Under design working loads, the lateral deflections are so small that the unit soil reaction "p" increases linearly with increasing lateral deflection "z" as expressed by $p = K_H z$, where K_H is the coefficient of horizontal subgrade reaction at the base.

(iii) The coefficient of horizontal subgrade reaction increases linearly with depth in the case of cohesionless soils.

(iv) The well is assumed to be a rigid body subjected to an external unidirectional horizontal force H and a moment M_0 at scour level.

Method of calculation

Step 1. Having determined the minimum grip length as explained in Sec. 9.7, calculate applied loads and moments as:

W = Total downward load consisting of dead load, live load etc., as acting on the base of well.

H = Total lateral load applied above the scour level.

M = Total external moment applied at the base of well due to eccentricity of live load, lateral loads, tilts and shift, etc.

Step 2. Using the dimensions of the well, calculate the following geometrical properties:

I_B = Moment of inertia of base section in the plane of bending, that is, about an axis perpendicular to the direction of horizontal force causing moment.

$$\left(I_B = \frac{\pi}{64} B^4 \text{ for a circular well} \right)$$

I_V = Moment of inertia of the vertical projected rectangle of the well below scour level

$$= L D_f^3 / 12 \qquad \qquad \text{... (9.25a)}$$

$$I = I_B + mI_v(1 + 2\mu' \alpha) \qquad \qquad \text{... (9.25b)}$$

where B = diameter of circular well, or width of base parallel to the direction of the horizontal force,

L = projected width of well in contact with soil offering passive resistance = $0.9 B$ for a circular well, the factor 0.9 being the shape factor. This factor is 1.0 for side of a rectangular or square well,

D_f = depth of well below scour level, that is, grip, and

m = ratio of horizontal subgrade modulus to vertical subgrade modulus at base level of well = K_H/K_v.

The value of m may be taken as 1.0 in most cases, unless determined by tests.

μ' = Coefficient of friction between the well sides and the soil = $\tan \delta$, where δ is the angle of wall friction taken as 2/3 ϕ but not more than 22.5°, where ϕ is the angle of internal friction of the soil on the sides.

$\alpha = [B/(\pi D_f)]$ for circular well and $[B/(2D_f)]$ for rectangular well.

Step 3. Check that the point of rotation of the well lies at the base by ensuring that the frictional force at the base is sufficient to restrain the movement of the well forward or backward. That is,

$$H > \frac{M}{r}(1 + \mu\mu') - \mu W \qquad \ldots (9.26a)$$

$$H < \frac{M}{r}(1 - \mu\mu') + \mu W \qquad \ldots (9.26b)$$

where $\qquad\qquad r = \frac{I}{mI_v} \cdot \frac{D_f}{2} \qquad\qquad \ldots (9.27)$

μ = coefficient of friction between base of well and soil. Due to roughness of plug concrete this is taken equal to tan ϕ.

Step 4. Check that the soil on the sides remain elastic by ensuring that the pressure parabola at the top remains below the passive pressure line. That is,

$$\frac{mM}{I} \not> \gamma (K_P - K_A) \qquad \ldots (9.28)$$

where γ = unit weight of soil (dry or submerged as the case may be) and K_P, K_A = passive and active earth pressure coefficients respectively. Coefficients K_P and K_A are to be calculated by Coulomb's theory using δ angle value as stated in the definition of μ' earlier. Although the shape of scour pit has an angle surcharge, it is usual and conservative to assume horizontal surface. For this condition K_A and K_P can be calculated from the following:

$$K_A = \left[\frac{\cos\phi}{\sqrt{\cos\delta} + \sqrt{\sin(\phi + \delta)\sin\phi}} \right]^2 \qquad \ldots (9.29a)$$

$$K_P = \left[\frac{\cos\phi}{\sqrt{\cos\delta} - \sqrt{\sin(\phi + \delta)\sin\phi}} \right]^2 \qquad \ldots (9.29b)$$

If the checks in steps 3 and 4 are not satisfied, increase the grip length and revise.

Step 5. Check the pressures at the base of well, the maximum compressive stress should be less than the allowable bearing pressure and minimum stress should not be tensile. That is,

$$\sigma_1 = \frac{W - \mu' P}{A} + \frac{MB}{2I} < q_a \qquad \ldots (9.30)$$

$$\sigma_2 = \frac{W - \mu' P}{A} - \frac{MB}{2I} > 0, \text{ i.e. no tension} \qquad \ldots (9.31)$$

where σ_1 and σ_2 = maximum and minimum base pressures, respectively,

$\qquad P$ = total horizontal reaction from the side = M/r, $\qquad\qquad \ldots (9.32)$

$\qquad A$ = area of base section of well,

$\qquad B$ = diameter or width of well in the plane of bending, and

$\qquad q_a$ = allowable bearing pressure.

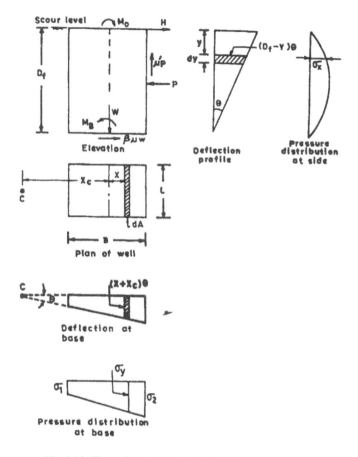

Fig. 9.13. Illustrating the Analysis of Well by Elastic Theory.

Derivation of equations used in elastic theory

(a) Total deflection at depth 'y' from scour level (Fig. 9.13)

$$= (D_f - y)\cdot\theta$$

Horizontal soil reaction

$$\sigma_x' = K_H \frac{y}{D_f}(D_f - y)\cdot\theta$$

or,
$$\sigma_x = m\frac{K_v\theta y}{D_f}(D_f - y)$$

where
$$K_H = mK_v$$

Total horizontal soil reaction

$$P = \int_0^{D_f} \sigma_x L d_y$$

$$= mK_v\theta \frac{L}{D_f} \int_0^{D_f} (yD_f - y^2)dy$$

or, $$P = \frac{2mK_v\theta I_v}{D_f} \qquad \qquad \text{... (9.33)}$$

where $$I_v = \frac{LD_f^3}{12} \qquad \qquad \text{... (9.25a)}$$

(b) Let M_P be the moment of P about base level

$$M_P = \int_0^{D_f} \sigma_x(D_f - y)Ldy$$

$$= m\frac{K_v\theta L}{D_f} \int_0^{D_f} y(D_f - y)^2 dy$$

$$M_P = mK_v\theta I_v \qquad \qquad \text{... (9.34)}$$

(c) Consider the soil reaction acting at the base. Vertical deflection at distance $(x + x_c)$

$$= (x + x_c)\theta$$

$$\sigma_y = K_v(x + 0 \ x_c)\theta$$

$$M_B = \int \sigma_y dAx = K_v\theta \int_{-B/2}^{B/2} (x + x_c)xdA$$

$$= K_v\theta \int_{-B/2}^{B/2} x^2 dA + K_v\theta x_c \int_{-B/2}^{B/2} xdA$$

Since origin is at the c.g. of base.

$$\int xdA = 0 \quad \text{and} \quad I_B = \int x^2 \ dA$$

$$M_B = K_v\theta I_B \qquad \qquad \text{... (9.35)}$$

(d) For equilibrium, $\Sigma H = 0$

or $$H + \beta\mu(W - \mu'P) = P$$

or $$P = \frac{H + \beta\mu W}{1 + \beta\mu\mu'} \qquad \qquad \text{... (9.36)}$$

(e) Taking moment about base

$$M = M_B + M_P + \mu'P\alpha D_f \qquad \qquad \text{... (9.37)}$$

where αD_f = distance from the axis passing the c.g. of base at which the resultant vertical frictional force on side acts normal to the direction of horizontal force,

   = $B/2$ for rectangular wells, and

   = 0.318 times diameter for circular wells.

Substituting equations (9.33), (9.34) and (9.35) we get

$$M = K_v \theta I_B + m K_v \theta I_v + \mu' \alpha 2 m K_v \theta I_v$$

or

$$\frac{M}{\left[I_B + m I_y (1 + 2\mu' \alpha) \right]} = K_v \theta$$

or

$$\frac{M}{I} = K_v \theta \qquad \qquad \dots (9.38)$$

where $I = I_B + m I_v (1 + 2\mu' \alpha)$ \qquad \dots (9.25b)

Equations (9.33), (9.34) and (9.35) can be rewritten as:

$$P = \frac{H + \beta \mu W}{1 + \beta \mu \mu'}$$

$$= 2 m K_v \theta \frac{I_v}{D_f}$$

$$= 2m \frac{M}{I} \frac{I_v}{D_f}$$

$$= \frac{M}{r} \qquad \qquad \dots (9.39)$$

where $r = \dfrac{D_f}{2} \dfrac{I}{m I_v}$ \qquad \dots (9.27)

or $H + \beta \mu w = \dfrac{M}{r}(1 + \beta \mu \mu')$

$$\frac{M}{r} - H = \beta \mu \left(W - \frac{M}{r} \mu' \right)$$

The value of β lies between the range,

   $\beta < 1$ or > -1. Thus.

(i) $\dfrac{M}{r} - H < \mu W - \mu \mu' \dfrac{M}{r}$

or $H > \dfrac{M}{r}(1 + \mu\mu') - \mu W$ ... (9.26a)

(ii) $\dfrac{M}{r} - H > \mu W + \mu\mu'\dfrac{M}{r}$

or $H < \dfrac{M}{r}(1 - \mu\mu') + \mu W$ ... (9.26b)

(f) The vertical reaction is given by

$$\sigma_y = K_v\theta(x + x_c)$$

or $W - \mu' P = \int \sigma_y dA = K_v\theta \int (x + x_c)dA$

or $\qquad\quad = K_v\theta x_c A \ (\because \int x dA = 0)$

or $x_c K_v\theta = (W - \mu'P)/A$

$\qquad\quad \sigma_y = K_v\theta x + K_v\theta x_c$

or, $\qquad\quad \sigma_y = K_v\theta x + \dfrac{W - \mu' P}{A}$

or, $\qquad\quad \sigma_1 = \dfrac{W - \mu' P}{A} + K_v\theta B/2$

or, $\qquad\quad \sigma_2 = \dfrac{W - \mu' P}{A} - K_v\theta B/2.$

Using equation 9.38 we get

$$\sigma_1 = \dfrac{W - \mu' P}{A} + \dfrac{MB}{2I}$$ ... (9.30)

$$\sigma_2 = \dfrac{W - \mu' P}{A} - \dfrac{MB}{2I}$$ ... (9.31)

(g) Condition of stability–The maximum soil reaction from the sides cannot exceed the maximum passive pressure at any depth provided the soil remains in an elastic state. Thus at any depth y:

$$\sigma_x \not> \gamma(K_P - K_A)$$

or, $$m\dfrac{K_v\theta}{D}y(D - y) \not> \gamma(K_P - K_A)$$

or, $$m\dfrac{K_v\theta}{D}(D - y) \not> \gamma(K_P - K_A).$$

Left hand side is maximum at $y = 0$

or $$mK_v\theta \not> \gamma(K_P - K_A)$$

or
$$\frac{mM}{I} \ngtr \gamma(K_P - K_A) \qquad \qquad \dots (9.28)$$

Ultimate resistance approach

For checking the ultimate load capacity of the well foundation, the applied loads are magnified by suitable load factors for various load combinations and the ultimate resistance is reduced by appropriate under-strength factors and then the two are compared. The steps to be taken are described below:

Step 1. Compute the applied loads W_u, H_u and M_u for various ultimate load combinations as follows:

(i) $1.1 D_L$
(ii) $1.1 D_L^{'} + 1.6 L_L$
(iii) $1.1 D_L + B + 1.4(L_L + W_C + E_P)$
(iv) $1.1 D_L + B + 1.4(W_C + E_P + W_L \text{ or } E_L)$
(v) $1.1 D_L + B + 1.25(L_L + W_C + E_P + W_L \text{ or } E_L)$

where D_L = dead load,

L_L = live load,

B = buoyancy,

W_C = water current force,

E_P = earth pressure force,

W_L = wind load, and

E_L = earthquake load.

The forces and loads will be considered as contributions to W_u, H_u and M_u as the case may be. In the ultimate resistance case, the point of rotation is assumed to occur at $0.2 D_f$ above base of well. The moment M_u is to be computed about this point instead of base. Moments due to shifts and tilts are to be included in the moment M_u.

Step 2. Check for maximum average pressure at base.

$$\frac{W_u}{A} \ngtr \frac{q_u}{2} \qquad \qquad \dots (9.40)$$

where q_u = the ultimate bearing capacity of the soil below base of well, and

W_u = ultimate vertical load acting at base.

If this condition is not satisfied increase the well diameter.

Step 3. Compute the ultimate moment of resistance of the base section

$$M_b = QW_u B \tan \phi \qquad \qquad \dots (9.41)$$

where Q = a constant depending on the shape of well and ratio D_f/B as given in Table 9.12.

Table 9.12. Values of Constant Q

D_f/B	0.5	1.0	1.5	2.0	2.5
Q for circular wells	0.25	0.27	0.30	0.34	0.38
Q for rectangular wells	0.41	0.45	0.50	0.56	0.64

Step 4. Compute the ultimate moment of resistance on the well sides. This has two parts, one due to passive ressistance and the other due to frictional resistance.

$$M_s = 0.10 \ \gamma(K_P - K_A) D_f^3 L \qquad\qquad \text{... (9.42)}$$

$$M_f = 0.18\gamma(K_P - K_A) \ D_f^2 \ LB \sin \delta \text{ for rectangular wells} \qquad \text{... (9.43)}$$

$$M_f = 0.11\gamma(K_P - K_A) D_f^2 \ B^2 \sin \delta \text{ for circular wells} \qquad\qquad \text{... (9.44)}$$

Step 5. Check the applied ultimate moment with the total ultimate moment of resistance.

$$M_u > 0.7(M_b + M_s + M_f) \qquad\qquad \text{... (9.45)}$$

where 0.7 is the strength reduction factor.

Derivation of equations used in ultimate resistance approach

(a) Base resisting moment (M_b)

The base resisting moment is the moment of the frictional force mobilized along the surface of rupture which is assumed to be cylindrical passing through the corners of the base for a square well. For circular wells, the surface of rupture corresponds to that of a part of sphere with its centre at the point of rotation and passing through the periphery of the base as shown in Fig. 9.14a.

If W_u is the total load modified by appropriate load factor, the load per unit width will be W_u/B, which will also be equal to the upward pressure as shown in Fig. 9.14b.

(i) Rectangular base–Consider the small arc of length $Rd\alpha$ at an angle of α from the vertical axis.

Its horizontal component $\quad = Rd\alpha \cos \alpha$

Vertical force on the element $\ = Rd\alpha \cos \alpha \ W/B$.

Due to this vertical force the normal force developed at the element is δF_n

where $\quad \delta F_n = \dfrac{W_u}{B} Rd\alpha \cos \alpha \cos \alpha$

$$= \dfrac{W_u R}{B} \cos^2 \alpha \ d\alpha.$$

Total normal force F_n is given by,

$$F_n = 2\int_0^\theta \dfrac{W_u R}{B} \cos^2 \alpha \ d\alpha$$

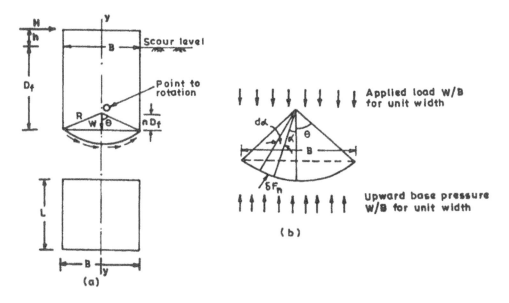

Fig. 9.14. Illustrating Computation of Base Resisting Moment.

$$= \frac{RW_u}{B}(\theta + \sin\theta\cos\theta)$$

where $\quad \sin\theta = \dfrac{B}{2R}, \quad \cos\theta = \dfrac{nD_f}{R}, \quad \tan\theta = \dfrac{B}{2nD_f}$

$$R = \sqrt{\frac{B^2 + 4n^2 D_f^2}{4}}$$

$$F_n = \frac{W_u}{2}\sqrt{1 + \frac{4n^2 D_f^2}{B^2}}\left[\tan^{-1}\frac{B}{2nD_f} + \frac{2nBD_f}{B^2 + 4n^2 D_f^2}\right]$$

Moment of resistance of the base about the point of rotation is

$$M_b = F_n \tan\phi\, R \qquad\qquad\qquad \dots (9.46)$$

Substituting the value of R in Eq. (9.47), we get on simplification

$$M_b = W_u B\mu\cdot\frac{1}{4}\left(1 + \frac{4n^2 D_f^2}{B^2}\right)\left[\tan^{-1}\frac{B}{2nD_f} + \frac{2n}{\dfrac{B}{D_f} + \dfrac{4n^2 D_f}{B}}\right]$$

or $\qquad M_b = W_u B\mu Q \qquad\qquad\qquad\qquad\qquad \dots (9.47)$

$$\text{where} \quad Q = \frac{1}{4}\left(1 + \frac{4n^2 D_f^2}{B^2}\right)\left[\tan^{-1}\frac{B}{2nD_f} + \frac{2n}{\dfrac{B}{D_f} + \dfrac{4n^2 D_f}{B}}\right] \quad \text{... (9.48)}$$

It may be noted that Q depends on n and D_f/B ratio.

(ii) Circular base—A multiplication factor of 0.6 is to be applied for the above expression of M_b in order to account for the surface of rupture being part of a sphere.

For both cases substituting the value 'n' equal to 0.2 for the point of rotation in Eq. 9.48, the values of Q for different values of D_f/B are obtained as given in Table 9.12.

(b) Side resisting moment (M_s)

Figure 9.15 shows the ultimate soil pressure distribution at the front and back faces of well. The point of rotation is assumed at 0.2 D_f above the base.

Let, $\gamma D_f(K_P - K_A) = X = BC;\ BF = Y.$

From triangles GEF and OEH

$$\frac{D_1}{X + Y} = \frac{D_1 - 0.2D_f}{Y}$$

or

$$\frac{X}{Y} = \frac{D_f}{5D_1 - D_f} \quad \text{... (9.49a)}$$

From triangles ABC, and AIE

$$\frac{D_f}{X} = \frac{D_f - D_1}{Y} \quad \text{... (9.49b)}$$

From Eqs. (9.49a) and (9.49b), we get

$$D_1 = D_f/3 \quad \text{... (9.49c)}$$

Moment of side resistance about 'O' is the algebraic sum of moments of triangles ABC and GEC.

$$M_s = \frac{1}{2} D_f X \frac{2}{15} D_f + \frac{1}{2}\frac{D_f}{3} 2X \frac{4D_f}{45}$$

$$= 0.096\, D_f^2\, X$$

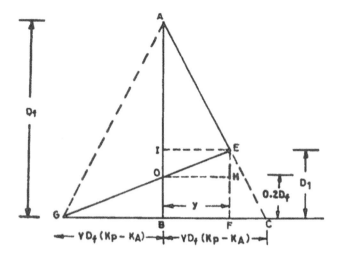

Fig. 9.15. Illustrating the Computation of Side Soil Resisting Moment.

Say $\approx 0.1\, D_f^2\, X.$

Substituting for X, $M_s = 0.1\, \gamma\, D_f^3\, (K_P - K_A)$ per unit length of well;

or   for a length of L, $M_s = 0.1\gamma\, D_f^3\, (K_P - K_A)L$ ... (9.50)

(c) Friction resisting moment (M_f)

Due to the passive pressure of soil as shown in Fig. 9.15, the frictional forces on the front and back faces of well will be acting in the vertical direction and will also produce resisting moment 'M_f'. In this analysis, the effect of the active earth pressure perpendicular to the direction of applied forces is neglected.

The vertical pressure due to friction at any level is sin δ times the pressure at that level, where δ is the angle of wall friction.

The friction force/unit width = (ΔAOE + ΔBOG) sin δ

Pressure at E $= \dfrac{2}{3}\gamma D_f (K_P - K_A)$

Area of triangle AOE $= \dfrac{2}{3}\gamma D_f (K_P - K_A)\dfrac{0.8D_f}{2}$

                                 $= \dfrac{0.8}{3}\gamma D_f^2 (K_P - K_A)$

Area of triangle BOG $= \dfrac{0.2D_f}{2}\gamma Df\, (K_P - K_A)$

                                 $= 0.1\gamma D_f^2 (K_P - K_A)$

Total friction force/unit width $= \left[\dfrac{0.8}{3} + 0.1\right]\gamma D_f^2 \, (K_P - K_A)\sin \delta$

$= \dfrac{1.1}{3}\gamma D_f^2 \, (K_P - K_A) \sin \delta.$

Moment about centre of rotation

(i) In case of rectangular wells of width B and length L

$$M_f = \dfrac{1.1D}{3}\gamma D_f^2 \, (K_P - K_A)\dfrac{B}{2} \sin \delta L$$

$$= 0.183\gamma(K_P - K_A)LB \, D_f^2 \, \sin \delta$$

Say $\qquad\qquad\qquad\qquad = 0.180\gamma(K_P - K_A)LB \, D_f^2 \, \sin \delta \qquad\qquad \dots (9.43)$

(ii) In case of circular wells

Lever arm $\qquad\qquad\qquad\qquad = \dfrac{B}{\pi}$

Therefore $\qquad\qquad M_f = \dfrac{1.1}{3}\gamma D_f^2 \, (K_P - K_A)\dfrac{B}{\pi}L \sin \delta$

Since $\qquad\qquad\qquad L = 0.9B$ in case of circular well

$$M_f = \dfrac{0.33}{\pi}\, \gamma(K_P - K_A)B^2 \, D_f^2 \sin \delta$$

$$= 0.105\gamma(K_P - K_A)B^2 \, D_f^2 \sin \delta$$

Say $\qquad\qquad\qquad\qquad = 0.11\gamma(K_P - K_A)B^2 \, D_f^2 \, \sin \delta \qquad\qquad \dots (9.44)$

9.14 DESIGN OF PIER CAP

The pier cap is the block resting over the top of the pier. It provides the immediate bearing surface for the support of the superstructure at the pier, and disperses the strip loads from the bearings to the substructure more evenly. Number of bearings on a pier cap depends on the width of the carriageway. Usually, for a two-lane bridge, load of the superstructure span is transferred to a pier cap through two bearings.

Pier cap is designed as an R.C.C. raft slab and should be constructed in M-20 grade. At the top of pier cap, any combination of the loads (i.e. as N-case or $N + T$ case or $N + T + S$ case) ultimately will give a vertical load V and moments M_{xx} and M_{yy}. The magnitude of the moment M_{xx} will be comparatively small and can be neglected in the design. Figure 9.16 shows the plan, the section and the loading diagram for a typical pier cap having two bearings. In this figure

$$R_1 = \dfrac{V}{2} + \dfrac{M_{yy}}{L_b} \qquad\qquad\qquad \dots (9.51)$$

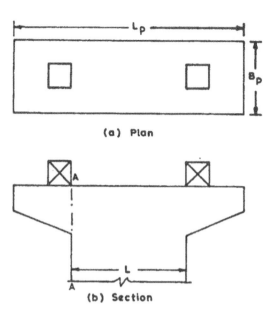

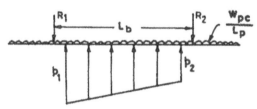

Fig. 9.16. Illustrating the Design of Pier Cap.

$$R_2 = \frac{V}{2} - \frac{M_{yy}}{L_b} \qquad \qquad \text{... (9.52)}$$

$$p_1 = \frac{V + W_{pc}}{L_p \times B_p}\left(1 + \frac{6e}{L}\right) \qquad \qquad \text{... (9.53)}$$

$$p_2 = \frac{V + W_{pc}}{L_p \times B_p}\left(1 - \frac{6e}{L}\right) \qquad \qquad \text{... (9.54)}$$

$$e = \frac{M_{yy}}{V + W_{pc}} \qquad \qquad \text{... (9.55)}$$

where W_{pc} = weight of pier cap,
L_b = distance between bearings, and
$L_p \times B_p$ = plan dimensions of pier section at its top.

The design of pier cap can now be done in the following steps:

(i) (a) Determine pressure on the concrete, q_c

$$q_c = \frac{R_1}{A_1}$$

... (9.56)

where A_1 = actual area of bearing.

(b) Obtain gross area of concrete, giving a spread of one horizontal to two vertical.

(c) Compute A/A_1. If the value of $\sqrt{A/A_1}$ is greater than 2.0, adopt 2.0.

(d) Allowable pressure on concrete

$$f_{ca} = f_{cc}\sqrt{\frac{A}{A_1}}$$

... (9.57)

where f_{cc} = permissible compressive stress of concrete in direct bearing.

(e) q_c should not exceed f_{ca}.

(ii) Based on the loading diagram given in Fig. 9.16c, determine the maximum bending moment. For example, section AA will be most critical for evaluating the bending moment in the example illustrated in Fig. 9.16. Determine the thickness of pier cap and amount of reinforcement with usual checking for nominal shear and bond length.

(iii) The pier cap should be of thickness of at least 225 mm up to a span of 25 m and 300 mm for longer spans. It should be reinforced atleast with two per cent mild steel or one and half per cent high yield strength deformed bars, distributed equally at top and bottom and provided in two directions at top and bottom. The reinforcement in the direction of the length of the pier should extend end to end of the pier, while the reinforcement at right angles to this should extend for the full width of pier cap. In addition to this, reinforcement provision should be made for two layers of mesh reinforcement each consisting of 6 mm bars at 75 mm c/c in both directions placed directly under the bearings.

9.15 DESIGN OF PIER

The shape of a pier depends on the type, size and dimensions of the superstructure. Piers can be solid, cellular, trestle or hammer-head types (Fig. 9.17). Solid piers are of masonry or mass concrete. Cellular, trestle and hammer-head types are constructed in reinforced concrete.

Solid and cellular piers are provided for river bridges. These are provided with cut waters to facilitate stream-lined flow and to reduce scour. The trestle type consists of columns with a bent cap at the top, and is suitable in flyovers and elevated roads. The hammer-head type provides slender substructure and is normally suitable for elevated roadways.

The top width of the pier depends on the size of bearing plates on which the superstructure rests. It is usually kept at a minimum of 300 mm more than out-to-out dimensions of the bearing plates, measured along the longitudinal axis of the superstructure. The length of the pier at the top should not be less than 1.2 m in excess of the out-to-out dimensions of the bearing plates

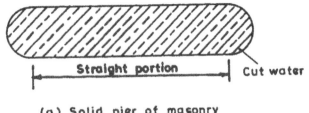

(a) Solid pier of masonry

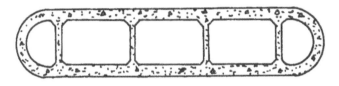

(b) Cellular R.C. Pier

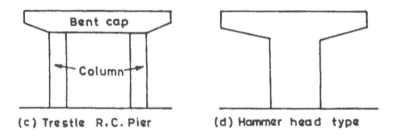

(c) Trestle R.C. Pier   (d) Hammer head type

Fig. 9.17. Types of Piers.

measured perpendicular to the axis of the superstructure. A pier is normally given a batter of 1 in 20 on all of its sides.

For design purposes, the section passing through the base of the pier is most critical. The design is carried out in the following steps:

(i) Considering the various forces acting on the superstructure and part of the substructure above the base of pier, obtain the net combined loading as V, M_{xx} and M_{yy}.

(ii) Determine area of cross-section of the pier base (A), its moment of inertias about x-x axis and y-y axis (I_{xx} and I_{yy}).

Then calculate
$$f_{1,2} = \frac{V}{A} \pm \frac{M_{xx}}{I_{xx}}.y \pm \frac{M_{yy}}{I_{yy}}.x \qquad ...(9.58)$$

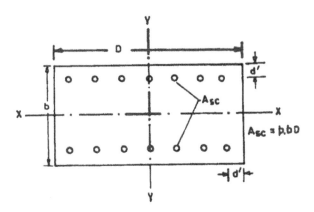

Fig. 9.18. Illustrating the Design of Pier.

If the whole base section of pier is in compression, provide nominal reinforcement. Otherwise, carry out the design as illustrated in the next step.

(iii) (a) Assume the trial section with reinforcement and effective cover, i.e. fix b, D, d' and p (Fig. 9.18).

(b) Compute vertical load capacity of the section using Eq. (9.59).

$$V_u = 0.45 f_{ck} A_c + 0.75 f_{sy} A_{sc} \qquad\qquad \text{... (9.59)}$$

where $A_c = b. D$, and

A_{sc} = amount of steel = pbD.

(c) Using charts of uniaxial bending (Figs. 5.6 to 5.10) obtain $\dfrac{M_{ux}}{f_{ck} Db^2}$ for given $\dfrac{V}{f_{ck} bD}$

and $\dfrac{p}{f_{ck}}$ and using charts for $\dfrac{d'}{b}$, and $\dfrac{M_{yy}}{f_{ck} bD^2}$ for given $\dfrac{V}{f_{ck} bD}$ and $\dfrac{p}{f_{ck}}$ using

charts for $\dfrac{d'}{D}$.

(d) Compute $\dfrac{V}{V_u}$, then

$$\alpha = 1.0 + \frac{\dfrac{V}{V_u} - 0.2}{0.8 - 0.2} \quad \text{for } 0.2 \leqslant \frac{V}{V_u} \leqslant 0.8 \qquad\qquad \text{... (9.60)}$$

when $\dfrac{V}{V_u} \leqslant 0.2$ then $\alpha = 1$

$\dfrac{V}{V_u} \geqslant 0.8$ then $\alpha = 2$

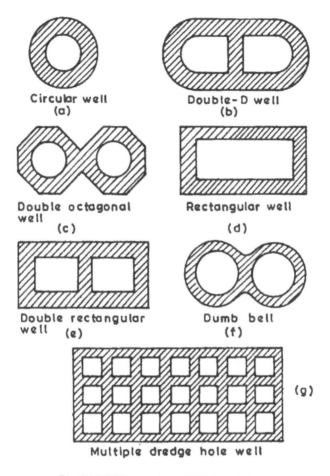

Fig. 9.19. Different Types of Well Foundations.

(e) For satisfactory section of pier, check

$$\left(\frac{M_{xx}}{M_{ux}}\right)^{\alpha} + \left(\frac{M_{yy}}{M_{uy}}\right)^{\alpha} > 1.0 \qquad \dots (9.61)$$

9.16 TYPES OF WELL FOUNDATION

Different types of wells in common use are shown in Fig. 9.19. The controlling factors in selecting the shape of the well foundation are the base dimensions of pier or abutment, the ease with which the well can be sunk, cost, considerations of tilt and shift during sinking and the magnitude of forces to be resisted.

Circular wells are used most commonly and the main points in their favour are their strength, simplicity in construction and ease in sinking. They are generally adopted for piers of bridges on narrow roads. In terms of the lateral stability for a given cross-sectional area, circular wells offer least resistance against tilting. They also have the disadvantage in accommodating a large oblong pier, the wells have to be correspondingly large which becomes uneconomical and increases obstruction to the flow of water.

Double D-well is common for the piers and abutments of bridges which are too long to be accommodated on circular wells. The dimensions of the well are so determined that the length and width of the dredge holes are almost equal. The main disadvantage is that the four corners at either end of the partition wall are far away from the centre of the dredge hole and these blind corners offer considerable resistance to sinking. Double octagonal well is free from this shortcoming of double D-well. It, however, offers great resistance against sinking on account of the increase in surface area.

Rectangular wells are generally adopted for bridge foundations having shallow depths. They can be adopted very conveniently where the bridge is designed for open foundations and change to well foundation becomes necessary during the course of construction on account of adverse conditions such as excessive inflow of water and silt into the excavation.

A dumb-bell shaped well is very similar to the double D-well except that the dredge holes are circular in shape. It has thus the advantages of twin circular wells in terms of easy accessibility to the dredging equipment and also has the advantage of double D-wells in terms of the monolithic structure.

For piers and abutments of very large sizes, wells with multiple dredge holes are used. Wells of this type have been used for the towers of Howrah Bridge.

As per Indian Standard (IS: 3955–1967) cross-section of a well should satisfy the following requirements:
(a) The dredge hole should be large enough to permit dredging.
(b) The steining thickness should be sufficient to transmit the load and also provide necessary weight for sinking and adequate strength against forces acting on the steining, both during sinking of the wells and service.
(c) It should accommodate the base of the structure and not cause undue obstruction to the flow of water.
(d) Overall size should be sufficient to transmit the loads to the soil.
(e) It should allow for the permissible tilt and shift of the well.

9.17 SINKING STRESSES IN WELLS

Skin friction—There is considerable skin friction between the well and the soil surrounding it. During the course of sinking the entire weight of the well is supported by skin friction when the soil below the cutting edge is removed by dredging. The skin friction at the time of sinking is much less than the normal skin friction which develops in the course of time. This is due to the fact that the structure of the soil gets disturbed during the course of sinking. If sinking is done through sand, a large quantity of sand keeps on flowing into the well and the quantity taken out by the dredger is generally two or three times greater than the volume displaced by the well. This results in loosening of the sand all round the well and also some subsidence at the bed level. This can be reduced to some extent if the static level of water inside the well is maintained higher than the outside by pumping water into it. This causes water to flow outside, below the cutting edge, which prevents the inflow of sand. Terzaghi has recommended that this should be done invariably when sinking is done by open dredging.

When sinking is done through clay it gets remoulded during the course of sinking and loses a considerable portion of its strength. The strength builds up again as the soil gets rest and considerable skin friction may develop between the soil and well in course of time. IS: 3955–1967 recommends the values of skin friction for designing the wells of bridges as given in Table 9.13.

Table 9.13. Skin Friction.

Soil	Skin Friction kN/m²
Silt and soft clay	7.30–29.30
Very stiff clay	48.80–195.30
Loose sand	12.20–34.20
Dense sand	34.20–68.40
Dense gravel	48.80–97.60

Skin friction may be neglected in designing the well foundation as the wells generally pass through loose and weak material before they are founded on good bearing stratum. The loose material is not capable of developing much skin friction. Vibrations produced by moving vehicles also reduce the skin friction. However, skin friction may be considered when the surrounding soil is good compact sand.

9.18 DESIGN OF WELL CAP

A well cap is needed to transfer the loads and moments from the pier to the well or wells below. The shape of the well cap is normally kept the same as of the well with a possible overhang all round of about 150 mm. If two or three wells support the pier the well cap will be extended to cover all the wells. The top of the well cap is usually kept at the bed level in case of rivers with seasonal flow or at about the low water level in case of perennial rivers.

The well cap is designed as a slab resting over the top of well.

For circular well, the design of the cap may be done in the following steps:

(a) Considering the forces on the superstructure and substructure, determine resultant vertical load V, moment M_{xx} and moment M_{yy} at the top of well cap.

$$\text{Compute } M = \sqrt{M_{xx}^2 + M_{yy}^2} \qquad \qquad ... (9.62)$$

$$e = \frac{M}{V}. \qquad \qquad ... (9.63)$$

(b) Figure 9.20 shows the plan and section of the well cap with the position of pier on it, along with the probable pressure diagram.

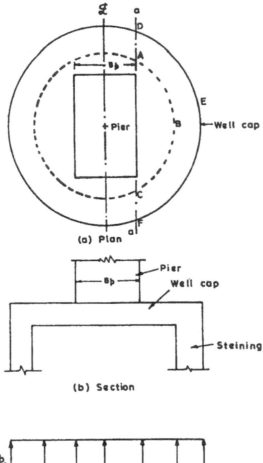

(a) Plan

(b) Section

Fig. 9.20. Illustrating the Design of Well Cap.

$$P_{1,2} = \frac{V}{A}\left(1 \pm \frac{6e}{B}\right) \qquad \qquad \text{... (9.64)}$$

where B = external diameter of well cap,

A = area of cross-section of well steining

$$= \pi\left(r_1^2 - r_2^2\right),$$

r_1 = outer radius of well, and

r_2 = inner radius of well.

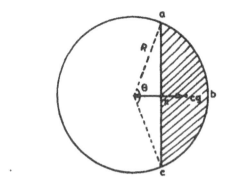

Fig. 9.21. Area and Centre of Gravity of a Segment of Circle.

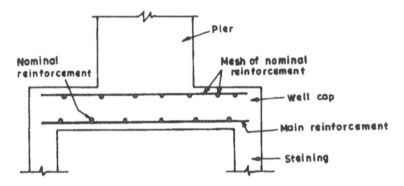

Fig. 9.22. Details of Reinforcement in a Well Cap.

(c) Critical section for finding B.M. will be a-a. Determine pressure intensity at section a-a, say p_3 (Fig. 9.20c). Assume that the shaded portion of the cap which is rested on steining is acted upon by a uniform pressure intensity of magnitude $(p_1 + p_3)/2 = p_4$.
From Fig. 9.21,

$$\text{Area of segment of a circle} = \frac{1}{2} R^2 (\theta - \sin \theta). \qquad \ldots (9.65)$$

Distance of c.g. of the segment from centre of circle

$$= \frac{4R \sin^3 (\theta / 2)}{3(\theta - \sin \theta)}. \qquad \ldots (9.66)$$

Using Eqs. 9.65 and 9.66 determine the areas of segments DEF and ABC with their centre of gravities (say A_1 and x_1; A_2 and x_2 respectively). Moment about secton a-a

$$= p_4 A_1 \left[x_1 - \frac{B_p}{2} \right] - p_4 A_2 \left[x_2 - \frac{B_p}{2} \right] - \frac{\text{Weight of well cap}}{\text{Area of well cap}} A_1 \left[x_1 - \frac{B_p}{2} \right] \qquad \ldots (9.67)$$

For this moment compute thickness of well cap and amount of reinforcement. Provide this reinforcement in the direction perpendicular to the length of pier. In other direction, provide nominal reinforcement (Fig. 9.22). Check for shear and bond in usual way.

9.19 DESIGN OF WELL STEINING

The well steining is the main body of the well. The thickness of steining is fixed based on the following considerations:

(i) It should be possible to sink the well without excessive kentledge.

   If the well of diameter d and steining thickness t is to be sunk under its own weight, then

$$\frac{\pi}{4}\left[d^2 - (d-2t)^2\right]\gamma_c \cdot h = \pi df \cdot h,$$

which gives

$$t = \frac{d}{2}\left[1 - \sqrt{\frac{4f}{\gamma_c d}}\right], \qquad \qquad \text{... (9.68)}$$

where f = skin friction,

   γ_c = density of concrete, and

   h = depth to which the well has progressed.

   If W_k is the weight of kentledge used for sinking, then the thickness of the steining can be obtained solving (Eq. 9.69).

$$\frac{\pi}{4}\left[d^2 - (d-2t)^2\right]\gamma_c h + W_k = \pi dfh \qquad \qquad \text{... (9.69)}$$

(ii) The wells should not get damaged during sinking.

   Consideration for the design of steining for wells which are to be sunk by open dredging are different from wells requiring pneumatic sinking. In India, pneumatic sinking is not resorted to directly and the common practice is to sink the wells by open sinking to the extent it is possible to do so and then to switch over to pneumatic sinking. Wells requiring pneumatic sinking are, therefore, designed to withstand forces acting on the steining during the course of open sinking and also which come into play when pneumatic sinking is resorted to.

   During the course of open dredging, soil is taken out with the help of grabs and there is only water inside the dredge holes, which exerts hydrostatic pressure on the well steining. Outside the well, the soil as well as water exert the pressure. This is partly cancelled by the water pressure acting from inside and the balance causes stresses in the steining in the horizontal direction. The net pressure diagram for which the steining should be designed is given in Fig. 9.23,

where γ_d = the dry unit weight of sand,

   γ_{sub} = submerged unit weight of sand,

   h = height of sand above water level, and

   h' = height of sand from cutting edge of spring water.

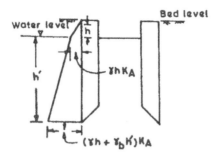

Fig. 9.23. Pressure Diagram for Well Steining during Open Sinking.

In a circular well this pressure will induce hoop compressive stresses which can be worked out using the following formulae applicable to thick shells:

$$\text{Stress along the inner face}, f_1 = \frac{2p_1 r_2^2}{r_2^2 - r_1^2} \qquad \text{... (9.70)}$$

$$\text{Stress along the outer face}, f_2 = p_1 \frac{r_1^2 + r_2^2}{r_2^2 - r_1^2} \qquad \text{... (9.71)}$$

where p_1 = net pressure outside the well,

r_1 = internal radius, and

r_2 = external radius.

Values of f_1 and f_2 should not exceed the permissible compressive stress of concrete.

The steining of a circular well which is to be sunk by pneumatic sinking acts like a thick shell subjected to air pressure from inside which causes hoop tension and longitudinal tension in the steining. Bending stresses will also be induced if the well has a shape other than circular. The air pressure is balanced by the hydraulic pressure and earth pressure from outside. They reduce the hoop tension as we go deeper below the bed level and there is hoop compression below a certain point where the air pressure is completely cancelled by the earth pressure and water pressure acting from outside. The pressure diagram for the horizontal forces is shown in Fig. 9.24. The portion of the steining which projects above the water level is subjected to the maximum hoop tension. In a circular well the hoop tension can be worked out easily using the following formulae which are applicable to thick shells.

$$\text{Stress along the inner face } f_1' = p_2 \frac{r_1^2 + r_2^2}{r_2^2 - r_1^2} \qquad \text{... (9.72)}$$

$$\text{Stress along the outer face } f_2' = \frac{2p_2 r_1^2}{r_2^2 - r_1^2} \qquad \text{... (9.73)}$$

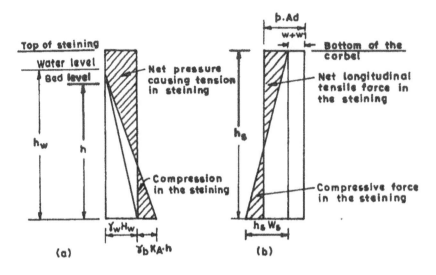

Fig. 9.24. Pressure Diagram for Well Steining during Pneumatic Sinking.

where p_2 = net pressure inside the well,

             r_1 = internal radius, and

             r_2 = external radius.

Hoop reinforcement per metre length

$$= \frac{(f_1' \text{ or } f_2'.t)}{\text{Permissible tensile strength of steel}} \qquad \qquad \text{... (9.74)}$$

where t = thickness of well steining.

(iii) If the well develops tilts and shifts during sinking it should be possible to rectify these within permissible limits without damaging the well.

(iv) The well should be able to resist safely earth pressure developed during sand blow or during other conditions like sudden drop that may be experienced during sinking.

(v) Stresses at various levels of the steining should be within permissible limits under all conditions for loads that may be transferred to the well either during sinking or during service.

It can be shown that in heavy well, section of zero shear force will be at a distance x from scour level which is given by

$$x = \left(\frac{2FH}{\gamma_b (K_p - K_A) B} \right)^{\frac{1}{2}} \qquad \qquad \text{..., (9.75)}$$

The maximum moment will be, $M_{\text{max}} = M_0 + \dfrac{2}{3} H \cdot x.$ \qquad \qquad \text{... (9.76)}

Stresses in the steining are: $\qquad f_{1,2} = \dfrac{V}{A} \pm \dfrac{M_{max}}{I} \cdot y,$ $\qquad$ ... (9.77)

where $\quad H$ = resultant horizontal force at scour level,

$\qquad = \sqrt{H_{xx}^2 + H_{yy}^2}$

$\quad H_{xx}$ = horizontal force at scour level across traffic,

$\quad H_{yy}$ = horizontal force at scour level along the traffic,

$\quad \gamma_b$ = submerged density of soil,

K_P, K_A = passive and active earth pressure coefficients,

$\quad M_{max}$ = maximum value of bending moment,

$\quad M_0$ = moment at the scour level,

$\qquad A$ = area of cross-section of well steining,

$\qquad = \pi(r_1^2 - r_2^2),$

$\qquad I$ = moment of inertia of well steining,

$\qquad = \dfrac{\pi}{64}(r_1^4 - r_2^4),$

r_1, r_2 = outer and inner radii of well,

$\qquad V$ = total vertical load acting up to the depth x below scour level, and

$\qquad F$ = factor of safety usually taken as 2.0.

First using Eq. (9.75), value of x is obtained which corresponds to the section of zero shear force below scour level. Eq. 9.76 will then give the value of maximum bending moment. Stresses in the well steining (f_1, f_2) corresponding to maximum bending moment can be computed using Eq. (9.77). These stresses should be within permissible limits.

In case, the steining section is in tension, it should be designed considering that it is subjected to a vertical load V and moment M applying the design principles of combined bending and thrust.

IRC Recommendations (IRC: 78–2000)

Thickness of steining: The thickness of the concrete steining should not be less than 500 mm and also not less than as given by Eq. 9.78.

$$t = Kd\sqrt{D_e} \qquad\qquad ... (9.78)$$

where $\quad t$ = minimum thickness of concrete steining in m,

$\qquad d$ = external diameter of circular well or dumb bell shaped well or small plan dimension of twin D well in m,

$\quad D_e$ = depth of well in m below LWL or GL whichever is greater, and

$\qquad K$ = a constant depending on the nature of subsoil and steining material (Table 9.14).

Table 9.14. Values of *K*.

S. No.	Type of Well	Material of Well	Surrounding Soil Type	Values of K
1.	Single circular or dumbbell	Cement Concrete	Sand	0.030
2.	Single circular or dumbbell	Cement Concrete	Clay	0.033
3.	Twin D	Cement Concrete	Sand	0.039
4.	Twin D	Cement Concrete	Clay	0.043
5.	Single circular or dumb	Brick Masonry	Sand	0.047
6.	Single circular or dumb	Brick Masonry	Clay	0.052
7.	Twin D	Brick Masonry	Sand	0.062
8.	Twin D	Brick Masonry	Clay	0.068

Reinforcement: For plain concrete wells, vertical reinforcements (either mild steel or deformed bars) in the steining should not be less than 0.12% of gross sectional area of the actual thickness provided. This should be equally distributed on both faces of the steining. The vertical reinforcement should be tied up with hoop steel not less than 0.04 per cent of the volume per unit length of the steining.

In case where the well steining is designed as a reinforced concrete element, it should be considered as a column section subjected to combined axial load and bending. However, the amount of vertical reinforcement provided in the steining should not be less than 0.2% (for either mild steel or deformed bars) of the actual cross sectional area of the steining. On the inner face a minimum of 0.06% (of gross area) steel should be provided. The transverse reinforcement in the steining should be provided in accordance with the provisions for a column but in no case should be less than 0.04% of the volume per unit length of the steining.

The vertical bond rods in brick masonry steining should not be less than 0.1% of the cross-sectional area and shall be encased into cement concrete of M15 mix of size 150 mm × 150 mm. These rods should be equally distributed along the circumference in the middle of the steining and should be tied up with hoop steel not less than 0.45% of the volume per unit length of the steining. The hoop steel should be provided in a concrete band at spacing of four times the thickness of the steining or three metres, whichever is less. The horizontal R.C.C. bands should not be less than 300 mm wide and 150 mm high, reinforced with bars of diameter not less than 10 mm placed at the corners and tied with 6 mm diameter stirrups at 300 mm centres.

9.20 DESIGN OF WELL CURB

The well curb is an R.C.C. ring beam which satisfies the following requirements:

(i) It should have a shape offering the minimum resistance during sinking, and should be strong enough to be able to transmit superimposed loads from the steining to the bottom plug. To

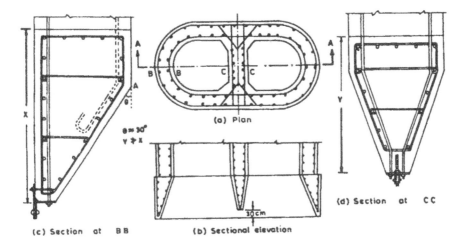

(a) Plon

(b) Sectional elevation

(c) Section at BB

(d) Section at CC

Fig. 9.25. Well Curb.

satisfy the above requirements, the shape and the outline dimensions of the curb should be as given in Fig. 9.25. The curb should invariably be constructed in reinforced concrete of mix not leaner than M20 with minimum reinforcement of 72 kg per cu.m. excluding bond roads. This quantity of steel should be suitably arranged as shown in Fig. 9.25 to prevent spreading and splitting of the curb during sinking and in service.

(ii) In case pneumatic sinking is indicated, the internal angle of the well curb should be made steep enough to provide easy access for the pneumatic tools.

(iii) In case blasting is anticipated, the outer faces of the curb should be protected with suitable steel plates of thickness not less than 6 mm up to half the height of the well curb, suitably reduced to 6 mm to a height of three metres above the top of the curb. The steel plates should be properly anchored to the curb and steining. The curb in such a case should be provided with additional hoop reinforcement of 10 mm dia mild steel or deformed bars at 150 mm centres. The latter reinforcement should also extend up to a height of three metres into the well steining above the curb, in which portion the mix of concrete in the well steining should not be leaner than M20.

Balwant Rao and Muthuswami (1963) have suggested two conditions to be considered for the design of curb.

a. Design of curb for sinking—During the process of sinking when the well is dredged the curb cuts through the soil by the dead weight of the steining including kentledge, if any. The stresses in the curb thus need to be considered. Fig. 9.26 shows the forces acting on the curb when the well has penetrated soil to a considerable depth.

Let d = mean diameter of curb in m,

 N = weight of steining in kN per m run,

 θ = angle in degrees of bevelling face with the horizontal,

 μ = coefficient of friction between soil and concrete of curb,

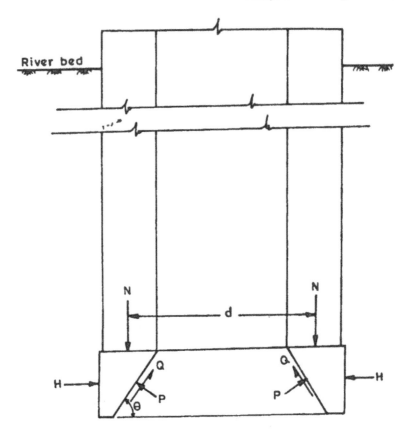

Fig. 9.26. Forces on Well Curb during Sinking.

P = force in kN per m run of curb acting normal to the bevel surface,
Q = force in kN per m length of curb acting tangentially to the bevel surface, and
H = horizontal resultant force in kN per m of curb.

Then $$Q = \mu P. \qquad \text{... (9.79)}$$

Resolving vertically, $\mu P \sin \theta + P \cos \theta = N.$

We get $$P = \left(\frac{N}{\mu \sin\theta + \cos\theta} \right) \qquad \text{... (9.80)}$$

Resolving horizontally $P \sin \theta - \mu P \cos \theta = H.$

We get · $$H = P (\sin \theta - \mu \cos \theta). \qquad \text{... (9.81)}$$

Substituting P from Eq. (9.81)

$$H \text{ per m run} = N \left(\frac{\sin\theta - \mu\cos\theta}{\mu\sin\theta + \cos\theta} \right) \qquad \text{... (9.82)}$$

Total hoop tension $= H\dfrac{d}{2}$

$$= 0.5\,N\left(\frac{\sin\theta - \mu\cos\theta}{\mu\sin\theta + \cos\theta}\right)d. \qquad \text{... (9.84)}$$

While sinking, active earth pressure of soil or external compression may not develop fully at the curb due to unsettled conditions.

Sometimes, during sinking, sand blow in case of a deep dredge may result in sudden descent of well. To account for these eventualities, hoop tension reinforcement is increased by 50% and vertical bond rods are provided. Tension in the curb will be given by

$$\text{Total hoop tension} = 0.75\,N\left(\frac{\sin\theta - \mu\cos\theta}{\mu\sin\theta + \cos\theta}\right)d. \qquad \text{... (9.85)}$$

b. Curb resting on bottom plug—Under conditions when the cutting edge is not able to move downwards due to reactions developed at the curb and bottom plug interface neglecting the effect of skin friction—reaction can be resolved into horizontal and vertical components. Assuming a parabolic arch within the thickness of the bottom plug, the weight of material filled in the well and that of the bottom plug will be transmitted directly to the bed through it. For the conditions assumed, the hoop tension H is given by the equation:

$$H = \frac{qd^2}{8r}\cdot\frac{d}{2}$$

$$= \frac{qd^3}{16r} \qquad \text{... (9.86)}$$

where $\qquad q = \dfrac{\text{total weight on the base}}{\text{area of plug}}$, and

$\qquad r$ = vertical height of imaginary inverted arch.

In granular soils, the hoop tension H is relieved by the active pressure around the curb (Fig. 9.27).

Hoop compression $\quad C = \dfrac{1}{2}(p_1 + p_2)\,b\dfrac{d}{2}$

$$= (p_1 + p_2)\frac{bd}{4} \qquad \text{... (9.87)}$$

where $\qquad\qquad p_1$ = active earth pressure at depth D_f,

$\qquad\qquad\quad = \gamma_b K_A D_f$

$\qquad\qquad p_2$ = active earth pressure at depth $(D_f - b)$,

$\qquad\qquad\quad = \gamma_b K_A (D_f - b)$

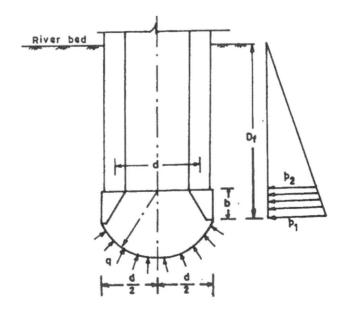

Fig. 9.27. Forces on Well Curb Resting on Bottom Plug.

K_A = coefficient of active earth pressure,

ϕ = angle of internal friction of soil, and

γ = submerged unit weight of soil.

Thus net hoop tension = $(H - C)$

$$= \frac{d}{4}\left(\frac{qd^2}{4r} - (p_1 + p_2)b \right) \qquad \qquad \text{... (9.88)}$$

At the junction of the curb and steining, a moment M_0 (Eq. 9.89) is developed due to the horizontal force H caused by bevelled action, that is

$$M_0 = H\frac{b}{2}. \qquad \qquad \text{... (9.89)}$$

Reinforcement provided at the corner A (Fig. 9.25) to take care of this moment should be taken along the bevel base and anchored well into the steining.

9.21 DESIGN OF CUTTING EDGE

The cutting edge usually consists of a mild steel equal angle of side 150 mm. The angle will have one side projecting downward from the curb as shown in Fig. 9.28a for soils where boulders are not expected. In soils mixed with boulders, the angle should have the vertical leg embedded in the curb in such a manner that the horizontal leg of the angle is flush with the bottom of the curb (Fig. 9.28b).

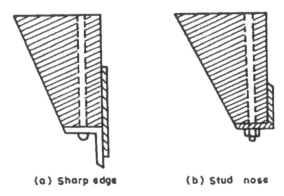

(a) Sharp edge   (b) Stud nose

Fig. 9.28. Shapes of Cutting Edge.

The cutting edge should be properly anchored to the well curb and should be strong enough not to fail in crushing while passing through the hardest strata expected to be met with (IS: 3955–1967).

9.22 DESIGN OF BOTTOM PLUG

Following points should be considered in the design and construction of the bottom plug:

(i) Bottom plug of a well usually consists of M20 cement concrete laid by means of tremie or skip boxes. Plugging done by concrete is also considered satisfactory. It has been noticed that the concrete laid for the bottom plug takes longer time to set and considerable portion of cement gets washed resulting in weak concrete and, therefore, 10% extra quantity of cement should be added.

(ii) Bottom plugging should always be done in one continuous operation when the water level is at its normal level. If there are two or more dredge holes, plugging should be done simultaneously and to equal heights in all dredge holes.

(iii) The bottom plug may be in the shape of a bulb. The advantage of this is that it will produce arch action, reduce hoop tension in curb and give greater bearing area.

(iv) While founding the well on the rock, it should be securely anchored to the bed by taking it 200 mm to 300 mm into the rock bed. Dowel bars when used should have twice the length in the plug to the length embedded in the rock to develop full bond (Fig. 9.29). This is necessary for the possibility of slips and tensile stresses that may exist between the base and rock surface.

(v) Before starting the plugging operation, the water in the well must be still and up to its normal level. Concrete must be of normal consistency as if being used on work out of water. Plugging of the well must be continuous and completed in a day.

(vi) It is essential that the bottom plug should be capable of receiving the load from the steining and transmitting it to the base in which case thickness of the bottom plug should be at least equal to half the diameter of dredge hole.

(vii) Sometimes forming of the bottom plug with a positive sump to get a bulb at the base may be difficult, especially in sandy strata. In such cases the transmission of the load by the

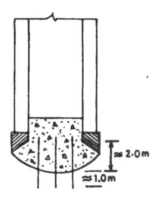

Fig. 9.29. Dowels in Rock and Bottom Plug.

plug by arch action may not materialize. To cover such contingencies it is better to fill the well above the bottom plug with sand up to top plug immediately below well cap.

(viii) In open caisson the top of the seal is carried to a level several metres above the bevelled portion of the cutting edge.

 (ix) Based on the theory of elasticity, the thickness of the seal t is given by the following relation:

For circular wells

$$t^2 = \frac{3W}{8\pi f_c}\,(3 + v).\qquad\qquad\qquad\qquad\text{... (9.90)}$$

If $\qquad\qquad\qquad\qquad W = q\pi r^2 \text{ and } v = 0.5$

Equation (9.90) reduces to

$$t^2 = 1.18 r^2 \frac{q}{f_c}.\qquad\qquad\qquad\qquad\qquad\text{... (9.91)}$$

For rectangular wells

$$t^2 = \frac{3qb^2}{4f_c(1+1.61\alpha)}\qquad\qquad\qquad\qquad\text{... (9.92)}$$

where W = total bearing pressure on the base of well,
 f_c = flexure strength of concrete seal,
 v = Poisson's ratio for concrete,
 R = radius of well at the base,
 q = unit bearing pressure against the base of well,
 b = short side of well, and
 α = width/length ratio or short side/long side of well.

9.23 TOP PLUG AND FILLING

After laying the concrete of the bottom plug and allowing it to remain undisturbed for at least 14 days, clean granular material (sand, sand and gravel, etc.) is filled in the dredge hole up to the desired height which extends up to the top concrete plug placed immediately below the well cap. The sand filling in a completely filled well serves the purpose of transmitting the load directly to the bottom plug and thus may give some relief to well steining. In seismic areas, the well is preferably filled only partly up to scour level and remaining portion may be filled with water. This is done to reduce the moment due to seismic force.

9.24 SINKING OF WELLS

The following steps are involved in well sinking:

(i) **Placing of curb:** The method adopted for correctly placing the curb depends on local configuration.

Following points are useful in the correct placement of curb:

(a) If the bed of the river is dry, the bed is excavated up to about 150 mm. Firstly, cutting edge is laid. The curb is then laid there and, after it has sufficiently hardened (in about five days), steining is built over it (Fig. 9.30).

(b) If the location of the well is at a place with water which cannot be drained, diverted or pumped out, sand island method which envisages the construction of temporary island by dumping sand on which curb and steining are built.

(c) At places where sand island cannot be made economically by just dumping sand there, the sand filling or island must be contained by suitable timbering or sheet piling or by rip rap forming coffer dams and must be protected against scour and erosion.

(d) In case of deep water it may be economical to build the curb on dry ground and float it to the site.

(ii) **Well steining:** Well steining should be built in initial short heights of about 2 m. The

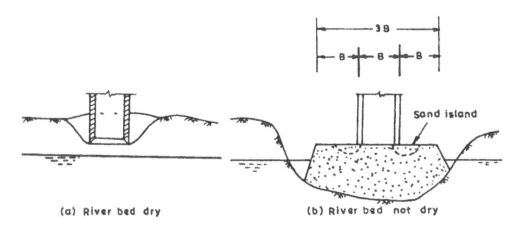

(a) River bed dry   (b) River bed not dry

Fig. 9.30. Laying of Well Curb.

steining should be built in one straight line from bottom to top. This can be ensured if the steining is built with straight edges preferably of angle iron.

(iii) Dredging, jetting and sinking: The process of dredging, jetting and sinking can be summarised as below:

(a) The removal of the material by dredging the wells is usually accommodated by clamshell buckets.

(b) When the caisson approaches very hard strata, special rock teeth should be added to the grab or very often heavy chiselling resorted to for breaking such strata.

(c) In those cases where due to the depth of water and peculiarities of soil, sinking by the use of grabs, etc. cannot be done, pneumatic sinking or in some cases sinking by divers with the use of steel helmet and compressed air is resorted to.

(d) The rate of sinking of a well is a function of the rate at which the bed material is removed from the well bottom.

(e) A well sinks by its own weight when the soil below its cutting edge is removed. If the side friction is so great as to retard sinking under its own weight, additional weight or kentledge will have to be added. When addition of kentledge does not make the well sink, it is best to suspend the work for some time and allow the water in the well to its normal level, then lower the water level by about 3 m to 6 m by pumping. The differential head causes increased flow in the well. This flow reduces surface friction and helps the well in sinking. In such cases the well sinks rapidly.

9.25 TILTS AND SHIFTS

Causes of tilts and shifts

The well should as far as possible be sunk true and vertical through all types of soils. A well may experience tilt and shift due to following reasons:

(i) Nonuniform bearing capacity (Fig. 9.31).

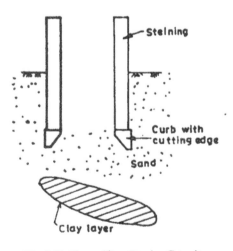

Fig. 9.31. Non-uniform Bearing Capacity.

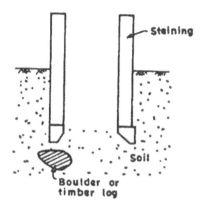

Fig. 9.32. Obstruction Below Cutting Edge.

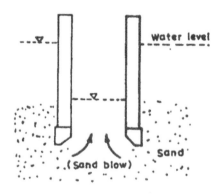

Fig. 9.33. Sand Blowing during Sinking.

(ii)   Obstruction on one side of the well (Fig. 9.32).

(iii) Sand blowing in wells during sinking (Fig. 9.33). It will cause sudden sinking of well.

(iv)   Method of sinking: Material should be removed from all sides equally otherwise the well may experience tilt.

(v)   Sudden sinking due to blasting may also cause tilting of well.

(vi)   Irregular casting of steining (Fig. 9.34). It will cause less friction on one side and, therefore, chances of tilting increases.

Rectification of tilt

The following measures may be useful in rectifying the tilt and shift during well sinking operation:

(i) Eccentric grabbing: Higher side is grabbed more by regulating the dredging suitably. This method is effective in initial stages of sinking but as the sinking progresses regulation of grabbing becomes more and more difficult. In case of tilted wells dredgers does not work

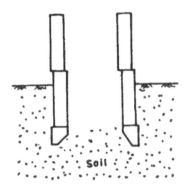

Fig. 9.34. Irregular Casting of Steining.

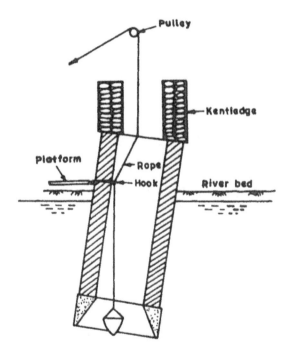

Fig. 9.35. Dredging with Hooking.

satisfactorily. Under such circumstances, a hole in the steining is made on the higher side and the rope of grab pulled through the hook (Fig. 9.35). Also, the well may be dewatered if possible, and open excavation on the higher side carried out.

(ii) Eccentric loading: To provide greater sinking effort on the higher side of the well, eccentric loading is necessary. The eccentric loading can be provided through a suitable platform (Fig. 9.36). As the sinking progresses, heavier kentledge with greater eccentricity is required to rectify the tilt.

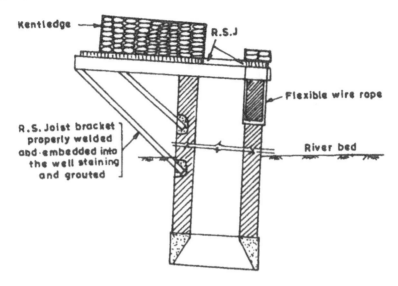

Fig. 9.36. Eccentric Placement of Kentledge.

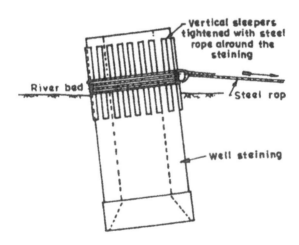

Fig. 9.37. Pulling the Tilted Well with Steel Rope.

(iii) **Water jetting:** Water jets are used on the outer face of the higher side. The jets will reduce friction on the higher side which will be helpful in rectifying tilt. The method, if used alone, is not very effective but is helpful if used with other methods.

(iv) **Arresting the cutting edge:** Temporary obstacles like widen sleepers or hooking below the cutting edge of the well on the lower side to avoid further tilt of the well could be useful.

(v) **Pulling the well:** Pulling the well towards the higher side by placing one or more steel ropes round the well with wooden sleepers aids the other methods of rectifying tilt (Fig. 9.37). This method is effective in early stages only.

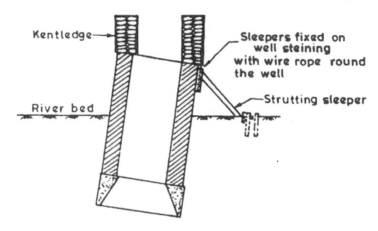

Fig. 9. 38. Strutting of Well on its Tilted Ride.

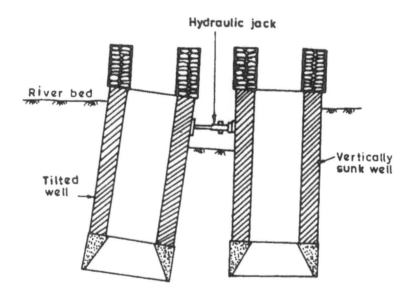

Fig. 9.39. Jacking the Tilted well.

(vi) Strutting the well: Strutting of well on the tilting side (Fig. 9.38) will avoid any further increase in the tilt.

(vii) Pushing the well by jacks: The wells may be pushed with suitable arrangements through mechanical or hydraulic jacks (Fig. 9.39).

A combination of some of the above-mentioned methods (Fig. 9.40) can-sometimes be usefully employed.

A tilt of 1 in 80 and a shift of 150 mm due to translation (both additive) in a direction which will cause most severe effect shall be considered in the design of well foundation (IRC : 78-2000).

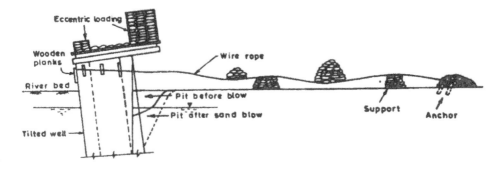

Fig. 9.40. Combination of Measures to Rectify Tilt and Shift of Well.

ILLUSTRATIVE EXAMPLES

Example 9.1

Design the pier and well foundation for a balance cantilever bridge for the following data:

Main span	= 30 m
Suspended span	= 15 m
Cantilever span	= 7.5 m
Loading	= IRC class A loading
Footpath load	= 2 kN/m²
Road width	= 8 m
Footpath width	= 1.5 m
Maximum design discharge	= 6000 m³/sec
Average velocity of flow	= 1.5 m/sec
Dead load of main span	= 4500 kN
Allowable soil pressure	= 450 kN/m² (static case)
Area of elevation	= 150 m²
Depth of centre of area of elevation above road level	= 1.75 m
Lacey's silt factor	= 1.0
Formation level of bridge	= 480.00 m
Bed level	= 462.50 m
H.F.L.	= 473.50 m
Level of bearing pins	= 475.45 m
Level of the base of the bearing	= 475.20 m
Width of bearing	= 500 mm
Length of outer edges of bearings	= 4.0 m.

Assume suitably any data not given.

Solution

(i) Depth of foundation

Effective linear waterway, L $= 4.8\sqrt{Q} = 4.8\sqrt{6000} = 371.8$ m

Discharge per metre width, D_b $\quad = \dfrac{6000}{371.8} = 16.14 \text{ m}^3/\text{sec/m}$

Normal scour depth, d $\quad = 1.34 \left[\dfrac{D_b^2}{K_{sf}} \right]^{\frac{1}{3}}$

$\quad = 1.34 \left[\dfrac{(16.14)^2}{1} \right]^{\frac{1}{3}} = 8.6 \text{ m}$

Maximum scour depth, D $\quad = 2d = 2 \times 8.6 = 17.2 \text{ m}$

Grip length $\quad = \dfrac{D}{3} = \dfrac{17.2}{3} = 5.73 \text{ m}$

Depth of the base of foundation below HFL
$\quad = 17.2 + 5.73 = 22.93 \text{ m, say 23 m}$

(ii) Preliminary dimensions of pier and well
(a) Top dimensions of pier

Width = Bearing width + clearance on either side
$\quad = 0.50 + 2 \times 0.25 = 1.0 \text{ m}$

Length = Length between the outer edges of bearings + 1.0 m + semi-circles
on either side (nose)
$\quad = 4.0 + 1.0 + 1.0 = 6.0 \text{ m.}$

(b) Dimensions of pier cap

Assuming 200 mm projections beyond the edge of the top of pier, then

Width = $1.0 + 2 \times 0.2 = 1.4 \text{ m}$
Length = $5.0 + 1.4 = 6.4 \text{ m}$
Thickness = 0.8 m (assumed)

(c) Bottom dimensions of pier

Height of pier = $475.20 - 462.50 - 0.8 = 11.9 \text{ m}$
Batter of the sides = 1 in 12

Width $\quad = 1.0 + 2 \times \dfrac{11.9}{12} = 3.0 \text{ m (approx.)}$

Length $\quad = 5.0 + \dfrac{2 \times 11.9}{12} + 3.0 = 10.0 \text{ m (approx.)}$

(d) Dimensions of well

Adopt a circular well for foundation with a single dredge hole.

External diameter $\quad = 10.0 \text{ m}$
Internal diameter $\quad = 7.0 \text{ m}$
Thickness of steining $\quad = 1.5 \text{ m}$
Thickness of well cap $\quad = 1.2 \text{ m}$

R.L. of base of foundation = $473.50 - 23.0 = 450.50 \text{ m}$
Depth of foundation below bed level = $462.50 - 450.50 = 12.0 \text{ m}$
Height of well = $12.0 - 1.2 = 10.8 \text{ m}$

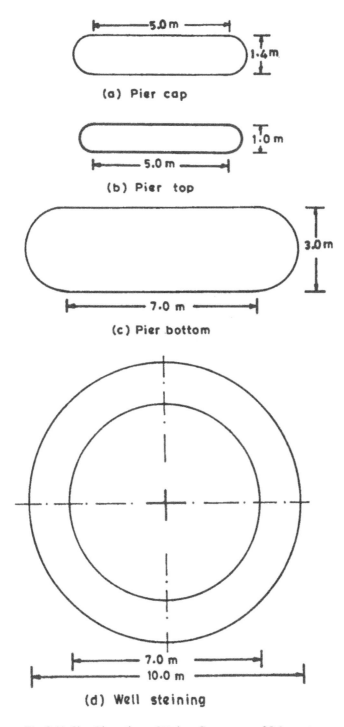

(a) Pier cap

(b) Pier top

(c) Pier bottom

(d) Well steining

Fig. 9.41. Plan Dimensions of Various Components of Substructure.

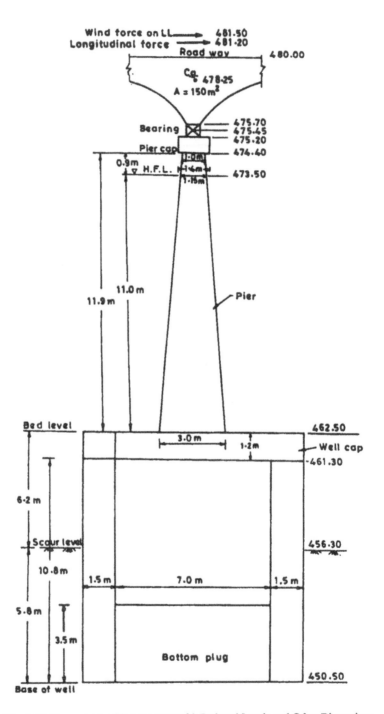

Fig. 9.42. Components of Substructure with Reduced Levels and Other Dimensions.

Cross-section of pier cap, pier at top and bottom, and of well are shown in Fig. 9.41. Trial dimensions and various reduce levels (RL) are shown in Fig. 9.42.

(iii) Load calculations

(a) Dead load

$$\text{From main span (30 m)} = 4500 \text{ kN}$$

$$\text{From suspended span (15 m)} = \frac{4500}{2} = 2250 \text{ kN}$$

$$\text{From cantilever span (7.5 m)} = \frac{4500}{4} = 1125 \text{ kN}$$

Total dead load reaction on a pier, E (Fig. 9.43)

$$= \frac{4500}{2} + \frac{2250}{2} + 1125 = 4500 \text{ kN}.$$

(b) Live load

Reaction on pier having rocker bearing (say at location E, Fig. 9.43) is obtained by drawing the influence line diagram. The magnitude of reaction is obtained by multiplying the ordinates of influence line diagram with the loads of IRC class A and adding together. Therefore, live load reaction.

$$= 27 \times 1.25 \left(\frac{10.7}{15} + \frac{11.8}{15} \right) + 114 \times 1.25 \left(1 + \frac{36.3}{37.5} \right) +$$

$$68 \times 1.25 \left(\frac{32}{37.5} + \frac{29}{37.5} + \frac{26}{37.5} + \frac{23}{37.5} \right) + 27 \times 1.25 \left(\frac{2.6}{37.5} + \frac{1.5}{37.5} \right) = 975 \text{ kN}.$$

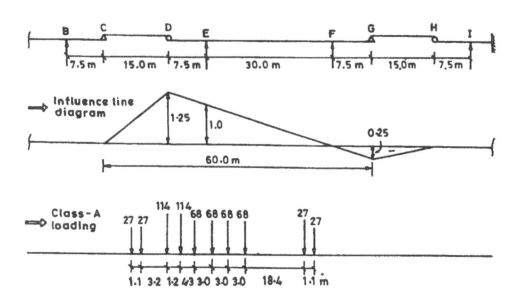

Fig. 9.43. Influence Line Diagram for Computing Live Load Reaction.

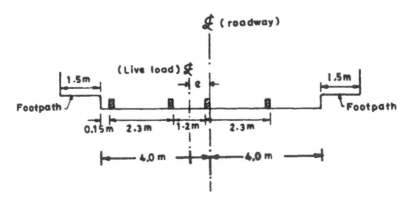

Fig. 9.44. Placement of Live Load on Roadway for Eccentricity Calculation.

It will act at an eccentricity of $(4 - 0.15 - 2.3 - 0.6)$ i.e. 0.95 m (Fig. 9.44).
  Reaction due to footpath loading

$$= \frac{1}{2} \times 52.5 \times 1.25 \times 2 = 66 \text{ kN.}$$

It will act at an eccentricity of 4.75 m from the centre of pier (Fig. 9.44).

(c) Impact load
  For IRC class A loading on R.C. bridges

$$\text{I.F.} \qquad\qquad = \frac{4.5}{6 + L}$$

$$\text{I.F. for main span} \qquad = \frac{4.5}{6 + 30} = 0.125$$

$$\text{I.F. for suspended span} = \frac{4.5}{6 + 15} = 0.214$$

$$\text{I.F. for cantilever span} \quad = \frac{4.5}{6 + 7.5} = 0.333$$

$$\text{Average I.F.} \qquad\qquad = 0.224$$

  Impact load for design of pier cap $= 975 \times 0.224 = 219$ kN.
  Impact load reduces to zero at a depth 3.0 m below the top of pier cap.
(d) Wind loads
(a) On dead load
Case I: No water condition i.e. no buoyancy and no water pressure.
  Height of centre of area in elevation above the bed level.
$$= 480.00 - 1.75 - 462.50 = 15.75 \text{ m.}$$

From Table 9.8:

Height, H (m)	p in kN/m^2	V in km/hr
15	1.07	128
20	1.19	136

At 15.75 m,

$$p = 1.07 + \frac{1.19 - 1.07}{20 - 15}(15.75 - 15) = 1.09 \text{ kN/m}^2$$

$$V = 128 + \frac{136 - 128}{20 - 15}(15.75 - 15) = 129.2 \text{ km/hr.}$$

Horizontal wind force = 1.09 × 150 = 163.50 kN.

Case II : Water at HFL (i.e. water current and buoyancy forces to be considered)

Height of centre of area in elevation above HFL

$$= 480.00 - 1.75 - 473.50 = 4.75 \text{ m}$$

From table, p = 0.63 kN/m^2

Wind force = 0.63 × 150 = 95 kN

Case III: If $V > 130$ km/hr, no live load is considered over the bridge. It is also assumed that high wind velocity and high flood are not likely to occur simultaneously. Hence, high wind on live load without water current forces are to be considered.

$p = 1.12$ kN/m^2 for V = 130 km/hr

Wind force on dead load = 1.12 × 150 = 168 kN

Wind force on live load = 4.5 × 2 × 18.8 = 169.2 kN.

Point of application of wind force on live load = 1.5 m above the road surface, i.e. at RL 481.50 m.

Summary

Dead load, Case I   163.50 kN acting at the centre of area in elevation

    Case II   95   kN acting at the centre of area in elevation

    Case III   168   kN acting at the centre of area in elevation

Live load   169.2   kN at 1.5 m above road surface

(e) Water current forces

Water pressure = 0.52 KV^2

Velocity component at the face of pier

(i) Along the pier axis = $V \cos (20 \pm \theta)$

(ii) Transverse to pier axis = $V \sin (20 \pm \theta)$

So, maximum water pressure

(i) Along the pier axis   = 0.52 × 0.66 ($\sqrt{2}$ × 1.5 cos 20)^2

                 = 1.36 kN/m^2

(ii) Transverse to pier axis = 0.52 × 1.5 ($\sqrt{2}$ × 1.5 sin 20)^2

                 = 0.41 kN/m^2.

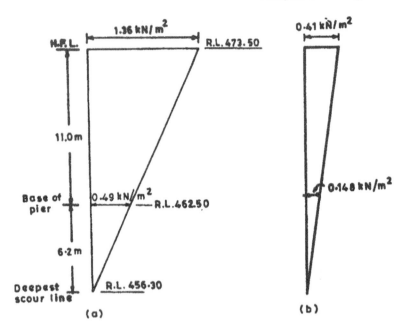

Fig. 9.45. Pressure Diagrams due to Water-current Forces: (a) Cosine Components, (b) Sine Component.

Refer Fig. 9.45
    Force due to water current

(i) Along the pier axis $\quad = \dfrac{1.36 + 0.46}{2} \times \dfrac{1.15 + 3}{2} \times 11$

$$= 21.2 \text{ kN}$$

It will act at $\dfrac{1.36 + 0.49 \times 2}{1.36 + 0.49} \times \dfrac{11}{3} = 4.64$ m below HFL

(ii) Transverse to pier axis $\quad = \dfrac{0.41 + 0.148}{2} \times \dfrac{5.16 + 7}{2} \times 11$

$$= 18.7 \text{ kN.}$$

It will act at $\dfrac{0.41 + 0.148 \times 2}{0.41 + 0.148} \times \dfrac{11}{3} = 4.64$ m below HFL

Force due to water current on well along the pier axis

$$= \frac{0.49}{2} \times 6 \times 10 = 14.7 \text{ kN}$$

This acts at $11 + \dfrac{6.2}{3} = 13.1$ m below HFL.

Force due to water current on the well transverse to pier axis

$$= \frac{0.148}{2} \times 6 \times 10 = 4.44 \text{ kN acting at 13.1 m below HFL}$$

(f) Dead weight of substructure

(i) Weight of pier cap $= [5.0 \times 1.4 \times 0.8 + \dfrac{\pi}{4} \times 1.4^2 \times 0.8] \times 25$

$\qquad\qquad\qquad\qquad\quad = 1171$ kN

(ii) Area of pier at top $\quad = 5.0 \times 1.0 + \pi/4 \, (1.0)^2 = 5.79$ m^2

Area of pier at bottom $= 3.0 \times 7.0 + \pi/4(3.0)^2 = 28.1$ m^2

Height of pier $\qquad\quad = 11.9$ m

Weight of pier $\qquad = \dfrac{5.79 + 28.1}{2} \times 11.9 \times 19$

$\qquad\qquad\qquad\qquad\quad = 3831$ kN.

(iii) Weight of well cap $\quad = \pi/4(10)^2 \times 1.2 \times 25$

$\qquad\qquad\qquad\qquad\quad = 2355$ kN.

(iv) Weight of well steining $= \pi/4(10^2 - 7^2) \times 10.8 \times 25$

$\qquad\qquad\qquad\qquad\quad = 10815$ kN.

(v) Weight of concrete plug $= \pi/4(7)^2 \times 3.5 \times 23$

$\qquad\qquad\qquad\qquad\quad = 3098$ kN.

(vi) Weight of sand fill $\quad = \pi/4(7)^2 \times (10.8 - 3.5) \times 18$

$\qquad\qquad\qquad\qquad\quad = 5057$ kN.

(g) Buoyancy

Width of pier at HFL $\qquad = 1.0 + 2 \times \dfrac{0.9}{11.9} = 1.15$ m

Length of pier at HFL $\qquad = 4.025$ m without noses

Area of pier at HFL $\qquad = 1.15 \times 4.025 + \pi/4(1.15)^2$

$\qquad\qquad\qquad\qquad\qquad = 4.95$ m^2

Area of pier at bottom $\qquad = 28.1$ m^2

Height of pier submerged

below HFL $\qquad\qquad\quad = 11.0$ m

Submerged volume of pier $= \dfrac{4.95 + 28.1}{2} \times 11 = 181.7$ m^3

Volume of well cap $\qquad\qquad = \pi/4 \times 10^2 \times 1.2 = 94.24$ m^3

Volume of well above scour level $= \pi/4 \times 10^2 \times 5.0 = 393$ m^3

Buoyancy on pier $\qquad\qquad = 10 \times 181.7 \times 0.15 = 248$ kN

Buoyancy on well cap $\qquad = 10 \times 94.24 \times 0.15 = 142$ kN

(h) Longitudinal forces

(i) On first train of vehicles

$= 0.2 \, (27 \times 2 + 114 \times 2 + 68 \times 4) = 110.8$ kN

(ii) On second train of vehicles

I for main span $\qquad = 0.1(27 \times 2 + 114 \times 2 + 68) = 35.0$ kN

II for cantilever span   = $0.1(68 \times 2) = 13.6$ kN
Total longitudinal force = $110.8 + 35.0 + 13.6 = 159.4$ kN.

It will act at a height of 1.2 m above roadway.
R.L. of the point of application = $480 + 1.2 = 481.2$ m.
Lever arm of longitudinal force with respect to bearing pins
$$= 481.2 - 475.45 = 5.75 \text{ m}.$$
Increase in reaction due to this force on rocker bearing
$$= \text{Decrease in reaction on roller (or free) bearing}$$
$$= \frac{159.4 \times 5.75}{30} = 31 \text{ kN}.$$
Loadings on the free bearings:

Dead load	=	4500 kN
Live load	=	$[27 \times 4 + 114 \times 4 + 68 \times 7) - 974.7 = 65.3$ kN
Impact load	=	$65.3 \times 0.224 = 14.6$ kN
Foot path reaction	=	$2 \times 52.5 - 66 = 39$ kN
Net vertical load on		
free bearing	=	$4500 + 65.3 + 14.6 + 39 - 31 = 4589$ kN

Coefficient of friction at roller bearing, $\mu = 0.03$
$$\mu R = 0.03 \times 4589 = 138 \text{ kN}.$$

Longitudinal force on fixed or rocker bearing will be $(159.4 - 138 = 21.4$ kN) or

$[(159.4/2) + 138 = 217.7$ kN] whichever is more, i.e. 217.7 kN
(i) Seismic force

$$\alpha_h = 0.1; \ \alpha_v = \frac{0.1}{2} = 0.05$$

(a) On dead load of superstructure
    $4500 \times 0.1 = 450$ kN (Horizontal) at R.L. 478.25 m
    $4500 \times 0.05 = \pm 225$ kN (Vertical)
(b) On live load
    $975 \times 0.1 = 97.5$ kN (Horizontal) at R.L. 481.50 m
    $975 \times 0.05 = \pm 49$ kN
(c) On pier cap
    $1171 \times 0.1 = 117.1$ kN (Horizontal) at R.L. 474.80 m
    $1171 \times 0.5 = \pm 58.6$ kN (Vertical)
(d) On pier
    $3831 \times 0.1 = 383.1$ kN (Horizontal). It will act at

$$\text{R.L.} \left(474.40 - \frac{2 \times 3 + 1.0}{3 + 1.0} \times \frac{11.9}{3} = 467.5 \text{m} \right)$$

    $3831 \times 0.05 = \pm 191.55$ (Vertical)

(e) On well cap

$$2355 \times 0.1 = 235.5 \text{ kN (Horizontal) at R.L. 461.90 m}$$
$$2355 \times 0.05 = \pm 118 \text{ kN (Vertical)}$$

(f) On well steining above scour line

$$10815 \times \frac{5}{10.8} \times 0.1 = 501 \text{ kN (Horizontal) at R.L. 458.80 m}$$

$$10815 \times \frac{5}{10.8} \times 0.05 = 251 \text{ kN (Vertical)}$$

(g) On sand filling above scour line

$$5057 \times \frac{5}{10.8} \times 0.1 = 234 \text{ kN (Horizontal) at R.L. 458.80 m}$$

$$5057 \times \frac{5}{10.8} \times 0.05 = 117 \text{ kN (Vertical)}$$

(j) Hydrodynamic pressures
Along the traffic (Fig. 9.46)
On pier:

Radius of enveloping cylinder, a $\quad = \dfrac{8.1}{2} = 4.05$ m

Height of submerged portion of pier, $H = 11.0$ m

From Table 9.10, for $\dfrac{H}{a}$ $\quad = \dfrac{11}{4.05} = 2.72$, $C = 0.644$

Therefore, $\quad F_{Hy} = C \cdot \alpha_h \cdot W = 0.644 \times 0.1 \times \pi \times 4.05^2 \times 11 \times 10$
$$= 365 \text{ kN}$$

It will act at R.L. $\quad = 462.50 + 5.5 = 468.00$ m
On well:

Radius of enveloping cylinder, $a = 5.0$ m
Height of submerged portion of well, $H = 6.2$ m

From Table 9.10 , for $\dfrac{H}{a} = \dfrac{6.2}{5.0} = 1.24$, $C = 0.434$

$$F_{Hy} = 0.434 \times 0.1 \times \pi \times 5.0^2 \times 6.2 \times 10$$
$$= 211 \text{ kN.}$$

It will act at R.L. $= 456.30 + 3.1 = 459.40$ m
Across traffic (Fig. 9.46)
On pier :

$$a = 2.1 \text{ m, } H = 11.0 \text{ m, } \frac{H}{a} = \frac{11}{2.1} = 5.23, C = 0.75 \text{ (Table 9.10)}$$
$$F_{Hy} = 0.75 \times 0.1 \times \pi \times (2.1)^2 \times 11 \times 10 = 114 \text{ kN.}$$

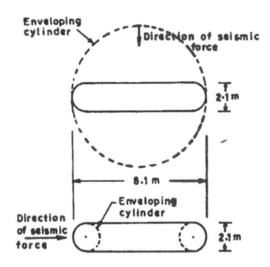

Fig. 9.46. Radii of Enveloping Cylinders for Computing Hydrodynamic Forces.

It will act at R.L. of 468.00 m.
On well:
$$F_{Hy} = 211 \text{ kN at R.L. } 459.40 \text{ m.}$$

(iv) Structural design

The design of the various components of the substructure requires the determination of combination of loads as given below:

(a) *N-Case*

$D_L + L_L + I_L + B + W_c + B_R + W_L$ (for no water condition, water at HFL and no live load condition separately).

(b) *(N + T)-Case*

All forces as in (a) + temperature effects

(c) *(N + T + S)-Case*

$D_L + L_L + I_L + B + W_C + B_R + E_L$ (considering earthquake load in longitudinal and transverse directions separately).

(v) Design of pier cap

Force Combinations

(a) *N-Case*

No Water condition:
$$V = 4500 + 975 + 66 + 219 + 31 = 5791 \text{ kN}$$
$$M_{xx} = 159.4 \ (475.45 - 475.20) = 40 \text{ kNm}$$
$$M_{yy} = [(975 + 219) \times 0.95 + 66 \times 4.75] + 163.50(478.25 - 475.20) +$$
$$169.2(481.5 - 475.20)$$
$$= 1448 + 499 + 1066 = 3013 \text{ kNm}$$

Water at HFL:
$$V = 4500 + 975 + 66 + 219 + 31 = 5791 \text{ kN}$$
$$M_{xx} = 40 \text{ kNm}$$
$$M_{yy} = 1448 + 95(478.25 - 475.20) + 1066$$
$$= 1448 + 290 + 1066 = 2804 \text{ kNm}.$$

No live load:
$$V = 4500 \text{ kN}$$
$$M_{xx} = 0$$
$$M_{yy} = 168(478.45 - 475.20) = 513 \text{ kNm}$$

(b) $N + T$ Case

No water condition:
$$V = 5791 \text{ kN} ,$$
$$M_{xx} = 217.7(475.45 - 475.20) = 55 \text{ kNm}$$
$$M_{yy} = 3013 \text{ kNm}$$

Water at HFL:
$$V = 5791 \text{ kN}$$
$$M_{xx} = 55 \text{ kNm}$$
$$M_{yy} = 2804 \text{ kNm}$$

No live load:
$$V = 4500 \text{ kN}$$
$$M_{xx} = 138(475.45 - 474.40) = 145 \text{ kNm}$$
$$M_{yy} = 513 \text{ kNm}.$$

(c) $N + T + S$ Case

In this case, wind forces will not be considered.
Seismic force in the direction of traffic:
Seismic force will not be considered on live load. Horizontal seismic force on dead load of superstructure
$$= 450 \text{ kN at R.L. } 478.25$$

Moment at the bearing level $= 450(478.25 - 475.45) = 1260 \text{ kNm}$

$$\text{Increase in reaction on pier} = \frac{1260}{30} = 42 \text{ kN}$$

$$\text{Reaction on roller end} = 4589 - 42 + 225 + 58.6 = 4831 \text{ kN}$$
$$\mu R = 0.03 \times 4831 = 145 \text{ kN}$$

$$\text{Longitudinal force at bearing level} = \frac{159.4 + 450}{2} + 145 = 450 \text{ kN.}$$

Moment M_{xx} at the top of pier $= 450(475.45 - 475.20) = 113 \text{ kNm}$

$$V = 4500 + 975 + 66 + 219 + 31 + 42 + 225 = 6058 \text{ kN}$$
$$M_{xx} = 113 \text{ kNm}$$
$$M_{yy} = 1448 \text{ kNm}$$

Seismic force across the traffic:

$$V = 6058 - 42 = 6016 \text{ kN}$$
$$M_{xx} = 55 \text{ kNm}$$
$$M_{yy} = 1448 + 450(478.25 - 475.20) + 117.1 \times 0.4 + 97.5(481.5 - 475.20)$$
$$= 3482 \text{ kNm.}$$

It may be noted that hydrodynamic pressures and water current forces will not come in the design of pier cap as the base of pier cap is located above the HFL.

Summary

	V (kN)	M_{xx} (kNm)	M_{yy} (kNm)
(i) N-Case			
No water condition	5791	40	3013
Water at HFL	5791	40	2804
No live load	4500	0	513
(ii) N + T-Case			
No water condition	5791	55	3013
Water at HFL	5791	55	2804
No live load	4500	145	513
(iii) N + T + S-Case			
Seismic along traffic	6058	113	1448
Seismic across traffic	6016	55	3482

As the permissible stresses in $(N + T)$ Case and $(N + T + S)$ case are increased by 15% and 25% respectively, N-Case (no water condition) seems to be most critical one for checking the section of pier cap. Hence, the further design of pier cap is done for this case only.

$$V = 5791 \text{ kN}, \quad M_{xx} = 40 \text{ kNm}, \quad M_y = 2804 \text{ kNm}$$

Assume that there are two bearings on the pier cap, then using Eqs. 9.51 to 9.56, we get

$$R_1 = \frac{V}{2} + \frac{M_{yy}}{L_b} = \frac{5791}{2} + \frac{2804}{3.5} = 3697 \text{ kN}$$
$$R_2 = 2094 \text{ kN}$$
$$A = (0.5 + 0.8)(0.5 + 0.8) = 1.69 \text{ m}^2$$

$$\frac{A}{A_1} = \frac{1.69}{0.5 \times 0.5} = 6.75, \quad \sqrt{\frac{A}{A_1}} = 2.6, \text{ adopt } 2.0.$$

Let the pier cap be constructed in M-20 grade

$$f_{ca} = f_{cc}\sqrt{\frac{A}{A_1}} = 0.45 f_{ck}\sqrt{\frac{A}{A_1}}$$
$$= 0.45 \times 20 \times 2.0 = 18 \text{ N/mm}^2$$

$$q_c = \frac{3697(1.5)}{0.5 \times 0.5} = 22182 \text{ kN/m}^2$$

$$= 22.2 \text{ N/mm}^2$$

Using the bearings of dimensions 0.6 m × 0.6 m.

$$q_c = \frac{3697(1.5)}{0.6 \times 0.6} = 15404 \text{ kN/m}^2$$

$$= 15.4 \text{ N/mm}^2 \ [< 18 \text{ N/mm}^2, \text{ safe}]$$

Dimensions of pier top $= B_p \times L_p = 1.0 \text{ m} \times 6.0 \text{ m}$ (with noses)

Weight of pier cap, $W_{pc} = 1171$ kN

$$e = \frac{M_{yy}}{V + W_{pc}} = \frac{2804}{5791 + 1171} = 0.403 \text{ m}$$

$$P_{1,2} = \frac{V + W_{pc}}{B_p \times L_p}\left(1 + \frac{6e}{L_p}\right)$$

$$= \frac{5791 + 1171}{1.0 \times 5.0}\left(1 \pm \frac{6 \times 0.403}{5.0}\right) \text{ [neglecting noses]}$$

$$= 2066 \text{ kN/m}^2, 719 \text{ kN/m}^2.$$

Refer Fig. 9.47. Critical section for bending moment will be a-a as beyond this the depth of pier will contribute to the thickness of pier cap if the pier is also constructed in R.C.C. If masonry pier is used, maximum bending moment will occur at the section of zero shear force. Let this section be at a distance of x from left hand side.

$$719 (x - 0.7) + \frac{1}{2}\left[\frac{1347}{5.0}(5 - x - 0.7) + 1347\right](x - 0.7)$$

$$= 3697 + 183 \ x$$

or, $\quad\quad x^2 - 14.38x + 38 = 0$

or $\quad\quad\quad x = 3.49 \text{ m}$

$$M_x = 3697 (3.49 - 1.45) + \frac{183 \times 3.49^2}{2} - \frac{2066 + 1314}{2} \times 2.79 \times \frac{2 \times 2066 + 1314}{2066 + 1314} \times \frac{2.79}{3}$$

$$= 7542 + 1115 - 7065 = 1592 \text{ kNm}.$$

Thickness of pier cap $= \sqrt{\dfrac{159\dot{2}(1.5) \times 10^6}{0.138 \times 20 \times 1400}} = 766 \text{ mm}.$

Provided overall thickness is 800 mm, hence safe. Reinforcement required for balanced section, $p = 0.96\%$.

Provide 0.96% reinforcement at top of the pier in the length direction and 0.375% in the width direction. Also provide a mesh of 0.75% reinforcement at the bottom. In addition to this, mesh of 6 mm dia. bars at 75 mm centres should also be provided directly under the bearings.

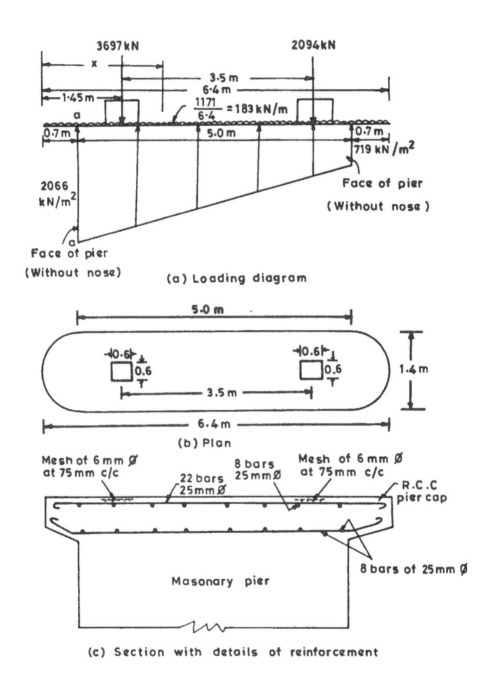

(a) Loading diagram

(b) Plan

(c) Section with details of reinforcement

Fig.9.47. Illustrating the Design of Pier Cap.

$$\text{Reinforcement at top} = \frac{0.96}{100} \times 800 \times 1400 = 10752 \text{ mm}^2$$

Using 25 mm dia. bars,

$$\text{Number of bars} = \frac{10752}{\dfrac{\pi}{4}(25)^2} = 22 \text{ bars.}$$

Therefore, provide 22 bars of 25 mm dia. at top in length direction and eight bars in width direction. At bottom, a mesh of 16 bars of 25 mm dia. may be provided. The details of reinforcement are shown in Fig. 9.47.

Design of pier

Combination of forces at the bottom of the pier (R.L. 462.50, Fig. 9.42) is given in Table 9.15.

Table 9.15. Combination of Forces at R.L. 462.50.

Load	Vertical	Along Traffic			Across Traffic		
	Force kN	Force kN	L.A. (m)	Moment kNm	Force kN	L.A. m	Moment kNm
1. Superstr. $D_L + L_L + B_L$							
(a) N-Case	5572	59.4	11.95	1905	975	0.95	926
					66	4.75	314
(b) N + T-Case	5572	217.7	11.95	2602	1041	—	1240
(c) N + T + S-Case	5839	450.0	11.95	5378	1041	—	1240
or	5797	217.7	11.95	2602	450.0	15.75	7088
					97.5	19.0	1852
					1041	—	1240
2. Wind	—	—	—	—	95.0	15.75	1497
3. Water curr.	—	18.7	6.34	119	21.2	6.34	135
4. Weight of substr.	1171	—	—	—	—	—	—
	3831	—	—	—	—	—	—
5. Buoyancy	– 248	—	—	—	—	—	—
6. Seismic	58.6	117.1	12.3	1440	—	—	—
	191.55	383.1	5.0	1915	—	—	—
or	—	365.0	5.50	2008	—	—	—
	58.6	—	—	—	117.1	12.3	1440
	191.55	—	—	—	383.1	5.0	1915
					114	5.5	627
Total							
N-Case (1a+2+3+4+5)	10326	178.1	—	2024	1157.2	—	2872
N + T-Case (1b+2+3+4+5)	10326	236.4	—	2721	1157.2	—	2872
N + T + S-Case (1c+3+4+5+6) or	10843	123.4	—	10860	21.2	—	1375
	10801	236.4	—	2721	1183	—	14297

The seismic condition gives forces and moments that are large and govern the design. More critical case is when seismic force is considered across the axis of bridge. Therefore,

$$V = 10801 \text{ kN}, \qquad M_{xx} = 2721 \text{ kNm}, \quad M_{yy} = 14297 \text{ kNm}$$

Refer Fig. 9.41 (c)

Area of pier cross-section

$$= 3.0 \times 7.0 + \frac{\pi}{4} \times (3.0)^2 = 28.065 \text{ m}^2$$

$$I_{xx} = \frac{7 \times 3^3}{12} + \frac{\pi}{64}(3)^4 = 19.72 \text{ m}^4$$

$$I_{yy} = \frac{3 \times 7^3}{12} + \frac{\pi}{64}(3)^4 + \frac{\pi}{4}(3)^2 \left[3.5 + \frac{4}{3\pi} \times \frac{3}{2} \right]^2$$
$$= 85.75 + 3.974 + 121 = 210.63 \text{ m}^4$$

$$f_{1,2} = \frac{V}{A} \pm \frac{M_{xx}}{I_{xx}} \cdot y \pm \frac{M_{yy}}{I_{yy}} \cdot x$$

$$= \frac{10801}{28.065} \pm \frac{2721}{19.72} \times 1.5 \pm \frac{14297}{210.63} \times 3.5$$

$$= 385 \pm 207 \pm 238$$
$$= 830 \text{ kN/m}^2 \text{ and } -60 \text{ kN/m}^2.$$

Allowable stress in compression $= 1.25 \times 1100 = 1375 \text{ kN/m}^2$

Allowable stress in tension $= 1.25 \times \dfrac{1100}{5} = 275 \text{ kN/m}^2.$

Therefore, stresses are within limits. It may be noted that the principles of limit design are not applied in this case as the pier is a solid masonry pier.

Design of well cap

Since the well is circular the horizontal forces and moments can be combined. Therefore,

$$V = 10801$$

$$H = \sqrt{236.4^2 + 1183^2} = 1206 \text{ kN}$$

$$M = \sqrt{2721^2 + 14297^2} = 14553 \text{ kNm}$$

$$e = \frac{14553}{10801} = 1.347 \text{ m}.$$

Refer Fig. 9.48a

Area of cross-section of well steining

$$= \pi(5^2 - 3.5^2) = 40 \text{ m}^2.$$

Using (Eq. 9.64)

$$f_{1,2} = \frac{V}{A}\left(1 \pm \frac{6e}{B}\right)$$

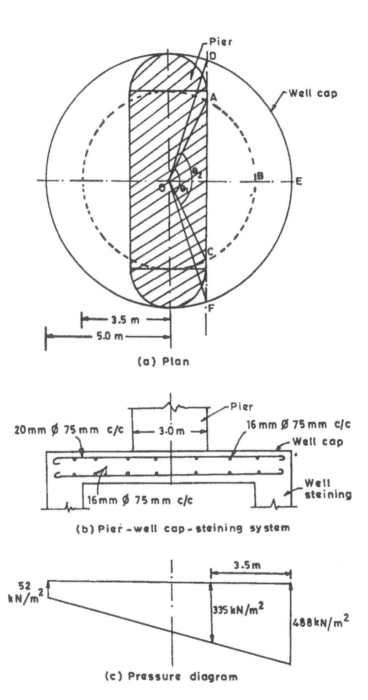

Fig. 9.48. Illustrating the Design of Well Cap.

$$= \frac{10801}{40}\left(1 \pm \frac{6 \times 1.347}{10}\right)$$

$$= 488 \text{ kNm}^2 \text{ and } 52 \text{ kNm}^2.$$

Area DEFCAD may be assumed to be acted by an average upward pressure of intensity

$$\frac{488+335}{2} = 412 \text{ kNm}^2 \text{ (Fig. 9.48c)}$$

Critical section for determining B.M. will be DF (Fig. 9.48a)

$$\angle \text{ DOF } = \theta_1 = 2 \cos^{-1} \frac{1.5}{5} = 145° = 2.53 \text{ radians}$$

$$\angle \text{ AOC } = \theta_2 = 2 \cos^{-1} \frac{1.5}{3.5} = 129° = 2.25 \text{ radians}$$

Therefore using Eqs. (9.65) and (9.66)

Area of segment DEF $\quad = \dfrac{1}{2} \times 5^2 \times (2.53 - \sin 145) = 24.45 \text{ m}^2$

Area of segment ABC $\quad = \dfrac{1}{2} \times 3.5^2 \times (2.25 - \sin 129) = 9.02 \text{ m}^2$

Distance of c.g. of segment DEF from O $= \dfrac{4 \times 5.0 \times \sin^3 \dfrac{145}{2}}{3(2.53 - \sin 145)} = 2.95 \text{ m}$

Distance of c.g. of segment ABC from O $= \dfrac{4 \times 3.5 \times \sin^3 \dfrac{129}{2}}{3(2.25 - \sin 129)} = 2.33 \text{ m}$

Weight of well cap = 2355 − 142 = 2213 kN
B.M. about section DF (Eq. 9.48)

$$= 412[24.45(2.95 - 1.5) - 9.02(2.33 - 1.5)] - \frac{2213}{\frac{\pi}{4}(10)^2} \times 24.45(2.95 - 1.5)$$

$$= 11522 - 999 = 10523 \text{ kNm}$$

Width of section DF = 9.52 m

Thickness of well cap $= \sqrt{\dfrac{10523(1.5) \times 10^6}{0.138 \times 20 \times 9537}} = 774 \text{ mm.}$

Provided thickness is 1200 mm. As the method of designing the well cap is approximate, let the total thickness of well cap remain as 1200 m.
Effective thickness of well cap = 1150 mm

$$1.5 \times 11440 \times 10^6 = 0.87 \times 415 A_{st} \left(1150 - \frac{415 A_{st}}{20 \times 9520 \times 1150}\right)$$

or
$$A_{st} = 4000 \text{ mm}^2.$$

Using 20 mm bars

$$\text{Spacing} = \frac{1000 \times \frac{\pi}{4} \times (20)^2}{4000} = 78.5 \text{ mm, say 75 mm c/c.}$$

Provide 16 mm dia. bars at spacing 75 mm c/c as nominal reinforcement. The details of reinforcement is shown in Fig. 9.48b.

Design of well steining

(i) As per IRC: 78-1973 minimum thickness of well steining can be obtained using Eq. (9.78) i.e.,

$$t = K d \sqrt{D_e}$$

$$= 0.033 \times 10.0 \sqrt{13.5} = 1.21 \text{ m } (< 1.5 \text{ m, o.k.})$$

(ii) Combination of forces at the scour level (R.L. 456.30) is given in Table 9.16.

The seismic condition gives forces and moments that are large and govern the design. More critical case is when seismic force is considered across the traffic. Therefore

$$V = 17800 \text{ kN}$$

$$H = \sqrt{2480.80^2 + 2147^2} = 2161 \text{ kN}$$

$$M = \sqrt{4204^2 + 24918^2} = 25270 \text{ kNm.}$$

Assuming $\quad \phi = 35° \text{ and } \delta = 17.5°$

$$K_A = \left[\frac{\cos \phi}{\sqrt{\cos \delta} + \sqrt{\sin(\phi + \delta)\sin \phi}}\right]^2$$

$$= \left[\frac{\cos 35}{\sqrt{\cos 17.5} + \sqrt{\sin(35 + 17.5)\sin 35}}\right]^2 = 0.246$$

$$K_p = \left[\frac{\cos 35}{\sqrt{\cos 17.5} - \sqrt{\sin(35 + 17.5)\sin 35}}\right]^2 = 7.35.$$

Adopting $\quad \gamma_b = 10 \text{ t/m}^3.$

Depth of zero shear force below scour level, x is given by:

$$x = \left[\frac{2FH}{\gamma_b(K_p - K_A)B}\right]^{\frac{1}{2}}$$

Table 9.16. Combination of Forces at Scour Level (R.L. 456.30).

Load	Vertical	Along Traffic			Across Traffic		
	Force kN	Force kN	L.A. (m)	Moment kNm	Force kN	L.A. m	Moment kNm
1. Superstr. $D_L + L_L + B_L$							
(a) N-Case	5572	159.4	18.15	2893	975	0.95	926
					66	4.75	314
(b) N + T-Case	5572	217.7	18.15	3951	1041	—	1240
(c) N + T + S-Case	5839	450.0	18.15	8168	1041	—	1240
or	5797	217.7	18.15	3951	450.0	21.95	9878
					97.5	25.2	2457
					1041	—	1240
2. Wind	—	—	—	—	95.0	21.95	2085
3. Water curr.	—	18.7	12.56	235	21.2	12.56	266
		4.44	4.10	18	14.7	4.10	60
4. Weight of substr.	1171	—	—	—	—	—	—
	3831	—	—	—	—	—	—
	2355	—	—	—	—	—	—
	5007	—	—	—	—	—	—
5. Buoyancy	−248	—	—	—	—	—	—
	−142	—	—	—	—	—	—
	−590	—	—	—	—	—	—
6. Seismic	58.6	117.1	18.5	2166			
	191.55	383.1	11.2	4291			
	118.00	235.5	5.6	1319			
	251.00	501.0	2.5	1253			
	619	1238	—	9029	—	—	—
		365	11.7	4271			
		211	3.1	654			
or	619	—	—	—	1238	—	9029
					114	11.7	1334
					211	3.1	654
Total N-Case (1a+2+3+4+5)	16956	182.54	—	3146	130.9	—	3651
N + T-Case (1b+2+3+4+5)	16956	240.84	—	4204	130.9	—	3651
N + T + S-Case (1c+3+4+5+6)	17842	2286	—	22375	35.9	—	1566
	17800	240.84	—	4204	2147	—	24918

$$x = \left[\frac{2 \times 2 \times 2161}{10(7.35 - 0.246) \times 10} \right]^{\frac{1}{2}}$$

$$= 2.46 \text{ m.}$$

If M_0 is the applied moment at the scour level

$$M_{max} = M_0 + \frac{2}{3} \cdot H \cdot x$$

$$= 25270 + \frac{2}{3} \times 2161 \times 2.46$$

$$= 28814 \text{ kNm}.$$

Additional weight of steining $= \frac{\pi}{4} (10^2 - 7^2) \times 2.46 \times 25 = 2462$ kN.

Buoyancy $= \frac{\pi}{4} \times 10^2 \times 2.46 \times 10 \times 0.15 = 291$ kN.

Net additional weight $= 2462 - 291 = 2171$ kN

Vertical weight up to the section of maximum B.M. $= 17800 + 2171$

$$= 19971 \text{ kN}$$

Eccentricity at the top of well due to permissible tilt $= \frac{10.8}{80} = 0.135$ m.

Eccentricity at depth of 2.46 m below scour level $= \frac{0.135}{10.8} \times 3.34$

$$= 0.0417 \text{ m}.$$

Moment due to tilt $= 19971 \times 0.0417 = 834$ kNm

Shift $= 0.150$ m

Moment due to shift $= 19971 \times 0.150 \text{ m} = 2995$ kNm.

Therefore   Total moment $= 28814 + 834 + 2995$

$$= 32643 \text{ kN/m}.$$

Area cross-section of well steining

$$= \frac{\pi}{4} (10^2 - 7^2) = 40 \text{ m}^2$$

$$I_{xx} = I_{yy} = \frac{\pi}{64}(10^4 - 7^4) = 373.0 \text{ m}^4$$

$$f_{1,2} = 1.5 \left(\frac{19971}{40} \pm \frac{32643}{373.0} \times 5.0 \right)$$

$$= 1.5[499 \pm 437] = 1404 \text{ kN/m}^2 \text{ and } 93 \text{ kN/m}^2$$

$$f_{cc} = 0.45 f_{ck} = 0.45 \times 20 = 9 \text{ N/mm}^2 = 9000 \text{ kN/m}^2$$

$$f_{1,2} \ll f_{cc}$$

Hence provide minimum reinforcement.

Assuming that the open dredging method has been used in sinking the well, then maximum pressure intensity acting on the well from outside (Fig. 9.23)

$$= 10 \times 12 \times 0.246 = 29.52 \text{ kN/m}^2$$

Maximum hoop stress (Eq. 9.71)

$$= \frac{29.52(5^2 + 3.5^2)}{(5^2 - 3.5^2)} = 86 \text{ kN/m}^2 = 0.086 \text{ N/mm}^2$$

$$\ll 20 \text{ N/mm}^2 \text{ [safe].}$$

Minimum area of vertical reinforcement

$$= \frac{0.2}{100} \times 40 \times 10^6 = 80000 \text{ mm}^2.$$

Using 25 mm dia. bars,

$$\text{Number of bars} = \frac{80000}{\frac{\pi}{4}(25)^2} = 163.$$

Use 110 bars of 25 mm dia. on the outer face and 60 bars on the inner face.

Transverse reinforcement $= \dfrac{0.04}{100} \times 40 \times 10^6 \text{ mm}^2/\text{m} = 16000 \text{ mm}^2/\text{m}.$

Provide 14 mm diameter bars at every 500 mm height on each face.

Design of well curb
(a) Stresses during sinking

Weight of steining per m run

$$\frac{\pi}{4}(10^2 - 7^2) \times 1.0 \times 25 = 1001 \text{ kN/m.}$$

Assuming angle of bevel face with vertical as 30° and $\mu = 0.4$
Total hoop tension (Eq. 9.85)

$$=0.75 \times 1001 \left(\frac{\sin 30 - 0.4\cos 30}{0.4\sin 30 + \cos 30} \right) \times \frac{10 + 7}{2}$$

$$= 919 \text{ kN.}$$

(b) Stresses in curb when resting on bottom plug

Net vertical weight

$$= \ 1.05(4500 + 975) + 66 + 31 + 42 + (1171 + 3831 + 2355) \times 1.05$$

$$+ \frac{\pi}{4}(10^2 - 7^2) \times 5.0 \times 25 \times 1.05 + \frac{\pi}{4}(10^2 - 7^2) \times 5.8 \times 25$$

$$+ 3098 + 5054 \times 1.05 \times \frac{10.8 - 5.8}{10.8 - 3.5} + 5054 \times \frac{5.8 - 3.5}{10.8 - 3.5}$$

$$- 10 \times 165.3 - 10 \times \frac{\pi}{4} \times 10^2 \times 12$$

$$= 5749 + 139 + 7725 + 1050 + 5800 + 3098 + 3635 + 1592 - 1650 - 9420 = 17718 \text{ kN}$$

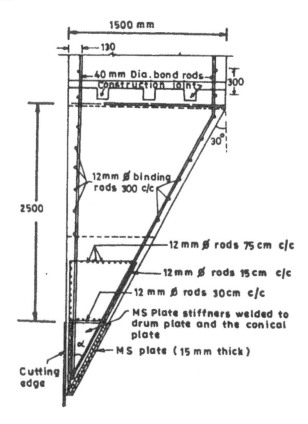

1500 mm

130

40 mm Dia.bond rods
Construction joint

300

30°

12mm ∅ binding
rods 300 c/c

2500

12 mm ∅ rods 75 cm c/c

12 mm ∅ rods 15 cm c/c

12 mm ∅ rods 30 cm c/c

MS Plate stiffners welded to
drum plate and the conical
plate

MS plate (15 mm thick)

α

Cutting
edge

Fig. 9.49. Details of Reinforcement in Well Curb.

$$q = \frac{17718}{\frac{\pi}{4} \times 10^2} = 226 \text{ kN/m}^2.$$

Using Eq. (9.86), assuming height of imaginary arch = 4.0 m.

Hoop tension $= \dfrac{qd^3}{16r} = \dfrac{226 \times 8.5^3}{16 \times 4.0} = 2169$ kN.

Assuming the height of well curb, as 2.5 m, using Eq. (9.87)

Hoop compression $\qquad = \gamma_b K_A (2D_f - b) \cdot \dfrac{bd}{4}$

$$= 10 \times 0.264 \times (2 \times 5.8 - 2.5) \times \frac{2.5 \times 8.5}{4}$$

$$= 128 \text{ kN.}$$

Net hoop tension $\qquad = 2169 - 128 = 2041$ kN

Design hoop tension $\qquad = 2041 \times 1.5 = 3062$ kN

Hoop reinforcement $= \dfrac{3062 \times 1000}{415} = 7379 \text{ mm}^2$

Minimum reinforcement $= 0.72 \text{ kN/m}^2$ excluding bond rods.
The arrangement of the reinforcement is shown in Fig. 9.49.

Design of bottom plug

The thickness of bottom plug may be obtained using Eq. (9.91).

$$t^2 = 1.18 \, r^2 \frac{q}{f_c}$$

$$= 1.18 \times 5.0^2 \times \frac{226 \times (1.5)}{1000 \times 10 \times 0.45}$$

$$t = 1.47 \text{ m.}$$

Provided thickness is 3.5 m, hence safe.

Stability analysis of well

Elastic theory

(i) W = Total downward load acting on the base of well
          = 17718 kN (considering full buoyancy)
          H = Total lateral load applied at scour level.
          = 2161 kN
          M = Total external moment applied at the base of well
          = 25270 + 2161 × 5.8 + 17718 × 0.15
          = 40461 kNm

          Diameter of well, $B = 10.0$ m
          Length of well, $L = 0.9\,B = 0.9 \times 10.0 = 9$ m
          $\mu = \tan 35° = 0.7$; $\mu' = \tan 17.5° = 0.315$
          $\gamma_b = 10 \text{ kN/m}^3$; $K_P = 7.35$, $K_A = 0.246$, $D_f = 5.8$ m.
Assume $m = 1.0$

(ii) $I_B = \dfrac{\pi}{64} \cdot B^4 = \dfrac{\pi}{64} \times 10^4 = 490.62 \text{ m}^4$

          $I_V = \dfrac{LD_f^3}{12} = \dfrac{9 \times 5.8^3}{12} = 146.34 \text{ m}^4$

          $\alpha = \dfrac{B}{\pi D_f} = \dfrac{10}{\pi \times 5.8} = 0.549$

          $I = I_B + mI_v(1 + 2\mu'\alpha)$
          $= 490.62 + 1.0 \times 146.34(1 + 2 \times 0.315 \times 0.549)$
          $= 687.60$ m.

(iii) $$r = \frac{I}{mI_v} \cdot \frac{D_f}{2} = \frac{687.60}{1.0 \times 146.34} \times \frac{5.8}{2} = 13.63 \text{ m}$$

$$\frac{M}{r}(1 + \mu\mu') - \mu W = \frac{40461}{13.63}(1 + 0.7 \times 0.315) - 0.7 \times 17718$$
$$= -8850 \text{kN}$$

$$\frac{M}{r}(1 - \mu\mu') + \mu W = \frac{40461}{13.63}(1 - 0.7 \times 0.315) + 0.7 \times 17718$$
$$= 14680 \text{kN}.$$

Therefore,

$$(H = 2161 \text{ kN}) > \frac{M}{r}(1 + \mu\mu') - \mu W.$$

and

$$(H = 2161 \text{ kN}) < \frac{M}{r}(1 - \mu\mu') + \mu W.$$

Therefore, point of rotation of well lies at the base.

(iv) $$\frac{mM}{I} = \frac{1.0 \times 40461}{687.60} = 58.84 \text{ kN/m}^3$$

$$\gamma(K_P - K_A) = 10.0(7.35 - 0.246) = 71.04$$

Hence $\dfrac{mM}{I} < \gamma(K_P - K_A)$, safe

(v) $$\sigma_1 = \frac{W - \dfrac{\mu' M}{r}}{A} + \frac{MB}{2I}$$

$$= \frac{17718 - \dfrac{0.315 \times 39718}{13.63}}{78.5} + \frac{40461 \times 10}{2 \times 687.60}$$

$$= 214 + 294 = 508 \text{ kN/m}^2$$

$$\sigma_1 < (q_a = 1.25 \times 450 = 562.5 \text{ kN/m}^2), \text{ safe}$$

$$\sigma_2 = 214 - 294 = -80 \text{ kN/m}^2.$$

As σ_2 is less than $\dfrac{q_a}{5}$, therefore well cross-section may be considered safe.

Ultimate resistance approach

(i) Assuming following ultimate load combination:

$$1.1D_L + B + 1.25(L_L + W_C + E_P + E_t)$$
$$W_u = 1.1 \times 4500 + 1.25(975 + 66 + 31) + 1.1(1171 + 3831 +$$

$$2355 + 10815 + 3098 + 5057) - 10(165.3 + \frac{\pi}{4} \times 10^2 \times 12) +$$

$$1.25 \times 0.05(4500 + 975 + 1171 + 3831 + 2355 + 5000 + 3461)$$

$$= 4950 + 1340 + 28957 - 11073 + 1331$$

$$= 25505 \text{ kN}$$

$$H_u = 1.25 \times 2161 = 2701 \text{ kN}$$

$$M_u = 1.25[25270 + 2161 \times 0.8 \times 5.8] + 25505 \times 0.15 +$$

$$\frac{0.135}{10.8} \times 0.2 \times 5.8 \times 25505$$

$$= 52141 \text{ kNm.}$$

(ii)
$$\frac{W_u}{A} = \frac{25505}{78.5} = 325 \text{ kN/m}^2$$

$$\frac{q_u}{2} = \frac{562.5 \times 3}{2} = 844 \text{ kN/m}^2$$

$$\frac{W_u}{A} < \frac{q_u}{2} \text{ (hence, safe).}$$

(iii)
$$M_b = Q W_u B \tan \phi; \text{ assuming } \phi \text{ of base soil as } 40°.$$

$$\frac{D_f}{B} = \frac{5.8}{10} = 0.58$$

From Table 9.12, for $\dfrac{D_f}{B} = 0.58$, $Q = 0.253$

$$M_b = 0.253 \times 25505 \times 10 \times \tan 40$$
$$= 54145 \text{ kN/m}$$

(iv)
$$M_s = 0.10 \, \gamma_b (K_P - K_A) \, D_f^3 L$$
$$= 0.10 \times 10 \times (7.35 - 0.246) \times 5.83^3 \times 9$$
$$= 12475 \text{ kN/m}$$

$$M_f = 0.11 \, \gamma (K_P - K_A) D_f^2 B^2 \sin \delta$$
$$= 0.11 \times 10 \times (7.35 - 0.246) \times 5.8^2 \times 10^2 \sin 17.5$$
$$= 7905 \text{ kN/m}$$

(v)
$$0.7(M_b + M_s + M_f) = 0.7(54145 + 12475 + 7905)$$
$$= 52168 \text{ kN/m.}$$

Therefore $0.7(M_b + M_s + M_f) > M_u$ [hence, safe].

Example 9.2

At one end of the bridge given in Example 9.1, the abutment carries the suspended span of

15.0 m. Assuming that the abutment carries fixed bearings of 500 mm total height and height of the concrete girder is 1.8 m, design the abutment.

Solution
(i) Trial dimensions: The trial dimensions are selected as shown in Fig. 9.50. It is assumed that at the ends, good bearing stratum is available at RL. 467.00 m. Open foundation is provided with embedment depth of 2.8 m. Therefore, the RL of base level of footing will be 467.20 m.

The stability of the abutment is checked here under low water level condition with earthquake force. When water is up to HFL, the water pressure in front of wall stabilizes the wall. If desired, calculations may similarly be made for considering water up to HFL, and without earthquake force.

The length of abutment is kept as 5.5 m.

(ii) Loads

(a) Dead load $= \dfrac{4500}{2} \times \dfrac{1}{2} = 1125$ kN

It will act at a distance of 0.7 m from point A (Fig. 9.50).

(b) Influence line diagram for reaction on the abutment will be as shown in Fig. 9.51.

$$\text{Live load reaction} = 114 \left(1 + \frac{13.8}{15}\right) + 68 \left(\frac{9.5}{15} + \frac{6.5}{15} + \frac{3.5}{15} + \frac{0.5}{15}\right) = 310 \text{ kN}.$$

$$\text{Reaction due to footpath loading} = \frac{1}{2} \times 1.0 \times 15 \times 200 = 1500 \text{ kg}$$

$$= 15 \text{ kN}$$

These reactions will act at a distance of 0.7 m from point A.

(c) Impact load

$$\text{Impact factor} = \frac{4.5}{6 + 15} = 0.214$$

$$\text{Impact load} = 0.214 \times 310 = 66 \text{ kN}.$$

(d) Braking force

$$B_R = 0.2 \, (27 \times 2 + 114 \times 2 + 68 \times 4) = 110.8 \text{ kN}.$$

It will act at a height of 1.2 m above roadway i.e. at R.L. of 481.20 m.

Increase in reaction due to this horizontal force on fixed bearing = decrease in reaction on roller bearing

$$= \frac{110.8 \times (481.20 - 477.95)}{15} = 24 \text{ kN}.$$

(e) Dead weight of abutment and earth fill
    Firstly considering up to section AB (Fig. 9.50)

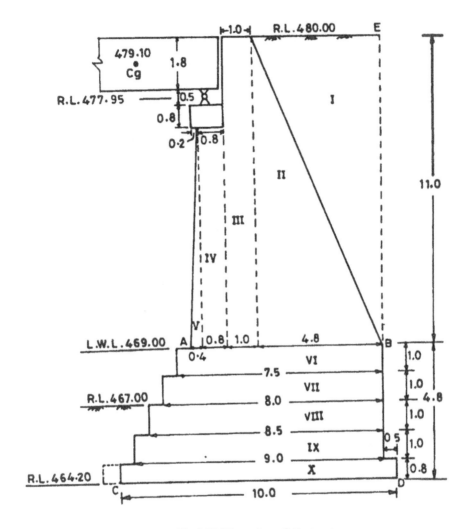

Fig. 9.50. Dimensions of Abutment.

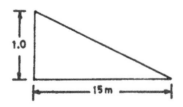

Fig. 9.51. Influence Line Diagram for the Reaction at the Abutment.

	Item	Weight (kN)	L.A. with Respect to Point A (m)	Moment about Point A (kN/m)
1	Weight of earth fill I	$\frac{1}{2} \times 4.8 \times 11 \times 18$ $= 475$	$7.0 - \frac{2}{3} \times 4.8$ $= 3.8$	1810
2	Weight of part II	$\frac{1}{2} \times 4.8 \times 11 \times 23$ $= 607$	$7.0 - \frac{4.8}{3}$ $= 5.4$	3280
3	Weight of part III	$1.0 \times 11 \times 23 = 253$	1.7	430
4	Weight of part IV	$0.8 \times (11 - 3.1) \times 23$ $= 145.4$	0.8	120
5	Weight of part V	$\frac{1}{2} \times (11 - 3.1) \times 0.4 \times 23$ $= 36$	0.267	10
6	Weight of bridgeseat	$0.8 \times 1.0 \times 25 = 20$	0.7	14
	Total	1536.4	—	5664

(f) Seismic force on superstructure

The critical case will be when seismic force acts in the direction of traffic.

Seismic force will not be considered on the live load.

Horizontal seismic load on the dead load

$$= 1125 \times 0.1 = 112.50 \text{ kN}.$$

It will act at R.L. 479.10.

Moment at the bearing level = 11250(479.10 − 477.95) = 129.375 m.

Increase in reaction on the fixed bearing

$$= \text{decrease in reaction on the roller bearing}$$

$$= \frac{112.5 \times 1.15}{15} = 8.6 \text{ kN}.$$

Vertical seismic force = 56.25 kN.

It will act at a distance of 0.7 m from point A.

(g) Temperature effects

Reaction on roller end = 1125 + (554 −310) + (30 − 15) + 0.214 × (554 − 310) − 24 − 8.6 + 56.25

= 1469 kN.

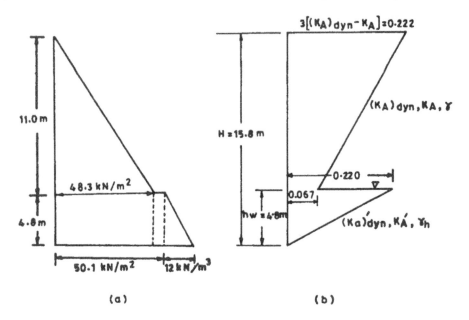

Fig. 9.52. Earth Pressure Diagram (*a*) Static Condition, (*b*) Dynamic Increment.

Frictional force at roller end = $0.03 \times 1469 = 44$ kN.

Therefore, longitudinal horizontal force on the fixed bearing

$$= \frac{110.8 + 112.50}{2} + 44 = 156 \text{ kN}.$$

It will act at R.L. 477.95 m.

(h) Seismic force on the abutment and earth fill

These are shown below in tabular form.

Item	Vertical Seismic Force			Horizontal Seismic Force		
	Force kN	L.A. (m)	Moment (kN/m)	Force kN	L.A. (m)	Moment (kN/m)
1. Part I	24	3.8	91.2	48	7.33	352
2. Part II	30	5.4	16.2'	60	3.66	220
3. Part III	13	1.7	22.1	26	5.50	143
4. Part IV	7.3	0.8	5.8	14.6	3.95	58
5. Part V	2.0	0.267	0.54	4.0	5.27	21
6. Bridge seat	1.0	0.7	0.7	2.0	8.3	17
Total	77.3	–	282.3	154.6	–	811

Earth pressure

When the live load is on the bridge or the abutment is having approach slab, surcharge due to live load on the backfill will not be considered.

The design is illustrated for this condition. The earth pressure is obtained along the face BE.

$$\phi = 35°, \delta = 23.3° \text{ (assumed on safer side)}$$

$$K_A = \left[\frac{\cos\phi}{\sqrt{\cos\delta} + \sqrt{\sin(\phi+\delta)\sin\phi}} \right]^2$$

$$= \left[\frac{\cos 35}{\sqrt{\cos 23.3} + \sqrt{\sin(35+23.3)\sin 35}} \right]^2 = 0.244$$

$$(K_A)_{dyn} = \frac{1 \pm \alpha_v}{\cos\psi} \left[\frac{\cos(\phi-\psi)}{\sqrt{\cos(\delta+\psi)} + \sqrt{\sin(\phi+\delta)\sin(\phi-\psi)}} \right]^2$$

where

$$\psi = \tan^{-1} \left[\frac{\alpha_v}{1 \pm \alpha_v} \right] = \tan^{-1} \left[\frac{0.1}{1 \pm 0.05} \right] = 5.44°$$

$$(K_A)_{dyn} = \frac{(1+0.05)}{\cos 5.44} \times \left[\frac{\cos(35-5.44)}{\sqrt{\cos(23.3+5.44)} + \sqrt{\sin(35+23.3)\sin(35-5.44)}} \right]^2$$

$$= 0.318$$

Static earth pressure $= \dfrac{1}{2} \gamma K_A H^2$

$$= \frac{1}{2} \times 18 \times 0.244 \times 11^2 = 266 \text{ kN.}$$

It will act at height of $\dfrac{11}{3}$ m above point A.

Dynamic increment $= \dfrac{1}{2} \times 18 (0.318 - 0.244) \times 11^2 = 80.6$ kN.

It will act at a height of $\dfrac{11}{2}$ m above point A.

Total vertical load $= 1125 + 310 + 15 + 24 + 1536.4 + 56.25 + 8.6 +$
$77.3 + 266 \sin 23.3 + 80.6 \sin 23.3$
$= 3347$ kN.

Total stablising moment $= (1125 + 310 + 15 + 24 + 8.6 + 56.25) \times 0.7 + 5664 +$
$282.3 + (266 + 80.6) \sin 23.3 \times 7.0$
$= 7967$ kN/m.

$$\text{Total overturning moment} = 811 + \left(266 \times \frac{11}{3} + 80.6 \times \frac{11}{2}\right) \times \cos 23.3 +$$

$$156(477.95 - 469.00)$$

$$= 3233 \text{ kN/m.}$$

Checking of stresses:

$$V = 3347 \text{ kN}$$

$$\text{Net moment} = 7967 - 3233 = 4734 \text{ kN/m}$$

$$e = 3.5 - \frac{4734}{3347} = 2.085$$

$$f_{1,2} = \frac{V}{A}\left(1 \pm \frac{6e}{B}\right)$$

$$= \frac{3347}{5.5 \times 7}\left(1 \pm \frac{6 \times 2.085}{7}\right)$$

$$= 242 \text{ kN/m}^2 \text{ and } -68 \text{ kN/m}^2.$$

The allowable stresses are 1650 kN/m^2 (compression) and $\dfrac{1650}{5} =$

330 kN/m^2 (tension), hence o.k.

Checking the stability at the base section of footing.

Net additional vertical weights will be as given below:

Weight of Item (kN)	L.A. with Respect to Point C (m)	Moment about Point C (kN/m)
a. Weight of part VI = 13 × 7.5 × 1.0 = 97.5	5.75	561
b. Weight of part VII = 13 × 8.0 × 1.0 = 104	5.5	572
c. Weight of part VIII = 13 × 8.5 × 1.0 = 110.5	5.25	580
d. Weight of part IX = 13 × 9.0 × 1.0 = 117	5.0	585
e. Weight of part X = 13 × 10.0 × 0.8 = 104	5.0	520
Total 533	—	2818

Total additional vetical force due to earthquake

$$= 0.05 \times 533 \times \frac{23}{13} = 47 \text{ kN.}$$

Total additional moment of vertical seismic force

$$= 0.05 \times 2818 \times \frac{23}{13} = 250 \text{ kNm.}$$

Total additional moment of seismic horizontal force

$$= 0.1(97.5 \times 4.3 + 104 \times 33 + 110.5 \times 2.3 +$$
$$117 \times 1.3 + 104 \times 0.4)$$
$$= 121 \text{ kNm.}$$

For submerged backfill

$$\delta = \frac{35}{2} = 17.5°, \gamma_b = 10 \text{ kN/m}^3$$

$$\psi = \tan^{-1}\left(\frac{\alpha_h}{1 \pm \alpha_v} \cdot \frac{\gamma_s}{\gamma_s - 1}\right) = \tan^{-1}\left(\frac{0.1}{1 + 0.05}\right)\left(\frac{20}{20 - 10}\right)$$

$$= 10.8°$$

$$K'_A = \left(\frac{\cos\phi}{\sqrt{\cos\delta} + \sqrt{\sin(\phi + \delta)\sin\phi}}\right)^2$$

$$= \left(\frac{\cos 35}{\sqrt{\cos 17.5} + \sqrt{\sin(35 + 17.5)\sin 35}}\right)^2$$

$$= 0.253$$

$$(K_A)'_{dyn} = \frac{1 \pm \alpha_v}{\cos\psi}\left(\frac{\cos(\phi - \psi)}{\sqrt{\cos 17.5} + \sqrt{\sin(35 - 17.5)\sin 35}}\right)^2$$

$$= \frac{1 + 0.05}{\cos 10.8}\left(\frac{\cos(35 - 10.8)}{\sqrt{\cos(17.5 + 10.8)} + \sqrt{\sin(35 + 17.5)\sin(35 - 17.5)}}\right)^2$$

$$= 0.495.$$

Earth pressure distribution in static case and pressure distribution of dynamic increment are shown in Figs. 9.52a and b respectively.

Static earth pressure $= 266 + 18 \times 0.253 \times 11 \times 4.8 + \frac{1}{2} \times 10 \times 0.253 \times 4.8^2$

$$= 266 + 241 + 29 = 536.$$

It will act at a height above point C

$$= \frac{266\left(4.8 + \frac{11}{3}\right) + 241\left(\frac{4.8}{2}\right) + 29\left(\frac{4.8}{3}\right)}{536}$$

$$= 5.37 \text{ m.}$$

Lateral dynamic increment

$$= \frac{1}{2} [(K_A)_{dyn} - K_a] \frac{\gamma}{H} [H - h_w]^2 [H + 2h_w] + \frac{1}{2} \times$$

$$[(K_A)'_{dyn} - K'_A] \frac{h_w^2}{H} [3\gamma_b(H - h_w) + \gamma_b h_w]$$

$$= \frac{1}{2}(0.318 - 0.244) \times \frac{18}{15.8} [15.8 - 4.8]^2 [15.8 + 2 \times 4.8] +$$

$$\frac{1}{2}(0.495 - 0.253) \times \frac{4.8^2}{15.8} [3 \times 10 (15.8 - 4.8) + 10 \times 4.8]$$

$$= 114 + 67 = 181 \text{ kN.}$$

It will act at a height of $\frac{15.8}{2} = 7.9$ m from the point C.

Checking of stresses

$V = 1125 + 310 + 15 + 66 + 24 + 1536.4 + 56.25 + 8.6 + 77.3 + (536 + 181) \sin 17.5 + 47 + 533 = 4014$ kN.

Total stabilising moment about point C

$= (1125 + 310 + 15 + 8.6 + 56.25) \times 3.2 + [475 \times 6.3 + 607 \times 7.9 + 253 \times 4.2 + 145.4 \times 3.3 + 36 \times 2.767 + 20 \times 3.2] + 2818 \times [24 \times 6.3 + 30 \times 7.9 + 13 \times 4.2 + 7.3 \times 3.3 + 2.0 \times 2.767 + 1.0 \times 3.2] + 250 + (536 + 181) \sin 17.5 \times 9.5$
$= 4848 + 9494 + 476 + 250 + 2048 = 17116$ kNm.

Total overturning moment about point C

$= [48 \times 12.13 + 60 \times 8.46 + 26 \times 10.3 + 14.6 \times 8.75 + 4.0 \times 10.07 + 2.0 \times 13.10] + 250 + 156[477.95 - 464.20] + 536 \times 5.37 \cos 17.5 + 181 \times 7.9 \cos 17.5$
$= 1552 + 250 + 2145 + 2745 + 1364 = 8056$ kNm.

Net moment $= 17116 - 8056 = 9060$ kNm

$$e = 5.0 - \frac{9060}{4014} = 2.74 \text{ m}$$

$$f_{1,2} = \frac{4014}{10 \times 5.5} \left[1 \pm \frac{6 \times 2.74}{10} \right]$$

$$= 192 \text{ kN/m}^2 \text{ and } -47 \text{ kN/m}^2.$$

Maximum base pressure is much less than the allowable soil pressure (i.e. 562.5 kN/m²). To bring the tensile stress (−47 kN/m²) to zero, the abutment base may be extended by 0.5 m as shown in Fig. 9.50.

REFERENCES

1. Arya, A.S. and S. Saran (1973), "Design of Abutments, Piers and Wells of Bhuntar Bridge," University of Roorkee, Roorkee.
2. Balwant Rao, B. and C. Muthuswami (1963), "Considerations in the Design and Sinking of Well Foundations for Bridge Piers," *Journal of the Indian Roads Congress*. New Delhi, Vol XXVII.
3. IRC: 6–2000— "Standard Specifications and Code of Practice for Road Bridges," Section II–Loads and Stresses.
4. IRC: 5–1998— "Standard Specifications and Code of Practice for Road Bridges," Section I–General Features of Design.
5. IRC: 45–1972— "Recommendations for Estimating the Resistance of Soil Below the Maximum Scour Level in the Design of Well Foundations of Bridges," Indian Roads Congress, New Delhi.
6. I.R.C.: 78–2000— "Standard Specification and Code of Practice for Road Bridges", Section VII – Foundations and Substructures."
7. IS: 3955–1967— Indian Standard Code of Practice of Design and Construction of Well Foundations, I.S.I., New Delhi.
8. IS: 1893–1984, 2002— Indian Standard Criteria for Earthquake Resistant Design of Structures, I.S.I, New Delhi.
9. Melville, B.W. (1997), " Pier and Abutment Scour: Integrated Approach", Journal of Hydraulic Engineering, A.S.C.E. Vol. 123, No. 2, pp.125–136.
10. Teng, Wyne, C. (1962), *Foundation Design*, Prentice-Hall, Englewood Cliffs.

PRACTICE PROBLEMS

1. What is a bridge ? Mention the factors which are responsible for putting the subject of bridge engineering on scientific footing. What are the factors to be considered while selecting the site for a proposed bridge?
2. How is maximum flood discharge of a river determined from the study of catchment area?
3. What are the substructures of a bridge ? Mention functions of each of them.
4. What is meant by 'economical span'? Derive the condition for an economical span, stating clearly the assumptions made in the derivation.
5. State how you would determine the depth of foundation for a pier of a bridge located near the mouth of a river.
6. What are the loads and forces to be considered in the design of a bridge pier?
7. Write a brief note on sinking of wells for a bridge across a shallow tidal river.
8. Sketch a typical cross-section of a masonry abutment and indicate the forces acting on the abutment.
9. Explain the "elastic theory" and ultimate resistance approaches for checking the lateral stability of a well.
10. Give the salient features of the design of
    (a) Pier cap
    (b) Pier
    (c) Well cap
    (d) Well steining
    (e) Well curb
    Describe your answer with neat sketches showing typical reinforcement details.
11. Describe the procedure of obtaining the following for the design of substructures of a bridge:
    (a) Live load reaction
    (b) Longitudinal forces
    (c) Water current forces
    (d) Wind forces
12. A reinforced concrete balanced cantilever bridge is to be supported on masonry piers open foundation. The data is as follows:

Formation level of the bridge	= 479.750 m
Max. discharge for linear waterway	= 5600 m³/sec
Max. discharge for foundation depth	= 7100 m³/sec
Mean velocity of flow during flood	= 4.5 m/sec
Low water level, L.W. L.	= 470.000 m
Design flood H.F.L.	= 473.660 m

Minimum clearance	$= 2.4$ m
Nature of bed strata at surface—boulder, shingle and sand silt factor f	$= 6.5$
Allowable pressure on soil at scour level	$= 450$ kN/m^2

This may be increased by 1 t/m^2 for every 1 m depth below scour depth.

Seismic force coefficient	$= 0.10$
Main spans adopted (c/c on piers)	$= 40$ m
Suspended span (c/c of bearings)	$= 20$ m
Cantilever span	$= 10$ m
Dead load of suspended span	$= 221$ kN
Dead load of each cantilever	$= 145$ kN
Dead load of main span	$= 511$ kn
Maximum L.L. reaction on fixed bearing	$= 150$ kN
Three bearings with base dimensions 50 cm × 85 cm	
Distance between centres of outer bearings	$= 4$ m
Level of bearing pins	$= 475.25$ m
Height of bearing below pin	$= 0.30$ m
Coefficient of friction in rollers	$= 0.03$
Piers are with semi-circular ends.	
Unit weight of concrete	$= 24$ kN/m^3
Unit weight of masonry	$= 22.40$ kN/m^3
Allowable compressive stress on masonry	$= 1100$ kN/m^2
Saturated unit weight of soil	$= 20$ kN/m^3

At one end of the bridge given the abutment carries the suspended span of 20 m span. The dead load of the span is 2200 kN. The maximum live load reaction is 750 kN. Assuming that the abutment carries fixed bearings of 0.30 m total height and height of the concrete girder is 2 m, design piers, well and abutments.

Marine Substructures

10.1 INTRODUCTION

In marine works, where the engineer pits his knowledge and skill against the forces of the sea, that most unrelenting of nature's forces, success brings satisfaction, for in excess of that attainable when derived from conquering forces, the magnitude and frequency of which can be assessed beforehand.

In earlier days the engineer had to rely almost entirely upon his own experience and judgment, amplified to some small extent of knowledge gained as a result of frequent personal contacts and discussions with a necessarily small circle of colleagues. It is to his credit that so few of his early works have in design, proved inadequate for their purpose.

Economic considerations might not have weighed quite so heavily then as now, and for this reason it behoves the engineer at all times to keep himself abreast of developments taking place throughout the world, in order that proved advances in design or construction might be reflected in his own works.

10.2 TYPES OF MARINE STRUCTURES

Following are the important types of marine structures:
1. Breakwaters
2. Wharves
3. Piers
4. Sea walls
5. Docks
6. Quay walls
7. Locks
8. Moorings

Figure 10.1 shows a plan of a harbour illustrating the relative positions of the above-mentioned structures.

10.3 BREAKWATERS

A breakwater is, as the name suggests, a structure meant to break large waves, preventing them from exerting their destructive influence upon the area sheltered for the reception of ships.

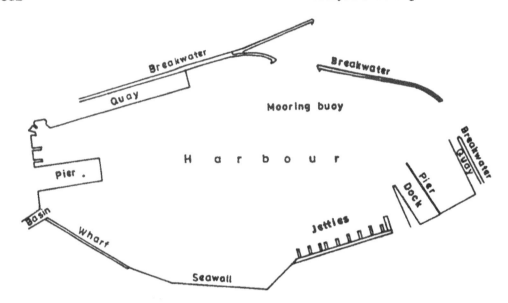

Fig. 10.1. Layout of a Harbour.

Breakwaters are constructed in the domain of sea and most of its portion remains submerged in sea. The breakwaters must have great strength and stability to withstand the enormous pressures developed due to the generation of large waves.

Following two methods are used in practice for destroying water waves or reducing their size.

Method I: By constructing a wall of sufficient height, the waves will be totally reflected, or in shallow water partially reflected and partly destroyed by the wave breaking. The reversal, in a comparatively short time, of the momentum of waves and their breaking causes great forces to be developed and the breakwater must, therefore, be of substantial construction.

Method II: The waves can be made to run up a long sloping beach, their energy then being absorbed gravitationally in surges and tumult of broken water, by movement of loose beach material, and by percolation of water through the material. In this way the dissipation of energy is spread over a longer period of time, and for the same size of waves the destructive forces developed are much less than in the first method.

Practically all breakwaters fall within the limits of the above two methods of reducing wave action.

Breakwaters are classified mainly into three types, namely, (i) mound or heap breakwater, (ii) vertical wall breakwater, and (iii) composite breakwater.

Mound or rubble stone type breakwater

Mound or rubble stone breakwater falls under the category of method II of reducing wave action. It consists of a heterogeneous mass of natural and, frequently, undressed stones. This is the simplest type of breakwater from the construction point of view as the stones are generally deposited at random and without any regard for bonding. The dumping of the stones is done till the heap or mound emerges out of water, the heap being consolidated and its side slopes regulated by the action of the waves. Usually the slope of breakwater gets flattened due to the drag action of the waves. Such flattening of slopes should be avoided.

According to most of the specifications, heavier stones should be used in the seaward slope. Usually, the weight of stone in this side should not be less than 30 kN. In the harbour side smaller stones may be permitted.

The disturbing action of waves is most keenly felt between high water level and low water level. It is in this region that the breakwaters should be protected from flattening. Protection may be either by (a) dumping heavy blocks of concrete on top and on front face. These blocks weigh 250 kN to 300 kN and are either deposited at random as shown in Fig. 10.2 or laid in courses as shown in Fig. 10.3, or (b) paving the upper part up to low water level by deep granite blocks as shown in Fig. 10.4, or (c) tetrapod method: the tetrapod is a four-pronged precast block (Fig. 10.5), the prongs radiating at 120° to each other from a fixed centre. Blocks of required weight depending on the height of waves are simply heaped on the exposed face of the breakwater at a slope of 1.5:1.

Vertical wall breakwater

A wall type breakwater falls under the category of method I of reducing wave action. It involves the construction of a masonry or concrete wall with vertical or nearly vertical faces. In Fig. 10.6, a concrete block breakwater is shown. It consisted of precast concrete sections of

Fig. 10.2. Mound Breakwater with At-Random Deposit of Stones.

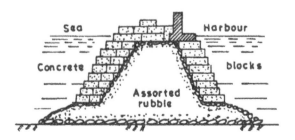

Fig. 10.3. Mound Breakwater with Concrete Blocks Laid in Courses.

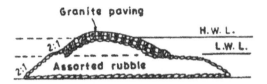

Fig. 10.4. Mound Breakwater with Top Protection.

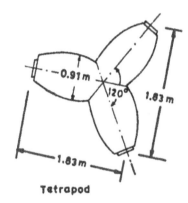

Tetrapod

Fig. 10.5. Tetrapod.

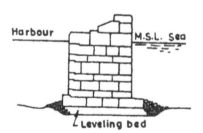

Fig. 10.6. Concrete Block Breakwater.

considerable weights which are arranged in such a manner that all joints stagger and key-way prevent any horizontal sliding of blocks. Usually the top of the bottom layer is made almost level with the bottom of the sea and extended to some distance on the sea-side.

A modified vertical face breakwater is shown in Fig. 10.7. The lower portion is a concrete caission type structure, built of prefabricated units and sunk into position on a prepared sea bed. After sinking, the interior is rapidly filled to water level with sand or gravel, and covered with a protective stone blanket. After initial settlements have occurred, openings between caissons are filled with concrete and a monolithic cap structure is cast on top.

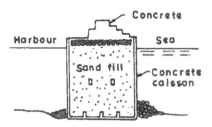

Fig. 10.7. Caisson Type Breakwater.

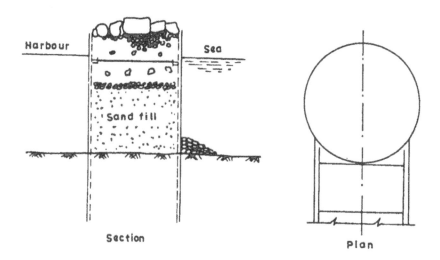

Section

Plan

Fig. 10.8. Straight Wall, Sheet Pile Type Breakwater.

Figures 10.8 and 10.9 illustrate a type of steel sheet pile breakwater adopted for moderate seasonal wave disturbances. The structure is vulnerable to storm damage before filling of the cells during construction, but this can be minimised by proper sequence of building operations.

Composite breakwaters

For deep water sites and at locations having large tidal variations, the quantity of stone required for a full height rubble mound is not economically feasible. Such a condition gives rise to a combination of rubble bases and various types of superstructures. Here the rubble mound provides the base which accommodates itself to the irregularities of the sea-bed, and may be deposited in deep water and allowed to stand for the purpose of obtaining a large part of total settlement before placing the superstructure (Fig. 10.10). Composite breakwater of this type may be divided into two classes, namely, those with superstructure founded at low water level, and those where the superstructure extends sufficiently far below low water to avoid the breaking of storm waves and disturbance to the rubble mound.

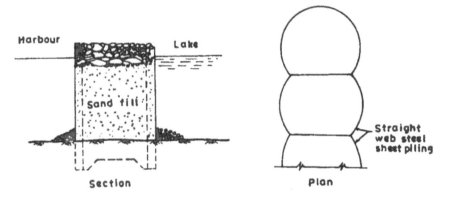

Fig. 10.9. Circular Sheet Pile Type Breakwater.

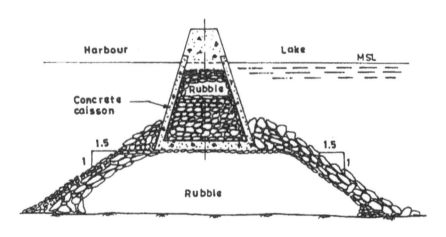

Fig. 10.10. Composite Breakwater.

Choice of type of breakwater

The vertical wall breakwaters should be constructed, when:

(a) Depth of water is sufficient to prevent the breaking of waves and in general the depth of water should be more than twice the wave height.

(b) Sea-bed is resistant to erosion.

(c) Foundations are not subjected to uneven settlement.

Mound or heap breakwaters are suitable in the following situations:

(a) When the breakwaters are to be constructed on a muddy bottom or a bottom disturbed by heavy wave action.

(b) When stone is available in plentiful and quarries lie conveniently adjacent to the site.

(c) When the alignment of the breakwaters is such that its axis forms with the direction of advance of the storm waves normal to the crests and troughs at an angle less than 45°.

In addition to above points, the following points may be looked for selection of breakwaters:

(a) Wall type breakwaters need special care and costly methods of construction. But after construction they do not require any further attention towards the variation of wave pressures. The mound breakwaters, on the other hand, are peculiarly susceptible to the constant fretting and atritional action of the waves. The preservation of mound breakwater necessitates, therefore, a constant replenishment of material.

(b) With the same or equal disposition of the axis of the breakwater, wall type breakwaters provide more navigable sheltered area.

(c) Wall-type breakwaters provide berthing facilities also.

As mentioned earlier, composite breakwaters are more suitable at deep water sites having large tidal variations.

10.4 WHARVES

Wharves are structures located in or at the edge of water deep enough to permit vessels to tie up against them. They provide direct connection between the water and land carriers. Wharf is also defined as the landing place for the vessels and these structures have water only on one side.

The main feature of a wharf is the vertical drop in elevation at its water edge. Therefore, the main forces, which must be resisted by a wharf, are horizontal ones due to lateral earth and water pressures and to the pulls and thrusts of the docking vessels.

Wharf structures may be classified in the four different groups, namely, (i) massive retaining wall, (ii) anchored bulk heads, (iii) piled structures, and (iv) relieving platform type.

Massive retaining wall type wharf structure

It is the usual type of the retaining wall to be constructed inside a coffer dam or as a floating caisson (Fig. 10.11a). For achieving economy, the wall may be of varying cross-section or by building it of precast concrete blocks.

Anchored bulk head type wharf structure

Similar to the massive retaining wall type, this also confines the earth behind a vertical wall, but the enclosing element is sheet piling wall usually of steel, although other materials can also be used (Fig. 10.11b). Stability of this wall is achieved by driving it down deep below the sea bottom and supporting it by a wale girder tied by horizontal steel rods to some anchorage that is well behind the wall. This anchorage may be a row of vertical sheet piles, combinations of vertical and batter piles or concrete slab which, by engaging large earth masses, resist the horizontal movement of anchor rods.

Piled wharf structure

In this type of wharf structure, a different principle for retaining the earth is utilized instead of retaining the earth by a vertical structure. This type of structure as shown in Fig. 10.11c can be supported on piles, cylindrical caissons or even solid piers spaced at certain intervals. This structure is made to bridge the natural slope. The chief characteristics of these types of structures are the absence of any lateral earth pressure. These structures have to be designed for the pull exerted by ships drawing themselves towards the wharf by means of rope and moving posts, and thrust transmitted to the deck.

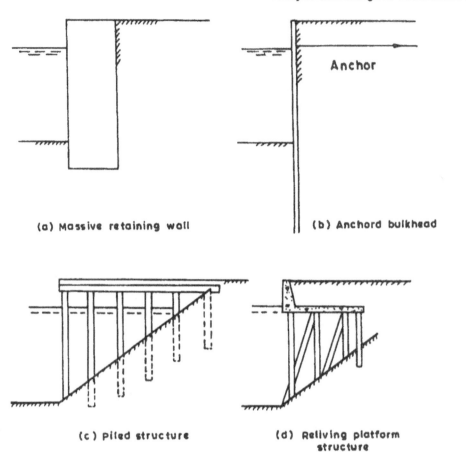

(a) Massive retaining wall

(b) Anchord bulkhead

(c) Piled structure

(d) Relieving platform
structure

Fig. 10.11. Types of Wharf.

The main advantage of piled wharf structures is that it is economical and quick in construction, as compared with filled structures which usually require considerable time to settle. Another advantage is that a pile structure presents a smaller obstruction to tidal prism. It does not affect materially the amount of water which flows in and out of a harbour with the changes of tide. A fourth advantage is the greater elasticity of pile structures, they absorb more readily the kinetic energy of moving vessels brought in contact with wharf.

Relieving platform type wharf structure

This consists of a combination of massive retaining wall and piled wharf structures (Fig. 10.11d). It is known as the relieving platform structure because, by allowing the natural slope to extend under it, the platform will materially relieve the total lateral earth pressure.

The structural analysis of a relieving platform may be made by breaking up it into two parts. Firstly, the sheet piling is analysed in the ordinary manner as a beam supported by the platform and at some depth below its intersection with the outside ground surface where equality exists between active and passive earth pressures. Finally, the pile group under the platform can be analyzed for the forces acting upon it, including the reaction from the sheet piling.

10.5 PIERS

A pier is a structure that projects from the shore-line out into the water. It has water on both the sides. Pier construction becomes economical in the case when the length of the berthing space is the deciding factor. Width of the pier depends on the length. The longer the pier, the wider it must be. It is due to the fact that a longer pier must handle more traffic at the shore end.

   Few typical structural arrangements of the piers are shown in Fig. 10.12. Pier shown in Fig. 10.12a consists of a fill placed between parallel walls of sheet piling tied together by anchor rods above low water level. Pier shown in Fig. 10.12b is simply a bridge built into the dredged area and consists of a deck supported on piles or cylindrical piers. In this case lateral earth pressures will not be acting on the piles, and this type of structure is suitable for large depths of water. Arrangement shown in Fig. 10.12c is applicable to a wide pier situated in deep water. Piles are used on both sides of pier for distances sufficient to bring the natural ground slope up to elevations where a small amount of sheet piling will retain the centre portion of the fill.

10.6 SEA WALLS

A sea wall is a shore line structure built for protecting and stabilizing the shore against erosion resulting from wave action. Figure 10.13 illustrates few different structural arrangements of sea walls. Figure 10.13a shows a gravity section type sea wall. These types of walls have been constructed up to 16 m height. Figure 10.13b shows a sheet pile type wall. Sheet pile protection serves two purposes, namely, (i) prevention of erosion, and (ii) support of upper cantilever wall with continuity of bending strength.

   The wall in Fig. 10.13c has curved surface. It is founded on piles as required by foundation conditions, and protected against erosion at the toe by sheet piles and riprap cover. This wall is suitable for location having wide beaches with relatively flat slope of sea shore. Figure 10.13d is an example of stepped face sea wall supported on piles and protected at the toe by sheet piling and riprap. This wall is suitable for moderate wave pressures.

10.7 DOCKS

Docks are enclosed areas for berthing ships to keep them afloat at a uniform level, to facilitate loading and unloading cargo. A dock may be either a wharf or pier. Broadly docks are classified into three types:

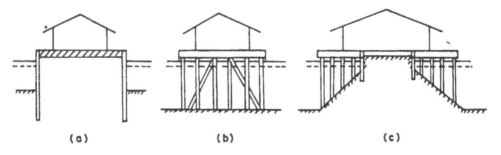

(a)   (b)   (c)

Fig. 10.12. Pier Structures.

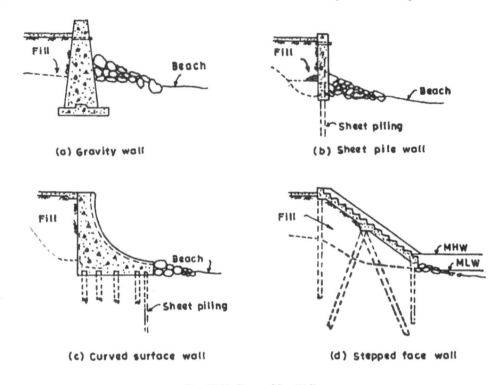

Fig. 10.13. Types of Sea Walls.

(a) *Dry docks:* As the name suggests, the arrangement in a dry dock is to take in a vessel, close the door, and pump out the water.

(b) *Floating docks:* It is a hollow structure, made of steel or R.C.C. To receive a vessel for repairs, the floating dock is sunk to required depth by allowing buoyancy chambers to be filled with water. After the vessel is berthed, the water is pumped out of buoyancy chambers, thus causing the dock to rise bodily, and in so doing lift the vessel above the water line.

(c) *Slipways:* In this, rails are laid on a firm ground at a uniform slope, from some distance under water to a point at which the longest vessel to be accommodated is completely out of range of tide.

Dock walls are designed as gravity retaining walls as shown in Fig. 10.14. The wall should be tested for the following two conditions:

1. Dock empty to withstand pressure of backfill.
2. Dock fill with backfill removed.

10.8 QUAY WALLS

Wharves along and parallel to the shore are generally called quays and their protection walls are called quay walls (Fig. 10.15). Various types of quay walls are: (i) masonry or concrete gravity walls, (ii) masonry or concrete walls on piles or wells, (iii) walls built on open pontoons, (iv) walls built on crib work, and (v) R.C.C. walls.

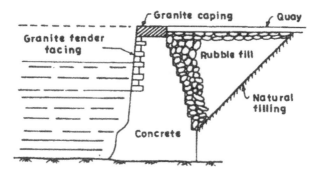

Fig. 10.14. Section of Dock Wall.

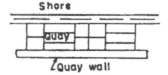

Fig. 10.15. Quay Wall.

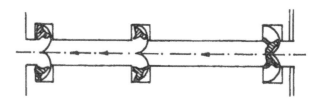

Fig. 10.16. A Typical Lock.

Quay walls are designed similar to retaining walls but on the water side they are subjected to varying water pressure, and on the land side, earth and continued water pressure, with proper allowance for surcharge.

10.9 LOCKS

A ship at berth expects to remain afloat and this, sometimes, becomes difficult at low water in a harbour with a big tidal range. Also, with big tidal range, there is difficulty in working cargo from the ships, and a more or less steady water level is advantageous.

For these reasons arrangements are, sometimes, made to close the entrance at docks by gates or caissons. Gates or caissons of dry docks are used to exclude water whereas the latter are for retaining water.

Figure 10.16 shows an horizontal section of a typical lock with three pairs of gates. For the lock operation as such only two pairs are needed. But in order to conserve water while handling smaller vessels, an additional intermediate pair of gates is sometime provided.

The dimensions of a lock are determined by the largest vessel utilizing the dock. The lock walls are like ordinary dock walls, except that they are supported by the floor against the lateral movement. The floor is an inverted arch, or if there is a flat floor it has curved launches tangential to the side walls, or otherwise it is so thick to admit of the existence of a vertical arc within itself.

10.10 MOORINGS

In addition to the facilities afforded by posts, rings, and bollards on the quay for securing vessels, in large basins floating and fixed moorings are provided. The former consist of buoys of various shapes anchored to the bottom of the basin, and the latter of piled stages suitably braced. A floating mooring has three principal components:

(i) Buoy, (ii) Cables, and (iii) Anchors.

Buoys are of diverse designs. The important types are cylinderical, drum, pear shaped and spherical (Fig. 10.17).

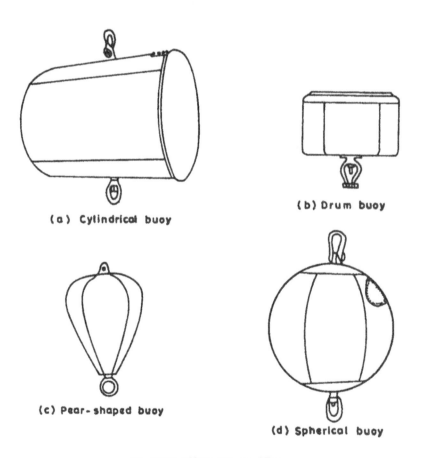

(a) Cylindrical buoy

(b) Drum buoy

(c) Pear-shaped buoy

(d) Spherical buoy

Fig. 10.17. Different Types of Buoy.

10.11 DESIGN LOADS

A waterfront structure should be designed to resist both vertical and horizontal forces. Following forces are considered in the design:

1. Dead load

The dead load (a vertical force) is caused by the weight of structure itself, including fill and any permanent construction which may be supported by it, such as transit sheds, warehouses, and cargo transfer equipment.

2. Live load

The live load (also a vertical force) is due to the transfer and storage of cargoes before they are loaded on sea and land carriers.

When the live loads act on the fill behind the structure, such as in a sheet pile wharf so that the loads are transmitted to the structure through increased earth pressure the retaining structure may be designed for uniformly distributed surcharge as given below in Table 10.1.

Table 10.1. Uniform Vertical Live Loading

Function of Berth	Uniform Vertical Live Loading kN/m^2
Passenger berth	5.0
Bulk unloading and loading berth	5.0 to 7.5
Container berth	15 to 25
Cargo berth	12.5 to 17.5
Heavy cargo berth	25 or more
Small boat berth	2.5
Fishing berth	5.0

If truck cranes are to be used in cargo handling, or if the backfill in a retaining structure is proposed to be placed with earth moving equipment of the crawler type, the uppermost portion of the waterfront structures, including the upper anchorage system should be designed according to the following loadings, whichever of the two is more unfavourable:

(a) Live load of 60 kN/m^2 from the back edge of the coping inboard for a 1.5 m width.
(b) Live load of 40 kN/m^2 from the back edge of the coping inboard for a 3.5 m width.

3. Lateral earth pressure

Lateral earth pressure due to retained fill or natural soil is an important item. Coulomb's theory is used for evaluating the pressure intensities. Following points may be considered in evaluating the earth pressures (IS : 4651 Pt. II–1969):

(a) In waterfront structures, wharves constructed as gravity retaining walls or sheet-pile retaining walls have sufficient outward movement to mobilize active pressure.
   In highly plastic clays, pressures approaching at rest condition may develop unless wall movement can continue with time.

(b) Gravity wharf walls founded on cohesionless subgrade may not yield forward sufficiently as to mobilize full passive resistance. A suitable factor of safety (usually 2 to 2.5) is used. Gravity wharf walls founded on clayey subgrade may eventually sufficiently yield as to fully mobilize the passive resistance.

(c) Usually the backfill in a gravity wharf wall settles more than the wall and the angle of wall friction for active earth pressure is positive. The cases of gravity wharf walls on clayey foundations require to be individually examined for settlement and if the wall settles more with reference to the backfill, negative angle of wall friction will result.
Angle of wall friction for sheet-pile walls should be judged with reference to the deposition of the backfill, and the relative movements of the wall and the backfill.

(d) When the wall moves up relative to the backfill, as it may happen in a sheet-pile wall which is strongly held back by anchor at the top and when, deflecting, moves out and upward in the embedded portion, negative angle of wall friction will result.

(e) For computing active earth pressure, the value of angle of wall friction is taken as $2\phi/3$. For computing passive earth pressure, $\delta = \phi/3$.

4. Hydrostatic pressure

Hydrostatic pressure due to water is not an important item, as it is usually counter-balanced by being present on both sides of the structure. For massive structures the hydrostatic uplift or the buoyancy is a very important factor to include.

5. Berthing load

When an approaching vessel strikes a berth a horizontal force acts on the berth. The magnitude of this force depends on the kinetic energy that can be absorbed by the fendering system. The reaction force for which the berth is to be designed can be obtained and deflection-reaction diagrams of the fendering system chosen. The kinetic energy, E, imparted to a fendering system, by a vessel moving with velocity V m/s is given by:

$$E = \frac{W_D V^2}{2g} \times C_m \times C_e \times C_s \qquad \qquad \text{... (10.1)}$$

where W_D = displacement tonnage (DT) of the vessel in tonnes,
V = velocity of vessel in m/s, normal to the berth (Table 10.2),
g = acceleration due to gravity in m/s^2,
C_m = mass coefficient, and is given by

$$C_m = 1 + \frac{2D}{B} \qquad \qquad \text{... (10.2)}$$

D = draught of vessel in m,
B = beam of vessel in m,
C_e = eccentricity coefficient, and is given by

Table 10.2. Normal Velocities of Vessels.

S.No.	Site Condition	Berthing Condition	Berthing Velocities Normal to Berth in m/s			
			Up to 5000 DT	Up to 10000 DT	Up to 100000 DT	More than 100000 DT*
(i)	Strong wind and swells	Difficult	0.75	0.55	0.40	0.20
(ii)	Strong wind and swells	Favourable	0.60	0.45	0.30	0.20
(iii)	Moderate wind and swells	Moderate	0.45	0.35	0.20	0.15
(iv)	Sheltered	Difficult	0.25	0.20	0.15	0.10
(v)	Sheltered	Favourable	0.20	0.15	0.15	0.10

*DT = Displacement Tonnage (IS : 4651 Pt. III–1974)

$$C_e = \frac{1 + \left(\dfrac{l}{r}\right)^2 \sin^2 \theta}{1 + \left(\dfrac{l}{r}\right)^2} \qquad \ldots (10.3)$$

l = distance from the centre of gravity of the vessel to the point of contact projected along the water line of the berth in m,

r = radius of gyration or rotational radius on the plane on the vessel from its centre of gravity in m (Fig. 10.18),

= 0.25 L

θ = approach angle usually equal to 10°, for smaller vessels, θ is 20°, and

C_s = softness coefficient = 0.9 to 0.95.

Sometimes, $\dfrac{l}{r}$ will more commonly be approximately equal to 1.25. The value of

coefficient C_e for $\dfrac{l}{r}$ = 1.0 varies from 0.50 to 0.56 for θ varying from 0° to 20°. For $\dfrac{l}{r}$ = 1.25,

it varies from 0.39 to 0.46 for θ varying from 0° to 20°.

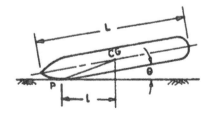

Fig. 10.18. Vessel Approaching Berth at an Angle.

Collision between a ship and a structure of the gravity-wall type will not, as a rule, result in any injury to the wharf, since the energy will be dissipated in the larger mass of concrete and fill. However, in the case of a structure supported on piles the energy of the moving vessel must be absorbed by the elasticity of the structure itself; stresses which have been set up must be accounted for.

6. Mooring loads

The mooring loads are the lateral loads caused by the mooring lines when they pull the ship into or along the dock or hold it against the forces of wind or current.

The maximum mooring loads are due to the wind forces on exposed area on the broad side of the ship in light condition:

$$F = C_w A_w P \qquad \qquad \text{... (10.4)}$$

where F = force due to wind in kN,
         C_w = shape factor = 1.3 to 1.6,
         A_w = windage area in m²,

$$= 1.175 \, L_P \, (D_M - D_L) \qquad \qquad \text{... (10.5)}$$

L_P = length between perpendicular in m,

D_M = mould depth in m,

D_L = average light draft in m, and

P = wind pressure in kN/m² to be taken in accordance with IS : 875–1964.

When the ships are berthed on both sides of a pier, the total wind force acting on the pier, should be increased by 50 per cent to allow for wind against the second ship.

7. Differential water table

In the case of waterfront structures with backfill, the pressure caused by difference in water level at the fillside and the waterside has to be taken into account in design. The magnitude of this hydrostatic pressure is influenced by the tidal range, free water fluctuations, the ground water influx, the permeability of the foundation soil and the structure as well as the efficiency of backfill drainage.

In the case of good and poor conditions of the backfill the differential water pressure may be calculated on the guidelines given in Fig. 10.19.

8. Earthquake forces

In areas susceptible to seismic disturbance, horizontal force equal to a fraction of the acceleration of gravity times the weight applied at its centre of gravity should be taken. The fraction will depend upon the likely seismic intensity of the area, and shall be taken in accordance with IS: 1893–1970. The weight to be used is the total dead load plus one-half of the live load.

9. Wind forces

Wind forces on structures should be taken in accordance with IS: 875–1964 as applicable.

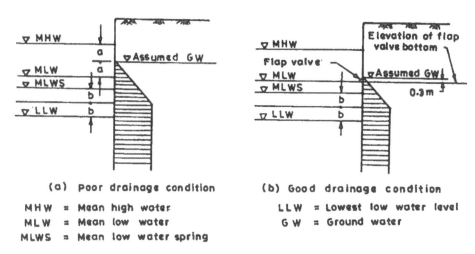

(a) Poor drainage condition

(b) Good drainage condition

MHW = Mean high water.
MLW = Mean low water
MLWS = Mean low water spring

LLW = Lowest low water level
GW = Ground water

Fig. 10.19. Guide for Calculating Differential Water Pressure.

10. Waves forces

Forces exerted by the sea waves are most important and are dealt in detail subsequently in Sec. 10.13.

10.12 COMBINED LOADS

The combination of loadings for design is dead load, vertical live loads, earth pressure plus either sea wave force, or berthing load, or line pull, or earthquake. The worst combination should be taken for design.

10.13 WAVE ACTION

Marine structures in general are exposed to the action of waves generated by high winds, sweeping over a considerable expanse of open water. Size of a wave depends on the four factors, namely (i) velocity of wind, (ii) duration of storm, (iii) depth of water, and (iv) the maximum distance over which wind can act which is usually known as "fetch" of the wave. Fetch of the wave can never exceed the distance from the marine structure to the windward shore, it usually can be assumed to be less than 550 km, even though the distance to the windward shore may be greater.

Wave disturbance is felt to a considerable depth; therefore, wave length and the depth of water has an effect on the character of the wave. Deep water waves are those which occur in water having a depth greater than one half the wave length ($d > L/2$), at which depth the bottom does not have any significant influence on the motion of water particles. Shallow water waves are those which occur in water having a depth less than one-half the wave length ($d < L/2$), and the influence of the bottom changes the form of orbital motion from circular to elliptical or near elliptical.

Waves are classified in the following manner:

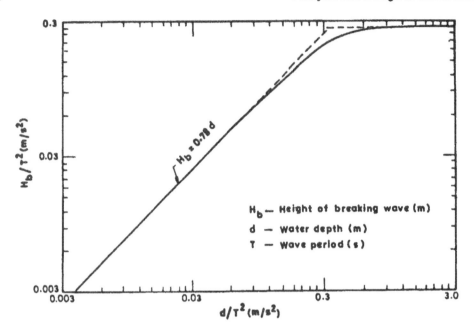

Fig. 10.20. Broken Waves.

(i) Broken waves: Waves break when the forward velocity of the crest particles exceeds the velocity of propagation of wave itself. In deep water this normally occurs when the wave height exceeds one-seventh of the wave length. When the wave reaches shallow water where the depth is equal to about 1.25 *h* it will usually break, although it may break in somewhat deeper water, depending upon the strength of the wind and condition of the bottom. Figure 10.20 gives the maximum height of wave as a function of the water depth.

(ii) Unbroken wave: An unbroken wave is a wave of oscillation, and even after breaking in deep water the wave will usually reform into an oscillatory wave of reduced height. Such waves travel without change of form at a constant speed, with a net displacement of water in the direction of wave travel.

Figure 10.21 shows the principal dimensions of an unbroken wave. The height *h* of the wave is measured from trough to crest. *d* is the depth of water from undisturbed level. d_0 is the depth of water from undulated sea level. *L* is the length of wave.

Various properties of the wave are discussed in the following heads:

1. Height of wave: For a given wind velocity, *V* in km/hour and a fetch *D* in km, the height *h* in metres can be computed for any wave from the following formulae:

According to Gaillard and Molitore (1935)

(a) For $D < 32$ km

$$h = 0.0322 \sqrt{VD} + 0.76 - 0.89 (D)^{\frac{1}{4}} \qquad\qquad \dots (10.6)$$

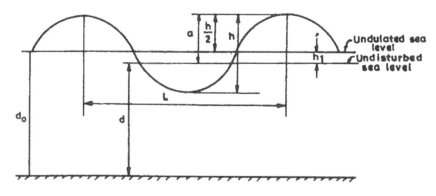

Fig. 10.21. Dimension of an Unbroken Wave.

(b) For values of $D > 32$ km

$$h = 0.0322 \sqrt{VD}. \qquad \text{... (10.7)}$$

Thomas Stevenson (1994) suggested the following formulae for the computation of the height of waves. These formulae are independent of the velocity of wind.

(a) For $D < 80$ km

$$h = 0.36 \sqrt{D} + 0.76 - 0.89(D)^{\frac{1}{4}}. \qquad \text{... (10.8)}$$

(b) For $D > 80$ km

$$h = 0.36 \sqrt{D}. \qquad \text{... (10.9)}$$

2. Length of wave: The highest wave has the minimum length height ratio (L/h). As the wind subsides the height of wave (h) diminishes and the ratio L/h increases.

Gaillard (1904) proposed the following formula for evaluating the value of L/h ratio in terms of wind velocity V. This formula is based on the basis of extensive model tests conducted by him

$$\frac{L}{h} = \frac{1350}{V}. \qquad \text{... (10.10)}$$

where V is the wind velocity in km/hour.

Cornick (1968) has given the values of the ratio L/h as listed in the Table 10.3.

Table 10.3. Values of L/h

Description	Wind Velocity (V) km/hour	L/h
Moderate wind	45.0	39
Strong wind	56.5	21
Storm	88.5	19

3. *Period of wave* : The time period of wave can be computed using Eq. 10.11.

$$T = \sqrt{\frac{2\pi L}{g}} \qquad \qquad \dots (10.11)$$

where T = time period of wave, s, and
          g = acceleration due to gravity, m/s^2.

4. *Height of wave above still-water level (a)* : Since the crest of the wave is above the still water level and the trough is below that level, it is sometimes erroneously assumed that the still-water level represents the mean of the two. However, many observations by various investigators show that the crest is about 2/3h above the still-water level in deep waters, and rising still higher in shallow water, a fact which must not be disregarded when examining the safety of the structure.

According to Gaillard (1904), the height of wave crest above still-water level (a) is given by the following formulae:

(a) For shallow water waves, with $d < 1.84h$

$$a = \frac{h}{2} + \frac{2h^2}{L} \qquad \qquad \dots (10.12)$$

(b) For deep water waves, with $d > 1.84h$

$$a = \frac{h}{2} + \frac{h^2}{L} \qquad \qquad \dots (10.13)$$

(c) For deep water ocean waves

$$a = \frac{h}{2} + 0.785\frac{h^2}{L}. \qquad \qquad \dots (10.14)$$

10.14 WAVE PRESSURE ON VERTICAL WALL

Vertical walls may be mainly classified in two groups, breakwaters and seawalls. Breakwaters are normally built in water deep enough to keep attacking waves from breaking. Seawalls, on the other hand, are generally built at the top of a beach and are subjected to the action of breaking waves. There is a considerable difference in wave pressure of the two types, breaking waves exerting a much greater pressure. The energy of the breaking wave is destroyed at the wall and, therefore, transmits a much greater energy to the wall than the nonbreaking wave, whose energy is mostly reflected by the wall.

If the wall is not built high enough, part of the wave will overtop the wall and cause disturbance on its leeside. Overtopping also may not be desirable for reasons of structural safety of the wall. Therefore, it will be assumed that the wall is high enough to obstruct totally the attacking waves.

Pressure due to nonbreaking wave
Following two approaches are popularly in use:

(a) Molitore-Gaillard approach

The maximum intensity of wave pressure is given by:

$$p_{max} = \frac{K.\gamma_w}{2g}(V + V_0)^2 \qquad \text{... (10.14a)}$$

where p_{max} = pressure in N/m^2,

γ_w = unit weight of water, 10000 N/m^3 for fresh water and 10200 N/m^3 for salt water,

V = velocity of propagation of wave in m/s,

$$V = 1.243 \sqrt{L \tanh\frac{2\pi d_0}{L}} \qquad \text{... (10.15)}$$

V_0 = maximum velocity of wave particle in its orbit in m/s,

$$V_0 = \sqrt{\frac{3.91h}{L \tanh\frac{2\pi d_0}{L}}} \qquad \text{... (10.16)}$$

K = empirical constant evaluated from Gaillard's observation as 1.20 to 1.71 for great lakes for winds of 48 to 112 km/hour. For ocean waves K is recommended as 1.8,

g = gravity acceleration, 9.81 m/s^2,

d = depth of water in m,

$$d_0 = d + a - \frac{h}{2} \qquad \text{... (10.17)}$$

a = height of crest of wave above undulated water level in m, and

h = height of wave in m.

This maximum wave pressure acts at a height h_1, above still-water level, the theoretical value for which is

$$h_1 = 0.785\frac{h^2}{L} \qquad \text{... (10.18)}$$

although better results are obtained by

$$h_1 = 0.12h. \qquad \text{... (10.19)}$$

When a wave is completely obstructed by a vertical surface of sufficient height, the wave crest is raised to a height equal to $h_m = 2a$ above the still-water level and at this height the pressure becomes zero. With the foregoing data, it is possible to construct the entire wave pressure diagram and thus evaluate the total force to be resisted by a proposed structure. In this method pressure diagram is taken as shown in Fig. 10.22.

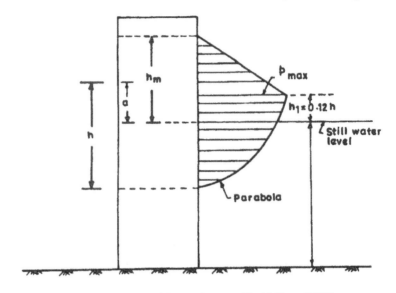

Fig. 10.22. Pressure Diagram Suggested by Molitore (1935).

(b) Sainflou approach

Sainflou (1928) used the concept of saint-venant and Flamant theory according to which a water particle at the surface of a deep-water wave oscillates about a point whose height above still-water level is given as

$$h_0 = \frac{\pi h^2}{4L} \coth \frac{2\pi d}{L} \qquad \qquad \dots (10.20)$$

Therefore, the crest height above still-water level, $a = h_0 + \dfrac{h}{2}$. When this wave is reflected from a vertical wall, a clapotis is created. The height of the centre of oscillation above still-water level is raised to

$$h_{oc} = \frac{\pi h^2}{L} \coth \frac{2\pi d}{L}. \qquad \qquad \dots (10.21)$$

The wave height of a clapotis is $2h$ (Fig. 10.23).

Sainflou gave a general formula for the pressure on a vertical wall, from which the net pressure diagram shown in Fig. 10.24 can be constructed. The pressure at the bottom is

$$P_2 = \frac{\gamma_w h}{\cosh\left(\dfrac{2\pi d}{L}\right)} \qquad \qquad \dots (10.22)$$

where γ_w = unit weight of water.

For the wave at crest position which presents the most critical condition, the wave pressure at still-water level, by simple proportion, is

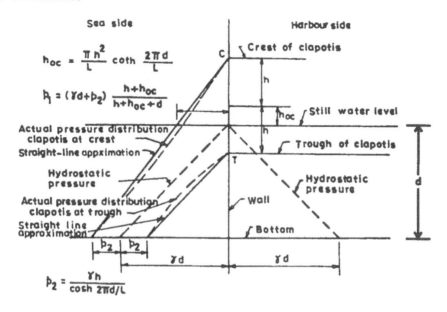

$$h_{oc} = \frac{\pi h^2}{L} \coth \frac{2\pi d}{L}$$

$$p_1 = (\gamma d + p_2) \frac{h + h_{oc}}{h + h_{oc} + d}$$

Crest of clapotis

C

h

Still water level

h_{oc}

h

Trough of clapotis

Actual pressure distribution clapotis at crest

Straight-line appximation

Hydrostatic pressure

Hydrostatic pressure

Actual pressure distribution clapotis at trough

T

Wall

Straight line approximation

Bottom

p_2 | p_2

γd

γd

$$p_2 = \frac{\gamma h}{\cosh 2\pi d/L}$$

Fig. 10.23. Clapotis.

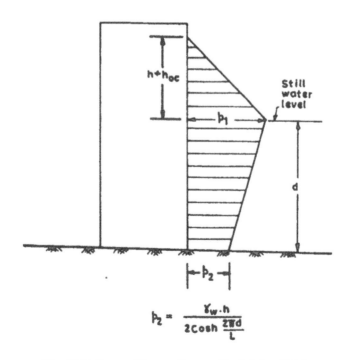

$h + h_{oc}$

Still water level

p_1

d

p_2

$$p_2 = \frac{\gamma_w \cdot h}{2 \cosh \frac{2\pi d}{L}}$$

Fig. 10.24. Pressure Diagram Suggested by Sainflou (1928).

$$p_1 = (\gamma_w d + p_2) \frac{h + h_{oc}}{h + h_{oc} + d}. \qquad\qquad \text{... (10.23)}$$

Sainflou method has been incorporated in IS code (IS: 4651 pt. III–1974).

Pressure due to breaking wave

Seawalls and inshore end of breakwaters and jetties are usually subjected to breaking waves. According to Minikin (1963) the total pressure on the wall from breaking waves is a combination of dynamic and hydrostatic pressures.

The dynamic pressure is concentrated at still-water level and is given by

$$p_m = 101 \frac{\gamma_w d h_b}{L d'} (d + d') \qquad\qquad \text{... (10.24)}$$

where p_m = dynamic pressure intensity in kN/m^2,
  h_b = height of wave just breaking on the structure in m,
  γ_w = unit weight of water in kN/m^3,
  d = depth of water below still-water level in m,
  d' = deep-water depth in m, and
  L = length of wave in m.

The hydrostatic pressure p_s on seaward side at still-water level and the pressure p_d at the depth d, are given by

$$p_s = \frac{\gamma_w h_b}{2} \qquad\qquad \text{... (10.25)}$$

$$p_d = \frac{\gamma_w (2d + h_s)}{2}. \qquad\qquad \text{... (10.26)}$$

The pressure diagram is shown in Fig. 10.25.

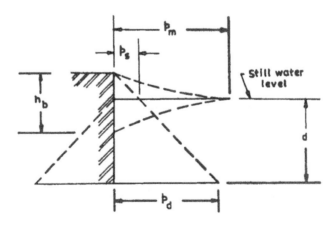

Fig. 10.25. Pressure Diagram Suggested by Minikin (1963).

Force and moment with water on landside

The resultant wave thrust R on structure per linear metre of structure is determined from the area of pressure diagram and is

$$R = \frac{p_m h_b}{3} + p_s \left(d + \frac{h_b}{4} \right). \qquad \qquad \dots (10.27)$$

The resultant overturning moment M about the base is given by

$$M = \frac{p_m h_b}{3} \cdot d + \frac{p_s d^2}{2} + \frac{p_s h_b}{4} \left(d + \frac{h_b}{6} \right). \qquad \qquad \dots (10.28)$$

With no water on landside

Thrust R per linear metre is

$$R = \frac{p_m h_b}{3} + \frac{p_d}{2} \left(d + \frac{h_b}{2} \right). \qquad \qquad \dots (10.29)$$

Moment M about the ground line is

$$M = \frac{p_m h_b}{3} \cdot d + \frac{p_d}{6} \left(d + \frac{h_b}{2} \right)^2. \qquad \qquad \dots (10.30)$$

10.15 SHIP IMPACT ON PILED WHARF STRUCTURE

Ship impact is an important force to be considered in the design of marine structures, e.g. wharves and piers. Expressions for the ship impact may be obtained by considering the impact of ship on a beam. Different expressions can be derived for different structures. In this section expressions for the ship impact are developed for a piled wharf structure from the first fundamental principles (Anderson, 1956).

Figure 10.26 shows a beam whose axis is fixed against rotation at both ends. If the top of the beam is permitted to move horizontally while the axis remains fixed in the vertical direction, then the deflection of the top due to the application of the load P is

$$\Delta = \frac{PL^3}{12EI} \qquad \qquad \dots (10.31)$$

where I is the moment of inertia of the cross-section of pile and E is the modulus of elasticity.

The strain energy which is assumed by the beam in producing this deflection is

$$= \frac{1}{2} P\Delta = \frac{P^2 L^3}{24EI}. \qquad \qquad \dots (10.32)$$

A moving ship contacting a piled wharf (Fig. 10.27) will cause horizontal deflections of several pile bents and will come to rest when its total kinetic energy has been converted into

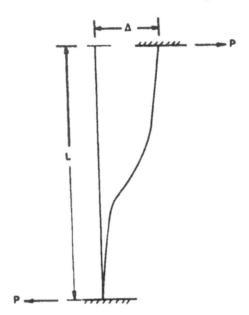

Fig. 10.26. Beam Fixed against Rotation at Both Ends.

strain energy of the deflected bents. The total strain energy of a bent can be written as the sum of energies stored in each pile. If the piles are assumed fixed at points below the ground line and fixed free at the pile cap, then

Total energy of the bent

$$= \frac{P_1^2 L_1^2}{24EI} + \frac{P_2^2 L_2^2}{24EI} \qquad \qquad ... (10.33)$$

where $L_1, L_2...$ are the lengths of the various piles in a bent. Piles are assumed to have same value of EI.

Energy imparted from the impact of ship per bent

$$= \frac{W_1}{2g} V^2 \frac{1}{N} \qquad \qquad ... (10.34)$$

$$W_1 = WC_m C_e C_s$$

where N = number of bents sharing the impact,

W = displacement tonnage (DT) of the vessel,

g = gravity acceleration,

V = velocity of ship,

C_m = mass coefficient (Eq. 10.2),

C_e = eccentricity coefficient (Eq. 10.3), and

C_s = softness coefficient = 0.9 to 0.95.

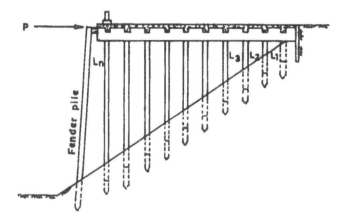

Fig. 10.27. Piles Wharf Structure.

Therefore

$$\frac{P_1^2 L_1^3}{24EI} + \frac{P_2^2 L_2^3}{24EI} + \dots = \frac{1}{N}\frac{W_1}{2g}V^2. \qquad \dots (10.35)$$

Considering the equal deflection of top of the piles, then

$$\frac{P_1 L_1^3}{12EI} = \frac{P_2 L_2^3}{12EI} = \frac{P_3 L_3^3}{12EI} = \dots \qquad \dots (10.36)$$

From Eq. 10.36

$$P_2 = P_1\left(\frac{L_1}{L_2}\right)^3 \; ; P_3 = P_1\left(\frac{L_1}{L_3}\right)^3 ;\dots \qquad \dots (10.37)$$

Substituting Eq. 10.37 in Eq. 10.35

$$P_1^2 L_1^3 + P_1^2\frac{L_1^6}{L_2^3} + P_1^2\frac{L_1^6}{L_3^3} + \dots = \frac{12EIW_1V_1^2}{Ng} \qquad \dots (10.38)$$

or

$$\frac{P_1^2 L_1^2}{K_1^4}\sum K^3 = \frac{12EW_1}{Ng}V^2 \qquad \dots (10.39)$$

where $K = \dfrac{I}{L}$ denotes the stiffness of the piles. Putting

$$C = V\sqrt{\frac{3W_1 E}{Ng\sum K^3}} \qquad \dots (10.40)$$

Therefore, $P_1 L_1 = 2C K_1^2$, similarly $P_2 L_2 = 2C K_2^2$, ..., $P_n L_n = 2C K_n^2$ ... (10.41)

Maximum bending moment in the piles are:

$$M_1 = \frac{1}{2} P_1 L_1; \quad M_2 = \frac{1}{2} P_2 L_2; \quad ... \qquad\qquad ... (10.42)$$

Solving the equations (10.39), (10.40) and (10.41), expressions for the bending moments may be written as:

$$M_1 = C K_1^2; \quad M_2 = C K_2^2; ... \qquad\qquad ... (10.43)$$

Equation (10.43) indicates that the maximum bending moment will occur at the top and bottom of the pile having smallest length i.e. in the rear pile. These moments are directly proportional to the velocity of the moving ship and also directly proportional to the square of the stiffness of pile.

Deflection of the pile bent is given by

$$\Delta = \frac{P_1 L_1^3}{12 EI} = \frac{CI}{6E}. \qquad\qquad ...(10.44)$$

If the pile top is considered as the hinged one, then proceeding exactly in similar manner, it can be shown that:

$$\frac{P_1^2 L_1^2}{K_1^4} \sum K^3 = \frac{1}{N} \frac{3 E W_1}{g} V^2 \qquad\qquad ... (10.45)$$

and the maximum bending moments are:

$$M_1 = P_1 L_1 = C K_1^2; \quad M_2 = P_2 L_2 = C K_2^2 ; ... \qquad\qquad ... (10.46)$$

where C is given by the equation (10.40).

The deflection of the pile in this case is:

$$\Delta = \frac{P_1 L_1^3}{3EI} = \frac{CI}{3E}. \qquad\qquad ... (10.47)$$

Usually the pile top is considered fixed free in the case of R.C.C. platforms. When the platform is a wooden one, then the pile top is considered as hinged. The pile bottom may be considered as the fixed one when it is embedded to 1.5 m in good soil and 3.0 m in soft or loose soil. The number of pile bents that can be assumed to share the impact of moving ship, parallel to its front face, can safely be taken as the number of bents in one half length of the ship.

10.16 DESIGN OF BREAKWATERS

Site considerations

The primary purpose of breakwaters is to protect the harbour areas from the action of heavy seawaves. These are constructed at locations such that they offer the desired protection at the least possible cost.

Since rock is one of the main materials used in the construction of breakwaters, its availability should be investigated. In this respect it will not only be necessary to determine that it is economically feasible to produce and deliver to the site a sufficient quantity of rock, but its density, soundness, and ability to break into large pieces when quarried will be important factors in determining its use.

Foundation investigation

After the tentative selection of site, an intensive investigation of local conditions must be made prior to any construction. Complete and detailed hydrographic surveys are necessary, not only of the tentative site but of adjacent areas. This phase of the investigation should be complete to the extent of supplementing standard survey procedure with bottom samplings, measurement of existing current and if possible, of amounts and directions of materials transported by these currents.

In addition to the hydrographic surveys, topographic surveys are needed showing land details in the vicinity of the breakwaters and along the shore for a considerable distance on either side of breakwater.

Conditions of the base soil/rock should be explored accurately for proper choice of the type of breakwater to be used at a particular site. The foundation analysis can be made by means of probings, wash borings, or drilling. Foundation materials vary from solid rock to soft mud and each gradation must be dealt differently. The quantities which need to be evaluated are compressibility, homogeneity, durability and scourability.

10.17 DESIGN OF RUBBLE-MOUND BREAKWATERS

General

In case of a rubble mound breakwater, the waves expend their energy on the structure, and the disturbing influence of waves is most keenly felt between high water and low water levels. It is in this region that the structure is most severely acted upon. A rubble mound breakwater consists of a central portion called the core, and protective layers called the armour. The core can consist of small pieces of stones, but the armour layer should have stone of bigger size. In between the core and the armour layer, graded stones should be provided, both for the better dissipation of energy and also to protect the finer material from being sucked out on the return wave. These intermediate layers of rock of smaller sizes, usually termed the 'filter course', may separate the inner core and the outside envelop of large armour rock.

Figure 10.28 illustrates in cross-section of the layers and weights of individual pieces of rock making up a typical rock-mound breakwater.

The design of a rock-mound breakwater must be based to a large extent on experience, supplemented in some cases by information obtained from model tests. There is no exact analytical analysis which can be made, but there are certain approximate formulae and design criteria which can be used by the designer to arrive at a safe design.

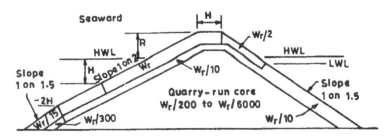

Fig. 10.28. A Rubble-mound Breakwater in Section.

Size of stones

The stability of a mound breakwater is dependent mainly upon the weight and shape of the individual pieces of armour rock on the slope of which they are laid, provided they are properly placed on the slope and with respect to each other so as to form a stable and reasonably close-fitting envelop around the core. It can be proved analytically and by model tests that the weight of individual pieces of armour rock will vary with the degree of slope on which they are laid, i.e., steeper slopes requiring heavier rock and flatter slopes lighter rock.

Following two correlations are popular in use to determine the weight of the stones in the armour:

(i) Iribarren (1965) correlation

$$W = \frac{K\gamma_r h^3 \gamma_w^3}{(\cos\beta - \sin\beta)^3 (\gamma_r - \gamma_w)^3} \qquad \qquad \dots (10.48)$$

where W = weight of individual armour stone in kN,

   K = a coefficient combining all unevaluated variables

   = 0.015 for natural rock, and 0.019 for artificial concrete blocks,

   h = height of wave in m,

   γ_r = unit weight of armour stone in kN/m³,

   γ_w = unit weight of sea water in kN/m³, and

   β = angle of slope with horizontal.

There are some limitations to above correlation. It is apparent that for slopes between 45° and 90° the $(\cos\beta - \sin\beta)^3$ term in the denominator is negative and therefore the stone weight also is negative. For a slope of 45° this term approaches zero and the stone size becomes infinite. Neither of these statements has physical meaning. In general, seaward slopes are less than 45° varying only between $1V : 3/2H$ and $1V : 2H$, and in this range of slopes, the Eq. 10.48 may be used.

The Eq. 10.48 is intended to be applied for the determination of stone sizes from the top of the breakwater (fixed by Iribarren, 1950 at $5/4h$ above still-water level) to a distance h below that level.

For lower positions on the slope, Eq. 10.48 can be used with h', substituted for h, where

$$h' = \frac{4\pi r^2}{LK'},$$ ... (10.49)

$$r = \frac{h}{2\sin(\pi h / L)},$$ ... (10.50)

L = length of wave,

K' = coth $(2\pi d_g / L)$, and   ... (10.51)

d_g = depth below still-water level of the lowest point on the slope for the use of stones of weight W.

(ii) Hudson (1952) correlation

$$W = \frac{\gamma_r h^3}{K_\Delta (S_r - 1)^3 \cot\beta}$$ ... (10.52)

where W, h, γ_r and β have the same meaning as explained for Eq. 10.48. S_r is the specific gravity of stone in armour portion and may be taken as 2.72.

K_Δ = damage coefficient. This may be taken as 3.2 for no damage and no overtopping criteria.

On the basis of model experiments, Hudson (1952) has suggested the following correlation for computing the slope angle β of the exposed side of mound-type breakwater:

$$\tan\beta = \frac{8}{T}\sqrt{\frac{h}{2g}}$$ ... (10.53)

where T = period of wave in s,

$$= \sqrt{\frac{2\pi L}{g}}$$ ... (10.54)

L = length of wave in m,

g = acceleration due to gravity in m/s², and

h = height of wave in m.

Crest elevation

The height to which a breakwater should be built will be somewhat dependent upon the purpose it is required to fulfil, the extent of enclosed water area, and the nature of the shipping work there. In the case of rubble-mound structures where vessels are not likely to be moored at, it is not always necessary to completely obstruct the waves. However, the volume of water passing over the top should not be such that it disturbs the functioning of the harbour. By and large, the crest elevation is kept $1.2h$ to $1.25h$ above the still water level, where h is the height of the wave (U.S. Corps of Engineers, 1957).

Crest width

The crest width of rubble-mound breakwater is selected basically keeping the following points in view:

(i) The weight of the stones are decreased by the buoyant force due to submergence,
(ii) The surge through the breakwater will give rise to a force to dislodge the stones, and
(iii) The portion of the wave passing over the crest will tend to dislodge the stones.

In addition to the above points, method of construction of the breakwater and its use other than the dispersing the waves also play role in deciding the crest width.

The minimum value of the crest width may be obtained using following correlation proposed by Hudson (1953):

$$B_c = 3 K_\Delta' \left(\frac{W}{\gamma_r} \right)^{\frac{1}{3}}$$

... (10.55)

where B_c = minimum crest width in m,

K_Δ' = layer coefficient = 1.0,

W = weight of stone in armour unit in kN, and

γ_r = unit weight of stones in armour unit in kN/m³.

In no case, the crest width should be smaller than 2.0 m.

Design steps
1. Using proper equations (Sec. 10.13) determine the height (h) and length (L) of the wave for the given data (Fetch of wave, D; Velocity of wind, V; Depth of still-water level, d).
2. Determine the time period of wave (Eq. 10.11) and slope β (Eq. 10.53):
3. Determine the weight of stone in the armour side (using Eqs. 10.48 to 10.52).
4. Decide the crest width (Eq. 10.55 or 2.0 m whichever is more) and the crest elevation (usually at a height of 1.20h to 1.25h) above still-water level.
5. Prepare the section of the breakwater.

10.18 DESIGN OF WALL TYPE BREAKWATER

The design of vertical wall type breakwater is checked for two combinations of the loads, namely
(i) Dead load plus wave pressure
(ii) Dead load plus seismic force

The design is carried out in a conventional way by checking the safety against sliding, overturning and base pressure as follows:

(a) F_s = Factor of safety against sliding $\not< 2.0$
(b) F_o = Factor of safety against overturning $\not< 2.0$
(c) Maximum base pressure intensity $< q_a$

where q_a is allowable soil/rock pressure.

10.19 DESIGN OF GRAVITY WALL AND ANCHORED BULK HEAD WHARF STRUCTURES

The design of gravity wall and anchored bulkhead wharf structures are checked for the following combination of the loads:

(i) Dead load plus vertical live loads plus earth pressure plus berthing load (i.e. ship impact).
(ii) Dead load plus vertical live loads plus earth pressure plus line pull.
(iii) Dead load plus vertical live loads plus earth pressure plus seismic force.

As mentioned in Sec. 10.11, impact of ship on these structures will not cause any damage to the wharf, since the energy will be dissipated in the large mass of the concrete or steel and fill.

The detailed designs of gravity type walls and anchored bulk heads have been discussed in Chapter 11.

10.20 DESIGN OF PILED WHARF STRUCTURE

In wharf structures supported on piles, the energy of the moving vessel must be absorbed by the elasticity of the structure itself, stresses which have been set up must be accounted for. As the piles are constructed on natural slope, force due to earth pressure will be zero. In general, these types of structures are designed for dead load plus live loads plus ship impact, this being the most critical.

Various steps involved in design have been illustrated through examples 10.4 and 10.5.

ILLUSTRATIVE EXAMPLES

Example 10.1

Design a wall breakwater for the following data:

$$\text{Depth of water} = 16.0 \text{ m}$$
$$\text{Wind velocity} = 130 \text{ km/hour}$$
$$\text{Fetch of wave} = 220 \text{ km}$$

Use Molitore's wave pressure diagram. The allowable rock pressure is 800 kN/m².

Solution

(i) $h = 0.0322 \sqrt{V \cdot D} = 0.0322 \sqrt{130 \times 220} = 5.45$ m

(ii) $\dfrac{L}{h} = \dfrac{1350}{V} = \dfrac{1350}{130} = 10.4$

$$L = 5.45 \times 10.4 = 56.6 \text{ m}$$

(iii) $1.84 \times h = 1.84 \times 5.45 = 10.0$ m. Therefore, $d > 1.84h$, hence

$$a = \frac{h}{2} + \frac{h^2}{L} = \frac{5.45}{2} + \frac{5.45^2}{56.6} = 3.249 \text{ m}$$

(iv)
$$d_0 = d + a - \frac{h}{2} = 16.0 + 3.249 - 2.725$$
$$= 16.524 \text{ m}$$

(v)
$$\frac{2\pi d_0}{L} = \frac{2\pi \times 16.524}{56.6} = 1.83$$

$$V = 1.243 \sqrt{L \tanh\left(\frac{2\pi d_0}{L}\right)}$$

$$= 1.243 \sqrt{56.6 \tanh(1.83)} = 9.18 \text{ m/s}$$

$$V_0 = \frac{3.91h}{\sqrt{56.6 \tanh(1.83)}} = 2.88 \text{ m/s}$$

$$p_{max} = \frac{K \cdot \gamma_w}{2g} (V + V_0)^2$$

$$= \frac{1.8 \times 10.20}{2 \times 9.81} (9.18 + 2.88)^2$$

$$= 136 \text{ kN/m}^2.$$

This maximum pressure intensity occurs at a height of $h_1 = 0.12\ h = 0.12 \times 5.45 = 0.654$ m above still-water level.

This pressure diagram is shown in Fig. 10.29.

Height of breakwater $= 16 + 6.498 = 22.498$ m, say 23.0 m

(vi) Let B be width of breakwater, refer Fig. 10.29.

Moment of wave pressure about the mid-point A of the base of breakwater

$$= \frac{1}{2} \times 136 \times 5.844 \left(16.654 + \frac{5.844}{3}\right) + \frac{2}{3} \times 2.855 \times 136 \left(13.799 + \frac{5}{8} \times 2.855\right)$$

$$= 11426 \text{ kNm.}$$

Weight of breakwater in terms of B

$$= 16 \times B \times (24 - 10.2) + 7.0 \times B \times 24 = 388.8B.$$

For no tension, the resultant of the weight of breakwater and horizontal wave pressure should be within the middle third.

Therefore,
$$388.8\ B \times \frac{B}{6} = 11426$$

or
$$B = 13.3 \text{ m, say } 13.5 \text{ m}$$

(vii) Horizontal wave pressure

$$= \frac{1}{2} \times 136 \times 5.844 + \frac{2}{3} \times 2.855 \times 136 = 656 \text{ kN.}$$

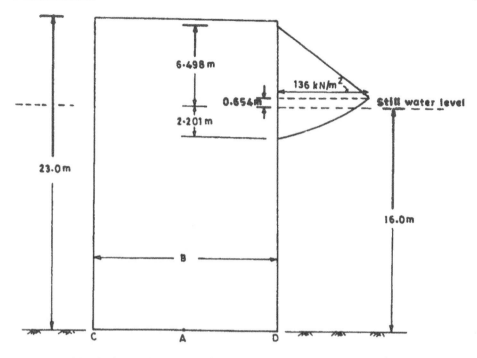

Fig. 10.29. Wall Breakwaters with Pressure Diagram Proposed by Molitore.

Assuming coefficient of base friction as 0.4.
Factor of safety against sliding

$$= \frac{0.4 \text{ (Weight of breakwater)}}{\text{Horizontal force exerted by wave pressure}}$$

$$= \frac{0.4 \times 388.8 \times 13.5}{656} = 3.2 \; [> 2.0, \text{ safe}].$$

Factor of safety against overturning

$$= \frac{\text{Stabilising moment about point } B}{\text{Overturning moment about point } B}$$

$$= \frac{388.8 \times 13.5 \times \dfrac{13.5}{2}}{11426}$$

$$= 3.10 \; [> 2.0, \text{ safe}].$$

As the eccentricity of the resultant with respect to point C is $\dfrac{B}{6}$, the maximum pressure intensity at the base

$$= \frac{2 \times \text{Weight of breakwater}}{\text{Area of breakwater}}$$

$$= \frac{2 \times 388.8 \times 13.5}{13.5 \times 1.0} = 777.6 \text{ kN/m}^2 \quad (< 800 \text{ kN/m}^2, \text{ safe})$$

Example 10.2

Design a wall breakwater for the data mentioned in Example 10.1 using the wave pressure diagram proposed by Sainflou.

Solution

(i) From the solution of Example 10.1

$$h = 5.45 \text{ m}, \ L = 56.6 \text{ m and } d = 16.0 \text{ m}$$

(ii)
$$\frac{2\pi d}{L} = \frac{2\pi \times 16.0}{56.6} = 1.775$$

$$h_o = \frac{\pi h^2}{4L} \coth\left(\frac{2\pi d}{L}\right)$$

$$= \frac{\pi \times 5.45^2}{4 \times 56.6} \coth(1.775) = 0.436 \text{ m}$$

$$a = h_o + \frac{h}{2} = 0.436 + \frac{5.45}{2} = 3.161 \text{ m}$$

$$h_{oc} = \frac{\pi h^2}{L} \coth\left(\frac{2\pi d}{L}\right) = 4 \times 0.436 = 1.744 \text{ m}.$$

Height of crest above still-water level

$$= h + h_{oc} = 5.45 + 1.744 = 7.194 \text{ m}.$$

Height of breakwater $= 16.0 + 7.194 = 23.194$ m, say 23.3 m

(iii)
$$p_2 = \frac{\gamma_w h}{\cosh(1.775)} = \frac{10.2 \times 5.45}{\cosh(1.775)} = 18.32 \text{ kN/m}^2$$

$$p_1 = (\gamma_w d + p_2) \frac{h + h_{oc}}{h + h_{oc} + d}$$

$$= (10.2 \times 16 + 18.32) \frac{5.45 + 1.744}{5.45 + 1.744 + 16.0} = 64.2 \text{ kN/m}^2.$$

Pressure distribution is shown in Fig. 10.30.

(iv) Moment of wave pressure diagram about the middle point of the base, A

$$= \frac{1}{2}(p_2 + p_1) \cdot d \cdot \frac{2p_1 + p_2}{p_2 + p_1} \cdot \frac{d}{3} + \frac{1}{2}(h + h_{oc}) \cdot p_1 \left(d + \frac{h + h_{oc}}{3}\right)$$

$$= \frac{1}{2}(18.32 + 64.2) \times 16.0 \times \frac{2 \times 64.2 + 18.32}{18.32 + 64.2} \times \frac{16.0}{3} +$$

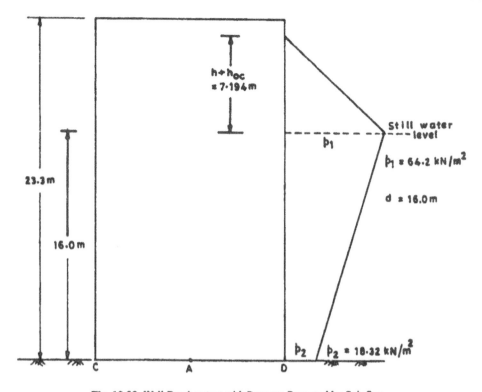

Fig. 10.30. Wall Breakwaters with Pressure Proposed by Sainflou.

$$\frac{1}{2}(5.45 + 1.744) \times 64.2\left(16 + \frac{5.45 + 1.744}{3}\right)$$

$$= 6260 + 4248 = 10508 \text{ kNm.}$$

For no tension:

$$388.8 \times \frac{B}{6} = 10508 \text{ or } B = 12.7 \text{ m, say (13.5 m)}$$

(v) Total wave pressure

$$= \frac{1}{2}(p_2 + p_1) \cdot d + \frac{1}{2}(h + h_{oc}) \cdot p_1$$

$$= \frac{1}{2}(18.32 + 64.2) \times 16 + \frac{1}{2}(5.45 + 1.744) \times 64.2$$

$$= 660 + 231 = 891 \text{ kN.}$$

(vi) Factor of safety against sliding

$$= \frac{0.4 \times 388.3 \times 13.5}{891} = 2.35 \ [> 2.0, \text{ safe}]$$

Factor of safety against overturning

$$= \frac{388.8 \times 13.5 \times \dfrac{13.5}{2}}{10508} = 3.37 \ [> 2.0, \ \text{safe}]$$

Maximum pressure intensity at the base

$$= \frac{2 \times 388.8 \times 13.5}{13.5 \times 1.0} = 777.6 \ \text{kN/m}^2 \ [< 800 \ \text{kN/m}^2]$$

It may be noted that the design of wall breakwater by the two methods, namely (i) using wave pressure diagram suggested by Molitore, and (ii) using wave pressure diagram suggested by Sainflou are almost same. However, IS 4651 (Part III-1974) recommends the use of Sainflou method, and the same may be adopted in designs.

Example 10.3

Design a rubble-mound breakwater for the data given in Example 10.1.

Solution
(i) From the solution of Example 10.1
$$h = 5.45 \ \text{m}, \ L = 56.6 \ \text{m and} \ d = 16.0 \ \text{m}$$

(ii) $T = \sqrt{\dfrac{2\pi L}{g}} = \sqrt{\dfrac{2 \times \pi \times 56.6}{9.81}} = 6.0 \ \text{s}$

Slope angle β of the exposed side

$$\tan \beta = \frac{8}{T} \sqrt{\frac{h}{2g}} = \frac{8}{6.0} \times \sqrt{\frac{5.45}{2 \times 9.81}} \ \text{or} \ \beta = 35°.$$

Provide the slope in top portion of the exposed side as 1 on 2, i.e.

$$\tan \beta = \frac{1}{2} = 0.5 \ \text{or} \ \beta = 26.5°.$$

In the lower portion of the exposed side, and on the harbour side, provide a slope of 1 on 1.5, i.e. 33° with horizontal.

(iii) Weight of stone in the armour portion is given by:

$$W = \frac{K\gamma_r h^3 \gamma_w^3}{10(\cos\beta - \sin\beta)^3 (\gamma_r - \gamma_w)^3}$$

$$= \frac{0.015 \times 24.0 \times 5.45^3 \times 10.2^3}{(\cos 26.5 - \sin 26.5)^3 (24.0 - 10.2)^3} = 344 \ \text{kN}$$

or
$$W = \frac{\gamma_r h^3}{K_\Delta (S_r - 1)^3 \cot\beta} = \frac{24.0 \times 5.45^3}{3.2(2.74 - 1)^3 \cot 26.5} \ \text{(assuming} \ S_r = 2.74)$$

$$= 115 \ \text{kN}.$$

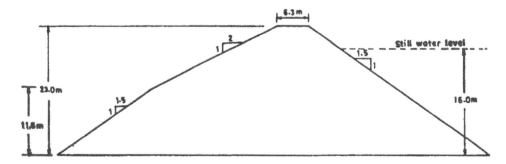

Fig. 10.31. Rubble-mound Breakwater.

The two formulae give different answers.
Use the weight of individual stone in the top armour portion as 225 kN.
(iv) Crest width

$$B_c = 3 \, K'_\Delta \left(\frac{W}{\gamma_r}\right)^{\frac{1}{3}} = 3.0 \times 1.0 \left(\frac{225}{24}\right)^{\frac{1}{3}} = 6.3 \text{ m.}$$

Height of the crest above still-water level = 1.25 h
$$= 1.25 \times 5.45$$
$$= 6.81 \text{ m, say } 7.0 \text{ m.}$$
Therefore, height of breakwater $\qquad = 16.0 + 7.0 = 23 \text{ m}$
(v) Section of the breakwater is shown in Fig. 10.31.

Example 10.4

Figure 10.32 shows a cross-section through a relieving platform type wharf which consists of a concrete platform of 14 m wide carrying sand fill of depth 1.35 m with 0.15 m thick concrete pavement. The whole structure is supported by eight vertical concrete piles and four batter piles. A concrete facial wall retains the sand fill at the out-shore edge.

Reduced level of dredged bottom is 10.00 m with respect to mean low water level (R.L. 0.00 m). R.L. of original ground is –4.3m. A line of reinforced concrete sheet piles were driven along the rear of the platform to retain the fill and to establish a wharf grade of +4.5 m.

A fender system is there to protect the wharf against ship impact. The fender system is of wood-wale spring type.

The platform and surroundings is subjected to a surcharge of 29 kN/m². Due to the fluctuation in tide from MLW at R.L. 0.00 m daily to +3.0 m the groundwater level behind the sheet pile will also fluctuate with low order. It is assumed that at MLW level, groundwater level will be R.L. +1.80 which will produce differential water pressure.

Following are the other data required in the design:

(1) γ_1 (sand dry) = 16.0 kN/m³
$$\phi_1 = 33°41'$$

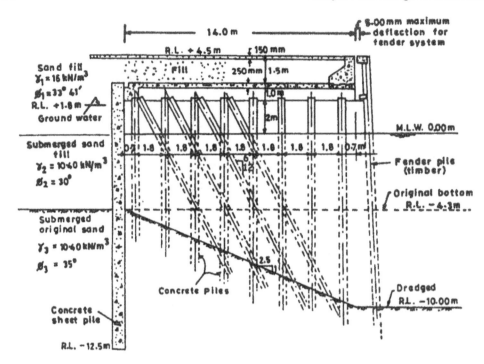

Fig. 10.32. Typical Section of a Relieving Platform Type Wharf.

(2) γ_2 (submerged sand) = 10.4 kN/m³

$$\phi_2 = 30°$$

(3) γ_3 (submerged original sand) = 10.4 kN/m³

$$\phi_3 = 35°$$

(4) Coefficient of permeability of fill =1 × 10⁻⁴ m per second

Coefficient of permeability of ground = 5 × 10⁻⁵ m per second

(5) Bent spacing 3 m c/c

Solution

(i) Determination of differential water pressure

Differential water pressure at R.L. 0.00 m = 1.8 × 10

$$= 18 \text{ kN/m}^2.$$

Hydraulic gradient will set up due to differential water pressure, and reduction of pressure will also take place to neutralize the gradient as water will flow through the bottom of the sheet pile.

Distance travelled by water from A to B = 4.3 + 8.2 + 8.2 = 20.7 m.

We know $v = K \cdot S$.

where v = velocity of flow,

K = coefficient of permeability, and

S = hydraulic gradient.

Velocity of flow through both the material will be same

$$K_{fill} \times S_{fill} = K_{ground} \times S_{ground}$$

$$S_{ground} = \frac{K_{fill}}{K_{ground}} \times S_{fill}$$

$$= \frac{1 \times 10^{-4}}{5 \times 10^{-5}} \times S_{fill}$$

$$S_{ground} = 2 S_{fill}$$

If h denotes head loss in kN/m² per metre run, then

$$h_{fill} \times 4.3 + 2 \times h_{fill} \times 16.4 = 18$$

$$h_{fill} = \frac{18}{4.3 + 32.8} = 0.485 \text{ kN/m}^2 \text{ per metre run}$$

$$h_{ground} = 2 \times 0.485 = 0.97 \text{ kN/m}^2 \text{ per metre run}$$

(a) Pressure at R.L. 4.3 m

$$= 18 - 0.485 \times 4.3$$
$$= 15.91 \text{ kN/m}^2.$$

(b) Pressure at R.L. 12.5 m

$$= 15.91 - 8.2 \times 0.97$$
$$= 7.955 \text{ kN/m}^2.$$

(ii) *Analysis of sheet pile*

Earth pressure distribution on sheet pile

Sheet pile is subjected to active earth pressure and supported by platform at top and supported by passive earth pressure at bottom.

Earth pressure is determined by Rankine's theory

For $\phi_1 = 33°41' = 33.7°$; $K_{A1} = 0.286$

$\phi_2 = 30°$; $K_{A2} = 0.334$, $K_{p2} = 3.0$

$\phi_3 = 35°$; $K_{A3} = 0.272$; $K_{P3} = 3.69$

$$K_{P3\ slope} = \left[\frac{\cos\phi}{1 - \sqrt{\sin\phi\,(\sin\phi - \cos\phi\,\tan i)}} \right]^2 = 1.72.$$

where, i = slope angle

$$= \tan^{-1} \frac{1}{2.5} = 21.8°$$

Preparation of pressure diagram with surcharge and active earth pressure

Active pressure—(Fig. 10.33c)

(i) $AB = K_{A1} \times q = 0.286 \times 29 = 8.30 \text{ kN/m}^2$

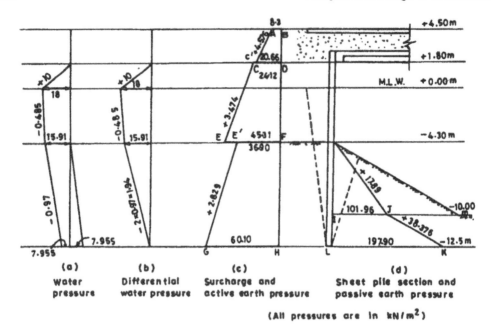

Fig. 10.33. Determination of Soil and Water Pressure.

(ii) Increment $= \gamma_1 \times K_{A1} = 16 \times 0.286 = + 4.576$ kN/m^3
     $CD = 4.57(4.5 - 1.0) + 8.3 = 20.66$ kN/m^2

(iii) $C'D = [\gamma_1(4.5 - 1.8) + 29] \times K_{A2}$
           $= [16 \times 2.7 + 29] \times 0.334 = 24.12$ kN/m^2
     Increment $= \gamma_2 \times K_{A2}$
               $= 10.4 \times 0.334 = 3.474$ kN/m^3
         $EF = 24.12 + 3.474(1.8 + 4.3) = 45.31$ kN/m^2

(iv) $E'F$ $= [72.20 + \gamma_3 \times h] \times K_{A3}$
            $= [72.20 + 6.1 \times 10.40] \times 0.272 = 36.90$ kN/m^2
     Increment $= \gamma_3 \times K_{A3}$
               $= 10.40 \times 0.272 = 2.829$ kN/m^3
     GH $= 36.90 + 2.829 \times (12.5 - 4.3) = 60.10$ kN/m^2

(v) Passive pressure
     Increment $= 10.40 \times 1.72 = 17.89$ kN/m^3
     IJ $= 10.40(10 - 4.3) \times 1.72 = 101.96$ kN/m^2

(vi) Increment $= \gamma_3 \times K_{p3}$
               $= 10.40 \times 3.69 = + 38.376$ kN/m^3
     LK $= 101.96 + 38.37(12.5 - 10) = 197.90$ kN/m^2

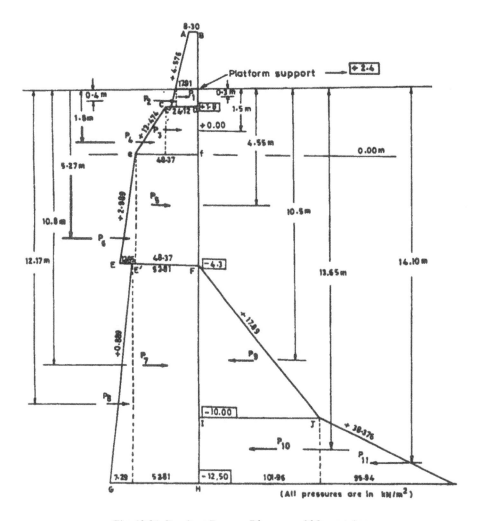

Fig. 10.34. Resultant Pressure Diagram and Moment Arms.

Determination of resultant pressure distribution diagram
    Refer Fig. 10.34

> *AB*— Same as before (water table below it)
> *CD*— Same as before
> *C'D*— Same as before

Increment = 3.474 + 10 (due to differential water pressure)
          = 13.474 kN/m³
at low water level 0.00 m R.L.

$$ef = 24.12 + 1.8 \times 13.474 \qquad = 48.37 \text{ kN/m}^2$$
$$EF = 45.31 + 15.91 \text{ (water pressure)} = 61.22 \text{ kN/m}^2$$

$$E'F = 36.90 + 15.91 \text{ (water pressure)} = 52.81 \text{ kN/m}^2$$
$$GH = \text{Same as before} \qquad\qquad = 60.10 \text{ kN/m}^2$$

To check the depth of embedment

Depth of embedment is safe if moment due to passive pressure (resisting) > moment due to active pressure (actuating) about top support.

Active pressure:

Force (kN)	Lever arm (m)	Moment (kNm)
P_1: 17.91×0.6	0.3	3.22
P_2 : $2.75 \times \dfrac{0.6}{2}$	0.4	0.33
P_3 : 24.12×1.8	1.5	65.12
P_4 : $24.25 + \dfrac{1.8}{2}$	1.8	39.28
P_5 : 48.37×4.3	4.55	946.36
P_6: $12.85 \times \dfrac{4.3}{2}$	5.27	145.60
P_7: 52.81×8.2	10.80	4676.85
P_8: $7.29 \times \dfrac{8.2}{2}$	12.17	363.75
	Total	6240.51 kNm

Passive pressure:

Force (kN)	Lever arm (m)	Moment (kNm)
P_9 : $101.96 \times \dfrac{5.7}{2}$	10.5	3051.15
P_{10} : 101.96×2.5	13.65	3479.39
P_{11}: $95.94 \times \dfrac{2.5}{2}$	14.1	1690.94
	Total	8221.48 kNm

$$\text{Factor of safety} = \frac{8221.48}{6240.51} = 1.317 \text{ (hence safe)}$$

(iii) Design of sheet pile

Design of sheet pile is to be carried out by equivalent beam method.

(a) Determination of point of zero shear

Let the point of zero pressure intensity lie at a depth y below the ground surface, then (Fig. 10.34):

$$\text{Intensity of active pressure} = \text{Intensity of passive pressure}$$

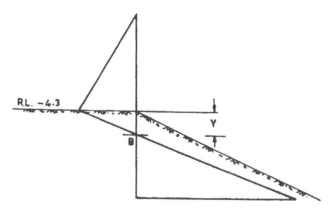

Fig. 10.35. Approximate Nature of Resultant Pressure Diagram.

$$52.81 + 0.889y = 17.89y$$
$$17y = 52.81 \text{ or } y = 3.10 \text{ m}$$

Resultant pressure diagram will be approximately as shown in Fig. 10.35.

(b) Determination of reactions

   Taking moment about R_1 (Fig. 10.36)

   Moment upto P_6 is same as before

Force (kN)	Lever arm (m)	Moment (kNm)
P_1: 10.746	0.3	3.22
P_2: 0.825	0.4	0.33
P_3: 43.416	1.5	65.12
P_4: 21.825	1.8	39.28
P_5: 207.991	4.55	946.36
P_6: 27.630	5.27	145.60
P_7: 81.855	7.73	632.74
Total force 394.29 kN		1832.65 kNm

$$R_2 \times 9.8 = 1832.65$$

$$R_2 = \frac{1832.65}{9.8} = 187 \text{ kN per metre run}$$

$$R_1 = (394.29 - 187) = 207.29 \text{ kN}$$

(c) Checking of the adequacy of depth of embedment the sheet pile may be treated as a simple beam.

   It can be checked conservatively by the following formula:

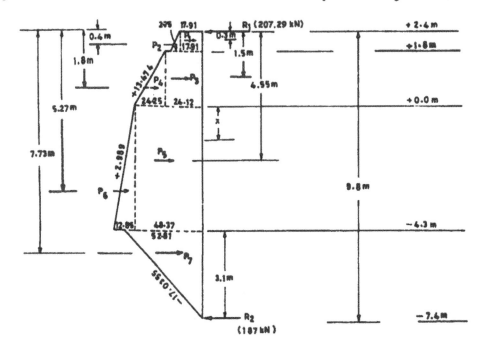

Fig. 10.36. Force Diagram to Determine Supports of Sheet Pile.

$$x = K_2 \left(y + \sqrt{\frac{6R_2}{\gamma(2K_P - K_A)}} \right)$$

where x = depth of penetration below original G.L.,

   $K_2 = 1.1$,

   R_2 = reaction = 187 kN,

   $\gamma K_P = 38.376$ kN/m^2 (Fig. 10.34),

   $\gamma K_A = 0.889$ kN/m^2 (Fig. 10.34), and

        y = depth where pressure diagram changes side from original G.L.

$$x = 1.1 \left(3.1 + \sqrt{\frac{6 \times 187}{2 \times 38.376 - 0.889}} \right) = 7.64 \text{ m.}$$

   Sheet pile is driven $(12.5 - 4.3) = 8.2$ from original G.L. which is more than 7.64 m, the required depth.

(d) Determination of moment

   Moment will be maximum at the section of zero shear force.

   Referring Fig. 10.36,

$$R_1 - P_1 - P_2 - P_3 - P_4 - 48.37x - 2.989 \times x \times \frac{x}{2} = 0$$

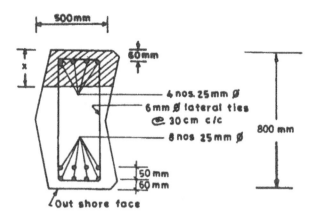

Fig. 10.37. Cross-section of a Sheet Pile.

or
$$48.37 \, x + 2.989 \, \frac{x^2}{2} = R_1 - P_1 - P_2 - P_3 - P_4$$
$$= 207.29 - 10.746 - 0.825 - 43.416 - 21.825$$
$$= 130.81$$
or
$$x = 2.5 \text{ m from R.L. } 0.00 \text{ m}$$

Now,

Moment about x
$$M_x = 207.29 \times 4.9 - 10.746(4.9 - 0.3) - 0.825(4.9 - 0.4) - 43.416\,(4.9 - 1.5)$$
$$- 21.825(4.9 - 1.8) - 48.37 \, \frac{2.5 \times 2.5}{2} - 2.985 \times 2.5 \times \frac{2.5}{2} \times \frac{2.5}{3}$$
$$= 1015.72 - 49.43 - 3.71 - 147.61 - 67.66 - 151.16 - 7.77$$
$$= 588.38 \text{ kNm.}$$

(e) Design of sheet pile section

Assuming the width of section as 500 mm.
$$0.138 f_{ck} b d^2 = 588.38 \times 1.5 \times 10^6$$

$$d = \sqrt{\frac{588.38 \times 1.5 \times 10^6}{0.138 \times 20 \times 500}}$$

$$= 799 \text{ mm.}$$

Provide the depth of section as 800 mm.

Tension reinforcement = 8 bars of 25 mm ϕ
$$= 3925 \text{ mm}^2$$
$$> 0.96\% \text{ i.e. } 3840 \text{ mm}^2.$$

Adopt the section of sheet pile as shown in Fig. 10.37.

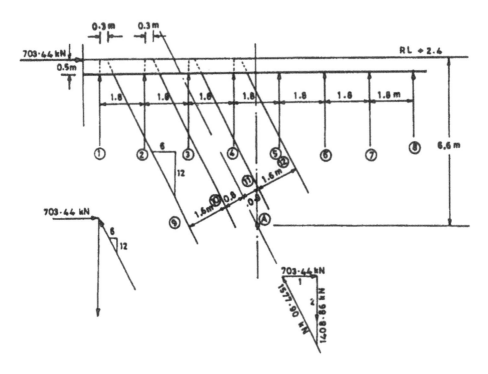

Fig. 10.38. Analysis of Pile Bent for Horizontal Load.

(iv) Analysis of pile bent

    (a) Loading:

Reaction from sheet pile = 207.29 kN per metre run

$$\text{Lateral pressure from fill} = \frac{(4.5 - 2.4) \times (8.30 + 17.91)}{2}$$

$$= 2.1 \times \frac{26.21}{2} = 27.52 \text{ kg}$$

Total = 207.29 + 27.52 = 234.81 kN.

As bent spacing is 3.0 m centre to centre,
Total thrust per bent = 234.81 × 3.0 = 703.44 kN

    (b) The weight of platform, fill and paving, etc.

$$
\begin{array}{llll}
\text{Sand fill} & = & 1.35 \times 16 = & 21.60 \text{ kN/m}^2 \\
\text{Top slab} & = & 0.15 \times 24 = & 3.60 \text{ kN/m}^2 \\
\text{Dock slab} & = & 0.25 \times 24 = & 6.00 \text{ kN/m}^2 \\
\hline
& & & 31.20 \text{ kN/m}^2 \\
\hline
\end{array}
$$

(c) Transverse girder which distributes the load on pile is assumed as stiff member.

Dead load on girder $= 31.20 \times 14 \times 3 = 1310$ kN
Weight of girder (assumed) $= 136$ kN
Total dead load $= 1310 + 136 = 1446$ kN
Live load @ 29 kN/m² $= 29 \times 14 \times 3 = 1230$ kN.

Assuming that vertical pile carries only the vertical load. The horizontal force can be resolved according to the batter angle of the pile as shown in Fig. 10.38.

From stress diagram

Load on batter pile $= 1577.90$ kN
Uplift on vertical pile $= 1408.86$ kN

This uplift load will be resisted by the dead load and pullout values of the vertical pile.

Therefore, Dead load on each pile $= \dfrac{1446}{8} = 180.75$ kN (P_1)

Live load on each pile $= \dfrac{1230}{8} = 154$ kN (P_2)

Uplift force $= \dfrac{1408.86}{8} = (-)\ 176$ kN (P_3) on each pile

Direct load on each batter pile $= \dfrac{1577.9}{4} = 395$ kN (P_4)

(d) From Fig. 10.38 it can be seen that the resultant of vertical and batter pile intersects at A, 6.6 m from the line of action of horizontal force. Now A is the point of rotation of the pile group.

Moment produced by the horizontal force about A will reduce or increase the force in vertical pile by the formula:

$$P_i + \frac{Md_i}{\Sigma d_i^2}$$

$M = 703.44 \times 6.6 = 4650$ kNm

Pile No.	d_i^2 in m²	$Md_i/\Sigma d_i^2$ (kN)
1, 8	$6.3^2 \times 2 = 79.38$	196.6
2, 7	$4.5^2 \times 2 = 40.50$	140.4
3, 6	$2.7^2 \times 2 = 14.58$	84.2
4, 5	$0.9^2 \times 2 = 1.62$	28.1
9, 12	$2.4^2 \times 2 = 11.52$	74.9
10, 11	$0.8^2 \times 2 = 1.28$	25.0
	$\Sigma d^2 = 148.88$	

Table: Summary of Pile Loads.

Pile No.	P_1	P_3 or P_4	$\dfrac{Md_i}{\Sigma d_i^2}$	Case I (without live load) (kN)	P_2	Case II Case I + P_2 (with live load)
1	180.75	−176	− 196.6	− 192.6	+154	− 38.6
2	180.75	−176	− 140.4	− 136.4	+154	+ 17.6
3	180.75	−176	− 84.2	− 80.2	+15wwq4	+ 73.8
4	180.75	−176	− 28.1	+ 24.1	+154	+ 129.9
5	180.75	−176	+ 28.1	+ 32.1	+154	+ 186.1
6	180.75	−176	+ 84.2	+ 88.2	+154	+ 242.2
7	180.75	−176	+ 140.4	+ 144.4	+154	+ 298.4
8	180.75	−176	+ 196.6	+ 200.6	+154	+ 354.6
9		+395	− 74.9	+ 320.0		+ 320.0
10		+395	− 25.0	+ 370.0		+ 370.0
11		+395	+ 25.0	+ 420.0		+ 420.0
12		+395	+ 74.9	+ 469.9		+ 469.9

(e) Design of pile section
     Design loads:

	Compressive	Tensile
Vertical pile	354.6 kN	−192.6 kN
Batter pile	469.9 kN	

Length of pile = Length above L.W.L + Length upto dredged bottom + Embedment in soil

$$= 2 + 10 + 1.5 = 13.5 \text{ m}$$

Pile is considered fixed at top and hinged at bottom, then
Effective length of pile = 0.75 × 13.5 = 10.1 m, say 10 m
     The pile section may be designed in usual way for the forces mentioned in the above table.
A section of pile shown in Fig. 10.39 is found adequate.
Check the safety against uplift.

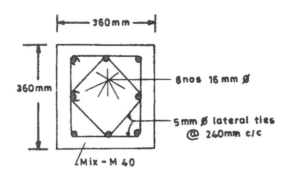

Fig. 10.39. Cross-section of Pile.

Vertical inshore pile (pile No. 1) in each bent must resist maximum uplift force of 192.6 kN.

Assuming average frictional resistance = 39 kN/m², and

$$\text{factor of safety} = 2$$

$$\text{Perimeter} \times f_s = \text{F.O.S} \times \text{uplift force}$$

$$4 \times 0.36 \times L \times 39 = 2 \times 192.6$$

$$L = \frac{2 \times 192.6}{4 \times 0.36 \times 39} = 6.85 \text{ m.}$$

To resist the pullout force, pile No. 1 is to be given embedment of 7 m. The embedment length for the other can be calculated in the same way.

(v) Design of wood springing top fender

Say a ship of 140000 kN displaced weight is moving to dock with a maximum speed 0.0447 metre per second normal to the dock. The dock is protected by a wood fender system of type shown in Fig. 10.40.

It is desired to determine the size of wood-wale to be used.

$$E = \frac{WV^2}{2g}, \text{ where } W = \text{weight of ship and } V = \text{velocity of ship.}$$

$$= \frac{1400 \times (0.0447)^2 \times 1000}{2 \times 9.81} = 14.27 \text{ kNm.}$$

Energy to be absorbed by the fender system = E = 14.27 kNm.

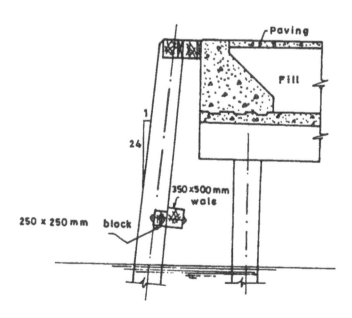

Fig. 10.40. Typical Springing Type Wood Fender.

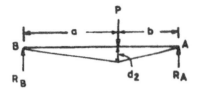

Fig. 10.41. Reaction on Beam BA.

Work done = $F \times d/2$, where d = displacement

$F \cdot d/2 = E/2$ or $Fd = 14.27$ kNm.

Assuming 350 mm × 500 mm wale supported on 75 mm thick block located each at 4.5 m c/c. Refer Fig. 10.41,

$$d_2 = \frac{Pa^2b^2}{3EIL}$$

Let

$a = 0.75\,L = 0.7 \times 4.5 = 3.375$ m

$b = 0.25\,L = 0.25 \times 4.5 = 1.125$ m

$E_{wood} = 124 \times 10^5$ kN/m²

$I = \dfrac{1}{12} \times 0.5 \times (0.35)^3 = 1.786 \times 10^{-3}$ m⁴

$$d_2 = \frac{P \times 3.375^2 \times 1.125^2}{3 \times 124 \times 10^5 \times 1.786 \times 10^{-3} \times 4.5}$$

$$= 4.82 \times 10^{-5}\,P.$$

Let, d_1 be the elastic compression, and displacement of the structure = 0.03 m.

$$\frac{P(d_2 + d_1)}{2} = \frac{1}{2} \text{ (kinetic energy imparted by the ship)}$$

(being two fender piles in adjacent span)

$P[4.82 \times 10^{-5}\,P + 0.03] = 14.27$

It gives $P = 315$ kN

$d_2 = 4.82 \times 10^{-5} \times 315 = 0.015$ m

Reactions on the wale:

$R_A = 0.75P = 0.75 \times 315 = 236$ kN

$R_B = 79$ kN.

Assuming following specification of timber

	Normal load (kN/m^2)	Impact load (kN/m^2)
Bending	16900	33800
Allowable stresses { Compression perpendicular to the grain	3200	6400
Horizontal shear	1000	2000

Assuming bearing width between fender pile and wale as 150 mm,

$$\text{Compressive stress in wale} = \frac{236}{0.5 \times 0.15} = 3146 \text{ kN/m}^2$$

$$[< 6400 \text{ kN/m}^2, \text{ safe}]$$

$$\text{Bending moment} = 236 \times b = 236 \times 1.125 = 265.5 \text{ kNm}$$

$$f_b = \frac{M}{I} y = \frac{265.5 \times 0.150}{1.786 \times 10^{-3}} = 22298 \text{ kN/m}^2$$

$$[< 33800 \text{ kN/m}^2, \text{ safe}]$$

$$\text{Horizontal shear} = \frac{V}{b' h} \times 1.5$$

where V = shear force,

b' = height of wale, 0.5 m,

h = depth of wale, 0.35 m, and

$$\text{Horizontal shear} = \frac{236 \times 1.5}{0.5 \times 0.35} = 2022 \text{ kN/m}^2$$

$$[\approx 2000 \text{ kN/m}^2, \text{ safe}].$$

Hence the assumed section of the wale is safe.

Batter pile for opposite direction for ship impact

Maximum docking force from springing type fender pile system

$$= 2 \times R_A$$
$$= 2 \times 236 \text{ kN}.$$

Assuming that this force is to be distributed by lower slab in five bents

$$\text{Load per bent} = \frac{2 \times 236}{5} = 94.4 \text{ kN} << 703.44 \text{ kN}.$$

So batter pile in the opposite direction is not required for ship impact.

(vi) Design of platform:

Design of platform consists of four elements

(a) Design of slab

(b) Design of inshore longitudinal girder

(c) Design of outshore longitudinal girder

(d) Design of transverse pile cap girder

(a) Design of slab

Assuming clear distance between two pile cap girders = 2.4 m

$$\text{Span} \quad = 2.4 \quad m$$

Dead load = 31.20 kN/m^2 (as calculated before)

Live load = 29   kN/m^2

Total load = 60.2 kN/m^2

$$\text{Moment} = \frac{wl^2}{12} \text{ (for continuous slab)} = \frac{60.2 \times (2.4)^2}{12}$$

$$= 28.9 \text{ kNm}$$

$$d = \sqrt{\frac{28.9 \times 10^6 \times 1.5}{0.138 \times 20 \times 1000}} = 125.3$$

Provide overall depth = 200 mm and effective depth as 150 mm.

$$0.87 \times 415 \times A_{st} \left[d - \frac{415 A_{st}}{20 \times 1000} \right] = 28.9 \times 10^6$$

$$A_{st} = 580 \text{ mm}^2.$$

Provide 12 mm dia. bars at 100 mm c.c both at top and bottom. Distribution steel @ 0.15% of gross cross-sectional area

$$= \frac{0.15}{100} \times 1000 \times 200 = 300 \text{ mm}^2.$$

Provide 6 mm dia. at 70 mm c/c.

Details of reinforcement are shown in Fig. 10.42.

(b) Design of inshore longitudinal girder

Load = sheet pile reaction R_1 per running metre

= 206.96 kN/metre run

$$M = \frac{206.96 \times 3^2}{12} = 155.22 \text{ kNm}.$$

Assuming the width of the beam (as 600 mm)

$$d = \sqrt{\frac{155.22 \times 10^6 \times 1.5}{0.138 \times 600 \times 20}} = 375 \text{ mm}.$$

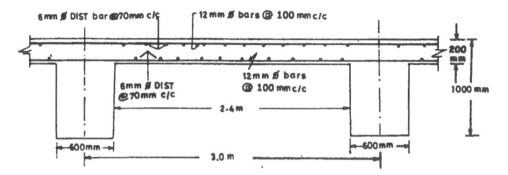

Fig. 10.42. Section of Deck Slab.

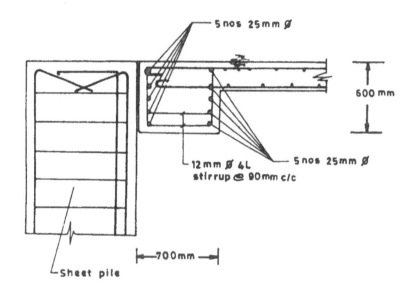

Fig. 10.43. Section of Inshore Longitudinal Girder.

Provide overall depth as 550 mm.

Provide five bars of 25 mm diameter on both sides.

Maximum shear force at support next to end support

$$= 0.6 \times (206.96 \times 3) = 372.5 \text{ kN}.$$

$$\text{Shear stress} = \frac{372.5 \times 1.5 \times 10000}{500 \times 600} = 1.86 \text{ N/mm}^2.$$

Provide 12 mm dia. four-legged stirrup 90 mm c/c.

The details are shown in Fig. 10.43.

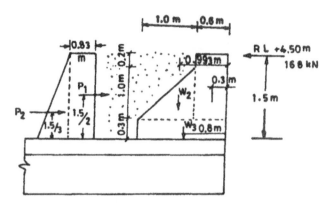

Fig. 10.44. Forces on Outshore Longitudinal Girder.

(c) Design of outshore longitudinal girder
This girder is resisting the earth pressure from fill and surcharge and reaction from fender system.
Refer Fig. 10.44

$$P_1 = 8.3 \times 1.5 \times 1 = 12.45 \text{ kN per metre run}$$

$$P_2 = 2.75 \times \frac{1.5}{2} \times 1 = 2.07 \text{ kN per metre run}$$

Overturning moment:

$$M_1 = 12.45 \times \frac{1.5}{2} = 9.34 \text{ kNm}$$

$$M_2 = 2.07 \times \frac{1.5}{3} = \underline{1.04 \text{ kNm}}$$

$$\underline{10.38 \text{ kNm}}$$

Resisting moment (due to weight of girder):

$$W_1 \times 0.3 = 0.6 \times 1.2 \times 24 \times 0.3 = 5.18 \text{ kNm}$$
$$W_2 \times 0.933 = 1/2 \times 1 \times 1 \times 24 \times 0.933 = 11.20 \text{ kNm}$$
$$W_2 \times 0.8 = 1.6 \times 0.3 \times 24 \times 0.80 = \underline{9.22 \text{ kNm}}$$

$$\underline{25.60 \text{ kNm}}$$

$$\text{Factor of safety against overturning} = \frac{\text{Resisting moment}}{\text{Overturning moment}}$$

$$= \frac{25.60}{10.38} = 2.46.$$

Reaction of fender system will produce a concentrated load at top

$$= 2R_A = 2 \times 236 = 472 \text{ kN}.$$

It may be at the transverse pile cap girder or at the mid span according to the arrangement of fender pile.

If it is designed as gravity type retaining wall it will overturn.

Moment due to reaction $= 472 \times 1.5 = 708$ kNm

Resisting moment for a pile bent of $3m = 3[W_1 \times 1.3 + W_2 \times 0.66 + W_3 \times 0.8]$

$$= 3[17.28 \times 1.3 + 12 \times 0.66 + 11.52 \times 0.8] = 118.80 \text{ kNm}$$

The balance moment $= 708 - 118.80 = 589.2$ kNm

$$d = \sqrt{\frac{1.5 \times 589.2 \times 10^6}{0.138 \times 20 \times 300}} = 1033 \text{ mm}.$$

Overall depth provided $= 1600$ mm

Provide three bars of 20 mm dia.

$$\text{Pull-out force} \qquad = \frac{589.2}{1.5} = 392.8 \text{ kN}$$

$$\text{Downward reaction on pile} = 200 \text{ kN} < 392.8 \text{ kN}$$

So uplifting of pile is possible.

Reinforcement of pile is to be taken up to the top of longitudinal girder (Fig. 10.45). Provide three bars of 20 mm diam.

When the load acts at the centre of the girder, it produces horizontal moment.

$$\text{Moment} = \frac{WL}{4} = \frac{472 \times 3}{4} = 354 \text{ kNm}$$

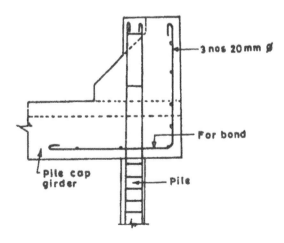

Fig. 10.45. Reinforcement Details when Impact at the Transverse Pile Cap Girder.

7 nos 12 mm ⌀
bars

Fig. 10.46. Reinforcement Details when Impact at the Centre of Girder.

$$0.87 \times 415 A_{st} \left[1600 - \frac{415 A_{st}}{20 \times 300} \right] = 354 \times 10^6$$

$$A_{st} = 630 \text{ mm}^2.$$

Provide seven bars of 12 mm dia.

This will be supported by extruded pile length above girder. So there must be some fixing effect at the support where moment will be opposite in nature to the span moment. Therefore, seven bars of 12 mm dia. are also provided at support on sea-side (Fig. 10.46).

(d) Design of transverse pile cap girder
Refer Fig. 10.47

$$\text{Dead load} = 31.20 \text{ kN/m}^2$$
$$\text{Live load } = 29 \text{ kN/m}^2$$

Load on girder per linear metre = 31.20 × 3 =   93.60 kN/m
Weight of girder @ 9.5 kN/m   =   9.5   kN/m
                                                 _____
                                                 103.10 kN/m

Live load per metre run = 29 × 3   = 87 kN/m
Dead load + live load = 103.10 + 87   = 190.10 kN/m.
Case 1—Girder without live load

$$M_n = M_{n-1} + V_{n-1}\, a + P_m a_m$$

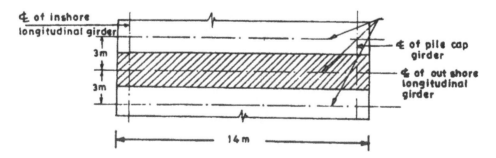

Fig. 10.47. Loading Area of the Pile Cap Girder.

where M_n = moment at section n,
          M_{n-1} = moment at previous section,
          V_{n-1} = shear at previous section,
          a = distance from previous section,
          P_m = loads and reaction between previous section and section 'n', and
          a_m = distance of P_m from section n.
Determination of moment (see Fig. 10.48):

$$M_a = -103 \times 0.7 \times \frac{0.7}{2} = -25.2 \text{ kNm}$$

$$M_b = M_{n-1} + V_{n-1} \times a + P_m \times a_m$$

$$= -25.2 - 265 \times 0.3 - 103 \times 0.3 \times \frac{0.3}{2}$$

$$= -25.2 - 79.5 - 4.6 = -109.3 \text{ kNm}$$

$$M_c = -109.3 - 10 \times 1.5 - 103 \times 1.5 \times \frac{1.5}{2}$$

$$= -240.3 \text{ kNm}$$

$$M_d = -240.3 - 301 \times 0.3 - 4.6$$
$$= -335 \text{ kNm}$$

$$M_e = -335 - 2 \times 1.5 - 116$$
$$= -454 \text{ kNm}$$

$$M_f = -454 - 237 \times 0.3 - 4.6$$
$$= -529.7 \text{ kNm}$$

$$M_g = -529.7 + 112 \times 1.5 - 116$$
$$= -476.9 \text{ kNm}$$

$$M_h = -476.9 - 67 \times 0.3 - 4.6$$
$$= -501.6 \text{ kNm}$$

$$M_i = -501.6 + 322 \times 1.5 - 116$$
$$= -134.6 \text{ kNm}$$

$$M_j = -134.6 + 197 \times 1.8 - 116$$
$$= +54 \text{ kNm}$$

$$M_k = +54 + 100 \times 1.8 - 166$$
$$= +68 \text{ kNm.}$$

Maximum moment in span jk will occur where S.F. = 0. If x is the distance measured from j towards k then,

$$100 = 103x \text{ or } x = 0.97 \text{ m}$$

$$M_{jk} = +54 + 100 \times 0.97 - 103 \times 0.97 \times \frac{0.97}{2} = 139 + 97 - 48.5$$

$$= 102.54 \text{ kNm}$$

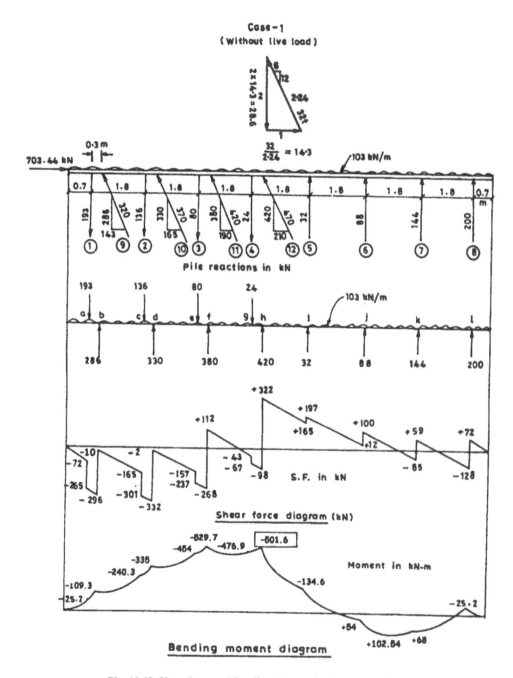

Fig. 10.48. Shear Force and Bending Moment for Transverse Cap Girder.

Similarly, shear force and bending moment diagrams are drawn for case II, i.e. considering live load (Fig. 10.49).

Out of the two cases, design moments and shear force are as given below:

Maximum hogging (tension at top) moment = 529.7 kNm
Maximum sagging (tension at bottom) moment = 220.5 kNm
Maximum shear force = 383 kN

Assume width = 600 mm

$$d = \sqrt{\frac{529.7 \times 10^6 \times 1.5}{0.138 \times 20 \times 600}} = 692.3 \text{ mm}$$

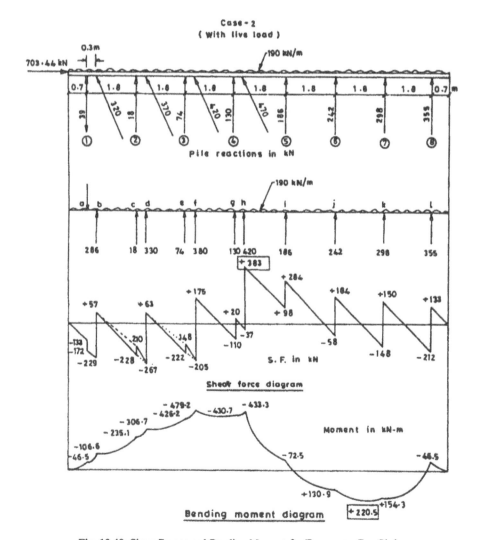

Fig. 10.49. Shear Forces and Bending Moment for Transverse Cap Girder.

Provide overall depth = 1100 mm and
Effective depth   = 1000 mm.

Provide nine bars of 25 mm dia. at top and four bars of 25 mm dia. at bottom. Also provide four-legged stirrups of 10 mm dia. at 100 mm c/c.

Example 10.5

The wharf consists of a series of sheet pile circular cells connected by sheet pile arcs. The piling is cut off at R.L. 1.2 m from M.L.W. 0.00 m (Fig. 10.50). A concrete wall on the bulkhead line provides a straight dock wall for the support of the fender system and to protect the sheet piling.

The soil formation at the location of the wharf consists of silty clay over-lying a sand stratum R.L. +10.00. The silty clay will be removed completely to 3 m in back of cells and then on a 1 in 5 slope. The area behind the cell will be filled with coarse sand and gravel.

The diameter of a cell is 16 m and other construction details shown in plan (Fig. 10.51). The angle of internal friction and unit weight of different soils are written in sectional elevation in Fig. 10.50. The cell is subjected to a pull of 19.80 kN/m and surcharge 24.40 kN/m^2 as shown. Design the cellular sheet pile wharf. The allowable soil pressure is 500 kN/m^2.

Solution

(i) General

Cell diameter = 16 m

Equivalent width, $B = 0.9 \times 16 = 14.4$ m

Weight of sand gravel fill inside the cell

$$= (\gamma_1 \times 3.7 + \gamma_2 \times 16.7) \times 14.4$$
$$= (17 \times 3.7 + 10.40 \times 16.7) \times 14.4 = 3406.75 \text{ kN}$$

For
$$\phi_1 = 33° \, 41', K_{A1} = 0.287$$
$$\phi_2 = 26°34', K_{A2} = 0.382$$
$$\phi_3 = 38°40', K_{A3} = 0.231 \text{ and } K_p = 4.33$$

(ii) Computation of earth pressure with surcharge

$$ab = 24.40 \text{ (surcharge)} \times K_{A1}$$
$$= 24.40 \times 0.287 = 7 \text{ kN/m}^2$$

Increment $= \gamma_1 \times K_{A1}$
$$= 17 \times 0.287 = + 4.88 \text{ kN/m}^3$$
$$bc = 4.88 \times 3.7 = 18.06 \text{ kN/m}^2$$
$$ac' = (24.40 + \gamma_1 \times 3.7) \times K_{A2}$$
$$= (24.40 + 17 \times 3.7) \times 0.382 = 33.35 \text{ kN/m}^2$$

Increment $= \gamma_2 \times K_{A2}$
$$= 10.40 \times 0.382 = +3.97 \text{ kN/m}^3$$
$$c'd = 3.97 \times 10.5 = 41.68 \text{ kN/m}^2$$
$$ad' = (24.40 + 17 \times 3.7 + 10.40 \times 10.5) \times K_{A3}$$
$$= (24.40 + 62.90 + 109.20) \times 0.231 = 45.39 \text{ kN/m}^2$$

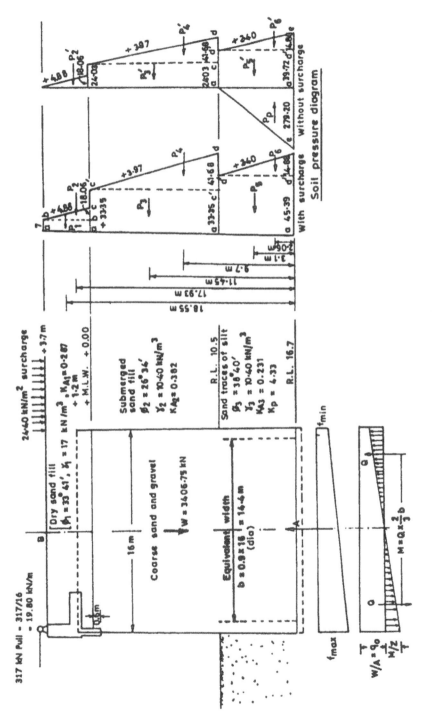

Fig. 10.50. Cellular Sheet Pile Wharf—Typical Section and Soil Pressure Diagram.

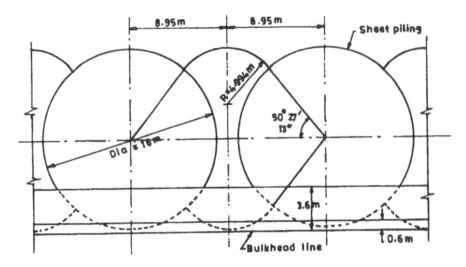

Fig. 10.51. Plan of Cells.

Increment $= \gamma_3 \times K_{A3}$
$= 10.40 \times 0.231 = +2.40 \ \text{kN/m}^3$
$d'e = 2.40 \times 62 = 14.88 \ \text{kN/m}^2$

(iii) Computation of earth pressure without surcharge

Increment $= \gamma_1 \times K_{A1} = 17 \times 0.287 = +4.88 \ \text{kN/m}^3$
$ab = 4.88 \times 3.7 = 18.06 \ \text{kN/m}^2$

$ac = \gamma_1 \times 3.7 \times K_{A2}$
$= 17 \times 3.7 \times 0.382 = 24.03 \ \text{kN/m}^2$

Increment $= \gamma_2 \times K_{A2}$
$= 10.40 \times 0.382 = +3.97 \ \text{kN/m}^3$
$cd = 3.97 \times 10.5 = 41.68 \ \text{kN/m}^2$
$ad' = (17 \times 3.7 + 10.40 \times 10.5) \ K_{A3}$
$= (62.90 + 109.20) \times 0.231 = 39.76 \ \text{kN/m}^2$
$de = 2.4 \times 6.2 = 14.88 \ \text{kN/m}^2.$

Passive pressure $= 10.40 \times 4.33 \times 6.2 = 279.20 \ \text{kN/m}^2.$

(iv) Determination of unbalanced moment about base due to active and passive soil pressure

From pressure diagram (without surcharge)

P'_2 $= 33.41 \ \text{kN}$

$\left. \begin{array}{l} P'_3 = 24.03 \times 10.5 = 252.31 \ \text{kN} \\ P'_4 \qquad\qquad = 218.82 \ \text{kN} \end{array} \right\} 471.13 \ \text{kN}$

Force (kN)			L.A. (m)	Moment about A (kNm)
Pull		19.80	21.11	417.97
P_1	7.00×3.7 =	25.90	18.55	480.44
P_2	$18.06 \times \dfrac{3.7}{2}$ =	33.41	17.93	599.05
P_3	33.35×10.5 =	350.18	11.45	4009.50
P_4	$41.68 \times \dfrac{10.5}{2}$ =	218.82	9.70	2122.54
P_5	45.39×6.2 =	281.42	3.10	872.39
P_6	$14.88 \times \dfrac{6.2}{2}$ =	46.13	2.06	95.02
		975.65 kN		8596.90 kNm
P_P	$279.20 \times \dfrac{6.2}{2}$ =	865.52 kN		1782.97 kNm
Sliding force		110.13 kN		Unbalanced moment 6813.93 kNm

$$\left.\begin{array}{l} P_5' = 39.72 \times 6.2 = 246.26 \text{ kN} \\ P_6' \qquad\qquad\quad = 46.13 \text{ kN} \end{array}\right\} 292.39 \text{ kN}$$

(v) Determination of base pressure

$$\text{Eccentricity } e = \frac{M}{W} = \frac{6813.93}{3406.75}$$

$$= 2.0 \text{ m} < \left\{\frac{B}{6} = \frac{14.4}{6} = 2.4 \text{ m}\right\}$$

$$f_{1,2} = \frac{W}{A} \pm \frac{M}{Z}; \quad Z = \frac{1}{6} \times bB^2 = 1/6 \times 1 \times (14.4)^2 = 34.56 \text{ m}^3$$

$$f_{max} = \frac{3406.75}{1 \times 14.4} + \frac{6813.93}{34.56}$$

$$= 235.68 + 197.16$$

$$= 433.74 \text{ kN/m}^2 < 500 \text{ kN/m}^2$$

$$f_{min} = 39.42 \text{ kN/m}^2 > 0 \text{ hence o.k. (no overturning)}$$

(vi) Check against sliding

The sliding resistance at the base of the cell

$$= W \tan \phi = 3406.75 \tan 38.68 = 2725.45 \text{ kN}$$

Horizontal sliding force $= 110.13$ kN $<<$ 2725.45 kN, hence safe

(vii) Shear stress due to overturning moment
From base pressure diagram (Fig. 10.50)

Overturning moment, $M = Q \times (2/3) B$

Shear force, $Q = \dfrac{3M}{2D} = \dfrac{3 \times 6813.93}{2 \times 14.4} = 709.78$ kN.

This vertical force will be resisted by any vertical plane say AB. Maximum allowable shear on the plane is $V \tan \phi$. But sheet pile will also contribute some frictional resistance due to roughness. A very close approximation of the influence of sheet piling is obtained by adding a coefficient of interlock friction usually $f = 0.3$.

Now $V_R = \Sigma V (\tan \phi + 0.30)$

At each level this V is equal to the horizontal active
Rankine force (without surcharge) as given below–

V	ϕ	$\tan \phi$	f	$(\tan \phi + f)$	V_R
33.41 (P_2')	33°41'	0.667	0.30	0.967	32.30
471.13 $(P_3' + P_4')$	26°34'	0.550	0.30	0.850	400.45
292.39 $(P_5' + P_6')$	38°40'	0.800	0.30	1.10	321.61
					Σ 754.36 kN

Factor of safety $= \dfrac{754.36}{709.78} = 1.06$. Hence safe.

(viii) Interlocking stress
Tension in the interlock is determined by the formula for hoop tension in a pipe subjected to internal pressure.

$$T = p\frac{D}{2}$$

where $T =$ tension in kN per metre interlocking,
            $p =$ horizontal earth pressure in kN/m², and
            $D =$ diameter of cell in metre.

Maximum earth pressure from earth pressure diagram at R.L. –10.5 m (Fig. 10.5) is

$$(33.35 + 41.68) = 75.03 \text{ kN/m}^2$$
$$T = 75.03 \times 16/2 = 600.24 \text{ kN/m}.$$

The recommended maximum working stress in the interlocks for the flat web sheet pile weighing 1.36 kN/m² or 1.50 kN/m² is 1400 kN per m.

600.24 kN/m is less than 1400 kN/m, hence O.K.

Cathodic protection is to be given for sheet piling below the concrete.

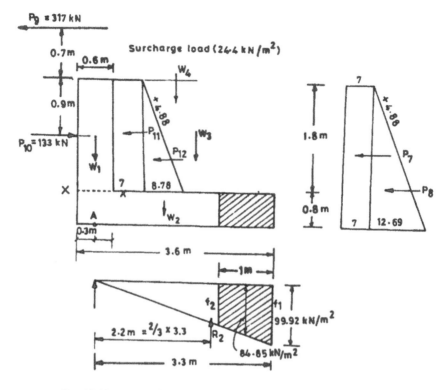

Fig. 10.52. Cross-section and Soil Pressure Diagram for Dock Wall.

(ix) Analysis and design of dock wall

Stem of the wall is to be designed for

(a) Outward pressures for earth and surcharge
(b) 317 kN pull of a mooring line
(c) For impact from vessel (say, 133 kN in this case)

A cross-section of the dock wall is assumed as shown in Fig. 10.52.

(b) Computation of forces

$$W_1 = 1.8 \times 0.6 \times 24 = 25.92 \text{ kN}$$
$$W_2 = 3.6 \times 0.8 \times 24 = 69.12 \text{ kN}$$
$$W_3 = 3.0 \times 1.8 \times 17 = 91.80 \text{ kN}$$
$$W_4 = 24.40 \times 3.6 \quad = 87.84 \text{ kN}$$
$$P_7 = 7 \times 2.6 \quad\quad\quad = 18.20 \text{ kN}$$

$$P_8 = 12.69 \times \frac{2.6}{2} \quad = 16.50 \text{ kN}$$

$$P_9 = 317 \quad\quad\quad\quad = 317 \text{ kN}$$
$$P_{10} = 133 \quad\quad\quad\quad = 133 \text{ kN}$$

(assumed horizontal impact from fender system)

$$P_{11} = 7 \times 1.8 \qquad = 12.60 \text{ kN}$$

$$P_{12} = 8.78 \times \frac{1.8}{2} \qquad = 7.90 \text{ kN}.$$

The soil pressure will vary from zero at sheet piling to a maximum at the interior edge of the footing on the centre line of cell. The increase in soil pressure will be assumed to follow a straight line variation in both directions as shown in Fig. 10.52.

Maximum soil pressure for permanent load and surcharge can be obtained from the condition that the moment about A will be zero (Fig. 10.52). Moment about A

$$R_2 \times 2.2 = W_2 \left(\frac{3.6}{2} - 0.3 \right) + W_3 \left(\frac{3}{2} + 0.3 \right) + W_4 \left(\frac{3.6}{2} - 0.3 \right) - P_7 \times \frac{2.6}{2} - P_8 \times \frac{2.6}{3}$$

$$= 69.12 \times 1.5 + 91.80 \times 1.8 + 87.84 \times 1.5 - 18.20 \times 1.3 - 16.50 \times 0.886$$

$$= 362.73$$

$$R_2 = \frac{362.73}{2.2} = 164.88 \text{ kN}.$$

If f_1 be the stress at the extreme edge then

$$\frac{1}{2} \times 3.3 \times f_1 = 164.88 \text{ or } f_1 = 99.92 \text{ kN/m}^2$$

The innermost 1 m wide ship of the bottom slab indicated by the hatching in Fig. 10.52 carries the load shown in Fig. 10.53.

Strees at the outer edge strip

$$f_2 = \frac{99.92}{3.3} \times 2.3 = 69.70 \text{ kN/m}^2$$

$$\text{Average stress} = \frac{69.70 + 99.92}{2} = 84.85 \text{ kN/m}^2$$

Direct load per metre run including surcharge

$$= 24.40 + 17 \times 1.8 + 0.8 \times 1.0 \times 24$$

$$= 24.40 + 30.60 + 19.20 = 74.20 \text{ kN/m}$$

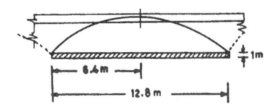

Fig. 10.53. Plan of Slab on Cellular Sheet Pile.

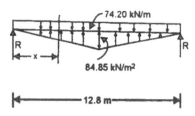

Fig. 10.54. Loading Diagram for Design of Slab.

$$\text{Reaction } R = 74.20 \times 6.4 - 84.85 \times \frac{6.4}{2}$$

$$= 474.88 - 271.52 = 203.36 \text{ kN.}$$

Design of base slab

Maximum moment will occur where S.F. = 0

Let at a distance x from left support S.F. = 0 (Fig. 10.54)

$$203.36 - 74.20x + \frac{84.85}{6.4} \times x \times x/2 = 0$$

It gives

$$x_1 = 4.79 \text{ m and } x_2 = 6.39 \text{ m}$$

$$M_{4.79} = 203.36 \times 4.79 - \frac{84.85}{6.4} \times \frac{(4.79)^3}{6} - \frac{74.20(4.79)^2}{2}$$

$$= 974.09 + 242.84 - 851.22$$

$$= 365.71 \text{ kNm (tension at bottom)}$$

$$M_{6.39} = 203.36 \times 6.39 - \frac{84.85}{6.4} \times \frac{(6.39)^3}{6} - \frac{74.20(6.39)^2}{2}.$$

$$= 1299.47 + 576.53 - 1514.87$$

$$= 361.13 \text{ kN (tension at bottom)}$$

Provide 25 mm ϕ bars @ 100 mm c/c.

Design of stem

The stem of the wall will resist the earth pressure and either the outward moving force or the inward docking force. The docking force is assumed to be distributed over a length of 5 m.

$$\text{Force/metre} = \frac{317}{5} = 63.40 \text{ kN.}$$

$$\text{Maximum moment at section } xx = 790 \times \frac{1.8}{3} + 12.60 \times \frac{1.8}{2} + 6.34 \times 2.5$$

$$= 4.74 + 11.34 + 158.50 = 174.58 \text{ kNm}$$

$$\text{Maximum inward moment} = 133 \times 0.9 - 12.60 \times 0.9 - 7.9 \times 0.6$$

$$= 119.70 - 11.34 - 4.74 = 103.62 \text{ kNm}$$

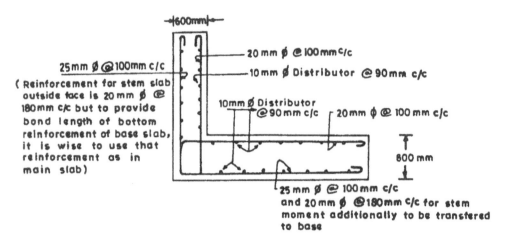

Fig. 10.55. Arrangement of Reinforcement for Dock Slab.

Provide 20 mm ϕ @ 100 mm c/c (see Fig. 10.55).

Distribution steel @ 0.15% gross cross-sectional area of concrete

$$= 1000 \times 600 \times \frac{0.15}{100} = 900 \text{ mm}^2.$$

Provide 10 mm ϕ distribution steel @ 90 mm c/c.

Provide 20 mm ϕ bars @ 180 mm c/c on the outside face.

Same moment will be transferred to bottom slab also, where they require 20 mm ϕ bars @ 150 mm c/c.

REFERENCES

1. Anderson, P. (1956), *Substructure Analysis and Design*, The Ronald Press Company, New York.
2. Cornick, H.F. (1968), *Dock and Harbour Engineering*, Vols. 1 and 2, Charles Griffin and Company Limited, London.
3. Du-Plat-Taylor, F.M. (1949), *The Design, Construction and Maintenance of Docks, Wharves and Piers*, Eyre and Spottiswoode Publishers Ltd., London.
4. Gaillard, D.D. (1904), "Wave Action in Relation to Engineering Structures," *Corps of Engineers*, U.S. Army, Paper No. 31.
5. Gaillard, D.D. and D.A. Molitore (1935), "Wave Pressures," *Trans. ASCE*, pp. 984-1017.
6. Hudson, R.Y. (1952), "Wave forces on breakwaters", *Proc. ASCE*, Vol. 78, No. 113.
7. Irribarren, R.C. (1942), "Wave Action in Relation to Harbour Protection Works," Dock and Harbour Authority, Vol. 23. Translated from Revista de Obras Publicas, Jan. 1941.
8. Irribarren, R.C. (1965), "Formulas pour calcul des digues en encroachements naturals on elements artificials," 21st Navigation Congress, Stockholm.
9. IS: 4651 (Part I)–1974, Code of Practice for Planning and Design of Ports and Harbours (Site Investigation).
10. IS: 4651 (Part II)–1989, Code of Practice for Design and Construction of Dock and Harbour Structures (Earth Pressures).
11. IS: 4651 (Part III)–1974, Code of Practice for Planning and Design of Ports and Harbours (Loading).
12. IS: 4651 (Part IV)–1989, Code of Practice for Planning and Design of Ports and Harbours (General Design Considerations).
13. IS: 4651 (Part V)–1980, Code of Practice for Planning and Design of Ports and Harbours (Layout and Founctional Requirements).

14. IS: 9527 (Part I)– 1981, Code of Practice for Design and Construction of Port and Harbour Structures (Concrete Monoliths).
15. IS: 9527 (Part VI)–1989, Code of Practice for Design and Construction of Port and Harbour Structures (Block Work).
16. IS: 9527 (Part III)–1983, Code of Practice for Design and Construction of Port and Harbour Structures (Sheet Pile Walls).
17. IS: 9527 (Part IV)–1980, Code of Practice for Design and Construction of Port and Harbour Structures (Cellular Sheet Pile Structures).
18. IS: 875–1987, Code of Practice for Design Loads (other than earthquake) for Buildings and Structures: Part 1 (Dead Load) ; Part 2 (Imposed Load); Part 3 (Wind Load); Part 4 (Snow Load) ; Part 5 (Special Loads).
19. IS: 1893–1984, Recommendations for Earthquake Resistant Design of the Structures, I.S.I., New Delhi.
20. Minikin, R.R. (1963), *Winds, Waves Maritime Structures*, Charles Griffin & Co. Ltd., London, Second Edition.
21. Oza, H.P. and G.H. Oza (1985), *Dock and Harbour Engineering*, Charotar Publishing House, New Delhi.
22. Quinn, A.D. (1972), *Design and Construction of Ports and Marine Structures*, McGraw Hill Book Company, USA.
23. Sainflou, G. (1928), "Essais sur Les Digues maritimes Verticales," Annales des Ponts et Chaussees, 98, Part 4, pp. 5–48.
24. Stevenson, T, "Design and Construction of Harbours".

PRACTICE PROBLEMS

1. Enumerate the various components of a harbour to be considered in the layout.
2. Describe the characteristics of seawaves.
3. What are the different types of breakwaters? Under what conditions will you suggest a vertical breakwater?
4. What is a rubble-mound breakwater? Discuss the points in favour and against.
5. Describe with sketches a composite breakwater.
6. Describe the various methods of getting wave pressures on a vertical wall.
7. Write a short note on 'ship impact'.
8. Define a wharf. What are the types of construction and discuss their merits.
9. Design a (i) wall type breakwater, and (ii) mound type breakwater for the following data—

   Depth of water = 20.0
   Wind velocity = 150 km/hour
   Fetch of wave = 200 km.
10. Design a piled wharf structure for the following data—

   Weight of ship = 180000 kN
   Number of piles in a bent = 14
   Number of bents sharing impact = 15
   Velocity of ship = 10 mm/sec
   Permissible value of deflection = 4 mm

Rigid Retaining Wall

11.1 INTRODUCTION

Retaining walls are structures used to provide stability for earth or other material where conditions disallow the mass to assume its natural slope. These are commonly used to hold back or support soil banks, coal or ore piles, and water. They differ from other types of retaining structures because they do not require external bracing for stability. For this reason, retaining walls have been widely used for variety of purposes as illustrated in Fig. 11.1.

On the basis of method of achieving stability, retaining walls are classified into the following types:

(i) **Gravity wall:** A gravity wall is usually constructed in stone masonry, brick or plain concrete (Fig. 11.2a). The weight of the wall provides the required stability against the effects of retained soil. The practical use of gravity retaining walls is controlled by height limitations (< 6.0 m), whereby the required wall cross-section starts to increase significantly with tall heights because of effects of approximately triangular soil pressure distribution behind the wall.

(ii) **Semi-gravity wall:** This type of wall is constructed in concrete, and it also derives stability from its weight. A small amount of reinforcement is provided for reducing mass of the concrete (Fig. 11.2b).

(iii) **Cantilever wall:** A cantilever wall is a reinforced-concrete wall that utilizes cantilever action to retain the mass behind the wall from assuming a natural slope. Stability of this wall is partially achieved from the weight of the soil on the heel portion of the base slab (Fig. 11.2c). These walls are suitable for retaining backfills of moderate heights (5.0 m to 7.0 m).

(iv) **Counterfort retaining wall:** A counterfort retaining wall is similar to a cantilever retaining wall, except that it is used where the cantilever is long or for very high pressures behind the wall and has counterforts, which tie the wall and base together, built at intervals along the wall to reduce bending moments and shears. As shown in Fig. 11.2d, the counterfort is behind the wall and subjected to tensile forces. Such walls are used when height of fill exceeds 7.0 m.

(v) **Buttressed retaining wall:** A buttressed retaining wall is similar to a counterfort wall, except that the bracings are in front of the wall and is in compression instead of tension.

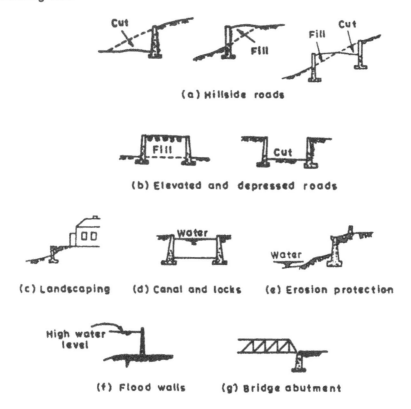

Fig. 11.1. Common Uses of Retaining Walls (Teng, 1977).

(vi) Crib wall: A crib wall is formed of timber, precast concrete, or prefabricated steel members, and filled with granular materials (Fig. 11.2e). Except for the exposed front face, crib walls often are completely covered with soil so that the cribbing is not visible. These are economical for small to moderate heights (4.0 m to 6.0 m).

(vii) Bridge abutment: Bridge abutments are often retaining walls with wing wall extensions to retain the approach fill and provide protection against erosion (Fig. 11.2f). They differ in two major respects from the usual retaining wall, namely: (a) they carry end reactions from the bridge span, and (b) they are restrained from the top so that active earth pressure is unlikely to develop.

(viii) Basement walls: Foundation walls of buildings including residential construction are retaining walls whose function is to contain the earth out of basements.

    Retaining walls are long structures. The usual design procedure to analyse a section of one unit in length except in counterfort or buttressed wall where a section that extends centre to centre distance between counterforts or buttresses is considered. The height of soil to be retained will usually vary along the length of the wall, and with homogeneous backfill and foundation conditions the most severe design loading occurs where the height is greatest. The wall cross-section required for this crucial location is also assumed for adjacent locations, although the actual height constructed may be lower.

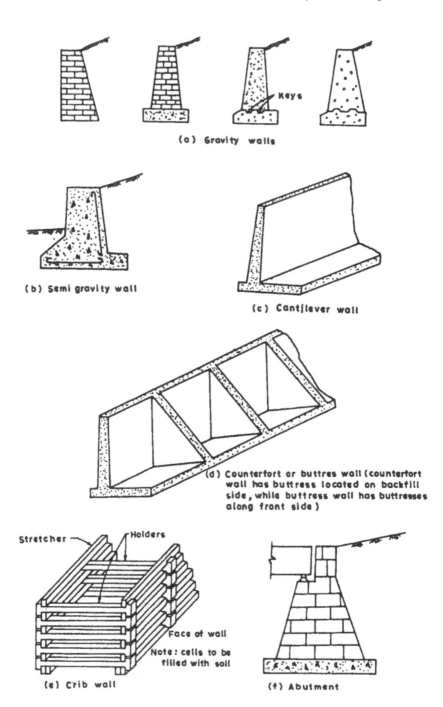

Fig. 11.2. Types of Retaining Walls.

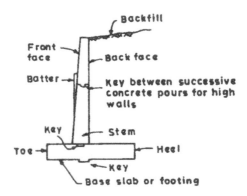

Fig. 11.3. Terms Used with Retaining Walls.

Figure 11.3 shows a typical section of a retaining wall illustrating the various terms used in the design.

11.2 COMMON PROPORTIONING OF RETAINING WALLS

The design of a retaining wall proceeds with the selection of tentative dimensions, which are then analysed for stability and structural requirements and are revised if necessary.

The tentative dimensions of a retaining wall may be selected on the basis of following informations:

(a) Gravity and semi-gravity walls

These walls, generally, are trapezoidal-shaped but also may be built with broken backs. The base and other dimensions should be such that the resultant of the forces falls within the middle one-third of the base. The top width of the wall should not be made less than 200 mm for the proper placement of the material. The base width ranges generally from 50 per cent to 70 per cent of the height of the wall. Small projections of the base both beyond the face and back of the wall are provided to reduce the pressure. The trial section is shown in Fig. 11.4.

(b) Cantilever wall

The tentative dimensions of a cantilever wall are shown in Fig. 11.5. Krishna and Jain (1966) suggested the following expressions for their proportioning on the concept that the resultant force remains within the middle third of the base:

(i) Horizontal backfill (Fig. 11.5a)

$$B = 0.95 \, H \sqrt{\frac{K_A}{(1-m)(1+3m)}} \qquad \qquad ... \ (11.1)$$

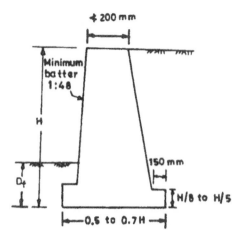

Fig. 11.4. Tentative Dimensions for a Gravity or Semigravity Wall.

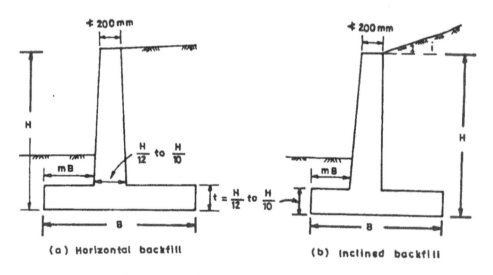

Fig. 11.5. Tentative Dimensions for a Cantilever Wall.

$$m = \frac{\text{Toe projection}}{B} = 1 - \frac{4}{9\beta} \qquad \ldots (11.2)$$

$$\beta = \frac{\gamma(H-t)}{q_a} \qquad \ldots (11.3)$$

where $\quad$ H = height of wall,

$\quad\quad\quad\quad$ B = base width of wall,

γ = density of backfill,
q_a = allowable soil pressure,
K_A = coefficient of active earth pressure, and
t = thickness of base slab ($H/12$ to $H/10$).

(ii) Inclined backfill (Fig. 11.5b)

$$B = H\sqrt{\frac{K_A \cos i}{(1-m)(1+3m)}} \qquad \qquad \text{... (11.4)}$$

$$m = 1 - \frac{3}{8\beta}. \qquad \qquad \text{... (11.5)}$$

Value of β is the same as given by Eq. 11.3.

The above equations are applicable for cohesionless backfill only. For $c - \phi$ soil, the base width may be taken 50 per cent to 70 per cent of the height of wall. The toe projection (mB) is taken 30 to 45 per cent of the base width.

(c) Counterfort or buttressed wall

The proportioning of toe and heel portion of the slab may be done as discussed above for cantilever wall. For economical design, the clear spacing between counterforts or buttresses is given by:

$$l^2 = 11.8\sqrt{\frac{H}{\gamma}} \qquad \qquad \text{... (11.6)}$$

where l = clear spacing between counterforts or buttresses, m,
          H = height of wall, m, and
          γ = density of soil, kN/m^3.

Typical proportions of counterfort retaining walls are shown in Fig. 11.6.

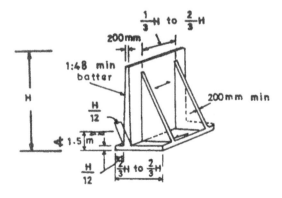

Fig. 11.6. Tentative Dimensions for a Counterfort Retaining Wall.

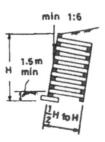

Fig. 11.7. Tentative Dimensions for a Crib Wall.

(d) Crib walls

Crib walls are usually constructed with a minimum batter of 1 to 6. The base width of wall ranges from 50 per cent to 90 per cent of height of wall. Figure 11.7 shows the common proportion of crib walls.

11.3 STABILITY OF RETAINING WALLS

The forces acting on a retaining wall are customarily taken per unit length for both gravity and cantilever walls. Counterfort walls are considered as a unit between joints, or as a unit centred on two counterforts.

Gravity and cantilever walls are acted upon by the forces shown in Figs. 11.8a and 11.8b respectively. The active earth pressure in gravity walls is computed using Coulomb's wedge theory taking angle of wall friction into account. In cantilever retaining walls it is convenient to determine the active earth pressure using Rankine's theory along the vertical plane BC passing through the heel of the wall (Fig. 11.8b). Weight of the soil above the heel slab is considered along with the weight of wall for stability analysis. Passive earth pressure is usually neglected.

The retaining walls are checked for structural stability and foundation stability by the procedure given below:

(a) Structural stability

The factor of safety against sliding, F_s is given by

$$F_s = \frac{\text{Horizontal resistance}}{\text{Horizontal force}} \qquad \qquad \text{... (11.7)}$$

where   Horizontal resistance = μ (total vertical force in granular soils),   ... (11.8a)

$\qquad\qquad\qquad\qquad = c_a B$ (in cohesive soils),   ... (11.8b)

μ = coefficient of base friction,

$\quad$ = 0.55 for coarse-grained soils without silt,

$\quad$ = 0.45 for coarse-grained soils with silt,

$\quad$ = 0.35 for silt,

$\quad$ = 0.60 for sound rock with rough surface,

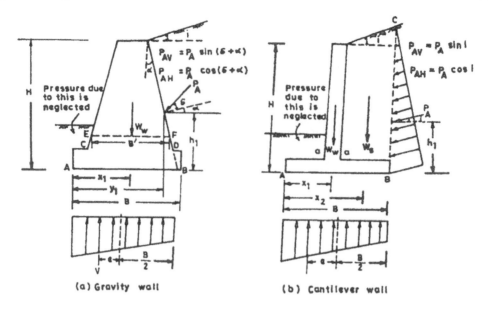

Fig. 11.8. Structural Stabilitiy of Retaining Wall.

c_a = base cohesion,
  = 0.5 c to 0.75 c,
c = unit cohesion of base soil, and
B = base width of wall.

In gravity walls (Fig. 11.8a)

$$F_s = \frac{\mu(W_w + P_{AV})}{P_{AH}}. \qquad \text{.. (11.9a)}$$

In cantilever walls (Fig. 11.8b)

$$F_s = \frac{\mu(W_w + P_{AV} + W_s)}{P_{AH}} \qquad \text{... (11.9b)}$$

where W_w = weight of retaining wall,
  P_{AV} = vertical component of earth pressure
    = $P_A \sin(\delta + \alpha)$ in gravity walls,
    = $P_A \sin i$ in cantilever walls,
  P_{AH} = horizontal component of earth pressure
    = $P_A \cos(\delta + \alpha)$ in gravity walls,
    = $P_A \cos i$ in cantilever walls, and
  W_s = weight of soil above heel slab.

The safety factor against sliding should be atleast 1.5 for cohesionless backfill and about 2.0 for cohesive backfill.

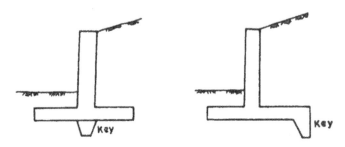

Fig. 11.9. Base Key in Retaining Wall.

When the recommended safety factor against sliding is difficult to attain, additional sliding stability may be derived from the use of a key beneath the base. The key is generally located either under the stem or at extreme end of the heel slab (Fig. 11.9). The benefit of provision of key is usually small unless it is embedded in hard strata or the key is sufficiently deep.

The safety factor against overturning, F_o is given by:

$$F_o = \frac{\text{Stabilising moment}}{\text{Overturning moment}}. \qquad \dots (11.10)$$

The stabilising and overturning moments are determined about point A (Figs. 11.8a and 11.8b). The expressions of F_o for gravity and cantilever walls are given below in Eqs. 11.11a and 11.11b respectively.

$$F_o = \frac{W_W \cdot x_1}{P_{AH} \cdot h_1 - P_{AV} \cdot y_1} \qquad \dots (11.11a)$$

$$F_o = \frac{W_w \cdot x_1 + W_s \cdot x_2}{P_{AH} \cdot h_1 - P_{AV} \cdot B}. \qquad \dots (11.11b)$$

Equations (11.11) may also be written as follows:

$$F_o = \frac{W_w \cdot x_1 + P_{AV} \cdot y_1}{P_{AH} \cdot h_1} \qquad \dots (11.12a)$$

$$F_o = \frac{W_w \cdot x_1 + W_s \cdot x_2 + P_{AV} B}{P_{AH} \cdot h_1} \qquad \dots (11.12b)$$

It may be noted that expressions (11.11) and (11.12) will give different value of F_o. Value of F_o may be obtained using both Eqs. 11.11 and 11.12 and lesser of the two adopted for design.

The factors of safety against overturning usually adopted are 1.5 and 2.0 respectively for granular and cohesive soils.

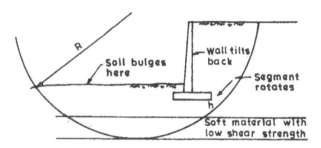

Fig. 11.10. Soil Shear Failure.

(b) Foundation stability

A retaining wall must also be proportioned to have sufficient factor of safety against failure of the foundation soil.

In general, foundation of a retaining wall is subjected to a vertical force, a horizontal force and moment. Therefore, the resultant load amounts to an eccentric-inclined load. The procedure of proportioning a footing subjected to eccentric-inclined load had already been dealt in Chapter 7.

Amount of tilt of a retaining wall can be estimated by using Eqs. 7.61 to 7.67 of Chapter 7. For adequate design the tilt should be less than the amount so that the wall at any point does not displace more than 0.1 per cent of the height of wall (Terzaghi, 1934; Rowe, 1952).

Where there is a layer of weak cohesive soil, the bearing capacity of this soil must be reliably evaluated. A retaining wall should not be attempted if weight of backfill will exceed the allowable bearing value of the underlying soil. In such cases lightweight backfill should be used instead of backfill, or replace the soft underlying soil with compacted granular soil. It will also help in reducing tilt if any. If the layer of weak soil is located within a depth of about 1.5 times the height of retaining wall, the stability of foundation soil should be investigated with the possibility of a sliding surface passing through this weak layer. This may be analysed by Swedish circle method (Fig. 11.10). The retaining wall should be proportioned so as to provide a factor of safety against deep foundation failure, at least equal to 2.0.

11.4 STRUCTURAL DESIGN OF RETAINING WALLS

Gravity walls

After satisfying the stability conditions, a gravity wall should be checked for shear, compressive and tensile stresses at different horizontal sections. The procedure of computing these stresses at any horizontal section say EF (Fig. 11.8a) is given below.

(i) At any section say EF (Fig. 11.8a), the shear stress is computed using Eq. (11.13).

$$\text{Shear stress} = \frac{P_{AH}}{B' \times 1} \qquad \qquad \dots (11.13)$$

where P_{AH} = total horizontal component of forces acting at section EF, and

B' = width of wall at EF.

The above shear stress should be less than the permissible shear stress.

(ii) At any section say EF (Fig. 11.8a) the maximum compressive stress, q_1 is given by the equation

$$q_1 = \frac{V}{B'}\left(1 + \frac{6e}{B'}\right)$$ ... (11.14)

where V = vertical component of the resultant force acting at section EF, and

   e = eccentricity with respect to the centre line of the section under consideration.

For satisfactory design q_1 should be less than the permissible compressive stress of the material of wall.

(iii) At any section, say EF (Fig. 11.8a), the maximum tensile stress q_2 (if $e > B'/6$) is given by:

$$q_2 = \frac{V}{B'}\left(1 - \frac{6e}{B'}\right).$$ ... (11.15)

The stress q_2 should be less than the permissible tensile stress of material of the wall.

Usually the gravity wall is checked at two sections, (i) CD, and (ii) AB. At section AB, as the soil is in contact with the base of the wall, for satisfactory design q_1 should be less than the allowable soil pressure and q_2 should not be less than zero.

Cantilever walls

In cantilever walls, design of upright, toe and heel slabs (Fig. 11.11a) are carried out separately. Forces on these three slabs which are taken into design are respectively shown in Figs. 11.11b, 11.11d and 11.11e.

The design of upright slab is carried out in the following steps.

(i) Maximum moment at section BC (Fig. 11.11b) is obtained. For this moment the thickness of stem and the amount of reinforcement are computed in the usual way.

(ii) The stem may be tapered to minimum (i.e. 200 mm) towards the top as the moment decreases to zero at the top of the wall.

(iii) The reinforcement may be curtailed towards top with the following approach (Fig. 11.11b).

Amount of steel, A_{sty} required at any section aa is proportional to

$$\frac{\text{Moment at aa}}{\text{Thickness of the stem at aa}}$$

Moment, M_y at depth y from the top of upright slab will be proportional to y^3.
Therefore, if d_y is the thickness of upright slab at section aa, then

$$A_{sty} \propto \frac{y^3}{d_y}.$$ ... (11.16)

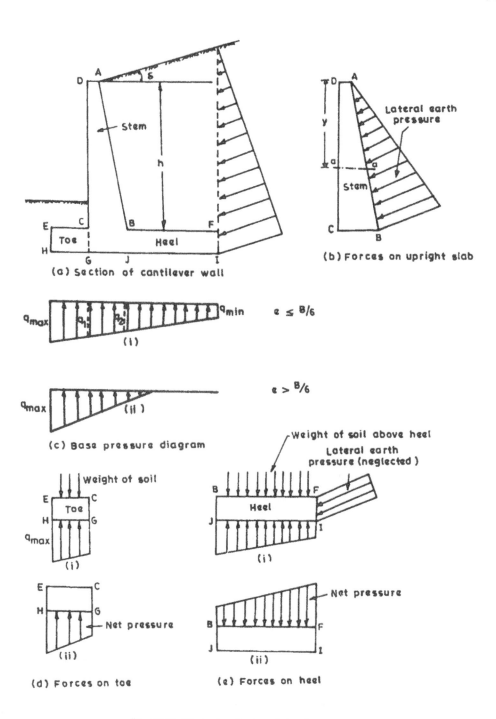

(a) Section of cantilever wall

(b) Forces on upright slab

(c) Base pressure diagram

(d) Forces on toe

(e) Forces on heel

Fig. 11.11. Illustrating Design of Cantilever Wall.

Equation (11.16) may also be written as:

$$\frac{A_{sty}}{A_{st}} = \frac{y^3}{h^3} \frac{d}{d_y}$$ ... (11.17)

where A_{st} = amount of steel at depth h determined in step (i), and
 d = thickness of the stem at depth h determined in step (i).

At any depth y, d_y is computed first. As the amount of reinforcement is known at the bottom of upright of slab, i.e. at depth h, A_{sty} is computed using Eq. 11.17. This amount of reinforcement should be safe in bond stress. Every reinforcing bar should be extended twelve times the diameter of bars or 230 mm whichever is more beyond the point it is no longer required for bond.

(iv) Temperature reinforcement equal to 0.15 per cent of section area of stem is provided in the longitudinal direction. Out of this reinforcement, 2/3rd is provided on the exposed side, while the remaining 1/3rd is provided on the backfill side. Nominal vertical reinforcement is provided on the exposed side to support the temperature reinforcement. A typical reinforcement diagram is shown in Fig. 11.12.

The toe will be acted on by the forces shown in Fig. 11.11d–i. The toe and weight of the soil above the toe acts in downward direction. The net pressure diagram will be a trapezoidal as shown in Fig. 11.11d–ii. For this pressure, the toe is designed as the cantilever with the reinforcement provided at the bottom (Fig. 11.12).

The heel is subjected to the weight of earth above it, the vertical component of earth pressure and the upward pressure from the soil below (Fig. 11.11e–i). The net result is downward pressure (Fig. 11.11e–ii) for which the heel is designed as cantilever with reinforcement at the top (Fig. 11.12).

In the base slab, reinforcement equal to 0.15 per cent of the concrete area is provided in the direction perpendicular to the main reinforcement at the top of the heel slab and at the bottom of toe slab (Fig. 11.12). This reinforcement covers the bending moments developed due to the fixity in upright slab, heel and toe.

In practice, cantilever walls are built in two stages: (i) the base of the wall, and (ii) the upright slab. A key is usually provided at the joint which is checked in shear. The shear stress is computed using the equation

$$\text{Shear stress} = \frac{P_H}{\text{width of key}}$$ ... (11.18)

where P_H = horizontal component of total pressure at key.

The width of key is usually kept about 1/4th the thickness of the wall, because the tensile zone of the wall is not so effective to resist shear as that in compression zone.

Counterfort wall

In counterfort wall, upright and heel slabs do not act as cantilever but act as continuous slab supported by the counterforts (Fig. 11.2d).

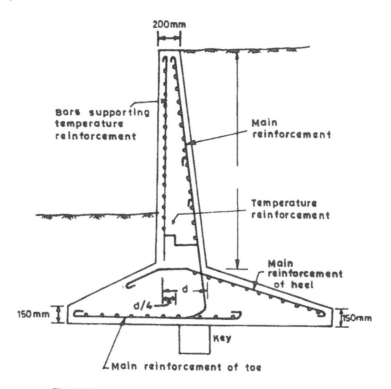

200mm

Bars supporting
temperature
reinforcement

Main
reinforcement

Temperature
reinforcement

Main
reinforcement
of heel

150mm

d/4

d

150mm

Key

Main reinforcement of toe

Fig. 11.12. Details of Reinforcement in a Cantilever Wall.

As the intermediate panels of slab are countinuous over counterforts, a positive bending moment equal to $pl^2/16$ and a negative bending moment equal to $pl^2/12$ are allowed in the design of both upright and heel slabs. Here p is the maximum pressure intensity (Figs. 11.11b, 11.11e–ii) and l is the clear spacing between the counterforts.

Sometimes, counterforts are also provided in the toe portions. If these are provided, design of the toe slab should be carried out in the same way as that of heel slab considering it continuous over the counterforts. Otherwise, the design is similar to that of cantilever wall.

The design of counterfort is discussed for two cases: (i) counterfort in toe portion slab, and (ii) no counterfort in toe portion. For the first case the back counterfort is acted on by the forces shown in Fig. 11.13b. The critical section for computing the bending moment and shear force is EF. Thickness of the counterfort at EF, amount of reinforcement, shear stress and bond stress are computed using the following expressions:

$$M_R = 0.255 f_{ck} \tan \beta d^3 \qquad \qquad \qquad \text{... (11.19)}$$

$$M_R = 0.87 f_y A_{st} \left[d - \sqrt{\frac{f_y A_{st}}{f_{ck} \tan \beta}} \right] \qquad \text{... (11.20)}$$

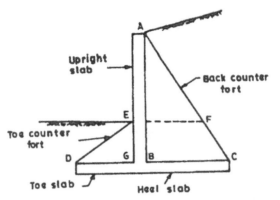

(a) Section of a counterfort wall

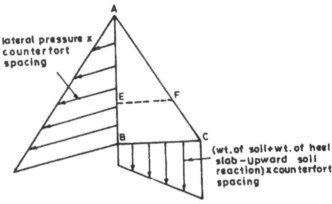

(b) Forces on the back counterfort

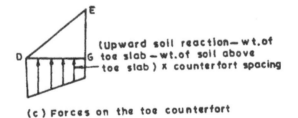

(c) Forces on the toe counterfort

Fig. 11.13. Illustrating the Design of Counterforts.

$$\tau_V = \frac{V_u \pm \dfrac{M_u}{d} \tan\beta}{bd} \qquad \dots (11.21)$$

where β is the angle which the inclined face of the counterfort makes with the vertical. Equations (11.19 to 11.21) are utilized for designing sections of linearly varying width.

The reinforcement is curtailed on the basis of the concept discussed for cantilever wall.

In case, there is no counterfort in toe portion, the bending moment and shear forces are computed at the section BC (Fig. 11.13b).

The toe counterfort is subjected to the forces shown in Fig. 11.13c. The bending moment and shear force for carrying out the design will be obtained at section EG. The main reinforcement will be provided at the bottom.

The upright slab has a net horizontal pressure and has a tendency to separate out from the counterfort. Horizontal ties are provided to connect together. Amount of the steel required to be provided in the form of horizontal ties is given by

$$\text{Steel required} = \frac{\text{Net horizontal force}}{\text{Permissible tensile stress of steel}} \qquad \text{... (11.22)}$$

Similarly, the heel receives a net vertical downward load and has a tendency to separate from the counterfort. Vertical ties are provided to connect the two together. The steel required in the form of vertical ties is given by

$$\text{Steel required} = \frac{\text{Net vertical force}}{\text{Permissible tensile stress of steel}} \qquad \text{... (11.23)}$$

The reinforcement in the counterfort is usually provided as shown in Fig. 11.14.

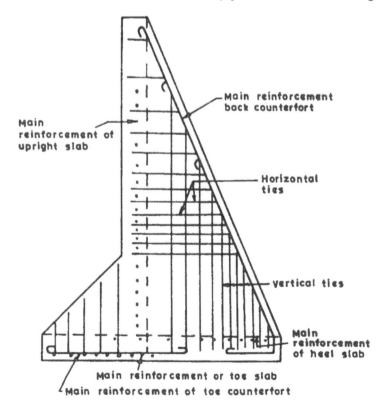

Fig. 11.14. Details of Reinforcement in a Counterfort Wall.

11.5 BACKFILL DRAINAGE

Fine grained soils cause large earth pressure against retaining walls; therefore, it is not economical to use such soils as backfill material. Ideal backfill materials are purely granular soils (clean sand, gravel, or sand and gravel) containing less than about five per cent of very fine sand, silt, or clay particles. If such a free-draining material is expensive in the locality, it should be preferably used in a wedge bounded by the back of the retaining wall and a plane rising at an angle of not more than 60 degrees with the horizontal. With this wedge of granular material against the wall, the earth pressure may be computed as if the entire backfill consists of the granular soil.

It is desirable to provide a suitable drainage system in the backfill to avoid the larger lateral pressure which will be induced if the backfill does not drain readily. The amount of drainage work depends upon the permeability of the backfill material. In pervious backfill, weep holes, Fig. 11.15a or a line of drain pipe, Fig. 11.15b, will suffice. Semipervious backfill (soils containing a small amount of fine sand, silt, or clay particles) require strips of filter material in addition to the drain pipes or weep holes, Fig. 11.15c. To provide adequate drainage in fine-grained backfill, a drainage blanket or double blankets are necessary, Fig. 11.15d and e (Teng, 1977).

All drain pipes or tiles should be provided with adequate clean-outs for periodical cleaning. The drain tiles or pipes should be embedded in selected filter material so as to prevent clogging the tile or pipes and carrying away soil particles by water.

Criteria for selecting the filter material are as follows (Terzaghi and Peck, 1981)

$$\frac{(D_{15})_{\text{filter}}}{(D_{85})_{\text{protected soil}}} \leq 5 \leq \frac{(D_{15})_{\text{filter}}}{(D_{15})_{\text{protected soil}}} \qquad \dots (11.24)$$

11.6 BACK ANCHORAGE OF RETAINING WALLS

If height of the backfill is large, it is more economical to anchor the wall back into the earth. Usually anchors are provided at 1/3rd height of the wall from base so that the force in the anchor becomes equal to the lateral earth pressure (Fig. 11.16). Due to this system, the bending moment in the upright slab decreases in the portion BC and becomes very small at the point C. It not only makes the thickness of the upright slab considerably smaller but also helps in keeping the base pressures within limits.

Anchoring is done by means of steel tie rods encased in concrete and tied with an anchor block made of concrete. If l is the centre to centre spacing of ties, then the area of steel in each tie will be given by the following equation:

$$\text{Area of steel in each tie} = \frac{\text{Total earth pressure} / l}{\text{Permissible tensile stress of steel}}.$$

The design of anchor block has been discussed in the next chapter.

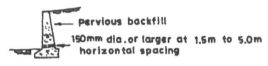

(a) Weep holes

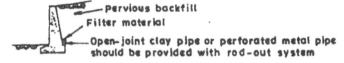

(b) Longitudinal drain pipe

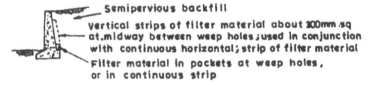

(c) Weep holes with filter strips

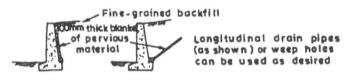

(d) Blanket drain

(e) Double blanket drain

Fig. 11.15. Common Types of Retaining Wall Drainage.

ILLUSTRATIVE EXAMPLES

Example 11.1

Design a solid gravity retaining wall to retain a 4.0-m embankment. The backfill soil is cohesionless having γ = 19 kN/m³ and ϕ = 33° and slopes 15° to the horizontal. Angle of wall friction may be taken as 20°. The allowable soil pressure at 1.0 depth below ground surface is 300 kN/m² and coefficient of base friction is 0.6. The wall is located in a seismic region having α_h = 0.08 and α_v = 0.04.

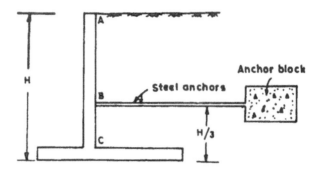

Fig. 11.16. Back Anchorage of Retaining Wall.

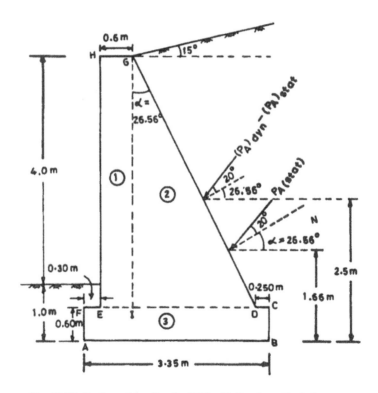

Fig. 11.17. Section of the Gravity Wall with Resultant Earth Pressure.

Solution

1. Trial section of retaining wall

   The trial section is as shown in Fig. 11.17

$$\gamma = 19 \ kN/m^3$$
$$\phi = 33°$$
$$\delta = 20° \ (assumed)$$

$$\alpha = \tan^{-1}\frac{2.2}{4.4} = 26.56°$$

$$i = 15°.$$

2. Earth pressure

By Mononobe and Okabe (1992), the dynamic active earth pressure coefficient is given by

$$(K_A)_{dyn} = \frac{(1 \pm \alpha_v)\cos^2(\phi - \alpha - \psi)}{\cos\psi \cos^2\alpha \cos(\delta + \alpha + \psi)} \times$$

$$\left[\frac{1}{1 + \left\{\dfrac{\sin(\phi + \delta)\sin(\phi - i - \psi)}{\cos(\alpha - i)\cos(\delta + \alpha + \psi)}\right\}^{\frac{1}{2}}}\right]^2$$

$$\psi = \tan^{-1}\left(\frac{\alpha_h}{1 \pm \alpha_v}\right)$$

Taking +ve sign

$$\psi = \tan^{-1}\left(\frac{\alpha_h}{1 + \alpha_v}\right) = \tan^{-1}\left(\frac{0.08}{1 + 0.04}\right) = 4.39°$$

$$(K_A)_{dyn} = \frac{(1 + 0.04)\cos^2(33 - 26.56 - 4.39)}{\cos 4.39 \cos^2 26.56 \cos(20 + 26.56 + 4.39)} \times$$

$$\left[\frac{1}{1 + \left\{\dfrac{\sin(33 + 20)\sin(33 - 15 - 4.39)}{\cos(26.56 - 15)\cos(20 + 26.56 + 4.39)}\right\}^{\frac{1}{2}}}\right]^2 = 0.858$$

Taking −ve sign

$$\psi = \tan^{-1}\left(\frac{0.08}{1 - 0.04}\right) = 4.76°$$

$$(K_A)_{dyn} = \frac{(1 - 0.04)\cos^2(33 - 26.56 - 4.76)}{\cos 4.76 \cos^2 26.56 \cos(20 + 26.56 + 4.76)} \times$$

$$\left[\cfrac{1}{1 + \left[\cfrac{\sin(33+20)\sin(33-15-4.76)}{\cos(26.56-15)\cos(20+26.56+4.76)} \right]^{\frac{1}{2}}} \right]^2 = 0.804.$$

Adopt the higher value of $(K_A)_{dyn}$, i.e. 0.858.

Dynamic earth pressure

$$(P_A)_{dyn} = \frac{1}{2} \gamma K_{A(dyn)} H^2$$

$$= \frac{1}{2} \times 19 \times 0.858 \times 5.0^2$$

$$= 203.77 \text{ kN/m.}$$

Static earth pressure coeff.

$$(K_A)_{stat} = \frac{\cos^2(\phi-\alpha)}{\cos^2 \alpha \cos(\alpha+\delta)} \left[\cfrac{1}{1 + \left\{ \cfrac{\sin(\phi+\delta)\sin(\phi-i)}{\cos(\alpha-i)\cos(\alpha+\delta)} \right\}^{\frac{1}{2}}} \right]^2$$

$$(K_A)_{stat} = \frac{\cos^2(33-26.56)}{\cos^2 26.56 \cos(26.56+20)} \times$$

$$\left[\cfrac{1}{1 + \left\{ \cfrac{\sin(33+20)\sin(33-15)}{\cos(26.56-15)\cos(26.56+20)} \right\}^{\frac{1}{2}}} \right]^2$$

$$= 0.696.$$

Static earth pressure

$$P_{A(stat)} = 1/2 \times 19 \times 0.696 \times 5^2 = 165.42 \text{ kN/m.}$$

Dynamic increment

$$= 203.77 - 165.42 = 38.35 \text{ kN/m.}$$

The static earth pressure will act at $h/3$ from the base and dynamic increment at $h/2$ from the base.

3. Structural stability analysis
Refer Fig. 11.17

  Weight of retaining wall

$$= \text{Weight of HGIE + Weight of GID + Weight of ABCF}$$
$$= 23 \times 0.6 \times 4.4 + 1/2(23 \times 2.2 \times 4.4) + 23 \times 3.35 \times 0.6$$
$$= 60.72 + 111.32 + 46.23$$
$$= 218.27 \text{ kN.}$$

Horizontal component of earth pressure

$$= 203.77 \cos 46.56 = 140.11 \text{ kN/m.}$$

Vertical component of earth pressure

$$= 203.77 \sin 46.56 = 147.95 \text{ kN/m.}$$

Safety factor against sliding

$$= \frac{(218.27 + 147.95) \times 0.60}{140.11} = 1.57 \text{ (safe).}$$

Stabilizing moment about point A

$$= 60.72 \times (0.3 + 0.3) + 111.32 \times \left(0.9 + \frac{2.2}{3}\right) + 46.23 \times \frac{3.35}{2} +$$
$$165.42 \sin 46.56[3.1 - 1.06 \tan 26.56] + 38.35 \sin 46.56 \times$$
$$[3.1 - 1.9 \tan 26.56]$$
$$= 664.24 \text{ kNm.}$$

Overturning moment about point A

$$= 165.42 \cos 46.56 \times 5/3 + 38.35 \cos 46.56 \times 2.5$$
$$= 255.47 \text{ kNm.}$$

Safety factor against overturning

$$= \frac{664.24}{255.47} = 2.6 \text{ (safe)}$$

4. Foundation stability

$$\text{Total moment about point A} = 664.24 - 255.47 = 408.77 \text{ kNm}$$
$$\text{Total vertical load} = 218.27 + 147.95 = 366.22 \text{ kN}$$

Eccentricity of vertical load $= \dfrac{3.35}{2} - \dfrac{408.77}{366.22} = 0.559$ m.

As eccentricity is less than $B/6$, maximum base pressure on section AB

$$= \frac{V}{B}\left(1 + \frac{6e}{B}\right)$$

$$= \frac{366.22}{3.35}\left[1 + \frac{6 \times 0.559}{3.35}\right]$$

$$= 218.76 \text{ kN/m}^2 < 300 \text{ kN/m}^2, \text{ o.k. safe.}$$

Minimum base pressure on section AB

$$= \frac{366.22}{3.35}\left[1+\frac{6\times0.559}{3.35}\right]$$

$$= 0.0 \text{ kN/m}^2 \text{ (safe)}.$$

Section ED of the wall (Fig. 11.17) may also be checked in the similar manner as illustrated above. In this case, the projections of the base slab on either side of ED are very small; therefore, the wall will also be safe considering its stability at section ED.

Example 11.2

A retaining wall has to be constructed to retain a soil embankment 6.5 m high above the ground surface. The top of the earth retained makes an angle of 10° with horizontal. The following properties of the backfill soil are obtained from tests:

$$\phi = 32°, \gamma = 16 \text{ kN/m}^3, q = 25 \text{ kN/m}^2.$$

The soil below ground surface was explored up to 15 m depth and it was found that the soil is medium dense sand with following properties:

$$\gamma = 18 \text{ kN/m}^3, \phi = 38°.$$

A plate load test was conducted at the site at 2.0 m depth below the ground surface. Load intensity versus settlement curve is shown in Fig. 11.18. Permissible settlement = 50 mm.

Design a cantilever retaining wall and show the details of reinforcement.

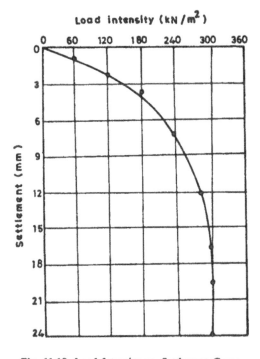

Fig. 11.18. Load Intensity vs. Settlement Curve.

Solution

1. Selection of type of foundation

As the foundation soil is medium dense sand up to the significant depth, provide open foundation at 2.0 m depth below ground level.

Total height of wall from base level = 6.5 + 2.0 = 8.5 m.

2. Proportioning of wall

Average thickness of the upright step and base slab is taken equal to 600 mm

$$K_A = \cos i \left(\frac{\cos i - \sqrt{\cos^2 i - \cos^2 \phi}}{\cos i - \sqrt{\cos^2 i - \cos^2 \phi}} \right)$$

For backfill soil:

$$K_{A1} = \cos 10 \left(\frac{\cos 10 - \sqrt{\cos^2 10 - \cos^2 32}}{\cos 10 + \sqrt{\cos^2 10 - \cos^2 32}} \right) = 0.32.$$

For soil below natural ground surface:

$$K_{A2} = \cos 10 \left(\frac{\cos 10 - \sqrt{\cos^2 10 - \cos^2 38}}{\cos 10 + \sqrt{\cos^2 10 - \cos^2 38}} \right) = 0.246.$$

According to Krishna and Jain (1966):

Base width $\qquad B = \sqrt{\dfrac{K_A \cos i}{(1-m)(1+3m)}}.$

where $\qquad m = 1 - \dfrac{3}{8\beta}$

and $\qquad \beta = \dfrac{\gamma_2(H-t)}{q_a} = \dfrac{18(8.5-0.6)}{300} = 0.474$

$$[q_a = 300 \text{ kN/m}^2, \text{ assumed}]$$

$$m = 1 - \frac{3}{8 \times 0.474} = 0.209$$

$$B = 8.5\sqrt{\frac{0.32\cos 10}{(1-0.209)(1+3\times 0.209)}} = 4.21\text{m}.$$

Adopt the base width as 4.6 m.

The trial proportioning is as shown in Fig. 11.19a.

3. Earth pressure computation

Total earth pressure (Refer Fig. 11.18b)

$$P = P_1 + P_2 + P_3 + P_4$$

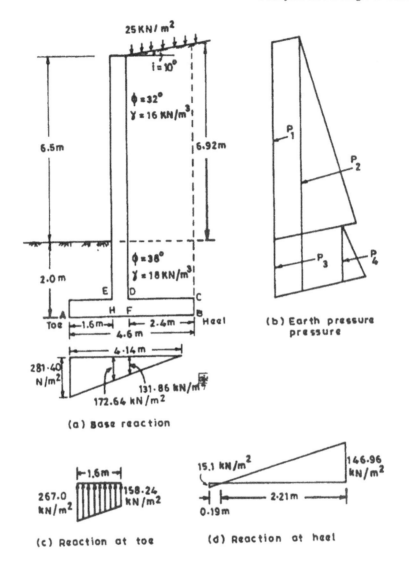

Fig. 11.19. Section of Cantilever Wall with Earth Pressure and Reaction Diagrams.

. $P_1 = K_{A1} \times 25 \times 6.92$

  $= 0.32 \times 25 \times 6.92 = 55.38$ kN acting at height of 5.46 m above B.

$P_2 = \dfrac{1}{2} K_{A1} \gamma_1 (6.92)^2$

  $= \dfrac{1}{2} \times 0.32 \times 16 \times (6.92)^2 = 122.59$ kN acting at a height of 4.30 m above B.

$P_3 = 0.246 \times (16 \times 6.92 + 25) \times 2.0 = 66.77$ kN acting at a height of 1.0 m above B.

$P_4 = \dfrac{1}{2} \times K_{A2}\gamma_2 \times (2.0)^2 = \dfrac{1}{2} \times 0.246 \times 18 \times 4 = 8.856$ kN acting at a height of 0.667 m above B.

$P = P_1 + P_2 + P_3 + P_4 = 253.6$ kN

It will act at a height of $\bar{h}$ from point B, where

$$\bar{h} = \frac{55.38 \times 5.46 + 122.59 \times 4.3 + 66.77 \times 1.0 + 8.856 \times 0.667}{253.6}$$

$= 3.55$ m

$P_H = 253.6 \cos 10 = 249.74$ kN

$P_V = 253.6 \sin 10 = 44.03$ kN.

4. Forces on wall

Sl. No.	Force	Description (kN)	Vertical Component	Lever Arm w.r.t Point B (m)	Moment kNm
1.	Wt. of upright slab	7.9 × 1 × 24 × 0.6	113.76	2.7	307.15
2.	Wt. of base slab	4.6 × 0.6 × 1 × 24	66.24	2.3	152.35
3.	Wt. of soil above heel above G.L.	2.4 × 6.71 × 1 × 16	257.66	1.18	304.00
4.	Wt. of soil above heel below G.L.	2.4 × 1.4 × 1 × 18	60.48	1.2	72.57
5.	Wt. of soil above toe	1.6 × 1.4 × 1 × 18	40.32	3.8	153.21
6.	Earth pressure	253.6			
	(i) Horizontal component	249.74		3.55	886.57
	(ii) Vertical component		44.03		
			582.49		1875.85

5. Bearing capacity computation

Total moment = 1875.85 kN

Total vertical load = 582.49 kN

Eccentricity with respect to point B $= \dfrac{1875.85}{582.49} = 3.22$ m.

Eccentricity e with respect to centre = 3.22 − 2.3 = 0.92 m

$$\frac{e}{B} = \frac{0.92}{4.6} = 0.2$$

The bearing capacity computations are carried out as proposed by Saran (1969) and Prakash and Saran (1971).

For $\qquad \frac{e}{B} = 0.20$ and $\phi = 38°$

$$N_\gamma = 25, \; N_q = 36$$

$$Q_{de} = \frac{4.6 \times 1.0}{3} \left(\frac{1}{2} \times 18 \times 4.6 \times 25 + 18 \times 2.0 \times 36 \right)$$

$$= 2851.2 \text{ kN} \gg 582.49 \text{ kN, hence safe}$$

$$\frac{S_e}{S_0} = 1.0 - 1.63 \frac{e}{B} - 2.63 \left(\frac{e}{B}\right)^2 + 5.83 \left(\frac{e}{B}\right)^3$$

$$= 1.0 - 1.63(0.2) - 2.63(0.2)^2 + 5.83(0.2)^3 = 0.615$$

$$\frac{S_m}{S_0} = 1.0 + 2.31 \left(\frac{e}{B}\right) - 22.61 \left(\frac{e}{B}\right)^2 + 31.54 \left(\frac{e}{B}\right)^3$$

$$= 1.0 + 2.31(0.2) - 22.61(0.2)^2 + 31.54(0.2)^3 = 0.81.$$

As the value of permissible settlement is 50 mm.

$$S_m = 50 \text{ mm}$$

$$S_0 = \frac{50}{0.81} = 61.70 \text{ mm}$$

$$\frac{S_p}{61.70} = \left(\frac{30(460 + 30)}{460(30 + 30)}\right)^2 = 0.28$$

$$S_p = 17.28 \text{ mm.}$$

Pressure corresponding to 17.28 mm settlement ≈ 300 kN/m² (Fig. 11.18).

Actual pressure on the foundation $= \dfrac{582.49}{4.6 \times 1} = 126.62$ kN/m², hence safe.

6. Safety factor against sliding and overturning

$$F_s = \frac{582.49 \tan(2/3 \times 38°)}{249.74} = 1.1 < 1.5 \text{ (unsafe).}$$

Shear key should be provided to check sliding.

Total moments of vertical loads about B = 989.28 kNm.

Distance of C.G. of vertical loads from point B $= \dfrac{989.28}{582.49} = 1.69$ m

Distance of C.G. of vertical loads from point A $= 4.6 - 1.69 = 2.91$ m

$$F_o = \frac{582.49 \times 2.91}{249.74 \times 3.55} = 1.91 (\text{safe}).$$

7. Upward reaction

As the contact pressure in eccentrically loaded foundation is still unknown, the design is carried for linear pressure distribution.

$$\text{As } \frac{e}{B} = 0.20 > \frac{1}{6}.$$

Effective width $= 3 \times (2.3 - 0.92) = 4.14$ m.

$$\text{Maximum base pressure} = \frac{2 \times 582.49}{4.41 \times 1} = 281.40 \text{ kN/m}^2.$$

Base pressure diagram is shown in Fig. 11.19a.

8. Design of toe slab

To be on the safe side, weight of the soil above toe has been neglected.
Weight of toe slab $= 0.6 \times 24 = 14.4$ kN/m^2.
Pressure distribution on toe is shown in Fig. 11.19c.

$$\text{Moment about EH} = \left[\frac{267.0 + 158.24}{2} \times 1.6 \times \left\{ \frac{2 \times 267.0 + 158.24}{267.0 + 158.24} \right\} \times \frac{1.6}{3} \right]$$

$$= 295.35 \text{ kNm.}$$

Maximum factored moment $= 295.35 \times 1.5 = 443.0$ kNm.
Using M-20 concrete

$$\text{Thickness of slab, } d = \sqrt{\frac{443.0 \times 10^6}{0.138 \times 20 \times 1000}}$$

Provide thickness of slab as 450 mm effective and 500 mm overall.
Area of tension steel

$$M_u = 0.87 \times \sigma_y A_{st} \left(d - \frac{\sigma_y A_{st}}{f_{ck} b} \right)$$

$$443.0 \times 10^6 = 0.87 \times 415 A_{st} \left(450 - \frac{415 A_{st}}{20 \times 1000} \right)$$

$$A_{st}^2 - 21686.75 \, A_{st} + 59.13 \times 10^6 = 0$$

$$A_{st} = 3198.28 \text{ mm}^2.$$

Provide 25 mm dia bars at 150 mm c/c spacing.

Per cent of tension steel $p = \dfrac{100 \times 3272.49}{1000 \times 450} = 0.73\%$.

Shear strength of concrete $= 0.55$ N/mm^2.

Nominal shear stress

$$\tau_v = \frac{V_u}{bd} = \frac{1.5 \times \left(\dfrac{267 + 158.24}{2} \right) \times 1.6}{1000 \times 450} \times 1000$$

$$= 1.13 \text{ N/mm}^2 > 0.55 \text{ N/mm}^2 \text{ (unsafe)}.$$

So adopt the effective thickness of slab equal to 700 mm and overall thickness equal to 750 mm.

$$\text{Nominal shear stress } \tau_v = \frac{1.5 \times \left(\dfrac{267 + 158.24}{2} \right) \times 1.6}{1000 \times 700} = 0.72 \text{ N/mm}^2.$$

Provide a reinforcement, $A_{st} = 5040.0$ mm^2.

Use 28 mm diameter bars at 120 mm c/c spacing.

Percentage of steel $\qquad = 1.5\%$

Shear strength of concrete $\quad = 0.72$ N/mm$^2 = \tau_v$ (hence safe).

9. Design of heel slab

    Downward load on heel slab

$$= 24 \times 0.6 + 18.0 \times 1.4 \times 16 \left(\frac{6.5 + 6.92}{2} \right)$$

$$= 146.96 \text{ KN/m}^2.$$

Net pressure distribution on the heel slab will be as shown in Fig. 11.19d.

Maximum moment

$$\frac{1}{2} \times 146.96 \times 2.21 \times \left(0.19 + \frac{2}{3} \times 2.21 \right) - \frac{1}{2} \times 15.1 \times 0.91 \times \frac{0.19}{3} = 269.78 \text{ kNm}.$$

$$\text{Maximum factored moment} = 1.5 \times 269.78 = 404.67 \text{ kNm}.$$

$$\text{Thickness of heel, } d = \sqrt{\frac{404.67 \times 10^6}{0.138 \times 20 \times 1000}} = 382.9 \text{ mm}.$$

Provide 700 mm effective and 750 mm overall depth as for toe slab.

Area of tensile reinforcement is given by:

$$404.67 \times 10^6 = 0.87 \times 415 \times A_{st} \left[700 - \frac{415 A_{st}}{20 \times 1000} \right]$$

or, $\qquad\qquad\qquad\qquad A_{st} = 1685.36$ mm.

Provide 16 mm diameter bars at 100 mm spacing c/c. Therefore,

$$\text{Provided } A_{st} = 2010.0 \text{ mm}^2.$$

Percentage of tension steel $= \dfrac{100 \times 2010.0}{1000 \times 700} = 0.29\%.$

Shear strength of concrete, $\tau_c = 0.37$ N/mm^2.

Nominal shear stress

$$= \dfrac{1.5 \times \left(\dfrac{1}{2} \times 146.96 \times 2.21 - \dfrac{1}{2} \times 0.19 \times 15.1\right)}{1000 \times 700} \times 1000$$

$$= 0.34 \text{ N/mm}^2 < \tau_c \text{ (hence safe).}$$

10. Design of upright slab

Maximum factored moment

$$= 1.5 \times 249.74 \times (3.55 - 0.60)$$
$$= 1105.10 \text{ kNm.}$$

Thickness of upright slab $= \sqrt{\dfrac{1105.10 \times 10^6}{0.138 \times 20 \times 1000}} = 632.77$ mm.

Provide 650 mm effective thickness and 700 mm overall thickness.
Area of tension steel

$$1105.10 \times 10^6 = 0.87 \times 415 A_{st} \left(650 - \dfrac{415 A_{st}}{20 \times 1000}\right)$$

$$A_{st}^2 - 31325.3 A_{st} + 1.475 \times 10^8 = 0$$

$$A_{st} = 5772.7 \text{ mm.}$$

Provide 28 mm diameter bars at 100 mm c/c spacing.

$(A_{st} = 6157.5 \text{ mm}^2 > 5772.7 \text{ mm}^2)$ o.k.

Percentage of tensile steel $= \dfrac{100 \times 6157.5}{1000 \times 700} = 0.88\%.$

Shear strength of concrete $\tau_c = 0.59$ N/mm^2.

Nominal shear stress $\tau_v = \dfrac{1.5 \times 249.74 \times 1000}{1000 \times 700}$

$$= 0.54 \text{ N/mm}^2 \ [< \tau_c, \text{ safe}]$$

11. Curtailment of reinforcement

Amount of steel, A_{ty} required at any section aa is proportional to

$$\dfrac{\text{Moment at aa}}{\text{Thickness of the stem at a}}.$$

Therefore,

$$A_{sty} \propto \dfrac{y^3}{d_y} \text{ or } \dfrac{A_{sty}}{A_{st}} = \dfrac{y^3}{h^3} \cdot \dfrac{d}{d_y}$$

where y is the depth of section from the top of upright slab. At a section 2.5 m above the bottom of upright slab

$$\frac{A_{sty}}{6157.5} = \left(\frac{5.25}{7.75}\right)^3 \times \frac{750}{572.58} \text{ or } A_{sty} = 2507.27 \text{ mm}^2.$$

So curtail the alternate bars at this section and extend the curtailed bar up to 340 mm above the theoretical section. Similarly, at a section 5 m above the bottom of upright slab

$$A_{sty} = 573.64 \text{ mm}^2.$$

Now further curtail the alternate bars and extend the curtailed bar up to 340 mm above the theoretical section.

12. Temperature reinforcement:

Temperature reinforcement required

$$= \frac{200 + 700}{2} \times 7.75 \times 1000 \times \frac{0.15}{100} = 5231.25 \text{ mm}^2.$$

Number of 10 mm diameter bar required for this $= \dfrac{5231.25}{78.5} = 66.64$.

Provide 44 bars on the exposed face at 175 mm c/c spacing, while on inner face 23 bars at 330 mm c/c spacing.

13. Design of shear key

The effect of shear key is to develop the passive resistance over a depth AG as shown in Fig. 11.20. Thus the fialure would occur along the surface G'G instead of AA'.

For computing the passive resistance below the toe, top overburden of 300 mm is usually neglected.

$$h_1 = 1.7 \text{ m}$$

$$\phi = 38°$$

$$b = 1.9 \tan 38 = 1.48 \text{ m}$$

$$K_P = \frac{1 + \sin\phi}{1 - \sin\phi} = 4.20$$

$$P_P = \frac{1}{2}\gamma K_p(h_1 + a + b)^2 - \frac{1}{2}\gamma K_p h_1^2$$

$$= \frac{1}{2} \times 18 \times 4.2 \times (1.7 + a + 1.48)^2 - \frac{1}{2} \times 18 \times 4.2 \times 1.7^2$$

$$= 37.8a^2 + 240.4a + 272.9.$$

Equilibrium of forces using factor of safety of 1.5,

$$1.5 \ (P_H) = 0.9 \ (P_P + \mu R_o)$$

$$1.5 \times 249.74 = 0.9 \left[37.8a^2 + 240.4a + 272.9 + 0.45\left(\frac{152.25 + 0}{2}\right) \times 2.7\right]$$

or $a = 0.089 \text{ m} = 90 \text{ mm}.$

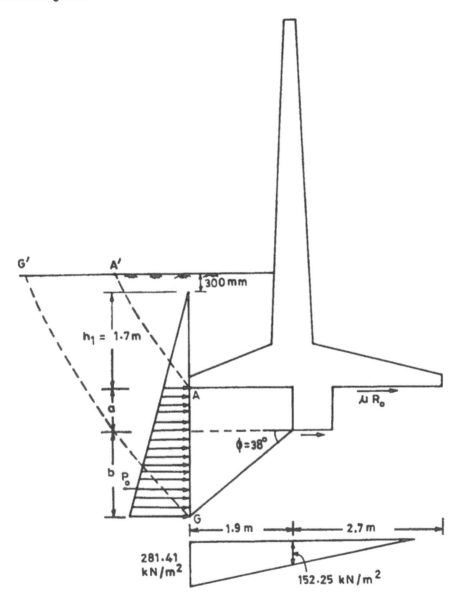

Fig. 11.20. Illustrating the Design of Base Key in Cantilever Wall.

Provide a shear key 400 mm × 400 mm. The steel from vertical wall should be embedded in the shear key.

Details of reinforcement are given in Fig. 11.21.

Example 11.3

Design a R.C.C. counterfort retaining wall to support a fill 7.0 m high. The top of fill is

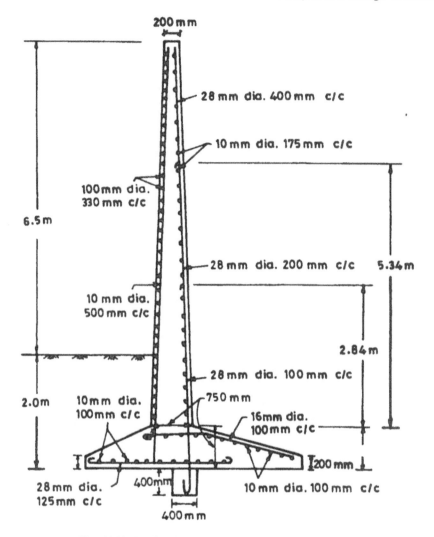

Fig. 11.21. Details of Reinforcement in the Cantilever Wall.

horizontal and has the characteristics: (i) angle of internal friction = 35°, and (ii) unit weight = 19 kN/m³. A competent stratum with an allowable soil presssure of 240 kN/m³ is available at a depth of 2.5 m. The coefficent of friction between the soil and concrete is 0.50.

Solution
1. Proportioning of wall

    Let the retaining wall is founded on stratum at 2.5 m below the ground surface, the total height of the wall will be (2.5 + 7.0) = 9.5 m.

    Assuming the thickness of base slab and stem as 350 mm, height of wall above point C = 9.5 – 0.35 = 9.15 m (Fig. 11.22).

Spacing between counterforts

$$l = \left[11.8 \left(\frac{h}{\gamma} \right)^{\frac{1}{2}} \right]^{\frac{1}{2}}, \text{ where } \gamma \text{ is in kN/m}^3$$

$$= \left[11.8 \left(\frac{9.15}{19} \right)^{\frac{1}{2}} \right]^{\frac{1}{2}} = 2.84 \text{ m, adopt 2.85 m.}$$

Base width, $\quad B = 0.95H \sqrt{\dfrac{K_A}{(1-m)(1+3m)}}$

$$K_A = \frac{1-\sin\phi}{1+\sin\phi} = 0.271$$

$$m = 1 - \frac{4}{9\beta}, \text{ where } \beta = \frac{\gamma h}{q_a} = \frac{19 \times 9.15}{240} = 0.724$$

Therefore, $\quad m = 1 - \dfrac{4}{9 \times 0.724} = 0.386$

$$B = 0.95 \times 9.5 \sqrt{\frac{0.271}{(1-0.386)(1+3 \times 0.386)}} = 4.1 \text{ m}$$

adopt 4.2 m.

2. Foundation stability analysis

Refer Fig. 11.22

Weight of base slab = 4.2 × 1 × 0.35 × 24 = 35.28 kN acting at 2.1 m from F.

Weight of vertical slab = 9.15 × 1× 0.35 × 24 = 76.86 kN
acting at (2.1 + 0.175) = 2.275 m from F.
Weight of fill material on heel = 2.1 × 9.15 × 1 × 19 = 365.08 kN
acting at 1.05 m from F.

Total vertical load per metre length = 477.225 kN.
Distance of C.G. of vertical load from F

$$= \frac{35.28 \times 2.1 + 76.86 \times 2.275 + 365.08 \times 1.05}{477.225}$$

$$= 1.325 \text{ m}$$

Total active pressure due to fill

$$P = \frac{1}{2} \times 0.271 \times 19 \times (9.5)^2$$

$$= 232.34 \text{ kN acting at 3.16 m above the base.}$$

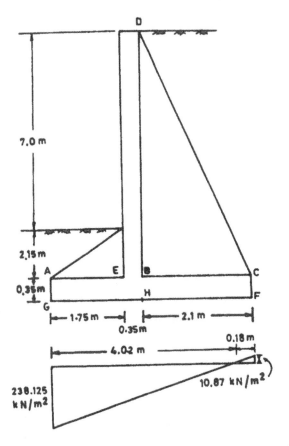

Fig. 11.22. Section of a Counterfort Retaining Wall with Base Pressure Diagram.

The resultant of horizontal and vertical forces strikes the base at a distance of

$$\left(\frac{232.34 \times 3.16}{477.225} + 1.325\right) = (1.54 + 1.325) = 2.867 \text{ m from F.}$$

The resultant strikes the base at $(4.2 - 2.867) = 1.333$ m from G.

Eccentricity of resultant with respect to centre = 0.767 m

Pressure under the toe and heel of retaining wall is given by

$$= \frac{W}{B \times 1}\left(1 \pm \frac{6e}{B}\right) = \frac{477.255}{4.2 \times 1}\left(1 \pm \frac{6 \times 0.767}{4.2}\right)$$

At toe = 238.125 kN/m² < 240 kN/m² (hence safe).

At heel = − 10.87 kN/m² (safe).

Soil reaction is shown in Fig. 11.22.

3. Structural stability analysis

Factor of safety against sliding $F_s = \dfrac{\mu W}{P}$

$$F_s = \frac{477.225 \times 0.50}{232.34} = 1.02 \text{ (unsafe).}$$

Hence a shear key is provided at the base below stem.
Stability against overturning
Resisting moment = 477.225 × (4.2 − 1.325) kNm
Overturning moment = 232.34 × 3.16 kNm

$$F_o = \frac{477.225 \times 2.875}{232.34 \times 3.16} = 1.86 \text{ (safe).}$$

4. Design of heel
Clear spacing between counterforts = 2.85 m.
Consider 1 m strip of heel near the outer edge F.
Net acting load of concrete and fill and due to soil reaction

$$= 0.35 \times 24 + 9.15 \times 19 + 10.87 = 193.12 \text{ kN/m}^2.$$

Net downward pressure = 193.12 kN/m^2.
Maximum factored negative bending moment at the counterfort

$$= 1.5 \times \frac{193.12 \times (2.85)^2}{12} = 196.065 \text{ kNm.}$$

$$\text{Depth of heel slab} = \sqrt{\frac{196.065 \times 10^6}{0.138 \times 20 \times 1000}} = 266.55 \text{ mm.}$$

Adopt 400 mm as effective depth and 450 mm as overall depth.

$$\text{Shear force} = \frac{1.5 \times 193.12 \times 2.85}{2} = 412.79 \text{ kN.}$$

$$\text{Nominal shear stress} = \frac{412.79 \times 1000}{1000 \times 450} = 0.91 \text{ N/mm}^2.$$

Top steel under the counterfort A_{st} is given by

$$196.065 \times 10^6 = 0.87 \times 415 \, A_{st} \left[400 - \frac{415 A_{st}}{20 \times 1000} \right]$$

or $\quad A_{st}^2 - 19277.10 A_{st} + 26.17 \times 10^6 = 0$

$$A_{st} = 1273.44 \text{ mm}^2.$$

Provide 16 mm diameter bars at a spacing of 150 mm c/c.
(Actual A_{st} = 1340.41 mm^2 > 1273.44 mm^2) o.k.

$$\text{Percentage of steel} = \frac{100 \times 1340.41}{1000 \times 450} = 0.297\%.$$

Shear strength of concrete $\tau_v = 0.38$ N/mm^2.

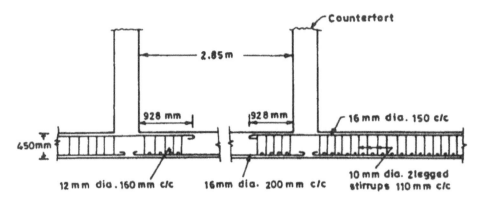

Fig. 11.23. Showing the Details of Reinforcement in the Section through Heel Slab.

So shear stirrups should be provided.

Design shear stress τ_{us} = 0.91 − 0.38 = 0.53 N/mm².

Provide 10 mm diameter two-legged stirrups at a spacing

$$S_v = \frac{0.87 \times 415 \times 400 \times 157}{0.53 \times 400 \times 1000} = 107 \text{ mm say } 110 \text{ mm.}$$

Bottom steel = $\dfrac{3}{4}$ × top steel = $\dfrac{3}{4}$ × 1273.44 = 955.08 mm².

Provide 16 mm diameter bars at a spacing of 200 mm c/c.

Heel near vertical slab

Upward soil reaction near the vertical slab

$$= \frac{238.125 \times 1.91}{4.02} = 113.23 \text{ kN/m}^2.$$

Downward pressure due to fill and concrete = 0.35 × 24 + 9.15 × 19

$$= 182.25 \text{ kN/m}^2.$$

Hence, net downward pressure = 182.25 − 113.23 = 69.02 kN/m².

It is nearly 40 per cent of that at F.

Therefore, steel required = 1340.41 × 0.4 = 536.16 mm².

Provide 16 mm diameter bars at 375 mm spacing c/c.

The top steel should extend up to a length at 58 ϕ or (x/5 + 230) mm whichever is more. The former is greater and equal to 928 mm.

The details of reinforcement are shown in Fig. 11.23.

5. Design of upright slab

Consider 1 m high strip at bottom.

Maximum earth pressure = $K_A \gamma H$ = 0.271 × 19 × 9.15 = 47.11 kN/m².

Factored negative bending moment in the slab near counterforts

$$= \frac{1.5 \times 47.11 \times 2.85^2}{12} = 47.83 \text{ kNm}$$

Depth of slab $\quad d = \sqrt{\dfrac{47.83 \times 10^6}{0.138 \times 20 \times 1000}} = 131.64 \text{ mm.}$

Adopt 200 mm effective and 230 mm overall depth.

Area of steel near counterfort

$$47.83 \times 10^6 = 0.87 \times 415 A_{st} \left[200 - \frac{415 A_{st}}{1000 \times 20} \right]$$

or $\quad A_{st}^2 - 9638.55\, A_{st} + 6.38 \times 10^6 = 0$

or $\qquad\qquad\qquad\qquad A_{st} = 715.59 \text{ mm}^2.$

Use 16 mm diameter bars at a spacing 250 mm c/c.

(Actual $A_{st} = 804.24 \text{ mm}^2 > 715.59 \text{ mm}^2$) o.k.

$$\text{Percentage of steel} = \frac{100 \times 804.24}{200 \times 1000} = 0.40\%$$

Shear strength of concrete $\tau_c = 0.43 \text{ N/mm}^2.$

$$\text{Factored shear force} = \frac{1.5 \times 47.11 \times 2.85}{2} = 100.69 \text{ kN.}$$

$$\text{Nominal shear stress } \tau_v = \frac{100.69 \times 10^3}{1000 \times 250} = 0.40 \text{ N/mm}^2.$$
$$< 0.43 \text{ N/mm}^2 \text{ (safe).}$$

Maximum bending moment on the front face $= \dfrac{3}{4}$ of negative B.M.

Therefore, the spacing 12 mm dia bars can be kept 330 mm c/c.

As the active earth pressure decreases towards top directly with height, area of steel A_{st} is proportional to the height. Hence the spacing of bars can gradually be increased to 500 mm near the top.

The details of reinforcement are shown in Fig. 11.24.

6. Design of toe

The upward soil reaction on toe near its outer end = 238.125 kN/mm².
Let the slab thickness be 500 mm.
Downward load due to slab = 0.5 × 1 × 24 = 12.0 kN/m².

Net upward reaction on the slab at point A = 238.125 − 12.0 (Fig. 11.21)
$$= 226.125 \text{ kN/m}^2.$$

$$\text{Maximum negative factored bending moment} = 1.5 \times \frac{226.125 \times 2.85^2}{12}$$
$$= 229.58 \text{ kNm.}$$

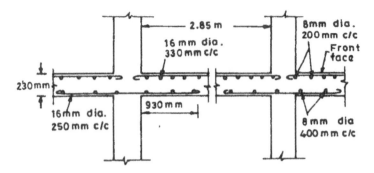

Fig. 11.24. Showing the Details of Reinforcement in the Section through Upright Slab.

Depth of slab, $d = \sqrt{\dfrac{229.58 \times 10^6}{0.138 \times 20 \times 1000}} = 288.41$ mm.

Adopt 300 mm as effective depth and 330 mm overall depth.
Area of steel required at the bottom near the counterfort is given by

$$229.58 \times 10^6 = 0.87 \times 415 A_{st} \left[300 - \frac{415 A_{st}}{20 \times 1000} \right]$$

or $A_{st}^2 - 14457.83 A_{st} + 30.64 \times 10^6 = 0$

$$A_{st} = 2579.48 \text{ mm}^2.$$

Hence provide 16 mm dia bars 75 mm spacing c/c ($A_{st} = 2680.8$ mm^2).

Factored shear force near counterfort $= 1.5 \times \dfrac{226.125 \times 2.85}{2}$

$$= 483.34 \text{ kN.}$$

Nominal shear stress $= \dfrac{483.34 \times 10^3}{1000 \times 300} = 1.38$ N/mm^2.

% of steel $= \dfrac{100 \times 2680.8}{1000 \times 300} = 0.89\%$.

Shear strength of concrete $\tau_c = 0.596$ N/mm^2 $< \tau_v$ (unsafe).

So, shear reinforcement is to be provided.
Design shear stress $\tau_v = 1.38 - 0.596 = 0.784$ N/mm^2.
Provide 10 mm diameter two-legged shear stirrups ($A_s = 157.0$ mm^2) at a spacing

$$S_v = \frac{0.87 \times 415 \times 300 \times 157.0}{0.784 \times 300 \times 1000} = 72.33 \text{ mm.}$$

But according to code, minimum spacing of stirrups should be limited to 100 mm in order to permit space for compaction of the concrete.

Therefore, revised area of stirrups

$$A_{sv} = \frac{V_{us}S_v}{0.87\sigma_v d}, \text{ where } V_{us} = \text{design shear force} = bd\tau_v$$

$$= \frac{0.784 \times 300 \times 1000 \times 100}{0.87 \times 415 \times 300} = 217.14 \text{ mm}^2.$$

Area of one leg = 108.57 mm².

Use 12 mm diameter two-legged vertical stirrups at 100 mm c/c.

Minimum area of shear reinforcement $A_0 > \dfrac{0.4bS_v}{\sigma_y}$

$$= \frac{0.4 \times 300 \times 100}{415} = 28.91 \text{ mm}^2 < 217.14 \text{ mm}^2, \text{ o.k.}$$

Top steel near the centre of span $= \dfrac{3}{4} \times 2680.8 = 2010.6 \text{ mm}^2.$

Provide 16 mm diameter bars at 100 mm c/c spacing.

The details of reinforcement are shown in Fig. 11.25.

7. Design of counterfort

Assume the thickness of counterfort as 500 mm.

c/c spacing of counterforts = 2.85 + 0.5 = 3.35 m

Active earth pressure at depth h below top $= K_A \gamma L h$

$$= 0.271 \times 19 \times 3.35\ h$$
$$= 17.25\ h \text{ kN/m}^2.$$

At the base, the pressure = 17.25 × 9.5 = 163.86 kN/m².

Front counterfort: This counterfort behaves as a horizontal cantilever bending about EF. Compression develops along AF and tension along AE (Fig. 11.26).

The inclination α of the counterfort is equal to

$$\tan^{-1}\left(\frac{2.5 - 0.33}{1.75}\right) = 51.11°.$$

Pressure at the counterfort at end A

$$= 226.125 \times 3.35 = 757.5 \text{ kN/m}.$$

Pressure at the counterfort at end E

$$= \frac{226.125}{4.02} \times 2.27 \times 3.35 = 427.75 \text{ kN/m}.$$

Moment of resistance of section with respect to compression

$$= 0.138\ f_{ck}\ bd^2\ \cos^2 \alpha$$

$$d = \sqrt{\frac{M_u}{0.138\ f_{ck}\ b\cos^2 \alpha}}$$

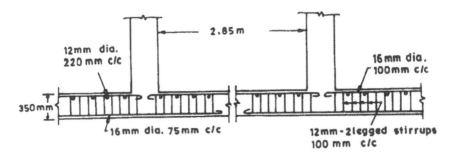

Fig. 11.25. Showing the Details of Reinforcement in the Section through Toe Slab.

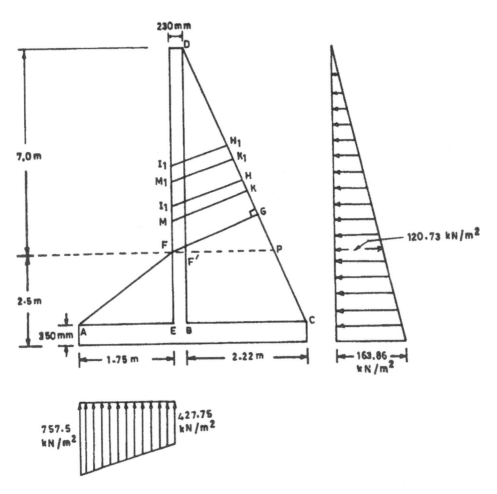

Fig. 11.26. Forces on the Counterforts.

$$M_u = 1.5 \left(\frac{757.5 + 427.75}{2} \right) 1.75 \left(\frac{2 \times 757.5 + 427.75}{757.5 + 427.75} \right) \times \frac{1.75}{3}$$

$$= 1487.43 \text{ kNm}$$

$$\cos^2 \alpha = 0.394$$

$$d = \sqrt{\frac{1487.43 \times 10^6}{0.138 \times 20 \times 500 \times 0.394}} = 1653.98 \text{ mm} = 1654 \text{ mm}.$$

So provide 1655 mm as effective and 1705 mm gross, which is already available. For balanced section provide 0.714% tensile steel.

$$A_{st} = \frac{0.714}{100} \times 1705 \times 500 \times 0.627 = 3820.68 \text{ mm}^2.$$

Use 10 bars of 28 mm diameter at the bottom of counterfort. These bars should extend beyond point E for at least (58 × 28 mm, i.e. 1624 mm).

Shear force along $\qquad$ EF $= \dfrac{757.5 + 427.75}{2} \times 1.75 = 1037.10$ kN.

Factored shear force $\qquad V_u = 1.5 \times 1037.10 = 1555.65$ kN.

Nominal shear force stress $\tau_v = \dfrac{V_u - (M_u / d) \tan \alpha}{bd}$

$$= \frac{1555.65 \times 10^3 - \dfrac{1487.43 \times 10^6}{1655} \times 1.239}{500 \times 1655}$$

$$\tau_v = 0.53 \text{ N/mm}^2$$

% of tension reinforcement $= \dfrac{100 \times 6154.4}{500 \times 1655} = 0.74\%.$

Hence shear strength of concrete $\tau_v = 0.55$ N/mm$^2 > 0.53$ N/mm^2. safe.
Hence shear reinforcement is not to be provided.

Back counterfort: Part DFP (Fig. 11.26) of the counterfort bends as a cantilever T-beam under the action of earth pressure due to fill. Below F, the depth of counterfort has upward bending. F is thus the critical point for the design of counterfort. The reinforcement found at point F is also used below it.

Intensity of earth pressure at E $= \dfrac{163.86}{9.5} \times (9.5 - 2.5)$

$$= 120.73 \text{ kN/m}.$$

Factored bending moment at F

$$M_u = 1.5 \times [1/6 \times 120.73 \, (7.0)]^2 = 1478.94 \text{ kNm}.$$

Factored shear force at F, $V_u = 1.5 [1/2 \times 120.73 \times 7.0]$
$$= 633.83 \text{ kN.}$$

Depth of counterfort

$$d = \sqrt{\frac{1478.94 \times 10^6}{0.1388 + 20 \times 500}} = 1035.22 \text{ mm.}$$

Depth available $= \dfrac{2.22}{(9.50 - 0.35)} \times 7.0 = 1.698 \text{ m} = 1698 \text{ mm (safe).}$

Effective depth = 1650 mm.

Here $\beta = 13.63°$ and $\cos \beta = 0.971$

$$1478.94 \times 10^6 = 0.87 \times 415 A_{st} \left[1650 \times 0.971 - \frac{415 A_t}{20 \times 500} \right]$$

$$A_{st}^2 - 38606.024 \, A_{st} + 98.70 \times 10^6 = 0$$

or $A_{st} = 2752.89 \text{ mm}^2.$

Provide seven bars of 24 mm diameter at the back counterfort. (Actual $A_{st} = 3166.72$ mm$^2 > 2752.89$ mm^2) o.k.

Shear stress in counterfort

$$\tau_v = \frac{V_u - (M_u / d) \tan \beta}{bd}$$

$$= \frac{633.83 \times 10^3 - \left(\dfrac{1478.94 \times 10^6}{1650} \right) \times 0.242}{500 \times 1650}$$

$\tau_v = 0.50 \text{ N/mm}^2.$

% of tension reinforcement $= \dfrac{100 \times 3166.72}{500 \times 1650} = 0.38\%.$

Shear strength of concrete $\tau_c = 0.39 \text{ N/mm}^2 < 0.50 \text{ N/mm}^2.$

Hence, provide 14 mm two-legged shear stirrups at a spacing 520 mm c/c. Along a line FG, draw perpendicular to DC, the effective lever arm and moment are nearly the same as those along a horizontal section through F so that the same reinforcement should be provided at G as at horizontal section through F.

The curtailment of reinforcement is done above FG section. Since area of steel at any depth h below top is proportional to h^2, number of bars at depth h will also be proportional to h^2.

Two bars can be curtailed out of seven bars at depth h, which is given by

$$h_1 = \sqrt{\frac{(7.0)^2 \times 5}{7}} = 5.91 \text{ m.}$$

In Fig. 11.26, MK represents the section beyond which two bars can be curtailed. Distance KG should, however, be not less than 58 diameters and HK not less than 12 dia where H is the point of actual curtailment.

The vertical distance of G above F =1650 sin β cos β
$$= 377.87 \text{ mm}$$

Distance

$$MF = (7.0 - 5.91) = 1.09 \text{ m}$$
$$KG = 1.09 \cos β = 1.05 \text{ m} = 1050 \text{ mm}$$
$$HG = 58 \text{ dia} = 58 \times 24 = 1392 \text{ mm}$$

Also $\qquad HG = 1050 + 12φ$
$$= 1050 + 12 \times 24 = 1338 \text{ mm}$$

Adopt $\qquad HG = 1340 \text{ mm}$

Hence vertical distance of H below top

$$= 7000 - 377.87 - 1340 \cos β = 5340.41 \text{ mm} = 5.34 \text{ m}$$
(first curtailment of bars)

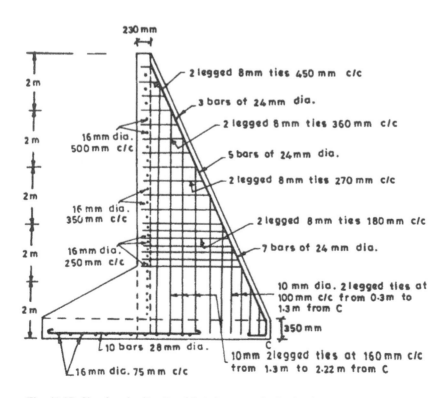

Fig. 11.27. Showing the Details of Reinforcement in the Section through Counterfort.

Total four bars can be curtailed at

$$h_2 = \sqrt{\frac{(7.0)^2 \times 3}{7}} = 4.58 \text{ m}$$

Vertical distance $M_1F = 7.0 - 4.58 = 2.42$ m
and $K_1G = 2.42 \cos \beta = 2.35$ m

GH_1 should not be less than 58ϕ and H_1K_1 should not be less than 12ϕ, where H_1 is the point of actual curtailment of two more bars.

$$58\phi = 58 \times 24 = 1392 \text{ mm}$$
$$GH_1 = K_1G + 12\phi = 2350 + 12 \times 24 = 2638 \text{ mm}$$
which being more should be adopted.

Hence vertical distance H_1 below top

$$= 7000 - 377.87 - 2638 \cos \beta = 4083.93 = 4.08 \text{ m}.$$

Therefore, the next two bars can be curtailed at 4.08 m from top.

The details of reinforcement are shown in Fig. 11.27.

8. Connection between counterforts and slabs

The vertical slab should be tied to the counterforts. The separating tendency will be present in top 7.0 m only.

Force causing separation at depth of 7.0 m

$$= 1/3 \times 19 \times 2.85 \times 7.0 = 126.35 \text{ kN/m}$$
$$\text{Factored force} = 1.5 \times 126.35 = 189.525 \text{ kN/m}$$

$$\text{Steel required} = \frac{189.525 \times 10^3}{0.87 \times 415} = 524.92 \text{ mm}^2.$$

Use two-legged 8 mm diameter ties at 180 mm c/c.

The spacing is gradually increased to 450 mm c/c towards the top.

The heel also tends to separate from counterfort. Vertical ties connect the two together. The end 300 mm strip of the heel may be assumed to be tied to the counterfort through the main reinforcement.

Net downward load on the next 1 m strip of heel

$$= (193.12 - 6.80) \times 1 \times 2.85 = 531.0 \text{ kN}.$$

$$\text{Area of steel} = \frac{531.0 \times 10^3}{0.87 \times 415} = 1470.91 \text{ mm}^2.$$

Provide 10 mm diameter 2-legged ties at spacing 100 mm c/c between 300 mm and 1300 mm.

In the balance $(2.22 - 1.30) = 0.92$ m strip of the heel.

The net downward load $= (193.12 - 66.34) \times 0.92 \times 2.85 = 332.40$ kN.

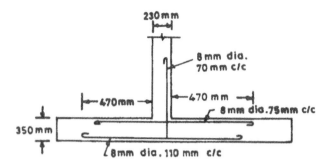

Fig. 11.28. Showing the Details of Reinforcement at the Junction of Vertical Slab and Heel.

$$\text{Area of steel} = \frac{332.40 \times 10^3}{0.87 \times 415} = 920.79 \text{ mm}^2.$$

Provide 10 mm diameter 2-legged ties at spacing 160 mm c/c in next 920 mm strip.

9. Reinforcement at junction of vertical slab and heel

To counter the bending of slabs near the junction in a direction perpendicular to the length of wall extra reinforcement is provided which is 0.3 per cent of sectional area of concrete and placed on the back of vertical slab. Reinforcement of magnitude 0.15% of sectional area is also provided on the top of heel and bottom of toe.

$$\text{Steel for upright slab} = \frac{0.3 \times 230 \times 1000}{100} = 690 \text{ mm}^2.$$

Provide 8 mm diameter bar at 70 mm c/c.
Steel for heel $= (0.15/100) \times 450 \times 1000 = 675 \text{ mm}^2$.
Provide 8 mm diameter bars at 75 mm c/c.
Steel for toe $= (0.15/100) \times 330 \times 1000 = 495 \text{ mm}^2$.
Provide 8 mm diameter bars at 110 mm c/c.
The details of reinforcement are shown in Fig. 11.28.

10. Temperature reinforcement

Provide 0.15 per cent of concrete area as temperature and shrinkage reinforcement. Vertical slab:

$$\text{Area of temperature reinforcement} = \frac{0.15}{100} \times 230 \times 1000 = 345 \text{ mm}^2.$$

Use 8 mm diameter bars at 200 mm c/c on the outside face and 400 mm c/c on the inside face of the vertical slab:

Heel slab:

$$\text{Area of temperature reinforcement} = \frac{0.15}{100} \times 450 \times 1000 = 675 \text{ mm}^2.$$

Provide 12 mm bars at 160 mm c/c.

Area of temp. reinforcement for toe slab $= \dfrac{0.15}{100} \times 330 \times 1000 = 495$ mm^2.

Provide 12 mm bars at 220 mm c/c.

REFERENCES

1. Krishna, J. and O.P. Jain (1986), *Plain Reinforced Concrete*, Vol. 1, Nem Chand and Bros., Roorkee.
2. Mononobe, N. and S. Okabe (1929), "Earthquake Proof Construction of Masonry Dam", Proc. World Engineering Congress, Vol. 9, p. 275.
3. Prakash, S. and S. Saran (1971), "Bearing Capacity of Eccentrically Loaded Footings", Proc. ASCE, Vol. 95, SM1, pp. 95-118.
4. Rowe, P.W. (1952), "Anchored Sheet Piles," *Proc. Inst. Civil Engineers*, Part I, Vol. 1, No. 1, London.
5. Saran, S. (1969), "Bearing Capacity of Footings subjected to Moments", Ph.D. Thesis, University of Roorkee, Roorkee, India.
6. Teng, W.C. (1977), *Foundation Design*, Prentice Hall of India Private Limited, New Delhi.
7. Terzaghi, K. (1934), "Large Scale Retaining Wall Test," *Engineering News Record*, Vol. 112, pp. 136–140.
8. Terzaghi, K. and R.B. Peck (1967), *Soil Mechanics in Engineering Practice*, John Wiley and Sons, New York.

PRACTICE PROBLEMS

1. Explain with neat sketches the situations where retaining walls are provided. List the various types of retaining walls giving the limitation of each with respect to height of backfill.
2. How is stability of a retaining wall checked ?
3. Describe stepwise the procedure of designing a cantilever wall. Give a neat sketch showing reinforcement in a cantilever wall.
4. What is the advantage of providing counterforts in a cantilever wall. Describe stepwise the design of counterfort. Give a neat sketch showing reinforcement in counterfort.
5. Design a gravity retaining wall to retain a 4.5-m embankment. The backfill is cohesionless having $\gamma = 19$ kN/m^3 and $\phi = 32°$ and slopes 15° to horizontal. Angle of wall friction is 21°. The wall is located in Zone III for earthquake considerations. The allowable soil pressure available at 1.25 m depth below ground surface is 300 kN/m^2.
6. Design a cantilever wall for retaining a soil embankment of 6.5 m high above ground surface. The backfill is cohesionless having $\gamma = 18$ kN/m^2 and $\phi = 30°$. The surface of the backfill is horizontal carrying a surcharge of 10 kN/m^2. The allowable soil pressure at 1.0 m depth below ground surface is 250 kN/m^2 and coefficient of friction between concrete and soil is 0.5.
7. Design a counterfort retaining wall for retaining an embankment of 8.5 m high above ground surface. The backfill is cohesionless having $\gamma = 17.5$ kN/m^3 and $\phi = 32°$ and slopes 10° to horizontal. The allowable soil pressure at 1.5 m depth below ground surface is 275 kN/m^3.

Sheet Pile Walls

12.1 INTRODUCTION

Sheet pile walls are also retaining walls constructed to retain earth, water or any other material. These walls are thinner in section as compared to gravity and cantilever walls. Sheet pile walls are generally used as waterfront structures, coffer dams, river bank protection and retaining the sides of cuts made in earth.

A sheet pile wall consists of a series of sheet piles driven side by side for retaining the fill material. Sheet pile may be of steel, timber or reinforced concrete. Timber sheet piling is used for short spans, light lateral loads, and commonly, for temporary structures in the form of braced sheeting. If it is used in permanent structures above water level, it requires preservative treatment, and even then, the usual life is relatively short.

Reinforced concrete sheet piles are precast concrete members. These piles are relatively heavy and bulky. They displace large volumes of soil during driving. The design of reinforced concrete sheet pile should account for handling and driving stresses.

Steel sheet piling is the most common type used because of the following advantages:

(i) It is of relatively light weight and may be reused several times.
(ii) It is easy to increase the pile length by either welding or bolting.
(iii) It has a long service life both above and below the water table with modest protection.
(iv) It is resistant to high driving stresses as developed in hard or rocky material.

12.2 TYPES OF SHEET PILE STRUTURES

Sheet piles may be used as:

(a) Cantilever sheet piles: When the height of the fill retained is small (< 5.0 m) sheet pile is simply driven into the ground (Fig. 12.1a). It acts as a cantilever and, therefore, termed as cantilever sheet pile.

(b) Anchored bulkheads: When the height of the fill retained is relatively large, the passive resistance of the soil in front of the embedded portion of the sheet pile may not be sufficient to keep the wall stable. The sheet pile wall is, therefore, held near the top by a tie bar suitably anchored at some distance behind the wall (Fig. 12.1b). This type of the sheet pile wall is termed as anchored sheet pile or anchored bulkhead, and are widely used for dock and

harbour strutures. The use of an anchor tends to decrease lateral deflection, the bending moment and the depth of penetration of the pile.

Anchored bulkhead may be either dredged type or backfilled type. In the former type, the sheet pile is first driven into ground and the material from the front of wall is dredged. In the latter type, the material at the back of the sheet pile is filled up and compacted.

(c) Braced sheeting in cuts: When the depth of excavation is more than 4 m, steel sheet piles are driven along the length of the cut. As excavation proceeds, wales and struts are inserted as shown in Fig. 12.1c. The wales are commonly of steel, and struts of steel or wood. Excavation then proceeds to a lower level, and another set of wales and struts is installed. This process continues until the excavation is complete. In most soils, it is advisable to drive the sheet piles several metres below the bottom of the excavation to prevent local heaves.

(d) Double wall cofferdam: The double wall cofferdam consists of two lines of sheeting tied to each other. The space between the walls being filled with soil (Fig. 12.1d). The sheeting in the cofferdam may be of steel with proper interlocking arrangement or timber suitably joined. The soil between the two lines of sheeting may be placed by end dumping from trucks, dredging or other similar arrangements. The two rows of sheeting are anchored to each other by a combination of wales and tie rods. Depending upon the height of the cofferdam, rows of tie rods are required.

(e) Cellular cofferdam: A cellular cofferdam is made by driving sheet piles (straight of web) to form a series of cells which are later filled with soil. These cells are interconnected for 'watertightness' and are self-stabilising against lateral pressure from soil and water. There are two types of cellular cofferdams, namely, (i) circular-cell type connected by arcs inside and outside (Fig. 12.1e), and (ii) diaphragm type in which the arcs inside and outside are connected by means of diaphragms (Fig. 12.1f).

There are other types, which may be considered as the modified forms of the two basic types. One type has large circular cells subdivided by straight diaphragms. These are called clover-leaf cofferdams (Fig. 12.1g). The other types are modified circular and diaphragm types and where small arcs on one side be omitted (Fig. 12.1h and i).

12.3 DESIGN OF CANTILEVER SHEET PILING WALL

General

Cantilever sheet piles are economical only for moderate wall heights, since the required section modulus increases rapidly with increase in wall heights; the bending moment increases with the cube of the cantilevered height of the wall. The lateral deflection of this type of wall, because of the cantilever action, will be relatively large. Since stability of wall depends primarily on the developed pressure in front of the wall, erosion and scour of wall, i.e. lowering of dredge line, should be controlled.

Figure 12.2a shows a section of sheet pile with its probable deflected shape. Saran and Vaish (1970) carried out a model study on cantilever sheet pile using sheet piles of different heights and thicknesses. Pressure distribution along the length of sheet pile was observed in most of the tests. Figure 12.2b shows a typical pressure distribution observed in a test. The conventional design of cantilever sheet piling is based on simplifying assumptions as

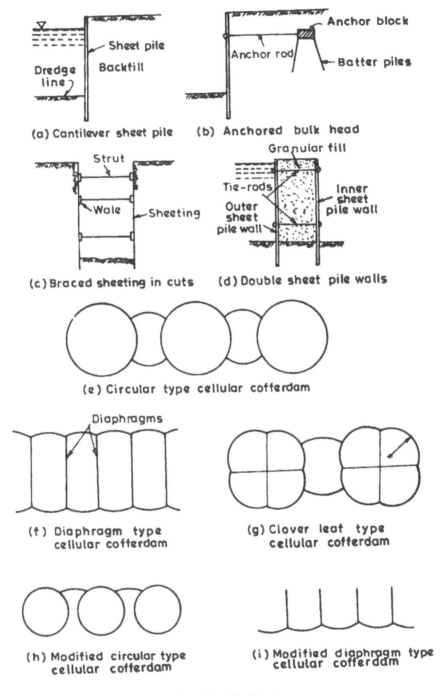

(a) Cantilever sheet pile

(b) Anchored bulk head

(c) Braced sheeting in cuts

(d) Double sheet pile walls

(e) Circular type cellular cofferdam

(f) Diaphragm type cellular cofferdam

(g) Clover leaf type cellular cofferdam

(h) Modified circular type cellular cofferdam

(i) Modified diaphragm type cellular cofferdam

Fig. 12.1. Sheet Pile Structures.

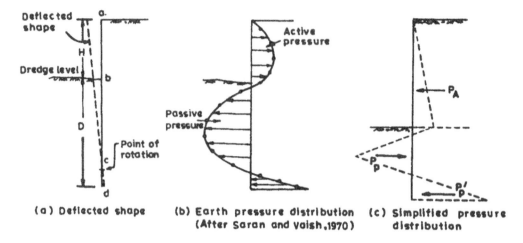

(a) Deflected shape   (b) Earth pressure distribution   (c) Simplified pressure
                             (After Saran and Vaish, 1970)   distribution

Fig. 12.2 Behaviour of Cantilever Sheet Pile.

indicated in Fig. 12.2c. Design procedures are discussed for both granular soils ($c = 0$) and cohesive soils ($\phi = 0$).

Cantilever sheet piling in granular soils

A simplified earth pressure diagram as shown in Fig. 12.3 is adopted for the design of cantilever sheet piling in granular soils. It is assumed that the soil below the dredge line has the same properties (γ, ϕ) as the backfill above the dredge line. The point of rotation, O located at a distance 'a' below the dredge line has zero earth pressure. If K_P and K_A are respectively the passive and active earth pressure coefficients, then in Fig. 12.3,

$$p_A = \gamma K_A H. \qquad \qquad \text{... (12.1)}$$

Slope of line DF $\qquad = \gamma (K_P - K_A). \qquad \qquad \text{... (12.2)}$

Therefore, $\qquad \gamma (K_P - K_A) a = p_A$

or $$a = \frac{p_A}{\gamma (K_P - K_A)} \qquad \qquad \text{... (12.3)}$$

$$p_{P1} = \gamma (K_P - K_A) y \qquad \qquad \text{... (12.4)}$$

$p_{P2} = $ Intensity of passive earth pressure due to backfill − Intensity of active earth pressure of the soil below the dredge line

$\qquad = \gamma K_P (H + D) - K_A \gamma D$

$\qquad = \gamma (K_P - K_A) D + \gamma K_P H. \qquad \qquad \text{... (12.5)}$

Let R_a represent the resultant of all the forces above point O, acting at distance $\bar{y}$ about point O (Fig. 12.3).

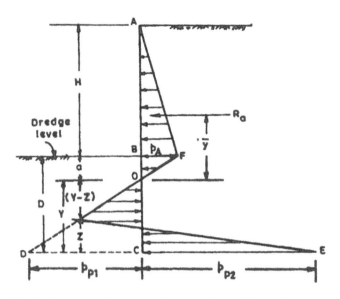

Fig. 12.3. Earth Pressure Diagram for Cantilever Sheet Piling in Granular Soil ($c = 0$).

The distance z can be found in terms of y by satisfying the condition of equilibrium $\Sigma F_H = 0$.

$$R_a + (p_{P1} + p_{P2}) \frac{z}{2} - p_{P1} \frac{y}{2} = 0$$

or
$$z = \frac{p_{P1}y - 2R_a}{p_{P1} + p_{P2}}. \qquad \text{... (12.6)}$$

Taking moments about the base of sheet pile and satisfying condition of equilibrium $\Sigma M = 0$.

$$R_a(y + \bar{y}) + (p_{P1} + p_{P2}) \frac{z}{2} \times \frac{z}{3} - p_{P1} \cdot \frac{y}{2} \times \frac{y}{3} = 0$$

or
$$6R_a(y + \bar{y}) + z^2 (p_{P1} + p_{P2}) - p_{P1}y^2 = 0. \qquad \text{... (12.7)}$$

Solving Eqs. 12.6 and 12.7, value of y can be obtained. Then

$$D = a + y \qquad \text{... (12.8)}$$

To cover the uncertainties in the evaluation of soil parameters and the difference in actual and assumed pressure distributions, it is recommended to use K_p/F instead of K_p in the above expressions, F being the factor of safety usually taken between 1.5 to 2.0. Alternatively, the depth D obtained from Eq. 12.8 may be increased by 20 to 40%. If D_a represents the value of D after using K_p/F or increasing it by some percentage, then

$$\text{Total length of sheet pile} = H + D_a. \qquad \text{... (12.9)}$$

Cantilever sheet piling in cohesive soils

In cohesive soils, consolidation may occur in the passive pressure zones. Tension cracks may develop in the active zone and become filled with water, thus increasing the lateral pressure considerably. Further, the clay may shrink and lose contact with wall. To account for this, the wall adhesion is not considered in the design.

Figure 12.4 shows the earth pressure diagram adopted for designing cantilever sheet pile in cohesive soil ($\phi = 0$), where

$$q_u = \text{unconfined compressive strength of clay}$$
$$= 2c$$
$$c = \text{undrained cohesion, and}$$

$$h_0 = \frac{2c}{\gamma}. \qquad \qquad \qquad \qquad ... \ (12.10)$$

Total resultant active earth pressure above the dredge line $P_A = 1/2(\gamma H - q_u)(H - h_0)$. It will act at a height of $\bar{y} = (H - h_0)/3$ above dredge level.

For statistical equilibrium, the sum of all the horizontal forces should be equal to zero.

$$P_A - (2q_u - \gamma H) D + \frac{1}{2}(2q_u - \gamma H + 2q_u + \gamma H) h = 0$$

or
$$h = \frac{(2q_u - \gamma H)D - P_A}{2q_u}. \qquad \qquad ... \ (12.11)$$

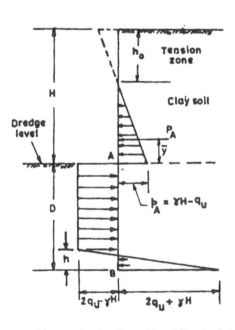

Fig. 12.4. Earth Pressure Diagram for Cantilever Sheet Piling in Cohesive Soil ($\phi = 0$).

Further, sum of the moments above base is zero

$$P_A[D + \bar{y}] - \frac{(2q_u - \gamma H)D^2}{2} + 4q_u \frac{h^2}{6} = 0. \qquad \text{... (12.12)}$$

Substituting the value of h from Eq. (12.11) in Eq. (12.12) and simplifying,

$$K_1 D^2 + K_2 D + K_3 = 0 \qquad \text{... (12.13)}$$

where
$$K_1 = 2q_u - \gamma H \qquad \text{... (12.14a)}$$

$$K_2 = -(\gamma H - q_u)(H - h_0) \qquad \text{... (12.14b)}$$

$$K_3 = \frac{P_A(6q_u\bar{y} + P_A)}{q_u + \gamma H} \qquad \text{... (12.14c)}$$

As discussed earlier, the depth of sheet pile computed from Eq. 12.13 may be increased by 20 to 40 per cent. Alternatively, the value of q_u may be divided by a factor 1.5 to 2.0.
It may be noted that the wall will be stable if

$$\frac{2q_u}{F} \geq \gamma H \qquad \text{... (12.15)}$$

F is the factor of safety usually taken between 1.5 and 2.0.

12.4 DESIGN OF ANCHORED BULKHEADS

General
Anchored bulkhead provides a vertical wall so that ships may tie up alongside, or to serve as a pier structure. The use of an anchor member tends to reduce the lateral deflection, the bending moment, and the depth of penetration of the pile. The use of more than one anchor may be desirable in high walls to control lateral deflections (Agarwal, Saran and Bansal, 1970).

The stability of an anchored bulkhead depends upon (a) relative stiffness of piling, (b) the depth of piling penetration, (c) the relative compressibility of the soil, and (d) the amount of anchor yield. The methods commonly used for the analysis and design of anchored bulkheads are (i) free earth support method, and (ii) fixed earth support method.

Free earth support method
The method is based on the assumptions that the sheet piling is preferably rigid as compared to the surrounding soil and is free to rotate having zero lateral movement at the tie rod (Fig. 12.5). The behaviour of an anchored bulkhead satisfies this assumption if the sheet pile is driven to a shallow depth.

(i) Granular soils
Figure 12.6 shows the earth pressure diagram for the anchored bulkhead in a granular soil. The problem basically is to estimate:

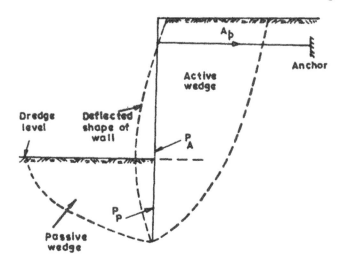

Fig. 12.5. Conditions for Free Earth Support of an Anchored Bulkhead.

(a) Minimum depth of penetration, D, necessary for stability
(b) Anchor pull, A_p
(c) Maximum moment in the sheet pile.

Let P_A = resultant active earth pressure acting at y_1 below the anchor rod level, and
           P_P = resultant passive earth pressure acting at y_2 below the anchor rod level.

For equilibrium $\Sigma H = 0$

$$A_p + P_P - P_A = 0.$$ ... (12.16)

As the slope of the line GH is $\gamma_b(K_P - K_A)$

$$a = \frac{P_{AE}}{\gamma_b(K_P - K_A)},$$ ... (12.17)

where P_{AE} = intensity of active earth pressure at point E.

Taking moments about the anchor rod level and satisfying the condition $\Sigma M = 0$

$$P_A y_1 = P_P \cdot Y_2.$$ ... (12.18)

From Fig. 12.6

$$P_P = \frac{1}{2}\gamma_b(K_P - K_A).y^2$$ ... (12.19)

$$y_2 = \left(h + a + \frac{2}{3}y\right)$$ ... (12.20)

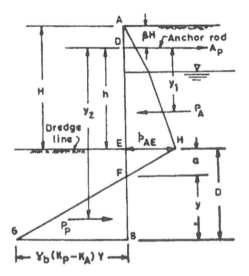

Fig. 12.6. Earth Pressures Diagram for Anchored Bulkhead in Granular Soil.

Substituting the values of P_P and y_2 in Eq. 12.18, we get

$$P_A \cdot y_1 = \frac{1}{2} \gamma_b (K_P - K_A) \, y^2 \left(h + a + \frac{2}{3} y \right). \qquad \qquad \text{... (12.21)}$$

On rearranging the terms,

$$\gamma_b \left(\frac{K_P - K_A}{3} \right) \cdot y^3 + \gamma_b \left(\frac{K_P - K_A}{2} \right) (h + a) \cdot y^2 - P_A \cdot y_1 = 0. \qquad \text{... (12.22)}$$

The solution of Eq. (12.22) will give the value of y and thus

$$D = y + a \qquad \qquad \text{... (12.23)}$$

$$A_P = P_A - P_P \qquad \qquad \text{... (12.24)}$$

Due to uncertainty in the development of passive resistance, the depth D calculated
.above is increased by 20 to 40 per cent. Alternatively the value of K_P may be reduced to
K_P/F, F being the factor of safety (1.5 to 2.0).

Maximum moment M_0 in the sheet pile is obtained at the section of zero shear force. Let
it occur at a distance x from the ground surface. If P_{ax} is the area of active earth pressure
diagram above this section, then for zero shear force

$$P_{ax} = A_p. \qquad \qquad \text{... (12.25)}$$

Solving Eq. 12.25, x can be obtained.

$$M_0 = P_{ax} \cdot \bar{x} - A_p (x - \beta H) \qquad \qquad \text{... (12.26)}$$

where $\bar{x}$ = distance of P_{ax} from the section at distance x from ground surface.

Rowe (1952) has proposed moment reductions for sheet piling designs based on the free-earth support piling method (Fig. 12.7). The Rowe moment reduction theory is based on the following factors:

(a) Relative density of soils
(b) Flexibility number

$$\rho = \frac{H_1^4}{EI} \times 10^{-10} \qquad \qquad \dots (12.27)$$

where H_1 = total length of pile in m,
        E = modulus of elasticity of piling material in kN/m², and
        I = moment of inertia of pile section in m⁴.

(c) Relative height of piling α and relative free board of the piling β as shown in Fig. 12.7.

After deciding D, the total length of the pile H_1 is obtained. Assuming the section of the sheet pile, flexibility factor ρ is computed. Using Fig. 12.7, the ratio M/M_0 is obtained which in turn will give actual moment M.

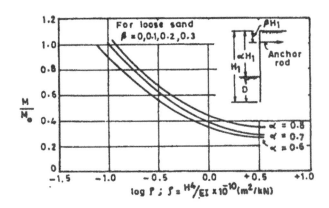

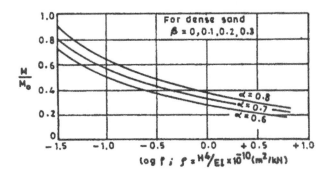

Fig. 12.7. Rowe's Moment Reduction Curves for Anchored Bulkheads in Sand.

For appropriate design,

$$\frac{M}{\text{Section modulus of sheet pile section}} \ngtr \text{Allowable steel stress} \qquad ...(12.28)$$

(ii) Cohesive soils

Figure 12.8 shows the earth pressure diagram for the anchored bulkhead when the soil below the dredge line is cohesive ($\phi = 0$) and the backfill is granular.

The surcharge at the dredge level due to backfill is

$$q = \gamma_t \cdot h_1 + \gamma_b \cdot h_3 = \gamma_e \cdot H \qquad ...(12.29)$$

For $\phi = 0, K_P = 1.0.$

Passive earth pressure may, therefore, be obtained by

$$p_P = q \cdot K_P + 2c_u \sqrt{K_P}$$

or $\quad p_P = q + q_u.$ $\qquad ...(12.30)$

Similarly, $\quad p_A = q - q_u.$ $\qquad ...(12.31)$

where $\quad q_u$ = uncontained compressive strength of clay = $2c_u$.

Net pressure acting below the dredge line

= Passive pressure acting towards right with zero surcharge above dredge line – Active pressure acting towards left considering surcharge q

$$= q_u - (q - q_u) = 2q_u - q. \qquad ...(12.32)$$

Let P_A and P_P represent total active earth pressure acting above the dredge line and net passive earth pressure acting below dredge line as shown in Fig. 12.8.

Taking moments of all forces at the anchor rod level and satisfying the condition $\Sigma M = 0$

$$P_A y_1 - D (2q_u - q) (h_2 + D/2) = 0 \qquad ...(12.33)$$

or

$$D^2 + 2Dh_2 - \frac{2P_A \cdot y_1}{2q_u - q} = 0. \qquad ...(12.34)$$

Equation (12.34) can be solved for D. This can be increased by 20 to 40 per cent to cover the uncertainties in the development of passive earth pressure. The force in the anchor rod, A_p can be found by summing the horizontal forces $\Sigma H = 0$

$$A_p = P_A - P_P. \qquad ...(12.35)$$

From Fig. 12.8, it can be seen that the wall will be unstable when

$$2q_u - q < 0$$

or

$$\frac{c_u}{\gamma_e H} < 0.25 \qquad ...(12.36)$$

where $\quad \gamma_e$ = effective unit weight of soil above dredge level.

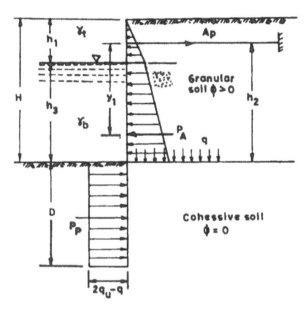

Fig. 12.8. Earth Pressure Diagram for Anchored Bulkhead in Cohesive Soil.

The term $\dfrac{c_u}{\gamma_e H}$ is known as stability number, S_n. This stability number is a function of the wall height H, but is relatively independent of the strength of material used in developing q. If wall adhesion c_a is taken into account, the stability number S_n becomes

$$S_n = \frac{c_u}{\gamma_e H}\sqrt{1+\frac{c_a}{c_u}} \qquad\qquad \text{... (12.37)}$$

At passive failure $\sqrt{1+c_a/c_u}$ is approximately equal to 1.25.

Rowe (1952) has proposed moment reductions for sheet piling designs based on the free earth support method (Fig. 12.9). The Rowe moment reduction theory is based on the following factors:

(a) Stability number, S_n

$$S_n = \frac{1.25 c_u}{\gamma_e H} \qquad\qquad \text{... (12.38)}$$

(b) Flexibility number ρ given by Eq. 12.27
(c) Relative height of piling α (Fig. 12.7).

Fixed earth support method

If the sheet piles are driven to a considerable depth, the deflected shape of the anchored bulkhead is likely to take the form shown in Fig. 12.10a. The lower end of the bulkhead is

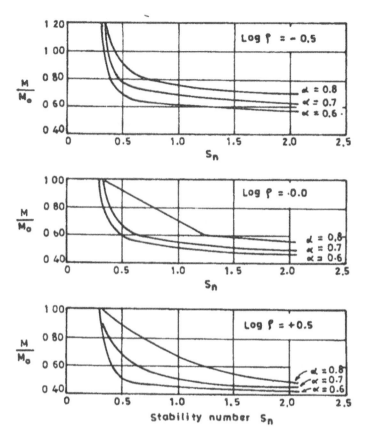

Fig. 12.9. Rowe's Moment Reduction Curves for Anchored Bulkheads in Cohesive Soil.

practically fixed in position due to the resistance offered by the adjacent soil. Figure 12.10b shows the probable bending moment diagram. The elastic line (Fig. 12.10a) changes its curvature at the point of contraflexure (Fig. 12.10b). Anchored sheet pile wall of this is called bulkhead with fixed earth support.

There are two approaches for solving the problem of bulkheads with fixed earth support, namely (i) elastic line method, and (ii) equivalent beam method. The elastic line method is relatively complicated and time consuming. The equivalent beam method represents a simplification of the elastic line method and it takes much less time and labour at a small sacrifice of accuracy. Keeping this in view only the equivalent beam method has been discussed here.

The procedure of evaluating the unknown quantities: (i) depth of penetration D, (ii) anchor pull A_p, and (iii) maximum moment in the sheet pile by equivalent beam method is discussed in the following steps:

(i) Active and passive earth pressures are assumed to develop, depending upon whether the pile deflects towards or away from the earth face. The pressure distribution is assumed to be linear as shown in Fig. 12.10c. The passive pressure developed near the bottom on

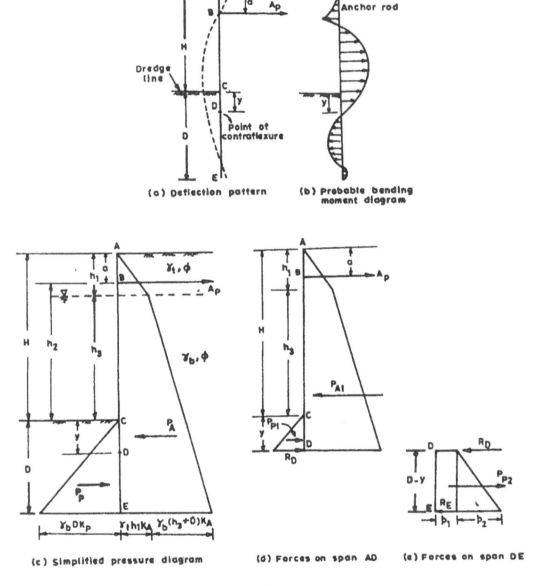

Fig. 12.10. Illustrating Fixed Earth Support Method.

the fill side is, however, assumed to be concentrated at the bottom of the pile as a reaction, R_E.

(ii) The position of the point of contraflexure is chosen. It depends on the relative density of granular soils and consistency of cohesive soils. For granular soils, it may be located as given in Table 12.1.

Table 12.1. Variation of y with ϕ.

ϕ	y (Fig. 12.10c)
20°	0.08 H
30°	0.08 H
40°	0.007 H

(iii) As there is no moment at the point of contraflexure, the span AE (Fig. 12.10c) is analysed as two independent spans AD and DE.

Forces acting on the span AD are shown in Fig. 12.10d. R_D represents the reaction at the point of contraflexure, and P_{A1} and P_{P1} are respectively the total active and passive earth pressures acting above the point D.

$$P_{A1} = \frac{1}{2}\gamma_t h_1^2 K_A + \gamma_t h_1(h_3 + y)K_A + \frac{1}{2}\gamma_b(h_3 + y)^2 K_A \qquad \text{... (12.39)}$$

$$P_{P1} = \frac{1}{2}\gamma_b y^2 K_P \qquad \text{... (12.40)}$$

where γ_t = density of fill soil above water level, and

γ_b = submerged density of fill soil.

Taking the moments of all the forces about the point B

$$R_D (H + y - a) = \frac{1}{2}\gamma_t h_1^2 K_A \left(\frac{2h_1}{3} - a\right) +$$

$$\gamma_t h_1 K_A(h_3 + y) \left(h_1 - a + \frac{h_3 + y}{2}\right) +$$

$$\frac{1}{2}\gamma_b(h_3 + y)^2 K_A \left(h_1 - a + \frac{2(h_3 + y)}{3}\right) -$$

$$\frac{1}{2}\gamma_b y^2 K_P \left(h_1 - a + h_3 + \frac{2}{3}y\right) \qquad \text{... (12.41)}$$

Solving Eq. (12.41), R_D is obtained.

(iv) Factors acting on span DE are shown in Fig. 12.10e.

$$p_1 = \gamma_b y K_p - \gamma_t h_1 K_A - \gamma_b(h_3 + y)K_A \qquad \text{... (12.42)}$$

$$p_2 = \gamma_b(D - y) (K_P - K_A) \qquad \text{... (12.43)}$$

Taking the moments of all the forces about the point E and putting it equal to zero, we get,

$$R_D(D - y) = p_1 \frac{(D-y)^2}{2} + \frac{1}{6} \gamma_b (D - y)^3 (K_P - K_A)$$

or

$$R_D = p_1 \frac{(D-y)}{2} + \frac{1}{6} \gamma_b (D - y)^2 (K_P - K_A) \qquad \qquad ... (12.44)$$

or

$$\gamma_b(K_P - K_A) D^2 - [2\gamma_b \, y(K_P - K_A) - 3p_1]D +$$
$$[\gamma_b \, y^2(K_P - K_A) - 3p_1 y - R_D] = 0 \qquad ... (12.45)$$

Equation 12.45 is quadratic in D and the solution of this will give the value of D. Values of R_D and p_1 are taken from Eqs. 12.41 and 12.42 and substituted in Eq. 12.45. As discussed earlier, the value of D is increased by 20 to 40%.

(v) The anchor pull is estimated by putting the algebraic sum of the forces acting on span AD (Fig. 12.10d) equal to zero.

$$A_P = P_{A1} - P_{P1} - R_D \qquad \qquad ... (12.46)$$

(vi) The maximum bending moment in the sheet pile is obtained by locating the section of zero shear force. Normally, it occurs at a point above the dredge level.

If x is the section of zero shear force from point A (Fig. 12.10c), then

$$\frac{1}{2} \gamma_t h_1^2 K_A + \gamma_t h_1 K_A (x - h_1) + \frac{1}{2} \gamma_b(x - h_1)^2 . \, K_A = A_P \qquad ... (12.47)$$

Solving Eq. (12.47), value of x can be obtained.

Hence

$$M_{max} = A_p(x - a) - \frac{1}{2} \gamma_t h_1^2 K_A \left(x - \frac{h_1}{3} \right) - \gamma_t \, h_1 K_A \times \frac{(x - h_1)^2}{2} -$$
$$\frac{1}{6} \gamma_b(x - h_1)^3 K_A. \qquad \qquad ... (12.48)$$

12.5 ANCHORAGE METHODS

Anchorages of sheet piling may be provided using any of the following:

(a) Anchoring to the existing structure (Fig. 12.11a),
(b) Batter piles (Fig. 12.11b),
(c) Concrete blocks or beams known as dead man (Fig. 12.11c).

Normally an existing structure is capable of taking the anchor pull. The design of batter piles has already been discussed in Chapter 8. The capacity of deadman is obtained for the following three cases:

(i) Continuous deadman near the ground surface, (ii) Short deadman near the surface, and (iii) Deadman at great depth.

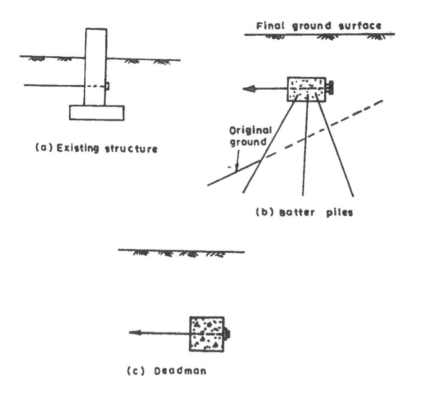

(a) Existing structure

Final ground surface

Original ground

(b) Batter piles

(c) Deadman

Fig. 12.11. Sheet Pile Anchorages.

A deadman whose length is considerably greater than its depth is termed as continuous deadman. In case when a continuous deadman is located near the surface (Fig. 12.12a, $h_1 <$ $0.6h_2$), it may be assumed to be acted upon by pressures as shown in Fig. 12.12b. The deadman capacity is, therefore:

$$D_c = \frac{(P_P - P_A)L}{F} \qquad \qquad ... (12.49)$$

where D_c = deadman capacity,
          P_P = passive earth pressure,
          P_A = active earth pressure,
          L = length of deadman, and
          F = factor of safety (1.5 to 2.0).

For a short deadman located near the ground surface, Teng (1976) has suggested the following equations for deadman capacity:
For cohesionless soils :

$$D_c = \left[L(P_P - P_A) + \frac{1}{3}K_0\gamma(\sqrt{K_P} + \sqrt{K_A})h_2^3 \tan\phi \right]\frac{1}{F} \qquad ... (12.50)$$

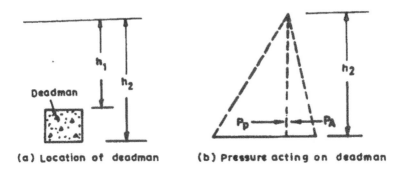

(a) Location of deadman   (b) Pressure acting on deadman

Fig. 12.12. Analysis of Deadman.

For cohesive soils:

$$D_c = [L(P_P - P_A) + q_u h_2^2] \frac{1}{F} \qquad \qquad ... (12.51)$$

where K_0 = coefficient of earth pressure at rest (0.4 to 0.55),

   q_u = unconfined compressive strength of soil, and

   ϕ = angle of internal friction.

The allowable capacity of a deadman at great depth below ground surface $h_1 > (h_2 - h_1)$ is approximately equal to the allowable bearing capacity of the footing of width $(h_2 - h_1)$ located at depth $(h_1 + h_2)/2$.

12.6 DESIGN OF BRACED SHEETING IN CUTS

Earth pressure diagrams

Sheet piles are used to retain the sides of the cuts in sands and clays. The sheet piles are kept in position by wales and struts (Fig. 12.1c). The first brace location should not exceed the depth of the potential tension cracks,

$$h_0 = \frac{2c}{\gamma} \tan (45 + \phi/2). \qquad \qquad ... (12.52)$$

Since the formation of cracks will increase the lateral pressure against the sheeting and if the cracks are filled with water, the pressure will be increased even more. The sheeting of a cut is flexible and is restrained against deflection at the first series of struts. The deflection, therefore, is likely to be as shown in Fig. 12.13. The actual earth pressure against the back of sheeting depends on the properties of the soil and the sequence of construction.

Based on the measurements on actual job sites (Peck, 1943; Spilker, 1937), empirical pressure diagrams have been proposed by Terzaghi and Peck (1967) depending on the type of soil. The pressure distribution on sheet pile walls to retain sandy soils is shown in Fig. 12.14 and to retain clay soils in Fig. 12.15.

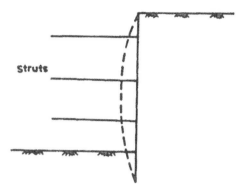

Fig. 12.13. Typical Deformation Pattern of a Braced Cut.

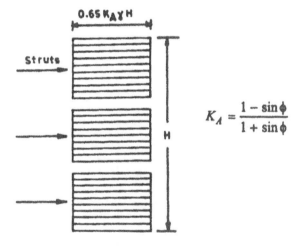

$$K_A = \frac{1 - \sin\phi}{1 + \sin\phi}$$

Fig. 12.14. Pressure Distribution on Strutted Excavations in Sands (Terzaghi and Peck, 1967).

Design

For the design of sheeting, it may be considered as the continuous beam supported at the wales, either cantilevered at the top, fixed, partially fixed, hinged, or cantilevered at the bottom, depending upon the amount of penetration below the excavation line. Bending moment and shear force diagrams of the sheeting may then be obtained, using the moment distribution method. Section of the sheet pile is then designed in the conventional way for the maximum bending moment. A rapid method of designing sheeting is to assume conditions as shown in Fig. 12.16. The top is treated as a simple cantilever beam, including the first two struts. The remaining spans between struts are considered as simple beams with a hinge or cantilever at the bottom.

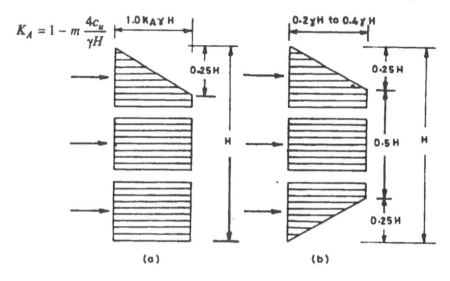

$$K_A = 1 - m\,\frac{4c_u}{\gamma H}$$

Fig. 12.15. Pressure Distribution on Strutted Excavation in (a) Soft to Medium Clays, $m = 1.0$ except when $\dfrac{\gamma H}{c_u} > 4$, then $m < 1.0$ and (b) Stiff Fissured Clays.

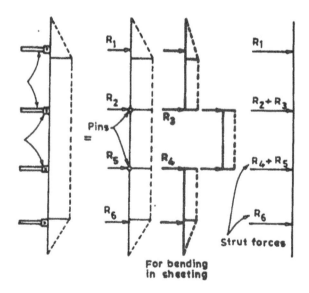

Fig. 12.16. Simplified Methods of Analysing Sheeting and Evaluating Strut Forces.

Struts are designed as columns subjected to an axial force. The wales may be treated as continuous members for the length of the wale or as simply supported members pinned at the struts. The later approach provides a conservative solution.

Modes of base failure

In sands

In the case of loose sands, the stability against 'piping' or 'boiling' in the base resulting from heavy upward seepage of ground water into the cut must be checked in cases where the sheet piles are not driven into an impermeable stratum. It is important either to prevent piping or to reduce it to an insignificant amount, since severe piping can lead to loss of ground outside the cut and even to undermining and collapse of the sheep pile wall. It can cause loss of passive resistance of the ground to the inward thrust of the sheet piling. Piping occurs when the exit gradient at the point of outflow into the cut reaches unity. At this stage the velocity of the upward flowing water within the cut is sufficient to lift and displace the soil through which it is flowing. Since velocity of flow is inversely proportional to the length of seepage path, the tendency to piping can be reduced by lengthening the seepage path. It may be done by driving the sheet pile to a sufficient depth or by providing a berm of permeable soil. The other effective method of preventing piping is to reduce the head of water causing piping. This can be done by pumping water from well points or deep wells placed at or below the level of the bottom of the sheet piles.

In soft clays

In the case of deep open cuts in soft clays, there is a possibility that the bottom may fail by heaving, because the weight of the blocks of clay beside the cut tends to displace the underlying clay toward the excavation (Fig. 12.17a). Figure 12.17b represents a cross section through a cut in soft clay. Bjerrum and Eide (1956) have studied the mechanics of bottom heave and gave the following formula for the critical depth of an excavation:

$$\text{Critical depth, } H_c = \frac{c \cdot N_c}{\gamma}. \qquad \qquad \cdots (12.53)$$

$$\text{Factor of safety against bottom heave, } F = \frac{c \cdot N_c}{\gamma H + q} \qquad \cdots (12.54)$$

where N_c = bearing capacity factor N_c (Fig. 12.17c),

c = undrained shear strength of clay in a zone immediately around the bottom of the excavation,

γ = density of soil,

H = depth of excavation, and

q = surface surcharge.

If sheet piles extend below the bottom of the cut, their stiffness reduces the tendency of the clay adjacent to the bottom to be displaced toward the excavation and, consequently, reduce the tendency toward heave. However, if the clay extends to a considerable depth below the cut, the beneficial effects of even relatively stiff sheeting have been found to be small. If the lower ends of the sheet piles are driven into a hard stratum, the effectiveness of sheet piles is increased appreciably.

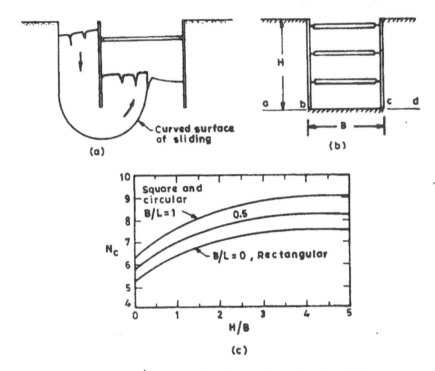

Fig. 12.17. (*a*) Base Heave in Strutted Excavation in Soft Clay, (*b*) Section through Open Cut in Deep Deposit of Clay, (*c*) Bearing Capacity Factor, N_c.

12.7 DESIGN OF CELLULAR COFFERDAMS

General

A cofferdam is a temporary structure used when for some purpose it is desired to exclude the water and expose the portion of the bottom of a river or lake or some other body of water. The cofferdam is usually constructed large enough to provide adequate working space. After the proposed area is enclosed by the cofferdam, the area is dewatered by pumping.

The most common types of cofferdams are: (i) Cantilever sheet piles, (ii) Braced cofferdams, (iii) Earth embankments, (iv) Double-wall cofferdams, and (v) Cellular cofferdams. The analysis and design of cantilever sheet piles, braced cofferdams and double-wall cofferdams have already been discussed. In this section, the analysis of cellular cofferdams is presented.

For economical design, approximate materials should be utilized for filling the cell. The material should be free draining having high angle of internal friction and should possess small amount of soil finer than 75 microns. Sand, gravel, crushed stone, or broken bricks are suitable material for filling.

The design of a cellular cofferdam requires consideration of the following factors:

(i)   Resistance to sliding where there is small penetration of the sheet piles (Fig. 12.18a)
(ii) Resistance to overturning (Fig. 12.18b)

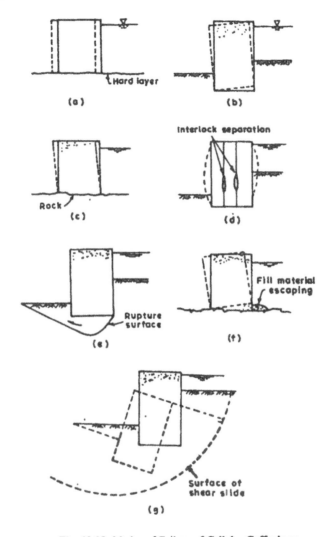

Fig. 12.18. Modes of Failure of Cellular Cofferdams.

(iii) Resistance to tilting (Fig. 12.18c)
(iv) Resistance to failure by interlock separation (Fig. 12.18d)
(v) Resistance to shear failure of base soil (Fig. 12.18e)
(vi) Resistance to tilting due to loss of contact of base with soil (F.g. 12.18f)
(vii) Resistance to general shear slide of the structure and soil retained by and beneath the structure (Fig. 12.18g)

Height and average width of cofferdams

The height of the cofferdam should be such that its top is always above the anticipated high water during the life-time of the cofferdam. Thus, the height can be decided by studying the flood records. Information regarding the elevation of the existing ground surface and the depth of scour should also be obtained from the site.

In case of high cofferdam berms are often used to reduce the relative height above the ground.

Instead of working with the actual cells having varying widths, usually one metre long section with an avergae uniform width is designed. The average theoretical width of a cellular cofferdam is that of a rectangular section having a section modulus equal to that of actual cofferdam from centre-to-centre of cells. A simple procedure may be used for design purposes whereby the average width is adopted such that the area in the rectangular section and that in the actual cofferdam are equal, that is

$$B = \frac{1}{L} \text{ (area of main cell + area of connecting cell)}$$

where B = average width, and

L = distance between centre-to-centre of cells.

It has been reported (TVA, 1957) that the two methods yield almost identical results and that the average widths by the methods of equal area vary upto six per cent higher than that by the method of equal section moduli. Thus, the method of equal area may be slightly on the unsafe side.

TVA engineers use the following relationships between the average width B and diameter D.

$$B = 0.875D \text{ with 60 degree Tee's.}$$
$$B = 0.785D \text{ with 90 degree Tee's.}$$

Stability analysis

There are three methods of stability analysis of cellular cofferdams namely:

1. Former Tennessee Valley Authority (TVA) method
2. Cumming's method
3. Hansen's method.

Of these methods the TVA (1966) is commonly used and the same is discussed here.

TVA method

Since a large portion of the cell-fill is saturated, the location of the line of saturation becomes the first step in the stability analysis. Based on observation of several cofferdams, TVA has arrived at the conclusion shown in Fig. 12.19. In case of perfectly free draining, the lower half may be assumed to be saturated (Fig. 12.19a) whereas for other types of fill, the saturation line at 2 : 1 slope should be analysed (Fig. 12.19b). In cases where earth berm is used, the saturation line slopes down to the top of the berm. In the earth berm, two locations of saturation line (Fig. 12.19c) should be used so as to make provision for the more critical condition.

Cofferdam on rock: For cellular cofferdams on rock, the following possibilities of failure should be analysed.

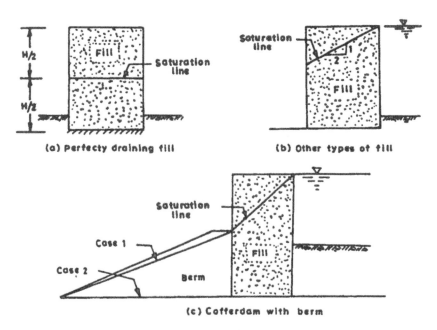

(a) Perfectly draining fill

(b) Other types of fill

(c) Cofferdam with berm

Fig. 12.19. Saturation Line for Analysis.

(a) Sliding: The cofferdam is subjected to horizontal forces due to the unbalanced hydrostatic pressure and the soil pressure. This horizontal force is resisted by the frictional resistance along the base of the dam and the passive earth pressure from the berm, if the berm is provided (Fig. 12.20a). The factor of safety against sliding, F_s is given by

$$F_s = \frac{W\mu + P_P}{P} \qquad \dots (12.55)$$

where W = effective weight of fill,
         P_P = passive resistance of berm,
         P = horizontal force due to soil and water, and
         μ = coefficient of base friction.

For satisfactory design, $F_s \not< 1.25$.

(b) Overturning: The cofferdam must be stable against overturning. Two types of analysis can be made when considering this type of stability. Firstly, the resultant of the forces should lie within the middle one-third of the base so that the soil does not come in tension. It means that (Fig. 12.20b)

$$e = \frac{P\bar{y}}{\gamma HB} \leq \frac{B}{6} \qquad \dots (12.56)$$

where P = horizontal pressure,
         $\bar{y}$ = height of point of application of P above base of dam,

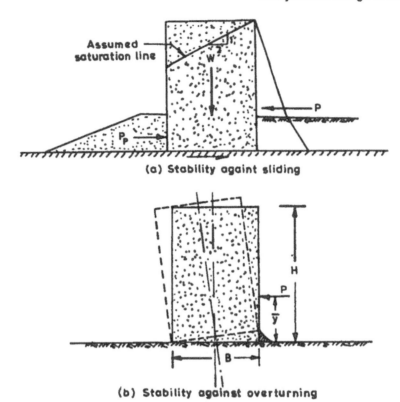

(a) Stability againt sliding

(b) Stability against overturning

Fig. 12.20. Stability of Cofferdam against Sliding and Overturning.

γ = density of soil, and
B = average width of cell.

Alternatively, it may be reasoned that as the cell tends to tip over due to the lateral thrust the fill runs out of the heel. For this to occur the frictional resistance of the sheet piling on the cell fill is developed. Assuming the lateral pressure between the fill and piling on the river side equal to external pressure, P and summing moments about the toe of the cell (Fig. 12.20b):

$$BP \tan \delta = P \bar{y} F$$

or $\qquad B = \dfrac{\bar{y}F}{\tan \delta}$ $\qquad\qquad\qquad\qquad$ ... (12.57)

where $\qquad \delta$ = angle of friction between cell fill and piling, and
$\qquad\qquad F$ = factor of safety usually taken as 1.25.

(c) Vertical shear
Shear along a plane through the centreline of the cell is another possible mode of failure

(Fig. 12.21a). For stability, the shearing resistance along this plane, which is the sum of soil shear resistance and resistance in the interlocks, must be equal to or greater than the shear due to overturning effects.

Assuming a linear pressure distribution at the base, the moment M due to external lateral pressure is given by Eq. 12.58 (Fig. 12.21a).

$$M = \frac{2}{3} B \cdot V \qquad \qquad ... (12.58)$$

where V = maximum vertical shear, and
             B = average width of cell.

Therefore $V = \frac{3}{2} \frac{M}{B}.$ \qquad \qquad ... (12.59)

The soil shear resistance is given by

$$S_1 = \frac{1}{2} \gamma H^2 K_A' \tan \phi \qquad \qquad ... (12.60)$$

where K_A' = coefficient of earth pressure.

The coefficient of earth pressure is computed from Mohr's circle. It is important to note that with shear on this plane, the lateral pressure is not a principal stress from the considerations of Mohr's circle. Krynine (1944) has derived the following expression for K'_A:

$$K_A' = \frac{\cos^2 \phi}{2 - \cos^2 \phi}. \qquad \qquad ... (12.61)$$

In addition, the friction on interlocks also offer resistance against vertical shear. This friction is equal to the interlock tension times coefficient of friction. The interlock tension force is computed using the conventional hoop-tension equation

$$T = q_a \cdot r. \qquad \qquad ... (12.62)$$

Since q_a increases with depth, the total hoop tension force for a cell of depth H is

$$T = \frac{1}{2} \gamma H^2 K_A r. \qquad \qquad ... (12.63)$$

Experiences at TVA indicated that the maximum pressure is developed at the height of $(3 H_c/4)$ above base of the cell of height H (Fig. 12.21c).

Hence $T = \frac{1}{2} \gamma H \,(3/4 H_c) \, K_A' \, r.$ \qquad \qquad ... (12.64)

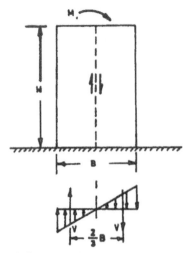

(a) Shear along centre line of cell

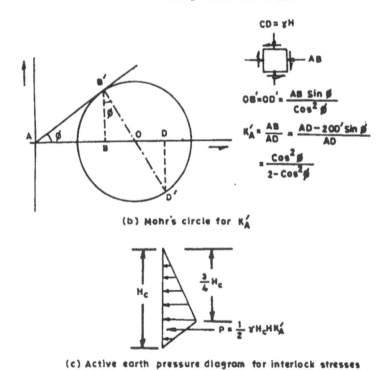

(b) Mohr's circle for K_A'

(c) Active earth pressure diagram for interlock stresses

Fig. 12.21. Vertical Shear in Cellular Cofferdam.

Thus, the friction force S_2 per unit width of cell is

$$S_2 = \frac{3}{8} \gamma \, H H_c \, K_A' \, \frac{r}{L} f \qquad\qquad \dots (12.65)$$

where f = coefficient of interlock friction (usually equal to 0.3), and

L = distance between cross walls for diaphragm cell

= r for circular cells.

The total shear resistance, S is

$$S = S_1 + S_2$$

$$= \frac{1}{2}\gamma H^2\left(\frac{\cos^2\phi}{2-\cos^2\phi}\right)\tan\phi + \frac{3}{8}\gamma H H_c\, K_A\, \frac{r}{L} f \qquad \ldots (12.66)$$

Factor of safety against vertical shear

$$= \frac{S}{V} \qquad \ldots (12.67)$$

$$\not< 1.25.$$

Cofferdam on soil: The stability analysis for cofferdams on rock is also applied to cofferdams on soil. In addition, the sheet piling must be driven to an adequate depth such that its bearing capacity is equal to at least 1.5 times the vertical force that acts on the piling.

In case of cofferdams on deep sand deposits, it should be examined that there is no sand boiling at the toe due to seepage of water. The danger of boil can be readily eliminated by providing inverted filters (Terzaghi, 1945), Fig. 12.22.

Cellular cofferdams on clay should be checked for the bearing capacity of clay and the tilting due to large compression at the toe of the cofferdam.

The ultimate bearing capacity of a clay stratum can be determined by

$$H = 5.7\frac{c}{\gamma}. \qquad \ldots (12.68)$$

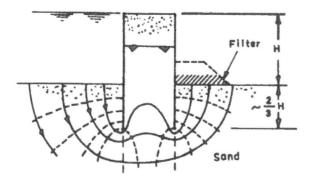

Fig. 12.22. Seepage in Cellular Cofferdam in Sand.

If a factor of safety of 1.5 is used, the maximum height of cofferdam on deep clay is

$$H = 3.8\frac{c}{\gamma}.$$ ... (12.69)

where H = maximum height of cofferdam above sheet piling,
             c = unit cohesion of clay, and
             γ = effective unit weight of deposit.

   Unequal base pressures in cofferdams on soft to medium clay may introduce a large amount of tilting. In such cases, the compressibility of the soil should be obtained by laboratory tests and estimation of tilt should be within permissible limits.

ILLUSTRATIVE EXAMPLES

Example 12.1
Determine the depth of embedment of the sheet piling shown in Fig. 12.23a. The soil has moist unit weight of 18 kN/m³ and submerged unit weight of 10 kN/m³. The soil possesses angle of internal friction of 32°.

Solution

(i) $$K_A = \frac{1-\sin 32}{1+\sin 32} = \frac{1-0.53}{1+0.53} = 0.307$$

$$K_P = \frac{1}{K_A} = \frac{1}{0.307} = 3.25$$

(ii) Earth pressure at the dredge line

$$p_A = (18 \times 2.5 + 10 \times 3.5) \times 0.307 = 24.56 \text{ kN/m}^2.$$

Hence, the distance of zero pressure below dredge line will be.

$$a = \frac{p_A}{\gamma_b(K_P - K_A)} = \frac{24.56}{10(3.25 - 0.307)} = 0.834 \text{ m.}$$

Resultant of all the forces above the point O,

$$R_a = \frac{1}{2} \times 0.307 \times 18 \times 2.5^2 + 18 \times 0.307 \times 2.5 \times 3.5 + \frac{1}{2} \times 0.307 \times$$

$$10 \times 3.5^2 + \frac{1}{2} [0.307 \times 18 \times 2.5 + 0.307 \times 10 \times 3.5] \times 0.834$$

$$= 17.27 + 48.35 + 18.80 + 10.24$$

$$= 94.66 \text{ kN/m.}$$

Taking moments of all the forces about the point O,

$$94.66 \, \bar{y} = 17.27 \left(0.834 + 3.5 + \frac{2.5}{3}\right) + 48.35 \left(0.834 + \frac{3.5}{2}\right) + 18.8 \times$$

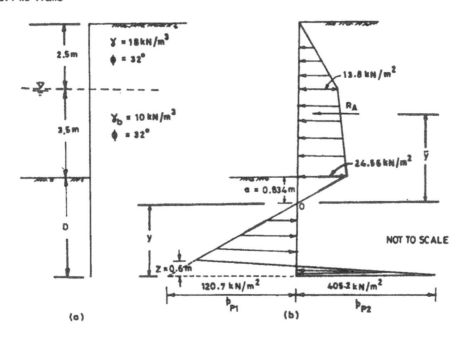

Fig. 12.23. Sheet Pile and Lateral Pressure Diagram.

$$\left(0.834 + \frac{3.5}{3}\right) + 10.24 \times \frac{2}{3} \times 0.834$$

$$= 89.24 + 124.94 + 37.61 + 5.69$$

$$\bar{y} = 2.72 \text{ m}$$

(iii) $p_{P1} = \gamma_b (K_p - K_A) y = 10(3.25 - 0.307) y = 29.43y \text{ kN/m}^2$
$D = 0.834 + y$
$p_{P2} = 18 \times 3.25 \times 2.5 + 10 \times 3.25 \times (3.5 + D) - 10 \times 0.307D$
    $= 260.00 + 29.43D = 243.75 + 29.43(0.834 + y)$

or $p_{P2} = 284.54 + 29.43y$

$$z = \frac{p_{p1} y - 2R_a}{p_{P1} + p_{P2.}} = \frac{29.43y^2 - 2 \times 94.64}{29.43y + 284.54 + 29.43y}$$

or $$z = \frac{y^2 - 6.43}{2y + 9.67}.$$

Now $$R_a(y + \bar{y}) + (p_{P1} + p_{P2}) \frac{z^2}{6} - p_{P1} \frac{y^2}{6} = 0$$

or $$94.66 (y + 2.72) + (29.43y + 284.54 + 29.43y) \times$$

$$\frac{(y^2 - 6.43)^2}{6(2y + 9.67)^2} - 29.43 \frac{y^3}{6} = 0$$

On simplification

$$y^4 + 9.11y^3 - 11.14y^2 - 142y - 190.05 = 0.$$

The solution of the equation by trial and error yield

$$y = 4.1 \text{ m.}$$

Therefore, $D = 4.1 + 0.834 = 4.934$ m.

Increase D by 30% and adopt depth of pile penetration as

$$1.3 \times 4.934 = 6.4 \text{ m, say } 6.5 \text{ m}$$

$$z = \frac{4.1^2 - 6.43}{2 \times 4.1 + 9.67} = 0.58 \text{ m}$$

$$p_{P1} = 29.43 \times 4.1 = 120.7 \text{ kN/m}^2$$
$$p_{P2} = 284.54 + 29.43 \times 4.1 = 405.2 \text{ kN/m}^2.$$

The pressure distribution is shown in Fig. 12.23. Total length of sheet pile is (6.0 + 6.5 = 13.5m).

Example 12.2
Determine the depth of embedment of the sheet piling shown in Fig. 12.24a.
    The soil is clay ($\phi = 0$) with unconfined compressive strength of 70 kN/m^2 and a unit weight of 17 kN/m^3.

Solution

(i) $$h_o = \frac{2c}{\gamma} = \frac{q_u}{\gamma} = \frac{70}{17} = 4.1 \text{ m}$$

$$P_A = \frac{1}{2}(\gamma H - q_u)(H - h_o)$$

$$= \frac{1}{2}(17 \times 6.0 - 70)(6.0 - 4.1) = 30.4 \text{ kN/m}$$

$$\bar{y} = \frac{6.0 - 4.1}{3} = 0.63 \text{ m}$$

$$2q_u - \gamma H = 2 \times 70 - 17 \times 6.0 = 38 \text{ kN/m}^2$$
$$2q_u + \gamma H = 2 \times 70 + 17 \times 6.0 = 242 \text{ kN/m}^2.$$

The pressure distribution will be as shown in Fig. 12.24.
(ii) For determination of h, equating summation of all horizontal forces:

$$30.4 - 38D + (38 + 242)\frac{h}{2} = 0$$

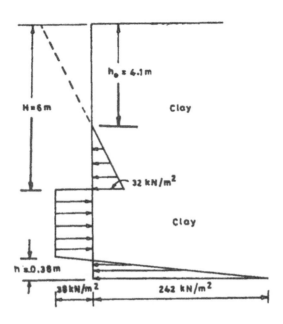

Fig. 12.24. Lateral Pressure Diagram.

or
$$h = \frac{38D - 30.4}{140}$$

(iii) For determining D, taking moments of all the forces about the base of sheet pile:

$$30.4 \, (D+0.63) - 38 \, \frac{D^2}{2} + (38 + 242) \, \frac{h^2}{6} = 0$$

or $\quad 30.4(D + 0.63) - 19 \, D^2 + \frac{280}{6} \times \frac{(38D - 30.4)^2}{140 \times 140} = 0$

or $\qquad\qquad\qquad\qquad D^2 - 1.6 \, D - 1.4 = 0.$

Solving for D, $D = 2.2$ m; $h = (38 \times 2.2 - 30.4) / 140 = 0.38$ m

(iv)   Increasing D by 40%, we have $D = 1.4 \times 2.2 = 3.1$ m
         Total height of sheet piling $= 6 + 3.1 = 9.1$ m.

Example 12.3
For the anchored bulkhead system shown in Fig. 12.25a, determine the following:
  (i)   Depth of embedment of sheet pile
  (ii) Anchor pull
  (iii) Maximum moment in the pile
  (iv) Capacity of the deadman

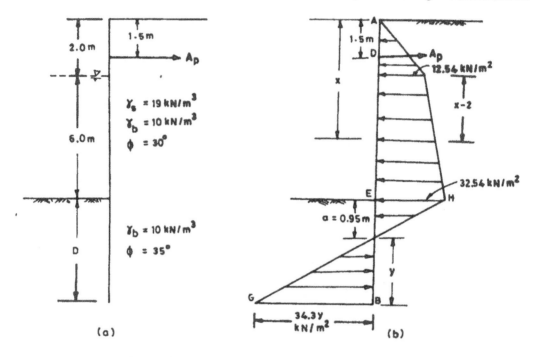

Fig. 12.25. Anchored Bulkhead and Lateral Pressure Diagram.

Solution

(i)
$$K_{A1} = \frac{1 - \sin 30}{1 + \sin 30} = 0.33$$

$$K_{P1} = \frac{1}{0.33} = 3.0$$

$$K_{A2} = \frac{1 - \sin 35}{1 + \sin 35} = 0.27$$

$$K_{P2} = \frac{1}{0.27} = 3.7$$

$$P_{AE} = (19 \times 2.0 + 10 \times 6.0) \times 0.33 = 12.54 + 20.0 = 32.54 \text{ kN/m}^2$$

$$a = \frac{32.54}{10(3.7 - 0.27)} = 0.95 \text{ m}$$

$$P_P = \gamma_b (K_{P2} - K_{A2}) y = 10 (3.7 - 0.27)y$$
$$= 34.3y, \text{ kN/m}^2$$

(ii) Taking the moments about anchor pull (Fig. 12.25b)

$$-\frac{1}{2} \times 2.0 \times 12.54 (0.67 - 0.5) + 12.54 \times 6.0(0.5 + 3.0) +$$

$$\frac{1}{2} \times 20.0 \times 6.0 (0.5 + 4.0) + \frac{1}{2} \times 32.54 \times 0.95 \times$$

$$\left(6.5 + \frac{0.95}{3}\right) - \frac{1}{2} \times 34.3y^2 (6.5 + 0.95 + 2/3y) = 0.$$

or $-2.13 + 263.34 + 270.0 + 105.36 - 17.15y^2(7.45 + 2/3y) = 0$

or $y^3 + 11.17y^2 - 55.64 = 0.$

By trial $y = 2.1$ m

$$D = y + a = 2.1 + 0.95$$
$$= 3.05 \text{ m}$$

(iii) $P_A = \frac{1}{2} \times 2.0 \times 12.54 + 12.54 \times 6.0 + \frac{1}{2} \times 20.0 \times 6.0 + \frac{1}{2} \times 32.54 \times 0.95$

$$= 12.54 + 75.24 + 60.0 + 15.45$$
$$= 163.23 \text{ kN/m}$$

$$P_P = \frac{1}{2} \times 34.3 \times 2.1 \times \frac{2.1}{2} = 37.81 \text{ kN/m}$$

$$A_P = P_A - P_P = 163.23 - 37.81 = 125.42 \text{ kN/m}$$

(iv) Let the section of zero S.F. occurs at a distance x from the ground surface, than

$$12.54 + 12.54 (x - 2) + \frac{1}{2} \times 10 \times 0.33 (x - 2)^2 = 125.42$$

or $(x - 2)^2 + 7.6 (x - 2) - 68.41 = 0.$

Therefore, $(x - 2) = \dfrac{-7.6 + \sqrt{57.76 + 273.64}}{2} = 5.3$

or $x = 7.3$ m.

Maximum moment in the sheet pile

$$= 12.54 \left(7.3 - \frac{2 \times 2}{3}\right) + 12.54 (7.3 - 2) \times \frac{5.3}{2} + \frac{1}{2} \times 10 \times 0.33 (5.3)^2 \times \frac{5.3}{3}$$

$$= 74.74 + 176.12 + 82.71 = 333.57 \text{ kNm}$$

(v) Total length of sheet pile

$$= 8 + 3.05 = 11.05 \text{ m.}$$

Assuming thickness of sheet pile as 6 mm

$$I = \frac{1 \times (6 \times 10^{-3})^3}{12} = 18 \times 10^{-9} \text{ m}^4.$$

Flexibility factor, $\rho = \dfrac{(11.05)^4 \times 10^{-10}}{200 \times 18 \times 10^{-9}} = 0.414$

$$\log \rho = -0.383$$

$$\beta = \frac{1.5}{11.05} = 0.136$$

$$\alpha = \frac{8}{11.05} = 0.724.$$

From Fig. 12.7, in loose sand, for $\log \rho = -0.383$

$$\alpha = 0.724 \text{ and } \beta = 0.136, \frac{M}{M_0} = 0.5.$$

Therefore,

Actual moment $= 0.5 \times 333.57 = 167$ kNm.

(vi) Let the deadman be a concrete block of $1.0 \text{ m} \times 1.0 \text{ m} \times 3.5 \text{ m}$ size.

$$h_1 = 1.5 - 0.5 = 1.0 \text{ m}$$
$$h_2 = 1.5 + 0.5 = 2.0 \text{ m}.$$

Therefore, $h_1 < 0.6\, h_2$.

Hence

$$\text{Deadman capacity} = \frac{(P_P - P_A)L}{F}$$

$$P_P = \frac{1}{2} \times 19 \times 3.0 \times 2.0^2 = 114 \text{ kN}$$

$$P_A = \frac{1}{2} \times 19 \times 0.33 \times 2.0^2 = 12.7 \text{ kN}.$$

$$\text{Deadman capacity} = \frac{(114 - 12.7) \times 3.5}{2.5} = 141 \text{ kN} > 125.42 \text{ kN (Hence o.k.)}$$

(vii) Provide actual depth of penetration of sheet piling as $1.2 \times 3.05 = 3.66$ m.
Hence total length of sheet piling $= 8 + 3.66 = 11.66$ m.

Example 12.4

Solve the problem of anchored bulkhead of Example 12.3 by fixed earth support method.

Solution

(i) The simplified pressure distribution is shown in Fig. 12.36.

(ii) The point of contraflexure is taken at depth $y = 0.08\, H$

$$= 0.08 \times 8 = 0.64 \text{ m (Table 12.1)}$$

(iii) Refer Figs. 12.10d and 12.26

$$P_{A1} = \frac{1}{2} \times 2.0 \times 12.54 + 12.54 \times 6.0 + \frac{1}{2} \times 20.0 \times 6.0 + 26.62 \times 0.64 +$$

$$\frac{1}{2} \times 2.7(0.64)^2$$

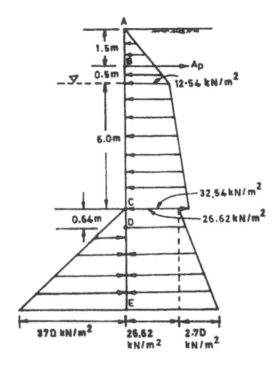

Fig. 12.26. Lateral Pressure Diagram.

$$= 12.54 + 75.24 + 60.0 + 17.04 + 0.55$$
$$= 165.37 \text{ kN/m}$$

$$P_{P_1} = \frac{1}{2} \times 37(0.64)^2 = 7.6 \text{ kN/m}.$$

Taking the moments of all the forces about the point having anchor rod:

$$R_D(6.5 + 0.64) = -12.54 (0.67 - 0.5) + 75.24(0.5 + 3.0) + 60 \times (0.5 + 4.0) +$$

$$17.04 \left(6.5 + \frac{0.64}{2}\right) + 0.55 \times \left(6.5 + \frac{2}{3} \times 0.64\right) - 7.6 (6.5 + 2/3 \times 0.64)$$

or $R_D \times 7.14 = -2.13 + 263.34 + 270.0 + 116.21 + 3.81 - 52.6$

or $R_D = 83.84 \text{ kN}$

(iv) Refer Figs. 12.10e and 12.26

$$p_1 = 37 \times 0.64 - 26.62 - 2.7 \times 0.64 = -4.67 \text{ kN/m}^2$$
$$p_2 = 37 (D - 0.64) - 2.7(D - 0.64) = 34.3 (D - 0.64) \text{ kN/m}^2.$$

Taking the moments of all the forces about point E

$$83.84(D - 0.64) = -4.67 \frac{(D-0.64)^2}{2} + \frac{34.3(D-0.64)^3}{6}$$

or
$$5.71(D - 0.64)^2 - 2.33(D - 0.64) - 83.84 = 0$$

or
$$D - 0.64 = \frac{2.33 + \sqrt{5.43 + 1915}}{2 \times 5.71} = 4.04 \text{ m}$$
$$D = 4.68 \text{ m}$$

(v) Anchor pull, $A_P = P_{A1} - P_{P1} - R_D$
$$= 165.37 - 7.6 - 83.84 - 74 \text{ kN/m}$$

(vi) Let the section of zero S.F. occur at a distance x from the ground surface, then

$$12.54 + 12.54(x - 2) + \frac{1}{2} \times 10 \times 0.33 (x - 2)^2 = 74$$

or
$$(x - 2)^2 + 7.6(x - 2) - 37.24 = 0$$

$$(x - 2) = \frac{-7.6 + \sqrt{57.76 + 149}}{2}$$

$$= 3.4$$

or
$$x = 5.4 \text{ m.}$$

Maximum moment in the sheet pile

$$= 12.54 \left(5.4 - 2 \times \frac{2}{3}\right) + 12.54(5.4 - 2) \times \frac{3.4}{2} + \frac{1}{2} \times 10 \times 0.33(3.4)^2 \times \frac{3.4}{3}$$
$$= 51.04 + 72.5 + 21.61 = 145.2 \text{ kNm}$$

(vii) Provide actual depth of penetration of sheet piling as $1.2 \times 4.68 = 5.6$ m
Therefore,
   Total length of sheet piling $= 8 + 5.6 = 13.6$ m.

Example 12.5
Solve the problem of anchored bulkhead of Example 12.3, if the soil below the dredge line is clay having $q_u = 80$ kN/m².

Solution
(i) Surcharge q at the dredge level
$$= 18 \times 2.0 + 10 \times 6.0 = 96 \text{ kN/m}^2.$$

Net pressure intensity below dredge level
$$P_P = 2q_u - q = 2 \times 80 - 96 = 64 \text{ kN/m}^2.$$

The earth pressure diagram will be as shown in Fig. 12.27.

(ii) Take the moments about the anchor rod level:

$$-\frac{1}{2} \times 2.0 \times 12.54(0.67 - 0.5) + 12.54 \times 6.0(0.5 + 3.0) + \frac{1}{2} \times 20.0 \times$$

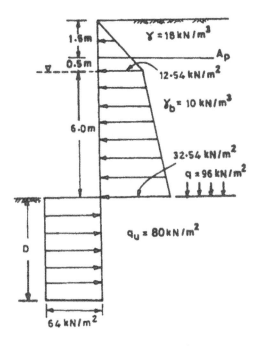

Fig. 12.27. Lateral Pressure Diagram.

$$6.0(0.5 + 4.0) - 64 \times D \left(6.5 + \frac{D}{2}\right) = 0$$

or $\qquad\qquad -2.13 + 263.34 + 270.0 - 32D^2 - 416D = 0$

or $\qquad\qquad D^2 + 13D - 16.6 = 0$

$$D = \frac{-13 + \sqrt{169 + 66.4}}{2} = 1.17 \text{ m}$$

(iii) Anchor pull,

$$A_p = \frac{1}{2} \times 2.0 \times 12.54 + 12.54 \times 6.0 + \frac{1}{2} \times 20.0 \times 6.0 - 64 \times 1.17$$

$$= 73 \text{ kN/m}$$

(iv) Let the section of zero S.F. occur at a distance x from the ground surface, then

$$12.54 + 12.54 (x - 2) + 1/2 \times 10 \times 0.33(x - 2)^2 = 73$$

or $\qquad\qquad x = 5.38$ m.

Maximum moment in the sheet pile

$$= 12.54 \left(5.38 - 2 \times \frac{2}{3}\right) + 12.54(5.38 - 2) \times \frac{3.38}{2} + \frac{1}{2} \times 10 \times 0.33 \times \frac{3.38^3}{6}$$

$$= 141 \text{ kNm.}$$

Assuming the thickness of sheet pile as 6 mm

$$I = \frac{1 \times (6 \times 10^{-3})^3}{12} = 18 \times 10^{-9} \text{ m}^4.$$

Flexibility factor, ρ

$$= \frac{H^4 \times 10^{-10}}{EI} = \frac{(8.0 + 1.17)^4 \times 10^{-10}}{200 \times 18 \times 10^{-9}} = 0.1964$$

$$\log \rho = -0.7068$$

$$S_n = \frac{c_u}{\gamma H} = \frac{40}{10 \times 9.17} = 0.436$$

$$\alpha = \frac{8}{9.14} = 0.872$$

$$\beta = \frac{1.5}{9.17} = 0.164.$$

By extrapolation, for $\log \rho = -0.7068$,

$\alpha = 0.872$ and $S_n = 0.436$ m, Fig. 12.9 gives

$$\frac{M}{M_0} > 1.0.$$

Hence actual moment may be adopted as 141 kNm.

(v) Provide actual depth of penetration as $1.3 \times 1.17 = 1.52$ m.
Total length of sheet pile $= 8.0 + 1.52 = 9.52$ m.

Example 12.6
A cut 3.0 m wide, 6.0 m deep is proposed in moist sand with $\phi = 38°$. Sketch the suitable scheme of sheeting, bracing and also determine the maximum sheet load. Assume the density of soil as 18 kN/m³.

Solution
(i) The scheme of sheeting and bracing as shown in Fig. 12.28a is proposed. It consists of vertical sheeting driven 0.5 m below the bottom elevation of the cut. Excavation is started from the top within sheeting. Soldier beams and struts are inserted at elevations shown in the Figure.

(ii)   Unit pressure $= 0.65\, K_A \gamma H$

$$= 0.65 \times \frac{1 - \sin 38}{1 + \sin 38} \times 18 \times 6.0 = 16.7 \text{ kN/m}^2.$$

The pressure diagram will be as shown in Fig. 12.28b.

(iii) Considering that struts have hinge joints at points B and C (Fig. 12.28c)
Taking moment about B,

$$R_1 \times 1.6 = 16.7 \times 2.2 \times 1.1$$

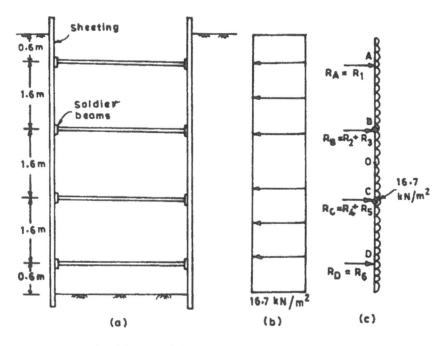

Fig. 12.28. Braced Open Cut with Earth Pressure Diagram.

$$R_1 = 25.26 \text{ kN/m}$$
$$R_2 = 16.7 \times 2.2 - 25.26 = 11.48 \text{ kN/m}$$
$$R_3 = R_4 = \frac{16.7 \times 1.6}{2} = 13.36 \text{ kN /m.}$$

Due to symmetrical loading diagram

$$R_5 = R_2 = 11.48 \text{ kN/m.}$$
$$R_6 = R_1 = 25.26 \text{ kN/m.}$$

Therefore,   Strut load at A, $R_A = 25.26$ kN/m
        Strut load at B, $R_B = 11.48 + 13.36 = 24.84$ kN/m
        Strut load at C, $R = 24.84$ kN/m
        Strut load at D, $R_D = 25.26$ kN/m

(iv) Cantilever moment $= \dfrac{16.7 \times 0.6^2}{2} = 3$ kNm/m

    Moment at the point $O = \dfrac{16.7 \times 3.0^2}{2} - 25.26 \times 2.4 - 24.84 \times 0.8$

$$= 75.15 - 60.62 - 19.87$$
$$= -5.34 \text{ kNm/m.}$$

Therefore, maximum moment in the sheet pile
        $= 5.34$ kNm/m.

Example 12.7

A strutted excavation, 1.75 m wide, is made in normally loaded clay of unit weight 19 kN/m^3. If the undrained shear strength of clay is 20 kN/m^2 and the cut is made up to a depth of 5.0 m, check the safety against base failure. Also show the pressure distribution and scheme of strutting.

Solution

(i) Critical depth, $H_c = \dfrac{cN_c}{\gamma}$

Let the cut be of 10.0 m length

Therefore, $\dfrac{B}{L} = \dfrac{1.75}{10.0} = 0.175$

$\dfrac{H}{B} = \dfrac{5.0}{1.75} = 2.85.$

For $\dfrac{B}{L} = 0.175$ and $\dfrac{H}{B} = 2.85$, Fig. 12.17c gives the value of N_c as 7.2.

Hence, $H_c = \dfrac{20 \times 7.2}{19} = 7.57$ m (> 5.0 m).

Therefore, the excavation pit is safe against base failure.

$$\dfrac{\gamma H}{C_u} = \dfrac{19 \times 5.0}{20} = 4.75 \ (> 4.0)$$

Therefore $m < 1.0$ Adopt $m = 0.4$

(ii) $K_A = 1 - \dfrac{m.4C_u}{\gamma H} = 1 - \dfrac{0.4 \times 4 \times 20}{19 \times 5.0} = 0.663.$

Unit pressure $= K_A \gamma H = 0.663 \times 19 \times 5.0 = 63$ kN/m^2.

The pressure distribution and scheme of sheeting is shown in Fig. 12.29a and b.

Example 12.8

The following data refer to a circular cellular cofferdam (Fig. 12.30).

Unit weight of fill	= 18 kN/m^3
Submerged unit weight of fill	= 10 kN/m^3
Angle of internal friction of fill material	= 32°
Coefficient of friction between fill and the rock	= 0.60
Coefficient of friction between fill and sheet pile	= 0.4
Coefficient of interlock friction	= 0.3
Permissible tensile stress in sheet piling	= 24 × 10^4 kN/m^2
Permissible tension in interlock	= 1500 kN/m.

Determine the diameter of cell.

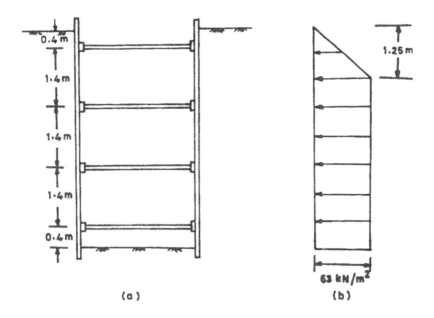

Fig. 12.29. Braced Open Cut with Earth Pressure Diagram.

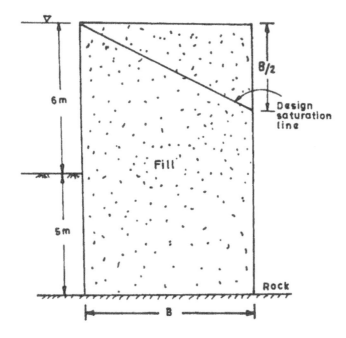

Fig. 12.30. Section of Cellular Cofferdam.

Solution

(i) Checking against sliding

Let B represent the average width of the circular cellular cofferdam.

Weight of cell fill $= 10 \times B \times 11 + (18 - 10) \times \dfrac{1}{2} \times B \times \dfrac{B}{2}$

$\qquad\qquad\qquad = (110 + 2B)B$

$\qquad K_A = \dfrac{1 - \sin 32}{1 + \sin 32} = 0.307.$

Lateral pressure after inside excavation

$$= \dfrac{1}{2} \times 10 \times 11^2 + \dfrac{1}{2} \times 10 \times 5^2 \times 0.307$$

$$= 605 + 76.75 = 681.75 \text{ kN.}$$

Frictional resistance $= W\mu = (110 + 2B)\, B \times 0.6$

$$F = \dfrac{(110 + 2B)\, B \times 0.6}{681.75}$$

Assuming $\qquad\qquad\qquad F = 1.25$

$\qquad (110 + 2B)\, B \times 0.6 = 1.25 \times 681.75$

or $\qquad\qquad 2B^2 + 110B - 1420.31 = 0$

$$B = \dfrac{-110 + \sqrt{4356 + 11363}}{4} = 7.7 \text{ m}$$

(ii) Checking against overturning

Moment of lateral pressure about the toe of the cofferdam

$$= 605 \times \dfrac{11}{3} + 76.75 \times \dfrac{5}{3} = 2347 \text{ kNm.}$$

Maximum allowable eccentricity $= \dfrac{B}{6}.$

For a factor of safety of 1.25

$$(110 + 2B) \times B \times \dfrac{B}{6} = 1.25 \times 2347$$

$$2B^3 + 110B^2 - 17602.5 = 0.$$

Solving by trial, $B = 11.5$ m.

Checking for overturning from shear of piling on cell fill:

Frictional resistance between piling and fill

$$= 681.75 \times 0.4 = 272.7 \text{ kN.}$$

Therefore, $\qquad B = \dfrac{2347 \times 1.25}{272.7} = 10.76$ m.

Therefore, adopt B as 11.5 m.

(iii) Checking against vertical shear

External moment of the lateral pressure about the centre of cofferdam = 2347 kNm.
Moment of vertical frictional resistance between piling and earth

$$= 76.75 \times 0.4 \times \frac{11.5}{2} = 176.5 \text{ kNm}$$

Net moment = 2347 – 176.5 = 2170.5 kNm.

Shear force on the central line of cell

$$= \frac{3}{2}\frac{M}{B} = \frac{3}{2} \times \frac{2170.5}{11.5} = 283 \text{ kN}.$$

Shear resistance offered by cell fill, $S = P \tan \phi$

S (on river side) = 681.75 tan 32 = 426 kN

$$K'_A = \frac{\cos^2 \phi}{2 - \cos^2 \phi} = \frac{\cos^2 32}{2 - \cos^2 32} = 0.561.$$

$$S \text{ (on dry side)} = \tan 32 \left[\frac{1}{2} \times 0.561 \times 18(11.0)^2 - \frac{1}{2} \times 0.561 \times 10 \left(\frac{11.0}{2} \right)^2 \right]$$

$$= 0.625(610.9 - 84.8)$$

$$= 328.8 \text{ kN}.$$

Average shear $= \frac{1}{2}$ (426.0 + 328.8) = 377.4 kN.

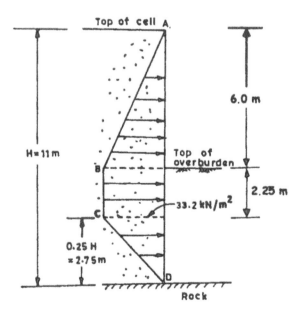

Fig. 12.31. Pressure Diagram for Computing Interlock Friction.

Computation of shear resistance offered by interlock friction: Assuming the cell to be filled with water at the top of overburden and until the cells have been filled, the inside overburden is not removed.

Pressure at the top of overburden

$$= 0.307 \times 18 \times 6 = 33.2 \text{ kN/m}^2$$

From Fig. 12.31,

Total pressure = area of figure ABCD

$$= \frac{1}{2} \times 33.2 \times 6 + 33.2(5 - 0.25 \times 11) + \frac{1}{2} \times 33.2 \times 0.25 \times 11$$

$$= 97.2 + 72.9 + 44.55 = 220 \text{ kN}.$$

Shear resistance due to interlock friction = $0.3 \times 220 = 66$ kN

Total shear   = $377.4 + 66 = 443.4$ kN

Safety factor $= \dfrac{443.4}{283} = 1.57 > 1.25$, safe.

Diameter of cell $= \dfrac{11.5}{0.875} = 13.15$ m.

REFERENCES

1. Agarwal, K.B., S. Saran and K.B. Bansal (1970), "Design of Anchored Bulkhead," *Journal of Concrete Construction and Architecture*, September Issue.
2. Bjerrum, L. and O. Eide (1956), "Stability of Strutted Excavations in Clay," *Geotechnique*, Vol. 6, No. 1, pp. 32–47.
3. Cummings, E.M. (1960), "Cellular Cofferdams and Docks", *Trans. ASCE*, Vol. 125, pp. 13–45.
4. Krynine, D.P. (1945), "Discussion on 'Stability and Stiffness of Cellular Cofferdam' by Karl Terzaghi" *Transactions ASCE*, Vol. 110, p. 1175.
5. Peck, R.B. (1943), "Earth Pressure Measurements in Open Cuts, Chicago (III) Subway," *Transactions ASCE*, Vol. 108, pp. 1008–1058.
6. Rowe, P.W. (1952), "Anchored Sheet Pile Walls" *PICE*, Vol. 1, part 1, pp. 27–70.
7. Saran, S. and Vaish, S.K. (1970), "Model Tests for Measurements of Earth Pressures behind Flexible Walls in Cohesionless Soils," *Journal on Indian National Society of Soil Mechanics and Foundation Engg.* Vol. 9, No. 4, p. 403.
8. Spilker, A. (1937), "Mitteilung iiber die Messung der Krafte in einer Baugrubenanssteifung," *Bautechnik*, Vols. 15, 16.
9. Teng, W.C. (1977), *Foundation Design*, Prentice Hall of India, New Delhi.
10. Terzaghi, K. and R.B. Peck (1967), *Soil Mechanics in Engineering Practice*, 2nd Edition, John Wiley and Sons, Inc., New York, p. 279.
11. T.V.A. (1957), *Cofferdam on Rock*, Technical Monograph 75, Tennessee Valley Authority, Knoxville, Tenn, p. 281.

PRACTICE PROBLEMS

1. Describe the analysis used to determine the depth of embedment of cantilever shear pile wall (a) in granular soils, and (b) in cohesive soils.
2. Describe the 'equivalent beam' and 'fixed earth support' methods of designing anchored bulkheads.

3. (a) Sketch a typical section of a braced open cut showing the position of struts. Also mark the deformation pattern of the side of such cuts.

(b) Describe a method of checking the stability of a cut in clays against base failure.

4. What is a cofferdam? Explain its purpose and list the different types of cofferdams.

5. Describe stepwise the procedure of designing cellular cofferdams.

6. Determine the depth of embedment for a sheet pile shown in Fig. 12.32.

7. Determine the depth of embedment, anchor pull and maximum moment in the anchored bulkhead shown in Fig. 12.33. Solve the problem by

(a) Equivalent beam method, and (b) Fixed earth support method.

8. A cut 4.0 m wide, 7.0 m deep is proposed in cohesionless soil with $\phi = 37°$. Sketch the suitable scheme of sheeting and bracing and also determine the maximum strut load. Assume the density of soil as 19 kN/m^3.

9. A strutted excavation, 2.0 m wide, is made in normally loaded clay of unit weight 18 kN/m^3. If the undrained shear strength of clay is 24 kN/m^2 and the cut is made up to a depth of 6.0 m, check the safety against base failure. Also show the pressure distribution and scheme of strutting. If the cut is made in stiff fissured clay, what will be the change in the strut load.

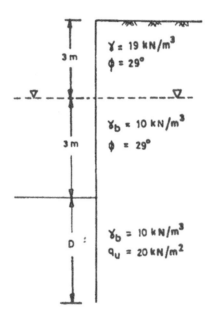

Fig. 12.32. Section of Sheet Pile.

10. The following data refer to a cellular cofferdam (Fig. 12.34)

Unit weight of fill	=	19 kN/m^3
Submerged unit weight of fill	=	10 kN/m^3
Angle of internal friction of fill material	=	35°
Coefficient of friction between fill and rock	=	0.62
Coefficient of friction between fill and sheet pile	=	0.43
Coefficient of interlock friction	=	0.32

Design: (a) Circular cellular cofferdam, and (b) Diaphragm type cellular cofferdam.

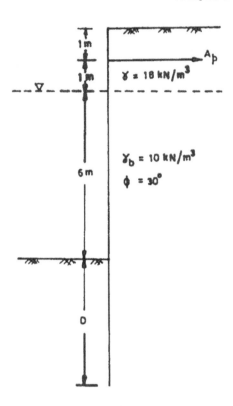

Fig. 12.33. Section of an Anchored Bulkhead.

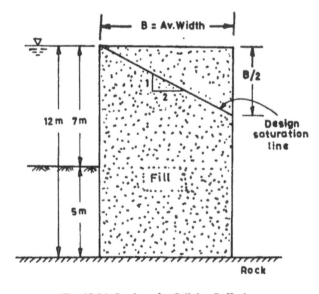

Fig. 12.34. Section of a Cellular Cofferdam.

Foundations in Expansive Soils

13.1 INTRODUCTION

Fine grained soils of less than 0.002 mm particle size are generally classified as clays. The important features of clay are plasticity and cohesion. In general, most of the particles in silt range and coarser than silt are approximately equidimensional, while the most common shape of almost all the clay size particles are platy. Like all other constituents of soil, clays are also products of weathering of rock. But, the important difference is that in case of clays, the effect of chemical weathering is most pronounced and original rock minerals undergo considerable changes to impart very different properties to clays. There are several minerals which are classed as clay minerals. However, the three main groups of minerals are kaolinite, illite and montmorillonite. The mineral montmorillonite is the main constituent of clays classed as expansive soils which are known to expand or swell when in contact with water and contract or shrink when dry. The typical swelling/shrinkage behaviour is due to the basic mineral composition of the montmorillonite.

The problem of damage to structure in expansive soil areas is worldwide. There is considerable damage to built up property. These soils occur in all the continents. In India about 20 per cent of total area is covered by expansive soils which are very often referred as black cotton soil. The damage to structures is mainly by differential heaving of the dried-up soils. Climatic environment is the most important for the swelling and shrinkage behaviour of these soils. Besides this local man-made conditions also cause changes in moisture and volume changes. From time to time various attempts have been made to modify the foundation practice to overcome the problem. The factors affecting volume changes are rather complex, attempts for quantification of the swelling and shrinkage have only limited success. However, certain types of foundations have ensured very reasonable trouble-free performance of the structures.

13.2 MINERAL STRUCTURE

The basic structural units of clay minerals are mainly of two types and held together by ionic bonds. These units are known as 'tetrahedral' units and 'octahedral' units. The central ion of the tetrahedron is generally, although not universally, silicon Si^{4+} surrounded by four ions of oxygen^{2-}. This unit is known as silica unit. The octahedral unit consists of an aluminium or magnesium, Al^{3+} or Mg^{2+} enclosed by six hydroxyl $(OH)^-$ ions. This unit is known as gibbsite. The clay minerals are for convenience represented by a combination of these symbols. The basic crystal layers of clay minerals consist of tetrahedral silica sheets arranged in hexagonal basal pattern combined with octahedral gibbsite unit.

Kaolinite minerals are the main constituents of China clay and kaolin. Their basic layers consist of a silica sheet combined with a gibbsite sheet which are strongly bonded to form a crystalline lattice. Montmorillonite minerals are mainly found in the expansive soils, bentonite and fullers earth. The basic layers consist of a gibbsite unit sandwiched between two tetrahedral units. The formation of typical montmorillonite mineral is shown in Fig. 13.1.

With such an arrangement of molecules, the outside surface consists of oxygen atoms only, resulting in very weak bonds between the basic layers. On hydration, these basal cleavages planes admit water molecules into the crystalline structure and cause swelling. This absorption of water can be 300% to 700% and this is the reason of high swelling of expansive clays. The crystalline structure of illite is basically the same as of montmorillonite, but for potassium ions present between the two units, i.e., it consists of two gibbsite and four silica units. The potassium ions compensate their out-of-balance electrostatic charge and strengthen the bond. This results in comparatively lesser absorption of water which is about half-way between the kaolinites and montmorillonites. Kaolinites absorb about 90% of their dry weight. The range of properties of the three minerals are summarized in Table 13.1.

Table 13.1. Properties of Clay Minerals (Gromoko, 1974).

S. No.	Property	Kaolinite	Illite	Montmorillonite
1.	Schematic structure			
2.	Particle thickness	0.5μ–2μ	0.003μ–0.1μ	9.5μ
3.	Particle diameter	0.5μ–4μ	0.5μ–10μ	0.5μ–10μ
4.	Specific surface m/gm	5–30	65–100	600–800
5.	Cation exchange Capacity me/100 gm	3–15	10–40	80–150
6.	Swelling %, under			
	10 kN/m²	negligible	350	1500
	20 kN/m²	negligible	150	350

Apart from the basic nature of the mineral, the overall behaviour of clays get modified by the minerals present in water. From engineering considerations, the overall behaviour is important. The expansive soils show large volume changes, low permeability, high compressibility and high sensitivity and these are the properties which are considered.

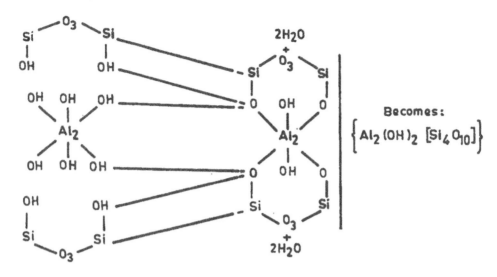

Fig. 13.1. Formation of Montmorillonite.

13.3 IDENTIFICATION OF EXPANSIVE SOILS

The purpose of identification of expansive soils is to qualitatively characterize the potential volume change and to take necessary precautions in selection and design of foundations.

There are a number of tests used in identifying the clay minerals and studying their detailed characteristics, namely (i) X-ray diffraction, (ii) Microscopic examination, and (iii) Differential thermal analysis (D.T.A.). There are some simpler tests that can be conveniently performed for determining the expansive characteristics important from engineering considerations. These tests are described below.

Free swell test

This test was suggested by Holtz and Gibbs (1956). In this test, 10 cubic centimetres of dry soil passing through a 425-micron sieve is slowly poured into a 100 cc graduated jar filled with water. Firstly, the volume of the settled soil (V_i) is read from the graduations of the cylinder. The soil is then left in the jar for 24 hours and then the volume of swelled soil (V_s) is measured. The free swell value in per cent (S_F), is then determined by

$$S_F = \frac{V_s - V_i}{V_i} \times 100. \qquad \text{... (13.1)}$$

In this test, bentonite may swell from 1200 to 2000 per cent, kaolinite about 80 per cent and illite from 30 to 80 per cent. According to Holtz and Gibbs (1956), soils having free swell values as low as 100 per cent may undergo considerable volume changes when wetted under light loads and should, therefore, be viewed with caution. However, the free swell test alone does not fully suffice to predict the swelling potential. It should therefore be supplemented by other tests.

Differential free swell test

In this test, two samples of dried soil weighing 10 g each, passing through 425-micron sieve are taken. One in put in 50 cc graduated jar containing kerosene oil (a non-polar liquid). The other sample is put in a similar jar containing distilled water. Both the jars are kept undisturbed for 24 hours and then their volumes are measured. The differential free swell in per cent (S_{DF}) is given by

$$S_{DF} = \frac{\text{Soil volume in water} - \text{Soil volume in kerosene}}{\text{Soil volume in kerosene}} \times 100.$$

The degree of expansiveness can be qualitatively assessed from the values of differential free swell (S_{DF}) from Table 13.2.

Table 13.2. Degree of Expansion Based on Differential Free Swell.

S_{DF} (%)	Degree of Expansion
Less than 20	Low
20 to 35	Moderate
35 to 50	High
Greater than 50	Very high

Colloidal content, plasticity index and shrinkage limit

Holtz and Gibbs (1956) on the basis of their extensive work on natural undisturbed soils indicated that colloidal content (fraction finer than 1), plasticity index and shrinkage limit can be qualitative indicators of expansive characteristics (Table 13.3).

Table 13.3. Degree of Expansiveness as a Function of Colloidal Content, Plasticity Index and Shrinkage Limit.

Classification Criteria	Identification Property	Approximate Range			
	Volume change* %	0–10	10–20	20–30	30
	Colloidal content %	0–15	10–25	20–35	25
	Plasticity index %	0–15	10–35	20–45	30
	Shrinkage limit %	12	8–18	6–12	10
Expansion Classification		Low	Medium	High	Very high

* From air dry to saturation under surcharge of 7 kN/m².

Table 13.4. Relationship between Swelling Potential of Soils and Plasticity Index.

Swelling Potential	Plasticity index per cent
Low	0–15
Medium	10–35
High	20–55
Very high	35 and above

Peck, Hanson and Thornburn (1974) have related plasticity index to the swelling potential. It is useful because of its simplicity (Table 13.4).

Gromko (1974) has related shrinkage limit to linear shrinkage and degree of expansion (Table 13.5).

Snethen (1979) based on field and laboratory studies on expansive soils of United States representing variety of conditions with respect to *in situ* water content, geology, topography climate, etc. proposed criteria for identifying expansive soils as given in Table 13.6.

Sridharan et al. (1985) in view of the negative free swell index values observed for soils containing kaolinite proposed that the free swell index be defined as the volume occupied by a unit weight of oven dry soil in water under no external restraint and accordingly suggested classification for expansive soils (Table 13.7).

Table 13.5. Correlation of Shrinkage Limit with Linear Shrinkage and Degree of Expansion.

Shrinkage Limit per cent	Linear Shrinkage per cent	Degree of Expansion
< 10	> 8	Critical
10–12	5–8	Marginal
> 12	0–5	Non-critical

Table 13.6. Criteria for Identifying Expansive Soils.

Liquid Limit %	Plasticity Index %	Natural Suction kN/m²	Potential Swell	Potential Swell Classification
60	35	400	1.5	High
50 to 60	25 to 35	150 to 400	0.5 to 1.5	Marginal
50	25	150	0.5	Low

Table 13.7. Classification of Expansive Soils Based on Free Swell Index.

Free Swell Index $(V_d/10)$, cc/gm	Swelling Potential
1.5	Negligible
1.5–2.0	Slight
2.0–5.0	Moderate
5.0–10.0	High
10.0	Very high

Besides this extensive research has given the following correlation of swelling potential with index properties and moisture content, etc.

(a) Seed, Mitchell and Cari (1962)

$$S = 0.00216 \, I_P^{2.44} \qquad \qquad \ldots (13.1)$$

(b) Ranganatham and Satyanarayana (1965)

$$S = 0.00413 \, I_S^{2.67} \qquad \qquad \qquad \text{... (13.2)}$$

(c) Nayak and Christensen (1971)

$$S = 0.0229 \, I_P^{1.45} \, c_l/w + 6.38 \qquad \qquad \text{... (13.3)}$$

(d) O'Neill and Ghazzaly (1977)

$$S = 2.27 + 0.131w_L - 0.27w \qquad \qquad \text{... (13.4)}$$

(e) Johnson and Snethen (1979)

$$\log S = 0.036w_L - 0.0833w + 0.458 \qquad \qquad \text{... (13.5)}$$

An estimate of the overburden (or footing) pressure necessary to restrain expansion to a tolerable quantity can be obtained by an equation by Komornik and David (1969) based on statistical analysis of some 200 soils. This equation is

$$\log \rho_v = \overline{2}.132 + 2.08w_L + 0.665\gamma d - 2.69w \qquad \qquad \text{... (13.6)}$$

Eq. (13.6) gives the value of σ_v in kg/cm^2.

The free swell obtained from Eqs. (13.1) to (13.5) may be reduced for confining pressure σ_v using Eq. (13.7) given by Bowles (1982).

$$S' = S\left(1 - 0.0735\sqrt{\sigma_v}\right) \qquad \qquad \text{... (13.7)}$$

In Eq. (13.7), value of σ_v is in kPa unit.

The notations used in Eqs. (13.1) to (13.7) are as follows:

S, S' = swelling potential,
w_L = liquid limit,
I_s = shrinkage index,
I_P = plasticity index,
w = initial water content,
c_l = per cent clay content, and
σ_v = overburden or footing pressure.

13.4 INDIAN EXPANSIVE SOILS

The area of major occurrence of Indian expansive soils is south of the Vindhyachal range covering almost the entire Deccan Plateau. But there are substantial areas of their occurrence in the other parts of the country as well. Although popularly known as black cotton soil, the expansive soils in India are not always necessarily of black colour.

The parent materials associated with expansive soils are either basic igneous, or sedimentary rocks. The formation of Indian expansive soils is usually associated with basalt.

However, these occur on granite, gneiss, shale, sandstone, slate or limestone and also as residual soils. Transported deposits up to 8 m depth have also been seen. The thickness of black cotton soil cover is highly variable from 0.30 m to 15 m. The composition of soil shows considerable variation with different depth horizons especially in their clay contents. The Indian expansive soils are classified as clay or silty having 30–70 per cent clay fractions, 17–45 per cent silt and 10–25 per cent sand. The clay fraction is very rich in silica. These consist of montmorillonite which in igneous rocks is formed by decomposition of feldspar and pyroxene and in sedimentary rocks it is a constituent of rock itself.

The range of chemical composition of black cotton soil is given in Table 13.8 (Katti, 1979).

The base exchange capacity of clay fraction is in the range of 100 to 130 m. eq/100 gm. The colour of soil is generally dark but these may be brown or yellowish brown at greater depths. As the depth increases, very often clay content decreases and relatively more carbonates in the form of nodules, 'kankar', are also found. The light coloured soils are sometimes mistaken as non- expansive soils, which may not be the case. Actually, colour is not a good guide to identify expansive soils.

The ranges of physical properties of Indian expansive soils are given in Table 13.9.

Table 13.8. Chemical Composition of Black Cotton Soils (Katti, 1979).

Description	Formula	Range
Silica	SiO_2	45–58
Alumina	Al_2O_3	13–18
Ferric oxide	Fe_2O_3	7–15
Lime	CaO	1–8
Magnesium oxide	MgO	2–5
Titanium oxide	TiO_2	0.5–2
Carbonate	CO_3	0.5–5
Sulphate	SO_3	1–2
Organic matter	—	0.5–4
Loss of ignition	—	5–17
pH	—	6–8

Table 13.9. Range of Physical Properties of Indian Black Cotton Soils.

Properties	Range
Liquid limit (w_l)	40–100%
Plasticity index (I_p)	20–60 %
Shrinkage limit (w_s)	8–18 %
Volumetric shrinkage	40–50 %
Differential free swell (Free swell index)	20–100 %

13.5 SWELL POTENTIAL AND SWELLING PRESSURE

When an expansive soil attracts and accumulates water, a pressure known as swelling pressure builds up in the soil and it is exerted on the overlying materials and structures, if there are any. Swelling pressure can be defined as the maximum force per unit area that needs to be placed over a swelling soil to prevent volume increase. The swelling pressure is not a unique property of soil, and it depends on several factors. The importance of estimating heave or swelling pressure is that if under given conditions these can be calculated, the loading on the foundation may be adjusted in such a way that the heave is neutralized by settlement. In other words, swelling pressure is counterbalanced by loading on foundations. The factors effecting swelling potential are:

1. Mineral type and its amount
2. Density—more compact strata swell more.
3. Loading conditions—surcharge loads reduce swelling. Also allowance for a small amount of swelling reduces swelling pressure to a large extent.
4. State of soil—stress history influence swell. Undisturbed soils show less swelling pressures. Static compaction allows more swell than dynamic compaction.
5. Duration of time—more time is conducive to greater swell.
6. Pore fluid—high salt concentration in water reduces swelling.
7. Initial moisture—dry soils swell more.

For the measurement of swelling pressure, the Indian standard IS: 2720 (part XLI) 1977 provides two methods, namely (i) the constant volume method in which volume change is prevented and the pressure is measured, and (ii) consolidometer method in which the volume change of the soil is permitted and the corresponding pressure required to bring back the soil to its original volume measured.

The first method requires continuous adjustment of pressure on the soil specimen taken in a consolidation cell, so that the soil volume at any time is equal to its initial volume. The second method consists of taking a few (more than three) initially identical soil specimens in consolidation cells of fixed ring type, subjecting them to different magnitudes of pressures and then allowing the soil to saturate. Under the different load intensities, some of the soil specimens would compress after saturation while some others would swell. After the volume change (compression or swelling) is complete and has been recorded, a load intensity versus volume change plot is obtained (Fig. 13.2a). From this plot, the pressure corresponding to zero volume change is read and is denoted as swelling pressure. It is convenient to plot the load intensity to logarithmic scale, as this would produce a straight line (Fig. 13.2b).

13.6 TRADITIONAL INDIAN PRACTICE

Traditionally the most common type of foundation in expansive soil has been the strip footing. Even now it is most prevalent type. The common characteristic features of this traditional Indian practice are:

1. The safe bearing capacity for the design is taken 50 to 75 kN/m^2. The widespread occurrence of cracking in lightly loaded structures with strip footing foundation is ascribed to the poor bearing capacity of the soil and, therefore, low values of safe bearing capacity are recommended. A value of 50 kN/m^2 seems to be commonly used value in design of foundations.

2. The minimum depth of foundation is taken as 2 m. This practice might have been adopted due to the reason that in most of cases, as the depth increases, the colour of black cotton soil

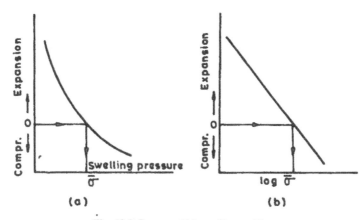

Fig. 13.2. Pressure-Volume Change Plots.

becomes lighter and at about 2 m depth, very often it is very light or completely different. It is wrongly concluded that soil, if not black in colour will not be expansive. Another reason for deeper footings might be that in these areas coarser fractions increase with depth. In some cases 'kankar' may be encountered and in such cases experience might show that apart from good bearing of soil, the expansive nature and consequent trouble due to cracking are really much smaller.

3. The bottom of the foundation trench is filled with well compacted sand or 'murrum' or broken stones. Also, back filling is specified by sand or nonclayey material. The obvious reason for this recommendation is to avoid the contact of the expansive soil with the foundation.

4. Plinth protection is sometimes suggested as an additional provision.

5. As the above provisions did not result in trouble-free foundations, in the late thirties, it was recommended to use reinforced concrete bands at foundation, plinth and lintel levels. The provision of reinforced concrete bands is an attempt to strengthen the superstructure and substructure.

It is known that the strip footing foundations, specially those of shallow depths, have a high risk of occurrence of cracks in the superstructure. But on account of the familiarity of an age-old practice, the use of strip footing foundation persists.

13.7. METHODS OF FOUNDATIONS IN EXPANSIVE SOILS

The well established reason for cracking of the buildings in expansive soils is the swelling and shrinkage of the soil. In the cycle of seasons, the foundations move differentially and the superstructure which does not take tensile stresses, suffers from cracks. As the maximum soil strata movements are at the ground level and diminish with depth, therefore, shallow foundations get affected by these movements. There are other factors also such as the nature of soil, initial moisture and density of soil, state of confinement of soil, external pressure, etc. which affect the volume changes in expansive soil.

From time to time several approaches have been followed to solve the problem. The various methods can be put under three categories:

1. Designing the structure to withstand the effects of swelling and shrinkage of strata.
2. Eliminating the swelling. This can be done by (a) stabilizing the moisture, (b) loading the soil more than the swelling pressure, (c) treating the soil so that its behaviour is changed, and (d) replacing the soil by suitable non-swelling soil.
3. Isolating the structure by taking down the foundations to a stratum which is not affected by swelling.

Strengthening of structures beyond a limit is not practical on account of economic reasons. Provision of reinforced concrete bands at various levels is an attempt in this direction. Heavily reinforced rafts have also been suggested. Cost is excessive. Very flexible structures also are not practical.

If there is no moisture change in the strata supporting foundations, there will be no volume changes. Prewetting has been tried. For isolated structures the method is not practical. Aprons and barriers can help in minimizing volume changes.

Greater loads on foundations have shown better performance. Lighter loading (conventional practice of not loading more than 5 kN/m^2) is not a safe approach. But quantification of a suitable loading is not practical as the factors affecting are many.

Stabilization of expansive soil by lime is well known but there are the problems of mixing and effective compaction. For building foundations it does not prove economical. Stabilization by lime (2% to 8%) decreases liquid limit, plasticity index and swelling, and increases optimum moisture content and strength of expansive clays. This method is more practical for roads. For deeper treatment lime slurry injection has also been tried. But it is not common.

13.8 REPLACEMENT OF SOILS AND 'CNS' CONCEPT

Replacement of soils by coarse-grained sand and murrum is an age-old practice in expansive soil area. Specially the back filling of the trenches after the construction of footings is a very common practice. These replacements and fillings are useful but are no guarantee for safety against cracks. The depth of such soil replacement is arbitrary and economics also is not favourable for handling large quantities.

Katti (1979) developed the concept of CNS (cohesive non-swelling soil) layer. It is based on the concept of providing similar environment existing at no volume change depth in saturated expansive soil by placing a soil system having cohesion but not having clay mineral of expanding type. It is further recommended that (i) the bearing capacity and settlement aspects of CNS be taken into consideration, and (ii) the CNS should be provided up to sufficient extent. The specifications of soil to be considered as CNS material are given in Table 13.10.

It is necessary to conduct large scale tests to arrive at optimum thickness of CNS layer with available CNS material.

The CNS has been tried for buildings at few places only and its success as building foundations is not established fully. On economy considerations also, the method may not be competitive because large-scale replacement (about a metre or more depth) of soil are needed and controlled compaction in layers is required. However, the method has been adopted on large scale for irrigation channels in Karnataka state.

Table 13.10. Specifications for Soil to be Considered as 'CNS' Material (Katti, 1979)

	Properties	Specifications Range
1.	Grain size analysis	
	Clay (< 0.002 mm) %	15–25
	Silt (0.06 to 0.002 mm) %	30–45
	Sand (2 to 0.06 mm) %	30–40
	Gravel (> 2 mm) %	10
2.	Consistency limit	
	Liquid limit %	30–50
	Plastic limit %	20–50
	Plasticity index %	10–25
	Shrinkage limit %	15 and above
3.	(a) Swelling pressure when compacted to standard proctor optimum condition with moisture content and at no volume change condition (kN/m^2)	less than 10
	(b) Swelling pressure when compacted to standard proctor optimum condition at no volume change condition (kN/m^2)	less than 5
4.	Clay minerals	Preferably kaolinite, illite
5.	Shear strength of samples compacted to standard proctor optimum condition after saturation	
	(a) 1/2 UCS (kN/m^2)	15–35
	(b) Consolidated direct shear test	
	@ 0.0125 mm/min, c_u (kN/m^2)	10–30
	ϕ_u (deg)	8–15
6.	Approximate thickness of CNS layer	
	Swelling pressure (kN/m^2)	thickness (m)
	100 to 150	0.75–0.85
	200 to 300	0.90–1.0
	350 to 500	1.05–1.15

13.9 UNDERREAMAD PILE FOUNDATIONS

Underreamed piles are bored cast *in situ* concrete piles having bulb(s) towards their toes. The underreamed pile, originally developed in Texas, USA, have been considerably experimented upon by Central Building Research Institute (CBRI), Roorkee, for use in Indian black cotton soils and appear to provide a good solution to foundation problems in such soils. The underreamed pile beam construction is the most common and appropriate type of foundation on swelling soils. The utility of this type of construction is to support the superstructure of the building over beams which are clear off the ground surface and span over piles anchored at a depth where the soil has a nearly stable water content during the various seasons of the year and is, thus, free from appreciable seasonal movements. This depth is about 3.5 m for black cotton soils in Central India.

The length of the pile may vary from 3.5 m to 4.0 m in deep deposits of black cotton soils. In shallow deposits, the pile is carried down into the nonexpansive layer to a minimum depth of 0.6 m. The spacing of piles depends upon the plan of the building, its loading and safe bearing capacity of piles. The spacing may vary from 1.5 m to 3 m. A pile is provided under every wall junction in the plan of the building so that a point load on the plinth beam is avoided.

Figure 13.3 shows the general configuration of underreamed piles. The diameter of the pile shaft (D) may vary from 200 to 500 mm and the ratio of the diameter of the enlarged base or bulb to that of shaft (D_u/D) may vary from 2 to 3. In case of multiunderreamed piles (i.e. having more than one bulb), the topmost bulb should be at a minimum depth of two times the bulb diameter. The centre to centre distance between the two bulbs may vary from 1.25 D_u to 1.5 D_u. Auger boring and an underreaming tool have been developed by CBRI, for making the hole before the concrete is cast. The pile is reinforced throughout its length to take care of the tensile stresses that are set up on account of the uplift forces due to swelling. During summer seasons, the soil shrinks and the pile acts as a column. Steel reinforcements should be sufficient to take care of this situation also.

The ultimate bearing capacity of underreamed piles can be calculated from soil properties (IS: 2911 (Part III)–1980). The soil parameters required are cohesion, angle of internal friction and soil density. If these properties are not available directly from laboratory and field tests, they may be indirectly obtained from *in situ* penetration test data. The factor of safety is normally 2.5 to 3.0 for calculating safe load.

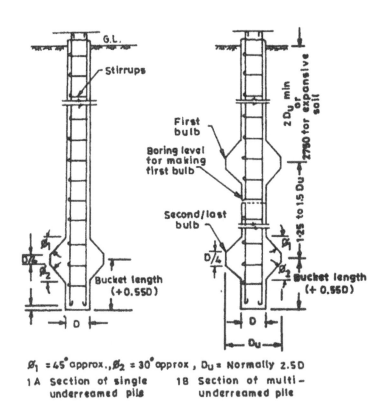

Fig. 13.3. Typical Details of Bored Cast-in-situ Underreamed Pile Foundations.

(a) Clayey soils

For clayey soils the ultimate load carrying capacity of an underreamed pile may be worked out from the following expressions:

$$Q_u = A_p \cdot N_c \cdot c_p + A_a \cdot N_c \cdot c'_a + c'_a \cdot A'_s + \alpha \cdot c_a \cdot A_s \qquad \ldots (13.8)$$

where Q_u (kN) = ultimate bearing capacity of pile,

A_p (m^2) = cross-sectional area of pile stem at toe level

$$= \frac{\pi}{4} D^2,$$

N_c = bearing capacity factor, usually taken as 9,

c_p (kN/m^2) = cohesion of the soil around toe,

A_a (m^2) = $\pi/4$ ($D_u^2 - D^2$), where D_u (m) and D (m) are the underreamed and stem diameter, respectively,

c'_a (kN/m^2) = average cohesion of soil around the underreamed bulbs,

α = reduction factor (usually taken 0.5 for clays),

c_a(kN/m^2) = average cohesion of soil along the pile stem,

A_s (m^2) = surface area of the stem, and

A'_s (m^2) = surface area of the cylinder circumscribing the underreamed bulbs.

The above expression holds for the usual spacing of underreamed bulbs placed at not more than one-and-a-half times their diameter. The following points are important in using Eq. 13.8:

(i) The first two terms in the formula are for bearing and the last two for friction components.
(ii) If the pile is with one bulb only, the third term will not occur.
(iii) For calculating uplift load first term will not occur in the formula.

(b) Sandy soils

For sandy soils the following expression be used:

$$Q_u = A_p \left(1/2 \cdot D \cdot \gamma \cdot N_\gamma + \gamma \cdot D_f \cdot N_q \right) +$$

$$A_a \left(1/2 \cdot D_u \cdot n \cdot \gamma \cdot N_\gamma + \gamma \cdot N_q \sum_{r=1}^{r=n} D_r \right) +$$

$$(1/2\pi \cdot D \cdot \gamma \cdot K \cdot \tan \delta) \left(D_1^2 - D_f^2 - D_n^2 \right) \qquad \ldots (13.9)$$

where A_p (m^2) = $\pi/4 \, D^2$, where D (m) is stem diameter,

A_a (m^2) = $\pi/4($ $D_u^2 - D^2$), where D_u (m) is the underreamed bulb diameter,

n = number of underreamed bulbs,

γ (kN/m^3) = average unit weight of soil (submerged unit weight in strata below water table),

N_γ and N_q = bearing capacity factors depending upon the angle of internal friction (Table 13.11),

D_r (m) = depth of the centre of different underreamed bulbs below ground level,

D_f (m) = total depth of pile below ground level,

K = earth pressure coefficient (usually taken 1.75 for sandy soils),

δ = angle of wall friction (may be taken equal to the angle of internal friction ϕ),

D_1 (m) = depth of the centre of the first underreamed bulb, and

D_n (m) = depth of the centre of the last underreamed bulb.

The following two points may be kept in view in using Eq. 13.9.

(i) The first two terms in the formula are for bearing component and the last one for friction component.

(ii) For uplift bearing on tip, A_p will not occur.

Table 13.11. Bearings Capacity Factors.

f	N_γ	N_q^*
20	3	10
22	5	12
24	7	15
26	9	18
28	13	22
30	18	29
32	25	37
34	34	50
36	47	65
38	65	90
40	85	128

* To be reduced to half for bored piles including underreamed piles.

(c) Soil strata having both cohesion and friction

In soil strata having both cohesion and friction or in layered strata having two types of soil, the bearing capacity may be estimated using both the equations 13.8 and 13.9. However, in such cases the load tests will be a better guide.

On the basis of a large amount of experimental data, CBRI has suggested bearing capacity values for typical dimensions of underreamed piles (CBRI Hand Book, 1978), and have been incorporated in IS: 2911 (Part III–1980). The values of the safe loads along with other details are given in Table 13.12.

After identifying the soil type and having standard penetration test data, the capacity of the underreamed piles (Table 13.12) can be obtained as outlined below:

(i) The safe bearing, uplift and lateral loads for underrreamed piles given in Table 13.12 apply to both medium compact ($10 < N < 30$) sandy soils and clayey soils of medium ($4 < N <$

Table 13.12. Safe Load for Vertical Bored Cast-in-situ Single Underreamed Piles of 3.5 m Length in Sandy and Clayey Soils including Black Cotton Soils

Size		Mild Steel Reinforcement		Rings Spacing of 6 mm Dia.	Compression			Uplift Resistance			Lateral Thrust
Dia. of pile	Under-reamed Dia.	Longitudinal Reinforcement			Safe Load	Increase Per 30 cm Length	Decrease Per 30 cm Length	Safe Load	Increase Per 30 cm Length	Decrease Per 30 cm Length	
(cm)	(cm)	(No)	(dia. mm)	(cm)	(t)	(t)	(t)	(t)	(t)	(t)	(t)
(1)	(2)	(3)	(4)	(5)	(6)	(7)	(8)	(9)	(10)	(11)	(12)
20	50	3	10	18	8	0.9	0.7	4	0.65	0.55	1.0
25	62.5	4	10	22	12	1.15	0.9	6	0.85	0.70	1.5
30	75	4	12	25	16	1.4	1.1	8	1.05	0.85	2.0
37.5	94	5	12	30	24	1.8	1.4	12	1.35	1.10	3.0
40	100	6	12	30	28	1.9	1.5	14	1.45	1.15	3.4
45	112.5	7	12	30	35	2.15	1.7	17.5	1.60	1.30	4.0
50	125	9	12	30	42	2.4	1.9	21	1.80	1.45	4.5

8) consistency including expansive soils. The values are for piles with bulb diameter equal to two-and-a-half times the shaft diameter.

The reinforcement shown is mild steel and it is adequate for loads in compression and lateral thrusts [columns (6) and (12)]. For uplift loads (column 9), requisite amount of steel should be provided. In expansive soils, the reinforcement shown in Table 13.12 is adequate to take upward drag due to heaving up of the soils.

The concrete considered is M 20. The minimum clear cover the longitudinal reinforcement should be 40 mm.

(ii) Safe load for piles of lengths different from those shown in Table 13.12 can be obtained considering the decrease or increase as from columns (7), (8), (10) and (11) for the specific case.

(iii) Safe loads for piles with more than one bulb can be worked out from Table 13.12 by adding 50 per cent of the loads shown in column (6) or (9) for each additional bulb in the values given in those columns. The additional capacity for increased length required to accommodate bulbs, should be obtained from columns (7) and (10).

(iv) Values given in column (12) for lateral thrust may not be increased or decreased for change in pile lengths. For longer and/or double and multi-underreamed piles higher lateral thrusts may be adopted after establishing from field load tests.

(v) For dense sand ($N > 30$) and stiff clay ($N > 8$) the safe loads in compression and uplift obtained from Table 13.12 may be increased by 25 per cent. The lateral thrust values should not be increased unless the stability and strength of top soil (strata up to a depth of about three times the pile shaft diameter) is ascertained and found adequate. For piles in loose ($4 < N < 10$) sandy and soft ($2 < N < 4$) clayey soils, the safe loads should be taken 0.75 times the values shown in Table 13.12. For very loose ($N < 4$) sandy and very soft ($N < 2$) clayey soils the values obtained from the table should be reduced by 50 per cent.

(vi) The safe loads obtained from Table 13.12 should be reduced by 25 per cent if the pile boreholes are full of sub-soil water or drilling mud during concreting. No such reduction may be necessary if the water is confined to the shaft portion below the bottom-most bulb.

(vii) The safe loads in uplift and compression given in Table 13.12 or obtained in accordance with (i) to (vi) should be reduced by 15 per cent for pile with bulb of twice the stem diameter. But no such reduction is required for lateral thrust shown in Table 13.12.

(viii) The safe loads for underreamed compaction piles can be worked out by increasing the safe load of equivalent bored cast-in-situ underreamed pile obtained from Table 13.12 by 1.5 times in case of medium ($10 < N < 30$) and 1.75 times in case of loose to very loose ($N < 10$) sandy soils. Depending upon the nature the initial increase may be up to a factor of 2 and initial load rests are suggested to arrive at the final safe load values for design in case of sizeable works. The values for lateral loads should not be increased by more than 1.5 times in all cases. In obtaining safe load of compaction pile the reduction for pile boreholes full of sub-soil water or drilling mud during concreting should be taken as 15 per cent instead of 25 per cent as given in (vi).

(ix) The reduction for piles with twice the bulb diameter is to be taken 10 per cent instead of 15 per cent as given in (vii). The provision of reinforcement in underreamed compaction piles will also be guided by driving considerations.

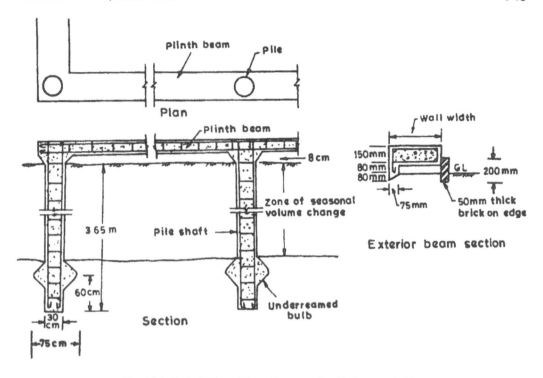

Fig. 13.4. Typical Pile and Beam Sections of an Underreamed Pile.

The plinth beams supporting masonry walls are designed for a maximum bending moment of $(wl^2/50)$ when the openings in the walls are near the supporting piles and $(wl^2/100)$ for blank walls when openings are in the centre with respect to the piles on either side of the opening, where w is the uniformly distributed load on the beam and l is the effective span. If the beams are not supported during construction till the masonry above it gains strength, the bending moment shall be increased to $wl^2/30$. The minimum overall depth of grade beam shall be 150 mm.

Figure 13.4 shows a typical pile and beam sections.

13.10 REMEDIAL MEASURES FOR CRACKED BUILDINGS

Unsatisfactory performance of foundations in expansive soils is too well known and the damage caused is very considerable. The major group of foundation failures is that of shallow spread footings. The reason is differential movement due to heaving and/or shrinkage. Other failures are of pile foundations which may be either due to the inadequacy of piles that get effected by the ground movements or the connecting beams at pile tops resting on the ground and other constructional defects.

The major thrust of remedial measures is generally towards moisture stabilization. One such remedy which has been tried successfully is an apron around buildings (Sharma, 1981) as shown in Fig. 13.5. The recommendations for laying are that the ground should be neither too

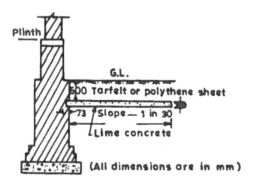

Fig. 13.5. Remedial Measure for Moisture Stabilisation.

wet nor too dry. Moisture stabilization for remedial measures by vertical cutoff walls around a building are also a potential method but not much used and reported. Underpinning and additional foundations are resorted to in few cases only.

ILLUSTRATIVE EXAMPLES

Example 13.1

Determine the capacity of 4.0 m long single underreamed pile of 500 mm stem diameter. Average cohesion value both within the strata of pile depth and below the toe is 100 kN/m².

Solution

$$Q_u = A_p N_c c_p' + A_a N_c c_a' + \alpha c_a A_s$$
$$A_p = \pi/4 \times (0.50)^2 = 0.196 \text{ m}^2$$
$$c_p = c_a' = c_a = 100 \text{ kN/m}^2$$
$$A_a = \pi/4 \times [(1.25)^2 - (0.50)^2] = 1.03 \text{ m}^2$$
$$A_s = \pi \times 0.50 \times 4.0 = 6.28 \text{ m}^2$$
$$N_c = 9.0$$
$$\alpha = 0.5$$
$$Q_u = 0.196 \times 9.0 \times 100 + 1.03 \times 9.0 \times 100 + 0.5 \times 100 \times 6.28$$
$$= 176.4 + 927 + 314 = 1417.4 \text{ kN.}$$

Safe capacity of pile $\quad = \dfrac{1417.4}{3} = 472 \text{ kN.}$

Example 13.2

Determine the length and capacity of tripple underreamed pile of stem diameter 400 mm. The stratum is a deep deposit of average cohesion of 70 kN/m².

Solution

(i) $D_u = 2.5 \times 0.4 = 1.0$ m

From the geometry of the pile (Fig. 13.3)

$$\text{Pile length} = 2D_u + 2 \times 1.25D_u + 0.55D + 0.55$$
$$= 2 \times 1.0 + 2 \times 1.25 \times 1.0 + 0.55 \times 0.4 + 0.55$$
$$= 5.27 \text{ m, say } 5.5 \text{ m}$$

(ii)

$$Q_u = A_p N_c \cdot c_p + A_a N_c c'_a + c'_a \cdot A'_s + \alpha c_a A_s$$
$$A_p = \pi/4 \times (0.4)^2 = 0.1256 \text{ m}^2$$
$$A_a = \pi/4 \times [(1.0)^2 - (0.4)^2] = 0.6594 \text{ m}^2$$
$$c_a = c'_a = c_a = 70 \text{ kN/m}^2$$
$$N_c = 9$$
$$A'_s = \pi \times 1.0 \times 2.5 = 7.85 \text{ m}^2$$
$$A_s = \pi \times 0.4 \times (2.0 + 0.22 + 0.55) = 3.48 \text{ m}^2$$
$$\alpha = 0.5.$$

Therefore,

$$Q_u = 0.1256 \times 9.0 \times 70 + 0.6594 \times 9.0 \times 70 +$$
$$70 \times 7.85 + 0.5 \times 70 \times 3.48$$
$$= 79 + 415 + 549 + 122 = 1165 \text{ kN}.$$

Safe capacity of pile $= \dfrac{1165}{3} = 388$ kN.

Example 13.3

Determine the capacity of a 5.0 m long single underreamed pile of stem diameter 400 mm. The stratum is a deep deposit of medium dense sand having $\gamma = 18$ kN/m^3 and $\phi = 30°$.

Solution

(i)

$$Q_u = A_p \left(1/2 \cdot D \cdot \gamma \cdot N_\gamma + \gamma D_f N_q\right) + A_a \left(1/2D_u \cdot n \cdot \gamma N_\gamma + \gamma N_q \sum_{r=1}^{r=n} D_r\right) +$$

$$1/2 \pi D \gamma K \tan\delta \left(D_1^2 + D_f^2 - D_n^2\right)$$
$$A_p = \pi/4 \times (0.4)^2 = 0.1256 \text{ m}^2$$
$$A_a = \pi/4 \times (1.0^2 - 0.4^2) = 0.6594 \text{ m}^2$$

For

$$\phi = 30°, N_\gamma = 18, N_q = 29/2 = 14.5$$
$$D_f = 5.0 \text{ m}; D_1 = 5.0 - (0.55D + 0.55) = 4.23 \text{ m}$$
$$k = 1.75, \delta = \phi = 30°; n = 1; D_\gamma = D_n = D_1 = 4.23 \text{ m}$$
$$Q_u = 0.1256(1/2 \times 0.4 \times 18 \times 18 + 18 \times 5.0 \times 14.5) + 0.6594\times$$
$$(1/2 \times 1.0 \times 1.0 \times 18 \times 18 + 18 \times 14.5 \times 1.0) + 1/2 \times \pi \times$$
$$0.40 \times 18 \times 1.75 \tan 30 (4.23^2 + 5^2 - 4.23^2)$$
$$= 172 + 279 + 285 = 736 \text{ kN}.$$

Safe load capacity of pile $= \dfrac{736}{3} = 245$ kN.

Example 13.4

Determine the length and capacity of a double underreamed pile of 0.5 m stem diameter. The strata is deep cohesionless soil with $\gamma = 19$ kN/m^3 and $\phi = 35°$.

Solution

(i) $D_u = 2.5 \times 0.5 = 1.25$ m

Length of pile $= 2 \times 1.0 + 1 \times 1.25 \times 1.25 + 0.55 \times 0.5 + 0.55$

                                                       $= 4.38$ m, say 4.5 m

(ii) $Q_u = A_p \left(1/2\ D\gamma N_\gamma + \gamma D_f N_q\right) + A_a[1/2\ D_u \cdot \pi \cdot \gamma \cdot N_\gamma + \gamma N_q \sum_{r=1}^{r=n} D_r] +$

                              $1/2\pi\ D\gamma K \tan \delta \left[D_1^2 + D_f^2 - D_n^2\right]$

                              $A_p = \pi/4(0.50)^2 = 0.196$ m^2

                              $A_a = \pi/4(1.25^2 - 0.50^2) = 1.03$ m^2

For $\phi = 35°,\ N_\gamma = 38$ and $N_q = 55/2 = 27.5$

                              $D_1 = 2.0$ m ; $D_2 = 3.56$ m, $D_n = 3.56$ m, $D_f = 4.5$

                              $\tan \delta = \tan 25 = 0.7$

Therefore, $Q_u = 0.196(1/2 \times 0.5 \times 19 \times 38 + 19 \times 4.5 \times 27.5)+$

                              $1.03[\ 1/2 \times 1.25 \times 2 \times 19 \times 38 + 19 \times 27.5(2.0 + 3.56\ )] +$

                              $1/2 \times \pi \times 0.5 \times 19 \times 1.75 \times 0.7[2^2 + 4.5^2 - 3.56^2]$

                              $= 496 + 3922 + 211 = 4629$ kN.

Safe capacity of pile $= \dfrac{4629}{3} = 1543$ kN.

REFERENCES

1. Bowles, J.E. (1982), "Foundation Analysis and Design". McGraw-Hill International Book Company, New Delhi.
2. Chen, F.H. (1975), *Foundations on Expansive Soils*, Elsevier, Scientific Publishing Co., Amsterdam.
3. Gromko, G.J. (1975), "Review of Expansive Soils," *Jr. Geotech. Engg. Div., Proc. ASCE*, GT6, Vol. 40.
4. Holtz, W.G. and Gibbs, H.J. (1956), "Engineering Properties of Expansive Clays", *Transactions ASCE*, Vol. 121, pp. 641–663.
5. Johnson, L.D., and Snethen, D.R. (1979), "Prediction of Potential Heave of Swelling Soil", *Geotechnical Testing Journal*, ASTM, Vol. 1, No. 3, pp. 117–124.
6. Katti, R.K. (1979), "Search for solutions to problems in Black Cotton Soils," First TGS Annual Lecture, *Indian Geotech. Journal*, No.1, Vol. 9.
7. Komornik, A. and David, K. (1969), "Prediction of Swelling Pressures of Clays," *Jr. Soil Mech. and Found. Division, ASCE*, SMI, Vol. 95.
8. Nayak, N.V. and Christensen, R.W. (1971), "Swelling Characteristics of Compacted Expansive oils."*Clay and Clay Minerals*, Vol. 19, pp. 251–261.
9. O' Neill, M.W. and Ghazzaly (1977), "Swell Potential Related to Building Performance," *Jr. Geot. Engg. Divn., Proc. ASCE*, GT 12, Vol. 103.
10. Peck, R.B., Hanson, W.E. and Thornburn, T.H. (1974), *Foundation Engineering*, John Wiley and Sons, New York.

11. Ranganatham, B.V. and Satyanarayna, B. (1965), "A Rational Method of Predicting Swelling Potential for Compacted Expansive Clays," *Proc. of 6th Int. Conf. on SMFE*, Montreal, Vol. 1, pp. 92–96.
12. Seed, H.B., Mitchell, J.K. and Cari, C.K. (1992), "Pressure Characteristics of Compacted Clays," *Highway Research Board, Bulletin 313*. pp. 12–39.
13. Sharma, D. (1981), "Apron—A Remedial Measure to Cracked Buildings in Expansive Soils," *Building Digest No. 141*, Central Building Research Institute, Roorkee.
14. Snethen, D.R. (1979), "Technical Guidelines for Expansive Soils in Highway Subgrades," FHWA-RD-79-51, Federal Highway Administration, Washington, U.S.A.
15. Sridharan, A., Sudhakar Rao, M. and Murthy, N.S. (1985), "Free Swell Index of Soils—A Need for Redefinition," *Indian Geotechnical Journal*, Vol. 15, pp. 94–95.
16. IS: 2720 (part XL)— 1977, Code of Practice for Methods of Test for Soils—Determination of free swell index of soils.
17. IS: 2720 (part XLI)—1977, Code of Practice for Methods of Test for Soils—Measurement of Swelling Pressure of Soil.
18. IS: 2911 (part III)—1980, Code of Practice for Design and Construction of Pile Foundations—Underreamed Piles.
19. IS: 11550–1985, Code of Practice for Field Instrumentation of Swelling Pressure in Expansive Soils.

PRACTICE PROBLEMS

1. Give the characteristics of expansive soils. Describe the procedure of their identification.
2. Describe the methods of obtaining 'swell pressure'.
3. Describe the salient features of underreamed piles. How are their capacities obtained in sand and clay ?
4. Determine the capacity of a 6.0 m long double underreamed pile (500 mm stem diameter) for the following data:
cohesion of soil at about 6.0 m depth = 75 kN/m^2
Average cohesion of the soil from ground surface to 6.0 m depth = 4.0 kN/m^2.
draw neatly the cross-section of the pile showing all dimensions and details of reinforcement.
5. For an overhead tank of capacity 1500 kilolitres, it was decided to provide underreamed piles. The soil strata up to 10.0 m depth is medium dense sand and water table is at 4.0 m depth. The standard penetration tests gave the following data:

Depth (m)	N-Value
1.5	8
3.0	15
4.5	9
6.0	18
7.5	25
9.0	29
10.5	36

Select suitable underreamed pile giving all its details. Give the number of piles and their arrangement. Illustrate your answer with neat sketches.

Foundations of Transmission Line Towers

14.1 INTRODUCTION

In the beginning, poles were the support system for the wires transmitting electric current. With the advancement in electrical engineering and need for supporting heavy conductors towers came in use. The most obvious feature of the present-day transmission towers is that these are tall structures, their height being much more than their lateral dimensions. These are space frames, almost exclusively built with steel sections, four-legged structures, having generally an independent foundation under each leg. The poles are also in use for transmission lines, but their use is mostly confined to urban and short distance transmission. The purpose of foundations of transmission towers, as in the case of foundations of other structures, is to safely transmit the loads to the substrata and to do it economically is another engineering consideration. The design of tower is done first, for load coming on to the foundations and so calculated that foundation design follows keeping in view the magnitudes and the nature of loads and the strata at site.

Tower foundations usually cost only 10 to 15 per cent of the cost of transmission towers; but due to a large number of foundations required in a transmission line, about 12 in a kilometre, and because of the present-day high and extra high voltage transmission requiring larger and heavier foundations, the total cost of foundations in a transmission project becomes quite substantial. Apart from the financial aspects, past records show that failures of tower foundations have more often been responsible for collapse of towers than other causes. These failures have usually been associated with certain deficiency in the design or construction of foundation or both. From the engineering point of view, the problem of design and construction of tower foundations becomes challenging because of the variety of soil conditions encountered and the remoteness of the construction sites. With the increased transmission voltage, the sizes of tower and loads on the foundation have increased and thus the problem of design and construction as well as the risk involved in case of failure of a tower has also increased.

14.2 NECESSARY INFORMATION

For the design and construction of foundation the following information is required (IS: 4091–1979):

(a) Route map showing the proposed layout of the towers with the general topography of the country and important towns, villages, etc., in the vicinity;

(b) Sections of trial borings or pits showing soil data at the site of work;

(c) Details of general layout of the towers;

(d) The nature, direction and magnitude of loads at the base of transmission tower both under normal condition and broken wire condition;

(e) Special information, for example, prevailing wind direction, depth of frost penetration and earthquake;

(f) A review of the performance of similar structures, if any, in the locality; and

(g) Maximum deformation allowed at the base of the tower or pole.

In the case of river crossings with towers or poles located in the river bed, the following additional information is needed:

(a) A site plan showing the details of the site selected for the crossing extending at least 90 m upstream and downstream from the central line of the crossing. The plan should normally include the following:

   (i) The approximate outlines of the bank,

   (ii) The direction of flow of water,

   (iii) The alignment of the crossing and the location of the towers, and

   (iv) The location of trial pits or borings taken in the river bed.

(b) A cross-section of the river at the site of the proposed crossing indicating the following:

   (i) The bed line up to the top of the banks and the ground line beyond the edges of the river, with levels at intervals sufficiently close to give a clear outline of marked features of the bed or ground, showing right and left bank and names of villages on each side;

   (ii) The nature of the surface soil in bed and banks with trial pits or borehole sections showing the levels and nature of the various strata down to the stratum suitable for founding the towers;

   (iii) The ordinary flood level;

   (iv) Low water level;

   (v) The highest flood level and years in which it occurred;

   (vi) The estimated depth of scour or if the scour depth has been observed, the depth of scour so observed.

(c) The maximum mean velocity of water current.

14.3 FORCES ON TOWER FOUNDATIONS

Figure 14.1 shows a sketch of typical transmission towers. The foundation of a tower is designed for the following forces:

(i) Weight of tower, cables insulators, clamps and other fixtures.

(ii) Wind pressure on the above component as mentioned in (i).

(iii) Unbalanced tension in cable when dead ended or broken on one side.

(iv) Loads due to earthquakes if located in seismic region.

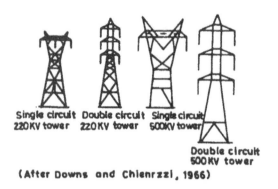

Single circuit   Double circuit   Single circuit
220KV tower   220KV tower   500KV tower

Double circuit
500KV tower

(After Downs and Chienrzzi, 1966)

Fig. 14.1. Types of Transmission Line Towers.

The tower foundations are checked for the two following cases:

Case I : Normal condition (N.C.): Considering the forces listed at serials (i) and (ii) or (iv),

Case II : Broken wire condition (B.W.C.): Considering the forces listed at serials (i), (ii) and (iii).

In thses two cases different multiplying factors are used for getting ultimate loads. The present-day practice is to use factors of 2 for NC and 1.5 for BWC.

A typical case of a four-legged transmission tower of rectangular base provided in a line running (say) North-South (Fig. 14.2) is dealt here:

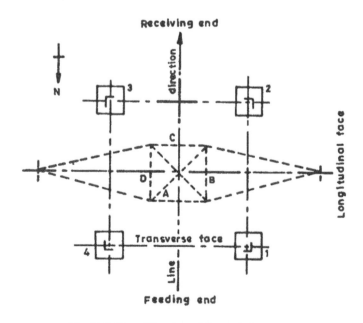

Fig. 14.2. Plan of Four-legged Transmission Tower.

a = Width of base along East-West.

b = Width of base along North-South.

W = Total weight of tower including one span of cables.

W' = Total weight of any unbalanced load such as due to a cable at off centre distance 'c'.

F = Resultant force of wind on tower and one span of cables acting at a distance 'd' above the ground and wind blowing (say) West to East.

P = Pull of any balanced force towards South applied at a height 'h' above ground and 'f' West of axis of tower. For example a dead ended or a broken cable on the North.

In the case of broken wire condition (BWC) all the cables are not considered broken on one side of the tower, as the design will not be economical. Only one at a time, ground wire, top cable or lower cable is considered broken. Then assuming that the forces divide equally among all the four foundations under each leg and that the torsional forces are in circumferential direction, the reactions can be summarized as in Table 14.1.

It may be noted that in Table 14.1 the transverse forces at No. 3 and No. 4 and torsional forces at No. 7 and No. 8 in actual behaviour do not get divided equally on the four foundations and generate new sets of forces but have been assumed so for simplifying the calculations.

Table 14.1. Forces on Foundations.

No.	Magnitude of Force	Direction of Each Foundation			
		N-E Corner No. 2	S-E Corner No. 1	S-W Corner No. 4	N-W Corner No. 3
1.	$W/4$	Down	Down	Down	Down
2.	$W' \cdot c/a$	Down	Down	Up	Up
3.	$F/4$	East	East	East	East
4.	$F \cdot d/a$	Down	Down	Up	Up
5.	$P/4$	South	South	South	South
6.	$P \cdot h/4$	Up	Down	Up	Down
7.	$P[af / 4(a^2 + b^2)]$	North	North	South	South
8.	$p[bf / 4(a^2 + b^2)]$	West	East	East	West

14.4 GENERAL DESIGN CRITERIA

For satisfactory performance, a tower foundation should be structurally adequate and be able to transmit the forces to the soil such that:

(i) There is a fair margin between the ultimate bearing capacity of the soil and the soil pressure actually developed.

(ii) The foundation should neither settle nor tilt beyond permissible limits. The limits of permissible settlement and deflection depend upon the configuration and the rigidity of the tower. Excessive settlement or deflection may cause serious secondary stresses in the members resulting in the collapse of the tower. Generally, the ratio of the base width and the tower height determines the permissible movement of the foundation. Many broad-base towers can perhaps tolerate a settlement of 20/25 mm without showing signs of any serious distress.

(iii) Against pull or uplift the foundation should have an adequate holding power or uplift resistance.

14.5 CHOICE AND TYPE OF FOUNDATION

The choice of tower foundation depends upon: (1) type of tower, (2) magnitude and type of loading, and (3) nature of strata (Bhandari et al., 1990). In a simple case of wires carried by poles, a portion of the pole buried in the ground serves as its foundation. In earlier stages at small size four-legged towers, their legs extended into the ground to provide support.

Later on with the use of heavy towers, the following types of foundations came in practice (Deb, 1971; Sharma, 1992).

(a) Concrete pad and chimney type (Fig. 14.3)

This is the most common type of footing used in India and in some countries of the continent. It consists of a plain concrete footing pad, the size and depth of which are decided either on the basis of bearing area necessary for transmission of the vertical downward load or from consideration of the amount of holding power required to resist the uplift force. The stub angle is taken inside and effectively anchored to the bottom pad by cleat angle and keying rods; the muff or the shaft with stub angle inside works as a composite member. The pad may be either pyramidal, Fig. 14.3a, or stepped, Fig. 14.3b. Stepped footings will   require less

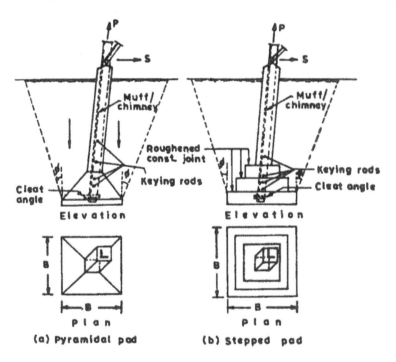

Fig. 14.3. Concrete Pad and Chimney Type Foundations.

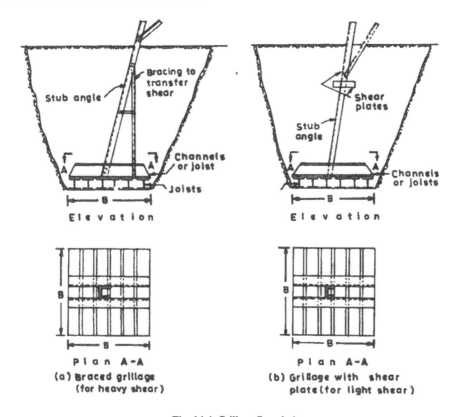

Fig. 14.4. Grillage Foundation.

shuttering materials but need more attention during construction to avoid cold-joints between the steps. The pyramidal footings, on the other hand, will require somewhat costlier form of works. In this type of footing, where the chimney is comparatively slender, the lateral load acting at the top of the chimney will cause bending moment. Therefore, the chimney/shaft should be checked for combined stress due to direct pull/thrust and bending.

This type of footing can be designed both for hard and comparatively soft soil and, when proportioned and designed correctly, gives satisfactory results.

(b) Grillage foundation (Fig. 14.4)

The grillage foundations are fabricated from steel sections and have the advantage of quicker construction. But they require much more steel and can be adopted for towers of moderate size only. For heavier towers grillage foundations are not an economical engineering alternative. Corrosion also is a problem. Grillages usually consists of I-beams or channel sections. For transference of horizontal thrust suitable shear plates are attached to the stub (Fig. 14.4). In case of towers at angles or at dead end a suitable bracing member is attached to the stub and grillage as shown in Fig. 14.4.

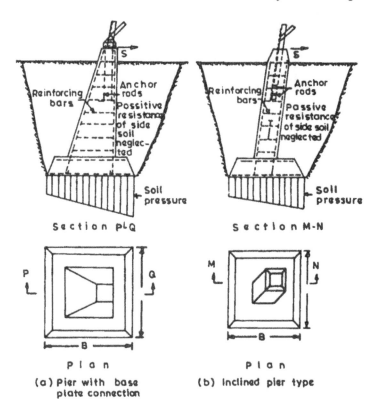

Fig. 14.5. Concrete Spread Footings.

(c) Reinforced concrete footings (Figs. 14.5 and 14.6)

In the case of large uplift forces or large compression loads increase in the spread footing area becomes necessary as going deeper is not always possible and in such cases pad and chimney type footing of mass concrete are not practical. Reinforced concrete footings are an answer. Chimneys also need reinforcement as lateral thrust causes moments in adddition to direct compression and uplift stresses. Concrete foundations have the advantage of adaptation of their geometry according to the needs of tower and strata. Two more common types of concrete foundations are shown in Fig. 14.5. The advantage of the one with vertical face of chimney (Fig. 14.5a) is that it resists better thrust. The large base footings may need suitable stepping and if adequate depth is not available to get the unreinforced base, the footing designed as cantilevers are required. A typically reinforced stepped footing is shown in Fig. 14.6. Adjustment of footing size and its depth to cater for the loads is an important consideration based on the constructional facility, type of strata, size of tower base and economy.

(d) Bored and cast-in-situ foundations (Fig. 14.7)

This type of foundations are being increasingly used all over the world on account of their better performance and economy over the traditional spread footings, as there is no backfilling, and construction time is reduced. As there is no need of compaction of the backfilled materials,

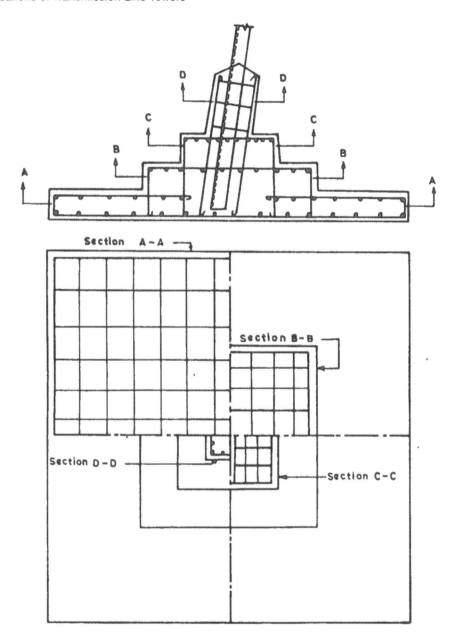

Fig. 14.6. Reinforced Stepped Footing.

better lateral resistance is assured. Holes are provided by augering and enlargement of toe is done by suitable tools. In India, boring of holes is done by spiral augers and bulb portion by underreaming tools. These are hand operated. In western countries, truck mounted machines are quite common. Boring at an angle in line with the stub is also possible in fairly stable strata. A

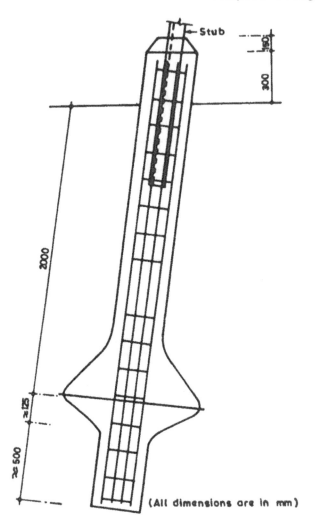

(All dimensions are in mm)

Fig. 14.7. Bored Underreamed Pile (Inclined).

typical underreamed pile is shown in Fig. 14.7. For unstable strata under water, more than one bulb and steep angles are not advisable. In such cases vertical piles should be preferred. In unstable strata the boreholes are stabilized by the circulation of drilling mud of bentonite in water.

Another version of underreamed piles popular in India is bored compaction piles which have the advantage of both bored and driven piles in sandy soils which show higher load capacity because the strata get compacted during piling process (Mohan et al., 1979). The design of piles in uplift is done by considering friction on pile stem and bearing on enlarged bulbs (Sharma et al., 1978; Prakash, 1980).

The bored piles can be provided in restricted spaces. These are very suitable for the foundations of railway track transmissions where space available is limited and minimum interruptions are required (Sharma et al., 1969). Besides this, underreamed piles are the best solutions to the foundation problems in expansive soils, such as black cotton soils of India. These soils swell and shrink with seasonal moisture changes and for stable foundations a minimum 3.5 m depth is necessary because movement of strata at this depth are negligible even in deep deposits of expansive soils (IS: 2911–Part lll, 1980).

For having desired capacity of piles, the available variables are the depth, diameter, number of bulbs and their diameter, number of piles in a group and their spacing. These can be suitably adjusted according to the nature of strata, type and magnitudes of loads. The design and construction of underreamed piles are dealt in detail in Chapter 13. Certain relevant aspects of these piles are contained in IS: 4091 (1979) and IS: 5613 (Part III/Section ll, 1989) also. Large number of piles will need a larger pile caps which are generally not less than 450 mm thick, and suitably reinforced both at top and bottom. The stub can be directly buried or held by plates with bolts in pile concrete. A typical design on a group of bored compaction piles is shown in Figs. 14.8 and 14.9.

(e) Precast concrete foundations

Due to difficulties inherent in getting good quality concrete in the isolated tower locations, attempts have been made to manufacture foundations either reinforced or prestressed in the factory. In Russia and some other European countries, precast reinforced spread footings of the type shown in Fig. 14.10 are reported to have been successfully used. Primary advantages of the

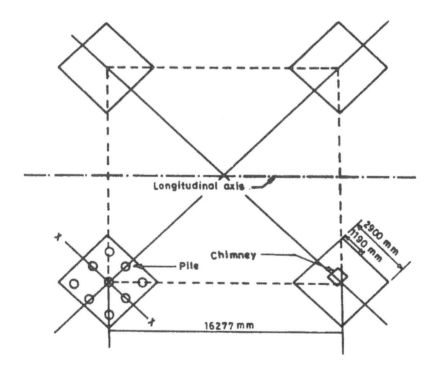

Fig. 14.8. Layout of Pile Groups.

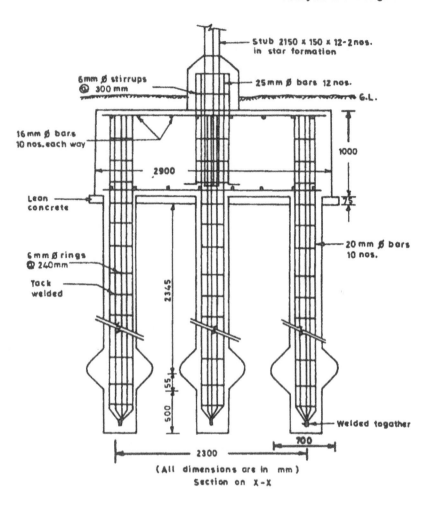

Fig. 14.9. Details of Pile Group.

prefabricted foundations are: (i) better control over quality of concrete and workmanship, (ii) saving in labour cost, and (iii) repeated use of form works and more economical use of materials of construction due to working under factory condition. Major handicaps in the use of prefabricated foundations are, however, the limitations of transport and handling available for installation of the completed footing at site. To partially overcome these difficulties, footings can be manufactured in component parts, transported, assembled and erected at site. This requires joining of the parts mechanically at site and involves careful detailing in the design stage. A precast footing more or less similar to the steel grillage is shown in Fig. 14.10b. In Fig. 14.10a is shown a circular base footing with prestressed concrete pedestal used in Finland (Gloyer et al., 1960). The precast footings are competitive when manufactured in bulk and where facilities exist for easy transportation of the tower foundations to site.

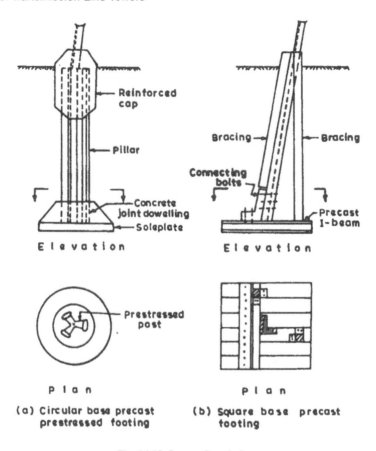

Fig. 14.10. Precast Foundations.

14.6 DESIGN PROCEDURE

There are two parts of the design: (i) stability analysis; and (ii) strength design. Stability analysis aims at eliminating the possibility of failure by overturning, uprooting, sliding and tilting of the foundation due to soil pressure being in excess of the ultimate capacity of the soil. Strength design means proportioning the components of the foundation to the respective maximum moment, shear, pull and thrust or combination of the same.

From design considerations, the foundations may be classified in two categories: (i) footing type, and (ii) piles. Therefore, the design procedure is discussed keeping two types of foundations in view.

Stability against base failure

The base of the footing or pile cap is usually subjected to a resultant vertical load and moments. If for one leg of tower,

W = Total vertical load including that of footing or pile cap

M_x = Moment about x-axis

M_y = Moment about y-axis.

For footings, base pressures are given by

$$q_{1,2} = \frac{W}{BL} + \frac{M_x}{I_x} \cdot \frac{L}{2} \pm \frac{M_y}{I_y} \cdot \frac{B}{2} \qquad \qquad \text{... (14.1)}$$

where B = width of base of the footing,

                L = length of base of the footing,

                I_x = moment of inertia of base of footing about x-axis, and

                I_y = moment of inertia of base of footing about y-axis.

The maximum pressure on the soil, q_1 should not exceed the allowable soil pressure. The allowable soil pressure, however, can be increased by 25% if the loading considered includes dead load and wind or earthquake load. The minimum base pressure, q_2 should not be less than zero to avoid tension and loss of contact of footing with soil.

In case of pile foundation, the maximum load in a pile can be obtained using the footing equation:

$$V_{pi} = \frac{W}{n} \pm \frac{M_x}{\sum x_i^2} y_i \pm \frac{M_y}{\sum y_i^2} x_i \qquad \qquad \text{... (14.2)}$$

where n = total number of piles in the group,

                x_i = distance of the ith pile from y-axis,

                y_i = distance of the ith pile from x-axis, and

                V_{pi} = load in the ith pile.

The maximum load in the pile should be less than the safe capacity of the pile.

Stability against uplift

In the case of footings, the uplift loads are assumed to be resisted by the weight of the footing plus the weight of an inverted frustum of a pyramid (or cone) of earth on the footing pad with sides inclined at an angle up to 30° with the vertical (Matsuo, 1967; Fig. 14.11a). A 30° pyramid or cone is taken for an average firm cohesive soil. In the case of non-cohesive materials such as sand and gravelly soils, a 20° pyramid or cone is considered. Interpolation can be done for in between classification.

A footing with an undercut generally develops uplift resistance that is higher than that of an identical footing without an undercut (Figs. 14.11b and c). However, for design purposes, the 20° and 30° cone or pyramid assumption shall be taken with a factor of safety of 1.0 for undercut footing, and 1.5 for footings without an undercut.

For footings below water table, submerged weight of the soil shall be taken.

For square footing, the volume of earth in frustum of pyramid is given by

$$V = \frac{D}{3} \ (3B^2 + 4D^2 \tan^2 \theta + 6BD \tan \theta). \qquad \qquad \text{... (14.3)}$$

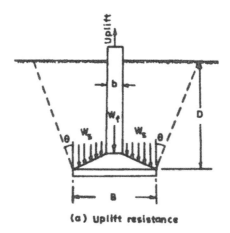

(a) Uplift resistance

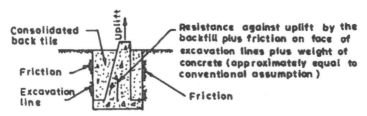

(b) Actual action without under-cut

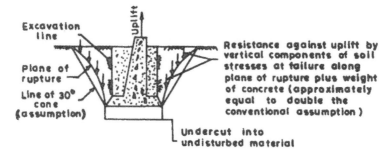

(c) Actual action with under-cut

Fig. 14.11. Soil Resistance to Uplift.

In case of circular slab for augered footing, the volume of soil is given by

$$V = \frac{\pi D}{12} (3B^2 + 4D^2 \tan^2 \theta + 6BD \tan \theta) \qquad ... (14.4)$$

where V = volume of earth,

               B = width or diameter of footing,

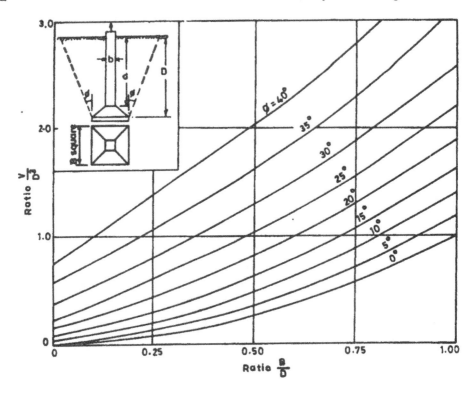

Fig. 14.12. For Volume of Frustum of Pyramid: Square Footing.

D = depth of base of footing below ground surface, and
θ = pyramid or cone angle with vertical.

Intercept charts given in Figs. 14.12 and 14.13 may be used to get the value of V. These charts are based on Eqs. (14.3) and (14.4) respectively. To obtain the weight of the soil, the volume of earth obtained should be multiplied by the effective unit weight of soil.

The piles in uplift should be designed by the usual considerations of the friction on stem and bearings on annular projections. The details of computations of uplift capacity of piles have already been given in Sec. 8.17 of Chapter 8. In case of heavy uplift forces and moments multiple underreamed piles may be used. In case of loose to medium sandy soils bored compaction underreamed piles may be used. The procedure of getting the uplift capacity of underreamed piles has already been dealt in Chapter 13.

Stability against overturning

Stability of the foundation against overturning in the case of uplift plus shear may be checked by the following criteria (Fig. 14.14).

(i) The foundation tilts about a point in its base at a distance of 1/6th of its width from the toe;

(ii) The weight of the footing acts at the centre of the base; and

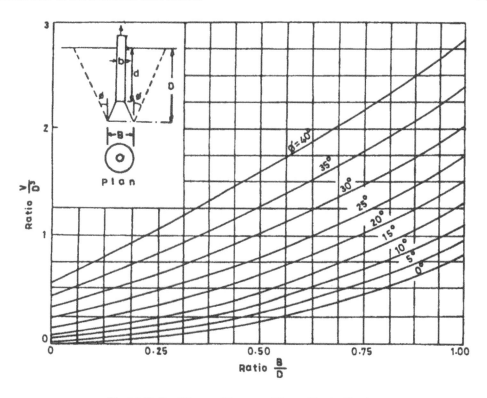

Fig. 14.13. For Volume of Frustum of Cone: Circular Footing.

(iii) Mainly that part of the cone which stands over the heel causes the stabilising moment. However, for design purposes, this may be taken equal to half the weight of the cone of earth acting on the base. It is assumed to act through the tip of the heel.

The maximum soil pressure below the base of the foundation (toe pressure) will depend on the vertical thrust on the footing and the moment at the base level due to the horizontal and other eccentric loadings, while the maximum passive pressure in the soil adjacent to the chimney/shaft will depend on the amount of the horizontal shear and the rotation of the shaft. The expression for checking stability against overturning

$$\frac{W'}{2} > \frac{U_p B/3 + H(D+h) - 1/3(BW_f)}{5/6B} \qquad \dots (14.5)$$

where W' = weight of soil in the cone/pyramid for stability,

U_p = uplift on the leg of tower,

H = maximum horizontal shear on tower leg, and

W_f = weight of tower.

Stability against lateral load

For footing type foundation (Fig.14.14): The analysis of footing type tower foundations subjected to lateral loads is based on the following assumptions (Deb, 1971; Wiggins, 1969):

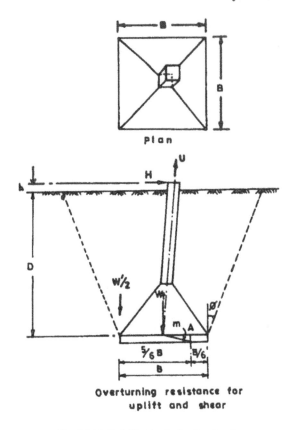

Fig. 14.14. Stability Study for Overturning.

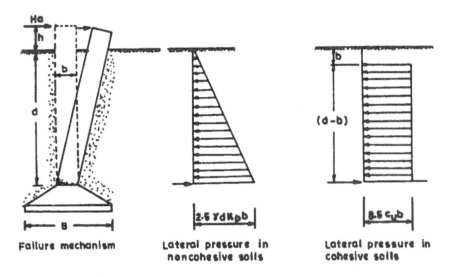

Fig. 14.15. Failure Mechanism and Soil Pressures for Shallow Chimney.

(i) The shaft subjected to lateral load is assumed as a cantilever beam on elastic soil support. In shallow shafts (Fig. 14.15), maximum moment will develop either at the junction of footing slab and shaft or at a point along the length of shaft. In deep shafts (Fig. 14.16), maximum moment will always occur at a point along the length of the shaft. As the lateral load increases, a hinge will develop at the point of maximum moment (Broms, 1965). After the formation of the hinge, the chimney is transformed into a mechanism and redistribution of moments takes place. The lateral load is then primarily resisted by the reaction of surrounding soil. Complete collapse of shaft will occur when soil reaction attains its ultimate value or when the redistributed moment in the shaft exceeds the ultimate resisting capacity of shaft section.

(ii) The ultimate lateral earth pressures in shallow shaft and deep shafts are taken as per Figs. 14.15 and 14.16 respectively for the case of non-cohesive and cohesive soils.
In these figures

$$K_p = \frac{1 + \sin\phi}{1 - \sin\phi} \qquad\qquad ... (14.6)$$

c_u = Undrained cohesion

On the basis of above assumptions and from the conditions of equilibrium, necessary equations relating to ultimate lateral load and ultimate moment resistance were derived. The results obtained from these equations have been plotted in non-dimensional form as shown in Figs. 14.17 and 14.18 for non-cohesive soils, and Figs. 14.19 and 14.20 for cohesive soils.

The charts of Figs. 14.17 to 14.20 may be used as given below:

(a) For the known values of d/b and h/b, obtain $H/\gamma\, K_p b^3$ or $H/c_u b^2$ from Fig. 14.17 or Fig.

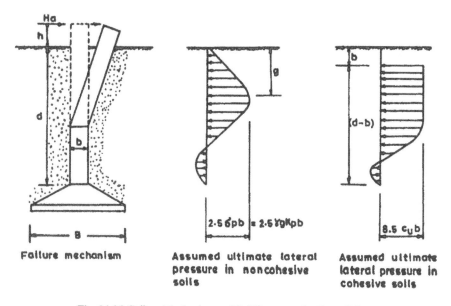

| Failure mechanism | Assumed ultimate lateral pressure in noncohesive soils | Assumed ultimate lateral pressure in cohesive soils |

Fig. 14.16. Failure Mechanism and Soil Pressures for Deep Chimney.

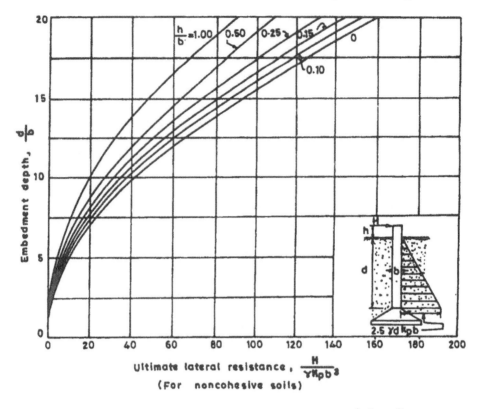

Fig. 14.17. Intercept Charts for Non-dimensional Parameters $\dfrac{d}{n}, \dfrac{h}{d}, \dfrac{H}{\gamma K_p b^3}$.

14.19 depending whether soil is non-cohesive or cohesive. Compute H by multiplying the obtained value with $\gamma K_p b^3$ or $c_u b^2$. If applied lateral load $H_a < H$, the shaft is regarded deep, otherwise shallow. The maximum moment in shaft in either case will occur along the length of shaft at a depth 'g' below ground level, where

$$g = 0.9 \frac{H}{\gamma K_p b} \text{ for non-cohesive soils} \qquad \text{... (14.7)}$$

or

$$g = \frac{H}{8.5 c_u b} + b \text{ for cohesive soils.} \qquad \text{.. (14.8)}$$

In case of shallow footing, however, the maximum moment in shaft will occur at its junction with the base slab if the depth of shaft d is less than g calculated above.

(b)   For the known value of $H/\gamma K_p b^3$, value of the maximum moment can be obtained by reading the corresponding value of $M/\gamma K_p b^4$ from Fig. 14.18 in non-cohesive soils. Similarly, in cohesive soils, maximum moment can be obtained by getting the value of $M/c_u b^3$ corresponding to $H/c_u b^2$ from Fig. 14.20.

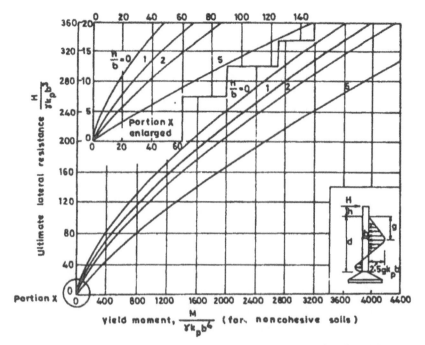

Fig. 14.18. Intercept Charts for Non-dimensional Parameters $\dfrac{H}{\gamma K_p b^3}, \dfrac{h}{b}, \dfrac{M}{\gamma K_p b^4}$.

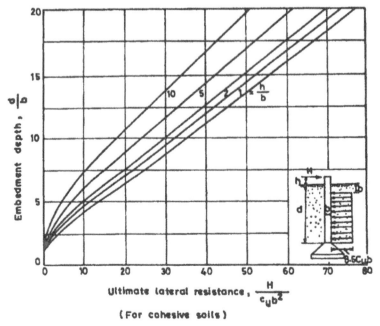

Fig. 14.19. Intercept Chart for Non-dimensional Parameters $\dfrac{d}{b}, \dfrac{h}{b}$ and $\dfrac{H}{C_u b^2}$.

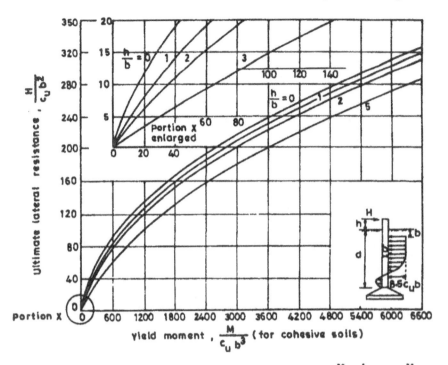

Portion X

yield moment , $\dfrac{M}{c_u b^3}$ (for cohesive soils)

Fig. 14.20. Intercept Chart for Non-dimensional Parameters $\dfrac{H}{c_u b^2} \cdot \dfrac{h}{b}$ and $\dfrac{H}{c_u b^3}$.

When the leg of tower is fixed at the top of the shaft by anchor bolts, Figs. 14.21 and 14.22 should be used for getting maximum bending moment in non-cohesive and cohesive soils respectively.

(iii) The section of the shaft is then designed for the combined pull and the maximum bending moment obtained in step (ii)b.

(iv) The moment at the junction of the shaft and the base slab should be taken into account in the design of base slab.

(v) In the case of long shaft, it is needed to compute its lateral deflection. A shaft can be regarded as long if

$$\eta d > 4 \text{ (in cohesionless soils), and}$$
$$\beta d > 2.25 \text{ (in cohesive soils)}$$

where

$$\eta = \left(\dfrac{\eta_h}{EI} \right)^{1/5} \qquad\qquad \text{... (14.9)}$$

$$\beta = \left(\dfrac{K_h b}{4EI} \right)^{1/4} \qquad\qquad \text{... (14.10)}$$

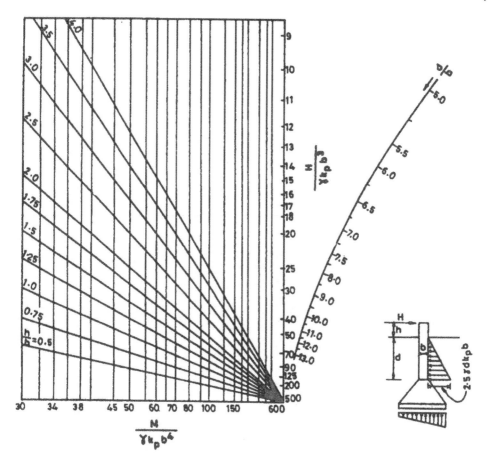

Fig. 14.21. Alignment Chart for Non-dimensional Parameters $\dfrac{M}{\gamma K_p b^4}$, $\dfrac{H}{\gamma K_p b^3}$, $\dfrac{d}{b}$ and $\dfrac{h}{b}$ for Shallow Foundation with Fixed Base in Non-dimensional Soils.

η_h = coefficient of lateral subgrade reaction of non-cohesive soils,

K_h = coefficient of lateral subgrade reaction of cohesive soils, and

EI = flexural rigidity of shaft member.

The deflection of the shaft at ground level, y_g is given by:

In non-cohesive soils

$$y_g = \frac{2.40 H_w}{\eta_h^{3/5}(EI)^{2/5}}. \qquad \dots (14.11)$$

In cohesive soils

$$y_g = \frac{2H_w(h\beta + 1)}{K_h b} \qquad \dots (14.12)$$

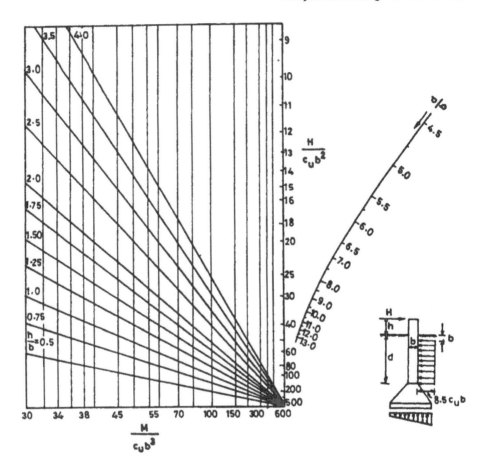

Fig. 14.22. Alignment Chart for Non-dimensional Parameters $\dfrac{M}{c_u b^3}$, $\dfrac{H}{c_u b^2}$, $\dfrac{d}{b}$ and $\dfrac{h}{b}$ for

Shallow Foundation with Fixed Base in Cohesive Soils.

where H_w = applied working lateral load.

Values of η_h and K_h may be taken as given in Sec. 8.14 dealing with laterally loaded pile.

The structural design of the foundation members is done in conventional way and explained through illustrative examples.

ILLUSTRATIVE EXAMPLES

Example 14.1

Design a suitable foundation for a 20° angle tower to be used in a double circuit 132 kV transmission line. The foundation is located in medium dense sand with $\phi = 30°$ and $\gamma = 17$ kN/m³. Depth of groundwater table is 5.0 m below the ground level. Use overload factors of 2 and 1.5 for normal and broken·wire conditions respectively. The foundation is subjected to the following loadings:

Table 14.2. Loadings on the Foundation.

Nature of Load	Load in kN under Condition	
	N.C.	B.W.C.
Downward	400	450
Uplift	300	380
Shear in transverse direction	3.3	25
Shear in longitudinal direction	—	16

Solution

(i) The design ulimate loads are obtained by using overload factors of 2.0 and 1.5 respectively for normal and broken wire condition. The design loads are given in Table 14.3.

(ii) Assuming the depth and base width of foundation to be 3.0 m and 1.7 m respectively. Ultimate bearing capacity of soil

$$q_u = 1.3cN_c + \gamma DN_q + 0.4\gamma BN_\gamma$$

For $\phi = 30°$, Table 7.2 gives

Table 14.3. Design Ultimate Loadings of Foundation.

Nature of Load	Design Ultimate Loading in kN with Overload Factor	
	N.C.	B.W.C.
Downward	800	675
Uplift	600	570
Shear in transverse direction	6.6	37.5
Shear in longitudinal direction	—	24

$N_q = 18.4$ and $N_\gamma = 22.4$.

Therefore $q_u = 17 \times 3 \times 18.4 + 0.4 \times 17 \times 1.7 \times 22.4$

$= 1197$ kN/m^2.

(iii) Stability analysis

(a) Design for uplift

Design uplift force = 600 kN (For normal condition)

Foundation proportioning is done as shown in Fig. 14.23(a).

$$B = 1.70 \text{ m}, D = 2.25 + 0.75 = 3.0 \text{ m}$$

$$\frac{B}{D} = \frac{1.7}{3.0} = 0.567.$$

From Fig. 14.12 for $B/D = 0.567$, $\dfrac{V}{D^3} = 1.32$.

Therefore, volume of soil $V = 1.32 \times (3.0)^3 = 35.64 \text{ m}^3$.

Weight of soil $W_s = 35.64 \times 17 = 605.88 \text{ kN}$.

Weight of footing $W_f = (25 - 17) \left\{ 0.40^2 \times 2.55 + \dfrac{0.75}{3}(1.70^2 + 0.40^2 + \right.$

$$1.7 \times 0.40) + 0.05 \times 1.7^2 + 25.0 \times 0.40^2 \times 0.25$$

$$= 12.46 \text{ kN}.$$

Therefore, $U_p = W_s + W_f = 605.88 + 12.46 = 618.34 \text{ kN} > 600 \text{ kN}$ (hence safe).

(b) Design for downward load

Design downward load $= 800 \text{ kN}$ (normal condition)

Weight of foundation $= 12.46 \text{ kN}$

Weight of earth directly on the base slab

$$= 17 \times \left[1.7^2 \times 3.0 - \left\{ 0.40^2 \times 2.25 + \dfrac{0.75}{3}(1.7^2 + 0.40^2 + 1.7 \times 0.40) \right\} \right]$$

$$= 17 \times (8.67 - 1.2925)$$

$$= 125.42 \text{ kN}.$$

Total downward load $= 800 + 12.46 + 125.42 = 937.88 \text{ kN}$.

Intensity of pressure $= \dfrac{937.88}{1.70^2} = 324.53 \text{ kN/m}^2$.

(c) Design for shear

Design shear $= 37.5 \text{ kN}$ (B.W.C.)

$$K_P = \dfrac{1 + \sin 30^\circ}{1 - \sin 30^\circ} = 3$$

$$\dfrac{d}{b} = \dfrac{2.25}{0.4} = 5.625, \quad \dfrac{h}{d} = \dfrac{0.32}{2.25} = 0.142.$$

From Fig. 14.17 for $d/b = 5.625$ and $h/d = 0.142$,

$$\dfrac{H}{\gamma K_p b^3} = 12.0.$$

Therefore, available $H = 12.0 \times 17 \times 3 \times 0.40^3$.

$$= 39.168 \text{ kN} > 37.5 \text{ kN}.$$

The muff, therefore, acts as a deep member.

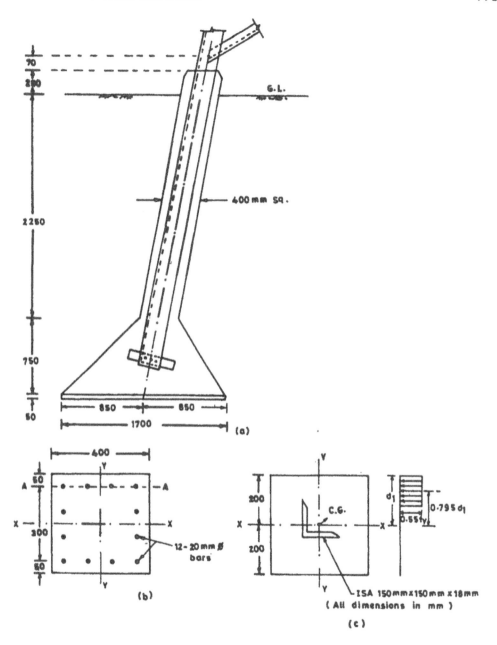

Fig. 14.23. Concrete Pad Foundation (Example 14.1).

(iv) Design for strength

(a) Design moment
Moment in the transverse direction

For $\qquad \dfrac{H}{\gamma K_p b^3} = \dfrac{37.5}{17 \times 3 \times 0.40^3} = 11.49$ and $\dfrac{h}{b} = \dfrac{0.25}{0.40} = 0.625.$

Fig. 14.18 gives $\dfrac{M}{\gamma K_p b^4} = 28.$

Therefore $M_{xx} = 28 \times 17 \times 3 \times 0.40^4 = 36.55$ kNm
Moment in longitudinal direction

For $\qquad \dfrac{H}{\gamma K_p b^3} = \dfrac{24.0}{17 \times 3 \times 0.40^3} = 7.39$ and $\dfrac{h}{b} = 0.625.$

Figure 14.18 gives $\dfrac{M}{\gamma K_p b^4} = 16.$

Therefore, $M_{yy} = 16 \times 17 \times 3 \times 0.40^4 = 20.89$ kN.

(b) Design of shaft
    Let the shaft be of reinforced concrete.

Tension combined with bending

    Design uplift   = 570   kN
    Design moments M_{xx} = 36.55 kNm
    and M_{yy} = 20.89 kNm

Use M20 concrete and Fe415 reinforcement of $f_y = 415$ N/mm².
Adopt the arrangement of reinforcement as shown in Fig. 14.23(b)
Neglecting tension in concrete the steel required to resist the direct pull

$$= \dfrac{570 \times 10^3}{0.87 \times 415} = 1578.72 \text{ mm}^2.$$

For bending, consider centroid of compression to coincide on line A-A, the lever arm, i.e.,
distance of centroid of tensile steel

$$\bar{x} = \dfrac{4 \times 300 + 2 \times 200 + 2 \times 100}{8} = 255 \text{ mm}.$$

Therefore, steel required to resist tension arising out of M_{xx}

$$= \dfrac{36.55 \times 10^6}{225 \times 0.87 \times 415} = 449.92 \text{ mm}^2.$$

Similarly, steel required to resist $M_{yy} = \dfrac{20.89 \times 10^6}{225 \times 0.87 \times 415}$

$$= 257.15 \text{ mm}^2.$$

Since M_{xx} and M_{yy} are reversible, total area of steel required

$$= 1578.72 + 2 \times (449.92 + 257.15) = 2992.86 \text{ mm}^2.$$

Provide 12 bars of 20 mm ϕ. It gives $A_{st} = 3769.91$ mm²

Thrust and bending moment

Design thrust $\qquad = 675$ kN (B.W.C.)

Design moments, $M_{xx} = 36.55$ kNm

and $\qquad\qquad M_{yy} = 20.89$ kNm

Percentage of steel $\quad p = \dfrac{3769.91}{400 \times 400} \times 100 = 2.3\%$

$$\frac{d'}{D} = \frac{50}{400} = 0.125$$

$$\frac{P_u}{f_{ck}bD} = \frac{675 \times 1000}{20 \times 400 \times 400} = 0.211$$

$$\frac{p}{f_{ck}} = \frac{2.3}{20} = 0.115$$

From design aids for $\dfrac{d'}{D} = 0.125$, $\dfrac{P_u}{f_{ck}bD} = 0.211$ and $\dfrac{p}{f_{ck}} = 0.115$

$$\frac{M_u}{f_{ck}bD^2} = 0.16$$

or $\qquad M_{ux1} = 0.16 \times 20 \times 400^3$

$\qquad\qquad\qquad = 2.048 \times 10^8$ Nmm $= 204.8$ kNm $= M_{uy1}$

$\qquad P_z = 0.45 f_{ck} A_c + 0.75 f_y A_{sc}$

$\qquad\qquad\quad = 0.45 \times 20 \times 4002 + 0.75 \times 415 \times 3769.91$

$\qquad\qquad\quad = 2613.3 \times 10^3$ $N = 2613.3$ kN

$$\frac{P_u}{P_z} = \frac{675}{2613.3} = 0.26$$

$$\frac{M_{xx}}{M_{ux1}} = \frac{36.55}{204.8} = 0.18$$

$$\frac{M_{yy}}{M_{uy1}} = \frac{20.89}{204.8} = 0.10$$

The value of α can be obtained by interpolation

$$\alpha = 1 \text{ for } \frac{P_u}{P_z} = 0.2$$

$$= 2 \text{ for } \frac{P_u}{P_z} = 0.8$$

$$\alpha = 1.0 + \left(\frac{0.26 - 0.20}{0.80 - 0.20}\right) \times 1.0 = 1.10$$

The strength of section is checked by the condition that:

$$\left(\frac{M_{xx}}{M_{ux1}}\right)^{\alpha} + \left(\frac{M_{yy}}{M_{uy1}}\right)^{\alpha} < 1$$

$$= (0.18)^{1.10} + (0.10)^{1.10} = 0.231 < 1, \text{ hence safe.}$$

Alternative design (Fig. 14.23d)

Consider the muff to have a stub angle and no other reinforcing bars and the stub angle reinforcement is coinciding with the centroid of the concrete section. It is assumed that the concrete in the tension side has cracked.

The tensile force due to bending will be

$$F_{txx} = \frac{36.55 \times 10^3}{0.795 \times 200} = 229.87 \text{ kN}$$

$$F_{tyy} = \frac{20.89 \times 10^3}{0.795 \times 200} = 131.38 \text{ kN}.$$

Therefore,

Total tension = $570 + 229.87 + 131.38 = 931.25$ kN.

Area of steel = $\dfrac{931.25 \times 10^3}{0.87 \times 250} = 4281.60 \text{ mm}^2.$

Use an angle section 150 mm × 150 mm × 18 mm.

Compressive stress on the section is not critical. The angle is assumed to act as a rigid reinforcement and being near the centroid of the concrete, has good concrete cover around it.

Example 14.2

Design a suitable tower foundation for a double circuit 144 kV transmission line without any deviation. The foundation is located in cohesive soil with allowable bearing pressure as 250 kN/m^2. Consider $c_u = 20$kN/m^2. $\gamma = 17$ kN/m^3 and $\phi = 35°$ for computation of uplift only. The foundation is subjected to the loadings given in Table 14.4.

Table 14.4. Loadings on the Foundation.

Nature of Load	Load in kN under	
	N.C.	B.W.C.
Downward	200.0	275.0
Uplift	150.0	225.0
Shear (transverse)	12.0	17.0
Shear (longitudinal)	—	8.0

Solution

(i)   In the design, ultimate loads are obtained by using overload factors of 2.0 and 1.5

respectively for normal and broken wire condition. The design loads are given in Table 14.5.

(ii) Assuming the proportioning of foundations dimensions as shown in Fig. 14.24(a).

(iii) Stability analysis

  (a) Check for deep or shallow foundation:

     Design shear = 25.5 kN (B.W.C.)

From Fig. 14.24(a), $d = 1.70$ m, $b = 0.30$ m and $h = 0.40$ m.

So, $\dfrac{d}{b} = \dfrac{1.70}{0.30} = 5.67$ and $\dfrac{h}{b} = \dfrac{0.40}{0.30} = 1.33$.

From Fig. 14.19

$$\frac{H}{c_w b^2} = 14.0.$$

Table 14.5. Design Ultimate Loadings on the Foundation.

Nature of Load	Design Ultimate Loading in kN with Overload Factors	
	N.C.	B.W.C.
Downward	400.0	412.5
Uplift	300.0	337.5
Shear in transverse direction	24.0	25.5
Shear in longitudinal direction	—	12.0

Therefore, available $H = 20 \times 14 \times 0.30^2$

$= 25.2$ kN < 25.5 kN (design shear).

Therefore, the shaft behaves as shallow member.

(b) Check for downward load:

  Buoyant weight of foundation

$$W_f = (25.0 - 17.0)\left[0.3^2 \times 1.7 + \frac{0.9}{3}(1.70^2 + 0.3^2 + 1.70 \times 0.30) + 0.05 \times 1.70^2\right]$$

$+ 25.0 \times 0.30^2 \times 0.25$

$= 11.32$ kN.

Weight of earth directly on base

$$W_s = 17.0\left[1.70^2 \times 2.6 - \left\{0.3^2 \times 1.7 + \frac{0.9}{3}(1.70^2 + 0.3^2 + 1.70 \times 0.30)\right\}\right]$$

$= 109.93$ kN.

Moment at base:

Actual $H = 25.5$ kN (B.W.C.)

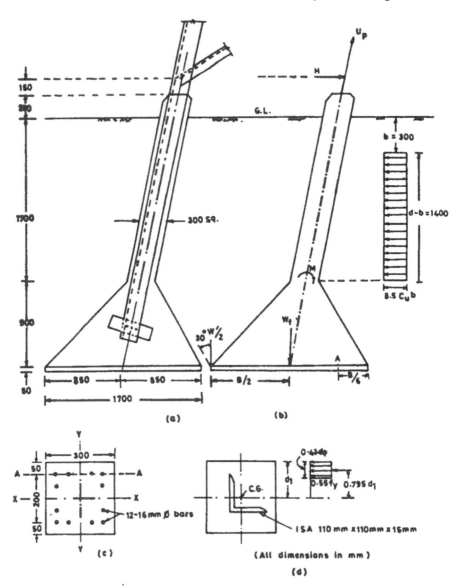

Fig. 14.24. Concrete Pad Foundation (Example 14.2.)

Actual $\quad \dfrac{H}{c_u b^2} = \dfrac{25.5}{20 \times 0.3^2} = 14.16, \dfrac{d}{b} = 5.67,$ and $\dfrac{H}{b} = 1.33.$

From Fig. 14.20, for shallow footings $\dfrac{M}{c_u b^3} = 32.0.$

or $\qquad\qquad M = 20 \times 0.3^3 \times 32 = 17.28$ kNm.

For checking stability consider vertical loads and moments for B.W.C., which will be critical. The shear in longitudinal direction will have no effect on the stability as actual $H = 12.0$ kN < 25.2 kN and in this direction the shaft will behave as deep foundation.

Now,   Design vertical load   = 412.5 kN
        Self weight of foundation = 11.32 kN
        Weight of earth on base   = 109.93 kN

      Total   533.75 kN

      Area of base $= 1.70 \times 1.70 = 2.89$ m^2

and   section modulus $z = \dfrac{1}{6} \times (1.70)^3 = 0.819$ m^3.

Therefore,   pressure at base $= \dfrac{533.75}{2.89} \pm \dfrac{17.28}{0.819}$

                  $= 205.79$ kN/m^2 and 163.58 kN/m^2.

From above the maximum toe pressure $= 205.79 < 250.0$ kN/m^2 (safe).

(c) Check for uplift

$$B = 1.70 \text{ m}, D = 2.6 \text{ m}, \frac{B}{D} = \frac{1.7}{2.6} = 0.654.$$

$$\text{From Fig. 14.12, for } \phi = 35°, \frac{V}{D^3} = 1.8.$$

Therefore, weight of earth in the frustum of pyramid

$W_s = 1.80 \times (2.6)^3 \times 17.0 = 537.82$ kN
             $W_f =$ 11.32 kN
          Total 549.14 kN

Design uplift $= 337.5$ kN < 549.14 kN, hence safe.

(d) Check for overturning

Considering the effect of 'H' and the available soil resistance, the moment at the base of chimney, $M = 17.28$ kNm.

For limiting condition, taking moments of all external forces at the assumed point of overturning at 'A' Fig. 14.24(b)

$$\frac{W'}{2} = \frac{(U_p - W_f) \times B / 3 + M}{5 / 6B}.$$

where W' is the weight of earth necessary for stability

$$\frac{W'}{2} = \frac{(549.14 - 11.32) \times 17 / 3 + 17.28}{5 / 6 \times 1.7} = 227.32 \text{ kN}$$

$$< \left(\frac{537.8}{2} = 268.9 \text{ kN} \right), \text{ hence safe.}$$

(iv) Design for strength:

(a) Design of shaft
Tension combined with bending
Depth of the point of maximum moment below G.L.

$$g = \frac{H}{8.5c_u b} + b$$

$$= \frac{25.5}{8.5 \times 20 \times 0.3} + 0.3 = 0.80 \text{ m.}$$

This is less than d, hence the maximum moment occurs along the length of chimney.

For $\dfrac{H}{c_u b^2} = 14.16$, $\dfrac{d}{b} = 5.67$, $\dfrac{h}{b} = 1.33$.

Figure 14.20 gives $\dfrac{M}{c_u b^3} = 18$.

Therefore, moment in transverse direction

$$M_{xx} = 20 \times 0.3^3 \times 45$$
$$= 24.30 \text{ kNm.}$$

Moment in longitudinal direction

For $\dfrac{H}{c_u b^2} = \dfrac{12.0}{20 \times 0.3^2} = 6.67$, $\dfrac{h}{b} = 1.33$.

Figure 14.20 gives $\dfrac{M}{c_u b^3} = 18$

$$M_{yy} = 9.72 \text{ kNm}$$
$$\text{Design uplift} = 337.5 \text{ kN}$$
$$\text{Design moments } M_{xx} = 24.30 \text{ kNm}$$
$$M_{yy} = 9.72 \text{ kNm}$$

Use M20 concrete and Fe415 reinforcement.
Adopt the arrangement of reinforcement as shown in Fig. 14.24(c). Neglecting tension in concrete, the steel required to resist the direct pull

$$= \frac{412.5 \times 10^3}{0.87 \times 415} = 1142.50 \text{ mm}^2.$$

For bending, consider centroid of compression to coincide on line A-A, the lever arm, i.e. distance of centroid of tensile steel

$$\bar{x} = \frac{4 \times 200 + 2 \times 50 + 2 \times 150}{8}$$

$$= 150 \text{ mm.}$$

Therefore, steel required to resist tension arising due to M_{xx} is

$$= \frac{24.30 \times 10^6}{0.87 \times 415 \times 150} = 448.69 \text{ mm}^2.$$

Similarly, steel required to resist M_{yy}

$$= \frac{9.72 \times 10^6}{0.87 \times 415 \times 150} = 179.47 \text{ mm}^2.$$

Since M_{xx} and M_{yy} are reversible, total area of steel

$$= 1142.50 + 2 \times (48.69 + 179.47)$$
$$= 2398.83 \text{ mm}^2.$$

Provide 12 bars of 16 mm ϕ, total area of steel = 2412.74 mm^2 > 2398.83 mm^2, o.k.

Direct compression with bending

Design thrust   = 412.5 kN

Design moment M_{xx} = 24.30 kNm

M_{yy} = 9.72 kNm

Percentage of steel $p = \dfrac{2412.74}{300 \times 300} \times 100 = 2.68\%$

$$\frac{d'}{D} = \frac{50}{300} = 0.16$$

$$\frac{P_u}{f_{ck}bd} = \frac{412.5 \times 1000}{20 \times 300 \times 300} = 0.229$$

$$\frac{p}{f_{ck}} = \frac{2.68}{20} = 0.134$$

From design aids, for $\dfrac{d'}{D} = 0.16$, $\dfrac{P_u}{f_{ck}bD} = 0.229$ and $\dfrac{p}{f_{ck}} = 0.134$

$$\frac{M_u}{f_{ck}bD^2} = 0.16$$

$$M_{ux1} = 0.16 \times 20 \times 300 \times 300^2$$
$$= 86.40 \times 10^6 \text{ Nmm} = M_{uy1}$$
$$P_z = 0.45f_{ck}A_c + 0.75f_y A_{sc}$$
$$= 0.45 \times 20 \times 300^2 + 0.75 \times 415 \times 2412.74$$
$$= 1560.9 \times 10^3 \text{ N}$$
$$= 1560.9 \text{ kN}$$

$$\frac{P_u}{P_z} = \frac{412.5}{1560.9} = 0.264$$

$$\frac{M_{xx}}{M_{ux1}} = \frac{24.30}{86.40} = 0.281$$

$$\frac{M_{yy}}{M_{uy1}} = \frac{9.72}{86.40} = 0.112$$

The value of α can be obtained by interpolation

$$\alpha = 1 \text{ for } \frac{P_u}{P_z} = 0.2$$

$$= 2 \text{ for } \frac{P_u}{P_z} = 0.80$$

$$\alpha = 1.0 + \left(\frac{0.264 - 0.20}{0.80 - 0.20}\right) \times 1.0$$

$$= 1.106$$

The strength of section is checked by the condition

$$\left(\frac{M_{xx}}{M_{ux1}}\right)^{\alpha} + \left(\frac{M_{yy}}{M_{uy1}}\right)^{\alpha} < 1$$

$$= (0.281)^{1.106} + (0.112)^{1.106}$$

$$= 0.334 < 1, \text{ hence safe.}$$

Alternative design (Fig. 14.24(d))

Use M20 concrete and mild steel reinforcement. Neglecting tension in concrete and considering the stub angle reinforcement to coincide with the centroid of concrete section.

The tensile force due to bending will be

$$F_{txx} = \frac{24.30 \times 10^3}{0.795 \times 150} = 203.77 \text{ kN}$$

$$F_{tyy} = \frac{9.72 \times 10^3}{0.795 \times 150} = 81.51 \text{ kN}$$

Therefore, total tension $= 337.5 + 203.77 + 81.51$
                               $= 622.78 \text{ kN}$

Area of steel $= \frac{622.78 \times 10^3}{0.87 \times 250} = 2863.35 \text{ mm}^2.$

Provide $110 \times 110 \times 15$ angle, area $= 3081 \text{ mm}^2 > 2863.35$, o.k.

Example 14.3
The foundation of a transmission tower with 30° deviation of a double circuit 220 kV line is subjected to the loadings given in Table 14.6. The overload factors are 2.00 and 1.50 for normal and broken wire conditions respectively. The foundation is located in non-cohesive soils with angle of internal friction $\phi = 30°$ and unit weight $\gamma = 16 \text{ kN/m}^3$. The ultimate bearing capacity of soil is 500 kN/m^2. Design a suitable grillage foundation.

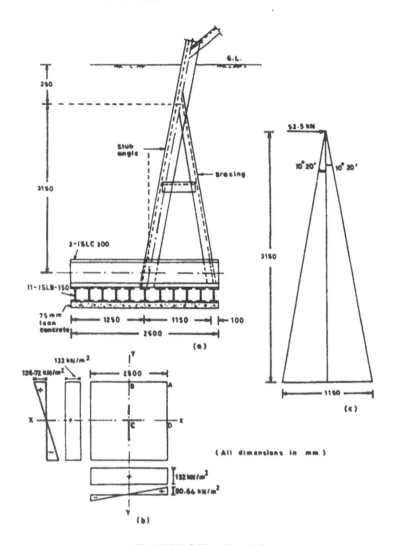

Fig. 14.25. Grillage Foundation.

Table 14.6. Loadings on the Foundation.

Nature of Load	Loads in kN under Condition	
	N.C.	*B.W.C.*
Downward thrust	450	550
Uplift	400	500
Shear in transverse direction	4.5	35
Shear in longitudinal direction	—	20

Solution

(i) The design ultimate loads are obtained by using overload factors of 2.0 and 1.5 respectively for normal and broken wire conditions. The design loads are given in Table 14.7.

(ii) Stability analysis

(a) Design for downward load: Adopt the foundation dimensions as shown in Fig. 14.25(a).

$$\text{Area of base} = 2.5 \text{ m} \times 2.5 \text{ m}$$
$$= 6.25 \text{ m}^2.$$

Soil pressure will be maximum with vertical and side thrust under B.W.C.

Table 14.7. Design Ultimate Loadings on the Foundation.

Nature of Load	Design Ultimate Loading in kN with Overload Factor	
	N.C.	*B.W.C.*
Downward thrust	900	825
Uplift	800	750
Shear in transverse direction	9	52.5
Shear in longitudinal direction	—	30.0

$$\text{Vertical load} \quad = 825.0 \text{ kN}$$
$$\text{Foundation load} = 12.0 \text{ kN (assumed).}$$
$$\text{Total} \quad\quad\quad = 837.0 \text{ kN}$$

Neglecting the passive resistance of soil, the moment at the base level due to side thrust,

$$M_{xx} = 52.5 \times 4.0 = 210.0 \text{ kNm}$$
$$M_{yy} = 30 \times 4.0 = 120.0 \text{ kNm.}$$

$$\text{Section modulus } z \text{ of bases} = \frac{(2.5)^3}{6} = 2.60 \text{ m}^3.$$

$$\text{Therefore, toe pressures} = \frac{837.0}{6.25} \pm \frac{210.0}{2.60} \pm \frac{120}{2.60}$$
$$= 133.92 \pm 80.77 \pm 46.15.$$

From above, the maximum pressure = 260.84 kN/m² < 500 kN/m² (hence safe).
and the minimum pressure = 7.0 kN/m²

(b) Check for uplift

$$B = 2.5 \text{ m}, D = 3.25 \text{ m} \quad \text{and} \quad \frac{B}{D} = \frac{2.5}{3.25} = 0.77$$

From Fig. 14.12, $\frac{V}{D^3} = 1.90.$

Therefore, weight of earth in the frustum of pyramid

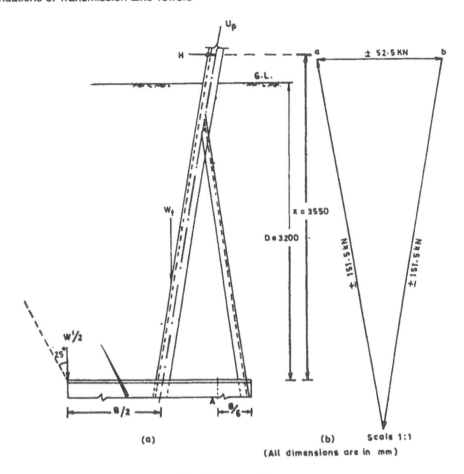

Fig. 14.26. Force Diagram.

$W_s = 1.90 \times 16 \times 3.25^3 \quad = 1043.57 \text{ kN}$

Weight of foundation, $W_f = \underline{\quad 12.0 \quad \text{kN}}$

$\qquad\qquad\qquad$ Total $\quad \underline{1055.57 \text{ kN}}$

The total downward load is thus greater than 800 kN, the maximum uplift. Hence the foundation is safe against uplift.

(c) Check for overturning

$\qquad$ Under normal condition

$\qquad$ For limiting condition, taking moment about point A (Fig. 14.26(a))

$$\frac{W'}{2} = \frac{(U_p - W_f) \times B/3 + H \times x}{5/6B}$$

where, W' is the weight of earth necessary for stability.

or
$$\frac{W'}{2} = \frac{(800 - 12.0) \times 2.5/3 + 90 \times 3.55}{5/6 \times 2.5}$$

$$= 330.53 \text{ kN} \left(< \frac{1043.57}{2} = 521.78 \text{ kN} \right).$$

Under B.W.C.
$$\frac{W'}{2} = \frac{(750.0 - 12.0) \times 2.5/3 + 52.5 \times 3.55}{5/6 \times 2.5}$$
$$= 384.66 \text{ kN} < 521.78 \text{ kN (hence safe)}.$$

(iii) Design for strength

(a) *Bottom tier*: Maximum soil pressures at A, B, C and D for B.W.C. (Fig. 14.25(b))

$$P_c = \frac{825.0}{6.25} = 132.0 \text{ kN/m}^2$$

$$P_d = 132.0 + \frac{52.5 \times 4.0 \times 6}{2.5^3}$$
$$= 132.0 + 80.64 = 212.64 \text{ kN/m}^2$$

$$P_b = 132.0 + \frac{30.0 \times 4.0 \times 6}{2.5^3}$$
$$= 132.0 + 46.08 = 178.08 \text{ kN/m}^2$$
$$P_a = 132.0 + 80.64 + 46.08 = 258.72 \text{ kN/m}^2.$$

Therefore, average pressure $= \dfrac{132.0 + 212.64 + 178.08 + 258.72}{4}$

$$= 195.36 \text{ kN/m}^2.$$

Ultimate moment about x-x $= 195.36 \times 2.5 \times \dfrac{1.25^2}{2}$

$$= 381.56 \text{ kNm}$$

Plastic modulus required $= \dfrac{381.56 \times 10^6}{415} = 919.42 \times 10^3 \text{ mm}^3$

Using ISLB 150, plastic modulus of each $= 91.8 \times 10^3 \text{ mm}^3$.

Therefore, number of joists in bottom tier $= \dfrac{919.42}{91.8} = 10.01$.

So, provide 11 nos. of above member.

(b) *Top tier:* Plastic modulus required $= 919.42 \times 10^3 \text{ mm}^3$.
Providing three numbers of ISLC 300, the available plastic modulus

$$= 3 \times 403.2 \times 10^3 = 1212.6 \times 10^3 \text{ mm}^3 > 919.42 \times 10^3 \text{ mm}^3 \text{ required}.$$

(c) *Shear* : Shear stress in both top and bottom tiers will be very small.

Stresses in the members (Fig. 14.26(b))

$$f_{bc} = \pm 151.5 \text{ kN} \quad f_{ca} = \pm 151.5 \text{ kN}$$

Therefore, total uplift = 750.0 + 151.5 = 901.5 kN
and total compression = 825.0 + 151.5 = 976.5 kN.

Provide angle 130 × 130 × 15, with three holes for 20 mm dia. bolt, net area for taking tension,

$$= 3681 - 3 \times 21.5 \times 15 = 2713.5 \text{ mm}^2.$$

Ultimate tensile strength = 2713.5 × 415 = 1126.1 × 10³ N
$$= 1126.10 \text{ kN} > 901.5 \text{ kN}.$$

Compressive strength

Maximum ultimate axial compression = 976.5 kN

Effective length of the member $$= \frac{3150}{2 \times \cos 10° 20'} = 1600.96 \text{ mm}.$$

Least radius of gyration r_{zz} = 25.3 mm.

Now $\dfrac{b}{t} = \dfrac{110}{16} = 6.9$ and $\dfrac{256}{\sqrt{f_y}} = \dfrac{256}{\sqrt{415}} = 12.56 > 6.9.$

Therefore, it is a compact section.

Slenderness ratio $= \dfrac{1600.96}{25.6} = 63.28.$

From IS: 802–1967
Maximum permissible stress in axial compression

$$f = f_y - \frac{(f_y - k)}{130} \left(\frac{l}{r} - 20 \right) \text{ for } 20 < \frac{l}{r} < 150.$$

where $\quad k = \left(\dfrac{\pi^2 E}{(150)^2} \right); f_y = 415 \text{ N/mm}^2$

$$E = 2.1 \times 10^5 \text{ N/mm}^2; k = \left(\frac{\pi^2 \times 2.1 \times 10^5}{(150)^2} \right) = 92.11$$

$$f = 415 - \frac{(415 - 92.11)}{130} (63.28 - 20)$$
$$= 307.50 \text{ N/mm}^2.$$

Therefore, total buckling load = 307.50 × 3681
$$= 1131.92 \times 10^3 \text{ N}$$
$$= 1131.92 \text{ kN} > 976.5 \text{ kN. Hence o.k.}$$

(d) Design of bracing

Adopt an angle 75 mm × 75 mm × 8 mm

Design for tension

Axial ultimate tension = 151.5 kN

With one 16 mm φ connecting bolt hole,

Effective area = $a + bk_1$

$$a = \frac{1138}{2} - 17.5 \times 8 = 429 \text{ mm}^2$$

$$b = \frac{1138}{2} = 569 \text{ mm}^2$$

$$K_1 = \frac{1}{1 + 1/3 \times b/a} = \frac{1}{1 + 1/3 \times 569/429}$$
$$= 0.693.$$

Therefore, effective area

$$= 429 + 569 \times 0.693$$
$$= 823.56 \text{ mm}^2.$$

Tension capacity = 823.56 × 415

$$= 341777.32 \text{ N}$$
$$= 341.78 \text{ kN} > 151.5 \text{ kN, o.k.}$$

Design for compression

$$\text{Effective length} = \frac{3150}{2 \times \cos 10°20'} = 1600.96 \text{ mm.}$$

Radius of gyration $r_{zz} = 14.6$ mm

and $\dfrac{b}{t} = \dfrac{75}{8} = 9.37$ and $\dfrac{256}{\sqrt{f_y}} = 12.56.$

Hence, it is a compact section.

$$\text{Slenderness ratio} = \frac{1600.96}{14.6} = 109.65.$$

Maximum permisible stress in axial compression

$$f = 415 - \frac{(415 - 92.11)}{130} (109.65 - 20).$$
$$= 192.31 \text{ N/mm}^2.$$

Therefore total buckling load

$$= 1138 \times 192.31$$
$$= 218858.0 \text{ N}$$
$$= 218.858 \text{ kN} > 151.5 \text{ kN, hence o.k.}$$

REFERENCES

1. Bhandari, R.K., Sharma, D. and Prakash, C. (1990), "Cost Effective Tower Foundations for Transmission Line Towers in Dubai," *Irrigation and Power Journal*, Vol. 47, No. 2.
2. Broms, B.R. (1965), "Design of Laterally Loaded Piles," *Journal of Soil Mechanics and Foundation Division*, ASCE, Vol. 91, No. SM3, Proc. Paper 4342.
3. Deb, K.R. (1971), "Foundations for Transmission Tower," *Jr. of Irrigation and Power.*
4. Downs, D.I. and Chieurzzi, R. (1966), "Transmission Tower Foundations," *Proc. ASCE, Jr. Power Division*, Vol. 92, No. PO2.
5. Gloyer, H. and Vogelsang, Th. (1960), "New Types of Foundation for Overhead Line Construction," C16RE, paper 232.
6. IS: 4091–1979, Indian Standard Code of Practice for Design and Construction of Foundations for Transmission Line Tower and Poles.
7. IS: 802: Part 1–1995, Code of Practice for Use of Structural Steel in Overhead Transmission Line Towers, Section 1 (Materials and Loads).
8. IS: 802: Part 1–1992, Code of Practice for Use of Structural Steel in Overhead Transmission Line Towers, Section 2 (Permissible Stress).
9. IS: 2911 (Part III)–1980, Code of Practice for Design and Construction of Pile Foundations (Underreamed Piles).
10. IS: 5613 (Part II, Section 2)-1985, Code of Practice for Design, Installation and Maintenance of Overhead Power Lines.
11. Matsu, M. (1967 and 1968), "Study on the Uplift Resistance of Footings (1 and 2)," *Soil and Foundation*, Vol. VII, No. 1 and Vol. VIII, No. 2.
12. Prakash, C. (1980), "Uplift Resistance of Underreamed Piles in Silty Sand," *Indian Geotechnical Journal*, Vol. 10, No. 1.
13. Sharma, D. (1992), "Foundation for Transmission Line Tower", Short term Course on Foundations for Structures, Department of Continuing Education, University of Roorkee, Roorkee, India.
14. Sharma, D., Jain, M.P. and Prakash, C. (1978), *Handbook on Underreamed and Bored Compaction Pile Foundation*, C.B.R.I., Roorkee.
15. Sharma, D. and Soneja, M.R. (1969), "Transmission Line Tower Foundations," Symp. Application of Soil Mech. and Found. Engg., Institute of Engineers, West Bengal Centre, Kolkatta.
16. Wiggins, R.L. (1969), "Analysis and Design of Tower Foundation," Proc. ASCE, *Jr. Power Division*, Vol. 95, No. PO1.

PRACTICE PROBLEMS

1. What are the various types of foundations used for transmission line towers? Explain the method of selecting a proper type of foundation. Illustrate your answer with neat sketches.
2. Give the necessary information required for the design and construction of transmission line tower foundations. What are the forces considered in design?
3. How is the safety of a tower foundation checked against—
   (i) Uplift,
   (ii) Overturning, and
   (iii) Lateral thrust?
4. Design a suitable foundation for a 25° angle tower to be used in a D.C. 132 kV transmission line. The foundation is located in dense sandy soil with $\phi = 35°$ and $\gamma = 18$ kN/m^3. Depth of groundwater table is 4 m below the ground level. Use overload factor of 2.0 and 1.5 for normal and broken wire condition respectively. The foundation is subjected to the loadings given in Table 14.8.

Table 14.8. Loadings on the Foundation.

Nature of Load	Loads in kN under Condition	
	N.C.	B.W.C.
Downward thrust	450	405
Uplift	340	400
Shear in transverse direction	3.5	26
Shear in longitudinal direction	—	18

5. Design a suitable tower foundation for a double circuit 144 kV transmission line without any deviation. The foundation is to be located in cohesive soil with $c_u = 17$ kN/m^2 and bearing capacity = 270 kN/m^2, consider $\gamma = 15$ kN/m^3 and (for computation of uplift only) $\phi = 38°$. The foundation is subjected to the loadings given in Table 14.9.

6. The foundation of a transmission tower with 25° deviation of a double circuit 220 kV line is subjected to the loadings given in Table 14.10. The overload factors are 2.00 and 1.50 for normal and broken wire conditions respectively. The foundation is located in non-cohesive soils with angle of internal friction $\phi = 32°$ and unit wt. $\gamma = 18$ kN/m^3. The ultimate bearing capacity of soil is 540 kN/m^2. Design a suitable grillage foundation.

Table 14.9. Loadings on the Foundation.

Nature of Load	Loads in kN under Condition	
	N.C.	B.W.C.
Downward	220.0	300.0
Uplift	165.0	240.0
Shear in transverse direction	14.0	18.0
Shear in longitudinal direction ,	—	8.5

Table 14.10. Loadings on the Foundation.

Nature of Load	Loads in kN under Condition	
	N.C.	B.W.C.
Downward thrust	470	565
Uplift	425	520
Shear in transverse direction	4.7	39
Shear in longitudinal direction	—	22

Reinforced Earth

15.1 INTRODUCTION

Reinforced earth is a comparatively new construction material which is formed by the association of soil, usually cohesionless, and tension resistant elements in the form of sheets, strips, nets or mats. The reinforcement is so arranged in the soil mass as to reduce or suppress the tensile strains which might develop under gravity and boundary forces. Most granular soils are strong in compression and shear but weak in tension. The performance of such soils can be substantially improved by introducing reinforcing elements in the direction of tensile strains, more or less in the same way, as in reinforced concrete. Figure 15.1 shows the three main component parts of a reinforced earth fill, namely (i) earth fill, (ii) reinforcement, and (iii) facing unit fixed to one end of the reinforcing strips. The facing unit is not a vital component; however, it is necessary to prevent surface erosion and to give an aesthetically acceptable external appearance.

Reinforced earth possesses many novel characteristics which render it eminently suitable for construction of geotechnical structures, particularly retaining structures. It employs prefabricated elements which can be easily handled, transported, stored and assembled. Ordinary frictional soil constitutes most of its bulk which is hauled and compacted in place, as for earthen embankments. Unlike rigid retaining walls, the stress concentration near the toe is small because of the relatively high base to height ratio, usually between 0.8 to 1.0 of the reinforced earth walls. This renders reinforced earth retaining structures eminently suitable for poor sub-soil conditions. As the soil forms over 99.9 per cent of the volume of the retaining wall (skin elements excluded), it is appreciably flexible and can withstand large differential settlement without distress. Consequently, reinforced earth is very economical in a variety of circumstances and can be adopted to great advantage for high retaining structures and for poor subsoil conditions.

Due to the novel behaviour of reinforced earth, it has been used in the construction of almost all engineering structures, namely, retaining walls, abutments, embankments, rail-road tracks, foundation slabs, earthen dams, etc. Figure 15.2 illustrates some of the engineering applications of reinforced earth.

15.2 BASIC MECHANISM OF REINFORCED EARTH

The basic mechanism of reinforced earth can be explained in a simple way, by the Rankine

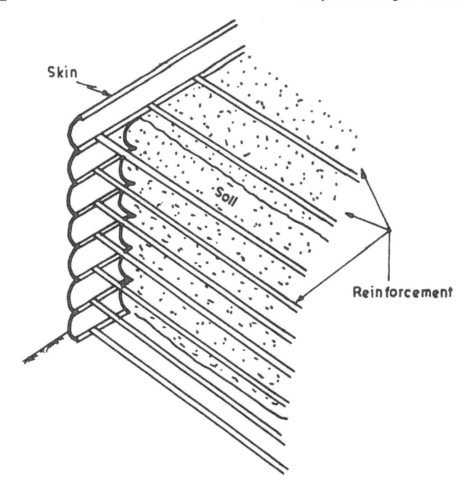

Fig. 15.1. Reinforced Earth.

state of stress theory. If a two-dimensional element of cohesionless soil is subjected to uniaxial stress, it will fail instantaneously as the Mohr circle of stress will cut the strength envelop (Fig. 15.3a). If the element is subjected to equal biaxial stresses, it will undergo uniform compression without failure. If the stresses are unequal, it will behave like a triaxial sample in axial compression with failure occurring at large lateral strains. At failure, the lateral stress σ_3 will be related to the major principal stress σ_1 as $\sigma_3 = K_A \sigma_1$, K_A being the coefficient of active earth pressure; at this instant the Mohr circle is tangential to the strength envelope (Fig. 15.3b). To hold the element without failure, the lateral stress σ_3 must be increased. If reinforcement is provided in the direction of σ_3, interaction between the soil and reinforcement will generate frictional forces along the interface. Tensile stresses will be produced in the reinforcement and a corresponding lateral compression $\Delta\sigma_3$ in the soil element, as long as there is no slippage between the two. It will be analogous to the existence of a pair of plates (Schlosser and Vidal, 1969) which prevents lateral expansion of the soil

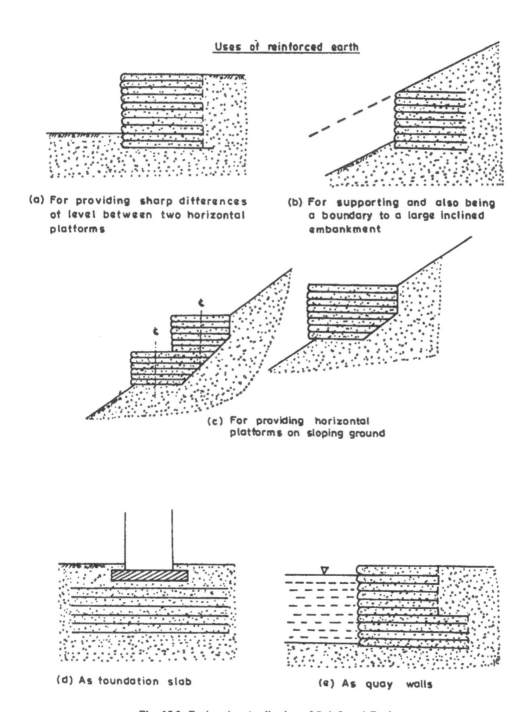

Uses of reinforced earth

(a) For providing sharp differences of level between two horizontal platforms

(b) For supporting and also being a boundary to a large inclined embankment

(c) For providing horizontal platforms on sloping ground

(d) As foundation slab

(e) As quay walls

Fig. 15.2. Engineering Application of Reinforced Earth.

(Contd).

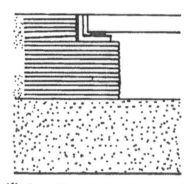

(f) As bridge abutment

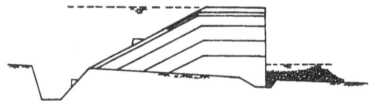

(g) As reinforced earth dam (after Cassard et al., 1979)

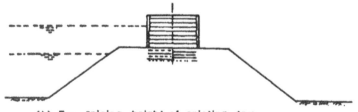

(h) For raising height of existing dam

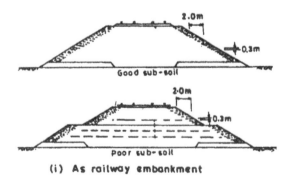

(i) As railway embankment

Fig. 15.2 Engineering Applications of Reinforced Earth.

(Contd).

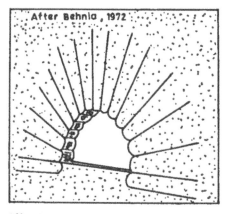

(j) As reinforced earth arch

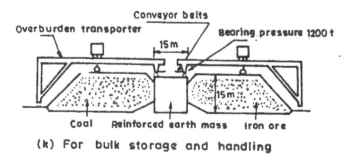

(k) For bulk storage and handling

Fig. 15.2. Engineering Applications of Reinforced Earth.

lement (Fig. 15.3c). The additional lateral stress will shift the Mohr circle to the right and away from the strength envelope. Thus soil reinforcement friction is fundamental to the concept of reinforced earth.

A different view of the influence of reinforcement on the behaviour of soil mass has been advanced by Bassett and Last (1978). The reinforcement is assumed to restrict the 'dilatancy' of soil leading to an increase in the mobilized shear strength. The reinforcement also causes a rotation of principal strain directions relative to the unreinforced case. The most effective directions for the reinforcement can be estimated by the zero extension characteristics which are thought to represent potential slip or rupture surfaces. Such a determination of the most effective alignment of reinforcement is likely to open up a much wider field for the application of reinforced earth technique.

15.3 CHOICE OF SOIL

Three principal considerations in the selection of the soil for reinforced earth are:

1. Long term stability of the complete structure,
2. Short term (or construction phase) stability, and
3. Physicochemical properties of materials.

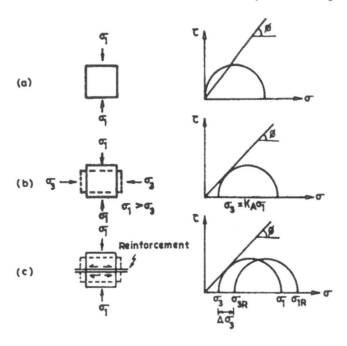

Fig. 15.3. Basic Mechanics of Reinforced Earth.

Granular soils compacted to densities that result in volumetric expansion during shear are ideally suited for use in reinforced earth structures. Especially, where these soils are well-drained, effective normal stress transfer between reinforcement and soil backfill will be immediate as each lift of backfill is placed, and shear strength increase will not lag behind the vertical loading. Further, granular soils behave as elastic material in the range of loading normally encountered in reinforced earth structures. Due to these two reasons, no post-construction movements will occur.

On the other hand, in fine-grained materials which are normally poorly drained, effective stress transfer will not be immediate. Therefore, there will be an extremely low factor of safety during construction. Further, due to elasto-plastic or plastic behaviour of such soils, there is a possibility of post-construction movements. The highly stressed reinforcement in cohesive backfill might be susceptible to creep (McKittrick, 1978) and to greater attack by corrosion in metal reinforcement (Murray and Boden, 1979). These three reasons make fine-grained soils unsuitable for use in reinforced earth construction.

The detailed specifications of an ideally suited soil are given below (McKittrick, 1978):

Sieve Size	% Passing
150 mm	100
75 mm	75–100
75 microns and P.I. < 6	0–25

If per cent finer than 75 microns is greater than 25% and per cent finer than 15 microns is less than 15%, material is acceptable if $\phi = 30°$ and P.I. < 6%.

In other soils the increase in strength by provision of reinforcement will be relatively less.

Department of Transport (1978), however, recommends the use of cohesive frictional fill also provided if it corresponds to the grading and plasticity characteristics given in Table 15.1

Table 15.1. Grading Limits of Cohesive Frictional Soil.

B.S.S.	Grading Limits Percentage Passing by Weight	Remarks
125 mm	100	The fill shall be finer than
90 mm	85	these grading limits
10 mm	25	
600 micron	10	
63 micron	10	
2 micron	10	
Liquid limit	≯ 45%	
Plasticity index	≯ 20%	
ϕ	≮ 20%	

In case of cohesive-frictional fills both the soil reinforcement friction and adhesion will have to be considered.

In certain cases it may be more economical to use more than one type of soil. A good quality free draining cohesionless soil near the face of wall (and at the top when carrying concentrated loads) and a relatively inferior frictional or cohesive-frictional soil elsewhere. This will ensure development of adequate soil-reinforcement friction in areas where it is needed most. It will, however, be necessary to ensure good drainge of the bottom and rear of the backfill when inferior backfill soil is used.

15.4 REINFORCEMENT

General

The materials which have been used successfully as reinforcement in various engineering structures are galvanised steel, Mg-Al alloy, stainless steel, wood, ferrocement, polymer and plastic. Reinforcement may take the form of strips, grids, anchors, continuous sheets, mats, nets and planks.

Metal reinforcement

The most popular form of reinforcement is strip, a few centimetres wide and a couple of millimetres thick and of length commensurate with the height of structure. Strip type reinforcements are usually of metal and used by virtue of their strength, stiffness, cheapness and resistance to creep deformation. The most commonly used metal for strip reinforcement is galvanised steel, sometimes aluminium magnesium alloy and 17% chrome stainless steel. The one difficulty in metal reinforcement is the accurate prediction of rates of reinforcement corrosion likely to occur during the design life of the structure. However, metal reinforcements are used by providing an additional thickness of 0.75–1.25 mm for galvanised steel and 0.1 to 0.2 mm for stainless steeel, depending upon the nature of soil, for making up the loss for corrosion. Strips may have several protrusions such as ribs or gloves to increase the friction between the reinforcement and soil.

Polymer reinforcement

Nonmetallic reinforcements are almost exclusively made of one, or a combination, of the many polymers available with the strength and deformation properties of the resulting reinforcement being largely governed by the specific polymer and the manufacturing processes used to form the end-product. In general nonmetallic reinforcements are less strong than their metallic counterparts. The main advantage of polymers is that they do not suffer from corrosion as such; however they are susceptible to attack by various other agencies. The degradation resistance of some of the more commonly used polymers is given in Table 15.2.

Table 15.2. Degradation Resistance of Various Synthetic Fibres.

| Resistance to Attack by | Types of Synthetic Fibres | | | | |
	Polyester	*Polymide*	*Polyethylene*	*Polypropylene*	*PVC*
Fungus	Poor	Good	Excellent	Good	Good
Insects	Fair	Fair	Excellent	Fair	Good
Vermin	Fair	Fair	Excellent	Fair	Good
Mineral Acids	Good	Fair	Excellent	Excellent	Good
Alkalis	Fair	Good	Excellent	Excellent	Good
Dry heat	Good	Fair	Fair	Fair	Fair
Moist heat	Fair	Good	Fair	Fair	Fair
Oxidizing agents	Good	Fair	Poor	Good	
Abrasion	Excellent	Excellent	Good	Good	Excellent
Ultraviolet light	Excellent	Good	Poor	Poor	Excellent

The main drawback in polymers is their tendency to creep. Creep, which is a time-dependent phenomenon, is manifested by strain, at constant load, in excess of that caused by initial loading. Therefore, the use of polymers as reinforcement should be made keeping this point in view.

Geotextiles

Geotextiles are porous fabrics manufactured from synthetic materials such as polypropylene, polyester, polyethylene, etc. They come in thicknesses ranging from 0.25 mm to 7.5 mm with permeablility comparable in range from coarse gravel to fine sand. They can

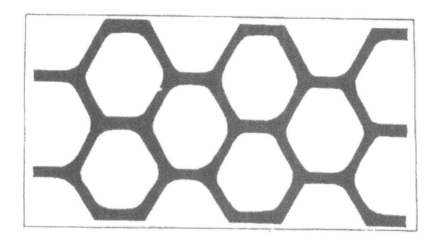

(a) Netlon grid

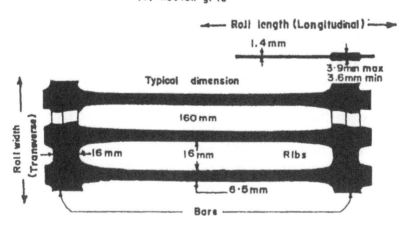

(b) Tensar SR80 Geogrid

Fig. 15.4. Typical Netlon and Tensar Grids.

be constructed in a variety of ways, the most common methods being: (i) woven, made from continuous monofilament fibres, and (ii) nonwoven, made from continuous or staple fibres laid down in a random pattern and then mechanically entangled into a relatively thick, felt-like fabric by means of punching with barbed needles.

In India, many firms are active in manufacturing geotextiles.

Geogrids

The manufacturing process of geogrids starts with an extruded sheet of polyethylene or polypropylene which is punched with a regular pattern of holes. Under the application of controlled heating, the sheet is stretched such that the randomly oriented long chain of molecules are drawn to an ordered and aligned state. This process increases the tensile strength and tensile stiffness of the polymer.

The uniaxially or biaxially oriented geogrid market is dominated by Netlon Ltd. of Blackburn, England. Their products are sold under the trade marks of Netlon and Tensar.

Figure 15.4 shows typical Netlon and Tensar geogrids.

Geomembranes

Geomembranes are impervious thin sheets of rubber or plastic material used primarily for linings and covers of liquid- or solid-storage impoundments.

Geocomposites

A geocomposite consists of a combination of geotextile and geogrid, or geogrid and geomembrane, or geogrid, geotextile and geomembrane, or any of these three materials with another material (e.g. with soil, styrofoam, deformed plastic sheets, steel anchors etc.). The application areas of geocomposites are numerous and growing steadily.

Figure 15.5 shows some typical geotextiles.

15.5 STRENGTH CHARACTERISTICS OF REINFORCED SOIL

The fact that reinforcement influences the stress-strain characteristics of soil and increases its strength can be demonstrated by the usual strength tests. Many researchers have studied the phenomenon of strength increase by subjecting reinforced samples to triaxial compression or plane-strain loading. For this purpose, the reinforced earth has been assumed to be an equivalent homogeneous material and the stress-strain and strength characteristics have been investigated for samples of cohesionless soil reinforced with discs, rings or fibres of different reinforcing materials (Long et al., Yang, 1972, Hausmann, 1976, Broms, 1977, Talwar, 1981, Singh, 1991). These studies have indicated appreciable increase in the deviator

Fig. 15.5. Typical Geotextiles.

(a) Rupture failure mode.

(b) Silippage failure mode.

Fig. 15.6. Shape of Samples in Two Failure Modes.

stress and revealed two different patterns of failure of reinforced earth samples—one in which the failure is caused by the rupture of reinforcement and the second in which the failure is due to slippage between the soil and reinforcement. Figure 15.6 shows the shape of samples in the two failure modes. The stress-strain characteristics for the two failure modes

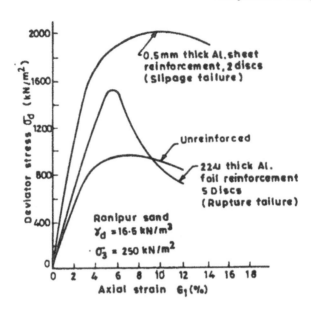

Fig. 15.7. Stress-Strain Behaviour of Reinforced and Plain Sand (Talwar, 1981).

are different (Fig. 15.7). The rupture failure mode is characterized by well-defined peak deviator stress at failure and diminished failure strain as compared to unreinforced soil. The stress-strain curves for the other failure mode (slippage failure) do not exhibit well-defined peaks, and failure strains indicate a ductile behaviour for the reinforced samples more or less like that of unreinforced samples. Obviously, the strength in the rupture failure mode is governed by the tensile strength of reinforcing elements, whereas in the slippage mode it is a function of the friction which develops at the soil-reinforcement interface, other conditions being same.

The mode of failure has a profound effect on the strength envelop of the reinforced earth. Rupture failure of reinforced earth leads to a strength envelop which is virtually parallel to that of unreinforced soil but exhibits a cohesion intercept which is a function of the rupture strength of reinforcement and its distribution in the sample (Fig. 15.8). Slippage failure, on the other hand, leads to an increase in the friction angle ϕ, with little or no cohesion intercept.

To illustrate the effect of reinforcement on the strength parameters of soil quantitatively, investigations of Talwar (1981) and Singh (1991) have been briefly summarised below.

Talwar (1981) conducted triaxial compression tests on samples of soil reinforced with discs of different materials as shown in Fig. 15.9. The soil used was dry Ranipur sand (SP, D_{10} = 0.13 mm, C_u = 1.85). The materials for reinforcing the triaxial samples were so chosen that failure could be achieved both by rupture of reinforcement and by slippage between reinforcement and soil. The following reinforcing materials arranged in increasing order of strength were employed:

(i)   Aluminium foil 22 micron thick,
(ii) Aluminium foil 50 micron thick,

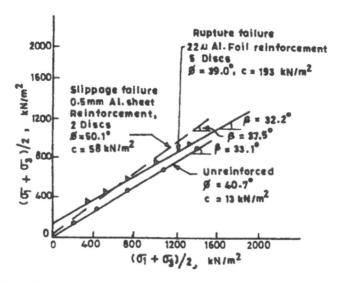

Fig. 15.8. Strength Envelopes of Reinforced and Plain Sand (Talwar, 1981).

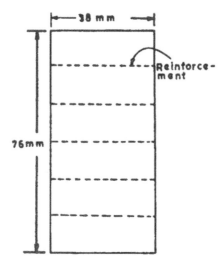

Fig. 15.9. Pattern of Reinforcement in Triaxial Sample (Talwar, 1981).

(iii) Fibre glass cloth 0.08 mm thick, and
(iv) Aluminium sheet 0.5 mm thick.

The reinforcements were used as discs of 35 mm diameter in five beds (Fig. 15.9). The tests were performed at two densities 16 kN/m³ and 16.5 kN/m³ which correspond to medium and dense states (D_R = 67.9% and 79.5%) respectively. The tests were performed at

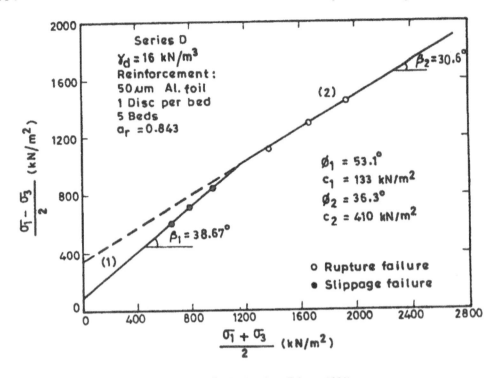

Fig. 15.10. Mohr's Envelop (Talwar, 1981).

different confining pressures varying from 20 kN/m² to 500 kN/m². A typical Mohr's envelop plot is shown in Fig. 15.10. The results obtained from this study are listed in Table 15.3.

Singh (1991) carried out triaxial compression tests on Amanatgarh Sand (SP, D_{10} = 0.19 mm, C_u = 1.58) placed at relative density of 75%. The samples were reinforced with: (i) Woven white geotextile (W), and (ii) Woven black geotextile (B). The properties of the two geotextiles are given in Table 15.4. The soil samples were reinforced as shown in Fig. 15.11. The tests were performed for different confining pressures varying from 50 kN/m² to 500 kN/m². The results obtained from this study are given in Table 15.5.

It may be noted from Tables 15.3 and 15.5 that at higher confining pressures, mode of failure changes from slippage to rupture. This is due to the fact that at higher confining pressures dilatation of sand is reduced so tensile force on reinforcing disc is not much. This leads to reduced frictional force at soil-reinforcement interface. Hence, friction angle φ is not much affected at higher confinement. But at higher confinement strength gain is consequence of utilization of tearing strength of the reinforcement. Since tearing strength of the reinforcement is not a function of confinement, strength of reinforced samples increases slowly with confining pressure leading to lesser friction angle at high confinement.

Table 15.3. Shear Strength Parameters of Reinforced Sand (Talwar, 1981).

Type of Sample	Cohesion c, kN/m²	Angle of Internal Friction φ (Deg)	Failure Mode	Remarks
		Medium dense sand, $D_R = 67.9\%$		
1. Unreinforced	10	38.5	Shear	(i) Sample was reinforced as shown in Fig. 15.9
2. Reinforced with				
(a) 22 micron Al foil	170	37.0	Rupture	
(b) 50 micron Al foil	133	53.0	Slippage $\sigma_3 < 100$ kN/m²	
	410	37.0	Rupture $\sigma_3 > 100$ kN/m²	(ii) σ_3 is confining pressure
(c) Fibre glass	166	46.0	Partly rupture	
		Dense sand, $D_R = 79.5\%$		
1. Unreinforced	13.2	41.0		Shear
2. Reinforced with				
(a) 22 micron Al foil		118	40.5	Rupture
(b) 50 micron Al foil		80	64.2	Slippage $\sigma_3 < 100$ kN/m²
		365	40.0	Rupture $\sigma_3 > 100$ kN/m²
(c) 0.5 mm Al sheet		58	50.1	Slippage

15.6 REINFORCED EARTH RETAINING WALLS

General

The current design methods make a number of simplifying assumptions. It is assumed that the retaining wall moves out sufficiently for the Rankine's state of stress to develop in the soil mass. The direction of principal stresses are assumed to coincide with the vertical and horizontal, and the vertical stress is assumed to be either uniform or varying linearly, at any depth z from top.

Based on the results of small-scale models and theoretical analysis, the following recommendations have been made:

(i) Failure due to slipping can be prevented by keeping width to height ratio of the wall greater than 0.8.

Table 15.4. Properties of Geotextiles.

	Bombay Dying (Woven) (Properties of Different Geotextiles)	
1. Quality no. (styles)	.0037	Black PD 380/B
2. Material	100% Polypropylene	
3. Specific gravity	0.91	0.91
4. Weight/Eq. meter in gms.	303.00	276.00
5. Thickness in mm (@ 100 gms/cm^2)	0.78	0.68
6. Breaking strength: (IS: 1969–1963) (5 × 10 cms)		
Warpway (kg)	390.1	245.7
Warpway (kg)	333.0	182.0
7. Elongation at break (%): (IS: 1969–1963)		
Wrap	38.0	46.9
Weft	31.8	27.8
8. Grab strength test: (3″ × 1″ strip) (ASTM-D-1682)		
Wrapway (kg)	266.9	214.8
Weftway (kg)	237.0	152.8
9. Elongation (%): (Grab test) (ASTM-D-1982)		
Wrapway	36.8	45.3
Weftway	33.1	30.3
10. Tear strength: (Single rip) (ASTM-D-1982)		
Wrapway (kg)	69.3	21.2
Weftway (kg)	71.6	18.0
11. Water permeability: (Litres/Sec/metre at 10 cm water head)	40.2	4.2
12. Pore size in Microns:		
Mean	174.0	25.0
Maximum	243.0	69.0

Properties common to all Qualities:

13. Good resistance to chemical

14. Resistant to biological degradation

15. Thermal stability - 0° to 120°

16. Width offered—as desired (Upto 330 cms).

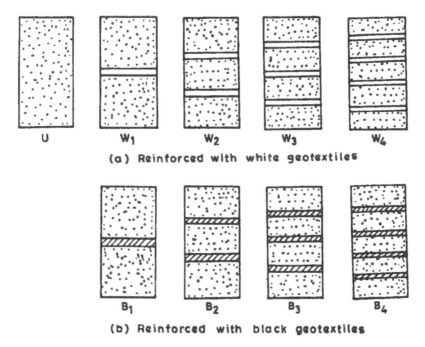

Fig. 15.11. Pattern of Reinforcement in Triaxial Sample (Singh, 1991).

(ii) The vertical pressure in the wall close to the face can be assumed to vary linearly with depth, with maximum value at the base.

(iii) The maximum traction force (tension) in strips is reached close to the face.

While designing a reinforced earth wall, one has to consider:

(i) External stability, and
(ii) Internal stability.

In the external stability analysis, it is assumed that the reinforced earth wall is an integral unit and behaves as a rigid gravity structure and conforms to the simple laws of statistics.

The internal stability deals with the design of reinforcement with regard to its length, cross-section against tension failure and ensuring that it has a sufficient anchorage length into the stable soil.

External stability

The possible external failure mechanisms are:

(a) Sliding,
(b) Overturning,
(c) Tilting/bearing failure, and
(d) Slip failure.

Table 15.5. Shear Strength Parameters of Reinforced Sand (Singh, 1991).

Type of Sample	Cohesion c, kN/m²	Angle of Internal Friction ϕ (Deg)	Failure Mode	Remarks
1. Unreinforced	15	39	Shear	σ_3 is the confining pressure
2. Reinforced with white geotextile as (Refer Fig. 15.11)				
(a) W_1	33	41	Rupture of reinforceemnt	
(b) W_2	50	46	Partly ruptured	
(c) W_3	34	54	Slippage $\sigma_3 < 100$ kN/m²	
	290	38	Rupture $\sigma_3 > 100$ kN/m²	
(d) W_4	20	60	Slippage $\sigma_3 < 100$ kNm²	
	350	38	Rupture $\sigma_3 > 100$ kN/m²	
3. Reinforced with black geotextile as (Refer Fig. 15.11)				
(a) B_1	20	41.5	Rupture	
(b) B_2	35	50	Slippage $\sigma_3 < 100$ kN/m²	
	240	39	Ruture $\sigma_3 > 100$ kN/m²	
(c) B_3	60	54	Partly rupture $\sigma_3 < 200$ kN/m²	
	280	40	Rupture $\sigma_3 > 200$ kN/m²	
(d) B_4	100	56	Slippage $\sigma_3 < 200$ kN/m²	
	450	38	Rupture $\sigma_3 > 200$ kN/m²	

These basic mechanisms are illustrated diagrammatically in Fig. 15.12.

Consider the external stability of the surcharged vertical wall shown in Fig. 15.13 which shows the externally applied forces assuming a Rankine distribution of lateral earth pressure and a trapezoidal distribution of ground bearing pressure.

(a) Sliding

$$\text{Factor of safety against sliding} = \frac{\text{Resisting force}}{\text{Sliding force}}$$

$$\text{Factor of safety} = \frac{\mu(\gamma_w H \cdot L \cdot + q \cdot L)}{\frac{1}{2}K_{Ab}\gamma_b H + K_{Ab}qH} \qquad \ldots (15.1)$$

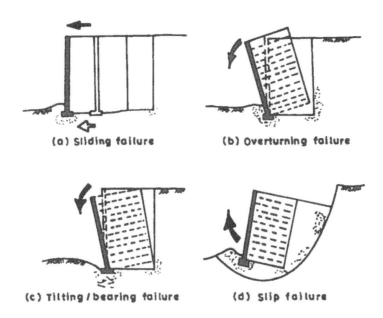

Fig. 15.12. Failure Mechanism Considered in External Stability of Reinforced Earth Wall.

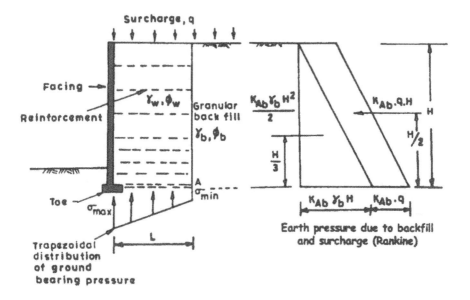

Fig. 15.13. Reinforced Earth Wall.

$$= \frac{2\mu(\gamma_w H + q)}{K_{Ab}(\gamma_b H + 2q)(H/L)}$$

where
μ = coefficient of reinforcement soil friction,
γ_b = desnsity of backfill,
K_{Ab} = coefficient of active pressure for backfill

$$= \frac{1 - \sin\phi_b}{1 + \sin\phi_b}$$

ϕ_b = angle of internal friction of backfill,
K_{Aw} = coefficient of active pressure for wall-fill,
ϕ_w = angle of internal friction for wall-fill,
γ_w = density of wall-fill,
H = height of wall, and
L = length of reinforcement.

The minimum factor of safety against sliding is usually taken as 2.

(b) Overturning

Overturning moment about the toe $= K_{Ab}\gamma_b \dfrac{H^3}{6} + K_{Ab}q \dfrac{H^2}{2}$

Restoring moment about the toe $= \gamma_w H \cdot \dfrac{L^2}{2} + q \dfrac{L^2}{2}$

Factor of safety against overturning $= \dfrac{\text{Restoring moment}}{\text{Overturning moment}}$

$$= \frac{\gamma_w HL^2/2 + qL^2/2}{K_{Ab}\gamma_b H^3/6 + K_{Ab}qH^2/2}$$

$$= \frac{3(\gamma_w H + q)}{K_{Ab}(\gamma_b H + 3q)(H/L)^2} \qquad \dots (15.2)$$

The minimum factor of safety against overturning is usually taken as 2.

(c) Tilting/bearing failure

Referring to Fig. 15.13, the maximum and minimum ground bearing pressures are given by

$$\sigma = \frac{V}{A} \pm \frac{M}{I} y = (\gamma_w H + q) \pm \frac{K_{Ab} \cdot \gamma_b H^3/6 + K_{Ab}q \cdot H^2/2}{(1 \times L^3/12)} \frac{L}{2}$$

$$\sigma_{max} = (\gamma_w H + q) + K_{Ab}(\gamma_b H + 3q)(H/L)^2 \qquad \dots (15.3)$$

$$\sigma_{min} = (\gamma_w H + q) - K_{Ab}(\gamma_b H + 3q)(H/L)^2 \qquad \dots (15.4)$$

For safe design

$$\sigma_{max} \ngtr / q_a$$

$$\sigma_{min} = \nless / 0.0$$

where q_a = allowable soil pressure.

The above equations do not take into account the vertical and horizontal line loads acting at the top of the wall (Fig. 15.14). The expressions of factors of safety against sliding and overturning, and of base pressures will be as given below :

Factor of safety against sliding

$$= \frac{2\mu\left[\gamma_w H + q + (V_1/L)\right]}{K_{Ab}\left[\gamma_b H + 2q\right](H/L) + (2H_1/L)} \qquad \dots (15.1a)$$

Factor of safety against overturning

$$= \frac{3(\gamma_w H + q) + \left[6V_1(d-e)/L^2\right]}{\left[K_{Ab}(\gamma_b H + 3q) + (6H_1/H)\right](H/L)^2} \cdot \qquad \dots (15.2a)$$

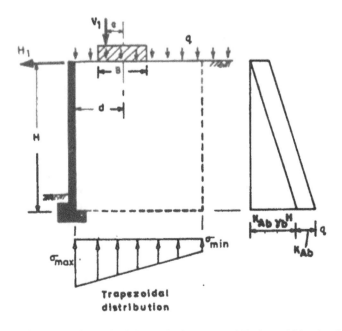

Fig. 15.14. Reinforced Earth Wall with Vertical and Horizontal Line Loads.

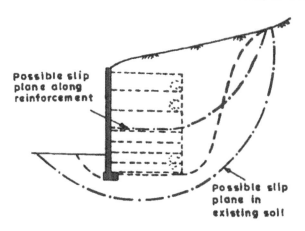

Fig. 15.15. Possible Slip Failures.

$$\sigma_{max} = (\gamma_w H + q + V_1/L) + K_{Ab}(\gamma_b H + 3q)(H/L)^2 + 3[2HH_1 + V_1(L - 2d + 2e)]/L^2.$$
$$... (15.3a)$$

$$\sigma_{min} = (\gamma_w H + q + V_1/L) + K_{Ab}(\gamma_b H + 3q)(H/L)^2 - 3[2HH_1 + V_1(L - 2d + 2e)]/L^2.$$

$$... (15.4a)$$

(d) Slip Failure

All potential slip surfaces should be investigated (Fig. 15.15). Where slip planes already exist, residual soil parameters should be adopted.

The factor of safety for reinforced earth structures against rotational slip is the same as for conventional retaining structures usually 1.5.

Internal stability

The internal stability is essentially associated with the tension and wedge pull-out failure mechanisms as shown in Fig. 15.16.

In general, two methods used in practical design utilise one of the following:

(a) Tie Back-Wedge Analysis, and
(b) Coherent Gravity Analysis.

The main differences between these two analyses originate from the basic assumptions made about (i) the shape of the failure zone, (ii) the rotation of the wall facing, and (iii) the lateral pressures acting within the reinforced soil. Out of the two methods, the Tie Back-Wedge Analysis is more popular in use and the same is discussed here.

(a) Tension failure

Consider the internal stability of a surcharged reinforced earth wall constructed with uniform frictional soil fill (Fig. 15.17).

The tensile force (T_i) per metre width in the reinforcement at depth h_i is given by

$$T_i = K_{aw}\sigma_{vi}S_{vi}S_{Hi} \qquad ... (15.5)$$

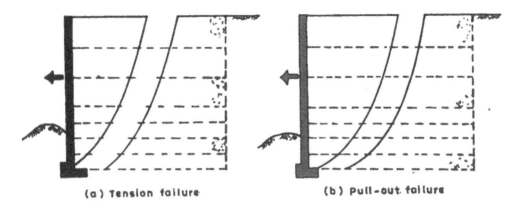

Fig. 15.16. Tension and Pull-Out Failure Mechanisms.

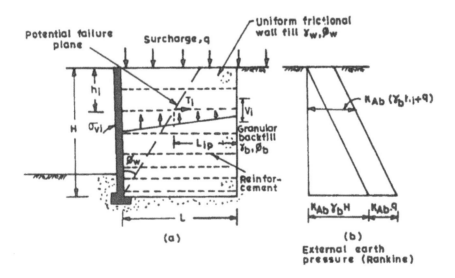

Fig. 15.17. Section of Reinforced Earth Wall Showing Tension in a Reinforcement.

where S_{Vi} and S_{Hi} are the vertical and horizontal spacing of reinforcements at depth h_i.

Using $H = h_i$ in Eq. 15.3

$$\sigma_{Vi} = (\gamma_w \cdot h_i + q) + K_{Ab} (\gamma_b \cdot h_i + 3q) (h_i/L)^2. \qquad \text{... (15.6)}$$

Substituting this value in Eq. 15.5

$$T_i = K_{aw}[\gamma_w h_i + q) + K_{Ab} (\gamma_b h_i + 3q) (h_i/L)^2] S_{Vi} \cdot S_{Hi} \qquad \text{... (15.7)}$$

In the case of a cohesive frictional soil Eq. 15.7 becomes

$$T_i = K_{aw}[\gamma_w h_i + q) - 2c/\sqrt{K_{aw}} + K_{Ab} (\gamma_b h_i + 3q) (h_i/L)^2]S_{Vi} \cdot S_{Hi} \qquad \text{... (15.7a)}$$

However, the effect of cohesion may be neglected. The design will become slightly on the conservative side. For no rupture of reinforcement

$$T_i \ngtr T_a \qquad \qquad \cdots (15.8)$$

where T_a = permissible tensile strength of the reinforcement.

The above equations do not take into account any of the tensions induced by vertical and horizontal line or point loads acting at the top of the wall. These tensions must be evaluated separately and added to the values given above. These additional tensions can be calculated as follows:

(i) Vertical line loads

A vertical load dispersal of 2:1 is usually assumed as indicated in Fig. 15.18a. A trapezoidal pressure distribution is assumed for calculating the vertical stress on each reinforcement; this stress can then be converted into an equivalent tension in the reinforcement using Eq. 15.5.

A conservative but simple approach is to calculate the equivalent eccentricity e' of the vertical load V_1 at depth h_i as shown in Fig. 15.18b.

Accordingly,

$$\sigma_H = \frac{V_1}{B_i} + \frac{6V_1 e'}{B_i^2}. \qquad \qquad \cdots (15.9)$$

When $h_i \leqslant (2d - B)$; $B_i = h_i + B$ and $e' = \dfrac{eB_i}{B}$.

Therefore, $$\sigma_H = \frac{V_i}{(h_i + B)}\left[1 + \frac{6e}{B}\right] \qquad \qquad \cdots (15.10)$$

and $$T_i = \frac{K_{Aw} \cdot V_i}{(h_i + B)}\left[1 + \frac{6e}{B}\right] S_H \cdot S_{HI} \qquad \qquad \cdots (15.11)$$

When $(2d - B) < h_i \leq 2(L - d) - B$; $B_i = d + \dfrac{(h_i + B)}{2}$ and $e' = \dfrac{B_i}{2} - d + \dfrac{e \cdot B_i}{B}$.

The equation of tension in the reinforcement becomes

$$T_i = \frac{6K_{Aw}V_i d}{B_i^2}\left[\frac{B_i(2B + 3e)}{3Bd} - 1\right]. \; S_H \cdot S_{HI}. \qquad \qquad \cdots (15.12)$$

Finally, if $h_i > 2(L - d) - B$, $e' = \dfrac{L}{2} - d + \dfrac{eL}{B}$

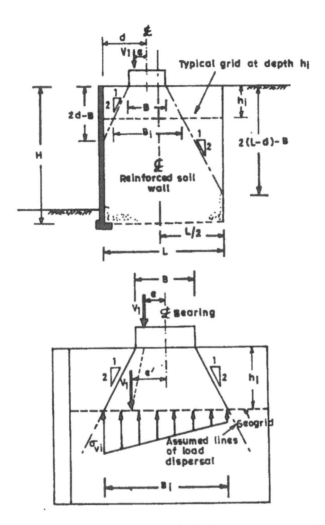

Fig.15.18. Illustrate the Procedure of Computing Tension in Reinforcement due to Vertical Line Load.

$$T_i = \frac{6K_{aw}V_1 d}{L^2}\left[\frac{L(2B+3e)}{3Bd}-1\right] S_{Vi}\cdot S_{Hi}. \qquad \text{... (15.13)}$$

(ii) Horizontal line loads

The dispersion of horizontal line load (H_1) may be taken as shown in Fig. 15.19 and depth of influence h_e given by

$$h_e = \frac{B+2d}{2\tan(45-\phi_w/2)}. \qquad \text{... (15.14)}$$

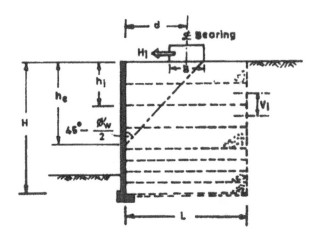

Fig. 15.19. Illustrate the Procedure of Computing Tension in Reinforcement due to Horizontal Line Load.

Consequently, when $h_e < H$, the tension induced into the reinforcement at depth h_i is given by

$$T_i = \frac{2H_1}{h_e} \left[1 - \left(\frac{h_i}{h_e}\right)\right] S_{Vi} \cdot S_{Hi} \qquad \qquad \dots (15.15)$$

If $h_e > H$ then it is convenient to make the simplifying assumption that $h_e = H$, therefore

$$T_i = \frac{2H_1}{H} \left[1 - \left(\frac{h_i}{H}\right)\right] S_{Vi} \cdot S_{Hi} \qquad \qquad \dots (15.15)$$

In the case when mat type reinforcement (i.e. geotextile or geogrid) is used, the tension in the reinforcement (T_i) is obtained per unit width of wall and, therefore, in the above expressions S_{Hi} will be unity.

(b) Wedge/pull-out failure

In addition to investigating the internal mechanisms of tension, it is also necessary to consider separately the possibility of inclined failure planes passing through the wall forming unstable wedges of soil bounded by the front face of the wall, the top ground surface and the potential failure plane.

The basic assumptions made in this analysis are:

(i) Each wedge behaves as a rigid body.
(ii) Friction between the facing and the fill is ignored.
(iii) No potential failure plane passes through the contact area under a bridge abutment bank seat or similar support.

Figure 15.20 indicates the loads and forces to be considered in the analysis, namely:

(i) Self weight of the fill in the wedge
(ii) Uniformly distributed surcharge (q)
(iii) Vertical loading (V_1)
(iv) Horizontal loading (H_1)
(v) The resultant force (R) on the potential failure plane.
(vi) Total tension (T) in the grids intercepted by the potential failure plane.

The analytical procedure is one of investigating potential failure planes passing through points such as a, b, c, etc. (Fig. 15.21) immediately behind the facing of the wall. A simple triangle of forces can be used to determine the value of T for several different values of β_1,

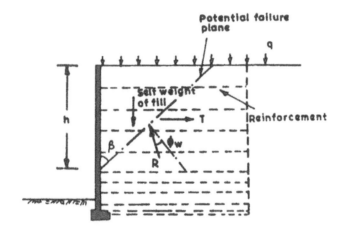

Fig. 15.20. Pull-out Failure Mechanism.

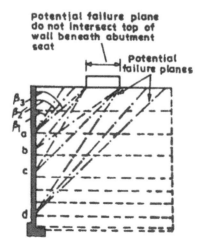

Fig. 15.21. Possible Failure Mechanism.

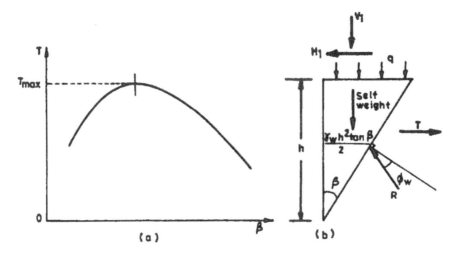

Fig. 15.22. Equilibrium of a Typical Wedge.

β_2, β_3, etc.). As β is varied for each point the value of T changes and reaches a maximum as indicated in Fig. 15.22.

Equilibrium of wedge (Fig. 15.22b) gives:

$$T = \left[\frac{h \tan\beta(\gamma_w h + 2q) + 2V_1}{2\tan(\phi_w + \beta)} \right] + H_1. \qquad \text{... (15.17)}$$

The maximum value of T is obtained by maximising it w.r.t. β. The tension taken by an individual reinforcement is considered equal to the total tension T divided by the number of reinforcing layers intersected by the failure wedge.

Anchorage

Once the maximum value of T has been found, the designer should ensure that all grids/strips have an adequate anchorage length into the resisting zone to prevent pull-out failure.

The anchorage length required for a grid/strip is calculated using a coefficient of interaction (α) which varies with the type of soil and form of grid/strip. This coefficient is usually obtained from a 300 mm direct shear box or suitable pull-out tests.

In addition, the calculation of the anchorage length should be based on the maximum tension which can occur in the grid/strip under consideration. Clearly this tension should not exceed the safe design strength of the grid/strip being used.

The anchorage length (L_{ip}) for the reinforcement at depth h_i (Fig. 15.17) is determined using the following equation:

$$L_{ip} = \frac{T_i \times \text{Factor of safety}}{2\alpha \tan\phi_w(\gamma_w h_i + q)}. \qquad \text{... (15.18)}$$

The coefficient of 2 in the denominator of the above equations is to account for interaction on both sides of grid/strip.

A factor of safety of 2 is usually taken for design purposes.

15.7 WALL WITH REINFORCED BACKFILL

The usefulness of the patented reinforced earth retaining wall of Vidal has been proved economical by thousands of such structures constructed all over the world. But situations can be met where reinforced earth walls may not provide ideal solution. This can be true for a location with limited space behind the wall (Fig. 15.23) or for narrow hill roads on unstable slopes, which may not permit the use of designed length of reinforcement (Fig. 15.24). In such circumstances, a rigid wall with reinforced backfill may appear more appropriate. Backfill is reinforced with unattached horizontal strips/mats/nets laid normal to the wall (Fig. 15.25).

Pasley (1822), based on field trials, demonstrated that substantial reduction occurs in the magnitude of lateral earth pressure on the wall by reinforcing its backfill. Broms (1977, 1987) and Hausmann and Lee (1978) performed model tests on rigid retaining walls with reinforced backfill and reported that considerable reduction in moments occur at the base of the wall. Saran et al. (1979), Talwar (1981) and Saran and Talwar (1983) developed theoretical analysis for obtaining earth pressure and its point of application in vertical wall with reinforced backfill. The results were presented in the form of non-dimensional charts. Garg (1988) and Saran et al. (1992) extended the work of Talwar (1981) for uniformly

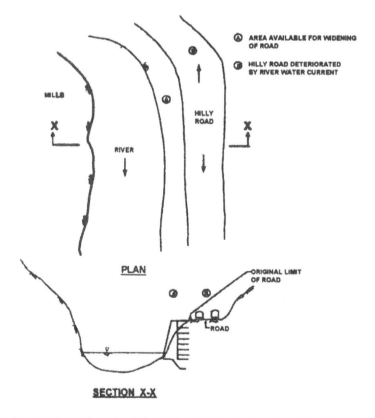

Fig. 15.23. An Example of Suitability of Wall with Reinforced Backfill.

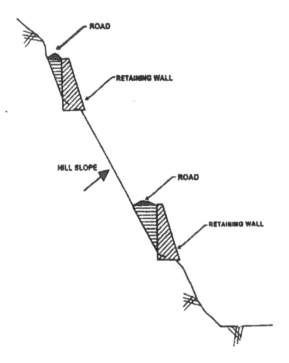

Fig. 15.24. Possible Situation for Application of Retaining Wall with Reinforced Backfill.

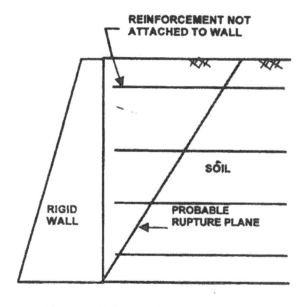

Fig. 15.25. Rigid Wall with Reinforced Backfill.

distributed surcharge on the backfill. Saran and Khan (1988) and Khan (1991) presented theoretical analysis for the inclined retaining wall with reinforced backfill and having a uniformly distributed and line load surcharge on the backfill. The analyses by Talwar (1981), Garg (1988) and Khan (1991) were based on the limit equilibrium approach and the soil in the backfill was considered as cohesionless. Mittal (1998) extended the analysis of Khan (1991) for cohesive-frictional soil. The results of all the above analyses were presented in the form of non-dimensional charts to determine the value of the total earth pressure and the height of the point of application from the base of wall. Further, the analytical findings were validated by carefully conducted model tests (Talwar, 1981; Garg, 1988; Garg and Saran, 1989; Khan, 1991; Mittal, 1998; Saran et al., 2001).

Pseudo-static analysis of retaining wall has been carried out to determine the seismic pressures and their point of application (Saran and Talwar, 1981, 1982; Saran, 1998). The results of these analyses indicated that provision of the reinforcement in the backfill reduces the seismic pressures significantly.

All the above studies illustrated the effectiveness of unattached reinforcement in reducing the earth pressure on a rigid wall. To highlight the salient features of the methodology adopted by above mentioned investigators, the work of Saran et al. (1992) has been described here briefly.

Analysis

The analysis was developed for a retaining wall of height H with vertical back face retaining cohesionless backfill having dry density γ_d and an angle of internal friction ϕ. The backfill which carries a uniform surcharge of intensity q is reinforced with unattached horizontal strips of length L and width W, placed at a vertical spacing of S_z and a horizontal spacing of S_x. A failure plane BC, making an angle θ with the vertical, passes through the heel of the retaining wall (Fig. 15.26).

The frictional resistance offered by a reinforcing strip will be located in the shorter portion of the strip, which moves relative to the failure plane. The shorter portion of the strip is referred as the effective length. For example, if strip DF is cut by failure plane at E, then the effective length will be either DE or EF. In the case in which the portion of the strip length within the wedge $DE < EF$, then EF will not move out of the soil mass. DE will come out of the wedge as the latter moved away from the stationary portion of the backfill. If $EF < DE$, the strip will move with the failure wedge, pulling length EF out of the stationary mass of backfill. Therefore, the effective length of the strip will be the smaller of DE or EF. A reinforcing strip, located completely within the moving wedge, will not contribute any frictional resistance to the movement of the wedge.

An element $IJKM$ (Fig. 15.26) of the failure wedge of thickness dy, located at a distance y from the top of the wedge, is in equilibrium under the following intensities of forces:

p_y = pressure intensity acting uniformly on IJ in the vertical direction due to the self-weight of the backfill lying above IJ and the uniform surcharge q;

$(p_y + dp_y)$ = uniform reaction intensity acting upward on KM in the vertical direction;

p_θ = reaction-intensity on JK acting at an angle ϕ to the normal on JK;

p = pressure-intensity on IM acting at an angle δ to the normal on IM;

σ_n = vertical stress due to the weight of an element $IJKM$ acting downward,

= $\gamma \cdot dy$; and

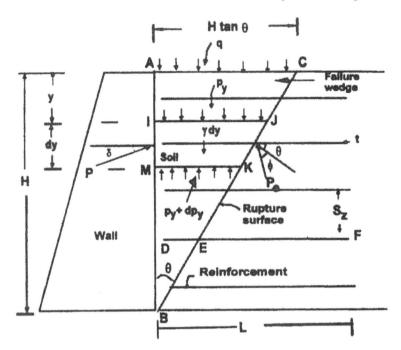

Fig. 15.26. Failure Wedge and Various Intensities of Forces Keeping Element *IJKM* in Equilibrium .

$t = (T/S_z)$ = intensity of tension in the reinforcing strip, which is assumed to be transmitted uniformly to the soil layers of thickness S_z encompassing the strip.

Neglecting second-order and higher order terms, the static equilibrium of an element *IJKM* ($\Sigma H = 0$, $\Sigma V = 0$ and $\Sigma M = 0$) of failure wedge *ABC* (Fig. 15.26) finally yields the following expression:

$$\frac{dp}{dy} = -C_1 \frac{p}{H-y} + C_2\gamma - C_3 \frac{dt}{dy} \qquad \qquad \text{... (15.19)}$$

where,

$$C_1 = \frac{2\sin\delta\cos(\theta+\phi)}{\sin(\theta+\phi-\delta)} \qquad \qquad \text{... (15.20)}$$

$$C_2 = \frac{\tan\theta\cos(\theta+\phi)}{\sin(\theta+\phi-\delta)} \qquad \qquad \text{... (15.21)}$$

$$C_3 = \frac{\sin(\theta+\phi)}{\sin(\theta+\phi-\delta)} \qquad \qquad \text{... (15.22)}$$

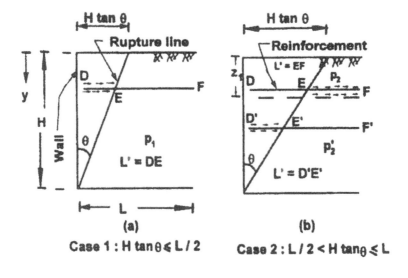

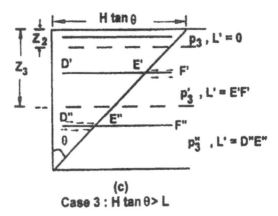

Fig. 15.27. Effective Length Criteria of Reinforcing Element.

Tension T at the limiting equilibrium can be taken as

$$T = \frac{2Wf^{*}\sigma_{v}l'}{S_{x}} \qquad \text{... (15.23)}$$

where l' = effective length of strip, f^{*} = coefficient of friction between reinforcing strip and soil, and

$$\sigma_{v} = \gamma\left(y + \frac{dy}{2}\right) + q \qquad \text{... (15.24)}$$

l' will vary for each reinforcing strip, depending on the wedge angle θ and the length L of the strip as shown in Fig. 15.27.

Case 1: $H \tan \theta < L/2$
$l' = (H - y) \tan \theta$

Case 2: $L/2 < H \tan \theta < L$
$l' = L - (H - y) \tan \theta$ for $y < Z_1$
$l' = (H - y) \tan \theta$ for $y > Z_1$
$Z_1 = H - L/2 \cot \theta$

Case 3: $H \tan \theta > L$
$l' = 0$ for $y < Z_2$
$l' = L - (H - y) \tan \theta$ for $Z_2 < y < Z_3$
$l' = (H - y) \tan \theta$ for $y > Z_3$
$Z_2 = H - L \cot \theta$
$Z_3 = H - L/2 \cot \theta$

The differential equation (Eq. 15.19) is solved for these three cases separately by substituting appropriate boundary conditions. For presenting the results in non-dimensional form, lateral earth pressure p is considered to consist of two parts: (i) lateral earth pressure due to backfill only, p_γ and (ii) lateral earth pressure due to only surcharge load, p_q i.e.

$$p = p_\gamma + p_q \qquad \qquad \qquad ...\ (15.25)$$

Expressions for pressure intensities $p_{1\gamma}$ and p_{1q} (Fig. 15.27a) for case 1, $p_{2\gamma}, p_{2\gamma}', p_{2q}$ and p_{2q}' (Fig. 15.27b) for Case 2 , and $p_{3\gamma}, p_{3\gamma}', p_{3\gamma}'', p_{3q}, p_{2q}', p_{3q}'$ (Fig. 15.27c) for Case 3 were obtained.

Expressions for pressure intensities are integrated over their respective domains to obtain the resultant earth pressure. The distance of the point of application of the resultant earth pressure is first obtained from the top of wall by integrating the moment of pressure intensity in each case and dividing it by the respective resultant earth pressure. The height of the point of application of the resultant earth pressure above the base of wall is obtained by substracting this distance from the total height of the wall.

The resultant earth pressure and the height of its point of application above the base of the wall in each case is expressed in non-dimensional form as follows:

For Case 1:

$$K_\gamma = \frac{P_{1\gamma}}{\left(\frac{1}{2}\right)\gamma H^2} = \frac{\int_o^H p_{1\gamma} dy}{\left(\frac{1}{2}\right)\gamma H^2} \qquad \qquad ...\ (15.26)$$

$$\frac{H_\gamma}{H} = 1 - \frac{\int_o^H p_{1\gamma} \cdot y \cdot dy}{\int_o^H p_{1\gamma} dy} \qquad \qquad ...\ (15.27)$$

$$K_q = \frac{P_{1q}}{qH} = \frac{\int_o^H p_{1q} dy}{qH} \qquad \qquad ...\ (15.28)$$

$$\frac{H_q}{H} = 1 - \frac{\int_0^H P_{1q} \cdot y \cdot dy}{H \int_0^H P_{1q} dy} \qquad \qquad \text{... (15.29)}$$

In Case 2:

$$K_\gamma = \frac{P_{2\gamma}}{\left(\frac{1}{2}\right)\gamma H^2} = \frac{\int_0^{Z_1} P_{2\gamma} dy + \int_{Z_1}^H p'_{2\gamma} dy}{\left(\frac{1}{2}\right)\gamma H^2} \qquad \qquad \text{... (15.30)}$$

$$\frac{H_\gamma}{H} = 1 - \frac{\int_0^{Z_1} P_{2\gamma} \cdot y \cdot dy + \int_{Z_1}^H p'_{2\gamma} dy}{H\left(\int_0^{Z_1} P_{2\gamma} dy + \int_{Z_1}^H p'_{2\gamma} \cdot dy\right)} \qquad \qquad \text{... (15.31)}$$

$$K_q = \frac{P_{2q}}{qH} = \frac{\int_0^{Z_1} P_{2q} \cdot y \cdot dy + \int_{Z_1}^H p'_{2q} dy}{qH} \qquad \qquad \text{... (15.32)}$$

$$\frac{H_q}{H} = 1 - \frac{\int_0^{Z_1} P_{2q} \cdot y \cdot dy + \int_{Z_1}^H p'_{2q} \cdot y \cdot dy}{H\left(\int_0^{Z_1} P_{2q} dy + \int_{Z_1}^H p'_{2q} dy\right)} \qquad \qquad \text{... (15.33)}$$

$$K_\gamma = \frac{P_{3\gamma}}{\left(\frac{1}{2}\right)\gamma H^2} = \frac{\int_0^{Z_2} P_{3\gamma} dy + \int_{Z_2}^{Z_3} p'_{3\gamma} dy + \int_{Z_3}^H p''_{3\gamma} dy}{\left(\frac{1}{2}\right)\gamma H^2} \qquad \qquad \text{... (15.34)}$$

$$\frac{H_y}{H} = 1 - \frac{\int_0^{Z_2} P_{3\gamma} \cdot y \cdot dy + \int_{Z_2}^{Z_3} p'_{3\gamma} \cdot y \cdot dy + \int_{Z_3}^H p''_{3\gamma} \cdot y \cdot dy}{H\left(\int_0^{Z_2} P_{3\gamma} dy + \int_{Z_2}^{Z_3} p'_{3\gamma} dy + \int_{Z_3}^H p''_{3\gamma} dy\right)} \qquad \qquad \text{... (15.35)}$$

$$K_q = \frac{P_{3q}}{qH} = \frac{\int_0^{Z_2} P_{3q} dy + \int_{Z_2}^{Z_3} p'_{3q} dy + \int_{Z_3}^H p''_{3q} dy}{qH} \qquad \qquad \text{... (15.36)}$$

$$\frac{H_q}{H} = 1 - \frac{\int_0^{Z_2} P_{3q} \cdot y \cdot dy + \int_{Z_2}^{Z_3} p'_{3q} \cdot y \cdot dy + \int_{Z_3}^H p''_{3q} \cdot y \cdot dy}{H\left(\int_0^{Z_2} P_{3q} dy + \int_{Z_2}^{Z_3} p'_{3q} dy + \int_{Z_3}^H p''_{3q} dy\right)} \qquad \qquad \text{... (15.37)}$$

It should be mentioned here that the closed-form solutions of these equations have been obtained. The details of the derivations are available elsewhere (Garg, 1988). The expressions of K_γ, K_q, $\dfrac{H_\gamma}{H}$ and $\dfrac{H_q}{H}$ have following two non-dimensional coefficients describing the reinforcement characteristics:

(i) Spacing coefficient, $D_p = \dfrac{f^* WH}{S_x S_z}$ for strip type reinforcement   ... (15.38)

$$= \dfrac{f^* H}{S_z}$$ for mat type reinforcement   ... (15.38)

(ii) Length coefficient $= \dfrac{L}{H}$ ... (15.39)

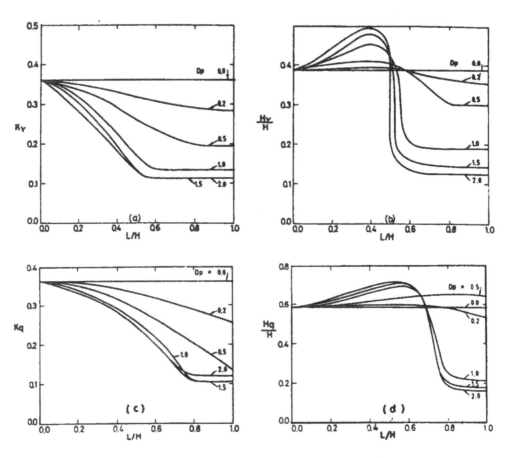

Fig. 15.28. Non-dimensional Chart for Resultant Pressure and Height of Point of Application — (a) and (b) Due to Backfill; (c) and (d) Due to Surcharge Loading ($\phi = 25°$).

Parametric study

The earth pressure coefficient K_γ and K_q, and the height of the point of application, $\dfrac{H_\gamma}{H}$ and $\dfrac{H_q}{H}$ were evaluated for ϕ equal to 20°, 25°, 30°, 35° and 40°, L/H ratio for a range 0.0 to 1.0 at interval of 0.2 and D_p with a range 0.2 to 2.0 at variable intervals. Three such typical sets of design charts for $\phi = 25°$, 30° and 35° are provided in Figs. 15.28, 15.29 and 15.30. The non-dimensional pressure coefficients, K_γ and K_q reduce with an increase in L/H upto 0.6 and thereafter these are almost constant. K_γ and K_q also reduce with an increase in non-dimensional coefficient D_p upto about 1.0 beyond which the reduction is insignificant.

Variation of $\dfrac{H_\gamma}{H}$ and $\dfrac{H_q}{H}$ with L/H shows that point of application of the resultant earth pressures (P_γ and P_q) moves towards the bottom of the wall for $L/H > 0.6$.

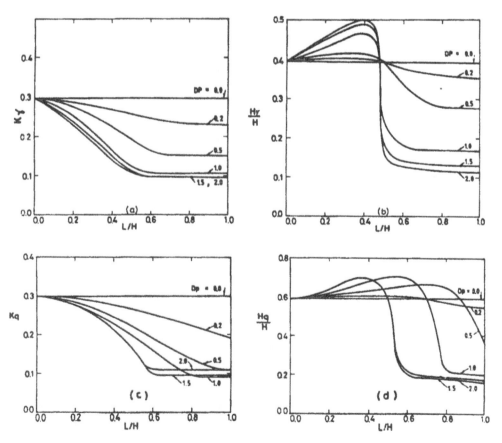

Fig. 15.29. Non-dimensional Chart for Resultant Pressure and Height of Point of Application – (a) and (b) Due to Backfill; (c) and (d) Due to Surcharge Loading ($\phi = 30°$).

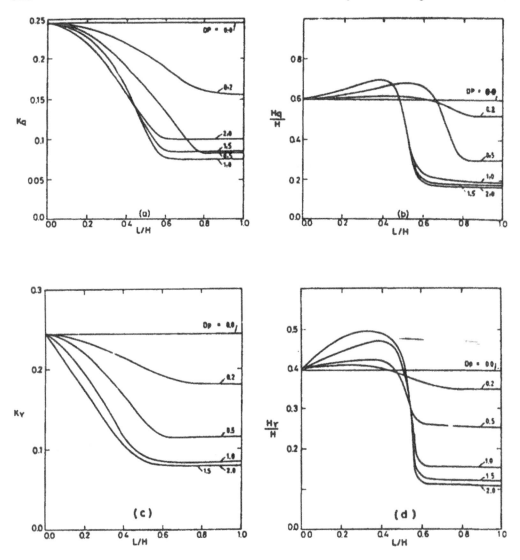

Fig. 15.30. Non-dimensional Chart for Resultant Pressure and Height of Point of
Application — (a) and (b) Due to Backfill; (c) and (d) Due to Surcharge Loading ($\phi = 35°$).

15.8 FOOTINGS ON REINFORCED SAND

Due to the meager availability of good construction sites, a foundation engineer today
frequently comes across the problem of erecting the structures on low bearing capacity
deposits. The traditional solutions to such situations have been—deep foundations placed
through the loose soil, excavation and replacement with suitable soil, stabilizing with injected
additives or applying the techniques for densification of soil. All of these methods have a
certain degree of applicability, but all suffer from being either expensive or time consuming.

The newly emerging alternate method is to remove the existing weak soil up to a shallow depth and replace it by the soil reinforced with horizontal layers of high tensile strength reinforcement. In recent years, geosynthetics like geotextiles or geogrids and metal strips have replaced the traditional materials in reinforcing the soils for improvement of bearing capacity and the settlement of soil beds (Saran, 1998).

In the following section, a method of analysis for calculating the pressure intensity corresponding to a given settlement of the footings resting on reinforced soil foundation has been presented for the following cases: (i) Isolated strip footings and (ii) Isolated rectangular footings.

Binquet and Lee (1975b) were perhaps the first who proposed an analytical approach for getting the pressure on an isolated strip footing resting on reinforced sand corresponding to a given settlement, which was further modified by Murthy et al. (1993). In the analyses presented here attempt has been made to overcome the shortcomings in the Binquet and Lee (1975b) approach for isolated strip footings, modifying the assumptions made by them to more realistic ones and further extending to isolated rectangular footings subjected to central-vertical loads (Kumar, 1997; Kumar and Saran, 2000; 2001; 2003).

Pressure ratio

For convenience in expressing and comparing the data, a pressure ratio term has been introduced and is defined as:

$$p_r = q/q_0,$$

where

q_0 = the average contact pressure of a footing on unreinforced soil, at a given settlement, and

q = the average contact pressure of the same footing on reinforced soil, at the same settlement.

It may be mentioned here that the pressure intensity of the footing on unreinforced soil at a given settlement can be obtained using standard penetration test data or standard plate load test data or using a method developed by Prakash et al. (1984) and Sharan (1977).

Analysis of isolated strip footings

Consider a strip footing of size, B. If the footing is loaded with uniformly distributed load of q, then normal and shear stresses can be calculated at any depth Z using the theory of elasticity (Poulos and Davis, 1974). The analysis is based on the following assumptions:

(1) The central zone of soil moves down with respect to outer zones. The boundary between the downward moving and outward moving zones has been assumed as the locus of points of maximum shear stress at every depth, Z. The location of separating planes can only be inferred from the location of broken ties (Binquet and Lee, 1975a) and deformation pattern of reinforcement after failure (Kumar, 1997; Kumar and Saran, 2001).

(2) At the plane separating the downward and lateral movements, the reinforcement is assumed to undergo two right angles bend around two frictionless rollers and T_D is vertically acting tensile force (Fig. 15.31). In a reinforcing layer embedded within the

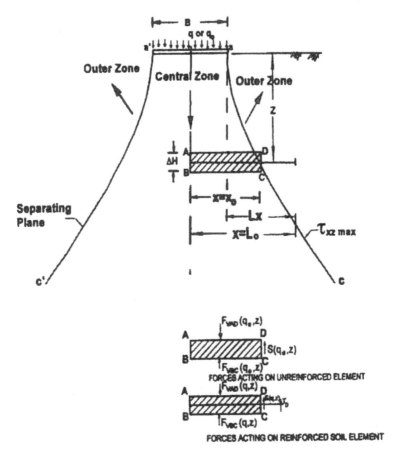

Fig. 15.31. Assumed Separating Plane and Components of Forces for Pressure Ratio Calculation of Isolated Strip Foundation on Reinforced Soil.

soil, the kink will form due to the relative movement along the plane separating the downward and lateral flow. The relative vertical movement decreases at larger depths, the right angle kink in reinforcement will not form. As the reinforcement placement at depths beyond the footing width will not be feasible from economic and construction considerations, assuming a right angle kink for reinforcements placed within this depth is reasonable. Further, from the basic mechanics, tension in reinforcement can be considered equally effective in the vertical direction at right angle bends.

(3) The mobilization of friction is dependent on the relative movement of soil and reinforcement. As the settlement of the footing at the surface causes a vertical settlement of different magnitude at different layer levels, settlement in this investigation has been assumed to vary in proportion to vertical stress at that point. This concept results in 30% of surface settlement at depth B and negligible settlement at depth $2B$ (Fig. 15.32). Therefore, soil-reinforcement friction coefficient, f^*, has been assumed to vary with depth as per the following equation:

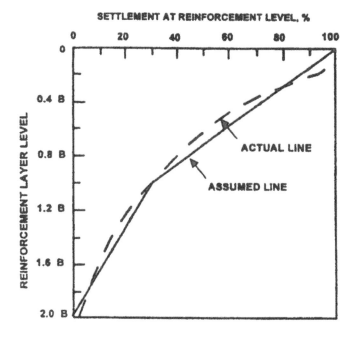

Fig. 15.32. Effective Settlement at Different Reinforcement Layer Level
(After Murthy et al., 1993).

$$f^* = m.f \qquad\qquad\qquad\qquad ... (15.40)$$

Where $\quad m = [1 - Z/B) \, 0.7 + 0.3]$ for $Z/B \leq 1.0 \qquad\qquad ... (15.41a)$

$\qquad\quad m = [(2 - Z/B) \, 0.3]$ for $Z/B > 1.0 \qquad\qquad\qquad ... (15.41b)$

$\qquad\quad f = \tan \phi_f,\ \phi_f$ is soil-reinforcement friction angle. $\qquad\qquad ... (15.41c)$

The variation of m with depth is shown in Fig. 15.32.

(4) As the normal force on the lower layers of reinforcement gets affected by the load carried by the upper layers, for N_R layers of reinforcement provided in the foundation soil, the normal force responsible for the development of reinforcement force has been assumed to vary in proportion of $r_1, r_2, r_3, \ldots\ldots, r_{NR}$, such that $r_1 + r_2 + r_3 + \ldots\ldots +$ $r_{NR} = 1$ and failure has been assumed for various combinations of reinforcement pull out and breakage at different layer levels.

(5) The forces evaluated in the analysis are for the same size of footing and for the same settlement for a footing on reinforced and unreinforced soil.

(6) Elastic theory has been applied to estimate the stress distribution inside the soil mass, as no stress equations are available for anisotropic non-homogeneous material like reinforced sand. However, it has been demonstrated later that results are not affected by this assumption.

Computation of force developed in reinforcement (T_D)

To evaluate the force developed in the reinforcement due to applied load on the footing, it was assumed that the plane separating the downward and lateral flow is the locus of points of maximum shear stress, $\tau_{xz\,max}$ at every depth, Z. In Fig.15.31, ac and $a'c'$ are assumed separating planes. Consider an element $ABCD$ at depth, Z (Fig. 15.31), which is the volume of soil lying between two vertically adjacent layers of reinforcement. The forces acting on the element are shown in Fig. 15.31 for unreinforced and reinforced foundation soil. The force developed in the reinforcement, T_D may be expressed in terms of pressure ratio as under

$$T_D = \left[J_z B - I_z \Delta H\right] q_0 \, (p_r - 1) \qquad \qquad ...\,(15.42)$$

in which

$$J_z = \frac{\displaystyle\int_0^{x_0} \sigma_z dx}{(q \text{ or } q_0)B} \qquad \qquad ...\,(15.42a)$$

$$I_z = \frac{\tau_{xz\,max}}{q \text{ or } q_0} \qquad \qquad ...\,(15.42b)$$

T_D is the force developed in the reinforcement, if there is only one layer of reinforcement in the foundation soil and that is placed at depth Z. ΔH is taken equal to vertical spacing between the horizontal layers of reinforcement, if layers are at equal vertical spacing and is equal to the average of two adjacent layers, if layers are at different vertical spacing. The stress equations have been solved by numerical integration.

Computation of reinforcement-pull-out frictional resistance (T_f)

The pull out frictional resistance shall be due to the vertical normal force on the length of the reinforcement, which is outside the assumed plane separating the downward and the outward flow (Fig. 15.31). The normal force consists of two components, one due to the applied bearing pressure and the other due to the normal overburden pressure of soil.

The reinforcement pull out frictional resistance per unit length of the strip footing is given by

$$T_f = 2f^*LDR \left[M_z \, Bq_0 \, p_r + \gamma \, (L_0 - X_0) \, (Z + D_f)\right] \qquad ...\,(15.43)$$

where T_f is the pull out frictional resistance per unit length of strip footing at depth Z, developed due to the reinforcing length beyond the assumed plane ac. D_f is the depth of footing below ground level.

$$L_0 = 0.5B + L_X \qquad \qquad ...\,(15.43a)$$

L_x = Extension of reinforcement beyond the either edge of footing.
LDR = Linear density of reinforcement
  = Plan area of reinforcement/Total area of reinforced soil layer
  = 1, for geogrid or other geosynthetic sheets covering the total soil layer.
f^* = Soil-reinforcement friction coefficient.

$$M_Z = \frac{\int_{x_0}^{L_0} \sigma_z \, dx}{qB} \qquad \qquad \cdots \ (15.43b)$$

Non-dimensional charts have been prepared for x_0/B, J_z, I_z and M_z for different Z/B values; M_z at different Z/B values for L_x/B values of 0.5, 1.0, 1.5, 2.0, 2.5 and 3.0 (Fig. 15.33).

Analysis of isolated rectangular footings

Computation of driving force (T_D)

Consider a rectangular footing of length, L and width, B. Figure 15.34 shows the plan and section of the assumed planes separating the downward and the outward flow, which

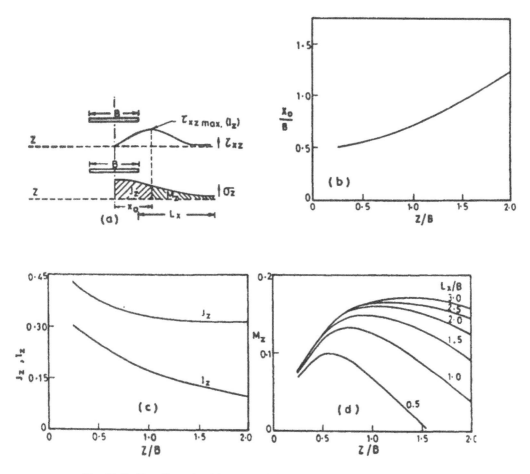

Fig. 15.33. Non-dimensional Length and Force Components for Pressure Ratio
Calculation of Isolated Strip Footing on Reinforced Soil.

are loci of points of maximum shear stress at every depth, Z. The location of the separating planes can only be inferred from the location of broken ties (Binquet and Lee, 1975a) and the deformation pattern of reinforcement after failure (Kumar, 1997).

The driving force in the x-direction is given by

$$T_{Dx} = [\, J_{xz}\, BL - I_{xz}\, L\Delta H\,]\, q_0 \{p_r - 1\} \qquad\qquad ...(15.44)$$

in which,
$$J_{xz} = \frac{\displaystyle\sum_{i=1}^{i=n} \int_0^{X_i} \sigma_Z(q,x,y,Z) dx \delta y}{qBL} \qquad\qquad ... (15.44a)$$

and
$$I_{xz} = \frac{\displaystyle\sum_{i=1}^{i=n} \tau_{xz}(q, X_i, Z) \delta y}{qL} \qquad\qquad ... (15.44b)$$

X_i is the horizontal position of peak shear stress at the ith element. Equations are solved by numerical integration and/or by summation. The interval, X_i is divided into small units of size 0.01B and the length of reinforcement $(L + 2L_y)$ is divided into 'n' parts, each of size $\delta y = 0.01B$. L_y represents extension of reinforcement beyond the edge of the footing in the y-direction.

Similarly, the driving force in the y-direction is given by

$$T_{Dy} = [\, J_{yz}\, BL - I_{yz}\, B\Delta H\,]\, q_0(p_r - 1) \qquad\qquad ... (15.45)$$

where J_{yz} is the total non-dimensional force due to applied pressure on area $g'\, h'\, k'\, j'\, g'$ marked as 'A_2', Fig. 15.34, and

I_{yz} is the total non-dimensional shear force along $g'\, h'$.

If the footing is loaded with a uniformly distributed load, q, then these components will be:

$$J_{yz} = \frac{\displaystyle\sum_{k=1}^{k=p} \int_0^{Y_k} \sigma_Z(q,x,Z) dy \delta x}{qBL} \qquad\qquad ... (15.45a)$$

$$I_{yz} = \frac{\displaystyle\sum_{k=1}^{k=p} \tau_{yz}(q, Y_k, Z) \delta x}{qB} \qquad\qquad ... (15.45b)$$

Y_k is the horizontal position of peak shear stress at the kth element. The stress equations are solved by numerical integration and/or by summation. The interval, Y_k is divided into small

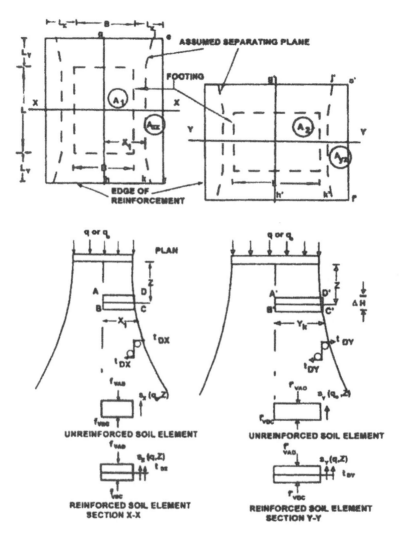

Fig. 15.34. Plan and Section of Assumed Separating Plane and
Components of Forces for Pressure Ratio Calculation.

units of size $0.01B$ and the length of reinforcement $(B + 2L_x)$ is divided into 'p' parts, each of size, $\delta x = 0.01B$. L_x represents extension of reinforcement beyond the edge of footing in x-direction.

Computation of reinforcement-pull-out frictional resistance (T_f)

Considering both components over the whole area, A_{xz}, outside the separating plane, we get reinforcement pull out frictional resistance in x-direction at depth, Z, for a footing placed at depth, D_f as:

$$T_{fx} = 2f_e LDR \ [M_{xz} \ q_0 \ p_r + \gamma \ A_{xz}(Z + D_f)] \ BL \qquad \qquad ... \ (15.46)$$

where,

$$M_{xz} = \frac{\displaystyle\sum_{i=1}^{i=n} \int_{X_i}^{(0.5B+L_x)} \delta_Z\,(q,x,y,Z)\,dx\delta y}{qBL}$$

...(15.46a)

$$A_{xz} = \frac{\displaystyle\sum_{i=1}^{i=n}(0.5B + L_x - X_i)\delta y}{BL}$$

...(15.46b)

Similarly,

$$T_{fy} = 2f_e L D R\ [M_{yz}\ q_0\ p_r + \gamma\ A_{yz}(Z+D_f)]\ BL \qquad ...\,(15.47)$$

where M_{yz} is a non-dimensional frictional resistance factor due to the vertical normal force on the plan area of reinforcement, $f\ e'\ f\ k'\ f$, marked as A_{xz}, which is outside the assumed plane separating the downward and the outward flow (Fig. 15.34) and A_{yz} in non-dimensional form is:

$$M_{yz} = \frac{\displaystyle\sum_{k=1}^{k=p} \int_{Y_k}^{(0.5L+L_y)} \sigma_Z(q,x,y,Z)\,dy\delta x}{qBL}$$

... (15.47a)

and

$$A_{yz} = \frac{\displaystyle\sum_{k=1}^{k=p}(0.5L + L_y - Y_k)\delta_x}{BL}$$

... (15.47b)

The forces T_{Dx}, T_{Dy}, T_{fx} and T_{fy} are the forces in x and y directions only. As the reinforcement would be pulled out simultaneously from all the directions with the application of load on the footing, the direction giving an average value of pressure ratio has been adopted. Thus, the direction making an angle of 45° with x or y direction gives the average value and has been adopted for the calculation of pressure ratio, which coincidentally gives the results obtained by adding the forces in the x and y directions. Total driving force is, therefore, given by

$$T_D = T_{Dx} + T_{Dy} = \left[\left(J_{xz} + J_{yz}\right)BL - \left(I_{xz}L + I_{yz}\,B\right)\Delta H\right]q_0(p_r - 1)$$

$$= \left[\left(J_z\right)BL - \left(I_{xz}L + I_{yz}\,B\right)\Delta H\right]q_0(p_r - 1) \qquad ... (15.48)$$

where

$$J_z = J_{xz} + J_{yz} \qquad ... (15.48a)$$

The total pull-out frictional resistance T_f is, therefore, given by:

$$T_f = T_{fx} + T_{fy}$$

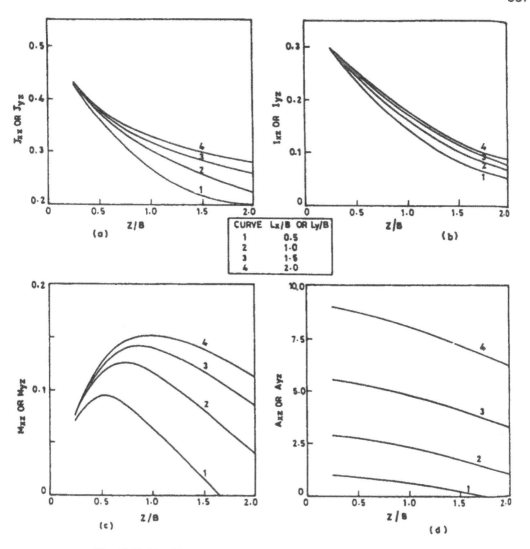

Fig. 15.35. Non-Dimensional Area and Force Components for Pressure Ratio
Calculation of Isolated Square Footing on Reinforced Soil.

$$= 2f_e \, LDR[\; (M_{xz} + M_{yz}) \, q_o p_r + \gamma(Z + D_f) \, (A_{xz} + A_{yz}) \, BL]$$

$$= 2f_e \, LDR \; [(M_z) \, q_o p_r + \gamma(Z + D_f) \, (A_z)BL] \qquad \qquad \ldots \; (15.49)$$

where $M_z = M_{xz} + M_{yz}$ and $A_z = A_{xz} + A_{yz}$

Charts have been prepared for J_{xz}, J_{yz}, I_{xz}, I_{yz}, M_{xz}, M_{yz}, A_{xz} and A_{yz} corresponding to different Z/B values for various values of L_x or L_y and for L/B ratio of 1, 2 and 3 (Figs. 15.35 to 15.39).

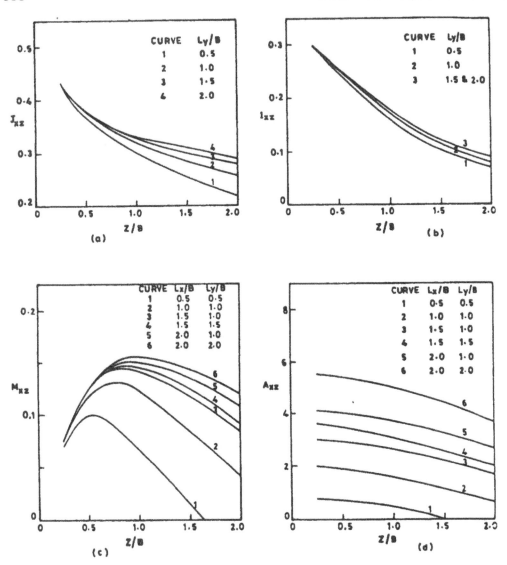

Fig. 15.36. Non-Dimensional Area and Force Components for Pressure Ratio Calculation of Isolated Rectangular Footing ($L/B = 2$) on Reinforced Soil (in X-Direction).

ILLUSTRATIVE EXAMPLES

Example 15.1

Design a suitable layout of geogrid reinforcement for the 4-m high vertical earth wall shown in Fig. 15.40. The other data is given as below:

Safe design strength of grids = 16.5 kN/m

Density of soil (wallfill and backfill) = 19 kN/m^3

Angle of internal friction of wallfill and backfill

$$\phi_w = \phi_b = 30°$$

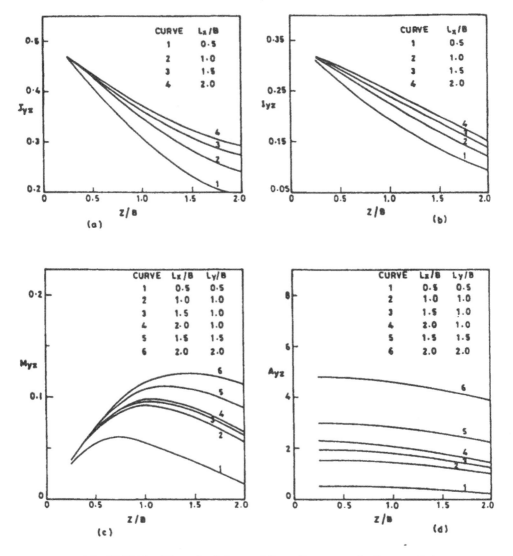

Fig. 15.37. Non-Dimensional Area and Force Components for Pressure Ratio Calculation of Isolated Rectangular Footing ($L/B = 2$) on Reinforced Soil (in Y-Direction).

Coefficient of base friction, $\mu = 0.5$

Allowable soil pressure = 140 kN/m²

Coefficient of interaction between grid and wallfill, $\alpha = 0.9$.

Solution

External stability

$$K_{Ab} = K_{AW} = \frac{1 - \sin 30°}{1 + \sin 30°} = 0.33$$

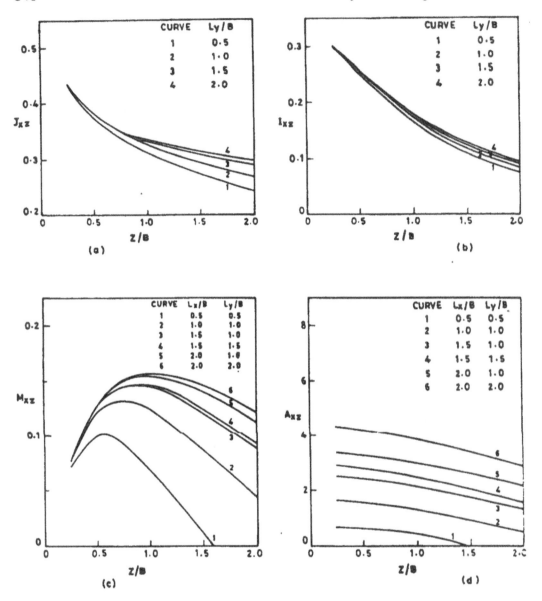

Fig. 15.38. Non-Dimensional Area and Force Components for Pressure Ratio
· Calculation of Isolated Rectangular Footing ($L/B = 3$) on Reinforced Soil (in X-Direction).

$\mu = 0.5$.

Factor of safety against sliding

$$= \frac{2\mu(\gamma_w H + q)}{K_{Ab}(\gamma_b H + 2q)(H/L)}$$

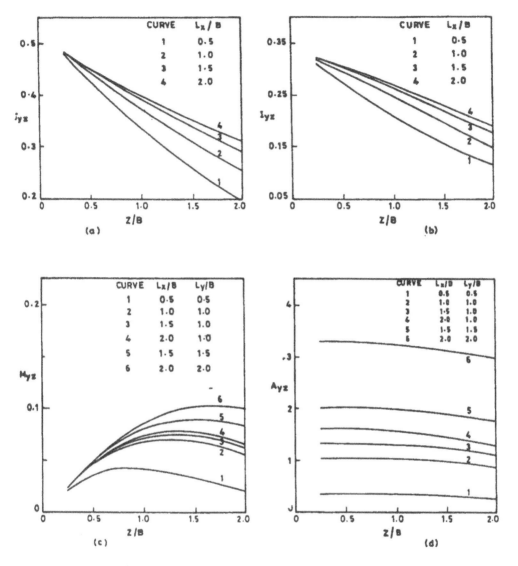

Fig. 15.39. Non-Dimensional Area and Force Components for Pressure Ratio Calculation of Isolated Rectangular Footing ($L/B = 3$) on Reinforced Soil (in Y-Direction).

$$= \frac{2 \times 0.5 \times (19 \times 4 + 12)}{0.33(19 \times 4 + 24)(H/L)} > 2 \text{ (say)}.$$

It gives

$$H/L < 1.33 \text{ and so } L > 3 \text{ m}.$$

Assume $L = 3.5$ m and determine factor of safety against overturning and tilting/bearing failure.

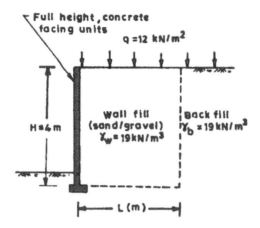

Fig. 15.40. Section of Reinforced Earthwall.

Factor of safety against overturning

$$= \frac{3(\gamma_w H + q)}{K_{ab}(\gamma_b H + 3q)(H/L)^2}$$

$$= \frac{3(19 \times 4 + 12)}{0.33(19 \times 4 + 36)(4/3.5)^2}$$

$$= 5.4 \; (> 2).$$

Ground bearing pressure under base of wall

$$\sigma_{max} = (\gamma_w H + q) + K_{ab} \, (\gamma_b H + 3q) \, (H/L)^2$$
$$\sigma_{min} = (\gamma_w H + q) - K_{ab} \cdot (\gamma_b H + 3q) \, (H/L)^2$$
$$\sigma_{max} = (19 \times 4 + 12) + 0.33 \, (19 \times 4 + 36)(4/3.5)^2$$
$$\phantom{\sigma_{max}} = 88 + 49 = 137 \; kN^2 \; (< 140 \; kN/m^2)$$
$$\sigma_{min} = 88 - 49 = 39 \; kN/m^2 \; (> 0/kN/m^2)$$

and so $L = 3.5$ m is adequate.

Internal stability:

Considering the tension failure

Tension in the reinforcement at depth h_i

$$T_i = K_{aw} \, [\gamma_w h_i + q + K_{ab} \, (\gamma_b h_i + 3q) \, (h_i/L)^2] S_{vi}$$

or $\qquad 16.5 = 0.33 \, [19h_i + 12 + 0.33 \, (19h_i + 36) \, (h_i/35)^2] S_{vi}$

Using this expression a plot of S_{vi} against h_i can be made to indicate the maximum allowable spacings for the grids by equating all values of T_i to safe design strength 16.5 kN/m (Fig. 15.41).

Clearly the tension in these grids should not exceed the safe design strength. Selected layout of reinforcement is shown in Fig. 15.42.

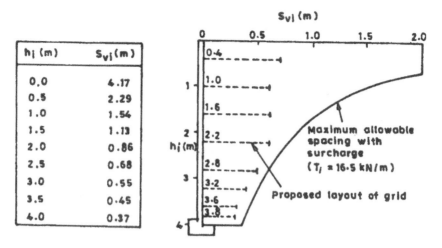

h_i (m)	S_{vi} (m)
0.0	4.17
0.5	2.29
1.0	1.54
1.5	1.13
2.0	0.86
2.5	0.68
3.0	0.55
3.5	0.45
4.0	0.37

Fig. 15.41. Proposed Layout of Grids.

Considering wedge failure

Using equation (15.17), the total tension is given by

$$T = \frac{h \tan\beta \,(19h + 24)}{2 \tan(\phi_w + \beta)}$$

It can be shown that maximum tension will occur at $\beta = 45 - \phi_w/2 = 45 - 30/2 = 30°$

Therefore, $\qquad T = \dfrac{h \tan 30° (19h + 24)}{2 \tan(30° + 30°)}$

or $\qquad T = 0.17h \,(19h + 24).$

Values of T corresponding to values $h = 1, 2, 3$ and 4 m are given in Table 15.6.

The effective numbers of grids (N) given in the third column of the Table were derived from the proposed layout of grids disregarding any grid that was less than a distance of 0.50 m (say) from the bottom of the wedge. All of the grid tensions (= T/N) are less than safe design strength and so the layout of grids is satisfactory.

Table 15.6. Tension in the Reinforcement Considering Wedge Failure.

h (m)	T (kNm)	Effective No. of Grids (N)	Tension in Each Grid kN/m
1	7.3	1	7.3
2	21.1	2	10.6
3	41.3	4	10.3
4	68.0	6	11.3

Anchorage length

The anchorage length required for each grid can now be calculated.

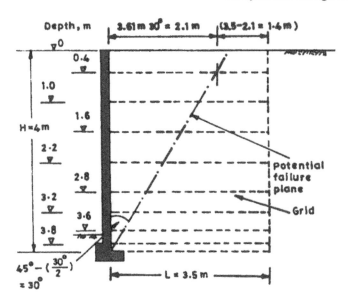

Fig. 15.42. Final Layout of Grids.

The tension (T_i) to be substituted into Equation (8.9) is then determined from either

(a) the tension failure analysis, or
(b) the wedge failure analysis

whichever gives the larger value.

Figure 15.42 clearly indicates that the proposed spacings of the grids towards the top of the wall are much close than required to satisfy the tension failure criteria.

Consequently, it is the wedge failure mechanism which governs in this case.

The previous calculations indicate that the potential wedge fai ire mechanism shown in Figure 15.42 produces a maximum tension of 11.3 kN/m when the surcharge is present.

When the surcharge is not present, this tension reduces to

$$T = \frac{4\tan 30^\circ (19 \times 4 + 0)}{2\tan 60^\circ \times 6}$$

$$= 8.44 \text{ kN/m.}$$

The required grip lengths (L_{ip}) will be:

$$L_{ip} = \frac{T_i \times \text{Factor of safety}}{2\alpha \tan\phi_w(\gamma_w h_i + q)} = \frac{T_i \times 2.0}{2 \times 0.9 \times \tan 30^\circ (19 \times h_i + q)}.$$

Putting T_i = 11.3 kN/m for surcharge case, q = 12 kN/m² and T_i = 8.44 kN/m for q = 0, anchorage length computations at various values of h_i are given in Table 15.7.

Table 15.7. Anchorage Length Computations.

Grid No.	h (m)	Anchorage 'L_{ip}' with Surcharge ($q = 12$ kN/m²) $T_i = 11.3$ kN/m	Anchorage 'L_{ip}' without Surcharge $T_i = 8.44$ kN/m	Grid Inside The Wedge (4−h) tan 30° (m)	Total Grid Length Col. 3 or col. 4 (Higher Value) + col. 5 (m)	Provided Grid Length (m)
1	2	3	4	5	6	7
1	0.4	1.1	2.2	2.1	4.3	4.3
2	1.0	0.73	0.90	1.67	2.57	3.5
3	1.6	0.53	0.55	1.33	1.88	3.5
4	2.2	0.41	0.44	1.0	1.41	3.5
5	2.8	0.35	0.32	0.67	1.02	3.5
ʼ6	3.2	0.32	0.28	0.44	0.76	3.5
7	3.6	0.29	0.25	0.22	0.57	3.5
8	3.8	0.27	0.23	0.0	0.27	3.5

Example 15.2

Design a reinforced earth abutment for retaining a 6.0-m high backfill as shown in Fig. 15.43. The vertical and horizontal bridge seat loads are 150 kN/m (V_1) and 25 kN/m (H_1) respectively. The allowable soil pressure is 250 kN/m² and the safe design strength of geogrid reinforcement is 17.5 kN/m. Assume μ and α as 0.55 and 0.95 respectively.

Solution

(i) External stability

Refer Fig. 15.30

$$W_1 = 25 \text{ kN/m (Self weight of bridge seat, assumed)}$$
$$W_2 = 20 \times 6 \times 4.2 = 504 \text{ kN/m}$$

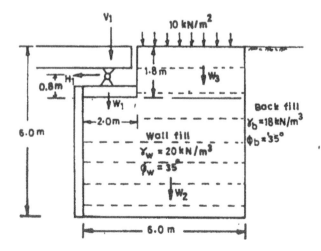

Fig. 15.43. Section of an Earth Abutment.

$$W_3 = 20 \times 4 \times 1.8 = 144 \text{ kN/m}$$

$$K_{Ab} = \frac{1 - \sin 35°}{1 + \sin 35°} = 0.271$$

$$P_1 = 0.271 \times 18 \times 6 \times 6/2 = 88 \text{ kN/m}$$

$$P_2 = 0.271 \times 10 \times 6 = 16.3 \text{ kN/m.}$$

In computing the factors of safety against sliding and overturning, it is assumed that the uniform surcharge load is not acting on the wall portion. It will present relatively more critical case.

Factor of safety against sliding

$$= \frac{(150 \times 25 + 504 + 144) \times 0.55}{(25 + 88 + 16.3)} = 3.50 \ (> 2.0, \text{ safe}).$$

Factor of safetry against overturning

$$= \frac{150 \times 1.0 \times 25 \times 1.0 + 504 \times 3.0 + 144 \times 4.0}{25 \times 5.0 + 88 \times 6/3 + 16.3 \times 3.0} = \frac{2263}{349.9}$$

$$= 6.46 \ (> 2.0, \text{ safe}).$$

Bearing pressures

· Ground bearing pressures under the base of wall are obtained considering the surcharge on the wall.

Total vertical force $= 150 + 25 + 504 + 144 + 10 \times 4 = 863 \text{ kN/m.}$

This vertical force is eccentric to the vertical centre line of wall by a distance of e given by

$$e = 3.0 - \frac{2263 + 40 \times 4}{863} + \frac{349.9}{863} = 3.0 - 2.80 + 0.40$$

$$= 0.6 \text{ m}$$

$$\sigma_{max} = \frac{863}{6 \times 1} + \frac{863 \times 0.60 \times 6}{1 \times 6^2}$$

$$= 143.8 + 86.3 = 230.1 \text{ kN/m}^2 \ (< 250 \text{ kN/m}^2, \text{ safe})$$

$$\sigma_{min} = 143.8 - 86.3 = 57.5 \text{ kN/m}^2$$

(ii) Internal stability

Simplifying the calculations by adopting idealisation shown in Fig. 15.44.

$$q = 10 + 20 \times 1.8 = 46 \text{ kN/m}^2.$$

Load H_1 may be taken equal to 25 kN/m plus the total earth pressure acting on the top 1.8 m abutment.

Tension in the reinforcement at any depth h_i without considering the effect of vertical and horizontal loads V_1 and H_1 given by Eq.

$$T_{i1} = 0.271[20h_i + 46 + 0.271 (18h_i + 138) (h_i/6)^2]S_H$$

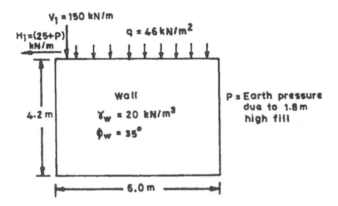

Fig. 15.44. Assumed Wall Section.

$$= [12.46 + 5.42h_i + 0.28\,h_i^2 + 0.036\ h_i^3]S_H.$$

Tension in reinforcement due to load V_1:

$$B = 2.0 \text{ m}, \ d = 1.0 \text{ m}, \ L = 6 \text{ m}$$
$$2\,(L - d) - B = 2(6 - 1.0) - 2.0 = 8.0 \text{ m} > H \ (= 4.2 \text{ m}).$$

When $h_i > 0$, Eq. 15.12 governs and $B_i = 1.0 + \dfrac{h_i + 2.0}{2}$

$$T_{i2} = \frac{6 \times 0.271 \times (150 + 25) \times 1}{(2 + 0.5h_i)^2} \left[\frac{(2 + 0.5h_i)(2 \times 2 + 3 \times 0)}{3 \times 2 \times 1} - 1\right] S_H$$

$$= \frac{95(1 + h_i)}{(2 + 0.5h_i)^2} \cdot S_H.$$

Tension in reinforcement due to load H_1:
Referring to Fig. 15.19, the depth of influence of horizontal load H_1 is

$$h_e = \frac{2.0}{\tan\left(45 - \dfrac{35}{2}\right)} = 3.84.$$

Eq. 15.15 gives

$$T_{i3} = \frac{2(32 + 0.271 \times 20 \times 1.8^2 / 2 + 0.271 \times 10 \times 1.8)}{3.84}\left[1 - \frac{h_i}{3.84}\right]S_H$$

$$= 23.8 \left[1 - \frac{h_i}{3.84}\right]S_H.$$

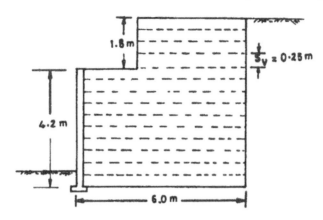

Fig. 15.45. Layout of Grids.

$$T_i = T_{i1} + T_{i2} + T_{i3}.$$

The above expression gives the following values of S_H for different values of h_i putting T_i equal to 17.5 kN/m

h_i (m)	S_H(m)
0	0.2916
0.5	0.2730
1.0	0.2640
1.5	0.260
2.0	0.258
2.5	0.2569
3.0	0.2556
3.5	0.2538
4.0	0.2512

The above example clearly shows the dramatic effects that vertical and horizontal line loads have on the spacing of the grids.

The layout of the grids is proposed as shown in Fig. 15.45.

Anchorage length:

Considering that the failure plane passes through the toe of the wall, then

$$T = \frac{4.2 \tan\beta \,(20 \times 4.2 + 2 \times 46) + 2 \times 150}{2 \tan(35 + \beta)} + 32 + 0.271 \times 20\frac{1.8^2}{2} +$$

$$0.271 \times 10 \times 1.8$$

$$= \frac{369 \tan\beta + 150}{\tan(35 + \beta)} + 45.66.$$

Varying β, T values are obtained as given below:

β (deg)	T (kN/m)
1	261
5	263
10	261
15	255
20	245

Reinforcements are provided at a uniform vertical spacing of 0.25 m (250 mm). Therefore, there will be 15 layers of grids in lower 4.2 m abutment.

$$\text{Tension per reinforcement} = \frac{263}{15} = 17.5 \text{ kN/m}$$

$$L_{ip} = \frac{17.5 \times 2.0}{2 \times 0.95 \times \tan 35 \,(20 \times 4.2 + 46)} = 0.2 \text{ m.}$$

Available embedment

$$= 6.0 - 4.2 \tan\left(45 - \frac{35}{2}\right)$$

$$= 3.81 \text{ m, hence safe.}$$

Example 15.3

Design a retaining wall of 8.0 m height retaining horizontal backfill carrying a uniformly distributed surcharge of 30 kN/m². The backfill and base soil have following properties:

$\gamma = 16.0$ kN/m², $c = 0$, $\phi = 30°$, $\delta = 20°$, $\mu = 0.5$, $q_a = 300$ kN/m²

Solution

(i) Let the backfill soil be reinforced with unattached geogrids (CE-121) layers of length equal to 0.6 times the height of wall. This geogrid has permissible tensile strength as 4.5 kN/m and coefficient of soil-geogrid friction (f^*) as 0.45. If vertical spacing between geogrid layers is adopted as 0.75 m, then

$$D_p = \frac{f^* H}{S_z} = \frac{0.45 \times 8.0}{0.75} = 4.8$$

From Fig. 15.29, for $\phi = 30°$, $L/H = 0.6$ and $D_p = 4.8$,

$$K_\gamma = 0.105; \qquad K_q = 0.160$$
$$H_\gamma = 0.180H; \qquad H_q = 0.7H$$
$$= 1.44 \text{ m} \qquad\qquad = 5.6 \text{ m}$$

(ii) Active earth pressure (P_A)

$$P_A = 0.5 \times 16 \times 0.105 \times 8^2 + 30 \times 0.160 \times 8$$
$$= 53.8 + 38.4 = 92.2 \ \text{kN/m}$$
$$P_A \cos \delta = 92.2 \cos 20 = 86.7 \ \text{kN/m}$$
$$P_A \sin \delta = 92.2 \sin 20 = 31.5 \ \text{kN/m}$$

(iii) Adopting the section as shown in Fig. 15.46

$$W = W_1 + W_2 + W_3$$
$$= 22.0 \times 0.5 \times 7.5 + 22.0 \times 0.5 \times 7.5 \times 2.0 + 22.0 \times 2.75 \times 0.5$$
$$= 82.5 + 165.0 + 30.0$$
$$= 277.5 \ \text{kN/m}$$

(iv) $$F_S = \frac{(277.5 + 31.5) \times 0.5}{86.7} = 1.78 \ (> 1.5, \text{safe})$$

Stabilising moment about point 'A' (Fig. 15.46)

$$= 82.5 \times 2.5 + 165.0 \times 1.58 + 30.0 \times 1.375 + 31.5 \times 2.75$$
$$= 594.9 \ \text{kN-m/m}$$

Overturning moment about point 'A'

$$= 53.8 \cos 20 \times 1.44 + 38.4 \cos 20 \times 5.6$$
$$= 275.0 \ \text{kN-m/m}$$

$$F_0 = \frac{594.9}{275.0} = 2.2 \ (> 2.0 \ \text{safe})$$

Net moment $= 594.9 - 275.0 = 319.9 \ \text{kN-m/m}$

$$e = 1.375 - \frac{319.9}{309}$$

$$= 0.34 \ \text{m} \ \left[< \frac{2.75}{6} = 0.458 \ \text{m}, \text{O.K.} \right]$$

$$p_{max} = \frac{309}{2.75} \left[1 + \frac{6 \times 0.34}{2.75} \right]$$

$$= 197.7 \ \text{kN/m}^2 \ [< 200 \ \text{kN/m}^2, \text{safe}]$$

(v) Force in a reinforcement layer at depth h_i is given by:

$$F = [\gamma \ (K_A - K_r) \ h_i + q \ (K_A - K_q)] \ S_z$$

K_A = Coefficient of active earth pressure in wall with unreinforced backfill = 0.3

For no rupture of reinforcement,

$$6.5 = [16.0 \ (0.3 - 0.105) \ h_i + 30 \ (0.3 - 0.16)] \ S_z$$

or $$6.5 = [3.12 \ h_i + 4.2] \ S_z$$

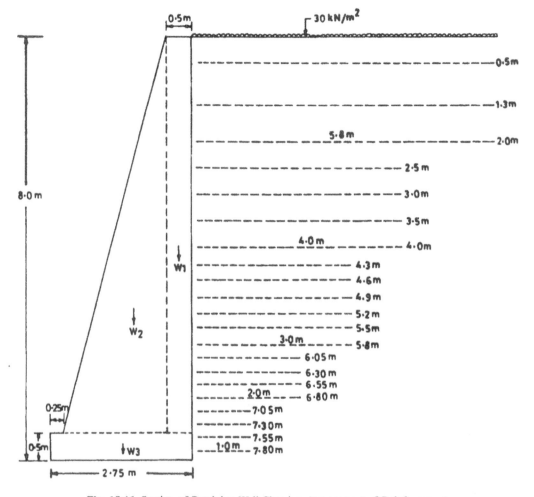

Fig. 15.46. Section of Retaining Wall Showing Arrangement of Reinforcement.

h_i (m)	S_z (m)
2.0	0.70
4.0	0.39
6.0	0.28
8.0	0.22

Provide the reinforcement as shown in Fig. 15.46.

Example 15.4

A plate load test was conducted on a 600 mm × 600 m square plate at a depth of 1 m below ground surface on a sandy soil, which extends upto a large depth. Pressure intensity versus settlement curve obtained from the test data is as shown in Fig. 15.47. The unit weight of the soil determined at the base of the test pit was 15.8 kN/m³. Design a rectangular

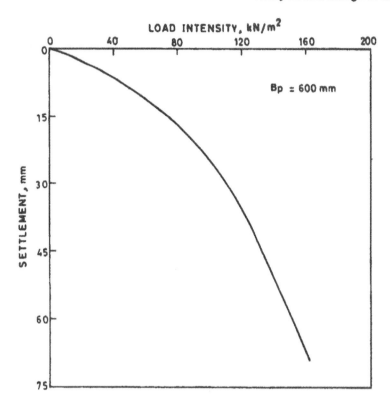

Fig. 15.47. Typical Plate Load Test Data.

footing of size 1 m × 2 m with its base at a depth of 1 m after reinforcing the foundation soil with geogrid having yield/rupture strength of 20 kN/m and soil-reinforcement friction angle from pullout test as 18°.

Take factor of safety of 3 for shear and 20 mm permissible settlement and factor of safety for geogrid as 2.

Solution

(a) Rectangular footing of size 1 m × 2 m placed at a depth of 1.0 m.

(i) Design considering un-reinforced sand

From Fig. 15.47, ultimate pressure of the plate (q_{up}) is obtained by double-tangent method. It gives 88 kN/m². As there is no overburden on the plate,

$$q_{up} = 0.4\gamma \, B_p \, N_\gamma$$

It gives, $$N_\gamma = \frac{88}{0.4 \times 15.8 \times 0.6} = 23.2$$

From Table 7.2, corresponding to $N_\gamma = 23.2$ values of ϕ and N_q are 31.4° and 27.2 respectively.

The ultimate bearing capacity of 1.0 m × 2.0 m footing located at 1.0 m depth is given by:

$$q_u = \gamma D_f \quad N_q + \frac{1}{2}\gamma \, B \, N\gamma \cdot S_\gamma$$

where S_γ = Shape factor.

The value of S_γ may be taken 0.8 for square footing ($L/B = 1$) and 1.0 for strip footing ($L/B \geqslant 8$). For rectangular footing, value of S_γ may be obtained by linear interpolation.

$$q_u = 15.8 \times 1.0 \times 27.2 + 0.5 \times 15.8 \times 1.0 \times 23.2 \times 0.9$$
$$= 597.71 \quad \text{kN/m}^2$$

Therefore,

Safe bearing capacity = q_s = 597.71/3 = 199.2 kN/m²

Settlement of the plate corresponding to 20 mm settlement (permissible) of actual footing can be computed from the following relation (Terzaghi and Peck, 1967):

$$S_p = S_f \left[\frac{B_p(B_f + 300)}{B_f(B_f + 300)} \right]^2$$

or
$$S_p = 20 \left[\frac{600(1000 + 300)}{1000(600 + 300)} \right]^2 = 15 \text{ mm}$$

From Fig. 15.47, pressure intensity corresponding to 15 mm settlement is 72 kN/m².
Hence allowable soil pressure is 72 kN/m².

(ii) Design considering reinforced sand

Use three layers of reinforcement with outer edge extension of 1.0B in both x and y directions keeping $U/B = 0.25$ and $S_v/B = 0.15$. Now, $f = \tan 18° = 0.32$.

$$\text{Rupture strength} = \frac{20(4B + 3B)}{2} = 70 \text{ kN}$$

$$L_x/B = L_y/B = 1.0 \text{ m}$$

Using Eq. 15.40 for getting the value of f_e, and Figs. 15.37 and 15.38 for obtaining A_{xz}, A_{yz}, J_{xz}, J_{yz}, I_{xz}, I_{yz}, M_{xz} and M_{yz}, values of T_D and T_f are obtained for each of three layers of reinforcement as given in Table 15.8.

Table 15.8. Calculation of Forces for Pressure Ratio Determination.

Z/B	$J_z =$ $J_{zz} + J_{yz}$	I_{xx}	I_{yx}	$M_z =$ $M_{xx} + M_{yx}$	$A_z =$ $A_{xx} + A_{yx}$	f_e	T_D	T_f
0.25	0.91	0.30	0.31	0.11	3.5	0.264	1.65	$0.12 q_0 p_r$ $q_0 (p_r - 1) + 73$
0.40	0.86	0.28	0.30	0.15	3.46	0.232	1.59	$0.136 q_0 p_r$ $q_0 (p_r - 1) + 71$
0.55	0.80	0.26	0.29	0.19	3.42	0.20	1.47	$0.152 q_0 p_r$ $q_0 (p_r - 1) + 69.66$

$$m_1 \ 1.65 \ q_0 \ (p_r - 1) = 0.12 \ q_0 p_r + 73 \text{ or } 70$$
$$m_2 \ 1.59 \ q_0 \ (p_r - 1) = 0.136 \ q_0 p_r + 71 \text{ or } 70$$
$$m_3 \ 1.47 \ q_0 \ (p_r - 1) = 0.1529 \ q_0 p_r + 69.66 \text{ or } 70$$
$$m_1 + m_2 + m_3 = 1$$
$$q_0 = 72 \text{ kN/m}^2$$

(i)   All layers fail in pullout, $p_r = 3.91$
(ii)   All layers fail in rupture, $p_r = 2.86$
(iii)   Layers I and II in pullout, layer III in rupture, $p_r = 3.43$
(iv)   Layers II and III in pullout, layer I in rupture, $p_r = 3.53$
(v)   Layers I and III in pullout, layer II in rupture, $p_r = 3.49$
(vi)   Layer I in pullout, layers II and III in rupture, $p_r = 3.1$
(vii)   Layer II in pullout, layers I and II in rupture, $p_r = 3.13$
(viii)   Layer III in pullout, layers I and II in rupture, $p_r = 3.19$

The Case (iii) is illustrated below:

$$m_1 \times 1.65 \times 72 \ (p_r - 1) = 0.12 \times 72 \ p_r + 73$$
$$m_2 \times 1.59 \times 72 \ (p_r - 1) = 0.136 \times 72 \ p_r + 71$$
$$m_3 \times 1.47 \times 72 \ (p_r - 1) = 70$$
$$m_1 + m_2 + m_3 = 1$$

or
$$m_1 \ (p_r - 1) = 0.07273 \ p_r + 0.6145$$
$$m_2 \ (p_r - 1) = 0.08553 \ p_r + 0.6202$$
$$m_3 \ (p_r - 1) = 0.6614$$
$$m_1 + m_2 + m_3 = 1$$

Adding the first three equations and putting $m_1 + m_2 + m_3 = 1$

$$p_r - 1 = 0.15826 \ p_r + 1.8961$$
or
$$p_r = 3.43$$

Similarly other seven cases were solved.
      Therefore, critical $p_r = 2.86$

∴ Pressure intensity of footing on reinforced soil at 20 mm settlement is

$$q = q_0\, p_r = 72 \times 2.86 = 205.92 \ \text{kN/m}^2$$

p_{ru} (for $q_0 = q_u$) = 594.71 is 1.23

∴

$$q_{ur} = 594.7 \times 1.23 + 15.8 \times 0.55 \times 1 \times 27.2$$

$$= 967.85 \ \text{kN/m}^2$$

$$FOS = 967.85 \ / \ 205.92 = 4.7 \ > \ 3, \quad \text{Safe.}$$

REFERENCES

1. Bassett, R.H. and Last, N.C. (1978), 'Reinforcing Earth Below Footing and Embankment'. Proc. Symposium on Earth Reinforcement, ASCE, Pittsburgh, pp. 202–231.

2. Binquet, J. (1978), "Analysis of Failure of Reinforced Earth Walls," Proc. Symposium on Earth Reinforcement, ASCE, Pittsburgh, pp. 232–251.

3. Binquet, J. and Lee, K.L. (1975a), "Bearing Capacity Tests on Reinforced Earth Slabs," Journal of the Geotechnical Engineering Division, ASCE, Vol. 101, No. GT12, pp. 1241–1255.

4. Binquet, J. and Lee, K.L. (1975b), "Bearing Capacity Analysis of Reinforced Earth Slabs," Journal of the Geotechnical Engineering Division, ASCE, Vol. 101, No. GT12, pp. 1257–1276.

5. Broms, B.B. (1977), "Triaxial Tests with Fabrics-Reinforced Soils". Proc. International Conferene on the Use of Fabrics, Geotechnics, Paris, Vol. III, pp. 129–133.

6. Broms, B.B. (1987), "Fabric reinforced soils", Proc. International Symposium on Geosynthetics and Geomembrane, Kyoto, Japan, pp. 13–54.

7. Department of Transport (D.O.T) Memorandum 1978, Technical memorandum (Bridges), B.E. 3/78, Reinforced Earth Retaining Walls and Bridge Abutments for Embankments.

8. Garg, K.G. (1988), "Earth Pressure on wall with reinforced backfill", Ph.D. Thesis, Indian Institute of Technology, Roorkee.

9. Garg, K.G. and Saran, S. (1989), "Earth Pressure on wall with reinforced backfill", International Workshop on Geotextiles, November, Vol. 1, Banglore.

10. Garg, K.G. (1998), "Retaining wall with reinforced backfill – a case study", Journal Geotextiles and Geomembrances, Elsevier, Vol. 16. 135–149.

11. Hausmann, M.R. (1976), "Strength of Reinforced Soil," Proc. Australian Road Research Board Conference, Vol. 8.

12. Hausmann, M.R. and Lee, K. (1978), "Rigid model wall with soil reinforcement", Proc. Symp. on Earth Reinforcement, ASCE, pp. 400–427.

13. Hoare, D.J. (1979), "Laboratory Study of Granular Soils Reinforced with Randomly Oriented Discrete Fibres," Proc. International Conference on Soil Reinforcement, Paris, pp 47–52.

14. Khan, I.N. (1991), "A study of reinforced earth wall and retaining wall with reinforced backfill", Ph.D. Thesis, Indian Institute of Technology, Roorkee.

15. Kumar, A. (1997), "Interaction of footings o reinforced earth slab", Ph.D. Thesis, University of Roorkee, Roorkee (UP), India.

16. Kumar, A. and Saran, S. (2000), "Analysis of square footing on reinforced soil foundations", Proc. Int. conf. on Construction Industry, Chandigarh, Vol. III, pp. 501–510.

17. Kumar, A. and Saran, S. (2001), "Isolated strip footings on reinforced sand", Journal of Geotechnical Engineering, SEAGS, Bangkok, Thailand, pp. 177–189.

18. Kumar, A. and Saran, S. (2003), "Bearing capacity of rectangular footing on reinforced soil", Journal of Geotechnical Engineering, Kluwer Academic Publishers, The Netherlands, vol. 21, pp. 201–224.

19. Long, N.T., Guegan, Y. and Legeay, G. (1972). "Etude de la Tenre Armee a l' Appareil Triazial," Rapport de Recherche du Laboratiore Central des Ponts et Chaussees, Paris, No. 17.

20. McGown, A. and Andrawes, K.Z. (1977), "The Influence of Non-woven Fabric Inclusions on the Stress-Strain Behaviour of a Soil Mass," Proc. International Conference on the Use of Fabrics in Geotechnics, Paris, Vol. 1, pp. 161–166.

21. Mckittrick, D.P. (1978), "Reinforced Earth, from Theory to Practice," Keynote Address, Symposium on Soil Reinforcing and Stabilising Techniques in Engineering Practices, Sydney.

22. Mittal, S. (1998), "Behaviour of retaining wall with reinforced c-φ soil backfill," Ph.D. Thesis, Indian Institute of Technology, Roorkee.

24. Murry, R.T. and Boden, J.B. (1979), "Reinforced Earth Wall Constructed with Cohesive Fill", Proc. International Conference on Soil Reinforcement, Paris.

25. Murthy, B.R.S., Sridharan, A. and Singh, H.R. (1993), "Analysis of reinforced soil beds", Indian Geotechnical Journal Vol 23(4), pp. 447–458.

26. Pasley, C.W. (1822), "Experiments on revetments", Vol. 2, Murry, London.

27. Poulos. H.G. and Davis, E.H. (1974). Elastic solutions for soil and rock mechanics", John Wiley and Sons. Inc., New York.

28. Prakash, S., Saran, S. and Sharan, U.N. (1984), "Footings and constitutive laws", Journal of Geotechnical Engg. Div., Vol. 110, No. 10, pp. 1473–1487.

29. Saran, S., Talwar, D.V. and Prakash, S. (1979), "Theoretical earth pressure distribution of retaining wall with reinforced earth backfill", International conference on Soil Reinforcement, Paris, March.

30. Saran, S. and Talwar, D.V. (1981), "Seismic pressure distribution on retaining walls with reinforced earth backfill", International Conference o Recent Advances in Geotechnical Eqrthquake Engineering and Soil Dynamics, St. Louis, MO, USA.

31. Saran, S. and Talwar, D.V. (1982), "Behaviour of reinforced earth wall during earthquake", VII Symp. on Earthquake Engg., University of Roorkee, Roorkee.

32. Saran, S. and Talwar, D.V. (1983), "Analysis and design of retaining walls with reinformced earth backfills", Journal I.E., Vol. 64, Pt. CI, July, pp. 20.

33. Saran, S. and Khan, I.N. (1988), "Analysis of inclined retaining wall having reinforced cohesionless backfill with uniformly distributed surcharge", Journal of India Geot. Engg., 28(3), pp. 250–269.

34. Saran, S., Khan, I.N (189), "A study on reinforced earth retaining wall", International Workshop on Geotextiles, November, Vol. 1, bangalore.

35. Saran, S., Garg, K.G. and Bhandari, R.K. (1992), "Retaining wall with reiforced cohesionless backfill", ASCE, Geot. Engg. Division, Vol. 118, No. 12, December issue, pp. 1869–1888.

36. Saran, S. (1998), "Behaviour of footings on reinforced sand", Invited Lecture, National Workshop on Reinforcement of Gravel and Slopes, Kanpur, August.

37. Saran, S., Garg, K.G. and Mittal, S. (2001), "Model studies on retaining wall with reinforced cohensive backfill", Proc. of Int. Conf. on Geotechnical and Environmental Challenges in Mountaineous Terrain, Nepal Engg. college, Nepal, Nov. 6–7, pp 110–120.

38. Saran, S.K. (1998), "Seismic earth pressures and displacement analysis of rigid retaining wall shaving reniforced backfill", Ph.D. Thesis, Indian Institute of Technology, Roorkee.

39. Schlosser, F. and Vidal, H. (1969), "Reinforced Earth", Bulletin de Liasion des Laboratoires Central des Ponts et Chaussees, No. 41, Paris.

40. Scholsser, F. and Long, N.T. (1974a), "Dimensionnement des Murs on Terre Armee," Dimensionement des Oulverages en Terre Armee-Murset Culees de Ponts, Laboratoire Central des Ponts et Chaussees, Paris.

41. Sharan, U.N. (1977), "Pressure Settlement Characteristics of Footing Using Constitutive Laws," Ph.D. Thesis, University of Roorkee, Roorkee, India.

42. Singh, V.K. (1991), "Strength and Deformation Characteristics of Sand Reinforced with Geotextiles," M.E. Thesis, University of Roorkee, Roorkee, India.

43. Talwar, D.V. (1981), "Engineering Behaviour of Reinfored Sand in Retaining Structures and Shallow Foundation," Ph.D. Thesis, University of Roorkee, Roorkee, India,

44. Vaishya, U.S. (1977), "Bearing Capacity and Settlement Characteristics of Footings on Reinforced Earth Slabs," M.E. Dissertation, Civil Engineering Department, University of Roorkee, Roorkee, India.

45. Vidal, H. (1966), "La Terre Armee," Annales des lInstitut Technique du Batiment et des Travaux Publics, France.

46. Vidal, H. (1978), "The Development and Future of Reinforced Earth," Keynote Address, Symposium on Earth Reinforcement, ASCE, Pittsburgh, pp. 1–61.

47. Yang, Z. (1972), "Strength and Deformation Characteristics of Reinforced Sand," Ph.D. Thesis, University of California, Los Angeles.

PRACTICE PROBLEMS

1. What is reinforced earth ? Explain the basic mechanism of reinforced earth.
2. How is soil selected in reinforced earth application? What are the popular reinforcing materials?
3. Give the salient feature of an investigation illustrating the increase in strength parameters of granular soil by providing reinforcement.
4. Describe stepwise the procedure of designing a reinforced earth wall.
5. Design a reinforced earth wall for retaining a 6.0-m high cohesionless. The soil in the wall and backfill has a density of 18 kN/m³ with angle of internal friction of 35°. The allowable soil pressure is 150 kN/m². Illustrate your taking
   (i) Galvanised strips as reinforcement,
   (ii) Geogrid as reinforcement.
6. Design a suitable layout of grid reinforcement for the wallfill forming the bridge abutment shown in Fig. 15.48. The other data is given below:
   (i) $V_1 = 180$ kN/m: $H_1 = 35$ kN/m, when dead load and live load are considered.
   (ii) $V_1 = 90$ kN/m: $H_1 = 0$, when only dead load is considered. Assume: $\mu = 0.55$m $\alpha = 0.92$ and $q_a = 250$ kN/ m².

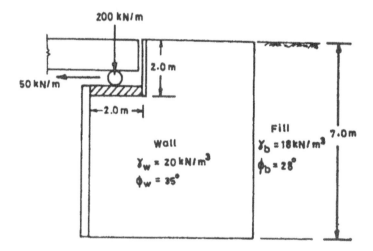

Fig. 15.48. Section of an Earth Abutment.

7. Design a retaining wall of 11.0 m height retaining a horizantal backfill carrying a uniformly distributed surcharge of 20 kN/m². The backfill and base sail have following properties:
   $\gamma = 15.5$ kN/m³, $c = 0$, $\phi = 35°$, $\mu = 0.55$, $q_a = 250$ kN/m².
   Give suitable layout of reinforcement in the backfill soil.
8. Considering the data of plate load test as shown in Fig. 15.47, determine the pressure corresponding to 25 mm settlement of the following footings resting on suitably reinforced earth slab
   (Geogrid, $R_T = 25$ kN/m):

   (a) Strip footing of width 1.5 m located at 1.0 m depth below ground surface.
   (b) Square footing of 1.5 m × 1.5 m size located at 1.0 m below ground surface, and
   (c) Rectangular footing of 1.5 m × 4.5 m size located at 1.0 m below ground surface.

   Assume suitably the reinforcement layout and any data not given.

APPENDIX A—SL UNITS

Table A-1 lists the basic SI units.

Table A-1 Basic SI units.

Physical Quantity	Unit	Symbol
Length	metre	m
Mass	kilogram	kg
Time	second	s
Electric current	ampere	A
Themodynamic temperature	kelvin	K
Luminous intensity	candela	cd

Table A-2 lists the derived SI units used in soil engineering.

Table A-2 Derived SI Units Used in Soil Engineering

Pysical Quantity	Unit	Symbol	Formula
Acceleration	metre per second squared	m/s^2	
Angular velocity	radian per second	rad/s	
Angular acceleration	radian per second squared	rad/s^2	
Area	square metred	m^2	
Area	hectare	ha	$hm^2 = 10^4 m^2$
Density	kilogram per cubic metre	kg/m^3	
Dynamic viscosity	Newton second per square metre	Ns/m^2	$kg. \, m/s. \, m^2$
Electric charge	Coulomb	C	$C = A.s$
Electric resistance	Ohm	Ω	$\Omega = V/A$
Force	Newton	N	$kg. \, m/s^2$
Frequency	Hertz	Hz	$1/s$
Kinematic viscosity	square metre per second	m^2/s	$kg. \, m^2/s^2$
Moment or torque	Newton metre	N.m	$kg. \, m^2/s^2$
Power	watt	W	j/s
Pressure	Pascal	Pa	N/m^2
Stress	Pascal	Pa	N/m^2
Unit weight	Newton per cubic metre	N/m^3	$kg/s^2. \, m^2$
Velocity	metre per second	m/s	
Voltage	volt	V	W/A
Volume	cubic metre	m^3	
Volume	litre	l	$dm^3 = 10^{-3} \, m^3$
Work/Energy	joule	J	N.m

CONVERSION FACTORS

To Convert from	To	Multiply by
in	mm	25.4
ft	mm	304.8
in^2	mm^2	645.16
ft^2	mm^2	92903
in^3	mm^3	16387
ft^3	m^3	0.028317
lb	kg	0.4536
ton (2240 lb)—long	kg	1016.05
ton (2000 lb)—short	kg	907.18
kg	lb	2.2046
kip	lb	1000
kip	kg	453.59
lb	N	4.4482
kg	N	9.8067
N	lb	0.22481
N	kg	0.1019
kip	kN	4.4482
Tonne—long	kN	9.8067
Ton–short	kN	8.8964
Tonne–long	MN	0.00981
MN	Tonne	101.64
kN	Tonne	0.1016
kip ft	kN m	1.3558
kg m	Nm	9.8067
Tonne m	kN m	9.8067
	MN m	0.00981
kN m	Tonne-m	0.1016
lb/ft	kg/m	1.4882
kip/ft	kN/m	14.5939
Tonne/m	kN/m	9.8067
lb/in^2	kN/m^2 (kPa)	6.8948
lb/in^3	kN/m^3	271.446
kip/ft^2	kN/m^2	47.88
kip/ft^3	kN/m^3	157.087
Tonne/m^2	kN/m^2	9.8067
Tonne/m^3	kN/m^3	9.8067
kg/cm^2	Tonne/m^2	10.00
	kN/m^2 (kPa)	98.0665
	MN/m^2 (MPa)	0.0980
kg/m^2	N/m^2 (Pascal)	9.8067
g/cm^3	kN/m^3	9.8067
kg/m^3	N/m^3	9.8067
Imperial gallon	Litre	4.546
U.S. Liquid gallon	Litre	3.785

Subject Index

Author Index

Printed and bound by CPI Group (UK) Ltd, Croydon, CR0 4YY

01/11/2024

01782612-0001